BIOLOGICAL OCEANOGRAPHY

BIOLOGICAL OCEANOGRAPHY

SECOND EDITION

Charles B. Miller

and

Patricia A. Wheeler

College of Earth, Ocean and Atmospheric Sciences
Oregon State University
Oregon, USA

⊗WILEY-BLACKWELL

A John Wiley & Sons, Ltd., Publication

Library of Congress Cataloging-in-Publication Data

Miller, Charles B., 1940–
    Biological oceanography / Charles B. Miller and Patricia A. Wheeler. – 2nd ed.
        p. cm.
    Includes bibliographical references and index.
    ISBN 978-1-4443-3301-5 (cloth) – ISBN 978-1-4443-3302-2 (pbk.)
    1. Marine ecology.   I. Wheeler, Patricia A., 1949–   II. Title.
    QH541.5.S3M55 2012
    577.7–dc23
                            2011046023

A catalogue record for this book is available from the British Library.

Set in 9.5/12 pt Classical GaramondBT by Toppan Best-set Premedia Limited
Printed and bound in Malaysia by Vivar Printing Sdn Bhd

1   2012

CONTENTS

Colour plates appear between pages 230 and 231

COMPANION WEBSITE

This book is accompanied by a companion website:

www.wiley.com/go/miller/oceanography

With figures and tables from the book for downloading

PREFACE TO THE SECOND EDITION

Scientific scholarship has changed markedly in the nine years since writing was finished on the first edition. Our journals are no longer printed on paper and studied in a library. Rather journals come to us as computer files. Only ten years back, papers printed adjacent to those suggested to us by colleagues or indexing services were often interesting and took us afield to exciting and divergent facts and ideas. Now the indices can find almost everything written about almost anything (metagenomics of arctic archaea, growth rings in whale teeth, swimming mechanics of amphipods, . . .); but the leaps into unrelated subjects are rarer. Moreover, the shear quantity of oceanographic (and biological!) literature is intimidating. There are more scientists everywhere and virtually all of them are publishing in English. Many long-standing journals now run to thousands and tens-of-thousands of "pages" yearly, and to that has been added a layer of new open access, on-line only journals. Thus, it is difficult to keep up with even sub-subspecialties (say, replacement of phosphorus in *Prochlorococcus* metabolism). We suspect that you prospective oceanographers, for whom this book is intended, will need to become extremely specialized, descending into the narrow wells you will drill into the expanding array of biological and oceanographic problems. Nevertheless, starting with a fairly wide perspective in your field will serve you well, and maintaining a usable introduction across the diverse aspects of biological oceanography has been our goal in revising *Biological Oceanography*.

Our perspective in this book is resolutely organismal: what are the prokaryotes, algae, protists and animals living in the seas and how do they make their way in the water, survive through generations, feed themselves and each other? How are these organisms distributed and why? What specific adaptations are required to live in blazing surface sunshine, dark deep waters, in bottom mud and near hydrothermal vents? How will those adaptations serve and change as the oceans warm and grow more acidic from dissolving carbon dioxide? We take this approach for the lighted upper layers, dim mesopelagic zone and the seafloor, emphasizing for each some aspects of ecological studies conducted in them. There are other perspectives, particularly an emphasis on biological-physical interactions. Those are mentioned many times here, but a more

directed treatment in that vein is provided in a third (2006) edition of *Dynamics of Marine Ecosystems* by K. H. Mann and J. R. N. Lazier, also published by Wiley-Blackwell. We recommend it when that emphasis is sought.

We have tried to select exemplary studies for many types of organism-organism and organism-habitat interaction. Some are recent. Many were left in from the first edition; not all topics have been studied recently and there is value in avoiding the impression that ocean ecology began in 2003. On the other hand, many topics have been "hot" in this decade, and for those we have chosen new examples. There is more molecular genetics, which has swept across all aspects of biology, including biological oceanography. A new first chapter considers some basic aspects of living in water, especially salt water. An updated discussion of spring blooms has been moved to Chapter 11 on pelagic biomes, and that discussion is greatly expanded. There is a new chapter on pelagic food chains, including identification of trophic levels from gut contents and from stable isotope ratios.

If you have been active in biological oceanographic research, your study may be shown or at least cited, but only a very small fraction of useful studies could be cited. Yours could be "on the cutting room floor", or likely we never found it at all. Not being included is not to be taken as judgment about your work. Some specialists found their fields under-represented in the first edition. That will happen again with the second edition, and we recommend to those who are teaching to fill gaps for their students with their own knowledge and distribution materials.

Like other scientists we are at least moderately opinionated about our subject matter and that of others. Much of that opinion has been allowed out on parade here. As you students and teachers find things to disagree with us about, perhaps something that can be explained better, let us know about it. For the moment we are still working and would like the communication to go both ways. Let us hear from you. Thanks.

Charles B. Miller and Patricia A. Wheeler
Corvallis, Oregon, USA
November 2011

ACKNOWLEDGEMENTS

Most warmly, we thank Martha Clemons and John Westall, our spouses, for helping us through the long sieges of study that revising this book has required. Thanks to Dean Mark Abbott for making the College of Earth, Ocean and Atmospheric Sciences (CEOAS, Oregon State University) facilities available to us deep into "retirement", and thanks to all our colleagues at CEOAS for sharing their expertise. Dave Reinert of CEOAS made improvements in many figures, and Guin Library staff Janet Webster, Susan Gilmont and Judy Mullen helped us with the mysteries of computerized literature. Thanks to the Oxford staff of Blackwell, particularly Ward Cooper for recruiting a second edition and Kelvin Matthews for seeing the manuscript through the press. Caroline Needham provided thoughtful copy editing, and Anne Bassett and Kevin Fung managed typesetting. Thanks to the one to three scientists who reviewed each chapter at the request of the publisher; Wiley-Blackwell resolutely kept your identities secret. All were helpful critics, offering useful suggestions. Even those disappointed in our chapters offered ways to improve them. We particularly thank Harold Batchelder, Michael Behrenfeld, Peter Franks, Bruce Frost, Erica Goetze, David Kirchman, Michael Landry, Tammi Richardson, Antoni Rosell Melé, Phillip Mundy, Evelyn Sherr, Barry Sherr, Suzanne Strom, Erik Thuesen, George Waldbusser and Jonathan Zehr, all of whom told us they provided reviews or looked at chapters at our request. Errors inevitably remain, all assignable to us not to the reviewers. Thanks also to the unhappy reviewer of Chapter 1, who felt it was too much an exercise in "cultural erudition". How else could we know we even approached that goal?

We thank all the scientists whose papers provided figures, over three hundred of you, some no longer alive. Dudley Chelton, Matthew Church, Erica Goetze, Nicolas Gruber, Christophe Menkes, Heidi Sosik, Taro Takahashi and Peter Thor provided modified figures or figure files with better resolution. Thanks to them.

Thanks to the following journals and organizations holding copyrights for permission to reprint figures: *Advances in Marine Biology* (Fig. 13.21); *American Naturalist* (7.19, 13.14, 13.16, 13.17); *Annual Review of Genetics* (Plate 2.1); *Applied Environmental Microbiology* (5.7, 15.6, Box Fig. 5.1.1); *Aquatic Microbial Ecology* (5.13, 7.10d&e, 9.6, 11.30); *Archiv für Hydrobiologie* (5.8); *Arctic* (Plate 11.3); *Biodiversity and Conservation* (13.20); *Biogeochemistry* (3.13); *Biological Bulletin* (12.18); *Biological Reviews* (12.10, 12.11); *British Journal of Experimental Biology* (Box Fig. 7.2); *Bulletin of the Ocean Research Institute, University of Tokyo* (10.10); (*Bulletin of Marine Science* (11.7, 17.12a&b); *Bulletin of the Scripps Institution of Oceanography* (10.5, 10.6, 10.7b, 10.8, 10.15); *CalCOFI Atlas* (6.7, 10.12, 10.24, 10.25); *CalCOFI Reports* (16.22a, 17.4, 17.21); *Canadian Journal of Zoology* (10.3, 15.4); *Current Opinion in Microbiology* (Plate 5.1); *Deep-Sea Research* (2.4, Box Fig. 2.3.2, 3.16); *Deep-Sea Research I* (13.9, 13.23a&b, 15.2, 15.8); *Deep-Sea Research II* (1.8b, 2.20, 3.9, 3.20, Plate 4.1, 4.10, 5.9, 8.5, 8.14, Plate 8.3, 10.14, 11.10, 11.11, 11.12, 11.15, 11.16, 11.18, 11.23, 11.25, 11.33, 11.34, 11.35, 13.26, 13.28, 13.29, Plate 13.1, 14.18, 14.29, Plate 16.4); *Discovery Reports* (6.5, 10.9); *Earth and Planetary Science Letters* (3.16); *Ecology* (Table 3.3, 11.14, 17.14); *Ecology Letters* (9.11); *Evolution* (10.22, 10.23); *Fisheries and Oceans Canada* (Plate 11.3); *Fisheries Investigations* (9.1); *Fishery Bulletin* (8.8, 10.16); *Geochimica et Cosmochimica Acta* (15.5); *GeoEye* (Plate 2.3); *Geophysical Research Letters* (11.17, 11.19, Plates 11.4, 11.6, 16.3); *Global Biogeochemistry* (Plate 16.2); *Global Change Biology* (16.21, 16.23); *Memoires of the Faculty of Fisheries, Hokkaido University* (7.16); *Hydrobiologia* (6.6); *IATTC Bulletin* (17.5a); *IATTC Annual Report* (17.15); *ICES Journal of Marine Science* (8.3, Plate 8.2, 9.9, 11.31, 11.32); *International North Pacific Fisheries Commission Bulletin* (17.6a) International Pacific Halibut Commission Technical Reports (17.8); *International Review of Cytology* (2.7a&b); *International Review of Hydrobiology* (7.11, 7.12); Isaacs Papers-Scripps Institution of Oceanography Library (13.1); *Izdatel'stvo Nauka, Moskow* (12.1); *Journal du Conseil* (11.9); *Journal of Crustacean Biology* (7.2, 15.7); *Journal of Experimental Biology* (7.9, Plate 7.1, 12.8); *Journal of Experimental Marine Biology and Ecology* (8.6, 8.7); *Journal of Fisheries Biology* (17.2); *Journal of Fluid Mechanics* (1.4); *Journal of Geophysical Research-Oceans* (Box Fig. 2.5.2, Table 3.2, 11.36, 11.39, 11.42, Plate 11.2); *Journal of Marine Research* (10.13, 14.19); *Journal of Marine Science* (3.5, 8.15); *Journal of Marine Systems* (4.11, 4.12, 16.22b&c, 17.20); *Journal of Phycology* (1.7, 2.6, 2.7c&d, 2.8, 2.9, 2.13, 2.14, 2.15, 3.2, 3.3, 3.7, 3.11, 3.12); *Journal of Plankton Research* (1.8a); *Journal of the Marine Biological Association, United Kingdom* (12.7, 12.9,

13.3a); *Limnology and Oceanography* (Table 2.3, 3.10, 3.14b, 3.17, 3.19, Box Fig. 3.1.1, Table 3.1, 7.1, 7.3, 7.4, 7.7, 7.8, 7.10a,b&c, 7.18, 7.20, 8.2, 8.4, 8.11, 8.12, 8.16, 8.19, 9.4, 9.5, 10.4a, 11.3, 11.26, 11.28, Plate 11.1, 14.21, 14.22, 14.26, 14.27, 14.28); *Limnology and Oceanography-Methods* (11.13); *Marine Biological Research* (11.20, 14.25a&b); *Marine Biology* (3.6c, 4.3, Plate 6.7, 7.5, 7.6, 7.13, 7.14, 7.15, Box Fig. 7.4, Box Fig. 7.5, 12.13, 12.14, 12.15, 12.16, 12.17, 14.1, 14.3, 14.4, 14.5, 14.5, 14.7, 14.8); *Marine Ecology Progress Series* 5.6, 5.10, 7.17, 9.2, 12.19, 14.20, 14.24, 16.19); Marine Geology (14.17c); *Memoires of the Geological Society of America* (1.6, 10.17, 10.19, 10.20, 10.21); *Microbiology and Molecular Biology Reviews* (5.3); Monterey Bay Aquarium (Plates 6.2, 6.3); Monterey Bay Aquarium Research Institute (13.6); *National Geographic* (Plates 6.6, 6.12); *Nature* (1.1, 3.15, 5.4, 8.17, 11.2, Plate 11.5, 14.16, 16.2, 16.15, 17.23); *Nature Geoscience* (16.13); *Nature Reviews* (5.1); *New Zealand Journal of Marine and Freshwater Research* (14.9, 14.10);

North Pacific Anadromous Fish Commission Bulletin (17.6b); *OCB News* (11.24); *Oceanography* (13.24, 13.25); *Oceanography and Marine Biology Annual Review* (13.7c&d, 14.12, 14.23); *Oceanologica Acrta* (16.17); *Philosophical Transactions, Royal Society of London* (8.1, 12.2, 12.4); *Physiologia Plantarum* (3.6a&b); *Proceedings of the National Academy of Sciences* (17.22); *Proceedings, Royal Society of London* (8.18); *Progress in Phycological Research* (2.5a); *Progress in Oceanography* (3.4, 3.8, 4.9, 4.8, 9.3, 9.7, 9.13, Plate 10.2, 11.37, 11.38, 11.40, 11.41, 16.24, 17.10); *Quaternary Science Review* (16.12); *Rapports et proces-verbaux, Conseil permanent, Exploration Mer* (17.9a,b&c); *Remote Sensing of Environment* (2.16); *Review of Fish Biology and Fisheries* (17.19); *San Francisco Estuary and Watershed Science* (14.11, Plates 14.1, 14.2, 14.3); *Science* (2.10, 3.1, 9.10, 10.18, 11.27, 12.3, 12.5, Plate 14.4); *South African Journal of Marine* Science (17.18); *The Veliger* (6.3); *Treatise on Geochemistry* (Elsevier) (15.1); *Trends in Biochemical Science* (15.3).

Chapter 1

Ocean ecology: some fundamental aspects

Biological oceanography could also be termed ocean ecology. The term encompasses the ecology of oceans just a short distance from the shore – perhaps from the lowest low-tide level onward, right out to the centers of the great oceanic gyres. Often, estuarine habitats are included in the study of the oceans. Oceanographers deal with questions like: what sorts of organisms inhabit different sectors and depths, and why? How is organic matter produced, by what types of "plants" (although we rarely say that word, as we will explain), and what controls their growth? Which animals constitute the herbivores and which the carnivores; and how do the carnivores locate their prey? How do the changing seasons affect the biota? What relationships prevail between organisms – from microbes to whales – and the chemical and physical character of seawater? How can worms and isopods make a living in mud beneath 4000 m of water in near-total darkness? What can we expect to harvest from the sea, and how can exploitation of fisheries or seafloor mines be achieved without damaging the resource or the habitat? How will ocean biota be affected by global climate change? Sometimes the key issues and answers to our questions come from marine biology, sometimes mostly from chemistry or physics. Fundamentally, biological oceanography straddles many disciplines, a fact which makes it a joy for the oceanographer.

Seawater

The root word in "ecology" is *oikos* (οικοσ), which is Greek for "house" or "habitat". It is the study of life in relation to its habitats, and obviously the key habitat in oceans is water – salt water. So, let us begin by considering water in some detail. The molecular structure of water, dihydrogen oxide, involves moderately strong covalent bonds between each of two hydrogen atoms sharing their single electrons with an oxygen atom. The water molecule is not linear; rather the hydrogen protons repel the overall electron shell to the far side of the oxygen atom and assume an angle of 105° from each other. Thus, the overall molecule is polar, being electropositive on the hydrogen side, and negative near the oxygen atom. This polarity creates a weaker bonding potential among the water molecules, especially in the liquid and solid phases. These *hydrogen bonds*, H-side to O-side, create a chaining effect, amounting in the liquid phase to arrays of "flickering clusters", and, in ice, to a weakly ordered crystal. As liquid water cools, the hydrogen bonds are less frequently disrupted by thermal motion, and the spatial array of more tightly bonded clusters progressively occupies less space. This means that water reaches its maximum density at 3.98°C (Caldwell 1978). However, the molecular ordering within ice is such that more space is filled by fewer molecules, so that the volume of ice is actually ~10% greater than the liquid phase at the density maximum, with the result that ice floats on water. Appropriately, much has been made of this unusual way in which water differs from comparable liquids. Lakes, for example, must cool entirely to ~4°C, becoming vertically homogeneous, before surface freezing can begin. Ocean salt water has a rather different equation of state (density being a function of temperature, salinity, and pressure), such that the temperature at which the maximum density occurs decreases with both salinity and pressure (see the data in Caldwell 1978), and overturning is not a necessary preliminary to freezing.

Biological Oceanography, Second Edition. Charles B. Miller, Patricia A. Wheeler.
© 2012 John Wiley & Sons, Ltd. Published 2012 by John Wiley & Sons, Ltd.

In addition, because of the hydrogen bonding, water has a very large *specific heat capacity* ("specific" means relative to the mass). The amount of heat required to warm a gram of water by 1°C (the specific heat) is defined as 1 calorie. The calorie is now considered to be an "archaic" unit equal to ~4.180 joules g^{-1} K^{-1}, varying somewhat with temperature and pressure (Why should anything be left as easy to remember?) which is a very large amount of energy when compared with the requirement for, say, ethanol with weaker hydrogen bonding at 0.58 calories g^{-1} °C^{-1}. This means that oceans are very slow to warm and very slow to cool, enabling currents headed poleward from the tropics to carry massive amounts of heat to high latitudes. In addition, very large amounts of heat must be added to water to force evaporation (2257 kJ kg^{-1} = 540 calories g^{-1}), and removed to allow ice formation (334 kJ kg^{-1} = 80 calories g^{-1}). For reasons that we will leave to the physical chemists, the temperature of liquid water remains fixed during freezing, at 0°C for pure water, and a few degrees lower for salt water (hence the salting of icy highways). Once frozen, ice can become even colder. Water also has a fixed boiling point at a given pressure, where the molecules escape explosively to the gaseous phase. This is 100°C at 1 atmosphere pressure. The effect of pressure on the phase transition is important in deep-sea hydrothermal vents, such that the boiling point of water at a depth of 2000 m is over 330°C. Thus, magma-heated water can emerge from the seafloor without exploding into steam. Water does evaporate into overlying air at sub-boiling temperatures, and this evaporation is more rapid when the temperature difference between the air and the water is greater. Thus, oceans, lakes, puddles, wet sand, and plant transpiration all pump water vapor into the atmosphere, leading to cloud formation, enhanced reflection of sunlight back to space, and rainfall that varies geographically, seasonally, and year to year. As is becoming obvious here, every aspect of the chemistry and physics of water is ecologically important.

The electrostatic polarity of water molecules also means that they will take on a preferential orientation adjacent to ionically bound molecules, to salts. For sodium chloride, for example, the oxygen atoms will tug on the sodium ion, and the hydrogen atoms will pull on the chloride. This tugging will be sufficient to dissociate the ionic bonds of many salts, and the water molecules will then encase the freed ions. Thus, dissolved salts will accumulate as the water flows over the land and rises through magma-heated rocks. These salts will then be transported into the sea. The sea is at the bottom of the hill, so to speak – an enormous evaporation basin in which the salts accumulate. Over sufficient time, a balance will emerge between the delivery rate and the processes that convey the salts into sedimentary structures (coastal salt-beds, manganese nodules, hydrothermal vent towers, etc.), such that the proportions of the different ions are relatively constant. Thus, the overall "salinity" is established by the remarkably constant proportions of the major dissolved ions (see Table 1.1).

All of those are termed *conservative* ions, and their proportions vary only slightly – a fact recognized by Forchhammer in 1864, but confirmed by the careful analytical work of William Dittmar (1884) with samples collected from the world's oceans on the Challenger Expedition (1873–1876). Calcium content does vary somewhat with depth, due to dissolution, under high pressure, of shells made from $CaCO_3$, and the bicarbonate content varies according to the amount of carbon dioxide in solution (the CO_2 content of the oceans is rising because seawater is absorbing the carbon dioxide generated by the burning of fossil fuels). Because of the near-constant proportions of major salts, the total salinity can be quite closely estimated by determining any one of the dissolved ions, e.g. chloride can be measured using a silver nitrate titration, or by measuring the overall electrical conductivity of the water. In modern practice, salinity of a sample is expressed as a ratio of its conductivity to that of a "standard" seawater, and is taken to have no units (the units in the ratio cancel), and is expressed on a "practical salinity scale". Salinity is often expressed simply as, say, $S = 35$, a number related to the grams of salt per kilogram of seawater, but no longer stated as such (parts per thousand). $S = 35$ is

Table 1.1 The proportions of the major dissolved ions in seawater. Total salts = 35.17 g kg^{-1} seawater.

CATIONS	IN SEAWATER (g kg^{-1})	ANIONS	IN SEAWATER (g kg^{-1})
Na$^+$	10.78	Cl$^-$	19.35
Mg^{2+}	1.28	SO$_4^{2-}$	2.71
Ca^{2+}	0.41	HCO$_3^-$	0.126
K$^+$	0.40	Br$^-$	0.067
Sr^{2+}	0.008	B(OH)$_4^{3-}$	0.026
		F$^-$	0.001

close to the overall average of ocean salinity. The upper range of S is ~40 in parts of the Red Sea. Unlike those conservative ions, others, like nitrate (NO_3^-), that are taken up by photosynthesizing algae and by bacteria, can vary *non-conservatively*. Nitrate varies from almost immeasurable amounts in the surface layers of oligotrophic central gyres to $45\,\mu M$ (micromolar) in the deep North Pacific. These μM quantities are not large enough to make the measurement of chloride or conductivity unreliable as an index of the overall mass of dissolved salts, S, although nitrate does make a measureable addition to seawater density in very deep waters in the Pacific.

Cell membranes mostly only pass salt ions through specific, energy-using, protein channels, but water passes through more freely, passing from the side with the lower solute (salt and everything else) concentration to the side with the higher solute concentration. This *osmotic* flow is actually down the gradient of water concentration. Cells and tissue fluids of much marine life, including algae and most invertebrates, are *isosmotic* with seawater. That is, solute and water concentrations are the same inside and outside their cells. Cells of freshwater plants and animals, on the other hand, must contain some salts and dissolved organic matter, so they have water pushing in through any porous cell surface. To avoid over-inflation, rupture, and death, they must steadily pump water back out. Protists have specialized organelles which do that, and metazoans have kidneys at several levels of complexity to perform this function for the body as a whole.

Fish evolved in fresh water. The impermeability of their skin and scales limits water influx to the gill membranes, which must be exposed to the water for oxygen exchange, and that lessened influx is pumped out by their efficient kidneys. When some fish colonized the estuaries and oceans (probably stepwise in that order), the problem was reversed, with water moving out through the gills. Several solutions evolved. Sharks and rays came to tolerate large tissue concentrations of urea, giving their tissues osmotic equivalence with the sea. Bony fishes developed a system of swallowing water and then excreting the salts both via the kidneys and from desalination glands on the gills. Fish that come and go between fresh water and salt water, including salmon, shad, eels, and others, must shift between these modes, in some cases (e.g. steelhead trout) back and forth many times. Many seabirds, although not impacted by the osmotic differential with seawater, must drink to replace water lost at their lungs; they eliminate the salt with glands in their nostrils. Marine mammals do not have much cell membrane exposed to water, and by and large they avoid drinking. They are very efficient at retaining water from their prey and water produced by their metabolic reactions. Their specialized kidneys manage the balance of tissue electrolytes (salts). Estuarine animals and plants living in brackish water have a variety of means for tolerating both the intermediate and highly variable osmolarity. Studies of osmoregulation support a minor research industry favored by university faculty members spending the summer at marine stations.

The covalent bonds of hydrogen to oxygen in water are labile enough that the oxygen side of one molecule occasionally pulls one of the hydrogen atoms off another, producing hydronium (H_3O^+) and hydroxyl ions (OH^-). In suitably pure water (actually rather difficult to obtain), the abundance of each is 10^{-7} molar. In solutions of acid, the acid protons form more H_3O^+, increasing its molarity to 10^{-6} or much less, and neutralizing an equivalent amount of the OH^-, reducing its molarity to 10^{-8}. In solutions of bases, the opposite happens. The balance in any given acid or base solution is given by the negative logarithm of the hydronium molarity, or pH value, which then is 7 at neutrality, 1.0 for 1 M acid and 14.0 for 1 M hydroxide. Seawater is buffered at pH values ranging in surface waters from 7.9 to 8.4 (the near-surface ocean average is ~8.1) by a combination of its carbonate and borate components, with the carbonate contributing about 95% of the buffering effect. The chemistry of the system is complex, primarily because it involves the multiple dissociations of the carbonic acid (H_2CO_3) that forms when carbon dioxide (CO_2) dissolves in water. A very large part of the total carbonate load is as bicarbonate (HCO_3^-), which can both dissociate further – acting as an acid, or take up a proton – i.e acting as a base, hence the strong buffering action. The entire system is under stress from increasing dissolution of carbon dioxide from fossil-fuel burning and other human activities, a topic to be considered later. However, the most important concern arises from the fact that the dissociation of more carbonic acid both reduces the stability of shells and coral skeletons and increases their formation costs. Organisms have some capability for internal pH management, but, as acidity increases, the energetic costs of regulation increase. The acid–base relations of seawater have been extensively and carefully studied, and therefore we will leave their description to the ocean chemists. A point to keep in mind is that the pH scale, so commonly used, is logarithmic to base 10. Thus, a change from pH 8.1 down to 7.8, which may come about, would represent a factor of two increase in hydronium-ion molarity – a very large shift indeed.

Pelagic autotrophs are small

In sharp contrast to the land, large complex plants are usually absent. Sargassum weed (*Sargassum* spp.) suspended from gas bladders in the subtropical gyre of the North Atlantic is a special and localized exception. However, it provides a model that it is a little surprising not to find everywhere; examples exist of large, floating plants, but they just are not typical. Instead, almost all of the photosynthetic organisms in the water itself, that is in *pelagic* habitats, as opposed to attached to the bottom, are small, unicellular algae known as *phytoplankton*. The word "plankton" comes from Greek (πλαγκτος) and implies a necessity to drift with the currents. Clytemnestra, in Aeschylus's *Agamemnon*, used it in denying that her thoughts were wandering (*planktos*). A classical scholar suggested the word to Victor Hensen, a founder of planktology, to describe relatively passive swimmers. Phytoplankton range in cell diameter from about 1 μm to about 70 μm, with a few representatives up to 1 mm. It is important to form a mental sense of this size range. Typical bacteria are 1 μm diameter; red blood-cells are 7 μm; an object of 50 μm is just visible to the naked eye if contrast is high. Most algal cells in the sea are at the lower end of this range. Definitions for the "size jargon" of biological oceanography are in found in Box 1.1.

Why are pelagic autotrophs so small? Biological oceanographic dogma, which will not be contradicted here, says they are small in order to provide a large surface area relative to their biomass in order to absorb nutrients like nitrate, phosphate, and iron from extremely dilute solution. Soil water in land habitats provides somewhat higher levels of nutrients (Table 1.2). The modest difference is augmented in the soil-water case, however, by rapid resupply from the closely adjacent mineral phase; nutrients do not become so thoroughly depleted in soil water. Thus, rootlets and root hairs over a small fraction of a plant's surface can supply nutrients for growth and maintenance of very large structures. In the sea, the rate of supply is limited by diffusion from dilute solution to the absorbing cell surface, so surface area must be maximized relative to cell volume. This is achieved by being small. For example, diatoms are an abundant group among the phytoplankton. Many of them are cylindrical, and if we fix the length/diameter ratio at 1, then the surface-area to volume ratio varies as 6/length, increasing strongly as size gets smaller. The surface area of a 30 μm diatom of this shape is 4241 μm^2, while that of a 15 μm one is a quarter of that, 1060 μm^2. However, the smaller one has twice the *surface area per unit volume*. Surface-to-volume (S/V) ratios of spheres vary similarly as 6/diameter. The effect of size on S/V is stronger for more elongate shapes (you can prove that to yourself by doing the calculations).

It is not surface *per se* that matters, since phytoplankton cells only cover a small fraction of their surface with transport enzymes to move nutrients from outside to inside. The importance of small size is to provide a large *relative* surface toward which diffusion can move nutrients; it is the rate of diffusion that is limiting at low concentrations. At the size scale of phytoplankton, the boundary layers (see below) next to cell surfaces in contact with the water are large relative to the cells, inhibiting fluid exchange next to the boundary. Turbulent shear is mostly at larger scales than the size of cells. Specifically, there is shearing mostly at dimensions larger than the Kolmogorov length scale, typically multiple centimeters at ocean rates of turbulent energy dissipation. Below such dimensions, viscosity dominates, and the impact of turbulence is small (Lazier & Mann 1989). Thus, effectively, the water next to a cell exchanges only slowly, and, although sinking and turbulence can increase nutrient availability at a distance from a cell,

---

Box 1.1 Plankton sizes

Several sets of prefixes have been proposed to distinguish size classes of plankton. We seem to have settled on those proposed by Sieburth *et al.* (1978).

CHARACTERISTIC LENGTH	TERM (EXAMPLES)
<0.2 μm	Femtoplankton (viruses)
0.2–2 μm	Picoplankton (bacteria, very small eukaryotes)
2–20 μm	Nanoplankton (diatoms, dinoflagellates, protozoa)
20–200 μm	Microplankton (diatoms, dinoflagellates, protozoa, copepod nauplii, etc.)
0.2–20 mm	Mesoplankton (mostly zooplankton)
2–20 cm	Macroplankton

Table 1.2 Relatively low values of major nutrient concentration in surface waters compared to natural (as opposed to fertilized) soil-water values. Units are micromoles liter^{-1} (μM).

UPPER-OCEAN CONCENTRATIONS IN WINTER	NO$_3^-$	PO$_4^{3-}$
North Atlantic subarctic	6	0.3
North Pacific subarctic	16–20	1.1
Natural soil water	5–100*	5–30**

*Soil and agricultural chemists use strange units like kg NO$_3^-$ hectare^{-1} to 20 cm soil depth. They rarely attempt to extract soil water *per se*, which is difficult because soil is relatively dry and much of the water is associated with organic matter.
**Also hard to characterize. This range came from a soil-science text, but do not put much faith in it (units were 0.05 to 3.0 ppm, a usual unit in that field). Most published data are measured in μg PO$_4^{3-}$ (g soil)$^{-1}$.

supply is effectively limited to molecular diffusion. The diffusive flux of a dissolved solute, such as nitrate, toward an absorbing surface of area A is given by Fick's Law, which Fick derived (Cussler 1984) by analogy to Fourier's Law for heat conduction:

$$\text{flux (amount arriving/time)} = -AD\, \delta C/\delta x,$$

where D is the substance-specific diffusion coefficient and $\delta C/\delta x$ is the gradient of concentration (amount/volume) away (hence the minus sign) from the surface. As stated, diffusion is slow enough that only a small fraction of the cell surface needs to be occupied by transport enzymes to acquire the specific molecules that the cell must absorb. Estimates by Berg and Purcell (1977), based on rates of diffusion and handling time per molecule, can be interpreted to imply that only a few percent of the cell surface needs to be devoted to transport enzymes for any required solute. More would not be useful, due to limitation of diffusive supply to the surface. In a sense, this is life-enabling, since *many* different solutes require a membrane transporter or at least a passage channel. Experimental data (Fig. 1.1) from Sunda and Huntsman (1997) show that, at growth-limiting concentrations of ferric iron, phytoplankton cells of all sizes have equal (the diffusion-limited maximum) rates of iron uptake per unit area. Because the iron requirement is general, and only met by the Fe^{3+} ion, sufficient areal density of transporters evolves in most (all?) species, such that uptake is, in fact, diffusion-limited, not transporter-limited. Smaller cells with less mass relative to surface area, however, receive enough iron to sustain growth when large cells are iron-limited. In addition, small, oceanic phytoplankton have evolved to require less iron per unit mass by substantial rearrangements of both photosynthetic and oxidative metabolism.

Because phytoplankton are small, they are also individually ephemeral compared to terrestrial plants or to algae attached along the shore. Grazing terrestrial animals typically take a bite from a plant, which then heals; pelagic grazers typically ingest the entire phytoplankton cell, so it is gone. Therefore, maintenance of a population of cells, a *phytoplankton stock*, depends upon their rapid reproduction. And reproduction can be rapid. Many (not all) phytoplankton can double in number one or more times per day. Thus, if grazers are few and growth conditions (light, nutrients, temperature) are good, then stocks can grow exponentially. Doubling once per day, they can increase 1000-fold in 10 days. Rapidly growing diatoms can increase twice that fast. This potentially rapid increase is the basis for phytoplankton population outbursts or "blooms", and also for harmful algal blooms. However, blooms generally do not develop at the rate that phytoplankton cells divide. There is always substantial grazing, and stock increase is generally limited to modest daily percentages. Blooms most commonly occur (where they occur) in the spring, and spring phytoplankton blooms have been and remain a central interest in biological oceanography. We will consider them in some detail (see Chapter 11), including explanations for those wide oceanic stretches where they generally do not occur.

Water is heavy and, for small particles, sticky

Water, fresh or salt, has mass, and the principal unit of mass, the gram (g) was chosen to approximate unity for a convenient volume of water, the cubic centimeter (cm^3). Thus, the density of water at 1 atmosphere pressure and 0°C is $1.0\,\text{g cm}^{-3}$. Redefinitions of units of measure have caused tiny deviations that for most purposes can be ignored. Like all substances, water expands and contracts with temperature changes, expands both above and below 4°C. Seawater, because of the changed intermolecular attractions due to the electrostatic forces from its constituent ions, does not have a similar temperature of minimum density. It contracts down to its freezing point, which is well below 0°C because of so-called colligative effects. Thus, the density of warm seawater is less than that of cold, over the global ocean temperature range from ~−2°C in the Antarctic to ~40°C. Moreover the density varies with the salinity. Finally, water is not incompressible (contrary to a commonly taught myth); at deep ocean pressures it contracts substantially.

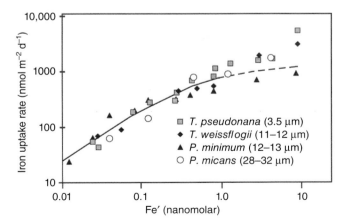

Fig. 1.1 Cellular uptake rate (per m^2 of cell surface area) of ferric iron from a culture medium (on a logarithmic scale) by phytoplankton cells of different mean diameters as listed, two species of *Thalassiosira* (diatoms) and two of *Prorocentrum* (dinoflagellates). Uptake was determined with radioactive iron as a tracer. Fe′ is a total of dissolved Fe(3+) species. Above [Fe(3+)] ≈ 0.75 nM at 20°C, ferric hydroxide precipitates, dashed curve, maintaining Fe′ at ≈0.75 nM. In the experiments, Fe′ was set by means of an iron chelating agent (After Sunda & Huntsman 1997).

Oceanographers use Greek symbols for different aspects of density, particularly σ_t for a measure of seawater's density if it is brought to the surface without heat exchange or salinity change, just decompression: $\sigma_t = 1000(\rho - 1)$, in which ρ is the actual density, usually a number like 1.02437 for which $\sigma_t = 24.37$. Thus, σ_t (sigma-t) is just shorthand for the modest but critically important variations of density due to salinity and temperature (not depth). A further refinement is often used, σ_θ (sigma-theta), accounting for adiabatic cooling from expansion (expanded, it will have the "potential temperature").

For rough calculations, the changes of density with temperature, T, and salinity, S, are:

$$\Delta\sigma_t \approx 0.20°C$$

(less useful than the approximate S effect because quite non-linear with T)

$$\Delta\sigma_t \approx 0.77 \text{ unit } S^{-1}$$

Actual density (ρ not σ_t) changes with depth (almost consistently symbolized as z), about

$$\Delta\rho = 0.0000044 \text{ g cm}^{-3}/\text{atmosphere pressure,}$$

and (again, for approximations) P increases 1 atmosphere for each 10 m of depth. Thus, at the bottom in the Marianas Trench, the density is ~1.069 g cm^{-3} (1069 kg m^{-3}). Just being in a stack adds to the stability of the ocean water column. It turns out that compression also affects the shape of organic molecules in deep-sea organisms, including bacteria, deep-diving seals, and whales. Enzymatic modulation of organic reaction rates depends upon very weak forces among atoms at the active sites of enzymes: hydrogen bonds and van der Waals' forces. Small distortions of an enzyme's shape can change the effectiveness of the bonding or bond release. Such effects become important at depth differentials around 1000 m (100 atmospheres). Thus, biochemistry and sometimes viability are affected by transfers of deep-sea fish, squid, shrimp, etc. to shipboard for experimentation. The biochemical reactions of deep-sea benthic bacteria must be studied in pressure chambers. On the whole, decompression does not tear enzymes apart, and they function again when placed back under pressure.

For precise calculation of density from conductivity (C, a measure of S), temperature (T) and pressure (D, because depth is proportional to pressure, hence "CTD") data, it is necessary to use empirical polynomial functions with extraordinary numbers of terms. For a current version, see Feistel's (2005) equation with 101 constants (many relating to sound speed, enthalpy, and other values of occasional interest) approximated to 15 decimal places.

Much of the significance of all this T–S–z detail is that the ocean is a vertical stack in which density increases downward, and the stacking is remarkably stable. Moreover, the stacking has major ecological consequences. Organization of the stack is created partly by sinking of cold, salty water near the poles: in the North Atlantic where the salty inflow of the Gulf Stream is refrigerated by frigid Arctic air, then sinks, whereas, in the Antarctic, exclusion of salt from forming sea ice into the water below adds to the density of extremely cold surface layers that also sink. These deep waters spread through the world ocean, making the deep waters cold everywhere. At the same time, the surface is heated by sunlight from above, decreasing the surface density, increasing the stability. Over the full range of depth, typically 4 km and in places 8 km or more, the compression of the water by pressure enhances the stability of the stacking. In order to open volume at depth for the sinking cold, salty water, the ocean everywhere is slowly being vertically mixed. This is most active in the upper layers driven by wind, tides, and internal waves, but must proceed at all depths. The deep limb of the circuit is (in large part) from the Norwegian and Irminger Seas to the vicinity of Drake Passage, then east across the South Atlantic and Indian Oceans and finally filling the deep Pacific. That full passage takes several thousand years. It is termed the "thermohaline circulation". Balancing the budget of sinking volume with that of upward mixing is not a simple set of measurements and is not yet accomplished. So-called internal tides provide only about half the necessary mixing energy. There has been recent interest in the possibility that stirring derived from the swimming motion of larger animals, from krill schools to whales, might provide nearly as much energy (Dewar *et al.* 2006; Visser 2007; Katija & Dabiri 2009).

Stable vertical stacking of the ocean "water column" (an essential bit of oceanographic jargon) is most significant ecologically because of the limits it sets on upward mixing of inorganic nutrients like nitrate, phosphate, and trace metals into the lighted surface layers where photosynthesis can support phytoplankton growth. The stability of stacking, the depths of the prominent pycnoclines (levels of strongest density change and, thus, most stable stratification), and the forces available to drive upward flow (upwelling) and vertical mixing, vary strongly over the world ocean, affecting the photosynthetic production potential of distinctive regions. This is a theme we shall return to repeatedly, a fundamental aspect of biological oceanography. Here, we give just one example of the density stacking and its variation with season. In discussing the variation of ocean biomes, we will consider the ecological consequences of different stacking patterns and mixing regimes. In the Atlantic north of the Gulf Stream, winter winds often mix the upper water column to below 300 m, making the profiles (Fig. 1.2; see also Fig. 11.23) of T, S, nutrients, and oxygen vertical to that depth. That is, those habitat conditions are homogeneous up and down. There is residual stratification below that depth, stratification that

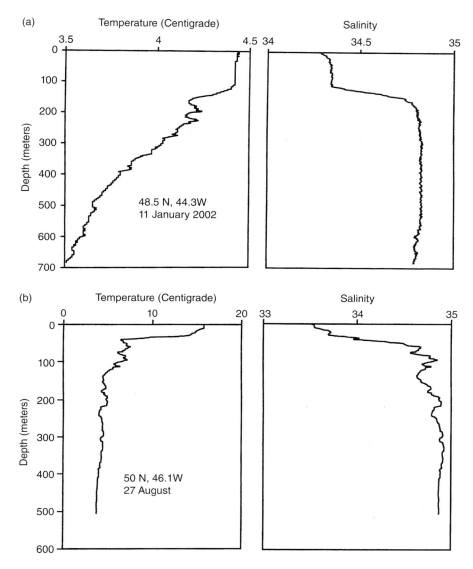

Fig. 1.2 (a) Winter and (b) summer profiles of temperature (*T*) and salinity (*S*) in the subarctic Atlantic Ocean south of Greenland. Note the differences between winter and summer scales. Summer is much more stratified than winter for both density-determining variables. Later mixing in winter from continued surface cooling and storms can homogenize the water column much deeper, to >300 m (see also Fig. 11.6). (Data from NOAA's World Ocean Database, WOD09: http://www.nodc.noaa.gov/OC5/WOD09/pr_wod09.html.)

the mixing does not overcome. Note that during mixing there is net temperature gain in the deeper reaches, net cooling above. Winds and mixing slow in the spring, and at some point solar heating warms and stabilizes an upper layer. This is set up and broken down several times by the spring alternation of calms and storms. By mid-summer there is strong stratification, primarily maintained by the elevated surface temperature above a gradient at variable depths around 35–45 m termed a seasonal thermocline. Blooms tend to occur after stratification is established at some level shallow enough to keep phytoplankton in the sunlit upper zone most of the time.

Atmospheric gases also dissolve in seawater

Nitrogen, oxygen, argon, and carbon dioxide all dissolve in seawater, and all of them obey Henry's Law: the equilibrium solubility is proportional to the partial pressure (in suitable units such as atmospheres) in the air above the water surface. The proportionality constant changes inversely with temperature, and for oxygen is approximately two-fold greater at −1°C than at 40°C. That is, saturation concentration *decreases* not quite linearly with rising temperature from 360 μmol kg⁻¹ at −1°C down to

$165\,\mu mol\,kg^{-1}$ at 40°C. Salinity reduces the saturation concentration; for example, at 0°C and salinity = 34, saturation is $351\,\mu mol\,kg^{-1}$ vs. $457\,\mu mol\,kg^{-1}$ in fresh water (the salt effect is slightly less strong at higher temperature). Units used in quantifying oxygen concentration most often now are $\mu mol\,kg^{-1}$ seawater; in an era not very long back they were $ml\,O_{2,\,STP}\,liter^{-1}$, referring to the gas volume at "standard temperature and pressure" of 0°C and 1 atmosphere. Both units (and also mass units, parts/million, partial pressures, . . .) appear in useful literature. Oxygen concentration can be measured with good precision and to low values by the Winkler titration, a series of redox reactions that are relatively simple to carry through to a sharp endpoint. It can also be measured with Clark electrodes and recently with optodes, in which luminescence from molecules embedded in an oxygen-permeable plastic is variably quenched depending upon adjacent oxygen concentration, and is measured by a light meter. Optodes have fairly long response times (many seconds), depending upon the permeability of the plastic, but the non-linear calibrations are stable over long periods and can cover the seawater concentration range.

If oxygen-depleted water is brought into the surface mixing layer and churns against the interface with the atmosphere, its oxygen content will rise toward saturation. However, the equilibration occurs relatively slowly, mostly because the mixing layer can be thick, but also because the oxygen itself has no particularly great "drive" around reaching saturation. The subjects of gas exchange coefficients and "piston velocities" are left to texts on chemical oceanography (e.g. Pilson 1998). Despite the slow oxygen-exchange rate, most deep water is "formed" at high latitudes where the water is very cold. Thus, the initial charge of oxygen for much of the subsurface ocean is close to the coldest possible saturation level.

Once oxygen is in solution, consumption by respiration of everything from bacteria to tiger sharks begins. Not all life depends upon oxygen (there are anaerobic microbes), but larger organisms with high activity levels and high metabolic rates depend upon oxidative respiration, a function primarily carried out in eukaryotic organisms by mitochondria – cell organelles specialized for this function. Down in the stratified layers well below vertical mixing from the surface, oxygen used is only replaced by the horizontal flow of water that left the surface some distance, often a very great distance, away. Thus, as organic matter sinks into these layers and is eaten and respired, the oxygen levels decrease. Additional decrease comes from animals that move between feeding near the surface and resting or hiding at depth, where those fish, squid, and plankton continue to use oxygen. Depletion generates oxygen-minimum zones in intermediate levels of the water column, particularly in the Arabian Sea and Pacific and most reduced from saturation in the Pacific toward the north (Plate 1.1).

There are also some nearly or completely anoxic layers beneath upwelling areas, particularly parts of the Peru Current and the Arabian Gulf. In the latter sites, dissolved oxygen as O_2 can be completely exhausted. Microbes in the anoxic layers continue to respire, at least to a point, using the oxygen atoms in nitrate and sulfate ions. The later activity releases sulfide, S^-, which is toxic to other life-forms and readily detected by odor in water collections. The limits of toleration for hypoxia among aerobic animals are variable among animal groups, species, and individuals. Mortality rates rise in bivalve mollusks and brittle stars at $[O_2] \sim 1\,ml\,liter^{-1}$ ($45\,\mu mol\,kg^{-1}$) with sharply greater mortality (often after emergence from the sediment) at $<0.5\,ml\,liter^{-1}$ (Diaz & Rosenberg 1995). Similarly low oxygen levels cause death or emigration in many other groups. Some pelagic animals that feed near the sea surface can migrate into very hypoxic, even anoxic, zones in order to rest and hide. An example is the Humboldt squid (or jumbo squid, *Dosidicus gigas*) of the eastern tropical Pacific and, lately, the California Current. It is equipped with very finely divided filaments in a very large gill for uptake of dilute oxygen. *Dosidicus* also has extreme capability for prolonged survival by anoxic metabolism, generating an "oxygen debt" that suitably adapted physiology (and the high exchange capability of the gills) relieves very quickly on return to the surface. In addition, the squids and midwater fishes that do spend time in such layers move very, very slowly. Movies of predator-attack and prey-escape events look like animal tai-chi exercises.

In recent decades, hypoxic and anoxic zones have appeared in many coastal areas, particularly offshore of major rivers like the Mississippi, Rhine, and Chang Jiang. This has been attributed to eutrophication near shore by agricultural nutrients washing into the coastal zone, greatly increasing algal production and subsequent oxygen-consuming decay above the seabed. Such coastal anoxia kills fish and benthos. Kills of continental shelf fauna by hypoxia have also been observed in recent times in areas where no obvious anthropogenic eutrophication has occurred. Off coastal Oregon, USA, recurring hypoxic episodes during the 2000s, some involving die-offs of fish and benthos, have been attributed to greater onshore transport of oxygen-depleted water from oceanic oxygen-minimum zones. Because more organic matter has been oxidized in such waters, they also contain more nutrients, which may have enhanced production of organic matter that then increases oxygen demand and depletion. In the Oregon case, a change in cycling between upwelling and relaxation events may have reduced the flushing of bottom layers with oxygenated water. Ocean ecology can be complex, and explanations of events can be both fuzzy and uncertain. Processes occurring on just one stretch of coast, or in just one fjord, may be more important there than the interactions that typically determine conditions and ecological relationships.

The types and importance of fluid drag

The mass density of seawater (any water) has other ecological effects, particularly the requirement for force to accelerate it aside during swimming. This force requirement produces what is known as *inertial drag*. It is the dominant resistance to sinking through water by large, dense objects and to swimming forward by larger, faster animals. There is another source of drag, which is the requirement for force to rearrange the intermolecular connections among water molecules in order to move through them. That is called *viscous drag*. The relative importance of inertial and viscous drag is expressed as a ratio, the Reynolds number, Re, which has the product of factors proportional to inertial drag in the numerator and the water viscosity in the denominator. Inertial drag is proportional to the linear size (l) of the moving body, often best chosen as the longest dimension perpendicular to the path, to the velocity (v) relative to the water, and to the water density (ρ): $lv\rho$. The viscosity (as discussed here, the *dynamic viscosity*) is the molecular resistance to shearing forces, symbolized μ (or often η), with SI units of Pascal·s (Pa·s) = Newton·s m^{-2}. Work with the units here. After some conversions, it will become apparent that those of the Reynold's number numerator (m, m s^{-1} and kg m^{-3}) and denominator cancel. Re is a dimensionless number.

Experimental work (also some theory) shows that at high Re, >~100, viscosity can be neglected in drag calculations, at least for processes like swimming, because inertial effects are so dominant. At Re less than ~1, inertial effects are small and viscous effects dominate. Algal cells, other protists, and many smaller metazoans like clam larvae or copepod nauplii, live in an apparently very viscous world, because both their l and v values are small. This has important effects on the mechanics of swimming and of approaching nearby food particles. The viscosity of water (and seawater, the effect of salt is small) varies not quite linearly with temperature, from ~0.65 mPa·s (milliPascal·s) at 40°C to ~1.8 mPa·s at 0°C. This difference approximately triples the work that ciliated or flagellated protists must do to move at 0°C compared to 40°C.

So, how does swimming work? When drag is principally inertial, so is the force exertion of an animal against water. A fin or tail sweeps through the water at an angle to the intended trajectory, and pushes a mass of water backward. There is an equal reaction on the mass of the fish or seal moving it forward. There are often elegant details. For example, a tuna that can swim at ~20 m s^{-1} has an ideal fusiform shape, minimizing the distance that water must be accelerated to the side and then back to the center line behind as it passes through the water. It has scales along its tail peduncle that lie flat during initial acceleration, and then extend out to initiate turbulence at intermediate speeds. That is useful because drag actually drops sub-

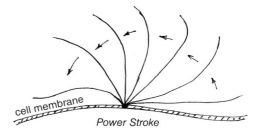

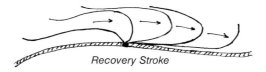

Fig. 1.3 Successive positions in the power and recovery strokes of a cilium.

stantially at the transition from smooth laminar flow along the skin surface to turbulent churning. To avoid drag from laterally extended pectoral and dorsal fins, tunas can pull them flat against the body into precisely fitted grooves.

Swimming by ciliated and flagellated cells is substantially different (Purcell 1977). Mass moved behind by a flagellum stroke is so small that the forward reaction is ineffective. But motion is achieved from the differential in the amount of intermolecular attractions that must be spatially rearranged between sliding a tube (a flagellum, say) lengthwise versus sideways through water. For reasons that may or may not be intuitively obvious (depending upon your brain wiring), the difference is ~1.7-fold. So, a cilium can be pushed backward perpendicular to the direction of motion, and then pulled back forward with most of the motion parallel to the trajectory (Fig. 1.3).

In the return stroke, the cilium's flexibility completes the motion as a rolling curl along the length. The viscous drag differential shoves the cell ahead. An alternate mode of exploiting the difference between along-shaft and across-shaft drag is use of flagellae wound into spirals. When rotated relative to the cell, often by molecular-scale rotary "motors", the vector component of the motion that is perpendicular to the axis of the spiral is sufficient to drive the cell steadily forward. There are many features of swimming (or filter feeding for that matter) in the viscous mode that are unexpected based on our experiences of swimming. When dominant, viscous drag is sufficiently powerful that there are no eddies shed aft or to the side. There is no inertial "carry" comparable to the long forward drift you experience after a swimming stroke. Rather, when force is not being exerted, there is remarkably close to zero continued forward motion; the stop is effectively instantaneous.

These two modes of swimming appear to be similar in some respects. Both ciliary motion and propulsion by lateral fins involve pushing back with a high drag element, then sliding it back forward by feathering or switching angles. The sources of the effective drag force, however, are substantially different. Some plankton, most particularly copepods, exploit the interface between drag dominated by inertia and drag dominated by viscosity. Because they are of intermediate size, with $l = 0.1$ to 1.0 cm, they can accelerate to sufficient v to enter the realm of inertial swimming (raising the Re), using paddle-like feet and tail-fan sweeps to accelerate to achieve extremely fast relative speeds (hundreds of body lengths s^{-1}). However, when power strokes are stopped, drag rapidly drops into the viscous range, with the advantage that at rest the sinking rate is very slow despite a moderate excess of density over the surrounding water. Little work is required to maintain a vertical location.

That brings us to sinking rates. A cannonball that missed its target (most of them did) would accelerate downward until inertial drag equaled the gravitational force attracting it, and then it proceeded to the bottom at that substantial terminal velocity (>100 m min^{-1}). The effective mass would, of course, be reduced by buoyancy from the water, the differential density determining the "effective" mass. Thus, a sufficiently hollow, perhaps aluminum, cannonball might move up not down after splashing in. The size of the cannonball makes only a miniscule difference. Sinking of a tiny fecal pellet from a zooplankter, partly filled with dense opal from diatom shells, will be affected primarily by viscous drag, and for a spherical fecal pellet the sinking velocity, V_s, is given by Stokes's Law:

$$V_s = \frac{2}{9} \frac{(\rho_p - \rho_f)}{\mu} g R^2,$$

in which g is gravitational acceleration, ρ_p and ρ_f are the densities respectively of the pellet and the fluid, R is the pellet radius, and μ is the dynamic viscosity. A modest difference from this depends upon the shape of the pellet, but use of an equivalent diameter of a sphere of the same volume will give a decent approximation. Care with units is required (!), but left to your attention. Notice that the larger the particle, the faster it sinks, with V_s varying with the square of the linear dimensions. If ρ_p is less than ρ_f, then the particle will rise. Consider the impact, mentioned above, of temperature on μ: a particle of ρ_p will sink about three times faster at 40°C than at 0°C, despite the effect of T on ρ_f. Stokes's Law is a simplified (viscous drag only) version of the Navier–Stokes's equation, the version of Newton's acceleration law $F = ma$, to which hydrodynamicists have given lifetimes of thought and a googol (10^{100}) of computer calculations.

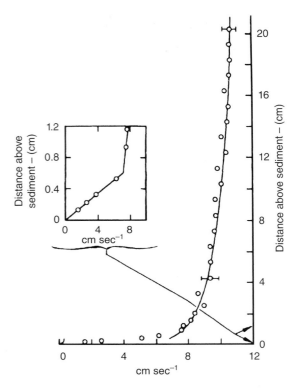

Fig. 1.4 A vertical profile of average velocity upward from a sandy-silt seabed at 199 m through the viscous boundary layer, in which velocity increases linearly with distance above bottom (inset), and then on upward (larger graph) through the "buffer layer". In the latter the velocity approaches that at a distance, ~10.5 cm s^{-1} here, within about 20 cm. Measurements were with a heated-thermistor velocity probe by Caldwell and Chriss (1979). Very mild turbulence (±0.5 cm s^{-1} at 20 cm, much less in the viscous layer; see Chriss & Caldwell 1984) has been averaged out.

Drag effects take on special characteristics at boundaries between water and solid surfaces, even soft ones like jellyfish skin or algal cell membranes. The fluid alongside, except in shear regimes strong enough to induce cavitation, remains stuck to the surface, so that exactly at the surface there is no relative motion. This is the "no-slip" condition. Velocity relative to the solid surface, the scales of a fish, say, increases away from the surface, reaching the full background relative velocity asymptotically at a considerable distance out. The zone of outward acceleration (Fig. 1.4) is called the fluid "boundary layer". As the figure shows, very close to a surface, up to about 1 cm but varying with the distant velocity, the local velocity increases linearly outward because viscous effects are dominant, and in this range viscosity damps turbulence in the flow. Away 0.5 to 1 cm, velocity increases more rapidly as inertial effects become more important, reaching an asymptote several dm out, or, in the case of the seabed, several dm up. Velocity in this range plots linearly vs. a logarithmic scale of distance from the surface. With distance from a surface, the potential for turbulence in the flow increases.

Boundary layers have many ecological effects that are well reviewed by Mann and Lazier (2006). They require that a swimming animal, particularly a small one, effectively must push along a mass somewhat larger than itself. It adds further dominance to the role of molecular diffusion for final transport of molecules to gills and cell surfaces. It means that small hairs side-by-side in a palisade, such as cilia on a ctenophore or setae on a euphausiid's (krill) leg, will have intersecting boundary layers and effectively form a solid paddle. Similarly, animals filter-feeding with setal or mucus meshes must generate substantial pressure to force water through their webs by narrowing the boundary layer of each strand. Boundary layers are less extensive at greater relative velocity. Because of boundary layers, drag tending to tip over benthic animals extending up into passing currents or to pull them out of the sediment is much reduced. As already stated, they mean that the supply of nutrients to algal cells depends upon molecular diffusivity, that is on the background concentrations, the potential cell-surface uptake rates and the solute-specific diffusivity constants. This list of boundary-layer effects is far from exhaustive.

For an extensive discussion of hydrodynamic effects on biological processes, refer to Steven Vogel's (1996) book *Life in Moving Fluids*.

Effects from having sun above, water below

Ocean water is held against the Earth by gravity, filling basins and with surfaces almost parallel to the so-called geoid, parallel apart from mild, long-range slopes created by the dynamics of flow on the curved and rotating form of the planet and, of course, except for surface waves. These sheets of water, thin relative to the Earth's diameter, are thus illuminated from above by sunlight and on some nights by moonlight. Light that is not reflected back into the atmosphere (and in part back into space) is progressively absorbed by the water and by both dissolved and particulate substances. Absorption increases with depth (z), following Beer's Law: $dE/dz = -kE$, for which the solution is $E_z = E_0 e^{-kz}$. That is, irradiance, E, declines exponentially with depth. The constant, k, an extinction rate for the overall spectrum of sunlight, has a value of $0.067\,m^{-1}$ for just seawater, and actual values in oligotrophic subtropical gyres are remarkably close to that when chlorophyll concentrations are on the order of $0.05\,\mu g\,liter^{-1}$ or less. More pigment-containing phytoplankton or more suspended sediment increase k, shoaling the levels reached by specific levels of irradiance. However, k also varies with the wavelengths of light. The wavelength of maximum transmission in pure water and in clear oceanic waters is around 435 nm (blue). Other wavelengths are more rapidly stripped out, eventually leaving only blue, with the only color vision

distinguishing shades of blue below about 100 m, and systems of photosynthetic and visual pigments must absorb near 435 nm. They do, mostly shifted toward the green at 465 nm (the extinction rate is almost constant from 410 to 475 nm). In neritic regions the inclusion of larger amounts of colored, dissolved organic substances (yellow transmitting, termed Gelbstoff or gilvin) and of phytoplankton (green transmitting) causes a shift in the wavelength of maximum transmissivity toward the green. Absorbance rapidly increases for longer wavelengths.

Actually all absorption of light by water has a minimum with respect to wavelength, a "window of clarity" (Yentsch 1980), right at the peak range of wavelengths of solar irradiance (Fig. 1.5) the wavelengths of visible and photosynthetically active light. This match of window and available light is one of the remarkable coincidences that make life on Earth possible. The coincidence of the solar spectrum to the window of water clarity allowed selective tuning of the light-absorbing pigments energizing photosynthesis by phytoplankton to the blue-dominated spectrum of light available at even moderate depths. The only light reaching depths below 100 m or so peaks very narrowly in the blue, so visual pigments of deep-sea fish and invertebrates (shrimp, squid) are adapted for generation of visual nerve impulses by absorption of those specific wavelengths.

The ocean layer that is sufficiently illuminated to support positive net photosynthesis, meaning more organic-matter generation than phytoplankton will respire themselves, is often considered to extend down to about the level receiving 1% of mid-day irradiance, and is termed the *euphotic zone*. It is not fully dark below that depth, and in clear tropical waters net photosynthesis may extend somewhat deeper, reaching to 120 m or so. In waters of a natural ecosystem, much additional absorbance comes from dispersed cells containing pigments, shoaling the euphotic zone depth. The effect of pigments is roughly proportional to chlorophyll concentration; chlorophyll will add about $0.02\,(m^{-1})$ to the absorbance coefficient in the blue (otherwise the absorbance minimum) for each $1\,\mu g\,liter^{-1}$. An extended treatment on the effects of pigments on the absorbance spectrum can be found in Morel (1991).

Most bioluminescence also has a narrow spectrum in the vicinity of 465 nm. In the deep sea, that makes possible matching by photophores on the undersides of fish, squid, and shrimp of downwelling irradiance to obliterate silhouettes that might be spotted by animals looking upward from below. In other applications, such as signaling between individuals, it would allow transmission of "messages" to the maximum possible distance. More will be said about the interaction of light with photosynthetic pigments as a function of depth when we discuss primary production. More will be said about deep vision and bioluminescence when we discuss mesopelagic habitats. The limits of water clarity (Fig. 1.5) are also important, both eliminating almost

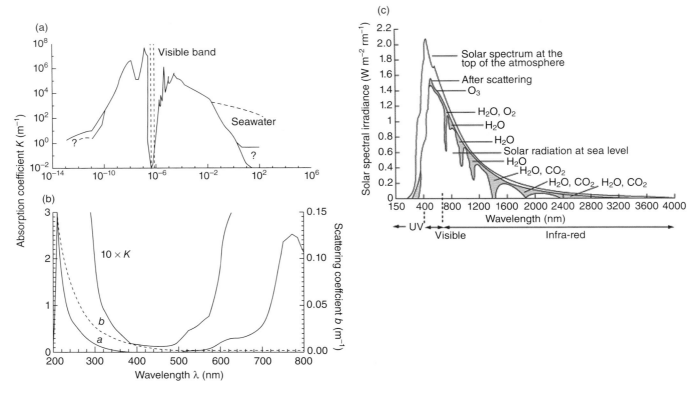

Fig. 1.5 (a) Light absorbance coefficients, k, of pure water (solid line) and seawater (dotted line) as a function of wavelength, showing a window of clarity around the visible band. (Data compiled from various sources.) (b) Detail of absorption (k) and scattering (b) spectra of seawater in the visible band plus near UV and near IR (data of Smith & Baker 1981). (c) Spectrum of radiative solar energy arriving at the surfaces of the atmosphere and ocean, with the differences labeled according to the principal absorbing gases in the atmosphere accounting for the difference. (Repeatedly published without attribution. Reproduced here from Falkowski & Raven 2007.)

all of the UV irradiance capable of damaging organic molecules (like DNA, for example) in the first few meters and reducing escape of warmth back to the atmosphere at infrared wavelengths.

Of course, the sun does not shine down from exactly the zenith on more than a very small part of the Earth at any time and there, in the tropics, only at local noon. The rest of the time, sunlight passes through the atmosphere to the surface at substantial angles that vary seasonally and through the day. Moreover, the familiar seasonal cycles of daylength, that are accompanied by changes in height of the sun above the horizon, change the depth of light penetration, duration of illumination for vision, the irradiance available in the day for photosynthesis, and the amount of surface layer heating. The lower the sun angle, the greater the area over which its light is spread, reducing the arriving photons per unit area. Many units are used in irradiance measurement. Those most commonly seen in oceanographic literature are watts m^{-2} (a measure of power, that is irradiance) moles of photons arriving at a one square meter surface per second (with energy or power depending upon wavelength). "Photosynthetically active radiation" (PAR), which will be discussed together with photosynthesis, is often determined in μmol photons s^{-1} m^{-2}. The equivalent unit μeinsteins s^{-1} m^{-2} is also in common use.

A key impact of this vertical arrangement, the sun heating the ocean from above, is the installation of relative buoyancy at the sea surface, adding stability to the stratification of the water column. Stable stratification limits vertical mixing by winds and tidal energy, and reduces the upward transfer of dissolved nutrients. Daily and seasonal variation in stratification and vertical exchange are key aspects of ecological processes in the sea.

Taking the individual or event-by-event viewpoint

A good deal of interest is expended in biological oceanography on bulk rates and quantities: the amount of photosynthesis occurring under a square meter of ocean surface, the biomass of zooplankton (mg m^{-3} or g m^{-2}) and its seasonal variations, the rate of downward "rain" of particulate organic matter into the deep sea, and more; the list is extensive. However, from the viewpoint of one predacious arrow worm (chaetognath), what matters are the potential for and rates of encounters with prey organisms (in their case mostly copepods) or with other chaetognaths of the same species that are also ready for mating. From the viewpoint (although it doesn't "look" *per se*) of a nitrogen-limited algal cell, the key to its growth potential is the likelihood that an ammonium or nitrate molecule will come adjacent to its cell membrane, that a ferric ion will diffuse near enough to bring it on board to act as a cofactor for nitrate reductase (to convert nitrate to ammonium). When the encounters needed for life processes are not occurring fast enough, there will be no photosynthesis, no food, no growth, no reproduction, no something, and ecosystem function will wind down. Looked at in this way, what matters are the event rates, and those depend in the first order upon the product of the concentrations of the two entities that must meet for an ecological interaction. Consider, for example, mating encounters by zooplankters. Copepods are the dominant small crustaceans in the sea, and they are dioecious (male and female functions in separate individuals). So, the probability of a mating encounter in an interval can be written

$$P_m = \beta \text{ [males][females]},$$

in which brackets indicate volume concentrations and β is termed an "encounter kernel". That terminology has been developed extensively, and with many examples, in a book by Thomas Kiørboe (2008). While he deals with plankton, as his title implies, the viewpoint can apply anywhere in marine (or any) ecology, including in the benthos.

A great deal of complexity can enter into establishing the value of β, by which we mean both establishing it in reality, for the organisms in the field, and in estimation of it by ecologists. We can think about, observe, and experiment upon the component factors affecting β, but in most cases we will not be able to measure every significant aspect of the encounter situation, especially not in the ocean where we inevitably remain rather clumsy observers. We are not even very good at measuring the effective concentrations. Of course, organisms from bacteria to whales are very good at raising their own concentrations at spots with high concentrations of the molecules, prey, or mates they need to encounter. Some of them are also very good at dispersing away from high concentrations of their predators, or only visiting those sites when the predators are somehow disabled, perhaps for example too dark for them to see. Suitable concentrating and avoidance behaviors are among the most obvious products of natural selection. Despite the difficulties, studying what matters to individual organisms in obtaining the encounters they require gener-

ates some of our best insights. As Kiørboe has shown, trying to find explicit functions to quantify β can be a fruitful mode of research. Among many variables, the relative motions of the encountering individuals (from molecules to whales) are often the most critical components of β. Signaling can also be important, as in the case of copepod mating when the female lays out tracks of attractant pheromones to alert males to a mating opportunity. Thus, in many cases β involves increasing apparent individual size, effectively a modification of volume concentration. The range of possibilities exploited in nature is wide. Sometimes attempts to apply this viewpoint will be explicit in the following chapters; just as often it will be an implicit alternative you can apply in thinking about what matters to life in the oceans.

General terminology for habitat partitions in the ocean

Habitats within the water column are termed *pelagic*, and seafloor habitats are termed *benthic*. Organisms living in pelagic zones are *plankton* (defined above) and *nekton* (from ηεκτος), large animals that swim well enough to move independently from displacements by horizontal currents. Successive pelagic layers downward are termed the *epipelagic*, *mesopelagic*, *bathypelagic* and *hadopelagic* zones. Respectively, they extend to about 200 m, to 1000 or 1200 m, to perhaps 4500 m and to the bottoms of the deepest trenches. The mesopelagic has enough light for useful vision, which strongly conditions life within it. Below that solar photons are uselessly rare and for some reason even bioluminescence is mostly removed from the adaptive repertoires of bathypelagic animals. Layers below the upper epipelagic, which is the euphotic zone, all depend upon downward transfer of food as sinking particles and vertically migrating animals.

Organisms living on the bottom are termed *benthos*. Benthic (the adjective) and benthos apparently are versions of the Greek "bathos" (βαθος), meaning depth. Benthic habitats share characteristics with both pelagic and terrestrial ones. They are (more-or-less) solid substrates, like the land, but they are continuously submerged in seawater. Thus, the basic physiological problems are the same as those for pelagic ocean life, but the two-dimensional aspect (at least relatively thin vertically) of a land habitat operates as well. Benthic habitats also grade downward in a series: *intertidal*, *subtidal* (shallow bottoms near shore), *bathyal* (continental slope depths), *abyssal* and *hadal* (trenches).

The solid Earth has two principal surfaces, the continental shields above sea level, and mostly at the level of steppe or lowland rainforest (~300 m elevation), and the abyssal

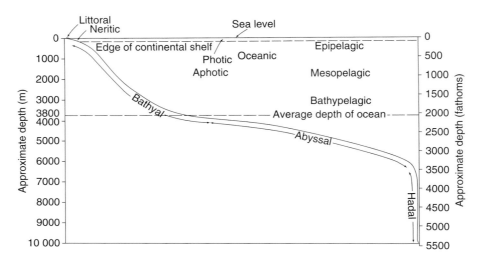

Fig. 1.6 Hypsographic curve for the world's oceans. Length along the abscissa is proportional to the area occupied by a given depth in the world's oceans. For example, depths in the abyssal category occupy about 60% of the seafloor (After Hedgpeth 1957).

plains at depths of about 4500 m. This deep-sea zone occupies about 60% of world ocean area (Fig. 1.6).

There are some rocky deep-sea sites, particularly at spreading centers, but most of the ocean is underlain by sediment-covered bottom 2000 to 5500 m below the surface productive layer. The extent to which both pelagic and benthic habitats have been studied with respect to the organisms living in them and their ecology declines rapidly as depth increases. While deep habitats are difficult to access and sample, they have been studied and much is known about them.

x	10^x	2^x
0	1	1
1	10	2
2	100	4
3	1,000	8
4	10,000	16
5	100,000	32

A few mathematical reminders

I The exponential function

This is, of course, basic mathematics, but, to refresh the concepts, we provide these notes. The exponential function appears repeatedly in biological oceanographic discourse. It appears repeatedly throughout science. It appears in analytical chemistry as Beer's Law, discussed above – the relationship between light absorption by a not quite fully transparent medium of transmission and the length of the absorbing column of medium. It appears in population dynamics, nuclear decay theory, everywhere.

Any function of the form $y = a^x$ can be termed exponential. We use the values of 10^x where x is an integer to give the place values in our usual number system:

However, when the term "exponential function" is invoked, the meaning frequently is the sequence of values for $y = e^x$, where "e" is the irrational number 2.71828. . . . This particular exponential function has the special property that the slope of the function (the change in y divided by the change in x, or dy/dx) at $x = 0$ is 1, *and* that the [slope of e^x] $= de^x/dx = e^x$ at all x. This function turns out to be (yes, lots of mathematics is hidden in that phrase) the exact relationship for any compound-interest problem when bank interest is compounded continually. The important thing isn't the compounding interval, but the interest rate (however much banks may try to convince you otherwise). Let's try an example. Let the interest rate, r, be 8% per year. If interest is compounded once per year, then the principal at T years $P_T = P_o(1 + r)^T$. If it is compounded n times per year, then it is $P_T = P_o(1 + r/n)^{nT}$. If it is compounded continuously, then the principal at T is $P_T = P_o e^{rT}$.

Let P_o be $1000 (or yen, or rubles, or euros):

	Values after T years of compounding interest N times per year				
T	$N=1$	2	3	4	∞
0	1,000	1,000	1,000	1,000	1,000
1	1,080	1,082	1,082	1,082.4	1,083.3
2	1,166	1,170	1,171	1,171.7	1,173.5
3	1,260	1,265	1,267	1,268.2	1,271.2
4	1,360	1,369	1,371	1,372.8	1,377.1

Even continuous compounding doesn't get you much. In fact, a change from 8% to 8.3%, i.e. a small change in the annual rate, is all it takes to cover all the possible effect of more frequent compounding. This continuous compounding is an excellent model for many processes such as the decline in concentration of phytoplankton when animals are filtering parts of the water and returning the water to the suspending volume.

Let us expand slightly on the mathematics for downward light extinction; the results are so important that immediate review is in order. The continuous-interest formula (or exponential function) turns out to be the solution (or integral) for an equation for the slope at any point of the curve of light intensity vs. depth. Such equations are differential equations, and their solutions when they have simple ones are always functions, such as the exponential. The absolute amount of change in the light between two depths depends upon: (i) the amount of light there is to be absorbed, E (principal); (ii) the fractional rate at which it is absorbed per meter, k (in this case a rate for negative interest); and (iii) the thickness of the absorbing layer, z ("time" at interest). We write:

$$dE/dz = -kE.$$

Differential equations of this sort are called "first order" (they involve first derivatives), "separable" differential equations (using this as an example, they can be rearranged to have dE and all functions of E on one side and dz and all functions of z on the other). They are "solved" by rearranging, then integrating:

$$\frac{dE}{dz} = -kE \Rightarrow \frac{dE}{E} = -kdz \Rightarrow \int_{surface}^{z} \frac{dE}{E} = -k\int_{0}^{z} dz$$

The integral of dE/E is natural log E, $\ln E$. The integral of dz is z. Thus, the integrals become:

$$\int_{surface}^{z} \frac{dE}{E} = -k\int_{0}^{z} dz \Rightarrow \ln E \Big|_{0}^{z} = -kz \Big|_{0}^{z}$$

And, finally, taking differences and antilogarithms, the "solution" is $E_z = E_0 e^{-kz}$, in which E_z is the intensity remaining at depth z relative to the just-below-surface intensity of E_0.

Populations above a reasonably small size grow exponentially, statistically exactly so if reproduction isn't synchronized in some way. When it is, then the exponential pattern appears for counts at equal intervals measured in reproductive cycles. If we use N for population numbers, then we get $N_t = N_0 e^{rt}$, and we talk of "r" as the rate of population increase (or decrease if its sign is negative). Both birth (b) and death (d) can occur as exponential functions, so we can write: $r = (b - d)$. If $N(t)$ and b are known, then you can solve for d, or conversely.

Examples of the exponential function will continue to appear in this book. Please practice with it, using a calculator. Get a very clear understanding of its characteristics.

Problem 1: It is desired to get an idea of the growth rate of some phytoplankton cells. A few are inoculated into a jar of sterilized seawater enriched with various fertilizer compounds (nitrate, phosphate, etc.). At the end of two days there were 200 cells per ml. At the end of four days there were 800. What is the exponential rate of growth? What is the doubling time? What is the formula relating doubling time to the exponential rate of growth?

Problem 2: A "simple" extension of the exponential function is the logistic equation, often used to characterize the increase of a population up to the limiting carrying capacity of its habitat. The logistic represents that by reducing the natural rate of increase, r, according to the fraction of the carrying capacity remaining:

Unlimited exponential increase	Logistic model
$dN/dt = rN$	$dN/dt = r(1 - N/K)N,$

where K is the carrying capacity, and N/K is the fraction of the "resource space" used. Find the solution to this equation. Serious mathematics students will determine the integral. Hint: it's a straightforward, first-order, separable differential equation. Review integration by parts. Others should not be embarrassed to use an integral table.

II Limiting factors

The notion of limiting factors is often traced to a German agricultural chemist named Baron Justus von Liebig (1803–1873). He was one of the early organic chemists, and he worked on the elemental content of plants in order to design effective fertilizers (Moulton 1942). One of his famous experiments was growing a plant in a pot using a known weight of soil. By later separating the plant and soil, he was able to show that the plant was made up of something other than constituents of the soil – invoking a conservation law of a sort. He was able to show that the plant was derived from water and air. His own statement of the concept of a limiting factor is now termed "Liebig's Law of the Minimum":

> ". . . growth of a plant is dependent upon the amount of the food stuff which is presented to it in minimum quantity."

We would add the qualifying phrase, "in proportion to its need for it". Note the singularity of this limiting factor; plural, interacting factors are not mentioned. The importance of interaction among potential limiting factors remains an issue of debate in ecology.

There are many examples of limiting factors as they affect living things. The characterizing signature of an analysis of a limiting factor is a hyperbolic function. Classic examples are the rates at which fish grow on different levels of feeding. In general, food eaten and growth both follow such hyperbolic patterns as food becomes more readily available. Thus, food availability is said to be a factor limiting growth. At the asymptote, other factors, including the intrinsic capacity for growth, become limiting. Hyperbolic relationships play a large part in marine ecology, and we use several functional forms to represent them in our models. Popular ones include:

1 Two linear segments meeting above the point of maximum curvature;
2 the Michaelis–Menten curve (Fig. 1.7) from enzyme kinetics (also known as the Monod function); and
3 the Ivlev equation.

In some instances there is little to choose among these representations, since the scatter of the data is usually great. The choice is made on the basis of convenience to the application. We will develop the list above so you have a reference. Two linear segments can usually be fitted by eye. These will represent the two basic parameters of the relation: the asymptotic growth rate and the slope of the initial response to increase of the limiting factor.

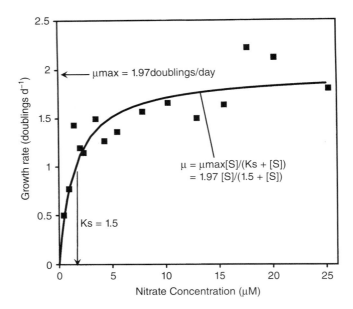

Fig. 1.7 Growth rate of the diatom *Asterionella japonica* as a function of available nitrate (squares), fitted by the hyperbolic Michaelis–Menten relationship. (Data and curve parameters from Eppley & Thomas 1969.)

The Michaelis–Menten equation is borrowed from biochemistry. Enzymes are often characterized in terms of their reaction kinetics. The data are measures of reaction velocity at various substrate concentrations. Reaction rate takes the form shown in Fig. 1.7.

Suppose there are a number of active sites on an enzyme to which a substrate can bind, and let the concentration of those sites be $[E]$. Binding is the slowest (limiting) step in conversion of substrate to product. The reaction is $E + S \leftrightarrow ES \rightarrow$ Product. The dissociation constant for the enzyme–substrate complex, ES, is $k_s = ([E] - [ES])[S]/[ES]$. Solving for $[ES]$:

$$[ES] = [E][S]/(k_s + [S])$$

Since ES will be transformed to product at a rate proportional to $[ES]$, we have

$$V = \text{Reaction Rate} = c[ES] = c[E][S]/(k_s + [S]),$$

where c is a proportionality constant. The maximal rate, V_{max}, will be attained when $[ES] = [E]$, that is when $V_{max} = c[E]$. Substituting, we have the Michaelis–Menten relation:

$$V = V_{max}[S]/(k_s + [S])$$

The graph of V vs. $[S]$ is hyperbolic with asymptote V_{max}. When $V/V_{max} = 0.5$, then $[S] = k_s$. Therefore, k_s is called the "half-saturation constant". It can readily be determined, and it is often used as a simple measure of enzyme substrate affinity. It can be used to characterize the slope of the initial portion of an ecological relationship of hyperbolic form. Note that k_s is a "backwards" variable: high k_s values denote *low* affinity, or slow approach to saturation.

The Ivlev (1945) approach is to suppose that with a great plethora of food, animals will eat some amount R_{max}, the maximum ration, and no more. At lesser food abundance, they will eat fractionally less. The maximum ration is approached asymptotically. The resulting equation is:

$$\text{Ration} = R_{max}(1 - e^{-\lambda \rho}),$$

where food density is ρ and λ is a constant, the Ivlev constant. To derive this, differentiate with respect to ρ, and then establish an argument for assumptions leading to the resulting differential equation. It is just as useful to simply examine the approach of Ration to R_{max} as ρ increases (the limit of $e^{-\lambda \rho} \to 0$ as $\rho \to \infty$).

Other functions for hyperbolic relationships are in use, and some will appear later in the text, for example the hyperbolic tangent function recommended by Jassby and Platt (1976) to characterize the increase of photosynthetic rate toward an asymptote with increasing available irradiance. When hyperbolic ecological or physiological data are strongly variable, it may not matter which deterministic function is used to represent the central tendency of responses to some forcing variable. The best choice may depend upon mathematical convenience, say in a numerical model, not on precision of fit.

Threshold effects are frequent in ecological relationships. It is sometimes found, for example, that animals won't feed at all unless there is more than some minimum of available food. This minimum is a threshold. Threshold effects are readily added to either the Michaelis–Menten (here restated in terms of food) or Ivlev equations:

$$R = R_{max}(\rho - \rho_t)/(D + \rho - \rho_t), \text{ and}$$
$$R = R_{max}(1 - \exp[-\lambda(\rho - \rho_t)]\}.$$

In both, ρ_t is a threshold food abundance for feeding. Both of these equations must be applied only where $R \geq 0$; that is, where $\rho \geq \rho_t$ use the equation, otherwise $R = 0$. Failure to follow this restriction (as in computer code for ecosystem models) will induce "negative" ingestion, nutrient uptake, . . . , all of which have unrealistic stabilizing effects on modeled ecological interactions.

Limiting factors are usually thought of as material or energetic requirements that an organism must draw from the habitat. However, the response to varying levels of limiting factor availability may be modified by other factors like temperature, salinity, ultraviolet radiation, frequency of large rocks in the path and endlessly on. For animals, growth rate (for copepods as an example, see Vidal 1980) varies not only with resources and conditions, but with life stage and growth already completed (body size). Growth rate does vary with food availability in the expected hyperbolic fashion, but, at least for small ectotherms, the height of the asymptote drops with increasing temperature. At higher temperatures they have greater metabolic costs, leaving less nutriment to support growth. In sum, control of most processes depends upon many variables in the habitat.

Deterministic functions (and models) vs. real data

Recall that "functions" defined mathematically assign *one* value of an output variable for each set of input variables supplied to it. They are very rigid things, said to be *deterministic*. There are also "relations" that assign a set or range of values for given inputs, but those are much harder to use, and they have not become popular in biological oceanographic representations. Consider Figs. 1.8a and 1.8b from Richardson and Verheye (1998) and from Hurtt and Armstrong (1999). The first shows measures of copepod egg production at different temperatures and chlorophyll concentrations (a measure of available food). The variations are more prominent than the trends, with just hints that there are relationships. Richardson and Verheye did not fit a function to the data at various temperatures, although the greatest production rates mostly were in the middle of the observed range. Some workers would have added a distribution function. They did fit an Ivlev curve to the chlorophyll data, a single value of the response variable (egg output) for each value of the phytoplankton abundance measure. Its predictive value is small.

The second shows the output of a modestly elaborate pelagic ecosystem model intended to represent the seasonal cycling of phytoplankton stocks. That is really just a function which issues one value of the quantity of phytoplankton for each set of system variables provided to it, including previous stock abundance. We will examine several such models in some detail. Hurtt and Armstrong were satisfied with this result, which does pass through the central tendency of the actual data from the model. Sometimes the best we can do in biological oceanography is to find rough approximations.

A note on biological terminology

Biological vocabulary has been undergoing a transformation, some of it driven by results from molecular genetics that have revised the understanding of phylogenetic relationships and, thus, taxonomy. Higher taxonomic categories within all subsets of the eukaryotes have recently been subject to recurring revision, as data revealing their phylogenetic relationships continue to accumulate. Stable systematics remain well ahead in the future. Mostly we will

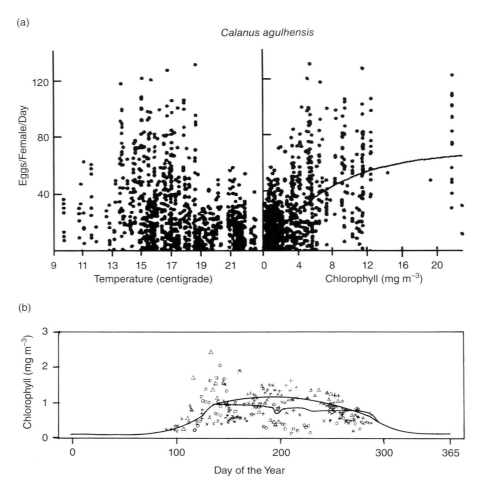

Fig. 1.8 Two examples of deterministic relationships fitted to marine ecological data affected by strong variation; the data were scattered by stochastic factors not considered by the deterministic models that are simplifying "best fits". (a) Egg production of *Calanus agulhensis* at different temperatures and chlorophyll concentrations in the Benguela coastal upwelling region. (After Richardson & Verheye 1998.) (b) Results from a pelagic ecosystem model similar to those described in Chapter 4. The two lines compare the chlorophyll concentration outputs from model versions generated by two modestly different fitting routines to actual chlorophyll concentration measures in several years (different symbols) at 59°N, 19°W in the Norwegian Sea compared with field time-series data (points) (After Hurtt & Armstrong 1999).

use quite classical category names that should remain recognizable, for example the usual names of zoological phyla. The vocabulary transformation has cast clouds of uncertainty around terms such as bacteria, protozoa, plant, and animal. We will only refer to multicellular, rooted, autotrophic organisms, not including large algae, as "plants". In common parlance, animals are heterotrophs, eating plants, fungi, and other animals. For protists making their living in that manner, we will sometimes use the word *protozoa*, meaning "simple animals", also "protozoan". We are aware that those are not considered taxonomic terms, but they are more direct than, say, "microheterotroph". Some protists function as both auto- and heterotrophs, that is as "mixotrophs". We will use that term when discussing their specific biology, but as they participate in community

photosynthesis or heterotrophy, they may be lumped with phytoplankton or protozoa. We intend to use "bacteria" only for eubacteria and "archaea" for that group quite recently recognized as profoundly distinct.

Conclusion

There are more fundamental aspects of seawater chemistry and physics, of the layout and motion of the oceans on the Earth, of "eco-math" that has roles in biological oceanography, but the introduction here should be sufficient preparation for studying the more interesting topics just ahead. The authors are excited to be getting on with it. Enjoy the ride.

The phycology of phytoplankton

Phytoplankon are the assemblage of photoautotrophic microorganisms making up the first trophic level of pelagic food chains. All belong to the botanical categories grouped as algae and studied by "phycologists". Compared to the roughly quarter million species of plants in terrestrial habitats, we find very few species of algae in the plankton, about 5000 (Tett & Barton 1995). Similar comparisons hold for zooplankton and fish, and an explanation will be sought below in the treatment of pelagic biogeography. Planktonic algae are classified into both ecological and botanical groups, sets associated functionally or taxonomically. We will mix those groupings in our study of what these algae are like. Representatives of most of the major algal *divisions* live as plankton. Divisions, the largest units of algal and plant classification and equivalent to the zoologist's phyla, are defined on both morphological and biochemical criteria. We will focus on a few ecologically dominant groups, some of them exactly parallel to the botanical ones, others not (Box 2.1).

The relative importance of phytoplankton groups varies with the ecological situation. The importance of the smaller phytoplankton has only been recognized recently, and, because of their size, some were not discovered until after 1979. We will examine each of these groups in considerable biological detail because of their large role in oceanic ecology. Often biological oceanographers must take off their ocean ecologist caps and put on marine lab T-shirts to masquerade as marine biologists. We'll do that now.

Evolution of phytoplankton

Globally, cyanobacteria and microalgae dominate marine photosynthesis. Cyanobacteria evolved about 2.85 billion years ago (Falkowski *et al.* 2004) and are simple prokaryotic cells without a membrane-bound nucleus or other cellular organelles. The microalgae are also single-celled organisms but have a more complex cell structure. They were actually formed as symbioses of photosynthetic prokaryotes or eukaryotes with heterotrophic eukaryotic hosts (Plate 2.1; Parker *et al.* 2008). Microalgae comprising the various taxonomic groups were formed by three types of these endosymbioses. In a primary endosymbiosis, a cyanobacterium was acquired by a heterotrophic eukaryote. In a secondary endosymbiosis, a eukaryotic heterotroph acquired a photosynthetic eukaryote. In a tertiary symbiosis, a dinoflagellate host engulfed a secondary endosymbiont, that provided its chloroplast to the dinoflagellate. The clue to this is that pigments and genes of the transferred chloroplasts match those of extant eukaryotic algae (Keeling 2010). In all of these symbioses, some or all genes from the chloroplasts were eventually transferred to the host nucleus.

Biological Oceanography, Second Edition. Charles B. Miller, Patricia A. Wheeler.
© 2012 John Wiley & Sons, Ltd. Published 2012 by John Wiley & Sons, Ltd.

Box 2.1 Ecologically distinct major groups of phytoplankton

GROUP	PHYCOLOGICAL TERMINOLOGY
Picoplankton	Photoautotrophs <2 μm
Cyanobacteria	Photosynthetic prokaryotes, size <2 μm, from division Cyanophyta (lately "cyanobacteria") *Synechococcus* and *Prochlorococcus*
Eukaryotic picoplankton	Very small, but structurally advanced forms
Microflagellates	An ecological assemblage of several divisions and classes: Cryptophyta, Haptophyta, Prasinophyceae, Prymnesiophyceae
Diatoms	Bacillariophyceae from division Heterokontophyta
Dinoflagellates	Dinophyceae from division Dinophyta

Main constituents of marine phytoplankton

Picoplankton – both prokaryotic and eukaryotic

Cyanobacteria

Bacteria are prokaryotes, organisms in which the macromolecules carrying genetic information, deoxyribonucleic acid (DNA), are not held in a nucleus ("karyon"), a specialized organelle surrounded by a membrane. The relatively simple DNA-bearing chromosomes disperse in the central region of the cell, which is the basis of the term "prokaryotic". Bacteria are more fully characterized in Chapter 5. One group of bacteria, called cyanobacteria, produce organic matter by photosynthesis. In that sense, they are algae as well as bacteria, so botanists classify them as the division Cyanophyta, the blue-green algae. Their photosynthetic pigments are arrayed in layers, thylakoids, around the cell periphery. Streams and ponds often support sizeable stocks of filamentous blue-green algae (*Nostoc*, *Anabaena*, and others), macroscopic forms familiar to many from introductory biology classes. It was not until surprisingly late that we realized cyanobacteria are important in marine, pelagic habitats.

Waterbury *et al.* and Johnson and Sieburth both reported in 1979 that very large numbers of photosynthetic bacteria can be counted in water samples from the ocean. These bacteria had been overlooked previously for several reasons, principally that they are small, all under 2 μm and most about 1 μm in diameter. Cyanobacterial thylakoids (Fig. 2.1) include alternate layers of phycobilisomes, particles of

protein binding photosynthetic accessory pigments including phycoerythrin. This pigment has a characteristic orange fluorescence in blue light that is strong enough to mask the red fluorescence of chlorophyll. This makes cyanobacteria readily countable with an epifluorescence microscope, an important tool in the study of this group's ecology (see Box 2.2).

Now that oceanographers watch for *Synechococcus*, we find that it frequently constitutes half or more of the photosynthetic biomass in coastal and oceanic areas throughout the euphotic zone. Small size doubtless accounts for this importance. Because sinking (or buoyant rising) rate is proportional to the square of cell diameter (Stokes's Law), picoplankton sink or rise very slowly. Even very modest reproductive rates compensate for any net losses to depth at the *Synechococcus* sinking speeds of less than a centimeter per day (Raven 1985). Minuscule size also maximizes relative surface area for nutrient absorption and reduces grazing loss to suspension feeders. *Synechococcus* possesses a variety of phycocyanin and phycoerythrin pigments that allow the different strains to utilize the wide range in light quality naturally occurring over both horizontal (coastal–oceanic) and vertical gradients (Scanlan *et al.* 2009). Phycoerythrin is also abundant in the Rhodophyta, red algae, which are progressively more dominant with increasing depth in subtidal habitats nearshore. In fact, rhodophyte chloroplasts resemble cyanobacteria in the form of their thylakoids and the presence of phycobilisomes. It is likely that these chloroplasts have descended from an early, intracellular symbiosis of cyanobacterial cells in the macrophytes (Plate 2.1).

One genus of filamentous cyanophytes, *Trichodesmium*, lives in tropical seas. "Tricho", as it is known to students of its biology, can generate gas vacuoles and float at the sea surface. It may also regulate buoyancy to move up and

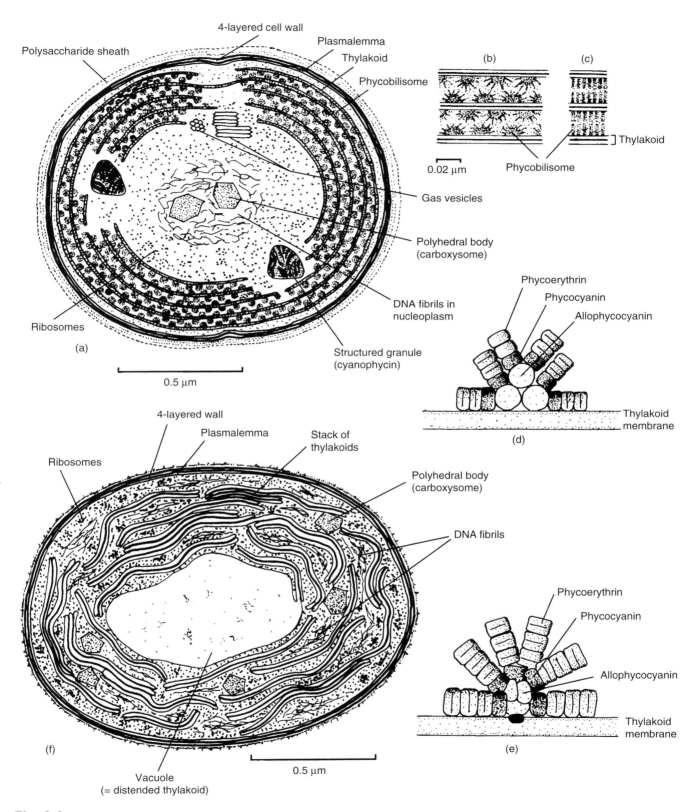

Fig. 2.1 Morphological plans of (above) single-celled cyanobacteria (*Synechococcus*) and (below) *Prochlorococcus marinus*. (After Van den Hoek *et al.* 1995.)

Box 2.2 Epifluorescence microscopy

Biologists, including biological oceanographers, have found increasing use for microscopes equipped so the user can see only light emitted as fluorescence by objects under examination, not light transmitted through or reflected off the particles. In the case of phytoplankton, this is accomplished by filtering cells on to a black membrane filter. Often they are fixed first with an aldehyde (e.g. glutaraldehyde), then filtered, covered with a drop of immersion oil and a cover slip, and examined in the microscope. Instead of illuminating them from below, they are illuminated with strong blue light from above at a small angle next to the objective or, most often, inserting the light with a prism through the objective lens (see diagram). Blue light excites fluorescence in green to red wavelengths. The portion of the optical pathway above the objective lens is equipped with a filter that removes all reflected blue light but passes the fluoresced light. The observer sees green, yellow, orange, and red fluorescence from cells and organelles against a black background. This is extremely useful for studies of picophytoplankton because cyanobacteria fluoresce at orange wavelengths while picoeukaryotes fluoresce in the red (Plate 2.2).

Thus, they can both be seen readily despite their small size and be separated into functional groups. Prochlorophytes only fluoresce weakly and are not readily seen microscopically. They can be distinguished by flow cytometry (see Box 2.3).

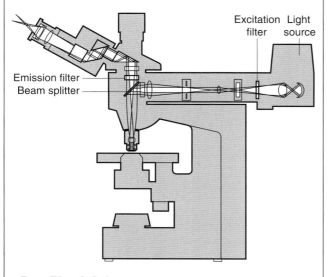

Box Fig. 2.2.1 Schematic of an epifluorescence microscope.

down, harvesting nutrients at depth then rising back toward brighter light. Like some genera of freshwater cyanophytes, it can fix gaseous nitrogen, N_2, into ammonium and nitrogenous organic molecules. Thus, given some phosphorus and trace elements, nitrogen limitation is alleviated, and the filaments can develop into extensive near-surface blooms. *Trichodesmium* blooms occur occasionally in all tropical waters, and are particularly common in the Arabian Sea and Red Sea probably due to abundant dust input (supplying iron), very warm waters and prolonged calm spells. When the Red Sea is red, the color usually comes from *Trichodesmium* mats. Another nitrogen-fixing, filamentous cyanophyte, *Richelia intracellularis*, is a symbiont of several diatom genera: *Rhizosolenia*, *Chaetoceros* and *Hemiaulus*. Nitrogen fixation by *Richelia* allows occasional blooms of these diatoms in the oligotrophic, central gyres. Moisander *et al.* (2010) studied the distribution of two unicellular N_2-fixing cyanobacteria, *Crocosphaera watsonii* and UCYN-A, by measuring the presence of the gene encoding the iron-protein in the nitrogenase enzyme. Results showed that UCYN-A has a broader latitudinal distribution than *Trichodesmium*, implying that significant N_2 fixation may occur in temperate, as well as, tropical waters. Interestingly, UCYN-A lacks the enzyme for assimilating CO_2 and releasing O_2 (Bothe *et al.* 2010). This protects the O_2-sensitive nitrogenase, but as a consequence UCYN-A is dependent on extra-cellular sources of organic carbon.

Prochlorococcus

The organisms most recently recognized as an important component of the phytoplankton are the picoplanktonic (<2 µm) cells of the genus *Prochlorococcus* (Chisholm *et al.* 1988). Genetic analysis revealed that *Prochlorococcus* is a cyanobacterium closely related to *Synechococcus*. *Prochlorococcus* can be the most abundant phytoplankton in oligotrophic, tropical waters, with its latitudinal distribution ranging from 45°N to 40°S, and it can account for a substantial fraction of primary productivity (Olson *et al.* 1990). Like *Synechococcus*, *Prochlorococcus* is a prokaryote without a membranous nuclear envelope (Fig. 2.1). These organisms have a layered cell wall braced with murein, the wall polymer of bacteria. They have several other distinguishing biochemical features, particularly dominance of the photosynthetic pigments by a "divinyl chlorophyll pigment", rather than the usual form of chlorophyll-*a* dominant in all other plants (Wu & Rebeiz 1988). They share with the Chlorophyta (green algae) and higher plants the inclusion of another molecular variant of chlorophyll, chlorophyll-*b*. Prochlorophytes were originally discovered as algal symbionts in some tropical algae by Lewin (Lewin & Withers 1975). They are also extracellular symbionts in some attached tropical ascidians (tunicates or sea squirts). Discovery of them in pelagic habitats came from work with automated cell-counting machines (see Box 2.3) in use for study of *Synechococcus*. In pelagic habitats,

Box 2.3 Flow cytometry

Biologists interested in blood-cell counts and in tissue-culture techniques have developed a suite of machines for automatically distinguishing, counting, and sorting large numbers of cells. The cells of interest are dispersed in isosmotic saline that is allowed to flow through a narrow tube fitted with a sequence of sensors, including electrical conductance meters, colorimeters, and fluorometers. If dispersed by sufficient fluid, the cells will pass through the sensors one at a time, allowing computer circuitry attached to the sensors to categorize them in respect to size, color, and fluorescent response to illumination in several wavelengths. If desired, the cells can be sorted into separate containers. To do that, the section of fluid in the tube containing a cell of a given category is tracked from flow velocity. The fluid is allowed to drip from an orifice at the end of the tube, which strips a few electrons from the drop surface, giving it a positive charge. Then the trajectory of its fall is steered into an appropriate container by surrounding electrodes that can be given positive (repulsive) or negative (attractive) charges under computer control. These devices have been adapted by oceanographers for identifying and counting very large numbers of cells, particularly picoplankton, according to their fluorescence and light-scattering signatures (Box Fig. 2.3.1). The positions of many individual cells on two-dimensional plots of light scattering versus fluorescence at different wavelengths generates cluster diagrams illustrating the relative importance of prochlorophytes and cyanobacteria in a given sample (Box Fig. 2.3.2).

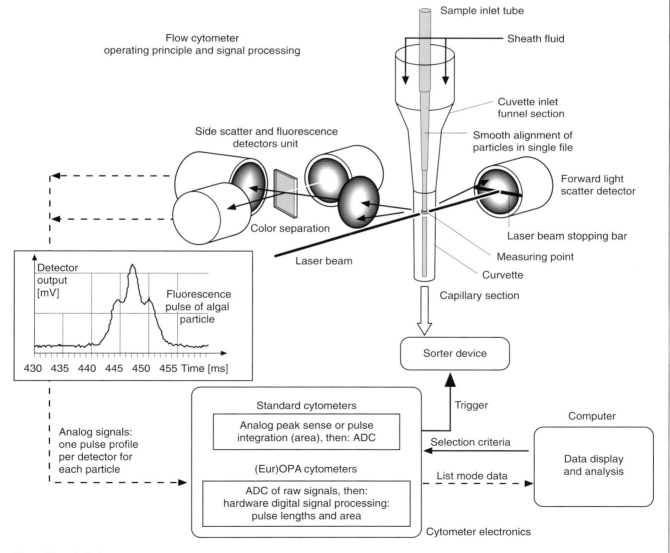

Box Fig. 2.3.1 Flow cytometer system layout (courtesy of Dr George Dubelaar, Cytobuoy, The Netherlands.)

(*Continued*)

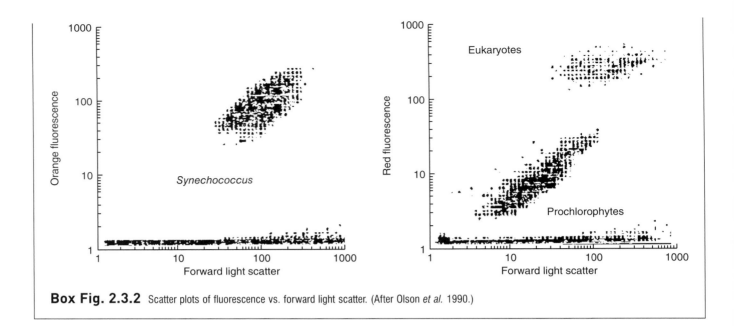

Box Fig. 2.3.2 Scatter plots of fluorescence vs. forward light scatter. (After Olson *et al.* 1990.)

the generic name, *Prochlorococcus*, recognizes their spherical (i.e. coccoid) form.

Prochlorococcus and *Synechococcus* occupy complementary but overlapping niches in the ocean. They form three ecological groups with widespread importance in the ocean: high-light-adapted *Prochlorococcus*, low-light-adapted *Prochlorococcus*, and diversely pigmented strains of *Synechococcus*. Comparisons of gene families in these three groups (Zhaxybayeva *et al.* 2009) suggest frequent gene transfers between *Synechococcus* and low-light-adapted *Prochlorococcus*. Such gene sharing (horizontal gene transfer) also occurs in bacteria, and can result in specific and presumably rapid ecological adaptation.

Eukaryotic picoplankton

In addition to cyanobacteria, the photosynthetic picoplankton often include numerous, eukaryotic forms smaller than 2 μm in cell diameter. The first report that cells of this type are abundant was again recent (Johnson & Sieburth 1982). Some of these forms are coccoid green algae, Chlorophyta, without flagellae but retaining basal bodies or even a rhizoplast structure showing relationship to a *Chlamydomonas*-like ancestor. The genome of one species, *Ostreococcus lucimarinus*, has been sequenced, showing a large number of selenoenzymes (Palenik *et al.* 2007). These enzymes are more catalytically active than enzymes without selenium and may be an adaptation to the organism's small size. Two other divisions, Heterokonta and Haptophyta, have marine species that are ecologically important eukaryotic picoplankton (Worden & Not 2008).

Liu *et al.* (2009) used a combination of genetic, pigment, and microscopy data to elucidate the abundance and diversity of very small haptophytes. These phytoplankton use 19'hexanoyloxyfucoxanthin (19-Hex) as an accessory photosynthetic pigment, and although this pigment is pervasive in the photic zone of the world oceans, haptophyte nuclear rRNA sequences (see Box 2.4) seemed scarce. Classical primers for small subunit ribosomal RNA (SSU rRNA) genes do not work well for the haptophytes, and when more specific primers were developed, the diversity and abundance of the haptophytes became apparent. These tiny haptophytes may contribute from 30 to 50% of the phytoplankton standing stock across the world ocean.

Microflagellates

Small, flagellated cells from a variety of algal groups account for a large fraction of marine primary production. Those groups are separable on the botanical grounds of differences in ultrastructure, pigment composition, details of biochemistry, and most recently by SSU rRNA comparisons. However, ecologists group them together as a convenience, since they are similar in size and general morphology. Their size ranges from 2 to about 30 μm, but most are less than 10 μm. All swim weakly by flagellar action, photosynthesize, and require special care to preserve. These common features mean that they must be studied by similar, specialized means, whatever their botanical affinities. Table 2.1 gives the salient features of the major groups based on features of the flagellae, cell wall, and abundant pigments. Details of cell division are also distinctive (Taylor 1976), but in a very complex fashion. The classification chosen follows Falkowski and Raven (2007) and borrows from group descriptions in Dodge (1979). In addition to the groups listed, a number of benthic algal groups have small,

Box 2.4 Molecular genetic classifications and phylogenetic reconstructions

In recent years, classifications of organisms and evaluation of their evolutionary relationships (phylogeny) have come to be based partly on similarities and differences in their DNA sequences. Moreover, extended sequences can be examined to determine what developmental and metabolic genes are present, or even to identify those genes active under different circumstances. To follow the arguments, you need a rudimentary understanding of molecular genetics. Those rudiments are provided here. For a more extended introduction, the Wikipedia article (http://en.wikipedia.org/wiki/DNA) is excellent and well illustrated.

The formulas, *genes*, for construction of proteins from 20 amino acids are stored in cells as sequences of four, small distinctive compounds, nucleotides (or "bases"): thymine (T), adenine (A), guanine (G), and cytosine (C). They are arrayed along and bonded to a chain of polymerized sugar molecules (deoxyribose). The chains are *deoxyribonucleic acid* (DNA). The code for each amino acid consists of three bases; for example, TGG codes for tryptophan. There are 64 codes (4^3) possible with four letters, so most amino acids are indicated in the code by several synonyms. The codes specifying particular amino acids vary in a few organisms. The construction of a specific protein involves enzymatic *transcription* of the DNA into similar molecules (with uracil, U, substituted for T) but on a ribose polymer: ribonucleic acid (RNA). Some of the codes indicate to transcription enzymes to "start transcribing here" or "stop transcribing". The transcription, termed messenger RNA or mRNA, is then *translated* into the protein by organelles (complex molecular machines) called ribosomes. In the ribosome, triplets of bases on the mRNA mesh with diffusing bits of RNA (transfer RNA, tRNA) linked to the specific amino acid appropriate to the triplet code, and the ribosome catalyzes the polymerization of the resulting amino acid sequence. The protein is then released, folds into its functional form, and additional complex processes incorporate it in the operating structure of the cell.

Genes are stored in cells as double helices of DNA, the two long polymers joined by hydrogen bonds between A and T and between G and C. It has to be duplicated at cell division, which involves decoupling that hydrogen bonding temporarily and forming complementary nucleotide sets along each single strand. A DNA polymerase enzyme complex unwinds and works along the chains, placing an A (and ribose) opposite to and hydrogen bonded to each T, Ts opposite As, Cs opposite Gs and Gs opposite Cs. One double helix becomes two double and identical helices. In bacteria and archaea (Chapter 5) the DNA chains (chromosomes) are

centrally located in the cells. In eukaryotes the chromosomes are also encapsulated in the cell's nuclear membrane ("karyon"). Chromosomes include both DNA and protein complexes called chromatin.

In the long process of revealing DNA structure, storage, replication and translation as proteins, molecular biologists learned how to "read" the code for long DNA sequences. Radically oversimplifying (see Sambrook *et al.* 2006), DNA from an organism is chemically extracted, and then selected portions are copied to generate readable quantities. There are two main ways: (i) The DNA is broken into bits that are installed in bacterial plasmids (closed loops of DNA), and those are multiplied through massive reproduction of carrier bacteria, usually *Escherichia coli*. DNA of specific interest is removed from the cloned plasmids and cleaned. Or, (ii) a sequence of interest can be selectively amplified by the polymerase chain reaction (PCR). As detailed in molecular biology books, PCR is an artificial amplification procedure done extracellularly. It allows amplification of very specific sequences, provided that reliably conserved portions of the sequences of interest are known to serve as "primers". The DNA produced in either way is purified and sequenced by Sanger's dideoxynucleotide chain-terminating method (see Sambrook *et al.* 2006), which has been automated using a version of PCR. The DNA code is read from the method's chromatogram. The result is literally spelled out:

. . . AGATTTCTGGTTTCTTAATGCCAGCTTTA . . .

Recent, automated techniques can obtain all or nearly all of an organism's genome by dicing up its DNA, amplifying (PCR) and sequencing all the pieces, then using computer comparisons to find matching overlaps and to reconstruct a probable whole.

Sequences for parts of genes, whole genes or greater lengths can be compared between individuals, species, or even phyla, for similarity. Levels of similarity or difference can indicate degrees of relationship. There are comparator algorithms based on a variety of distinct principles: neighbor-joining, evolutionary parsimony, and others. Similarities of a gene to others of known function can suggest its function. For that purpose, access to large libraries of code is important, to which end the US National Institutes of Health maintains GenBank, a massive web-accessible record of DNA sequences for organisms of all sorts. Together the mathematical techniques and computer operations of all these methods are termed "bioinformatics".

To examine relationships across wide ranges of relationship, it is required to use genes that are present in all

(Continued)

living things. Genes coding the RNA constituting ribosomes have been particularly important. This RNA comes in two subunits, termed large and small. Small subunit (SSU RNA) sequences have been particularly useful in identifying microbial relationships and very old phylogenetic connections generally. The terms 16S RNA and 18S RNA are commonly used, reflecting the molecular weights (in kilodaltons) of prokaryote and eukaryote SSU RNA, respectively. Conserved primers for SSU DNA have been known for ~30 years. Over the course of evolution to more complex life forms, ribosomal structure has become more complex and the code has gotten longer. This growing elaboration and the variations among elaborations on different phylogenetic branches are a basis for classification and reconstruction of evolutionary relationships. The conserved parts apparently are fundamental to ribosomal function, explaining their consistent sequences throughout the evolution of life on Earth. They provide starting places for both comparisons and for PCR amplification. Classification of life forms using SSU RNA was pioneered by C.R. Woese (e.g. Woese & Fox 1977), and then expanded as "exploratory systematics" by Norman Pace and colleagues (e.g.

Olsen *et al.* 1986). That work has profoundly reorganized thinking about microbial forms in particular.

Because mitochondria and chloroplasts were acquired originally as internal symbionts of ancestral cells, they retain a modest part of the DNA (and the ribosomes) required to construct their proteins. That DNA is clonally reproduced without sexual recombination (leading to simpler patterns of inheritance than those of nuclear or bacterial genes), evolves relatively rapidly, and can provide somewhat more direct tracing of ancestry. Mitochondrial DNA, mtDNA, is often favored for studies of phylogeny in eukaryotes, particularly animals. Gene variants are termed "haplotypes" because of the haploid character of mitochondrial genes. Sequences of particular mtDNA genes are widely used for species and strain identification using either short sequences from genes or species-specific, mtDNA PCR primers. The gene for cytochrome oxidase I (COI) is popular in that regard for animals, sometimes termed a DNA "barcode", and has been widely applied in the recent Census of Marine Life (COML).

Additional specific details of molecular genetics will be supplied as needed throughout the book.

Table 2.1 Salient features of taxonomic groups of microalgae. Several groups seldom abundant in marine habitats are left out: Euglenophyta, Raphidophyceae, Eustigmatophyceae. Principal pigments for the major groups are shown in Table 2.2.

	FLAGELLAE	CELL WALL
Prokaryotes (Cyanobacteria)		
Synechococcus	0	Murein
Prochlorococcus	0	Murein
Chlorophyta	Two (or four) apical, equal, smooth	Naked or with cellulose sheath, sometimes calcified
Prasinophyceae	One or two unequal, or four equal and scaly	Organic scales or naked
Cryptophyta	Two, apical, equal, sometimes hairy	Naked
Haptophyta	Two, equal or not, hairy + haptonema emerging between	Chlorophylls *c*1, *c*2 Organic scales
Prymnesiales	Two, unequal, hairy, haptonema vestigial	Organic or calcite scales
Heterokontophyta Bacillariophyceae (Diatoms)	One in male gamete, hairy, no microtubules	Naked gametes
Chrysophyceae and relatives*	Two, unequal, posterior short, smooth, anterior long	Naked or scales (some opal)
Dinophyta	Two, one girdling, one posterior	Cellulosic plates or often naked
Rhodophyta	0	Cellulose

*A number of groups related to the Chrysophyceae have recently been separated within a new phylum Heterokontophyta. These include Synurophyceae, Dictyochophyceae (including the Dictyocales or silicoflagellates that bear opal scales) and the very tiny Pelagophyceae (the principal eukaryotic picoplankton). Heterokontophyta are characterized by flagellae of unequal length (anterior "tinsel" and posterior smooth), chlorophyll-*c*, and chrysolaminarin as a storage product. The Chrysophyceae and Bacillariophyceae (listed above) are Heterokontophyta, as are the brown algae Pheophyceae and Xanthophyceae.

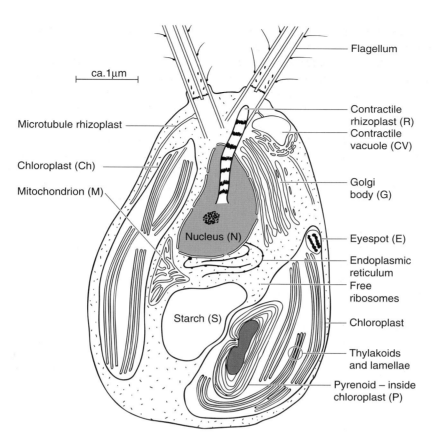

Fig. 2.2 General morphological plan of a number of microflagellate types.

flagellated, photosynthetic gametes that may be abundant in coastal phytoplankton from time to time. Those include the benthic diatoms, brown algae (Phaeophyceae), various Chlorophyta, Xanthophyceae and the Eustigmatophyta. Descriptions of the last two groups can be found in Van den Hoek *et al.* (1995).

Despite the diversity in details, a general pattern of cellular organization can be described that applies to most microflagellates. The cell is round to ovoid, occasionally spindle-shaped. It has an "anterior" end at which are located the insertions of the flagellae, most often two in number. Movement of the cell usually is toward the flagellae, which pull. The outer surface may be the naked cell membrane, or the cell may be covered with a secondary organic wall and bear organic, siliceous, or calcareous scales. Internally, the structure is as shown in Fig. 2.2.

The nucleus is pear-shaped, with its narrow end located anteriorly beneath the roots of the flagellae. The root of one flagellum (sometimes both flagellae) attaches to a spray of striated fibers, the rhizoplast, which extends posteriorly into the cytoplasm around the nucleus. Rhizoplast fibers can contract, pulling intermittently on the root to generate

the wave-like movements of the flagella. To one side of the nucleus and root structure of the flagellae sits a contractile vacuole. Located anteriorly is the Golgi body, a system of flattened membranous cisternae involved in secretory activity. In many forms it produces the scales that cover the cell externally.

There may be one or two chloroplasts. If there is just one, then it is cup-shaped and fills the posterior end of the cell. If two, they wrap the sides of the cell, one often extending farther posteriorly. Thylakoids, the layers of membrane inside the chloroplast bearing the photosynthetic pigment systems, are arranged parallel to the external surface of the cell. There may be two to many thylakoids, and thylakoids vary among groups in the number of membranes constituting them. In the center of the chloroplast cup is a *pyrenoid*, which organizes the formation of storage product, variously lipid or starch, in a mass below the nucleus. The anterior-most extension of the chloroplast often bears an eye-spot consisting of several layers of pigmented granules. This probably enables phototaxis or other responses to light and, thus, slight vertical migration. Mitochondria either scatter about between the chloroplast and the nucleus, or in some forms there is only a single mitochondrion of complex shape in this central position.

Microflagellates with scales (coccolithophores, for example) form them one at a time in a specialized organelle in the interior of the cell, and they are pressed out through the membrane on to the cell surface. In some forms this scale organelle is clearly related to the Golgi body. As the scales are formed and extruded they become arranged on the cell surface, possibly by successive rotation of the cytoplasm relative to the membrane (Brown *et al.* 1973). Some coccolithophorids are more morphologically complex and have modified "appendage" coccoliths that may serve to discourage grazers. See Young *et al.* (2009) for scanning electron micrographs of these appendage-bearing coccolithophores.

Variations in microflagellate morphology (Fig. 2.3) include two chloroplasts instead of one, absence of an eye-spot, and presence of additional flagellae or flagellum-like organelles. Details of reproduction vary among groups, adding characteristics useful in classification (Taylor 1976). However, most population increase in all groups occurs by simple division along a cleavage furrow. Sexual reproduction has been found in most microflagellates, with variation in the details and much left to be learned. *Pleurochrysis*, of the Haptophyta, exhibits alternation of a planktonic, scaled, diploid form with a benthic, filamentous, haploid form, both of which reproduce vegetatively (Gayral & Fresnel 1983). The benthic form releases naked swarmers that fuse to form zygotes that develop into the scaled form. Haploid–diploid alternation of two distinct forms also occurs in most coccolithophores (de Vargas *et al.* 2007). In chlorophycean forms like *Chlamydomonas*, the vegetative cells are haploid, with sex as fusion of vegetative cells modified to + and − forms with complementary cell structures to enable fusion and nuclear transfer (Goodenough & Weiss 1978). Meiosis takes place in the "zygote" (Triemer & Brown 1977).

The Prymnesiophyta are particularly prominent in the oceanic phytoplankton. *Emiliania huxleyi* and other coccolithophores bear calcareous plates. They sometimes bloom in sufficient density to reflect light strongly back into the sky, and appear in satellite images as creamy outlines of eddies (Fig. 2.4).

Their cousins, the Pavlovales, retain the full function of a structure between the two flagellae called a haptonema, constructed of three concentric sheaths of membrane surrounding a core of seven microtubules. The outer sheath may bear small scales. The haptonema can bend and coil through activity of the microtubules, allowing it to serve as a feeding organelle (Kawachi *et al.* 1991). Particles adhere to it, are moved to the base, stuck together into a mass, and then moved back out to the tip. The tip then twists around to the base of the cell, and the surface of the cell forms a food vacuole around it (phagocytosis). Mixing of autotrophy and phagotrophy is quite widespread, even in extremely small cells.

Diatoms – Bacillariophyceae

In nutrient-rich coastal waters and during oceanic spring blooms, diatoms usually dominate the phytoplankton. They range in size from the 2.0 mm *Ethmodiscus rex* of the warm, mid-ocean gyres down to about 2 μm, as for example *Nitzschia cylindroformis* common in the subarctic Pacific. Diatoms can divide more rapidly than other phytoplankton, hence their importance during blooms. Beside their importance as constituents of the plankton, diatoms migrate actively up and down in sand beaches, grow in the interstices on the bottom side of sea ice, and live on the surfaces of macroalgae. Their distinguishing feature is a hard mineral shell or frustule constructed of opal, that is, hydrated, polymerized silicic acid, $Si(OH)_4$. Opal has a hardness of seven, suggesting its value is as armor, and a density of $2.7 \, \mathrm{g \, cm^{-3}}$, suggesting problems with buoyancy maintenance. In fact, sinking after depletion of water-column nutrients may be of value, moving the cell downward toward a nutricline (gradient of increasing of nutrients at the base of the euphotic zone) or taking them rapidly away from grazers to deposit as resting spores in the sediment. All diatom shells are highly elaborate microscopically, and very beautiful.

Taxonomists define two main groups, primarily distinguished by frustule structure: the centrics (Centrales) and the pennates (Pennales). Centric diatoms derive from a radially symmetrical primitive form with frustules shaped like petri dishes (Fig. 2.5). The upper and lower valves (*epitheca* and *hypotheca*) each consist of a flat plate (the valve) and a cylindrical rim, the girdle band, that wraps the curved edge. The valve and the girdle are sometimes loosely connected, sometimes fused. The girdle of the lower valve fits inside the girdle of the upper valve. As the cell grows, this sliding joint can expand, providing room for increase in cell content. In some genera (e.g. *Rhizosolenia*) the joint is not actually sliding, but is a set of pieces or girdle rings, more of which can be added as the cell grows until it is tubular. Cytoplasm is located along the inner surface of the shell, forming a hollow lining around a large vacuole in the cell center. The cytoplasmic layer contains several of the cell organelles, most obviously the chloroplasts and mitochondria. "Cell sap", a fluid not unlike seawater but with variations in specific ion content, fills the central vacuole. The nucleus of the cell is usually located against the center of one valve, but prior to division it slides into a central cytoplasmic island suspended in the vacuole by strands from the sides. Many variant shapes appear among the centrics through extension of the shell in one axis or another and by adding spines.

Pennate diatoms are bilaterally symmetric, not radially symmetric. On each valve there are openings in the form of slits along the surface, termed raphes (Fig. 2.6). Typically, the shell is elongate parallel to the raphes. Cytoplasm streams along the raphes, generating cell movement. The

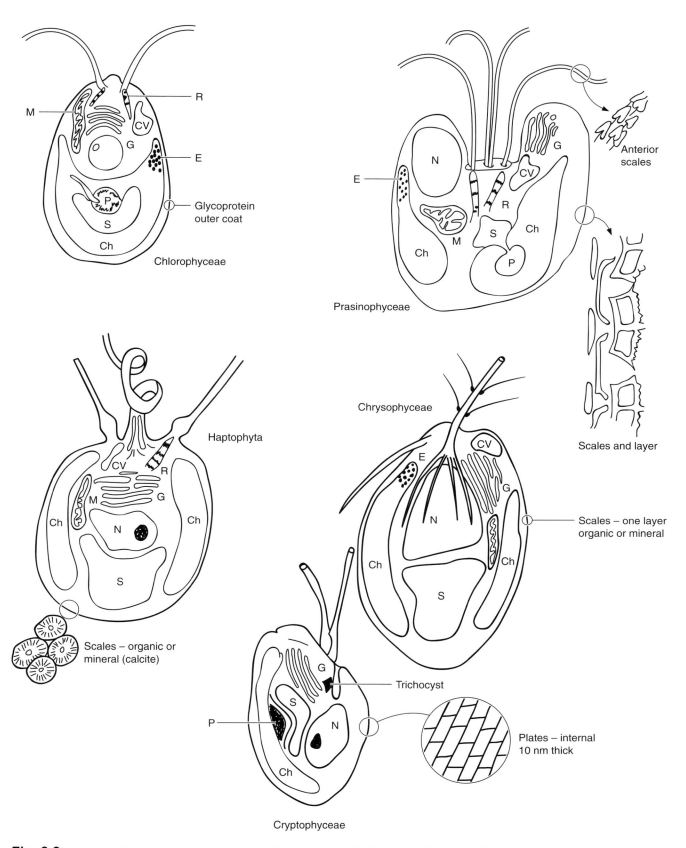

Fig. 2.3 Morphological variations among the more commonly occurring microflagellate groups. Refer to Fig. 2.2 for abbreviations.

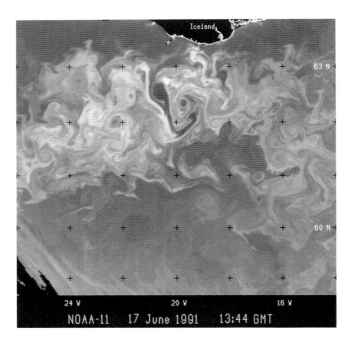

Fig. 2.4 Advanced very high-resolution radiometer (AVHRR) image of the visible reflections from an *Emiliana huxleyi* bloom in the Atlantic Ocean south of Iceland. Lighter colors are from higher reflectance from the plaques of calcite (coccoliths) on the cell surfaces. (Courtesy of Steve Groom, Plymouth Marine Laboratory, similar to fig. 2 in Robertson *et al*. 1994.)

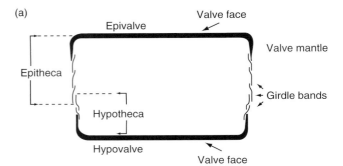

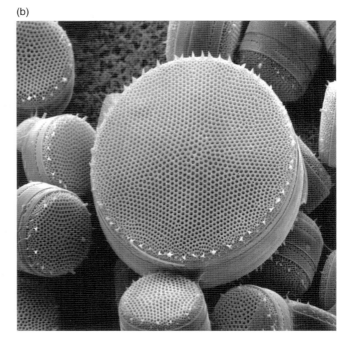

Fig. 2.5 (a) Diagrammatic section through a centric diatom frustule (shell) with standard terminology. (b) Scanning electron micrograph (SEM) of *Thalassiosira eccentrica*, a centric diatom, illustrating the simplest petri-dish shape and radial symmetry. The range of cell sizes shows the change from just before (small) to just after (large) auxospore formation. Cell diameters of *T. eccentrica* range from 12 to 100 μm. (Courtesy of Anne-Marie Schmid, Inst. Pflanzenphysiologie, University of Salzburg; also in Schmid (1984), with permission, Koetz Publ.)

Fig. 2.6 SEM of the pennate diatom *Navicula cuspidata*, showing the central raphe. (Courtesy of E.G. Vrieling, Department of Marine Biology, Centre for Ecological and Evolutionary Studies, University of Groningen, The Netherlands, with permission from the *Journal of Phycology*.)

cells always move along solid surfaces, and pelagic forms move only against each other in colonies. The best candidate mechanism for this locomotion (Edgar & Pickett-Heaps 1984) is based on ultrastructural analysis. Strands of mucus excreted through the raphe attach to the substrate while maintaining a connection to actin filaments through a complex of membrane-associated proteins (Edgar & Pickett-Heaps 1983). Actin is a contractile protein, important in practically all motile functions from cytoplasmic streaming to muscle contraction. Myosin is one element of the protein complex, and its movement against the actin filaments produces a rearward translocation of a membrane

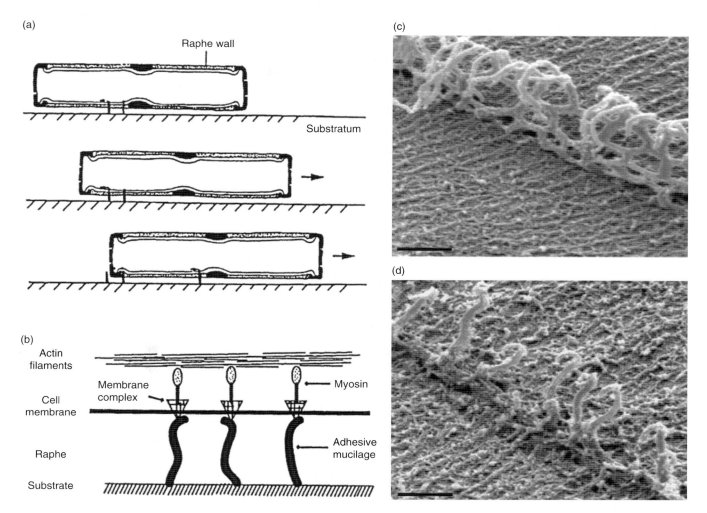

(a)

Raphe wall

Substratum

(b)

Actin filaments

Membrane complex

Cell membrane

Myosin

Raphe

Adhesive mucilage

Substrate

(c)

(d)

Fig. 2.7 Model of gliding locomotion in diatoms. (a) Strands of adhesive mucilage secreted by the diatom adhere to the substrate but also remain connected to components of the diatom cell membrane. These membrane elements are actively translocated rearward upon a framework of actin filaments, leading to the forward movement of the diatom relative to the substrate. At the rear of the cell, mucilage strands break and are deposited as a trail. (After Edgar & Pickett-Heaps 1983.) (b) Model for the organization of the motor apparatus in diatoms. Excreted musilage adheres to the substrate and to a membrane-associated complex that is linked to a diatom myosin. The myosin translocates the membrane complex and attached mucilage rearward along a tract of actin filaments leading to the forward gliding of the diatom. (After Heintzelman 2006.) (c and d) scanning electron micrographs (SEMs) of chemically fixed mucoid trails deposited by *Pinnularia viridis*. The adhesive strands protruding from most of the raphe may be fused together (c) or remain as individual strands (d). (c and d after Higgins *et al*. (2003), with permission from the *Journal of Phycology*.)

protein and its attached mucilage strand along the length of the raphe (Fig. 2.7a & b). At the rear of the cell the strands are broken off by the forward traction generated by the actin–myosin system and left behind as a mucus trail. The mucus strands can be fused together or remain as individual strands (Fig. 2.7c & d).

Diatom frustule sculpture is complex, producing a variety of contacts between the cell membrane and the water. There may be punctae, which are simply holes in the frustule. There are usually areolae, small boxes embedded in the shell wall with a large pore in the outer surface and a lacework with many small pores against the cell membrane. Some openings, such as the labiate process of both

centrics and pennates and the pore plate of the pennates, connect the water outside with specific cell organelles. Many forms have spines and processes (Fig. 2.8). These can be long, and they sometimes connect the cells into chains. In centrics this may be a slender siliceous thread linking the center of one valve to the center of the next (as in *Thalassiosira*), or a complete ring of elaborate, interlacing fencing around the valve edges (as in *Skeletonema*). In pennates the corners of elongate, nearly rectangular shells can be joined to form a star (*Thalassiothrix*), or cells can adhere to each other at the raphe, forming "rafts" in which the cells slide back and forth along each other's length (*Bacillaria*). It is often assumed that the very long side

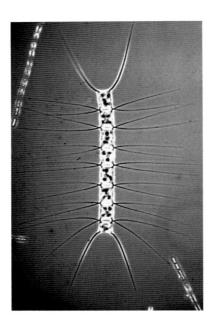

Fig. 2.8 Phase-contrast light micrograph of *Chaetoceros decipiens*, showing chaining of cells and fine siliceous spines (spines this fine are usually called setae). (Courtesy of J.D. Pickett-Heaps, University of Melbourne.)

spines either inhibit predation or enhance surface area to increase drag and inhibit sinking. However, for forms with siliceous spines, calculations show that the cost of spines in increased density exceeds the gain in increased drag. Moreover, Gifford *et al.* (1981) have shown that copepods more readily eat a form of *Thalassiosira weissflogii* with long, chitinous spines than they do a form without them, presumably because longer spines make the cell more readily detected and captured more effectively. The whole matter of spines is still not fully explained.

Diatom cells divide by an unusually elaborate process, since the cell not only must divide, but must fabricate new valves. Pickett-Heaps *et al.* (1990) have reviewed much of what is known about diatom mitosis. Division (Fig. 2.9) proceeds inside the old frustule until there are two protoplasts, one adjacent to each valve. Each has its own cell membrane at the central plane. New siliceous walls form inside and parallel to each of these membranes. Formation begins with the appearance of membranous vesicles, probably derived from one or more sets of Golgi apparatus (cell organelles involved in packaging of secretory products, particularly protein–carbohydrate complexes) in each cell. These vesicles aggregate in the center of the division plane, and coalesce to form a "silicalemma", a membrane surrounding a silica deposition vesicle (SDV). The two sides of the silicalemma may connect in a "donut hole" wherever there is to be a pore in the new frustule. There are other

elaborations equivalent to at least some of the sculpture of the new frustule, although it is not certain that all details are sculpted in that manner. The vesicle fills with silica with remarkable rapidity, in minutes.

Silicon for frustule building must be available in the form of silicic acid; in fact the entire process of cell division is placed under control of silicic acid availability. Enzymes requiring it as a cofactor control DNA replication in diatoms. Silicic acid is taken into the cell in most cases directly before frustule formation, not stored internally in any great quantity (Darley *et al.* 1976). Both uptake and deposition require energy (Lewin 1955). Opal is laid down as "nano" (less than 1 μm) spheres on a modified protein matrix (Kröger *et al.* 1999). A backbone of hydroxyl-group-bearing amino acids has paired, modified lysine residues spaced along it. One of those carries a chain of 5 to 10 repeats of N-methyl-propylamine, the other is *N,N*-dimethyl lysine (Fig. 2.10).

Vrieling *et al.* (1999) used vital dyes that fluoresce at low pH to show that precipitation of opal in the silica deposition vesicles occurs under acid conditions. At acid pH, both the hydroxyl groups and, particularly, the repeated N-methyl-propylamine chains catalyze silicic acid polymerization. The resulting opal frustules are well protected during the life of the diatom by organic coatings on the outer surfaces. Bacteria colonize and enzymatically degrade the organic matrix (Bidle & Azam 2001). Upon decay of the coating after cell death, weakly basic seawater slowly redissolves the opal. Thus, accumulation of diatomaceous sediments requires rapid burial rates, and often only species with the very thickest shells remain in the geological record.

In many, but not all, diatom species, cell size diminishes progressively as division proceeds (MacDonald 1869). Since the hypotheca becomes the epitheca for one of the daughter cells, that cell is smaller than its sister. This process terminates by formation of an *auxospore*. This is usually, if not always, coupled with sexual reproduction. Drebes (1977) provides a thorough discussion of sexual processes in diatoms. In centric forms (an example of a life-history sequence is shown in Fig. 2.11) many species produce four flagellated sperm by meiosis. A sperm fertilizes a single oocyte produced inside the shell of the reduced vegetative cell. The resulting zygote drops the frustule, grows to a large size, and forms a heavy cell membrane of organic material containing siliceous scales (Edlund & Stoermer 1991), and finally develops a shell of the vegetative type internally. This new frustule, surrounding the so-called "initial cell", is produced in association with several mitoses, indicating a relationship between the control of frustule formation and the control of mitosis *per se*. All of the daughter nuclei, save one, simply degenerate. Pennates exhibit similar sexual processes, although motile sperm are not involved. Rather the gametes are "isogametes", cells of the same size (sometimes amoeboid) produced by meiosis that fuse to form a zygote when exchanged between touching parent frustules.

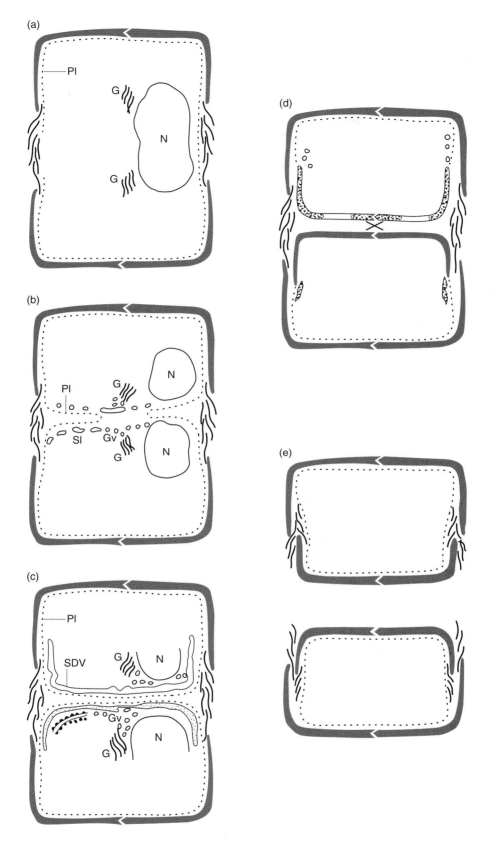

Fig. 2.9 Sequence of cell division and frustule formation events in the pennate diatom *Gomphonema parvulum*. (a) Cell elongates, additional volume covered by addition of more girdle bands, the nucleus enlarges with DNA duplication, and organelles, represented by Golgi body (**G**), are partitioned. (b) Nucleus divides, then protoplasts are separated by invagination of the plasmalemma (**Pl**) across the division plane, and vesicles (**Gv**) apparently derived from the Golgi body accumulate adjacent to it. (c) Formation is completed of silicalemma (**Sl**) around deposition vesicles (**SDV**) in which the valves will develop. (d) Opal is deposited in the SDV, then smaller SDV form at the sides for girdle band deposition (shown in lower cell). (e) Cells separate. (After Dawson 1973.)

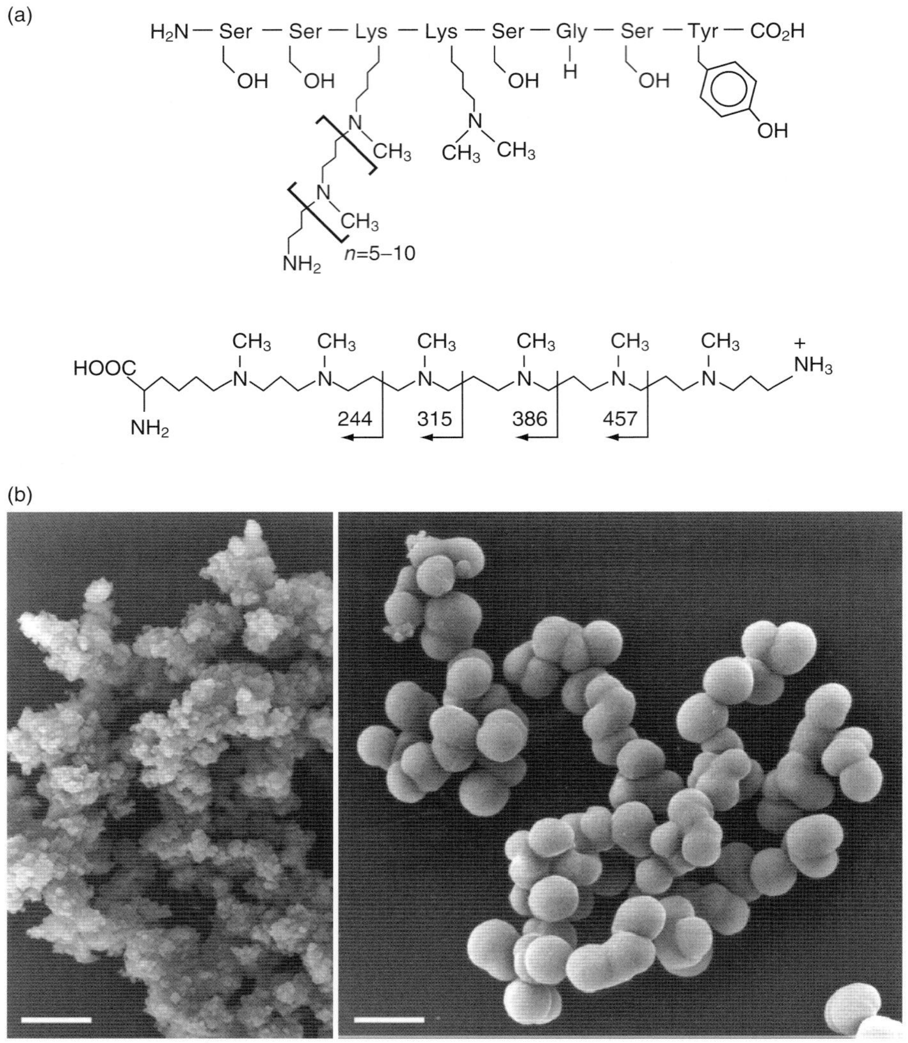

Fig. 2.10 (a) Amino acid sequence of silaffin, the protein promoting opal deposition from dilute silicic acid solutions and embedded in diatom opal; the side chain of poly-*N*-methyl-propylamine is shown expanded; the repeat number is variable. (b) Opal precipitated on two modifications of silaffin out of dilute silicic acid. (Courtesy of Nils Kröger, University of Regensburg, with permission from *Science*.)

The zygote produces a new frustule by an elaborate process (see Mann 1984; Pickett-Heaps *et al.* 1990).

D'Alelio *et al.* (2010) studied the life-cycle of a pennate diatom, *Pseudo-nitzschia multistriata*, by looking at a 10-year time series of cell size and abundances in the Gulf of Naples. They found that asexual and sexual stages occurred with remarkable regularity. Cell sizes ranged from 75 to 30 μm, and large cell sizes (indicating auxospore formation) occurred every 2 years (after ~200 generations) with a subsequent decrease in size over the following period. This is the first report of such regulation of cell division and sexual reproduction across a regional population in a diatom. The mechanism of this regulation is unknown.

Dinoflagellates – Dinophyceae (Pyrrophyta)

The biology of dinoflagellates has been reviewed at book length by a group of experts (Taylor 1987). Most of the flagellated phytoplankton have two flagellae. These may be similar in length, coiling, insertion, presence of scales or hairs, or they may be distinct. In dinoflagellates, one flagellum is structurally complex and wraps around the equator of the cell in a groove, the *cingulum*. Its wave-like movements serve to rotate the cell in the water as it swims, a characteristic from which the name of the group derives: διηοσ, "whirling". The other, simpler flagellum originates

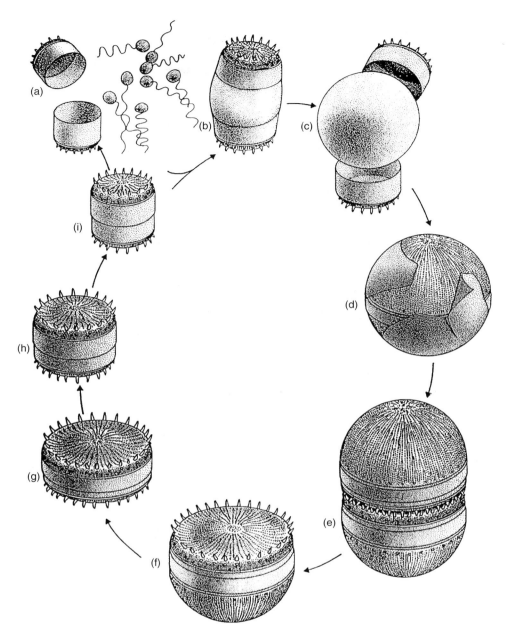

Fig. 2.11 Events in the life-cycle of *Stephanodiscus*. Cell reduced to small size (a) forms motile gametes. Gametes fuse with haploid oogonial protoplasts, which form enlarged auxospores (b). Auxospores expand (c), then break open revealing sculptured shell of the initial cell (d). (e) The initial cell divides, producing vegetative cells (f then g). Size reduction then proceeds again through many mitotic division cycles (h and i). Formation of motile gametes is only supposed for this genus, not observed directly. (After Round *et al.* 1990.)

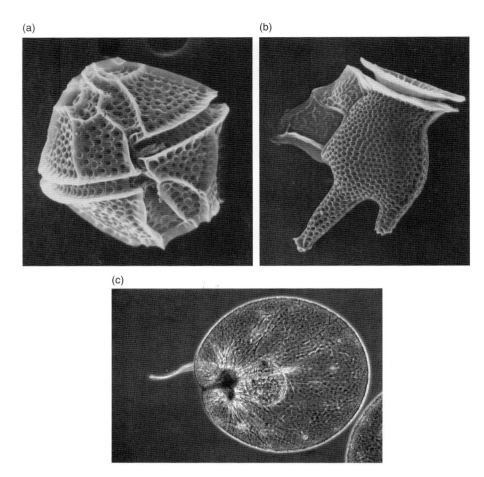

(a)

(b)

(c)

Fig. 2.12 Variation of form among dinoflagellates. (a) *Lingulodinium polyedrum*, a thecate dinoflagellate, ~30 μm. (b) *Dinophysis tripos*, a tropical flagellate with a "sail" ~100 μm long. (c) *Notiluca scintillans*, a large, naked, predatory dinoflagellate. The structure extending from the cell is a "tentacle", not a flagellum; diameter varies from 200 to 2000 μm (a and b – courtesy of J.D. Dodge, after Dodge (1985); c – courtesy of Jan Rines, University of Rhode Island.)

just behind the insertion of the first, passes "rearward" in a longitudinal groove, the *sulcus*, then extends beyond the cell. Waves propagate along this flagellum, pulling the cell through the water. Motion through the water at the scale of a flagellum is dominated entirely by viscous effects (see Chapter 1 for an explanation).

Dinoflagellates divide into three broad groups: the unarmored Gymnodiniales and the armored Peridiniales and Dinophysiales. Dodge and Crawford (1970), however, have shown that the species really form a gradual series. All forms have a pellicle constituted of a cell membrane externally and underlain by flattened vesicles. The vesicles are filled in armored forms with cross-linked cellulose, forming plates. The pellicle in both groups divides into an *epicone* ahead of the cingulum and a *hypocone* behind. Plates of armored forms are arranged in a heritable pattern over the epi- and hypocones, and the patterns distinguish subgroups and species. These morphological species designations do not always match up with SSU rRNA designations. A range of forms is illustrated in Fig. 2.12. Peridiniales are roughly

biconical, tapering from the cingulum toward rounded anterior and posterior ends. Plates in two rows surround both the epicone and hypocone. Some of the plates can bear expanded horns, as in *Ceratium*, a common genus. Plates may also bear spines or be otherwise elaborated. The Dinophysiales have much smaller individual plates, fused into anterior and posterior valves, with the cingulum and sulcus bordered by thin expansions or crests arising from the edges of the grooves. The pattern varies between species by differences in expansion and sculpture of crests. Many of the resulting forms are reminiscent of the intergalactic cruisers of science-fiction movies. Dinophysiales are mostly tropical and exclusively marine, rarely if ever a major constituent of the plankton.

Some members of both naked and armored groups can prey upon smaller organisms such as diatoms, ciliated protozoa even copepod nauplii. Some forms are not photosynthetic; they lack chloroplasts, and depend exclusively upon prey for nutrition. In *Protoperidinium* and the *Diplopsalis* group, a membranous sac, called a pallium, is produced by

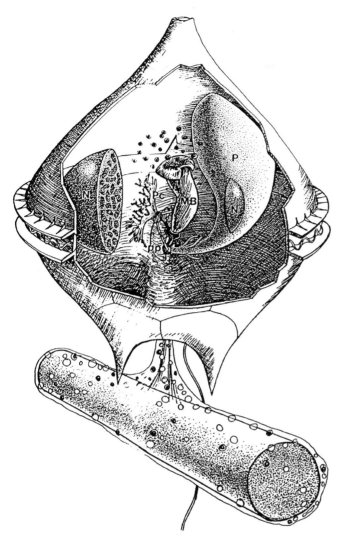

Fig. 2.13 A dinoflagellate, *Protoperidinium spinulosum*, externally digesting a cylindrical diatom in a pallial sac extruded from a microtubular basket (**MB**) through an aperture in the cell armor near the flagellar pore. Other abbreviations: **N** – nucleus, **pp** – pore plate, **Pc** – collecting pusule, **P** – sac pusule. (After Jacobson & Anderson 1992, with permission from the *Journal of Phycology*.)

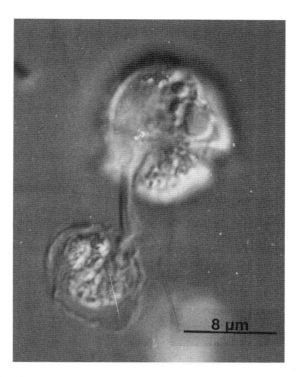

Fig. 2.14 The feeding peduncle of *Gymnodinium fungiforme* (above) extending into an unidentified food particle (below). (Courtesy of H. Spero, after Spero (1982), with permission from the *Journal of Phycology*.)

an elaborate apparatus (Fig. 2.13) internal to a "mouth", or pallial pore, located in the sulcus. In *Protoperidinium spinulosum*, prey become attached first to a slender pseudopod strengthened with protein microtubules called a tow filament. The pallial membranes are then extruded around it and digestion proceeds externally in a pallial sac (Jacobson & Anderson 1992).

The naked, non-photosynthetic *Gymnodinium fungiforme* attacks prey by insertion into their tissue of a tubular peduncle (Fig. 2.14; Spero 1982). From a cell about 15 μm in diameter, the peduncle can extend at 3 μm diameter for 12 μm, supported by microtubules derived from an internal microtubular "basket" similar to that seen in forms secreting pallial sacs. Cytoplasm from the prey is actively trans-

ported through the peduncle into the dinoflagellate cell, and forms a mass of food vacuoles. Many (if not all) photosynthetic dinoflagellates are also phagotrophs, so that mixotrophy is a hallmark of the group.

Some dinoflagellates from all groups are bioluminescent (hence an older division name, Pyrrophyta, meaning "fire plants"), although most groups also have forms that are not. Their light production accounts for a wide variety of dramatic marine phenomena, including shining boat wakes, night-sparkling beach sands, and entire bays glowing in the tropical night. One bay in Puerto Rico, Bahia Fosforente (also called Bahia Mosquito), is named for this glow, which is believed to arise from a complex scheme in which organic matter from mangrove trees along shore supports bacteria, which produce extracellular vitamin B_{12}, which then allows the vitamin-requiring and luminescent *Pyrodinium bahamense* to proliferate. Light is produced in some forms by a diffuse, soluble luciferin–luciferase system (a chemoluminescent compound and the enzyme moderating energy transfer to it), and in other forms by a similar but membrane-bound biochemistry. Dinoflagellates with membrane-bound luminescent organelles, called scintillons, are capable of remarkably intense, brief flashes, usually in response to mechanical stimulation of the cell. In many species, bright ambient light suppresses bioluminescence, presumably saving energy when luminescence could not be seen above the background. Experiments (Esaias & Curl 1972) make clear that luminescent flashing set off by motion of the feeding limbs of planktonic herbivores shuts down their feeding activity, reducing rates of grazing on dinoflagellates. The mechanism certainly exploits the fact that a predator with a mass of glowing particles in its vicinity will attract its own visual predators. Thus, selection soon leaves only those individuals that stop feeding when they stimulate luminescent flashing.

Internal organization is complex in the dinoflagellates. The large nucleus of the cell is centrally located. Distinct chromosomes (and lots of them) are present throughout the cell cycle in the form of thick rings of DNA strands that are wound into secondary helices like twist donuts (Oakley & Dodge 1976). There are almost no proteins in dinoflagellate chromosomes, and those that do exist are different from chromosomal proteins in most eukaryotes (Rizzo & Nooden 1973). Mitochondria and Golgi bodies are scattered through the cytoplasm. Photosynthesizing chloroplasts are located around the periphery of the cell. They are constructed in fairly typical fashion with many layers of stacked membranes. Additional organelles occur in some forms but not others. One or more sacs, called *pusules* and filled with a pink fluid, attach to the canal leading inward from the cell surface at the root of the longitudinal flagellum. Pusules are believed to be involved in osmoregulation or possibly buoyancy adjustment (Dodge 1972). Proteinaceous *pyrenoids* involved in starch formation may be present. Some forms have a reddish stigma or eye-spot, like those in a variety of protozoan forms, mediating changes in rate or direction of movement in response to light. In dinoflagellates, eye-spots certainly permit the vertical migratory behavior (down at dusk, up at dawn) exhibited by many forms (e.g. Eppley *et al.* 1968). *Trichocysts*, organelles also found in ciliated protozoa like *Paramecium* and in some Chlorophyta, are located beneath pores in the theca and can explosively project a small protein thread into the surrounding medium. It is believed the thrust thus generated pushes the cell suddenly away from the disturbance causing the discharge. Sudden displacements of only a few body diameters may be enough to prevent predation and may also be involved in prey capture.

Before asexual reproduction in dinoflagellates, the pellicle splits and the cell emerges, naked. Presumably a new cell membrane forms beneath the vesicular layer. The cell swells into a globular form and divides into two, and sometimes four or eight daughter cells. Flagellar replication takes place prior to division generally, and each daughter receives a complement of two flagellae. Some organelles – pusules are an example – appear to be absent during reproduction, reforming later in the cell cycle. Nuclear division differs in many details from that in most eukaryotic forms. The nuclear membrane is present throughout and divides by formation of a progressively constricted waist. There are no centrioles, but the chromosomes migrate into the daughter portions of the nucleus along the inner surface of the nuclear membrane, following microtubules that form and pass through the dividing nucleus from the cytoplasm. Sexual reproduction takes place in most, if not all, dinoflagellate species. Von Stosch (1973) has described this in detail for a species of *Gymnodinium*. Coupling involves fusion of two cells along their sulci, followed by pairing of the chromosomes from the usually haploid parent nuclei, release of diploid swarmers, and finally, meiotic production of cells that reacquire the usual vegetative cell form (Faust 1992). Many other schemes of sexual reproduction have been described as well (Beam & Himes 1979).

Dinoflagellates are responsible for the seasonally recurring phenomenon of *red tides*. Off California, and rarely Oregon, USA, red tides can be seen in summer from bluffs above the coastal sea as irregular patches of reddened water. The intensity of the color ranges from barely visible to an impression of a massive spill of tomato soup. These patches are formed by intense blooms of one or another dinoflagellate. Species commonly involved vary with location. Red tides of *Lingulodinium polyedrum* are the most usual off Southern California. Off Florida the commonly blooming species is *Karenia brevis*. Both forms contain potent neurotoxins, brevetoxins and yessotoxins, respectively. Some zooplankton avoid "toxic" cells, while others ingest them and are harmed or killed, and still others eat them and are unaffected. Red tides can kill fish in massive numbers, causing messy wash-ups on beaches, making vacationers (and hoteliers) unhappy. That is much more common in Florida than on the West Coast. Toxins can accumulate to lethal levels in clams and oysters, leading to neurotoxic shellfish poisoning in careless diners. Bona fide fatal cases are very few for the US West Coast. Worldwide, red tides and other toxic phytoplankton blooms appear to have been increasing in frequency, particularly at higher latitudes. It is uncertain whether the change is due to human impacts upon coastal environments (e.g. hog farm effluents), but it is extremely likely (Glibert *et al.* 2005). Global warming, also a human impact, may play a part in the increase. Intense scientific and public interest surrounded the discovery of a highly toxic dinoflagellate, *Pfiesteria piscicida*, which produces a potent, fish-killing neurotoxin that can be transferred from the water to the air, affecting people directly (Burkholder & Glasgow 1997). Details of the life-cycle stages of *P. piscicida* have been described by Litaker *et al.* (2002).

In addition to their several roles as phytoplankton, microheterotrophs, illuminators of white caps at night and toxic bloom culprits, dinoflagellates are algal partners in a diverse array of symbioses with animals. Called *zooxanthellae* when symbiotic, they reside intracellularly in their animal hosts that harvest photosynthate from them. Such partnerships with zooxanthellae are found in several groups of pelagic protozoa (foraminifera, radiolaria), coral polyps, sea anemones, the giant clam (*Tridachna* sp.) and sundry nudibranch snails. Volumes of information about these relationships have never quenched the curiosity they evoke, so research goes on.

Phytoplankton viruses

More than 50 species of phytoplankton have been shown to contain viruses or virus-like particles, and it is likely that viruses infect every major algal division (Munn 2006). Many of these viruses have now been characterized (Lawrence 2008), and they may play a major role in controlling the size and duration of phytoplankton blooms. This was well illustrated for *E. huxleyi* in mesocosms (1 m³ containers) (Martinez *et al.* 2007) where a bloom was followed by a rapid increase in virus particles. Exposure of the vegetative (coccolith-bearing diploid) stage to the virus induces the sexual stage of the life cycle (Frada *et al.* 2008). The haploid sexual stage of the coccolithophore is immune to the virus, thus allowing the population an escape mechanism.

Five viruses infecting diatom species have been studied, and these replicate either in the cytoplasm or in the nucleus. Eissler *et al.* (2009) studied the lytic cycle of an intranuclear virus infecting the diatom *Chaetoceros wighamii*. Following inoculation of cultures with the virus, rod-like arrays appear within the nucleus during the early stage of infection (Fig. 2.15).

These are replaced by virus-like particles over the following 24 hours, with about 20% of cells infected. Cell abundance then declines, and the free viral abundance increases.

Most algal viruses appear to be species-specific. Additional studies are under way to characterize the diversity of algal viruses, host specificity, and the ecological role of viruses. Eventually we may know how much phytoplankton mortality is due to viral lysis versus predation.

Phytoplankton standing stocks

Although many different types of phytoplankton can be present in seawater samples, it is often useful to have a measure of the total phytoplankton biomass or standing

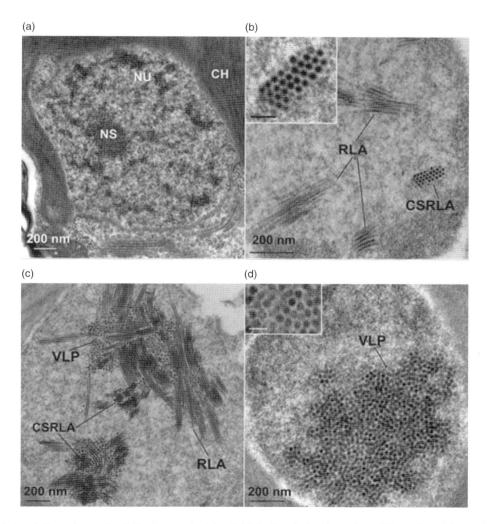

Fig. 2.15 Transmission electron micrographs of *Chaetoceros* cf. *wighamii*. (a) Section of a healthy nucleus: **NU** – nucleus, **NS** – nucleolus, **CH** – choroplast). (b) Section of a cell at early infection, showing rod-like arrays (**RLA**) and rod-like arrays in cross-section (**CSRLA**), including an insert for more detail (scale bar, 50 nm). (c) Thin section of a cell at mid-infection, showing virus-like particles (**VLP**), RLA, and CSRLA. (d) Thin section of a cell at late infection, showing VLP, including an insert for more detail (scale bar, 50 nm). (After Eissler *et al.* 2009, with permission from the *Journal of Phycology*.)

stock. Chlorophyll-*a* is present in almost all phytoplankton, and the amount of it in seawater is a reasonable, if imperfect, measure of phytoplankton standing stock. Moreover, chlorophyll concentration is relatively easy to determine. Phytoplankton are removed from a known volume of water with a suitably fine filter (glass-fiber mesh is most common), then chlorophyll is extracted with acetone and quantified by spectrometry, chromatography, or, since it shows red fluorescence in blue illumination, with a fluorometer. Parsons *et al.* (1984) give typical recipes for these techniques.

Greater spatial and temporal resolution of variations in phytoplankton stocks is often acquired with *in situ* fluorometers (Box 2.5). These can be deployed on moorings for time-series sampling, or lowered through the water column to determine vertical profiles of phytoplankton standing stocks. *In situ* fluorometric measurements are then calibrated with a set of extracted chlorophyll measurements.

Satellite-based estimates of chlorophyll

All of oceanography, particularly its physical and biological aspects, has been intensely challenged by satellite data. Snapshots from space of temperature distributions have challenged physical oceanographers, because earlier analyses of data taken from ships were blurred by widely spaced stations and the motion of patterns during sampling (low "synopticity"). Satellites swing across the Earth in minutes, gathering images from very wide swaths. Some sit in geostationary orbits and get instantaneous, nearly whole hemisphere images. In the mid-1970s, we were suddenly able to see the layout of variability on scales of a few kilometers. This reveals offshore jets and eddies, current meanders and surface ring structures. The dynamism of ocean processes was a major surprise. There are only vague hints in the literature that anyone anticipated that it would look as it does from this distant and instantaneous perspective. Physicists had dropped the necessary higher-order terms in their hydrodynamic models, thinking they were too small to have significant, observable effects. From the shipboard perspective, they had been right. From the instantaneous perspective they had blurred the picture dramatically. Biologists were surprised by similar data. The jets and swirls show up not only in satellite temperature maps (e.g. AVHRR), but in pictures based on ocean color, variation of which is mostly determined by chlorophyll content. Here, too, the swirls and jets, first seen in Coastal Zone Color Scanner images, are impressive. A picture (book cover) of the northwest Atlantic shows (albeit in "false" color) some of the dramatic features that changed the understanding of ocean processes.

Just as impressive are the short-term changes in quantities and distribution from one picture date or season to the next. Satellites provide new pictures on time scales of less than a week, the intervals required for records from orbital swaths to accumulate as regional or global images. Short-term and seasonal comparisons become possible, and the general correlation between production rates and chlorophyll standing stock makes possible more-or-less convincing estimates of regional and global primary production rates. We can add up approximations of how much carbon is being incorporated in organic matter and get at large-scale biogeochemical transformation rates with moderate precision. Recently and currently active color sensors on satellites include SeaWiFS (now dead), MODIS-AQUA (fading), MERIS (European Space Agency), Oceansat (India), and FY1-D (China). You can find an archive of Level 3 (i.e. elaborately corrected, averaged, and mapped) SeaWIFS images at http://seawifs.gsfc.nasa.gov/SEAWIFS.html. Monthly global averages available there show the dominant features of seasonal cycling of phytoplankton stock in all ocean regions, except for the most polar areas and the two subpolar zones in their respective winters.

SeaWiFS, operational from September 1997, recorded over swaths adding up to global coverage every two days. The Goddard Space Flight Center of NASA produced processed images rapidly, with weekly averages as global images appearing a few days after the data were in, monthly averages right at the end of the month, and so on. The global picture (Plate 2.3) shows the layout of blue (low chlorophyll) and green to red (artificial colors indicating high and higher chlorophyll) regions across the globe. Note that the current satellite array also generates color pictures of the land, which have their own uses.

Satellite-based estimates of chlorophyll fit ship-based data in a general fashion (Bailey & Werdell 2006) (Fig. 2.16) over a range from 0.02 to ~20 mg m^{-3}. The comparison in Fig. 2.16 is between SeaWiFS OC4v4 (Box 2.6) values and a matching global set of surface-ship values called NOMAD. A large fraction of satellite estimates, the output of Eqn. 3 in Box 2.6, are within the range one-third to three times the field estimates, even somewhat better in the oligotrophic range (<0.3 mg m^{-3}). The use of logarithmic scales in the plot accommodates both the wide range of chlorophyll in the oceans (0.01 to >20 mg m^{-3}) and the substantial variation (essentially uncertainty for biological purposes) in the relationship of the water-leaving spectral signal to variations in chlorophyll concentrations. It must be pointed out that not all of the variation comes from the satellite estimates. Chlorophyll is not measured by water sampling without error or variation, and modest variation is expected across the range of pixels included in the patches averaged for comparison to the ship data. Furthermore, roughly equal amounts of in-water light attenuation, and thus water-leaving spectral variation, derive from chlorophyll and from accessory pigments of phytoplankton. At any given concentration, there is significant variation in the ratio of their concentrations and thus their effects, but the ratios examined (on log–log

Box 2.5 *In situ* fluorometry

Because they contain chlorophyll, phytoplankton can be quantified *in situ* by exciting, then measuring, their fluorescence. Water adjacent to a small window on an *in situ fluorometer* (Box Fig. 2.5.1) is flashed with a xenon lamp filtered at 455 nm, and the resulting fluorescence due to chlorophyll at 685 nm is measured with a photomultiplier circuit. Corrections for light-source variation are made by reporting the ratio of fluorescence to light intensity measured internally in the housing. Neveux *et al.* (2003) made periodic calibrations of this signal by filtering phytoplankton from the vicinity of the sensor and extracting chlorophyll for determination by spectrofluorometry. Fluorescence measured in this way for near-surface water samples varies strongly with external illumination. Daytime fluorescence is reduced by non-photochemical quenching where the excitation energy is transferred to photoprotective pigments or dissipated as heat. Maximum fluorescence is seen at night. This variation decreases with depth. Thus, there is artifactual day–night and depth variation (Box Fig. 2.5.2). Careful accounting for such effects must be incorporated in field studies with fluorometers. These instruments are deployed as vertical water-column profilers and as moored recorders.

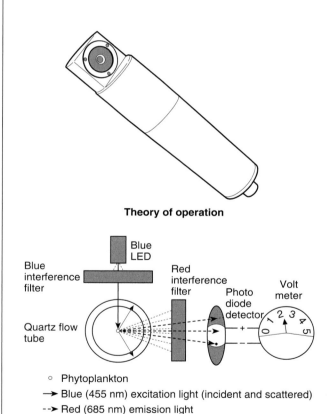

Theory of operation

- Phytoplankton
→ Blue (455 nm) excitation light (incident and scattered)
--→ Red (685 nm) emission light

Box Fig. 2.5.1 A commercial fluorometric chlorophyll *a* recording device with a diagram of its operating principle. Blue light is beamed into an observation space and the resulting red fluorescence is measured by a detector and recorded (Chelsea Instruments).

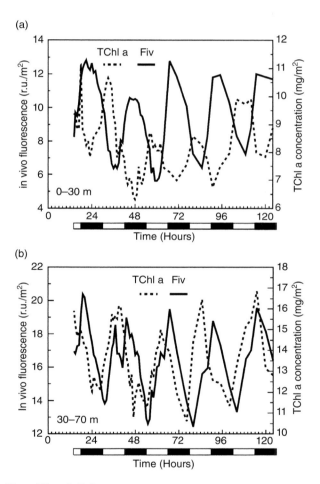

Box Fig. 2.5.2 Diel variations of integrated *in vivo* chlorophyll fluorescence (**Fiv** – solid line) in the (a) 0–30 m and (b) 30–70 m layers during a 5-day time series at the equator. Comparison with diel variations of integrated total extracted chlorophyll-*a* (**TChl** – a dashed line). (After Neveux *et al.* 2003.)

scales) over the entire range from severe oligotrophy to pea-soup green waters are well fitted by a 1:1 relationship (Trees *et al.* 2000). Those two aspects of chlorophyll and accessory pigment variation contribute both to the overall excellence of chlorophyll-*a* (C_a) estimates relative to *in situ* estimates and to the substantial variation around the general trend. While the satellite estimates of C_a are not always accurate for any given pixel, with sufficient areal averaging the data are very useful because the estimator shows no obvious bias. Regional and global totals benefit from massive averaging and should be reasonably accurate.

On the other hand, the NOMAD (or any sea-surface-"truth" data-set) can be dissected into regional subsets, and those often show substantial offsets from the matching surface data, particularly in coastal regions. For example, a plot of surface chlorophyll concentration from NOMAD vs. SeaWiFS C_a estimates from Eqn. 3 (see Box 2.6) from coastal sites in the Atlantic Ocean (Fig. 2.17) shows substantial overestimates from the satellite algorithm for both moderately rich North American waters (near $1\,mg\,Chl\,m^{-3}$) and for the very oligotrophic Mediterranean. The match is better, at least apparently unbiased, for estimates from near the southwest coast of Africa. In contrast, C_a is generally underestimated for waters around the Palmer Peninsula in Antarctica (not shown).

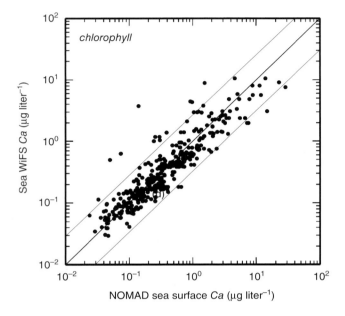

Fig. 2.16 Comparison of OC4v4 SeaWiFS chlorophyll estimates (text Eqn. 3) (ordinate) to the NOMAD shipboard measures (abscissa) of sea-surface chlorophyll from a wide range of ocean sites. The heavy line represents a 1:1 relationship, and the good fit of the central tendency implies biases are small. The thin lines represent offsets from one-third to three-fold, encompassing most of the comparisons. (After Bailey & Werdell 2006.)

Phytoplankton pigments

Photosynthesis underlies most biological energy and material conversions on Earth, so understanding of it is central to ecology and biological oceanography. Light-absorbing pigments make photosynthesis possible. Biological pigments generally are molecules carrying substantial systems of conjugated double bonds among carbon atoms. The resonant electrons of those unsaturated carbon chains can absorb a photon, shift to a new energy state, then pass the acquired energy into enzyme-regulated reactions. Chlorophyll-*a* (Fig. 2.18) acts as the key pigment in all photosynthetic organisms, except in *Prochlorococcus*, which has the very similar divinyl-chlorophyll. Functional

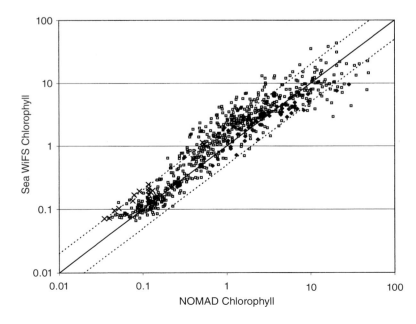

Fig. 2.17 Comparison of OC4v4 SeaWiFS chlorophyll estimates to NOMAD estimates from just Atlantic coastal sites. Lines represent the 1:1 relation and ½- to two-fold variation. Open squares are from along the North American coast; filled diamonds from off SW Africa; and **X**s from the Mediterranean. (Analysis and graph generously provided by Janet W. Campbell, University of New Hampshire.)

Box 2.6 Algorithms for satellite-based estimates of chlorophyll

The exact algorithms for SeaWiFS, MODIS, and MERIS data interpretation are not simple, and a sort of insiders-only technical haze covers the methodology. Outlines for many of the component algorithms are at:

http://oceancolor.gsfc.nasa.gov/DOCS/MSL12/master_prodlist.html/#prod11

Here is a general description for SeaWiFS; similar techniques are used for MODIS, MERIS and CZCS data. During daylight, a suite of spectral sensors looks down a telescope "folded" with prisms into a compact package. The image at the "eyepiece" of the telescope subtended a small (~1.2 × 1.2 km) sea-surface pixel. The telescope swung left and right across the path passing under the satellite, recording light from successive pixels. Dichroic beam-splitters and color filters divided light from each pixel into eight spectral bands, one for each recording sensor. Optical details differ among satellites; MERIS for example has a row of cameras looking down at different lateral angles. Sensors in all systems are charge-coupled devices similar to those in digital cameras. The spectral sensitivities in the chosen wavelength bands are calibrated before launch, and calibrations can be corrected in orbit by between-satellite comparisons. The spacecraft periodically transmit recorded results to ground stations by radio.

In operation, each sensor collects an amount of light (L_{total}) proportional to the water-leaving radiance from each pixel (L_W, which is the variable of interest) *plus* (i) light reflected from the sea surface in the pixel (L_r); *plus* (ii) light atmospherically scattered into the pixel-to-satellite path that left the sea outside the pixel (part of L^*); and *plus* (iii) light scattered directly from the atmosphere (another part of L^*). As a first processing step, many pixels are eliminated because a very broad spectrum (white light) indicates clouds, and because intense sun-glint from waves can overwhelm L_W. If those "flags" are not present, then L_{total} will be affected by the thickness of atmosphere between the sea and the sensor, which varies between pixels, thus requiring a pixel-specific transmittance fraction (T_A). So, we have:

$$L_{total} = L^* + T_A L_W + T_A L_r. \qquad (1)$$

L_r is taken to be modest and proportional to L^* and is lumped with it, giving $L_{total} = L^* + T_A L_W$. L^* is estimated from solar irradiance, which is approximated from a model (see just below). Transmittance can vary with atmospheric conditions, and T_A is estimated from some ratios of received irradiance at several wavelengths known not to vary from each other in their L_W. L_W is <10% of L_{total}, and a great deal of uncertainty in the L_W estimate is incurred in approximating T_A. All of the calculations, including the incoming irradiance model, must account for the zenith angle of the sun, the zenith angle of the satellite from observed pixels, and other angles including the curvature of the Earth (usually ignored by only using swaths near the satellite's nadir). These angles vary with time-of-day, season, and latitude, and the solar input varies with the distance from the sun (greater in summer, less in winter). Setting possible uncertainty aside for later evaluation, L_W values are calculated for each wavelength.

In most algorithms, L_W is next converted to $R_{rs} = L_W/E_t$, where E_t is the total downwelling irradiance of the same wavelength just below the sea surface. That is estimated from the model of solar irradiance for the pixel under study, corrected for atmospheric absorbance (T_A, again).

With estimates in hand of R_{rs} at several wavelengths, $R_{rs}(\lambda)$, a chlorophyll concentration (C_a) relationship is generated using C_a values of water samples collected from ships in the same pixels at close to the same time as the satellite overpass (in some data-sets within 3 hours). New sets of surface C_a measures are made on a recurring basis as new satellites recording at different wavelengths are launched, as old sensor calibrations shift and as new algorithms are tested.

Most such satellite estimators of chlorophyll in the sea surface in a pixel are functions of ratios of R_{rs} at several wavelengths. For example, one formulation used for SeaWiFS (termed Oc4v4) first determines R, the greatest of three possible ratios divided by $R_{rs}555$:

$$R = \max[R_{rs}443, R_{rs}490, R_{rs}510]/R_{rs}555, \qquad (2)$$

which is basically a ratio of blue light to green light coming up from the ocean (the numerical values refer to wavelengths in nanometers). Notice that smaller R corresponds to more chlorophyll (more green light relative to blue). Finally, chlorophyll ($mg\,m^{-3} = \mu g\,liter^{-1}$) is calculated from the fitted function:

$$C_a\ (mg\ m^{-3}) = anti\log_{10}(0.366 - 3.067R$$
$$+ 1.930R^2 + 0.649R^3 - 1.532R^4) \qquad (3)$$

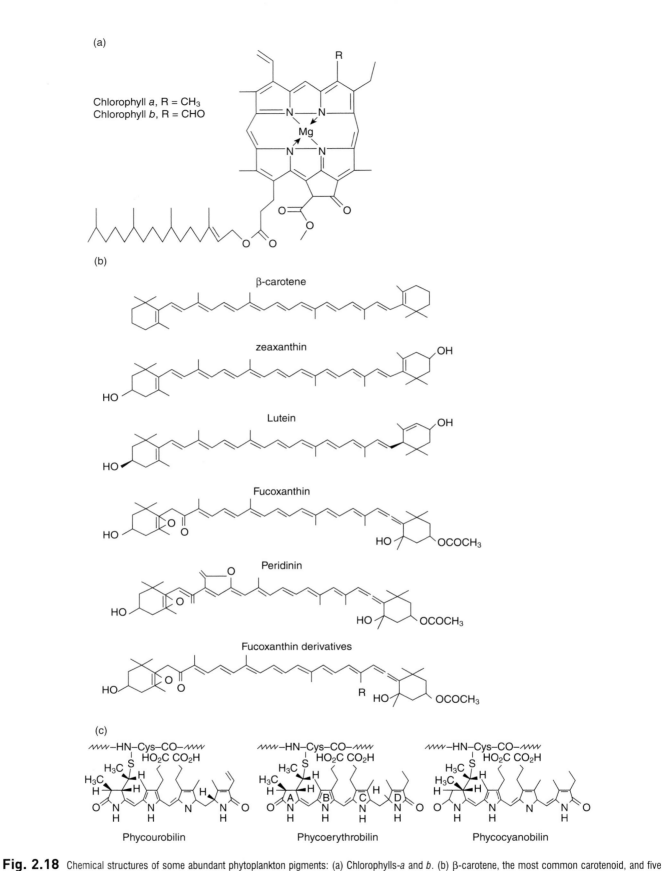

Fig. 2.18 Chemical structures of some abundant phytoplankton pigments: (a) Chlorophylls-*a* and *b*. (b) β-carotene, the most common carotenoid, and five xanthophyll derivatives. The R side chains for 19′ butanyloxyfucoxanthin and 19′-hexanyloxyfucoxanthin are: $CH_2-O-\overset{\overset{O}{\|}}{C}-C_3H_7$ and $CH_2-O-\overset{\overset{O}{\|}}{C}-C_5H_{11}$ respectively. (c) Several versions of phycoerythrin-like pigments based on open pyrrole structures.

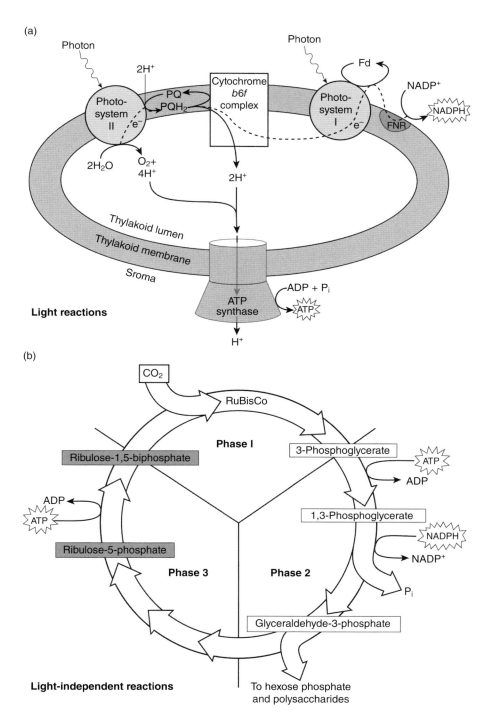

Fig. 2.19 Biochemical pathways of photosynthesis.

chlorophyll-*a* includes a large protein matrix that presents to the light a porphyrin ring with a magnesium atom held by central ligands. The ring has a tail, a linear carbon chain called phytol, by which it attaches to the protein portion of the system. Chlorophyll-*a* is associated with other protein-bound pigments (carotenoids, xanthophylls and several chlorophyll variants) in one of two types of "photosystem":

Photosystem I, or P700 for its wavelength of maximum absorption and Photosystem II, or P680. P680 applies absorbed energy (Fig. 2.19) to capture electrons from water, releasing protons ($4H^+$) that are actively transported into the thylakoid lumen, and oxygen (O_2) that diffuses out of the thylakoids. The electrons move in a complex series of transfers to P700. More photic energy is absorbed there, raising the energy level of the electrons, which then reduce $NADP^+$ (nicotinamide adenine dinucleotide phosphate) to NADPH, a moderately stable, diffusible molecule that carries reducing capacity (energy available for biosyntheses) into the cytoplasm. The protons pumped into the thylakoid generate an energetic gradient (up to 3.5 pH units) across the membrane. The only exit is through an enzyme complex, which uses the energy to esterify adenosine diphosphate (ADP) with additional phosphate ions to generate ATP. ATP is the principal, diffusible energetic molecule in cells. Along with NADPH, it is used to drive biosyntheses.

The remainder of the photosynthetic process occurs in "light-independent reactions", so called because they will proceed without illumination, given supplies of bicarbonate, ATP and NADPH. Sugars, lipids, and proteins are all produced by the dark reactions in the "stroma" layers of the thylakoids. ATP phosphorylates a five-carbon sugar, ribulose-5-phosphate (Ru5P) to ribulose-bisphosphate (RuBP), which thereby carries the energy to reduce and add a carbon atom from CO_2 in the presence of a key enzyme called ribulose bisphosphate decarboxylase ("RuBisCO"). The resulting six-carbon sugar splits into two three-carbon sugars that are both recycled to RuBP and forwarded to all the biosynthetic pathways of the cell. Those pathways, sequences of enzymatic reactions, apply the reducing power of NADPH to add complexity to the molecules and store more energy. Some of the free oxygen from P680 recycles within the cell to respiration, that is, oxidation of photosynthate, while the remainder diffuses from the cell. This photosynthetic oxygen production (about half of it globally comes from marine phytoplankton) drives the oxidative side of the ecological carbon cycle. Chemical steps in photosynthesis are complex but, thanks to ^{14}C-isotope labeling of substrates, are known in detail. Several biochemistry texts give good summaries (e.g. Mathews *et al.* 2000).

In addition to chlorophyll-*a*, all phytoplankton have accessory light-absorbing pigments. Some of these, termed "antenna" pigments, transfer electron excitation to chlorophyll-*a* in order to drive photosynthesis. In fact, a large fraction of the chlorophyll, including chlorophylls-*b*, -*c* and some -*a*, serves as antenna pigments, passing electron excitation to chlorophyll-*a* in the photosystems. β-carotene is another antenna pigment found in photosystems of all but one rare group of autotrophic plants (Chlorarachniophyta). Carotenes are hydrocarbon molecules with conjugated double bonds (Fig. 2.18) in chains long enough to retain and transfer photonic excitation. The Cyanophyta and Rhodophyta have distinctive pigments with this function, the phycobilins (Fig. 2.18), in which alternating double and single bonds occur in a tetrapyrrole resembling the porphyrin ring of chlorophyll, but not closed.

Other pigments, termed protective pigments, absorb photons to prevent photolytic damage to chlorophyll and the rest of the photosynthetic apparatus. Most of these belong to the class of pigments called xanthophylls.

Xanthophylls are close chemical relatives of β-carotene (Fig. 2.18), but all xanthophyll structures include one or more oxygen atoms. Different algal groups can be distinguished, even identified, by the kinds of pigments, particularly the xanthophylls, contained in their photosynthetic apparatus. There are about 30 classes of xanthophylls, but most phytoplankton groups have significant quantities of only one or a few (Table 2.2). Thus, these pigments can be extracted, identified by absorption spectra and quantified using chromatography (usually HPLC) to determine the presence or relative abundance of higher-order groups (divisions and classes) of the phytoplankton, as shown in Fig. 2.20.

Phytoplankton functional types

Phytoplankton functional types are groups of phytoplankton species that have in common a specific function of interest, e.g. calcification. Discriminating distinct groups of phytoplankton from satellite measures of ocean color is an area of active research. One approach is to use pigment-based size classes. Uitz *et al.* (2006) used satellite measures of surface chlorophyll and empirical relationships linking the amount of surface chlorophyll to the pattern of its vertical distribution and the relative proportions in three size categories: pico-, nano, and microplankton. To simplify the analysis, Uitz *et al.* (2010) used seven pigments to estimate how much of the chlorophyll-*a* is associated with each size group (Table 2.3).

There is some overlap of pigments among the various taxa (Table 2.2) and size groups, so the relationships are not exact. The results for June 2000 were calculated as the percentage and amount of vertically integrated chlorophyll associated with each group. Picoplankton form the dominant group in the subtropical gyres, with relative abundances ranging from 45 to 55%. However, picoplankton biomass is low everywhere. Microplankton are dominant in the subarctic zones and in the coastal upwelling zones, and are responsible for the highest vertically integrated chlorophyll values. Nanoplankton appear to be ubiquitous, with a relative contribution of 40–50%, with a significant enhancement of biomass in the equatorial regions and along the subantarctic convergence (Uitz *et al.* 2010). Other approaches for remote sensing of phytoplankton functional types include use of spectral measurements to define six phytoplankton groups (Alvain *et al.* 2008) and spectral retrievals of satellite-measured backscattering to define three size classes of phytoplankton (Kostadinov *et al.* 2010). Each of these approaches lacks the specificity of actual species identification and enumeration, but does provide a synoptic view of regional, seasonal (Fig. 2.20), and interannual variability of some phytoplankton groups in the ocean. Availability of newer spectral sensors for future satellite deployment may provide higher resolution for discriminating specific groups.

Table 2.2 Principal pigments in different phytoplankton groups reduced from Van den Hoek *et al.* (1995).

	SYNECHOCOCCUS	PROCHLOROCOCCUS	CHLOROPHYTA	PRASINOPHYTA	CRYPTOPHYTA	PRYMNESIOPHYTA	BACILLARIOPHYCEAE	CHRYSOPHYCEAE	XANTHOPHYCEAE	DINOPHYTA
Chlorophylls										
a	*	*	*	*	*	*	*	*	*	*
b		*	*	*						
*c*1						*	*	*	+	
*c*2					*	*	*	*	+	*
*c*3							*			
Phycobilins										
phycocyanin	*				*					
allophycocyanin	*									
phycoerythrin	*				*					
Carotenes										
α					*			+		
β	*	*	*	*	*	*	*	*	*	*
Xanthophylls										
zeaxanthin	*	*	+	+		+		+		
violaxanthin			*	*						
19′-hexanoyloxyfucoxanthin						+				
19′-butanoyloxyfucoxanthin						+				
diatoxanthin						*	*	+	*	
diadinoxanthin						*	*	+	*	*
alloxanthin					*					
peridinin										*
neoxanthin							+	+	+	

*, important pigment; +, pigment present.

Table 2.3 Pigments used as biomarkers to estimate abundance of phytoplankton functional types from satellite-derived estimates of surface chlorophyll (Uitz *et al.* 2010).

PIGMENT	ABBREVIATION	PHYTOPLANKTON GROUP	SIZE CLASS
Fucoxanthin	Fuco	Diatoms	Microplankton
Peridinin	Perid	Dinoflagellates	Microplankton
19′-hexanoyloxyfucoxanthin	Hex-fuco	Prymnesiophytes	Nanoplankton
19′-butanoyloxyfucoxanthin	But-fuco	Prymnesiophytes	Nanoplankton
Alloxanthin	Allo	Cryptophytes	Nanoplankton
Chlorophyll-*b* + divinyl Chl *b*	TChlb	*Prochlorococcus*	Picoplankton
Zeaxanthin	Zea	Cyanobacteria	Picoplankton

Equations for calculating the fraction (*f*) of chlorophyll attributed to each size group:

$f_{micro} = (1.41 [Fuco] + 1.41 [Peri]) / \Sigma DP_w$

$f_{nano} = (1.27 [Hex\text{-}fuco] + 0.35 [But\text{-}fuco] + 0.60 [Allo]) / \Sigma DP_w$

$f_{pico} = (1.01 [TChlb] + 0.86 [Zea]) / \Sigma DP_w$

where ΣDP_w represents the chlorophyll-*a* concentration reconstructed from the knowledge of the seven other pigments.

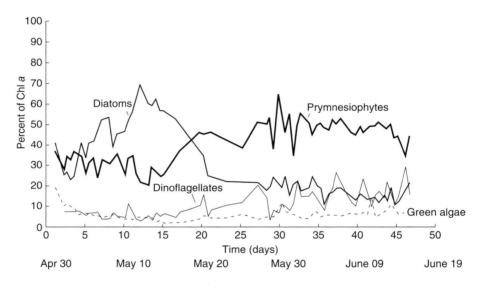

Fig. 2.20 Fractions of chlorophyll-*a* associated with accessory pigments from different phytoplankton groups (as labeled) during the course of the spring bloom in the northwest Atlantic (approximately 49°N, 20°W) in 1990. (After Barlow *et al.* 1993.)

In closing

This brief introduction covers just that part of phycology applicable to phytoplankton. Knowledge of algae is extensive and detailed. Because knowledge of algae is increasing rapidly with use of modern tools for microscopy, biochemistry, and molecular genetics, occasional hours in the library or on-line with recent issues of phycological journals are essential education maintenance for biological oceanographers. Refinements of techniques for *in situ* and remote sensing of phytoplankton distributions are providing important information for regional and global studies of variations in the standing stocks of the major phytoplankton groups. We address environmental factors affecting the algal growth rates and primary production in Chapter 3.

Chapter 3

Habitat determinants of primary production in the sea

In the trophic–dynamic approach to ecosystems, we try to measure the production at each link in the food chain or web, at each trophic level. "Production" is incorporation of new organic matter into cellular material, that is, an increase in biomass. For phytoplankton, this is done by photosynthesis, the dominant type of *primary production*. (Chemosynthesis is important in some environments.) *Gross* primary production is total photosynthate generated, and *net* primary production (i.e. growth) is gross production less respiration. Net production is available to herbivores. For herbivores, an increase in biomass can be expressed as a difference:

secondary production = phytoplankton eaten
− feces − respiration

As will be shown eventually, secondary and higher-level rates are much harder to measure than primary productivity. Production *rates* (biomass elaborated per unit time) are often termed *productivities*. Oceanographers speak of "the primary productivity", meaning the rate of phytoplankton production, usually measured as carbon newly incorporated in organic matter per unit area (or volume) per unit time.

The light-dependent reactions of photosynthesis (Fig. 3.1) comprise absorption of light energy; transfer of electrons through the photosynthetic reaction centers coupled with reduction of water to oxygen and production of adenosine triphosphate (ATP) and the reduced form of nicotinamide adenine dinucleotide phosphate (NADPH). The NADPH produced is then used as reducing power for the biosynthetic reactions in the Calvin cycle of photosynthesis. The light-independent reactions (formerly called "dark" reactions) of photosynthesis fix CO_2 into carbohydrates. Thus, photosynthesis is often described in terms of the biochemical elaboration of carbohydrate (sugars and their polymers). We will follow that tradition, but keep in mind that most unicellular algae direct more than half of their reduced carbon to protein synthesis, and that the principal store of high-energy molecules is often lipid. The overall photosynthetic reaction producing carbohydrates is:

$$2H_2O^* + CO_2 + light \xrightarrow{\text{Chl-}a} CH_2O + H_2O + O_2^*$$

The asterisks indicate that the oxygen produced is from the water reactant, not from the carbon dioxide. The water on the product side derives from a dehydroxylation step. This generalized reaction has two components, the light reactions of photosystem II (PSII) and photosystem I (PSI):

$$2H_2O^* + light \xrightarrow{\text{PSII}} 4H^+ + 4e^- + O_2^*$$

$$NADP^+ + H^+ + 2e^- \xrightarrow{\text{PSI}} NADPH$$

and the light-independent reaction:

$$CO_2 + 4H^+ + 4e^- \xrightarrow{\text{RuBisCO}} CH_2O + H_2O$$

where the free energy of cleavage of high-energy phosphate bonds of ATP and the reducing power of NADPH are used to fix and reduce CO_2 to form carbohydrate. This reaction is mediated by the enzyme ribulose bis-phosphate carboxylase (RuBisCO).

The pigment system (considered in Chapter 2) gives the phytoplankton cell access to energy from most of the visible spectrum. An example of the relative roles of different pigments in absorbing light is shown in Fig. 3.2a.

Biological Oceanography, Second Edition. Charles B. Miller, Patricia A. Wheeler.
© 2012 John Wiley & Sons, Ltd. Published 2012 by John Wiley & Sons, Ltd.

(a)

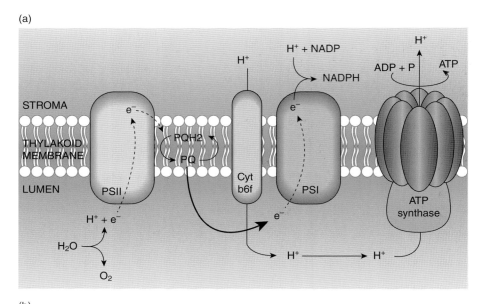

(b)

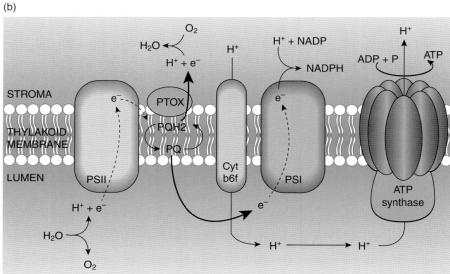

Fig. 3.1 Principal components and electron-flow pathways for the light reactions of photosynthesis. (a) Linear electron flow begins with excitation of the photosynthetic reaction centers, PSII and PSI, and with splitting of water by the oxygen-evolving complex of PSII. Electrons are passed from PSII to the plastoquinone (PQ) pool and then to PSI, where NADPH is formed. The cytochrome complex (cyt b_6f) pumps in protons that drive ATP production. (b) Under conditions of high light and low nutrients, plastoquinol terminal oxidase (PTOX) is activated and serves as an electron relief valve that uses excess electrons to reduce O_2 to water, so that less NADPH is formed than during linear electron transport. (After Zehr & Kudela 2009.)

Absorption by pigments is not 100% perfect, and some wavelengths are more effective than others at driving photosynthesis. This is shown in the "action spectrum" for photosynthesis (Fig. 3.2b). The highest overall effectiveness is found at those wavelengths centered at 465 nm, which are dominant in deeper ocean layers. This is true for phytoplankton generally. Lewis *et al.* (1985) have shown this by producing action spectra for natural phytoplankton assemblages at various ocean sites. All have their peak at 425–450 nm, a broad shoulder to 550 nm, and a rise at

675 nm (Fig. 3.3). Photosynthetic yield per quantum drops off rapidly below 425 nm and has low values from 600 to 650 nm – wavelengths not important in pelagic habitats except at the very surface.

There is a remarkable triple complementarity among the spectra of sunlight, the light transmissivity of water, and the absorption bands of various algal pigments. Absorption of light by water has a minimum with respect to wavelength, a window of clarity (Yentsch 1980), right at the peak range of wavelengths of solar irradiance (Fig. 1.5),

(a)

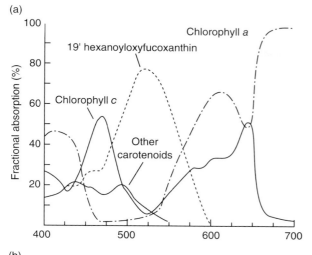

(b)

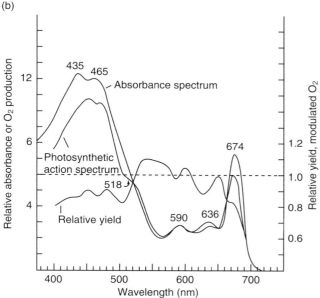

Fig. 3.2 (a) Fractional contribution to absorption of light by different pigments in live *Emiliania huxleyi*, a coccolithophorid. Contributions to absorbance add to 100% at each wavelength. Pigments are chlorophyll-*a*, chlorophyll-*c*, 19′ hexanoyloxyfucoxanthin, and other carotenoids. (b) Absorbance spectrum and photosynthetic action spectrum of oxygen production per quantum in *E. huxleyi*. The ratio [photosynthesis/absorbance], the relative yield, shows that absorbed quanta at all wavelengths are roughly equally effective. (After Haxo 1985.)

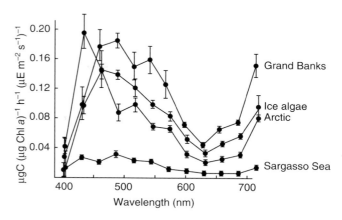

Fig. 3.3 Photosynthetic action spectra for natural phytoplankton assemblages from four regions: Grand Banks (showing ± 1 Standard Error), ice algae, Arctic open water, and Sargasso Sea. (After Lewis *et al*. 1985.)

The factors that control the overall rates of photosynthesis, primary productivity, in the sea are: (i) those that control the rates of reaction of the photosystems (PSI and PSII), and (ii) those that control the rates of the light-independent reactions. The former are light intensity and availability of water and carbon dioxide. Water and carbon dioxide are in abundant supply in seawater, although reaction rates can be forced somewhat by carbon dioxide loading. The latter include temperature and availability of *nutrients*. Nutrients are "fixed" nitrogen, phosphate, various metal ions, silicic acid (for diatoms and chrysophytes), and sometimes vitamins. Roles of a number of these are considered here.

Estimation of primary production

Measurement of productivity is most successfully done at the primary-producer level in pelagic habitats, because in some cases it is easy to separate phytoplankton from their herbivores and observe the rate of accumulation of phytoplankton cells. We have a number of methods that are simple in principle, more complex in practice. The basic method is to collect a bottle of seawater from a place and depth to be studied, check that no herbivores are present (use a coarse filter perhaps), and measure the increase in phytoplankton biomass during some time interval. This would most appropriately be the period of the natural illumination cycle, 24 hours, but adverse responses of the phytoplankton to confinement often force shorter incubations. Alternately, *any* photosynthetic product can be measured or the reduction of *any* photosynthetic resource can be determined. In practice only a few of

the wavelengths of visible and photosynthetically active light. The absorption peaks of phytoplankton pigments (Fig. 3.2b) tend to be centered over the deep trough in the absorption coefficient, *k*, spectrum of water (Fig. 1.5b). Most importantly, the absorbance peak for "antenna pigments" and chlorophyll-*a* acting together centers close to the deepest trough near 465 nm. This is not an accidental coincidence, but represents tuning by natural selection of the photosynthetic system to the properties of water.

these are useful. Look again at the overall reaction for photosynthesis:

$$2H_2O^* + CO_2 + light \rightarrow CH_2O + H_2O + O_2^*.$$

Oxygen production can be measured by the Winkler titration of samples before and after incubation. That only works easily in very rich environments with large populations of phytoplankton, because the usual change in oxygen concentration over a reasonable interval of measurement is about the same as the variability of the measurement. Photosynthetic O_2 evolution can also be measured with the tracer ^{18}O either directly or as the rate of ^{18}O and ^{16}O evolution from a mixture of $H_2^{18}O$ and $H_2^{16}O$ (Bender et al. 1987).

A long-favored method is the carbon-14 incorporation technique first introduced to ecology by Steeman-Nielsen (1952). A bottle of seawater, to which ^{14}C as sodium bicarbonate has been added, is incubated at its depth of collection in the sea (or more often in a simulated photic and thermal environment on deck). After an interval, the bottle is retrieved, the phytoplankton are filtered off, and the amount of ^{14}C incorporated in them is measured by scintillation counting. The net (since respiration goes on throughout the incubation as well as photosynthesis) primary production is obtained by multiplying the fraction of the ^{14}C taken up (e.g. counts per minute in filtered cells/counts per minute provided) by total carbonate in the bottle. Thus, a carbonate determination is required as well. The standard recipe for ^{14}C-uptake production measurement is given by Parsons et al. (1984). They give the following final equation for the calculation of the result:

$$\text{Photosynthesis (mg C m}^{-3}\text{ h}^{-1}) = [(R_s - R_b)W]/RT$$

(Eqn. 3.1)

where R is the counting rate to be expected for the entire addition of ^{14}C; R_s and R_b are counting rates for the filtered sample and a blank, respectively; W is the total weight of carbonate carbon in the water (mg C m^{-3}); and T is the duration of the incubation in hours. A dark-bottle uptake experiment is usually included in the observational design to account for carbon-isotope exchange processes other than those of primary production per se. Note that the volume of the incubation container is implicit in all of R_s, R_b, and R (say, ^{14}C counts per liter), so it cancels.

The ^{14}C method has been used all over the world since 1952. There is a large mass of data, and much of what we know about the vertical and geographical distribution of primary productivity comes from it. However, it has always been recognized that the method has some uncertainties associated with it, and during the 1980s even the general levels measured were questioned. The recognized problems fall into two classes:

1 Errors of the method itself, that is, procedural problems; and

2 Comparison of the results to field measures of variables that imply values different from those of the ^{14}C technique.

The simple outcome of the substantial reassessment, which started in about 1978 (Peterson 1980), is that some of the classical (pre-1980) ^{14}C-uptake values were too low. Fitzwater et al. (1982) introduced an elaborate cleaning of the bicarbonate inocula with ion-exchange medium (Chelex®), incubation in polycarbonate bottles, and maintenance of high cleanliness with respect to trace metals, particularly copper and zinc, in all aspects of the work. Rates compared to the standard inocula in glass bottles were two to eight times higher. The implication is that trace-metal contamination can poison phytoplankton, particularly in oceanic habitats. Trace-metal concentrations well above habitat levels were contaminants of early ^{14}C-bicarbonate used as tracer in productivity studies. The results from newer clean techniques are consonant with demonstrations of substantial sensitivity of phytoplankton to trace-metal poisoning. Methods closely akin to that of Fitzwater et al. (1982) have become standard. On the whole, new numbers are higher than old. For example, a subarctic Pacific study by Welschmeyer (1993) (Fig. 3.4) shows that clean methods in use after 1980 give higher results than those conventional in the 1960s and 1970s. The change for this comparison is almost a factor of two. In even more oligotrophic habitats, such as the Pacific central gyres, the increase is greater.

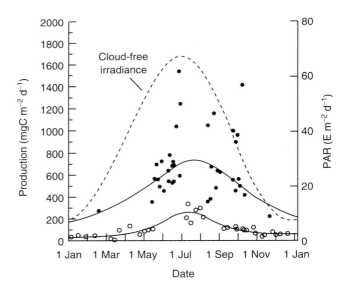

Fig. 3.4 Depth-integrated primary productivity in the subarctic Pacific compared to a generalized curve of PAR ("photosynthetically active radiation"). Filled circles are modern ^{14}C-uptake data from a trace-metal clean technique (1984–1988); open circles are data collected before 1980 with the then-standard ^{14}C-uptake method. (After Welschmeyer 1993.)

Another issue for ^{14}C measurements of primary production is whether the calculated rates represent gross or net primary production (Marra 2009). In short incubations with little respiratory loss, the ^{14}C technique estimates gross primary production. In longer incubation, some or much of the ^{14}C fixed and respired is refixed, so that the measurement is greater than net production, but less than gross production.

An array of studies showed that the microherbivores (protozoa, crustacean nauplii, rotifers) left in the incubation bottles of a ^{14}C-uptake study (because they cannot be filtered out without loss of phytoplankton) contribute a substantial fraction of grazing in the field, and they continue to graze during an incubation. Their contribution varies, but they can be eating one-third, even most, of the new photosynthate. Of course, at first they are just grazing on the phytoplankton that have no ^{14}C incorporated. As the experiment runs, their impact gets greater and greater, making short incubations preferable to long ones. This causes workers to use very high ^{14}C activity in their measurements. Landry and Hassett (1982) developed a dilution technique to account for microzooplankton grazing during primary-production studies, and we will consider it when we deal with microzooplankton grazing.

Bender et al. (1999) compared gross O_2 production rates to ^{14}C production in several ocean regions. Gross O_2 production tends to be about two to four times higher than net ^{14}C production. The ^{14}C estimates of primary production can be low if respiratory losses of [14]CO_2 and release of DO[14]C are not considered. Gross oxygen production rates will also exceed ^{14}C estimates when electrons are diverted from photosystem II (PSII) and used for reduction of oxygen rather than being moved to photosystem I (PSI) for synthesis of ATP (Fig. 3.1b). High levels of such oxygen reduction are common in oligotrophic waters with low nutrient availability and high light intensity (Mackey et al. 2008). Halsey et al. (2010) showed that in *Dunaliella tertiolecta* the difference between net and gross primary production and the allocation of carbon to different metabolites depended on the growth rate. They found that in slow-growing cells gross O_2 production and gross C-fixation were much greater than net production rates. Conversely, gross and net production rates were similar in short-term incubations for fast-growing cells. It is clear from field and laboratory studies that ^{14}C-based and O_2-based measures of photosynthetic activity measure different component processes.

The stable isotope ^{13}C can also be used to measure rates of photosynthesis when use of radioisotopes is impractical (Hama et al. 1983). Bicarbonate enriched with ^{13}C is added to seawater, and the incorporation of CO_2 into particulate carbon (PC) is followed by measuring changes in the [13]C:[12]C ratio of PC relative to the total CO_2 pool. The isotope ratio is usually measured by mass spectrometry. This method is less sensitive than the ^{14}C method; requires larger volumes of water for incubation; and is generally more expensive. Theoretically, ^{13}C and ^{15}N isotopes can be used together for the simultaneous measurement of carbon fixation and nitrogen uptake (Slawyk et al. 1977), but in practice separate incubations are usually done (Imai et al. 2002; Kudo et al. 2005, 2009).

A more recent approach to measuring primary production in the field is the use of fluorescence as a measure of the activity of PSII. Two common approaches are fast-repetition-rate fluorometry (FRRF) and pulse-amplitude-modulated (PAM) fluorescence. See Box 3.1 for details of the PAM method. These techniques are essentially instantaneous and avoid the requirement of containing water samples in incubation bottles. Cermeno et al. (2005) and Corno et al. (2006) found poor correspondence between near-surface (5 m) measurements of ^{14}C production and FRRF, but good correspondence between those measurements throughout the rest of the water column. Suggett et al. (2009) compared FRRF, [18]O production, and ^{14}C measurements. Their FRRF rates exceeded [14]CO_2 uptake by a factor of five to 10. Suggett et al. attributed the difference to the uncoupling of electron flow between PSII and PSI and stressed that use of FRRF to examine aquatic productivity needs to focus on "a systematic description of how electrons are coupled to C fixation in nature". In conclusion, the ^{14}C technique is still the "standard" measurement for primary production in the sea, and measures something between net and gross carbon fixation by phytoplankton. Other approaches are being developed, but most are based on the light reactions of photosynthesis; the exact relationship between total light energy captured during photosynthesis and net or gross carbon fixation remains to be determined.

We will return later to the measurement of primary productivity. We will compare rates in different ocean areas, try to explain the differences, sum up the global total and try to fit the oceans into the overall biogeochemical cycling of carbon. But, now we turn to the factors affecting the rates.

Effects of light intensity (also called irradiance, illumination, and photon flux)

The photosynthetic rate varies over the entire range of light intensity from darkness (negative net productivity due to respiration) to full sunlight at the sea surface (considerable "photoinhibition"). The relationship is called a photosynthesis vs. irradiance (P vs. E curve), where E refers to the flux of radiant energy in units of $mol\,quanta\,m^{-2}\,s^{-1}$ (Fig. 3.5). The symbol "I" used in older texts to denote

Box 3.1 Use of chlorophyll fluorescence to measure photosynthetic activity

Light energy absorbed by chlorophyll can be directed in three ways: energizing photosynthesis (photochemistry); dissipation as heat; or re-emission as fluorescence. Photochemistry includes the activity of PSI, PSII and the assimilation of carbon. The spectrum of emitted fluorescence has a peak at a longer wavelength than the absorbed light. To measure emitted fluorescence, a light source is switched on and off at high frequency and the detector is tuned to detect only fluorescence excited by the stimulating light.

Pulse-amplitude-modulated (PAM) fluorometers and fast-repetition-rate fluorometers (FRRF) are both used to measure the photosynthetic activity of phytoplankton and cyanobacteria. We provide a simplified explanation of PAM fluorometry here (Mackey *et al.* 2008) and refer the reader to Kolber *et al.* (1998) and Suggett *et al.* (2009) for details of the FRRF technique. Variable fluorescence (F_v from Box Fig. 3.1.1) is the maximum fluorescence to bright flashes of dark-adapted cells minus the fluorescence to standardized dim flashes, $F_v = (F_m - F_o)$. F_v provides an estimate of the maximum potential rate of electron transport through PSII for dark-adapted cells. The maximum *potential* relative photosynthetic efficiency is proportional to F_v. The relative fluorescence decreases with continuing exposure to light and reaches an asymptote (F_s) after about 5 minutes. Fluorescence emission (F_s) is smaller because some of the electron acceptors are reduced and no longer able to accept electrons. The fluorescence response of light-adapted cells to a saturating flash is F_m'. The *actual* relative photosynthetic efficiency of PSII in the light-adapted state is $(F_m' - F_s)/F_m'$, a ratio called Φ_{PSII}. Multiplying Φ_{PSII} by the intensity of photosynthetically effective light, I_A, provides a measure of the PSII electron transport rate at that intensity. Thus, $\Phi_{PSII} \cdot I_A$ is an estimate of gross photosynthetic oxygen production. See Mackey *et al.* (2008) for thorough derivation of those relationships.

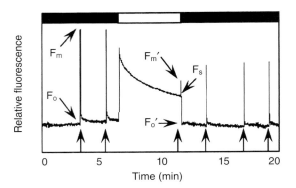

Box Fig. 3.1.1 The bar along the top shows periods of dark (black) and actinic light (white). Minimum fluorescence values (F_o and F_o') are determined using a low-intensity measuring light prior to delivery of the saturating light pulses. Arrows indicate the timing of saturating (3000 μmol quanta m^{-2} s^{-1}) actinic light pulses of 0.8 s duration. (After Mackey *et al.* 2008.)

PAM fluorometers generate data for F_o, F_m, F_s, F_o', and F_m' by the flashing and 'actinic' interval shown in Box Fig. 3.1.1. FRRF systems work on different principles, but are fast enough to generate roughly similar estimates as water column profiles. Both allow much higher spatial and temporal resolution of photosynthetic activity than is possible with the traditional ^{14}C technique. The fluorometric measures, however, have limited ecological relevance unless they can be converted to rates of carbon assimilation. When PSII, PSI and carbon assimilation activity are closely coupled, the fluorometric and ^{14}C measures give similar estimates of the rate of photosynthesis. However, the component photosynthetic processes are not always closely coupled, and that is a constraint for using PAM and FRRF for ecological studies. Some specific comparisons are described in the text.

irradiance is currently used to denote radiation intensity, which is the flux of radiant energy from a specified direction. The zero value of the photosynthetic rate scale cannot actually be at $E = 0$. Net photosynthesis is the result of "gross" photosynthesis minus respiration, which must have some small, positive value. The light intensity at which (Gross PS – Respiration) = 0.0 is termed the *compensation intensity*. However, in many regions (e.g. Fig. 3.5) the sensitivity to available light is so great that the positive intercept on the irradiance axis is not evident in the data. The response of the light-dependent (light-limited) portion of the relationship is linear and is represented by the initial

slope, dP/dE, often symbolized as α. At higher intensities, such that a great portion of the chlorophyll is at all times in an excited state, the processes limiting the overall rate are the light-independent reactions. Thus, the photosystems are light saturated, and no further increase in rate occurs in response to greater intensity. The P vs. E curve becomes horizontal.

At even higher intensities, the rate begins to fall off, an effect called *photoinhibition*. This effect has different causes in different phytoplankton. One important cause is "photorespiration". The intermediate products of the light-independent reactions include five-carbon, phosphate-

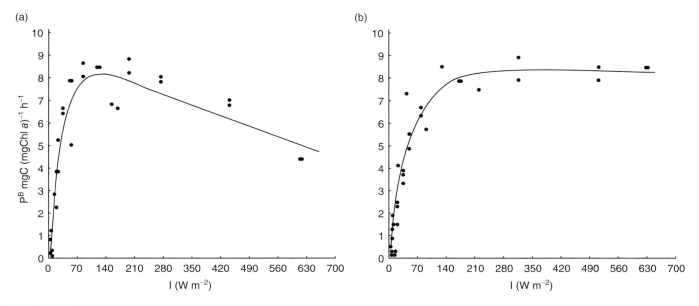

Fig. 3.5 Photosynthesis (per unit chlorophyll) vs. irradiance (*P* vs. *E*) curves developed in deck incubators for two different field situations: (a) a dense (13 mg Chl m^{-3}) diatom assemblage off Peru, exhibiting some photoinhibition; and (b) a predominantly flagellate assemblage of low density (0.3 mg Chl m^{-3}) off Nova Scotia with no evident photoinhibition (2 W m^{-2} = 10 µmol photons m^{-2} s^{-1}). (After Platt *et al.* 1980.)

esterified sugars, which are photolabile. Under intense light they break down into phosphoglycolic acid (C_2) and phosphoglyceric acid (C_3). The former is either excreted from the cell or metabolized to CO_2, but it cannot be returned to the photosynthetic pathway. The increase of this process (and probably other similar ones) reduces the overall rate of net photosynthesis.

Light-intensity response curves ("*P* vs. *E* curves") are all of this basic form, but differ in initial slope (α), maximum rate ("P_{max}"), intensity at onset of photoinhibition, and rate of decrease due to photoinhibition. When photosynthetic rates are normalized to (i.e., divided by) the levels of chlorophyll or cell carbon, the initial slope and maximum rates are denoted as α^B and $P_{max}{}^B$, where *B* denotes the normalization parameter used. The intersection of the initial slope of the *P* vs. *E* curve with P_{max} is K_E, the light-saturation parameter that indicates the irradiance where the shift in the controlling factor changes. Light responses may differ between clones of the same species cultured from different (or even the same!) habitats; between samples from different depths at the same station; and they differ substantially depending upon the history of light exposure of the phytoplankton. Various functional forms have been suggested to describe the *P* vs. *E* relationship. Several equations suggested by Platt and others are widely used. Platt and Jassby (1976) suggested a hyperbolic tangent function, $P = P_{max} \tanh(\alpha E/P_{max})$, which works well when photoinhibition is not obvious at higher intensities. This hyperbolic tangent function also has some theoretical attractions. Platt *et al.* (1980) fitted an array of *P* vs. *E* data for natural phytoplankton assemblages that did exhibit photoinhibition with the function:

$$P = P_{max}[1 - \exp^{(-\alpha E/P \max)}][\exp^{(-\beta E/P \max)}],$$

that has been applied by many field workers (e.g. Welschmeyer 1993). The parameter β is the intensity at onset of photoinhibition. This function applies to "normalized" photosynthetic rates, that is, rates per unit phytoplankton biomass. In fieldwork, the usual normalizing variable is chlorophyll-*a* concentration, and the ratio of the photosynthetic rate (carbon uptake) per unit volume to chlorophyll concentration is termed an *assimilation number*.

Different algal groups have significantly different light requirements for growth and photosynthesis (Richardson *et al.* 1983). Dinoflagellates and cyanobacteria photosynthesize and grow best at low light intensities. Diatoms can utilize low light intensities, but are much more tolerant of high light than most other groups. There are numerous reports and summaries of the photosynthetic parameters for *P* vs. *E* curves for most algal groups and many species of phytoplankton. We show some representative parameters in Table 3.1. Diatoms have the highest maximum rate of photosynthesis per unit chlorophyll ($P_{max}{}^{Chl}$), cyanobacteria are intermediate and the nanoflagellates (primarily prymnesiophytes) have the lowest $P_{max}{}^{Chl}$. Diatoms have the steepest initial slope for the rate of photosynthesis per unit chlorophyll (α^{Chl}), nanoflagellates have an intermediate slope, and cyanobacteria have the lowest slope.

In addition to genotypic differences, termed *photoadaptation*, in the ability of phytoplankton to utilize light, all phytoplankton exhibit short-term phenotypic adjustments, termed *photoacclimation*, in response to variations in light intensity (MacIntyre *et al.* 2002). The amount of

Table 3.1 Photosynthetic characteristics of dominant phytoplankton groups. α^{Chl} is the initial slope of the *P* vs. *E* curve with units of mg C (mg Chl-*a*)$^{-1}$ h^{-1} (µmol quanta m^{-2} $^{-1}$). P_{max}^{Chl} is the maximum rate of photosynthesis normalized to Chl-*a* with units of mg C (mg Chl-*a*)$^{-1}$ h^{-1}. The photosynthetic parameters (α^{Chl} and P_{max}^{Chl}) are calculated from a model of primary production calibrated with field data on specific phytoplankton pigments, irradiance as a function of depth, and specific absorption coefficients (Uitz *et al.* 2008)

GROUP	α^{Chl}	P_{max}^{Chl}
Diatoms	0.032 ± 0.007	4.26 ± 0.45
Prymnesiophytes	0.026 ± 0.005	2.94 ± 0.43
Cyanobacteria	0.007 ± 0.003	3.75 ± 0.37

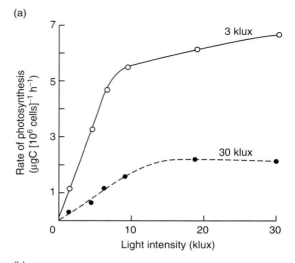

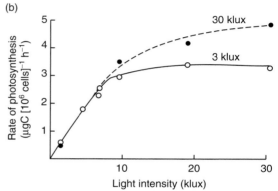

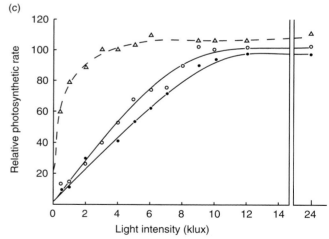

Fig. 3.6 Three types of acclimation to different light intensities, all in terms of photosynthesis per cell, not per unit chlorophyll. Changing chlorophyll per cell is part of the acclimation process. (a) *Chlamydomonas moevuss*. (b) *Cyclotella meneghiniana*. (c) *Phaeodactylum tricornutum*. Light intensities were 12 (dots), 5 (circles) and 0.7 (triangles) klux. (a after Jørgensen 1969, b after Jørgensen 1964, c after Beardall & Morris 1976.)

photosynthetically active pigment may increase at low intensities (more chlorophyll-*a*, more P700 in the *Chlamydomonas* and *Phaeodactylum*, causing an increase in the slope of the *P* vs. *I* curves in short-term measurements (Fig. 3.6 a & c). Or the activity of the dark-reaction system may be increased at high light intensities (driven by increased availability of light-energized reactants – ATP, NADPH), causing the P_{max} value to rise in short-term comparisons for *Cyclotella* (Fig. 3.6b).

Figure 3.7 shows *P* vs. *E* curves for low- and high-light acclimated *Skeletonema costatum*. Interpretation of the curve depends on the variable to which photosynthesis is normalized. When photosynthesis is normalized to Chl-*a* (Fig. 3.7a), low- and high-light-acclimated cells show similar rates of photosynthesis at low light intensities, but high-light-acclimated cells appear to outperform the low-light-acclimated cells at high light intensities. When the rates of photosynthesis are normalized to cell number (Fig. 3.7b), it is clear that low-light-acclimated cells have higher rates of photosynthesis at low light intensities, while high-light-acclimated cells have higher rates of photosynthesis at high light intensities. Finally, when photosynthesis is normalized to cell carbon (Fig. 3.6c), the low-light-acclimated cells have higher rates of photosynthesis than high-light-acclimated cells at all light intensities. This last approach has the advantage of showing the effect of variations in the *P* vs. *E* curve on the rate of incorporation of carbon into cell biomass under some conditions.

Physiological acclimation to variations in irradiance serves to minimize variations in growth rate when light varies. For phytoplankton, this entails a balance between the light and light-independent reactions of photosynthesis. At low irradiance, photosynthesis is limited by the rate of light absorption and photochemical energy conversion. At

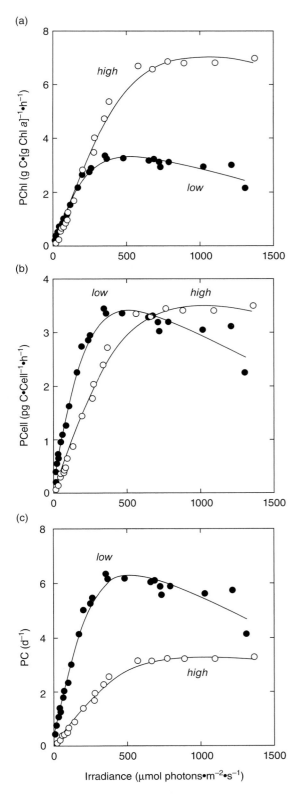

Fig. 3.7 *P* vs. *E* response curves for high-light (1200 μmol photons m^{-2} s^{-1}) and low-light (50 μmol photons m^{-2} s^{-1}) acclimated nutrient-replete cultures of the diatom *Skeletonema costatum*. (a) Chl-*a*-specific photosynthesis. (b) Cell-specific photosynthesis. (c) Carbon-specific photosynthesis. (After MacIntyre *et al*. 2002.)

high irradiance, photosynthesis is limited by the rate of electron transport, level of RuBisCO, or supply of ribulose 1,5 bisphosphate for the fixation of carbon.

Models for photosynthesis

The basic *P* vs. *E* curve suggests directly how photosynthetic rates per unit biomass of phytoplankton should vary downward in the sea. In the layers closest to the surface, productivity will either be at P_{max} or photoinhibited. As measures are made deeper in the water column, they can be expected to increase to P_{max} as irradiance decreases below photoinhibiting levels, then remain at P_{max} for some distance downward as light absorption by water and particles (including phytoplankton) progressively reduces available irradiance. From some depth not very far down, irradiance will be in the sloped part of the *P* vs. *E* curve, so that photosynthetic output gets progressively less. Since irradiance declines exponentially with depth (according to Beer's Law, $E_z = E_0 e^{-kz}$, where *k* is the extinction coefficient and *z* is depth), the *P* vs. *z* relation decreases exponentially downward from the depth where irradiance is less than that forcing photoinhibition, as shown by vertical profiles (Fig. 3.8a) of ^{14}C-uptake data. Of course, the primary production per unit volume is also a function of the amount of phytoplankton present, usually characterized by the chlorophyll profile. In temperate and high latitudes (Fig. 3.8b), that is maximal and fairly even through the upper mixing layers, then tapers off.

Extinction of light with depth means that at some depth, fairly close to the surface, irradiance will be the primary factor limiting productivity. That is true everywhere in the oceans. When it is said that phytoplankton growth is nutrient-limited, reference is to the upper water column or *euphotic* zone. The photosynthetic compensation depth varies with surface irradiance and the absorption coefficient (*k*), and is often assumed to be the depth at which irradiance is 1% of the surface value, maximally 70 m at lowest *k*(−0.067). However, subtropical gyres with minimal *k* have a strongly shade-adapted flora in the deeper euphotic zone forming a "deep chlorophyll maximum" (Fig. 3.9) around 100 m (~0.1% of the surface irradiance) in which net photosynthesis is still positive. This very widespread feature of the subtropical and tropical oceans is due to large amounts of chlorophyll per cell resulting from shade acclimation, although changes in species composition also contribute to some of the differences.

Bio-optical models of primary production

Bio-optical models of primary production (*P*) take the general form of:

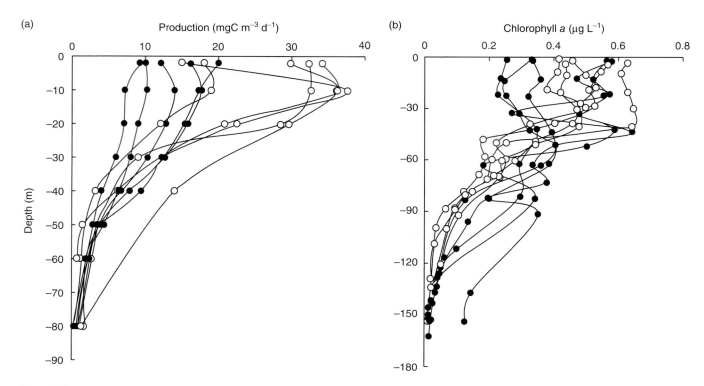

Fig. 3.8 (a) May (•) and September (○) profiles of primary productivity m^{-3} in the Gulf of Alaska (50°N, 145°W). (b) Chlorophyll profiles from the same site at various times are always maximal near the surface, and then taper off below 50 m. (After Welschmeyer 1993.)

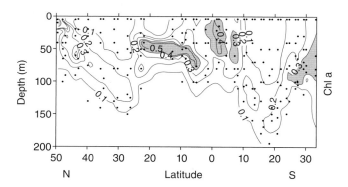

Fig. 3.9 A vertical chlorophyll section from 49°N to 33°S in the Atlantic Ocean. North of 45°N and south of 40° the chlorophyll maximum is in the upper 50 m. Deep chlorophyll maxima occur in the subtropical gyres between 100 and 175 m depth. Surface outcropping and a shallow (50 m) subsurface maximum occur within the equatorial band. (After Serret *et al.* 2006.)

$$P = PAR \times [\text{Chl-}a] \times a^* \times \Phi_c \qquad \text{(Eqn. 3.2)}$$

where PAR (mol quanta m^{-2} s^{-1}) is the photosynthetically available radiation, [Chl-*a*] is the chlorophyll-*a* concentration (mg m^{-3}), a^* is the Chl-*a* absorption coefficient (m^2 [mg Chl-a^{-1}]), and Φ_c is the quantum efficiency for carbon fixation (mol C [mol quanta]$^{-1}$).

Spectral irradiance (PAR) is measured as a function of depth, and variations with time of day can be calculated from models. Chl-*a* is measured fluorometrically or by HPLC of pigments extracted from phytoplankton collected on filters. The absorption coefficient, a^*, can be measured from samples collected on filters using a spectroradiometer. Bannister (1974) proposed an average value of 0.016 m^2 [mg Chl-a]$^{-1}$ for a^*, but more extensive data-sets (Bricaud *et al.* 1995) show a range of values (0.18 to 0.01 m^2 [mg Chl-a^{-1}]). The maximum quantum yield ($\Phi_{c\ max}$) of photosynthesis is determined as:

$$\Phi_{c\,max} = \alpha^B / \bar{a}^* \qquad \text{(Eqn. 3.3)}$$

where α^B is the slope of the P vs. E curve, mg C (mg Chl-a)$^{-1}$ h^{-1} (µmol quanta m^{-2} s^{-1})$^{-1}$ determined by the ^{14}C method and incubating samples collected from various depths under a range of light intensities, and $\bar{a}^*$ is the average chlorophyll-specific absorption coefficient of phytoplankton weighted by the spectral irradiance inside the incubation chambers used for determination of the P vs. E curves.

Claustre *et al.* (2005) used this particular bio-optical model combined with quantitative information on the composition of the phytoplankton community derived from HPLC pigment concentrations (Bricaud *et al.* 2004) to

Table 3.2 Physiological parameters and phytoplankton biomass variables for modeled photosynthesis derived from HPLC pigments. (After Claustre *et al.* 2005.)

	MICROPLANKTON	NANOPLANKTON	PICOPLANKTON
P_{max}^{Chl} (mg C [mg Chl-a]$^{-1}$ h^{-1})	6.27 ± 0.53	2.38 ± 0.23	0.13 ± 0.3
α^{Chl} (mg C [mg Chl-a]$^{-1}$ h^{-1})/(µmol quanta m^{-2} s^{-1})	0.093 ± 0.009	0.046 ± 0.004	0.014 ± 0.005
Absorption coefficient (m^2 mg Chl a^{-1})	0.021 ± 0.002	0.021 ± 0.001	0.038 ± 0.001
Quantum yield (mol C [mol quanta]$^{-1}$)	0.102	0.050	0.009

assess phytoplankton size-specific primary production in the North Atlantic Ocean. The results show that P_{max}^{B}, α^{B}, and $\Phi_{c\ max}$ all decrease as cell size decreases, but that a^*, the chlorophyll-a absorption coefficient, is highest for the smallest cells (Table 3.2). The latter result is expected due to the so-called "packaging" effects of chlorophyll in larger cells that decrease the effective cross-section for absorption of light. The bio-optical approach to estimating primary production requires simultaneous determination of P vs. E curves for discrete depths in the euphotic zone, but permits a continuous calculation of photosynthesis through the euphotic zone based on the rate of light absorption.

Estimating primary production from satellite-derived chlorophyll

In Chapter 2, we described how sensors on various satellites (SeaWiFS, MODIS, etc.) are used to provide regional and global images of surface chlorophyll concentrations in the ocean (Plate 2.3). A general correlation between production rates and chlorophyll standing stocks makes possible a more-or-less convincing estimate of regional and global primary production rates. Primary production (PP) is the product of the amount of chlorophyll present in the water column and the efficiency of light utilization (ε), i.e.:

$$PP = C_{sat} \times \varepsilon \qquad \text{(Eqn. 3.4)}$$

where C_{sat} is the satellite-derived estimate of surface chlorophyll, and ε includes the the Chl-a absorption coefficient (a^*) and the quantum yield (Φ_c) from Eqn. 3.3.

The simplest models of daily net primary production (NPP) incorporate the depth dependence as follows:

$$\Sigma NPP = P_{opt}^{b} \times C_{sat} \times DL \times Z_{eu} \times F \quad \text{(Eqn. 3.5)}$$

where P_{opt}^{b} is the maximum value of Chl-normalized photosynthesis in the water column (similar to P_{max}^{b}); DL is the duration of the photoperiod; Z_{eu} is the depth of the euphotic zone (depth of 1% surface PAR); and F describes the

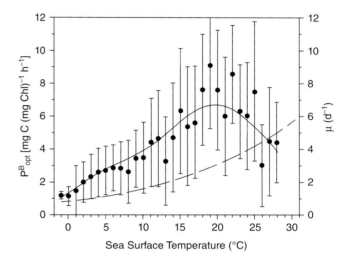

Fig. 3.10 Measured (•; ± SD) and modeled median value (solid curve) of the photoadaptive parameter, P_{opt}^{B}, as a function of sea-surface temperature. Dashed curve indicates the theoretical maximum specific growth rate (μ; d^{-1}) of phytoplankton described by Eppley (1972), which is used in a variety of productivity models. (After Behrenfeld & Falkowski 1997a.)

dependence of vertically integrated net primary production (ΣNPP) on the surface light intensity as it affects the depth of light-saturated photosynthesis.

Behrenfeld and Falkowski (1997a) compared measurements of P_{opt}^{b} with temperature (Fig. 3.10) and derived a single-factor empirical model for estimating the maximum value of Chl-normalized photosynthesis in the water column:

$$\begin{aligned} P_{opt}^{B} = &-3.27 \times 10^{-8} T^7 + 3.4132 \times 10^{-6} T^6 \\ &- 1.348 \times 10^{4} T^5 + 2.462 \times 10^{-3} T^4 - 0.0205 T^3 \\ &+ 0.0617 T^2 + 0.2749 T + 1.2956. \end{aligned}$$

$$\text{(Eqn. 3.6)}$$

The error bars based on measured values of P_{opt}^{b} are wide and certainly contribute to the uncertainty in the derived estimates of primary production, but the approximation appears to be adequate to represent regional differences in maps of global ocean productivity. Behrenfeld and Falkowski (1997b) also compared more elaborate time-integrated, wavelength-integrated and wavelength-resolved models for estimating primary production on global scales. The major difference among model outputs derives from the mode of estimation of P^b_{opt}, the "photoadaptive" variable. When the same chlorophyll fields are used in the models, global annual primary production in the ocean is about 44 Gt C yr^{-1}.

Measuring phytoplankton growth

Culture methods make possible the detailed study of the responses of phytoplankton species to environmental factors. Comparisons are made of rates of photosynthesis or cell division under various conditions. Reproduction rates are measured by counting cells before and after a period of growth and multiplication. Healthy, well-supplied algal cultures will increase exponentially for a considerable period, and responses are usually expressed as the exponential rate of increase,

$$\mu = (1/t)\ln(N/N_o), \qquad \text{(Eqn. 3.7)}$$

usually with units of d^{-1}, or as the number of cell divisions (doublings) per day,

$$\mu_2 = (1/t)\log_2(N/N_o). \qquad \text{(Eqn. 3.8)}$$

It is useful to memorize that if $\mu_2 = 1$, then $\mu = 0.69$.

If cells are growing under constant conditions, then the daily increase of any cellular component (e.g. carbon or nitrogen) can be measured and used to calculate a growth rate. Such conditions can be achieved in steady-state continuous cultures, but are not common for incubation times less than the generation period (doubling time) for phytoplankton. Growth rates can be determined as a function of nutrient concentration, using the Monod relationship (which is analogous to the more familiar Michaelis–Menten enzyme kinetic equation, see Chapter 1, pp. xx and Box 3.2):

$$\mu = \mu_{max}[S]/(K_s + [S]). \qquad \text{(Eqn. 3.9)}$$

Chan's (1978) data (Fig. 3.11) show that fully photoacclimated dinoflagellates typically reach maximal doubling rates at lower irradiance than diatoms, while diatoms have higher maximum growth rates. Actually, dinoflagellates and diatoms have the same photosynthetic rates per unit chlorophyll, but autotrophic dinoflagellates have much less chlorophyll relative to other cell constituents. When adapted to high irradiance, >200 μmol photons m^{-2} s^{-1}, typical dinoflagellates have 4 to 10 ng Chl (μg protein)$^{-1}$, whereas diatoms have 15 to 30 ng Chl (μg protein)$^{-1}$. Dinoflagellates also have a much more DNA than other algae, and consequently require more energy for cellular maintenance, giving them a lower growth efficiency (Tang 1996).

Illumination in the upper ocean is not continuous, but has a diel cycle (this is not news). It increases rapidly as the sun rises, peaks at noon, and drops to very low levels at night. Virtually all organisms, marine and terrestrial, have physiological cycles matched to the illumination cycle, and most have internal time-keeping processes allowing anticipation of events in the daily round. Algal growth and often multiplication vary with a daily periodicity, since photosynthesis must vary with the alternation of light and dark. Nelson and Brand (1979) have examined the phase relationships between cell division and the light–dark cycle for a variety of planktonic algae in culture. Timing of cell division tends to follow taxonomic lines (Fig. 3.12a & b).

Six species of diatom strongly favored daytime division. One species of dinoflagellate and six varied species of microflagellates had division maxima at night (Fig. 3.12c & d). There are counter-examples in the literature of dark-period division in diatoms (Eppley et al. 1971), but, on the whole, diatoms are diurnal dividers, while other forms divide at night. There are cycles in many other physiological parameters which parallel those of division.

Effects of nutrient availability

The organic matter of a typical phytoplankton assemblage in the sea has the following elemental composition in terms of atoms (or moles) relative to phosphorus:

$$C : N : P = 108 : 15.5 : 1$$

These numbers are known as the *Redfield ratios* after A.C. Redfield, who first measured them carefully in 1934 (see Redfield et al. 1963). Oxygen and hydrogen are also present in organic material, but clearly are never limiting in the marine environment. Carbon as carbonate (about 2 mM) is not limiting, either, although changes in abundance of the components of the carbonate system in seawater can alter the rate of photosynthesis. The quantities of fixed nitrogen and phosphorus in the ocean at great depth are in the rough ratio 16:1. The reason is that those elements are supplied to the deep sea primarily by decay of organic matter descending from above. Phosphorus and nitrogen are required for production of organic matter in roughly the Redfield ratios, and since

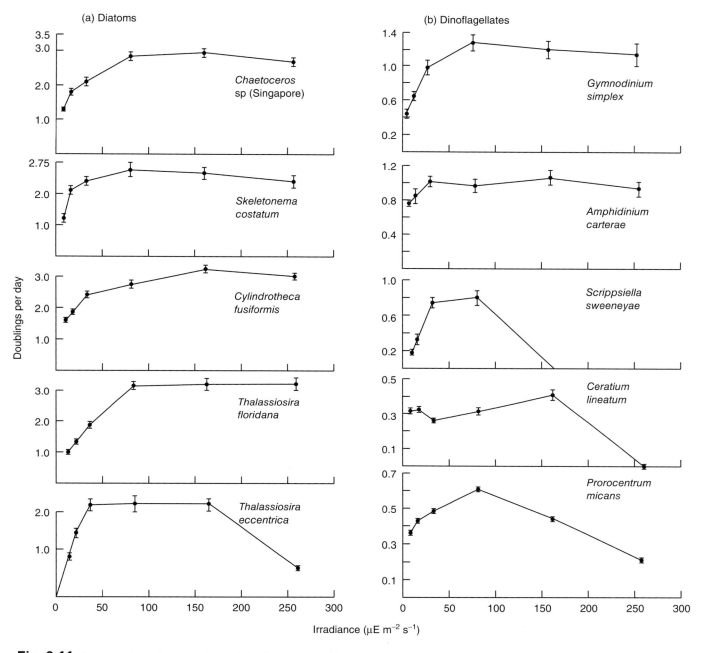

Fig. 3.11 Response of growth rates to changes in irradiance compared for diatoms (left) and dinoflagellates (right). These are exponential rates established after long acclimation at each irradiance. Bars are 95% confidence limits. (After Chan 1978.)

they are often in limited supply (rate of supply < (rate of demand for maximum growth rates), they partly control rates of phytoplankton production in the sea. They are known as "major nutrients". Silicon is also a major nutrient for diatoms and silicoflagellates. Each of these major nutrients shows characteristic vertical distributions in oceanic waters (Fig. 3.13) with low to non-detectable concentrations in the surface water, and increasing concentrations with depth, but with some differences in the loca-

tion of the nutriclines (steepest gradients in concentration) depending on the depth and rates of remineralization processes.

In addition, all phytoplankton have small, but important, requirements for a wide array of other elements. Particularly important are transition metals and common ions. Growth of phytoplankton could be limited by the supply of any of these. However, many (Na^+, Cl^-, SO_4^{2-}, Mg^{2+}, Ca^{2+}) are present far in excess of need. Thus,

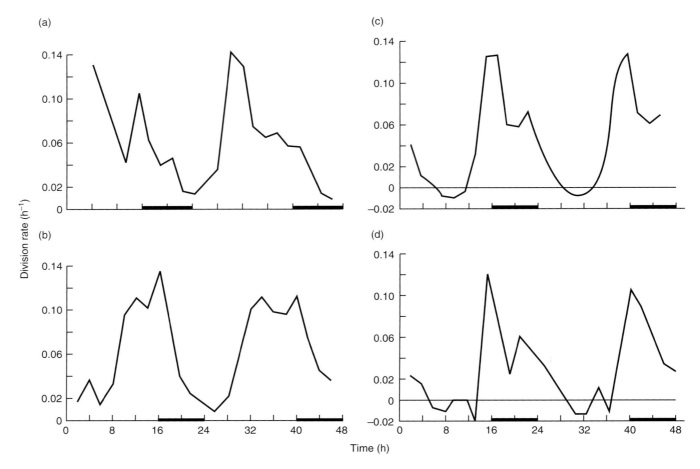

Fig. 3.12 (a) Division rate (h⁻¹) throughout the day–night cycle in two diatom clones (a) *Thalassiosira pseudonana* and (b) *Chaetoceros simplex*, and two flagellate clones (c) *Emiliania huxleyi* and (d) *Isochrysis galbana*. (After Nelson & Brand 1979.)

constituent elements important in minor quantities (trace elements) are Fe, Zn, Cu, Mn, Mo, and Co, termed "minor nutrients". There is evidence that at least some of these trace metals must be in organically chelated form before they can be used, but, on the whole, availability of many trace metals for uptake by cells is proportional to the free ion activity. Interference of trace metals with one another (some can occupy and block the uptake sites for the others) is another important aspect of phytoplankton nutrition. Finally, some of the vitamins needed for human nutrition are required as growth supplements by many species of phytoplankton. Such vitamins, particularly thiamin, biotin, and B_{12} (a protein-bound form of cobalamin), are present in small amounts in seawater, sometimes becoming reduced to limiting levels.

The kinetics of phytoplankton growth and uptake rates as a function of nutrient concentration are hyperbolic functions of various forms (Box 3.2). The response of phytoplankton growth to a range of concentrations of a single limiting nutrient (other nutrients in excess) is illustrated

(Fig. 1.7) for the case of nitrate. Growth rate increases rapidly over a small range of low values, then levels off. Once the nutrient requirement is met by adequate concentrations, something else becomes limiting (light or physiological potential at the prevailing temperature). In much of the current literature, K_s values are used to characterize the phytoplankton growth responses to nutrient availability (Table 3.3). Small K_s values indicate rapid response of the growth rate to increased nutrient availability. Large K_s indicates that relatively high concentration is required to achieve a near-maximal growth rate.

Determination of the responses of phytoplankton species and phytoplankton assemblages to variation in nutrient availability took much of the effort in phytoplankton ecology from 1970 to 1990. The data are complex and various. Note that the results in Fig. 1.7 are presented as doublings/day, a specific growth rate (growth/abundance/time). That is the ideal form for information on nutrient responses. However, many of the data available are as nutrient uptake rates combined with one determination of

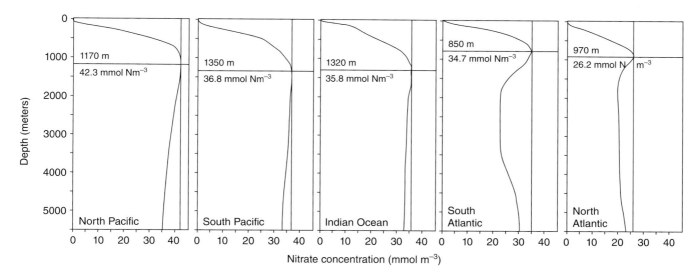

Fig. 3.13 Vertical nitrate profiles from various oceanic regions. (After Beckmann & Hense 2009.)

Box 3.2 Kinetics of nutrient uptake and growth

The Michaelis–Menten equation is the function most frequently used to represent cell growth as a function of the external nutrient concentration. We introduced it in Chapter 1. Here we present it with the symbols and parameters most often used for nutrient uptake kinetics, termed the Monod equation:

$$\rho = \rho_{max}[S]/(K_\rho + [S])$$

where ρ is the uptake rate normalized to biomass per unit time, ρ_{max} is the maximum uptake rate, K_ρ is the half-saturation constant and $[S]$ is the nutrient concentration. The Monod equation is used for short-term uptake rates (often in minutes or 1 hour). Phytoplankton can adapt to limiting nutrient concentrations by increasing ρ_{max} (increasing transport sites) or by reducing cellular needs for the nutrient in question. As a substrate (e.g. nutrient) affinity measure, K_ρ must be used with care, because it is affected by ρ_{max}. If their ρ_{max} values are different, two hyperbolae with the same initial slope will have different K_ρ values. Larger ρ_{max} forces a larger K_ρ. Kristiansen et al. (2000) give a nice example of this effect.

For steady-state conditions, the Monod equation can be used to describe growth rate as a function of substrate concentration:

$$\mu = \mu_{max}[S]/(K_\mu + [S]).$$

Another approach to examining kinetics of nutrient uptake is to grow cultures with a continuous nutrient supply and then assess cell growth rates as a function of the cellular nutrient content (Q). Under these conditions, the growth rate is calculated from the Droop equation (Droop 1968):

$$\mu = \mu_{max}[Q - Q_{min}]/Q$$

Under steady-state conditions, the nutrient uptake rate is equal to the product of the specific growth rate and the cellular quota:

$$\rho^{ss} = \mu Q$$

and this steady nutrient uptake is observed to follow a hyperbolic function of the external nutrient concentration (using ρ^{ss}_{max} and $k_{\mu Q}$ for distinction from ρ_{max} and k_ρ:

$$\rho^{ss} = \rho^{ss}_{max}[S]/(K_{\mu Q} + [S])$$

Morel (1987) describes the derivations and applications in more detail.

Both the Monod and Droop equations have been useful for the study of nutrient uptake and growth kinetics for cultured phytoplankton under steady-state conditions. The Monod model is simpler and preferred for steady-state conditions, while the Droop model is more complex and has the additional requirement of determining the variations in cell quota but it performs better under transient conditions.

Table 3.3 Half-saturation constants (K_s) ± 95% confidence intervals for nitrate uptake for eight clones of phytoplankton. (After Carpenter & Guillard 1971.)

SPECIES AND CLONE	SOURCE	K_S (µM)
Cyclotella nana		
3-H	Estuarine	1.87 ± 0.48
7-15	Shelf	1.19 ± 0.44
13-1	Oceanic	0.38 ± 0.17
Fragilaria pinnata		
0-12	Estuarine	1.64 ± 0.59
13-3	Oceanic	0.62 ± 0.17
Bellerochia sp.		
Say-7	Estuarine	6.87 ± 1.38
675D	Shelf	0.12 ± 0.08
SD	Oceanic	0.25 ± 0.18

the maximum growth rate at high nutrient concentration to produce a synthetic μ vs. [S] curve.

Equality of the uptake and growth curves is ensured (forced) in many studies by use of *chemostats* for determination of uptake rates. A chemostat is a container of well-mixed nutrient medium to which new nutrient solution is supplied at a volume rate (e.g. ml min^{-1}) μ_1, and from which culture is removed at a volume rate μ_2. The volume is maintained constant by choosing $\mu_1 = \mu_2$. Measurements of phytoplankton growth rate are made at system equilibrium by counting the number of cells per unit volume in the outflow and dividing by [flow rate/chamber volume]. Nutrient uptake rate per cell is equal to the difference in concentration between inflow and outflow divided by the flow rate and cell count in the chamber at steady-state. *Turbidostats*, used for example by Falkowski *et al.* (1985), have several advantages over chemostats. Turbidostat inflow and outflow are controlled by feedback from a particle-density sensor, allowing the stock to rise to equilibrium level without having to "fight" the outflow during "spin-up". Another (not used by Falkowski *et al.* 1985) is that the turbidostat can incorporate natural light cycling by reducing the outflow when growth or multiplication rates drop or rise. Rates then must be averaged over the diel cycle to compare effects of nutrient levels.

Affinity for nutrients in natural phytoplankton assemblages is extremely variable, but greater uptake capability (represented by lower K_s values) is generally observed as nutrient concentration goes down. Harrison *et al.* (1996) studied nitrate uptake kinetics from about 10°N to 63°N in the North Atlantic, including both upwelling and extremely oligotrophic conditions, finding a range of K_s values from 1 or 2 nM up to 1 µM, with most values in the 100-fold range from 5 nM to 0.5 µM. There was a very strong tendency toward lower K_s (again, greater affinity) when concentrations were below about 0.1 µM (100 nM). An especially sensitive, chemolumincescent nitrate analysis with a 2 nM detection limit (Garside 1982) enables such studies. Affinity for silicic acid in natural assemblages is similarly variable; in most of the oceans, including diatom-dominated coastal waters; K_s ranges from 0.5 to 5 µM (Nelson & Dortch 1996). However, south of the Antarctic polar front (~58°S), where silicic acid concentration is consistently above 20 and usually above 40 µM, the K_s for this diatom nutrient rises to about 20–40 µM (Nelson *et al.* 2001). Thus, when nutrients are abundant, less cellular machinery is provided to acquire them.

Nitrogen

Nitrogen is in a sense the most informative phytoplankton nutrient. That is because its oxidation states give information about the biological transformations to which given atoms have been subjected in the immediate past. Because of this information content, early studies of nutrient-uptake kinetics and regulation of phytoplankton growth focused on nitrate and ammonium utilization. Physiological and ecological studies have since broadened to include trace metals, co-limiting factors, and organic forms of nitrogen and phosphorus. Nitrogen is a fundamental constituent of proteins, nucleic acids, enzyme cofactors and at least one key, marine carbohydrate, chitin. Thus, it must be acquired and incorporated by phytoplankton for the initial synthesis of organic matter, and it remains essential at all steps in all food chains. However, very few organisms are capable of directly reducing and incorporating N_2, the form of nitrogen in the atmosphere that is abundant as dissolved gas in the ocean. This capability is restricted to an array of bacteria, some of which are photosynthetic cyanobacteria ("blue-green algae"). The biochemical transformations are referred to as "nitrogen fixation." They require a high energy input for reduction of N_2 to ammonium, which then can be incorporated in amino acids, purines, glucosamine, and so forth. Ammonium in marine systems can serve as an energy source for bacteria and archaea, which oxidize it to nitrite (NO_2^-), then nitrate (NO_3^-). Both NO_2^- and NO_3^- are available for uptake by most phytoplankton and many bacteria. Because they are biologically available, all these reduced and oxidized forms are referred to as "fixed" nitrogen.

Pelagic marine nitrogen fixation is estimated to be 121×10^9 kg N yr^{-1} (Galloway *et al.* 2004). Fixation is mediated by (i) the filamentous cyanophyte *Trichodesmium*; (ii) *Richelia*, a bacterium living endosymbiotically in several diatoms, most prominently *Rhizosolenia* and *Hemiaulus*; and (iii) other cyanobacteria and heterotrophic bacteria. Enzymes catalyzing nitrogen fixation require molybdenum and substantial iron as cofactors, and they may be limited by the low availability of iron (Falkowski 1997). *Trichodesmium* is generally restricted to oligotrophic, tropical waters. Surveys of its abundance and *in situ* fixation rate measurements suggest that it can account for more than half of the pelagic nitrogen fixation (Carpenter & Capone 2008). *Trichodesmium* is particularly abundant in the Red and Arabian Seas, where dust blown to sea from Africa provides enough iron to sustain nitrogen fixation. A unicellular cyanobacterium *Crocosphaera*, has a much wider temperature tolerance and a lower iron requirement than *Trichodesmium* (Fu *et al.* 2008; Moisander *et al.* 2010), and studies are under way to quantify its role in marine nitrogen fixation. We will return to the importance of oceanic nitrogen fixation in Chapter 11.

A key fact about the pelagic nitrogen cycle is that its redox transformations are vertically separated. When proteins or nucleic acids are metabolized, say by heterotrophs, the most usual form for excretion of free nitrogen is ammonium, but a fraction is in urea or uric acid – small organic molecules less toxic than ammonium in internal solution. Nitrogen in these forms is reduced. Eukaryotic heterotrophs do not use these reduced forms as fuel. Moreover, archaeal oxidation of ammonium to nitrite and bacterial oxidation of nitrite to nitrate appear to be inhibited by light, so both processes mostly occur below the euphotic zone. Because of the inhibition, "nitrifying" bacteria cannot build up a stock to take advantage of ammonium even at night. Finally, in euphotic zones, phytoplankton take up ammonium and urea in preference to oxidized forms of fixed nitrogen (Fig. 3.14). So, their concentrations are held at low levels, rarely exceeding $0.5\,\mu M$, except in estuaries.

Fixed nitrogen, like all nutrients, is eroded from the euphotic zone by the downward movement of photosynthetically generated organic matter. This occurs through the sinking of phytoplankton; vertical mixing of particles and dissolved organic matter; sinking of particulate wastes of grazers; and downward swimming of zooplankton and fish. Subsequent mineralization of organic matter introduces ammonium at depth, where nitrifiers can function. The second oxidation step, mediated by *Nitrobacter*, is more light-inhibited than the first, which is mediated in the ocean by archaea, so there is generally a thin layer with measurable concentration of NO_2^-. Below that, most fixed nitrogen is oxidized to NO_3^-. In most regions, the most important return supply of fixed nitrogen to the upper layer is the upward return of nitrate by vertical mixing, which is often seasonally accentuated, and by upwelling, which is geographically localized. Once in the euphotic zone, nitrate can be returned to the biological cycle through uptake by phytoplankton.

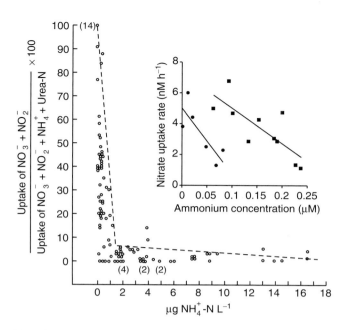

Fig. 3.14 Ratio of nitrate uptake to total fixed nitrogen uptake (the *f*-ratio) versus available ammonium concentration in Chesapeake Bay (from McCarthy *et al.* 1975). Inset: nitrate uptake rate vs. available ammonium estimated with $^{15}NO_3^-$ in the oceanic subarctic Pacific. (After Wheeler & Kokkinakis 1990.) In both habitats, ammonium availability suppresses nitrate utilization by phytoplankton.

Because production based on nitrate is using nutrient molecules newly arrived from outside the productive layer, it is termed *new production*. If the organic matter is then eaten, respired, and the nitrogen excreted as ammonium, its subsequent uptake and reincorporation in organic matter by phytoplankton is termed *recycled production*. This key distinction was spelled out by Dugdale & Goering (1967) in a seminal paper. At the time the paper was written, it was assumed that, by and large, fixed nitrogen was the usual limiting nutrient, so that ultimately the rates of the system would be set by the rate of supply of nitrate. Later it has become clear that availability of other nutrients, particularly iron, can set the rate of nitrate utilization. Iron is required for some components of the photosynthetic electron-transport systems and for nitrate reductase, an enzyme functioning to reduce nitrate to ammonium. Nonetheless, the rate of nitrate utilization remains a good measure of the new production. The rate of ammonium utilization is, in the same sense, a measure of recycled production. In ecosystems with very low nitrate, or high nitrate but low iron, the recycled production can be a much larger fraction of carbon fixation than new production. In systems with both nitrate and iron, new and recycled production are more equal. The relative importance is expressed as the "*f*-ratio":

$$f\text{-ratio} = \frac{\text{Uptake rate for nitrate}}{\text{Uptake rate for (ammonium + urea + nitrate)}}$$
$$(\text{Eqn. 3.10})$$

Usually, stable isotopes (^{15}N) of nitrogenous nutrients are supplied in incubation experiments to determine the uptake rates and the f-ratio (Box 3.3), which varies from about 0.5 (coastal upwelling systems with primary production dominated by diatoms) to 0.05 (highly oligotrophic, oceanic systems).

In a pelagic ecosystem at steady state (or on long-term average) the upward supply of inorganic nutrient elements

---

Box 3.3 Use of ^{15}N-tracers to determine new and regenerated production

An isotopic tracer method allows determination of these rates. It is based on labeling the fixed nitrogen supply, in bottle incubations of natural phytoplankton (much like a ^{14}C-uptake experiment) with nitrogen-15. After an interval of incubation, the ^{15}N incorporated from nitrate, or from a reduced form, into organic matter is determined by filtering the incubated water, and by recovering incorporated ^{15}N by destructive oxidation of the organic matter. The nitrogen is converted to N_2, then the ^{15}N/^{14}N ratios are determined by mass spectrometry or from emission spectra of the gas when excited by a high-voltage electric field. Each form of fixed nitrogen to be tested requires a separate labeling experiment. This technique is not perfect. Particularly, it is difficult to add sufficient ^{15}NH$_4^+$ (or labeled urea) to get a signal, that is a shift in the ^{15}N/^{14}N ratio of phytoplankton, without significantly increasing total NH$_4^+$. That quickly enhances NH$_4^+$ usage and reduces NO$_3^-$ uptake. In waters with very low nitrate, tracer additions of ^{15}NO$_3$ can increase productivity overall. Another problem (e.g. Bronk & Glibert 1994) is that recovering and accounting for all added tracer is difficult, because some incorporated nitrogen ends up as dissolved organic matter outside cells. Nevertheless, ^{15}N methods have shown the broad outlines of the pelagic nitrogen cycle. Roughly, new production is proportional to the ^{15}NO$_3^-$ uptake rate, and regenerated production is proportional to the ^{15}NH$_4^+$-uptake rate. Because the ^{14}C method is believed to estimate the total of new and regenerated production, the ratio of [New/(New + Regenerated)], measured by isotope uptake rates, is multiplied by the ^{14}C-uptake rate to determine the new production in terms of carbon.

---

by vertical mixing and advection into the euphotic zone should be equal to the downward flux at the base of the euphotic zone of those same elements incorporated in organic matter. Downward flux includes sinking of organic particles and vertical mixing in downward-decreasing gradients of dissolved and particulate organic matter. Thus, another way to estimate new production is to catch falling organic particles in traps moored at an appropriate depth in the water column and then determine the amount of carbon in the flux. Rates per unit area are determined by dividing the organic content after a trapping interval by trap area and time of deployment. Elskens *et al.* (2008) used tracer experiments and neutrally buoyant sediment traps to measure and compare new production and export production in the mesotrophic subarctic Pacific (47°N, 161°E). New production in the upper 50 m was about 20% of total primary production, while export production measured in traps at 150 m was about 10% of total primary production. Elskens *et al.* also estimated rates of remineralization and found that 80% of remineralization was in the upper 50 m, and 11% in the 50–150 m portion of the water column. Satellite measures of surface chlorophyll and deeper trap studies indicated that the Elskens *et al.* traps were deployed at the end of a diatom bloom. There was little change in phytoplankton biomass in the upper 150 m. The carbon and nitrogen assimilated during primary production were mostly consumed and remineralized by heterotrophs in the upper 150 m, and about 9% of the primary production sank into deeper water.

Elemental cycles are generally not in continuous balance. During seasons of more rapid primary production, dissolved, colloidal and particulate organic matter accumulates in the euphotic zone (Wheeler 1993). Recycling of this molecular detritus can require all the rest of the annual cycle. Thus, short-term trapping results will generally be less (apart from sampling problems) than short-term new production estimates.

New production should also be equivalent to the net oxygen production, [photosynthesis – respiration, PS – R] in the euphotic zone, since regenerated production is coupled to community respiration of photosynthate. There are severe difficulties in determining this in the field, particularly a requirement for precise estimation of oxygen-exchange rates at the sea surface. The concentrations of inert gases must also be measured to separate the effects of physical versus biological processes leading to supersaturation of gases in the surface mixed layer. Emerson *et al.* (1991) used argon and N_2 to estimate the effects of temperature and bubble processes on gas exchange rates between surface seawater and the atmosphere in the subarctic Pacific during the summer. The difference in saturation state between argon and oxygen indicates the biological component of O_2 supersaturation. Net oxygen production rates agreed reasonably well with ^{15}N estimates of new production for the summer samplings. To determine seasonal and annual estimates of net oxygen production, Emerson *et al.* have used *in situ* measurements of O_2 and N_2 in surface waters. Nitrogen gas can be used as an "inert" gas because rates of nitrogen fixation only change the concentration by ~0.1%. By measuring temperature, salinity, oxygen, and total dissolved gas pressure every two hours on a mooring at the Hawaii Ocean Time series, Emerson *et al.* (2008) determined a net biological oxygen production in the surface mixed layer of $4.8 \pm 2.7\,mol\,O_2\,m^{-2}\,yr^{-1}$. Emerson and Stump (2010) used the same method to measure net oxygen production in the subarctic Pacific. Measurements taken every three hours for nine months on a surface mooring indicated a mean summertime oxygen production of $24\,mmol\,O_2\,m^{-2}\,d^{-1}$, and very little net oxygen production during the winter. Net oxygen production can be scaled to carbon production using a photosynthetic quotient ($mol\,O_2$ evolved : $mol\,CO_2$ consumed) of 1.45. For both of these studies, net oxygen production rates agreed well with other estimates of new production, but lack of precision in the estimates of the physical processes results in an uncertainty of about 40% for the rates of net oxygen production.

Phosphate

Phosphorus, available in the ocean as phosphate (PO_4^{3-}), is incorporated in many biological molecules, for example, nucleic acids, and adenosine di- and triphosphates (ADP and ATP). Esterification of a phosphate group to a small molecule creates a diffusible, high-energy "currency" suitable for enzyme binding at sites adjacent to binding of substrates. Hydrolysis of the ester bond (dephosphorylation) then provides energy ($-62\,kJ\,mol^{-1}$) for substrate transformations. Phosphate in the ocean comes only from rocks, and it is ultimately removed from the ocean by incorporation in sediments. In the meantime, it cycles as a nutrient much as fixed nitrogen, but without an informative variety of oxidation states. In highly oligotrophic environments (central gyres), a significant fraction of dissolved phosphate is in small organophosphate compounds, and at least some phytoplankton in those regions bear surface enzymes which can remove phosphate from organophosphorus and incorporate it.

In general, and despite debates about the matter (Falkowski 1997; Cullen 1999; Tyrrell 1999), phosphate is the principal limiting nutrient for planktonic photosynthesis in freshwater systems but not in marine ones. That is because fresh water is always closely associated with land, and thus is better supplied with iron. Iron enables nitrogen fixation in cyanophytes, which can then carry on photosynthesis until phosphate is exhausted. In coastal marine systems, nitrate simply runs out first, which Tyrrell (1999) shows with a graph based on widely dispersed NO_3^- and PO_4^{3-} observations (Fig. 3.15). At low levels, phosphate is often positive when nitrate is at zero, usually not the opposite.

Nitrogen recycling is fast enough for production to continue at reduced levels without forcing the system to dependence on nitrogen fixation. In oceanic waters several factors work against nitrogen fixation. Subarctic, subantarctic and seaward antarctic ecosystems are colder than the adaptive range of nitrogen-fixing cyanobacteria. Moreover, except for the spring-blooming North Atlantic, nitrogen is usually not reduced to strongly limiting levels because of iron limitation. The huge oligotrophic sectors in equatorial belts and the subtropics do have significant levels of nitrogen fixation, which can be limited by either phosphorus or iron (Hutchins & Fu 2008). Van Mooy and Devol (2008) assessed nutrient limitation in the North Pacific subtropical gyre by measuring RNA synthesis (which requires a source of phosphorus) after additions of NH_4^+ and PO_4^{3-}. The NH_4^+ additions resulted in increased rates of RNA synthesis, while PO_4^{3-} had no effect, suggesting that the phytoplankton were N-limited. Low PO_4^{3-} concentrations in the Atlantic compared to the Pacific Ocean, have prompted the suggestion of phosphorus limitation there; however, Van Mooy *et al.* (2009) elucidated an intriguing adaptation of Atlantic phytoplankton to low phosphorus. Phytoplankton in the Sargasso Sea reduce their cellular phosphorus requirements by substituting non-phosphorus membrane lipids for phospholipids. The *Prochlorococcus* genome is radically minimized (Dufresne *et al.* 2003), further saving on phosphate.

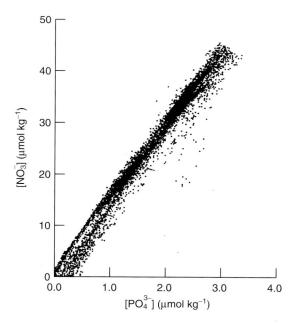

Fig. 3.15 A scatter plot from a series of global sampling sections (GEOSECS) of nitrate (NO_3^-) vs. phosphate (PO_4^{3-}). Some phosphate is usually left when nitrate is depleted below levels detected by standard techniques. (After Tyrrell 1999.)

Use of organic forms of nitrogen and phosphorus

Potential nutrients for phytoplankton growth (as well as heterotrophic organisms) include organic forms of nitrogen and phosphorus. These can be a significant component of the dissolved nutrient inventories in some regions, and net changes over annual cycles suggest that use of organic forms of nutrients may extend the productive season after inorganic forms are depleted (Banoub & Williams 1973). There are multiple sources and sinks for dissolved organic material, so generalizations about their significance are difficult, except to say that: (i) pool sizes are small; (ii) turnover rates can be high; and (iii) it can be difficult to separate the roles of autotrophic and heterotrophic microorganisms.

Trace metals and primary production

As mentioned above, phytoplankton require trace metals as cofactors for some enzymes and for components of electron-transfer chains in mitochondria and chloroplasts. These requirements are reflected in the vertical distributions of essential metals (e.g. zinc and iron) in the ocean. Most essential metals are depleted at the surface and increase with depth (Fig. 3.16).

Most trace metals are present at pico- to nanomolar levels in surface seawater, whereas cellular concentrations in phytoplankton are in the micromolar range. Many of the

trace metals are present in non-reactive chemical forms; they are bound to strong metal chelators. This complicates determination of the mechanism and kinetics of phytoplankton uptake, but some progress has been made. Four types of uptake systems have been identified for iron: (i) transporters specific to particular iron compounds, e.g. Fe-citrate and siderophores; (ii) FeII transporters that oxidize FeII as it is transferred across the membrane, e.g. oxidase–permease complexes; (iii) FeIII reductases; and (iv) unchelated FeIII transporters (Morel et al. 2008). These uptake systems can be studied in cultured phytoplankton, but it is extremely difficult to determine which systems are most important in natural populations of phytoplankton in seawater.

Due to the supply of iron from terrestrial environments, iron concentrations can be 100 to 1000 times higher in coastal than in oceanic waters. Early studies (Brand 1991; Sunda et al. 1991; see also Fig. 1.1) with cultured phytoplankton showed that the iron requirements of oceanic and neritic phytoplankton reflect this difference and that oceanic clones are able to grow at much lower iron concentration than their coastal counter parts. Sunda and Huntsman (1995) extended this work by studying iron-uptake rates of three coastal and three oceanic phytoplankton species ranging in diameter from 3 to 13 µm. They found that uptake rates normalized to cell surface area were similar in all six species. This similarity is explained if evolutionary pressures have driven all species to develop uptake at the maximum limits imposed by diffusion and ligand exchange rates. Thus, oceanic species have been forced to reduce their cell size and/or reduce their growth requirement for iron. They also found that coastal diatoms appear to accumulate 20–30 times more iron than is required to meet their metabolic needs, compared with a two- to three-fold excess uptake in a coastal dinoflagellate and an oceanic coccolithophore.

The relative requirements for trace metals to support phytoplankton growth have been determined by growing phytoplankton in cultures and then analyzing their metal and phosphorus content by high-resolution, inductively coupled, plasma mass spectrometry (HR–ICPMS) (Ho et al. 2003). Results for 15 different species can be presented as an expanded Redfield formula for those elements known to be required for growth:

$$C_{124}N_{16}P_1S_{1.3}\ K_{1.7}Mg_{0.56}Ca_{0.5}$$
$$(Fe_{7.5}Mn_{3.8}Zn_{0.80}Cu_{0.38}Co_{0.19}Mo_{0.03})/1000.$$

Although ratios varied among the species examined, in general: Fe > Mn > Zn > Cu > Co > Mo. The high cell quotas for Fe are due to its abundance and major role in electron transport in the photosynthetic systems. Over 90% of the phytoplankton metabolic requirement for Fe is for photosynthesis, with two Fe atoms/PSII

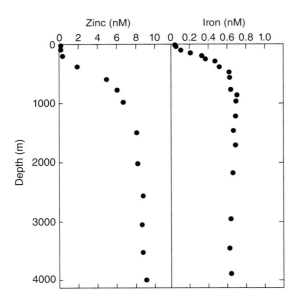

Fig. 3.16 Vertical profiles of dissolved zinc at (32°41′N, 144°59′W) and iron at (50°N, 145°W) in the North Pacific Ocean. (After Bruland 1980, and Martin et al. 1989.)

complex, 12 Fe atoms/PSI complex, and six Fe atoms/(Cyt) b_6f complex.

Falkowski *et al.* (2004) suggested that cyanobacteria and the green plastid algae (those derived from a primary endosymbiosis of a cyanobacterium; see Plate 2.1) evolved first in the Proterozoic ocean (2500 to 542 million years ago). At that time, oceanic Fe, Zn, and Cu concentrations were high, so the green plastid algae consequently have higher Fe, Zn, and Cu cell quotas than the later evolved "red plastid" algae (those derived from a secondary endosybiosis of a rhodophyte, see Fig. 2.1). The red plastid algae, including prymnesiophytes, diatoms, and some dinoflagellates, have higher Mn, Co, and Cd cell quotas than the green plastid algae. Quigg *et al.* (2003) attribute the high iron quota of "green plastid" algae to the high PSI : PSII ratio in this group, while the "red plastid" algae have a low iron requirement due to a low PSI : PSII ratio. For present-day conditions, however, the major differences in iron cell quotas are apparent in the comparison of oceanic and coastal phytoplankton, rather than between phylogenetic groups. For example, Strzepek and Harrison (2004) showed that the oceanic diatom *Thalassiosira oceanica* minimizes its iron demand by greatly reducing its level of PSI relative to PSII, yielding a PSI : PSII ratio of 0.1 compared with the PSI : PSII ratio of 0.5 found in the coastal diatom *Thalassiosira weissflogii*. Thus, the oceanic diatom has adapted to low iron by reducing its iron requirement, but at the cost of no longer being able to respond quickly to wide variations in light intensity using PSI.

Marchetti *et al.* (2006a & b) compared iron cell quotas for different diatom species grown in Fe-replete and low-Fe culture media. Oceanic isolates of *Pseudonitzschia*, a genus often found at sea during iron-enrichment experiments, accumulated 60 times more iron than needed for growth at low iron concentrations, while coastal isolates accumulated 25 times more than needed. Such accumulation ratios for oceanic and coastal isolates of *Thalassiosira* were 14 and 10 respectively.

Finkel *et al.* (2006) conducted a more comprehensive study to compare the genetic (phylogenetic) versus environmental (phenotypic) variations in iron demands of different phytoplankton. They examined Fe, Mn, Zn, Cu, Co, and Mo levels in five phytoplankton species grown at five different light intensities. Metal to phosphorus ratios varied by one to three orders of magnitude. The Fe : P for all species examined ranged from 2 to 1000 Fe : P (mmol : mol). Diatoms and cyanobacteria showed the widest ranges of Fe : P 2–251 and 7–1053, respectively, while the dinoflagellate and prasinophyte had much narrower ranges of 18–359 and 9–52 respectively. The results show that variations in light levels had as great or greater effect on metal : P ratios, than group or species differences.

Iron and light interactively affect the cell composition, the rates of primary production, and the growth of marine phytoplankton. The efficiency of the iron-uptake mechanism; variations in iron requirements for PSI and PSII; and wide ranges in iron cell quotas, all suggest the importance of successfully competing for limited iron supplies. Experiments with laboratory cultures suggest some of the strategies used by different phytoplankton. Mackey *et al.* (2008) examined variations in photosynthetic electron flow in natural phytoplankton communities dominated by either *Synechococcus* or *Prochlorococcus* in high-light, low-nutrient environments. Both groups have oceanic clones with low PSI : PSII ratios, presumably an adaptation to low iron availability. Low levels of PSI restrict the electron flow from PSII to PSI and expose PSII to photo-damage. In these picoplankters, the cell prevents photodamage with plastoquinol terminal oxidase (PTOX), an enzyme downstream of PSII that uses electrons to reduce oxygen and regenerate water. This pathway decouples oxygen cycling from CO_2 fixation in photosynthesis, and appears to be widespread in oceanic surface layers.

Effects of temperature variation on primary productivity

Increased temperature affects phytoplankton growth as it does other metabolic processes: it makes the reactions proceed faster. Since many reactions with a variety of kinetics are involved in photosynthesis, there is no simple relation between either photosynthetic rate or growth rate and temperature. Most importantly, different species respond differently to temperature variation. Furthermore, temperature interacts with other factors such as nutrient availability to determine rates, so that the relationships can be complex. However, at an ecological level the effect of temperature is to set an overall upper limit on growth rates. A data compilation by Smayda (1976, 1980) of maximum cell division rates for many species (Fig. 3.17) shows increasing exponents over ranges of 10°C or more, and then, for most species, rather steep drop-offs above a thermal limit. Eppley (1972) plotted a scatter diagram from a similar data compilation of rates from batch cultures, but without distinguishing the species. He then fitted an upper envelope to the data, using an exponential equation. Actual μ_{max} values for individual species do not look much like the envelope. The higher points, especially below 10°C are mostly for diatoms, which have higher growth potential than other phytoplankton, when only internal physiological interactions (rates set by temperature) are limiting. Flagellates are mostly slower. Bissinger *et al.* (2008) analyzed a larger data-set, completed a thorough statistical analysis and proposed this modification of Eppley's equation:

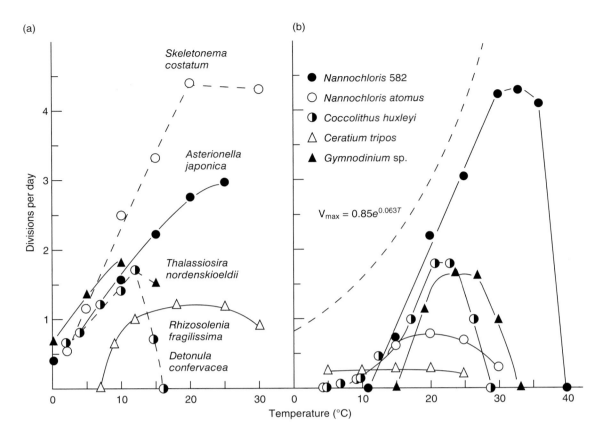

Fig. 3.17 Temperature effect on phytoplankton growth rate for several species: diatoms left, flagellates right. Dashed line on the right is the "Eppley curve", μ (doublings/day) = 0.85 exp(0.063T). (After Smayda 1976.)

$$\mu_{max} = 1.169\,\exp^{(0.0631T)}.$$

which suggests that growth at lower temperatures may be up to 30% faster than estimated from the original Eppley equation (Fig. 3.18).

Goldman and Carpenter (1974) examined the temperature responses of several species in chemostats. They used an Arrhenius plot (Fig. 3.19) to present the data, and fitted an Arrhenius-type equation to it, rate = $f(1/T$ in °K). Their description of results for a number of species is:

$$\mu_{max} = (1.8 \times 10^{10})\exp^{(-6842/T)}.$$

This curve is below that of Eppley, when that is plotted on the Arrhenius axes, but the actual quantitative difference is small (see also Goldman 1977). The difference obviously comes from Eppley's fitting of an upper envelope (which must be positioned subjectively) compared to the Goldman–Carpenter approach of fitting a relation to the central tendency of their data (probably more reliable). Both of these functions are commonly used in numerical models to characterize the response of phytoplankton to temperature.

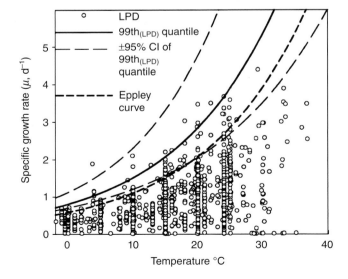

Fig. 3.18 Daily doubling rates of a wide variety of phytoplankton cultured at a range of temperatures. Fitted maximum rate curves are that due to Eppley (1972) and one fitted (with confidence limits) by Bissinger *et al.* (2008) to a larger set of species in the "Liverpool phytoplankton database" (LPD). (After Bissinger *et al.* 2008.)

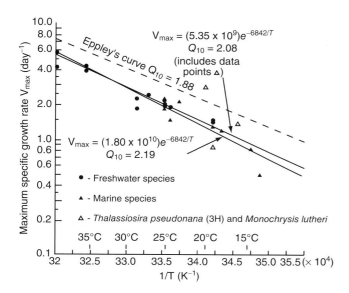

Fig. 3.19 An Arrhenius plot of V_{max} (nutrient and light saturated growth) vs. temperature. The form of the plot allows graphic determination of equation parameters. Data represented by points are from chemostat studies. (After Goldman & Carpenter 1974.)

Resting stages

Many diatoms and dinoflagellates form resting stages (cysts or spores) in response to unfavorable environmental conditions. Resting spores are more common in centric diatoms than in pennates (Hasle and Syvertsen 1997). Culture studies show that diatom resting stages can retain viability for at least two years. Field studies show that some benthic resting stages remain viable for only a few years, while others remain viable for decades (McQuoid & Hobson 1996; McQuoid et al. 2002). It is likely that these resting stages form the seed stock for pelagic phytoplankton blooms when they are resuspended from the sediment. Wetz et al. (2004) conducted experiments during the late winter to determine whether samples from the bottom boundary layer in coastal waters contained viable seed stocks that could serve to initiate a spring bloom. Growth occurred at light levels that were 40–50% of surface light, indicating that the resuspended cells could resume growth, but that winter mixing was deep enough to prevent growth.

Dinoflagellate cysts typically have a different cell wall from actively growing cells, and often require a period of cold and dark before being able to excyst. At the termination of a bloom, dinoflagellates (especially *Alexandrium* species) undergo sexual reproduction and the swimming zygotes form dormant cysts that can remain viable in the sediment for years. Recurrent toxic blooms in the Gulf of Maine have been a problem for a century or more. The main cyst seedbed is in the north-east part of the gulf near the mouth of the Bay of Fundy. One long-standing dilemma about these blooms is that the circulation patterns in the gulf are mainly to the west and south (Fig. 3.20) and should create, in effect, a one-way transport system with limited opportunity for cells to circulate back to the northeast. Anderson et al. (2005) and McGillicuddy et al. (2005) hypothesized a resolution based on models of the circulation patterns: cells accumulate in a retentive eddy near the mouth of the Bay of Fundy, allowing cysts to be deposited in that area to reseed future blooms. Cells that escape that retention zone into the eastern segment of the Maine Coastal Current bloom farther downstream and form a second seedbed offshore of the Androscoggin and Kennebec Rivers (Fig. 3.20). Germination of cysts in this second seedbed propagates the species farther along the Maine coast.

Cautions and future prospects

Phytoplankton ecologists have been a semi-independent subset among biological oceanographers. They have taken intense interest in factors controlling phytoplankton growth in the sea, and they have revealed many aspects of these relationships. However, several issues have become apparent. Cultured phytoplankton are not necessarily representative of the dominant phytoplankton in the ocean. The use of cultures inevitably involves unnatural conditions, including isolation of phytoplankton from their full set of natural associations. Production rate incubations disrupt normal grazing and regeneration processes. Thus, while light and nutrient conditions are important, phytoplankton stocks are as much controlled by grazing as by factors regulating cell growth. With exceptions like the onset of red tides, most of each day's phytoplankton growth is eaten *on the same day*. Increases such as spring blooms result from generally modest net differences between cellular multiplication and grazing. It is useful to examine phytoplankton in isolation, but the results must always be considered in light of the full set of interactions to which these small cells are subjected.

Molecular techniques and genomic analysis are elucidating both the phylogenetic relationships and previously unrecognized metabolic functions in the various phytoplankton groups. Genomic inventories suggest the existence of multiple enzymatic pathways for nutrient acquisition and metabolism, but do not indicate if and when these pathways are used. The challenge remaining is to combine these molecular and genomic techniques with *in vitro* and *in situ* experiments to determine key relationships of phytoplankton to both their physical–chemical environment and their co-existing microbes and zooplankton.

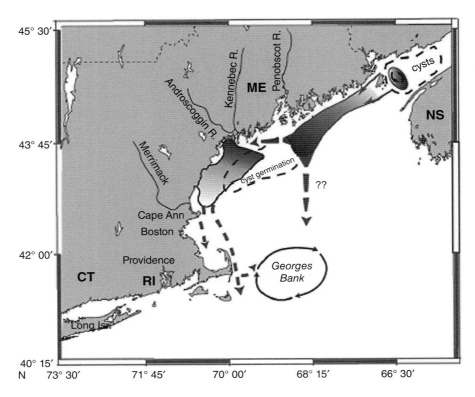

Fig. 3.20 Conceptual model of *A. fundyense* cysts and motile cell dynamics in the Gulf of Maine. Areas enclosed with dashed lines denote cyst seedbeds that provide inoculum cells. Major current systems are shown with shaded arrows. The shaded areas represent growth and transport of motile cells. (After Anderson *et al.* 2005.)

Ideally, we would like accurate estimations of primary production by phytoplankton and the subsequent use of the fixed carbon by other organisms. However, plankton communities are diverse. Several different organisms may serve similar functions, and a single organism can serve multiple functions. For example, *mixotrophs* have the capacity for both primary production and heterotrophic production (via use of dissolved organic material and/or phagotrophy of particulate organic material). This presents a problem in distinguishing between primary and secondary production processes. We will return to some of these complexities in Chapter 5 ("A Sea of Microbes") and Chapter 9 ("Pelagic Food Webs").

Chapter 4

Numerical models: the standard form of theory in pelagic ecology

Well before the advent of readily available computers for solving complex systems of difference equations, Riley (1946, also Riley *et al.* 1949) had introduced numerically solved rate-equation models of pelagic ecosystem dynamics representing the interactions among physical factors and several trophic levels. With the development of accessible computers in the 1960s and with improvements since, numerical models have become the basis of most theory in biological oceanography. Such theory is so pervasive that modeling is now a specialty within the field, much as theoretical and observational physics are done by different people. Models are indispensable in ocean ecology; almost whatever we do, we work back and forth between model populations or ecosystems and field observations. Just as important, models facilitate thought experiments about processes too widespread or too small to observe. We model effects of possible future events, like doubling atmospheric CO_2, that cannot be experimentally determined. There are benthic ecosystem models, fishery models, population-dynamical models for whales, and more. A standard problem, causal explanation of the spring phytoplankton bloom and other seasonal cycles, is a motivation for some of this theory, the part we will review. There are now models of every system and process that the imagination can isolate within the world's oceans, but very often the basic mechanics are those examined here.

Pelagic ecosystem models are often numerical approximations to the solutions of systems of differential equations. Not enough is known to include the full complexity of ecosystem processes, but what we do not know, we "parameterize" (guess, cover over by lumping categories, or some other strategy). The models often generate convincing simulations of our concepts of ecological processes. In what follows, we will examine several rather old, but easy to follow, pelagic ecosystem models. After reviewing seasonal cycling in temperate, coastal, pelagic ecosystems and a primer on difference-equation modeling, we will work toward simulating the processes generating and terminating (in places preventing) spring blooms.

Seasonality in phytoplankton

Because phytoplankton blooms in spring occur in temperate coastal areas and in the subarctic North Atlantic, where scientific ocean observations began, explanations of the seasonal cycles of productivity and standing stocks of phytoplankton and grazers became a central dogma of "classical" biological oceanography. In Chapter 11 we will consider spring blooms in more detail. Briefly, the sequence of events and their supposed causes are as follows:

1 Phytoplankton stock is low through the winter because strong vertical mixing keeps net losses from the photic zone greater than the possible net growth, despite sufficient nutrients. Low sun angles and short days contribute to the balance at low stock by keeping algal growth rates slow.

2 In spring, increased illumination and reduced winds generate some vertical stratification. This reduces the loss rate from the now better-lighted surface layer, and a

population of phytoplankton builds up. This is called the spring bloom.

3 As spring wears into summer, the growth of phytoplankton depletes the nutrients that are no longer rapidly supplied from depth because of density stratification. Productivity becomes nutrient limited and falls off. Simultaneously and subsequently, the increase in algae allows an increase in animals, which eat up the algae. Grazing, reduced algal growth rates, and sinking of algal cells due to nutrient starvation and agglutination produce a mid-summer low in algal stocks.

4 By fall, the grazing animals have declined, or have entered non-feeding resting stages for the forthcoming winter. The first, intermittent storms of the coming winter usually stir up some nutrients without completely mixing away the density stratification. The daylength is still moderately long, the sun is still high, and the gray cloud banks of winter hesitate on the horizon. The result is a brief but substantial fall bloom.

5 The fall bloom is mixed downward out of existence by the storms of early winter, which also resupply the surface with nutrients. The sea is plowed for the next spring's bloom.

The details of this scenario, and the variant processes in large ecosystems that work differently, are the principal subjects of many studies and, thus, of models in pelagic ecology.

Rate equation modeling

Rate equations are the inner machinery of most ecosystem models, so rudimentary understanding of these equations and their solution is essential. A.J. Lotka (1925), an American mathematical ecologist, and Vito Volterra (1926a & b), an Italian mathematician with interests in biological interactions and fisheries, developed a simple, differential-equation model of the interaction between a prey population and a predator population. The classic example of such an interaction is the snowshoe hare–Canadian lynx oscillation (MacLulich 1937; Stenseth *et al.* 1997). The model is usually called the Lotka–Volterra predator–prey model, and it provides a simple example of the steps in a process model. Let the size of the prey population be $N1$, that of the predator population $N2$. The following two equations might describe the *interdependent* changes in these two quantities:

$$dN1/dt = b * N1 - K1 * N1 * N2$$
$$= \text{prey increase} - \text{predation},$$
$$dN2/dt = K2 * N1 * N2 - d * N2$$
$$= \text{a fraction of predation} - \text{predator death}.$$

The * denotes multiplication (as in computer languages), b is the net [birth – death] rate of the prey in the absence of predation, d is the death rate of the predator, and K1 and K2 are constants. In modeling terminology, $N1$ and $N2$ are "state variables" of the system and b, K1, K2, and d are system "parameters".

Some assumptions behind this model are realistic, some are not:

1 In the absence of predation the prey species would increase exponentially at rate b (an unlimited habitat apart from predation is unrealistic).

2 The rate at which prey are eaten is proportional to the product of the densities of prey and predator, an "encounter" model (moderately realistic).

3 Time spent by predators consuming prey and converting them to new predators is negligible (no time lags, not realistic).

4 And still other assumptions.

Despite the unrealistic aspects, these equations are instructive in many ways and have been much studied. They are typical of differential equation systems describing ecological interactions in that they have *no explicit solutions*. That is, you cannot rearrange them (separate the variables), integrate and find functional representations for the time courses of $N1$ and $N2$ (which would be called a "solution" to such a system). So, we turn to other approaches to study them further.

Most often the interactions are approximated by finite-difference equations, and then some finite time increment is repeatedly applied to make sequential calculations for $N1$ and $N2$. The resulting "*implicit*" solution is a time-series of the variables, and it can be studied for its characteristics. This is called Euler's method, among other names. For the Lotka–Volterra equations the finite-difference version is:

$$\Delta N1/\Delta T = b' * N1 - K1' * N1 * N2,$$
$$\Delta N2/\Delta T = K2' * N1 * N2 - d' * N2,$$

in which the primes (') denote that the best constants might be a little different from the infinitesimal case. The Δ symbols imply changes over small, but finite, intervals. A solution is a very large number of sequential calculations of $N1$ and $N2$, a process greatly eased by use of a computer. The equations are programmed in some convenient computer language, then run through very large numbers of cycles of ΔT. We show a version called PROGRAM VOLTERRA as a Matlab script, a currently popular programming language, in Box 4.1.

With the program in hand, we need some preliminary values of the constants. We start with some that give a steady-state or equilibrium result, that is, $\Delta N1 = \Delta N2 = 0$ always. Thus, (dropping the primes for simplicity):

$$N1 * b * \Delta T = K1 * N1 * N2 * \Delta T \text{ and } K2 * N1 * N2 * \Delta T$$
$$= d * N2 * \Delta T.$$

So, $N1 = d/K2$ and $N2 = b/K1$.

Box 4.1 Computer programs for the Lotka–Volterra model

Instructions to the computer embodied in a program like this are simple and can be learned most quickly by working with an experienced programmer. The code here is a "Matlab" script, a popular commercial system for programming personal and other computers. It provides a variety of tools for mathematical manipulations, data handling and graphing. The "%" sign indicates a comment, an explanation to human readers of the program that computers will ignore. Here the whole comment statements are in bold.

To run this, type the statements (comments – i.e., % statements – not needed) into the editor window of a Matlab system. Use the semicolons; they stop the computer from listing every variable value in the command window at every step. Click the run symbol (a green triangle over a white rectangle), and the program will very quickly show a plot in a new window of the N2 vs. N1 phase diagram. For the stable state, this will be a single point at (1000, 10). For, say, a starting N2 value of 11 it will show a whorl of expanding population oscillations, ending when either N1 or N2 < 0 (Fig. 4.1).

```
%PROGRAM lotka_volterra.m
%1. Set up vectors of zeros for 1000 time steps
  N1=zeros(1,1000); N2=zeros(1,1000);
%2. Enter starting values N1=prey, N2=predators
  N1(1)=1000; N2(1)=11; %Stable point; change to experiment.
%3. Parameters; these are values for the 1000:10 stable point
  B1=0.1; K1=0.01; K2=0.0001; D2=0.1; %Change to experiment.
%4. Loop through 1000 time steps
  Dt=1.0; % Time step is 1 time unit
  for i=1:1000 % Approximate over 1000 time steps
      DN1=B1*N1(i)*Dt - K1*N1(i)*N2(i)*Dt;
      DN2=K2*N1(i)*N2(i)*Dt - D2*N2(i)*Dt;
      N1(i+1)=N1(i)+DN1;
      N2(i+1)=N2(i)+DN2;
  end
%5. Graph the results as a function of time
    subplot(3,2,1:2);
    plotyy((0:i), N1,(0:i), N2);
    xlabel('Time Steps')
    ylabel('Left No. N1; right No. N2')
    % As a phase diagram
    subplot(3,2,3:6);
    plot(N1,N2); xlabel('N1'); ylabel('N2');
```

Any values satisfying those equalities will be at equilibrium, but we will ignore the case $N1 = N2 = 0$. We try arbitrary values that meet the equilibrium conditions by running VOLTERRA on a computer with Matlab software. The screen interface of Matlab is convenient. Type the program into the editor window, hit the "run" button and the results will appear in a new window as a time-series plot. All the constants and starting values are in the program. To change any of them, just type in different numbers, push run again and the new results will appear. Thus, you can (and should) readily experiment. The program in Box 4.1 uses some equilibrium values: $N1 = 1000$, $N2 = 10, D = 0.1, K2 = 0.0001, B = 0.1, K1 = 0.01, \Delta T = 1$. Looking at the results every 10 time-steps, the boring output is:

TIME	N1	N2
10	1000	10
20	1000	10
30	1000	10
40	1000	10
:	:	:
:	:	:
90	1000	10

The model is at equilibrium, as predicted, and a single point appears in the plot. Now, edit the program, changing

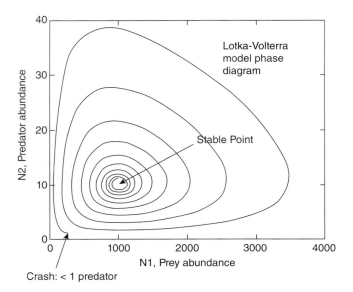

Fig. 4.1 Phase diagram for a Lotka–Volterra predator–prey model. Initial prey and predator numbers are 1000 and 11, just off the stable point (1000,10) for the parameters. The diagram shows an expanding oscillation that is said to "crash" when N2 becomes less than 1.

$N2 = 11$ and run it without changing any other state variable or parameter. The time-series of $N1$ and $N2$ values will start to oscillate, getting further and further from equilibrium until the model "crashes", with one of the species (and then the dependent other) going to extinction or "exploding". A graphical presentation of such models is often made in "phase space" (Fig. 4.1), a plot on axes scaled by $N1$ and $N2$. The time course is traced on this phase plane. By starting trials on various parts of the plane, the stability properties of the relationship between $N1$ and $N2$ can be investigated. As formulated, the model has only the one equilibrium point, which is unstable because even a slight displacement from it will result in a crash.

Predator–prey interactions with large regions of stability can be obtained by replacing the exponential increase of the prey by incorporating a habitat carrying-capacity (or limit) for the prey (a somewhat more realistic model). Let this maximum number of prey be M. An equation incorporating this feature is the "logistic equation":

$$\Delta N1/\Delta T = bN1 * ((M - N1)/M).$$

The rate of increase is decreased according to the fraction of the available space (or food, or nest sites) that is already occupied. The term $(M - N1)/M$ is called a "damper" term in differential equation terminology. Install a damper term in PROGRAM VOLTERRA and try its effects.

While this level of modeling might seem uselessly simple-minded, tests of important ecological hypotheses can be based on the Lotka–Volterra model. For example, Strom *et al.* (2000) pointed out a problem in understanding some nutrient-limited phytoplankton stock oscillations, namely

that the stocks are not eaten down to very low levels. One possible explanation is that where herbivorous protozoa control stocks, the grazers are small enough and varied enough that they can start eating themselves when phytoplankton stocks drop. That gives phytoplankton stocks some respite from grazing at the low end of the cycle. The feasibility of this as a stabilizing mechanism can be tested by slightly modifying the Lotka–Volterra equations as follows.

A program to solve a phytoplankton–grazer (P and Z) interaction posed as:

$$dP/dt = aP - 0.06PZ, (a = 0.69 \text{ d}^{-1}, \text{ one doubling/day})$$
$$\text{(Eqn. 4.1)}$$

and

$$dZ/dt = 0.02PZ - 0.5Z, \quad \text{(Eqn. 4.2)}$$

is provided in Box 4.2.

The solution settles immediately into boring and unrealistic limit cycles (Fig. 4.2a), regardless of initial conditions. However, with a density-dependent self-consumption term added to the grazer equation (2) (variant 2 in Box 4.2), there is a stabilization of the oscillation. The new equations are:

$$dP/dt = aP - 0.06PZ, (a = 0.69 \text{ d}^{-1}) \quad \text{(Eqn. 4.3)}$$

and

$$dZ/dt = 0.02PZ - 0.5Z - 0.03Z^2 \quad \text{(Eqn. 4.4)}$$

The self-limitation of grazers (by cannibalism, Fig. 4.2b), represented by the quadratic term in equation (4), is powerful and P and Z quickly become constants. A surprisingly realistic model can be produced by installing at each time step some random variation in the phytoplankton growth-rate constant (vary a from 0.52 to 0.86; variant 3 in Box 4.2). Now the variables sustain oscillations (Fig. 4.2c) something like those observed in real oligotrophic pelagic ecosystems. The simplicity of the model has the benefit that the proposed mechanism very likely causes the calculated effect, not some other aspect of the interaction. In the case of the interaction of real nano- and picophytoplankton with protozoan grazers, the self-limitation need not be actual intraspecific cannibalism, only a rising tendency for some microherbivorous species to feed on others as they become abundant and phytoplankton decrease. Notice, however, that the basic period of the oscillation without the grazer cannibalism (Fig. 4.2a) has approximately "re-emerged". Perhaps that, too, is how such interactions come to oscillate as they do. As in all such modeling, the success of the model only shows the *feasibility* of a concept. It does not prove that the same mechanism operates in the field. That requires tests in the field, or at least contained incubations involving real organisms from the field.

Box 4.2 Matlab script for Lotka–Volterra type model of nanophytoplankton–protozoan grazer interaction

```
%PROGRAM STROM.m
%1. Set up vectors of P, Z and T
% filled with NaN ("not a number") for
% 120 steps/day for 60 days:
  ndays = 60; nsteps = 120;
  P=ones(nsteps*ndays,1)*NaN; Z=P; T=P;

%2. Set starting values:
% PP=phytoplankton, ZZ=grazers
  PP=10.; ZZ=8.;

%3. Set Parameters
a=0.69; b=0.06; c=0.02; d=0.5; e=0.03;
%reduce parameters for small time steps
as=a/nsteps; bs=b/nsteps; cs=c/nsteps;
ds=d/nsteps; es=e/nsteps; ct = 0;

%4. Main daily loop
for i=1:ndays
%Variant 2: Remove next "%" to
% randomly vary a (0.52 to 0.86)
 %as= a*(1.+ 0.5*(rand-0.5))/nsteps;
%5. "Euler" loop to closely track the
% solution:
  for j=1:nsteps
    ct = ct+1;
    T(ct)=ct/nsteps;
    P(ct)=PP; Z(ct)=ZZ;
    dP=as*PP-bs*PP*ZZ;
    dZ=cs*PP*ZZ-ds*ZZ;
% Variant 2: Remove next "%" to add
% grazer cannibalism
    %dZ=dZ-es*ZZ*ZZ;
    PP=PP+dP; ZZ=ZZ+dZ;
  end
end

%Graph result
figure;
plot(T(1:end),P(1:end),'g'); hold on
plot(T(1:end),Z(1:end),'r');
axis([0 60 0 100]);
```

Same instructions as at bottom of Box 4.1.

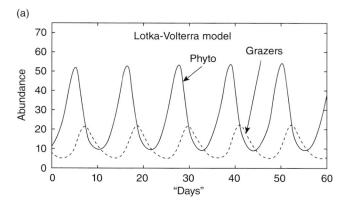

(a)

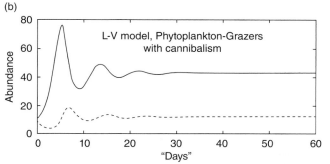

(b)

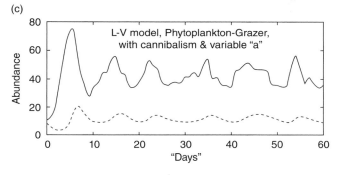

(c)

Fig. 4.2 (a) Persistent oscillations of a Lotka–Volterra model of phytoplankton–grazer interaction with no damping process, plotted as a time-series. (b) The same phytoplankton–grazer relationship, but with the grazers slightly self-limiting (a quadratic damper term added). (c) Exactly like (b), but with random variation of the phytoplankton growth rate.

A simple pelagic ecosystem model

One of our basic goals in pelagic ecology is to understand at whatever level possible the interactions among nutrient availability (a measure of the essential resource base), phytoplankton growth (dependent upon species, nutrients, illumination, and temperature), phytoplankton stock size (dependent upon growth, grazing, mixing, and sinking) and zooplankton stock size (dependent upon grazing and mortality). The most basic models are termed nutrient–phytoplankton–zooplankton, or NPZ, models.

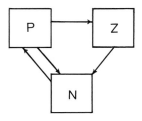

Fig. 4.3 A box diagram of the Franks–Wroblewski–Flierl NPZ model.

An unrealistically simple, but instructive, version of an NPZ model was provided by Franks *et al.* (1986). It examines the time course of quantities of dissolved nutrient and assimilated nutrient (i.e. algal and herbivore biomass) in a theoretical upper water column. Since there is no spatial variation, such a model is termed zero-dimensional (even though time is a "dimension"). Nutrients are taken up by phytoplankton and converted to phytoplankton stock. Phytoplankton are eaten by zooplankton and converted to tissue, with losses to metabolism. Those losses appear immediately as nutrient available to phytoplankton. Both phytoplankton and zooplankton die and decay at significant rates, the nutrients released appearing immediately as dissolved and available nutrients. The most important oversimplification is that the system is closed; nothing mixes in or sinks out. The flow diagram for these interactions is shown in Fig. 4.3.

The difference equations can be written out in words, often a useful modeling exercise:

Change in P = +Nutrient Uptake
           – Mortality (of P) – Grazing

Change in Z = +(Growth Efficiency) $*$ Grazing
           – Mortality(of Z)

Change in N = –Nutrient Uptake + (1 – Growth Efficiency)
           $*$ Grazing + Mortality(P) + Mortality(Z)

Next, these equations are written out as differential (or directly as difference) equations. The key challenge at this step is to find suitable and effective functional relations to represent the interactions accurately. The functions used here differ from those of Franks *et al.* (1986) only in letting γ, not $(1 - \gamma)$, be zooplankton growth efficiency (0.3 of ingested food). They are:

$$\frac{dP}{dt} = \frac{V_m NP}{K_s + N} - mP - ZR_m(1 - e^{-\Lambda P})$$

$$\frac{dZ}{dt} = \gamma ZR_m(1 - e^{-\Lambda P}) - kZ$$

$$\frac{dN}{dt} = -\frac{V_m NP}{K_s + N} + mP + kZ + (1 - \gamma)ZR_m(1 - e^{-\Lambda P})$$

A Matlab script with suitable code for solving these equations is shown in Box 4.3. Names of the state variables, parameters and their standard values are shown in Table 4.1.

It is imagined that the upper water column has just stratified at model initiation and illumination is great enough to sustain rapid phytoplankton growth, which is limited only by nutrient ("nitrogen") availability. Conditions are right for a spring bloom. The system starts with lots of nutrients and small quantities of phytoplankton and zooplankton.

Both phytoplankton, P, and zooplankton, Z, are just quantities, and their units represent their nutrient (as nitrogen) content. Neither has age nor size structure. Phytoplankton increase as a hyperbolic function of nutrient availability, a relation represented by the Michaelis–Menten function (Chapter 3), $dP/dt = (V_m NP/(K_s + N))$. They also have a proportional death rate, $-mP$, beside that from grazing. Zooplankton graze as a function of phytoplankton availability, P, with the rate increasing hyperbolically according to an Ivlev function, $dP/dt = -PZR_m(1 - e^{-\Lambda P})$, to an asymptotic value R_m. That is, planktonic grazers will eat more if more food is offered, but only up to a point. Beyond that amount, their ingestion rate levels off (e.g. Frost 1972). The model zooplankton die at a rate proportional to their abundance, $-kZ$ (k for "kill").

The initial parameter set (modified from Franks *et al.*1986) produces a strong, brief bloom that is reduced by grazers (Fig. 4.4a). Nutrients are partially regenerated, and then, after a few damping oscillations, proportions settle to a steady state. To a limited extent, the model can be modified to apply more realistic rate parameters and initial values. Phytoplankton seldom grow at 2.0 d^{-1}; a more realistic rate would be one doubling per day, that is: $V_m = 0.69$ d^{-1}. High-temperate North Atlantic waters before the spring bloom contain more nitrate: 10–12 μM. Substituting those values (Fig. 4.4b), produces a strong, more delayed bloom that is eaten down by grazers. Both P and Z then fall to very low (but not zero) values with almost all the nutrient in inorganic form. The cycle repeats after 75 days, and then settles into similar strong oscillations with an approximately 50-day period. A common modeling strategy can be applied to make the model appear somewhat more realistic: a threshold amount of phytoplankton is included, P_0, that must be present to induce grazers to eat. This is installed in the Ivlev function as:

$$dP/dt = -PZ(R_m - e^{-\Lambda(P-P_0)}),$$

along with a statement forcing dP/dt to zero when $P < P_0$. The grazing threshold enforces low, constant P and Z values after the initial bloom (Fig. 4.4c). The large amount of nutrient regenerated as the bloom is consumed is not realistic and is forced by the closure of the upper water column.

More realism can be obtained by letting some of the phytoplankton and zooplankton that die ($-m * P - k * Z$)

Box 4.3 Matlab script for an NPZ model similar to that of Franks *et al.* (1986)

```
%Program Franks
clear all
ndays=60; nsteps=120;
%Set up storage vectors for results
Pct=ones(nsteps*ndays,1); Tct=Pct;
Nct=Pct; Zct=Pct;
%Set values of all parameters
%for standard run. Daily rates
%are reduced for small time steps
Vm=2./nsteps; m=0.1/nsteps;
Rm=1.5/nsteps; d=0.2/nsteps;
Mix=0.02/nsteps;
%Try Vm=0.69; try other changes
Gamma=0.3; Ks=1.; Lmda=1.; NatZ=10.6;
%Set values of starting conditions
NIT=1.6; P=0.3; Z=0.1; P0=0.;
%P0 initially set to zero; try 1.0
ct=0;
%Main loop starts here,
%one cycle per model day
for i=1:ndays
  %To add autumn mixing use
%the following statements;
%make ndays=200.
%if i>120
  %Mix=Mix+0.02/nsteps; end
%Subloop to allow nstep
%time steps per day
```

```
for j=1:nsteps
  ct=ct+1; Tct(ct)=ct/nsteps;
  Nct(ct)=NIT; Pct(ct)=P; Zct(ct)=Z;
  UPTAKE=Vm*NIT/(Ks+NIT);
  if P>P0
    Ivlev=Rm*(1-exp(-Lmda*(P-P0)));
  else Ivlev=0.;
  end
  delP=UPTAKE*P-m*P-Z*Ivlev;
  delZ=Gamma*Z*Ivlev-d*Z;
  delN=-UPTAKE*P+m*P+...
      (1-Gamma)*Z*Ivlev+d*Z;
%To mix with deeper water and sink
%some organic matter use the
%following delN= statement instead:
  %delN=-UPTAKE*P+0.6*m*P+0.4
  % *(1-Gamma)*Z*Ivlev+0.4*d*Z
  % +Mix*(NatZ-NIT);
  %Calculate new values of P, Z, NIT:
  P=P+delP; Z=Z+delZ;
NIT=NIT+delN;
  end
end
figure
plot(Tct(1:end), Nct(1:end),'k');
hold on;
plot(Tct(1:end), Pct(1:end),'g');
plot(Tct(1:end), Zct(1:end),'r');
```

"sink", and by mixing nutrient-rich (10.6 μM) deep water into the upper layer with corresponding removal of upper-layer water (with nutrients and phytoplankton), both at a rate of 2% of upper-layer water daily. Also, an autumn (fall) bloom can be added by increasing the mixing rate from some arbitrary day onward, shown as day 120, at 2% per day. All of these changes are specified in program comments (Box 4.3). The result (Fig. 4.4d) has a large, if too brief, spring bloom, low summer standing stocks, and a fall bloom not quite half as strong as that in spring. Nutrients then rapidly return to initial levels and the fall bloom is removed by increased mixing and grazing.

A (somewhat) more complex NPZ model

That Franks *et al.* (1986) model has evoked a great deal of analysis (e.g. Busenberg *et al.* 1990; Edwards *et al.* 2000),

but it is not realistic in several respects, apart from the extreme simplification of its biological variables relative to the complexity of an actual pelagic community. Most importantly, we understand that blooms are initiated in spring when illumination becomes sufficient above the shallowest significant mixing barrier for net increase (growth – grazing) to exceed stock losses due to mixing. Thus, a somewhat more realistic model would simulate control of primary production by seasonally varying sunlight and include the decrease of light with depth, at least to the bottom of a mixing layer. Since phytoplankton (*P*) growth varies non-linearly with irradiance (the *P* vs. *E* relation), and irradiance decreases exponentially downward, the integration of production in the mixing layer should be by summing production stepwise down the water column, rather than by integrating the available light and then applying a *P* vs. *E* relation to the mean *E*. Mixing variation must be included in at least simplified fashion, say by varying the mixed-layer depth through simulated seasons. Phytoplankton can be considered to be evenly distributed

Table 4.1 Symbols and standard values or initial values for the Franks *et al.* (1986) model.

VARIABLES AND PARAMETERS	VALUES IN STANDARD RUNS
V_m = maximum phytoplankton growth rate	2 d^{-1}
N = nutrient concentration	Start at 1.6 µmol N liter^{-1}
K_s = half saturation constant for nutrients	1 µmol N liter^{-1}
P = phytoplankton stock size	Start at 0.3 µmol N liter^{-1}
m = phytoplankton mortality rate (apart from grazing)	0.1 d^{-1}
Z = zooplankton stock size	Start at 0.1 µmol N liter^{-1}
γ = zooplankton growth efficiency	0.3
R_m = maximum zooplankton ration	1.5 d^{-1}
Λ = Ivlev constant	1.0 (µmol N liter^{-1})$^{-1}$
d = zooplankton mortality rate	0.2 d^{-1}

through the mixed layer, absent below, with losses to depth when the layer shallows, dilution when it deepens. Nutrient limitation and grazing are independent factors that terminate spring blooms, so they must be included as alternative controls of the phytoplankton stock. Zooplankton can be taken to sustain their stock within the mixed layer by swimming, which may or may not be realistic for protozoan grazers.

Evans and Parslow (1985) developed a model quite similar to that of Franks *et al.*, with the added features just listed. It is modified somewhat here. They represented control of phytoplankton growth rate by multiplying the nutrient-limitation effect by the light-limitation effect. It is likely preferable, as represented in the equations below, and as applied by Denman and Peña (1999), to choose at each time step the lesser of the rates set by light or by nutrients; that is, to strictly apply Liebig's "law of the minimum".

The model has the following equation set:
Change of mixing depth,

$$\Delta M_Z/\Delta t = \zeta^+(t);$$

Change of nutrient,

$$\Delta N/\Delta t = -GP + (\zeta^+(t) + 0.025M_Z)(N_{deep} - N)$$
$$+ 0.5mP + 0.5\,grazing\,H$$

G = phytoplankton growth rate, d^{-1} = min{V_{max} (−exp(−$\alpha Ez/V_{max}$)), V_{max} $N/(K_s + N)$}

Change of phytoplankton,

$$\Delta P/\Delta t = GP - mP - grazing\,H - \zeta^+(t)P;$$
$$grazing = c(P - P_0)/(d + P - P_0)$$

Change of herbivores,

$$\Delta H/\Delta t = grazing\,fH - carnH - (M_z(i) - M_z(I - 1))$$

The $\Delta[M_z, N, P, \text{or } H]/\Delta t$ values, as implemented in the model, are whole-day (24 h) changes. State variables N, P, and H are in units of nutrient (as nitrogen) concentration, µmol liter^{-1}. The symbol $\zeta^+(t)$ means that change in concentrations (N and P) due to altered mixed-layer depth (Fig. 4.5) only occurs when the mixed layer deepens, not when it shoals. This is not applied to herbivores, H, which are assumed to swim up to stay above the limit to mixing; thus mixed-layer deepening dilutes them, shoaling concentrates them. Mixing also includes $0.025M_z$ exchange between the mixing layer and deeper water each day. Phytoplankton growth rate, when limited by available light, is related to it by a saturating P vs. E curve. Almost any suitable function will serve, and a simple one used by Denman and Peña (1999), V_{max} (1 − exp[−$\alpha E_z/V_{max}$]), is used here. The quantity 1 − exp(−$\alpha E_z/V_{max}$) varies from 0 to 1. Estimates of available sea-surface light, E_0, are from standard irradiance functions for the latitude and date (Brock 1981), ignoring variations due to clouds (although cloud effects can be added). Extinction down the water column is done here by a numerical integration from dawn to sunset on every model day, meter-by-meter down through the mixed layer. Both phytoplankton and zooplankton have daily mortality losses, mP and $carnH$, proportional to their abundance. Grazing has a hyperbolic relationship to P and is proportional to H, but with a threshold, P_0, below which grazing stops. Herbivores grow with an efficiency f, i.e. added $H = f$ $graz\,HP$. A program for this model is shown in Box 4.4. Initial variable and parameter values (changed from Evans & Parslow 1985) are those shown in the program.

The model was run until the cycle was stable (takes ~3 years; we used 6), and then restarted with the sixth year values for January 1 (as in Box 4.4). The result (Fig. 4.6a) shows a cycle (that recurs consistently) with a spring bloom in April, summer low in P, small fall bloom, and winter low. The nutrient cycles inversely and zooplankton bring down the phytoplankton bloom and peak toward the end of May. The timing of the spring bloom is about right for the Atlantic at the modeled latitude, 47°N. Nutrients (fixed nitrogen here) are drawn down to realistically low values for the North Atlantic. Phytoplankton growth is light limited until the peak of the spring bloom and again shortly after the fall bloom; it is nutrient limited from mid-April to September (Fig. 4.6b).

Note that we, like Evans and Parslow, provided some diffusion through the density barrier at the bottom of the

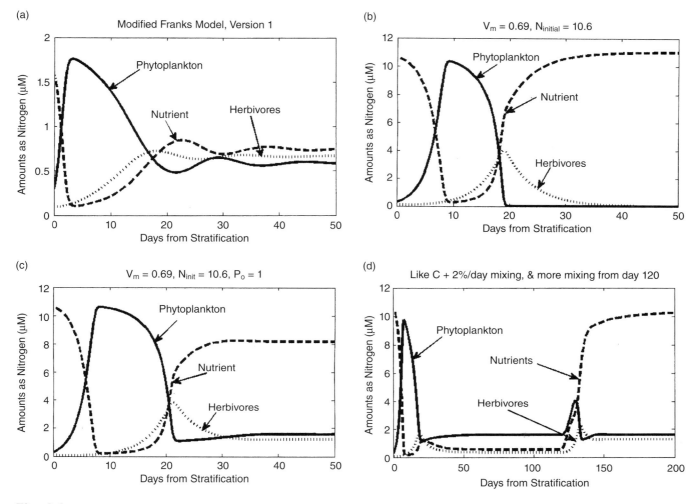

Fig. 4.4 Annual nutrient, phytoplankton and herbivore stock variation in a Franks–Wroblewski–Flierl model. (a) Similar to original model. (b) V_m reduced to $0.69\,d^{-1}$ and initial nutrient increased to $10.6\,\mu M$. (c) Like (b), but with a threshold phytoplankton abundance, P_0, for grazer activity. (d) Like (c) with mixing daily from a lower layer constantly at initial nutrient and with no phytoplankton.

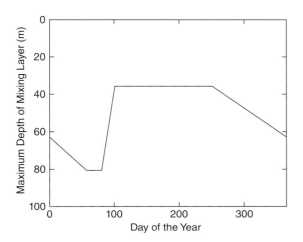

Fig. 4.5 Mixed-layer depth cycle imposed in the Evans–Parslow model.

mixed layer. Thus, there is always some nutrient input from depth. Nutrient is also recycled; as implemented here half of the phytoplankton mortality ($0.07\,d^{-1}$) is returned to the mixed layer as nutrient, and 15% of grazing reduction of P is removed from the mixed layer. Those can be taken as the "export production".

A "standard run" of a model like this (Box 4.4; Fig. 4.6a) applies parameters and starting values that produce output pleasing to the modeler in some sense (it fits his/her data; it fits his/her intuition of how the cycles should look; . . . ; it doesn't crash). The next step is varying parameters and input values to examine the sensitivity of the model to them. For example, in this model the small fall bloom starts on model day 250, exactly when mixing depth starts to descend. The model is very sensitive to shifts in mixing depth. As a test, the model can be run with some random variation (only $\pm 1\,m$) in mixed-layer depth. The necessary changes are shown as "comment" lines in Box 4.4 (i.e. lines starting "%", with instructions for making the changes). If

Box 4.4 Mid-latitude, Atlantic production cycle model, similar to that by Evans and Parslow (1985), driven by irradiance (W m⁻²) variation (modeled after Brock 1981), nutrient and herbivory cycling

```
%Mid-latitude, Atlantic production cycle driven by irradiance (W/m^2)
%variation (modeled after Brock, 1981), nutrient and herbivory cycling.
clear all %Location and run length:
Lat=47.; nyears=1; %Storage vectors:
Daystr=NaN*ones(365*nyears,1); Pstr=Daystr; Nutstr=Daystr; %Storage vectors:
Hstr=Daystr; MLZ=zeros(1,365); DayNstr=NaN*ones(2,365*nyears);
%Solar constant (W/m^2), atmospheric attenuation, PAR fraction):
SolarK=1373; AtmAtt=0.5; ParFrac=0.48;
%Light (water extinction, phyto extinction), from Fasham et al. (1990):
ex=.1; ey=0.12; %Nutrient Parameters, starting nutrients and mixing depths:
Ks=1.5; DeepNut=10.0; Nut=5.4541; Mprev=62;
%Phyto growth (alpha [h^-1/W m^-2], Vmax[d^-1], m=mortality [d^ -1]), and starting value:
alpha=0.04; Vmax=0.7; P=1.1228; m=0.07;
%Hebivore parameters - grazing rate (c&d), threshold P, growth efficiency,
%starting value, death rate:
c=0.35; d=1.0; Po=0.1; f=0.35; H=0.4048; carn=0.07;
%Generate yearly sequence of mixed layer depths (MLZ):
%for Yr=1:nyears %(remove % for random effects on MLZ)
  for i=1:365
      %X=1.-2.*rand; %(remove % for random effects on MLZ)
      X=1. ; %(add % for random effects on MLZ)
      if i<58; MLZ(i)=X+62+i*18/58;
      elseif i >= 58 && i <81; MLZ(i)= X+80;
      elseif i >80 && i <101; MLZ(i)=X+80-45*(i-80)/20;
      elseif i >100 && i <250; MLZ(i)=X+35;
      elseif i>=250; MLZ(i)=X+35+27*(i-250)/115.;
      end
  end
for Yr=1:nyears %Year-to-year loop (add
% for random effects on MLZ)
  for i=1:365 %Main daily loop
      Day=(Yr-1)*365+i; Daystr(Day)=Day;
      NL=Vmax*Nut/(Ks+Nut); diff=0.025*MLZ(i);
      ext=ex+ey*P;
% Surface PAR at latitude for times of day from dawn to noon and
% production rate integration down to MLD:
%I. Declination=angle of sun above the equator:
      D1=23.45*sind(360.*(284.+i)/365.);
%II. Angle(deg.) between south (i.e., noon) and setting sun:
      W1=acosd(-1.*(tand(Lat)*tand(D1)));
%III. One-half daylength, hours dawn to noon:
      L1=W1/15.; % Earth rotates 15=degrees/hour
%IV. Distance of Earth from sun relative to average, a minor effect:
      Rx=1./sqrt(1.+0.033*cosd(360.*i/365.)); %bookkeeping:
      VTofD=L1/40.; TofD=12.01-L1-VTofD; SGr=0.;
      for j=1:40 %summing production dawn to noon in 40 steps
        TofD=TofD+VTofD; %for the following see Brock (1981):
        W2=(TofD-12.)*15.;
        CosZen=sind(D1)*sind(Lat)+cosd(D1)*cosd(Lat)*cosd(W2);
```

```
            Isurf=SolarK*CosZen/(Rx*Rx);
            Io=Isurf*AtmAtt*ParFrac;
%Sum photosynthesis down to MLZ, meter by meter
          for k=1:MLZ(i) %progressive light extinction to MLZ(i):
            Iz=Io*exp(-1*ext*k);
              % Denman-Pena function for phyto growth, scaled (0-1):
              Gr=1-exp(-alpha*Iz/Vmax);
              SGr=SGr+Gr;
          end
      end
      AveGr=2*Vmax*SGr/(MLZ(i)*40.); %(2* to get dawn to dusk)
      %convert rate to daily growth multiplier:
      if NL < AveGr
        G=NL;
      else
        G=AveGr;
      end
      %mixing due to mixing layer deepening, zeta:
      if MLZ(i) > Mprev; zeta=MLZ(i)-Mprev;
      else zeta=0.; end
      %update state variables (P, Nut, H)
      graz=c*(P-Po)/(d+P-Po);
      if P < Po; graz=0.; end
      xmix=(diff+zeta)/MLZ(i);
      P=P+G*P-m*P-graz*H-xmix*P;
      Nut=Nut-G*P + 0.5*m*P + xmix*(DeepNut-Nut) + 0.5*graz*H;
      DelH=f*graz*H-carn*H-(MLZ(i)-Mprev)*H/MLZ(i);
      H=H+DelH; %change to new mixed layer depth:
      Mprev=MLZ(i); %store variables for plotting:
      Pstr(Day)=P; % For P as Chl, multiply P by *8*12/50;
      % 8=C/N, 12=mg C/mmoleC, 50=C/Chl
      Nutstr(Day)=Nut; DayNstr(1,Day)=Day; DayNstr(2,Day)=Nut;
      Hstr(Day)=H;
  end
end
%plot results vs. days from start on 1 January:
plotyy(Nutstr(1:end),'k',Pstr(1:end),'g'); %Chlorophyll, green line
%Nutrient, mmoles/m^3 fixed nitrogen, blue line
ylabel('Nutrient Units');
hold on %Herbivores as nitrogen, red line
plot(Daystr(1:end),Hstr(1:end),'r');
```

you run the model with those changes (and set *n*year = 3), the output will be much the same, but with somewhat realistic variability in the summer values of *N*, *P*, and *H*, both during summer and between years. Since the random variation in mixing is about the same magnitude as the onset of mixed-layer deepening in fall, the summer oscillations usually replace the fall bloom.

Having established a model that produced recurring seasonal cycles including a spring bloom, Evans and Parslow set out to discover what would eliminate blooms. Many oceanic ecosystems, mostly HNLC systems (see Chapter 11), consistently do *not* exhibit seasonal blooms. They tried eliminating variation in mixed-layer depth. That damps the cycle amplitude in our model, too, but cycles remain. Simplify the program appropriately and try that. If the model is realistic in its fundamental relationships, then it appears that blooms are affected by, but do not solely depend upon, mixed-layer variation. The principal drivers must then be the cycle of insolation and the response of grazers to it. Next, they asked whether the parameters describing grazing could be responsible for the absence of strong phytoplankton cycling seen in many oceanic areas, and found them strongly effective. In our version, increasing the maximum per capita grazing rate, c, from 0.35 to 0.6 flattens the cycles of phytoplankton stocks, eliminating the spring bloom (Fig. 4.6c), despite the ongoing cycling of illumination and mixing rate. Nutrients, while cycling, remain high in the mixed layer all year. That, in fact, remains the theory explaining the absence of blooms in the subarctic Pacific and subantarctic. As will be seen (Chapter 11), it is thought (and tested) that trace-metal limitation sets up phytoplankton–grazer relations appropriate

for grazing to consistently balance phytoplankton growth, but with oscillations somewhat like those of summer in our model.

When you try different parameters in this program, you will likely find some sets with strong day-to-day oscillations and some producing negative values of state variables. That is because of the whole-day time-steps. Changes of actual oceanic phytoplankton biomass do come in daily increments (with increase in the light, decrease in the dark), so there is a touch of realism in this choice. We have followed Evans and Parslow in sticking to quite a shallow (80 m) winter mixed layer, much shallower than usual North Atlantic winter mixing. A useful exercise to try is reprogramming the model to increase winter mixing, say, to 300 m, i.e. below the usually calculated critical depths (defined in Chapter 11). To limit bloom duration you may need stronger grazing. You will not get a classic North Atlantic bloom, at least not with the initial parameters.

An aside: problems in solving of differential equation systems numerically

So far we have only applied the so-called Euler method for approximating solutions to differential equations. It is a good scheme for simple equations, and the adequacy of a solution can be checked by rerunning the approximation with much shorter time steps. To do that, you also have to reduce the rate parameters in equivalent ratio. Eventually, at some extremely small time step, this will always work, reproducing the result for a somewhat longer step, but it may take an eon of computer time. When the time step is too large, the derivative for one or more of the curves being simulated can at some point force the sequence of state values far off the path of a correct solution. This will not stop the integration; it will just go off into misleading state variable space, sometimes producing temptingly wonderful but utterly wrong results. This happens when the real solution is strongly curved or it abruptly shifts direction within a time step, departing from the exactly linear jumps of the derivative sequence. An example is shown in Fig. 4.7.

There are a number of ways to check and correct the accuracy of an integration as it proceeds: Runge–Kutta methods of different orders (the Euler method is order 1) and other variable time-step or "vicinity checking" schemes. The Marquardt–Levenberg approach is still widely used. If you become serious about such modeling, spend time learning its complexities and pitfalls, all of them outside the scope of this book.

There are also programming pitfalls to avoid. Units and conversion factors must be consistent: state variables in

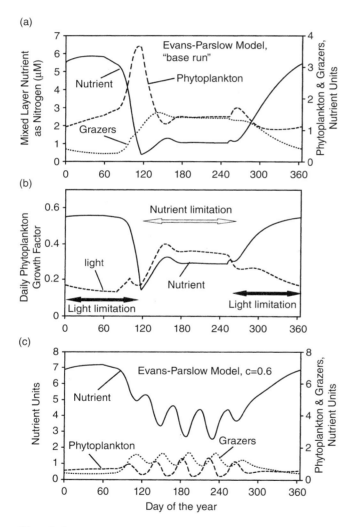

Fig. 4.6 (a) The annual cycle of the modified Evans–Parslow model. Right scale in nutrient-content units (μM) for phytoplankton (dashed lines) and grazers (thin lines); left scale for nutrient (μM, thick lines). (b) Comparison of the effects of light and nutrient limitation of phytoplankton growth rate in the mixed layer. The smaller of the light-limitation (solid line: rate = $V_{max} A_z\{1 - \exp[\alpha I_z/V_{max}]\}$, where A_z means average over depth) or the nutrient limitation (dashed line: rate = $V_{max} N/[K_s + N]$) operates each model day. (c) Effect on Evans–Parslow model output of increasing grazing rate; the spring bloom is suppressed and replaced by low-level summer oscillations.

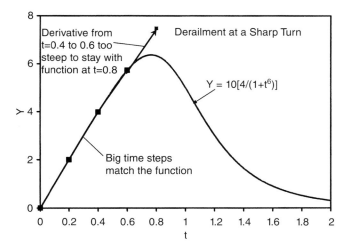

Fig. 4.7 Example of an implicit integration with overly long time steps (dark line from square to square) departing from the trajectory of the function (the gray curve) representing the correct solution.

mass or moles, phytoplankton growth rates appropriate to the irradiance units (watts/area, photons/[area × time)) and so forth. Thresholds like $(P - P_0)$ in a grazing function must not be applied if $P < P_0$. And there are others!

More sophisticated models, subarctic Pacific ecosystem dynamics

Several models, still NPZ models, but all with somewhat more sophistication, have been developed to examine the so-called "Subarctic Pacific Problem" (Frost 1993; Fasham 1995; Denman & Peña 1999; Denman et al. 2006). The oceanic Gulf of Alaska, in particular, has strong seasonal cycles of phytoplankton growth rates that, unlike those in the oceanic North Atlantic, do not result in cycling of phytoplankton stock, particularly as measured by chlorophyll concentration. The Frost paper, summarized here, has the advantage (much to be commended in modeling reports) of providing the mechanics of the model in sufficient detail that a person with moderate skill can program it from the equations provided. Some of the other models highlight additional features of the system.

The goal is to model the essential relations among phytoplankton, grazers and nutrients that characterize the subarctic Pacific (discussed in more detail in Chapter 11). In the oceanic sectors of the Gulf of Alaska and west almost to Japan, there are *no* phytoplankton blooms. Instead, there is a low amplitude oscillation of phytoplankton stocks keeping chlorophyll-*a* between about 0.15 and 0.65 µg liter^{-1} (Fig. 11.5) with occasional run-ups to almost 1.0 µg liter^{-1}, but very rarely more than that. Levels are somewhat higher

to the west, seaward of Sakhalin and Hokkaido. Just as important, nitrate in the surface mixed layer remains greater than 6 µM right through the year; it never drops to a value low enough to limit phytoplankton growth. There is an annual cycle in surface nitrate (Fig. 4.8), between 17 µM in March and about 7 µM in July, but it is not depleted. Pelagic ecosystems with both of these features are referred to as high-nitrate, low-chlorophyll (HNLC) systems. As discussed in Chapters 3 and 11, Martin et al. (1989) suggested that the key factor keeping phytoplankton stocks low is probably limitation of the growth rate of larger phytoplankton by the very low availability of iron. Their iron-enrichment experiments showed large phytoplankton eventually blooming in containers with added iron, but not in those without it. The phytoplankton that are present are growing rapidly, one doubling per day or more, but they are small. The effect of this (Miller et al. 1991a & b) is that phytoplankton are then susceptible to grazing by protozoans. Protozoans, in turn, can grow about as fast as phytoplankton, which enables them to hold the phytoplankton stocks within narrow limits.

Frost (1993) captured most of this in a one-dimensional model, with that one dimension being depth. It describes processes down the water column by installing separate NPZ models in each of numerous layers, in this case an upper mixing layer and a series of deeper layers through the progressive density stratification to the bottom of the euphotic zone. Exchange between adjacent layers is included as a process at each time step. Like the Evans–Parslow model already described, phytoplankton growth depends upon irradiance (but here according to the Jassby & Platt *P* vs. *E* function), modified somewhat when available fixed nitrogen concentration is reduced. To model the subarctic Pacific, it is assumed that the phytoplankton are very small, because big phytoplankton are strongly growth-limited by low iron availability. Since the phytoplankton are small, the grazers are characterized as protozoans with suitably high potential growth rates. The upper layer is assumed to be well mixed, so only one set of state variables is followed there. Below the mixed layer, however, the values vary with both biological processes and mixing between successive 1m layers at realistic rates. All variable and parameter names are given in Table 4.2.

Here is the equation set for the mixing layer at the surface:

$$\frac{\Delta P}{\Delta t} = \frac{1}{Z_m}\left\{\sum_{Z=0}^{Z_m} P * PMAX * \tanh\left[\frac{\alpha PAR_z}{PMAX^*}\right]\right\}$$
$$- \frac{e(P - P_o)H}{f + P - P_o} + \frac{K_v}{Z_m}(P_{Z_m+1} - P)$$

$$\frac{\Delta H}{\Delta t} = H\left[\frac{\gamma e(P - P_o)}{f + P - P_o} - \frac{mH}{b + H}\right] + \frac{K_v}{Z_m}(H_{Z_m+1} - H)$$

$$\frac{\Delta D}{\Delta t} = 0.3\left(H\frac{e(P - P_o)}{f + P - P_o}\right) - wD + \frac{K_v}{Z_m}(D_{Z_m+1} - D)$$

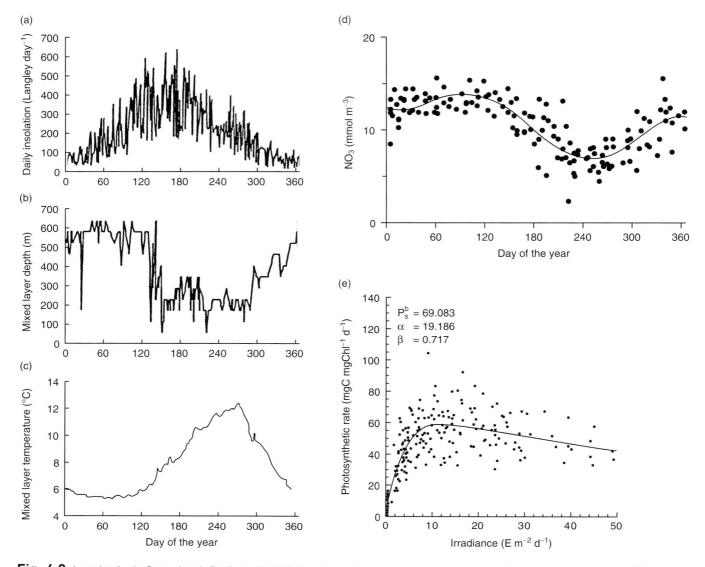

Fig. 4.8 Input data for the Frost subarctic Pacific model. (a) Daily irradiance from weather-ship radiometer data. (b) Mixed-layer depth from daily CTD casts. (c) Mixed-layer temperature. (d) Annual nitrate cycle based on data for many years, and used for the starting value on year day 1. (e) *P* vs. *E* relation used in the model with large initial slope. (After Frost 1993.)

$$\frac{\Delta NO_3}{\Delta t} = -\frac{\xi}{Z_m}\left\{\sum_{Z=0}^{Z_m} P * PMAX * \tanh\left[\frac{\alpha PAR_Z}{PMAX^*}\right]\right\}$$

$$\left(1 - \frac{NH_4}{d_{NH_4} + NH_4}\right) + \frac{K_v}{Z_m}(NO_{3Z_m+1} - NO_3)$$

$$\frac{\Delta NH_4}{\Delta t} = -\frac{\xi}{Z_m}\left\{\sum_{Z=0}^{Z_m} P * PMAX * \tanh\left[\frac{\alpha PAR_Z}{PMAX^*}\right]\right\}$$

$$\frac{NH_4}{d_{NH_4} + NH_4} + \xi H\left[\frac{0.4e(P - P_o)}{f + P - P_o} + \frac{0.4(mH)}{h + H}\right]$$

$$+ \xi(0.4wD) + \frac{K_v}{Z_m}(NO_{4Z_m+1} - NH_4)$$

Here is the very similar equation set for layers below the first stratification barrier to mixing. Terms involving K_v are implemented as the fractional mixing at each time step between a given vertical meter and those above and below:

$$\frac{\Delta P_z}{\Delta t} = \left\{P_z * PMAX * \tanh\left[\frac{\alpha PAR_z}{PMAX^*}\right]\right\}$$

$$- \frac{e(P_z - P_o)H_z}{f + P_z - P_o} + K_v(P_{z-1} - P_{z+1} - 2P_z)$$

$$\frac{\Delta H_z}{\Delta t} = H_z\left[\frac{\gamma e(P_z - P_o)}{f + P_z - P_o} - \frac{mH_z}{h + H_z}\right] + K_v(H_{z-1} + H_{z+1} - 2H_z)$$

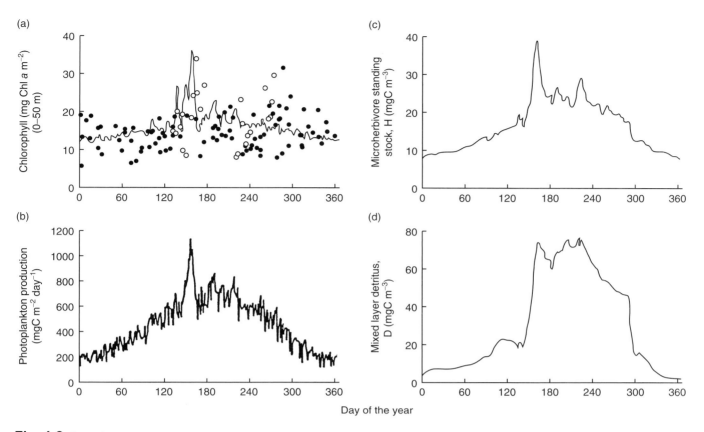

Fig. 4.9 Output from the subarctic Pacific model. (a) Integrated chlorophyll in the upper 50 m compared to field data; (b) phytoplankton production rate; (c) Microherbivore standing stock, *H*, in the mixed layer. The cycle of primary productivity appears in this second trophic level, not in the phytoplankton; (d) Mixed layer detritus, *D*. (After Frost 1993.)

$$\frac{\Delta D_z}{\Delta t} = 0.3 \left(H_z \frac{e(P_z - P_o)}{f + P_z - P_o} \right) - wD_z + K_v(D_{z-1} + D_{z+1} - 2D_z)$$

$$\frac{\Delta NO_{3z}}{\Delta t} = -\xi \left\{ P_z * PMAX * \tanh\left[\frac{\alpha PAR_z}{PMAX*} \right] \right\}$$

$$\left(1 - \frac{NH_{4z}}{d_{NH4} + NH_{4z}} \right) + K_v(NO_{3z-1} + NO_{3z+1} - 2NO_{3z})$$

$$\frac{\Delta NH_{4z}}{\Delta t} = -\xi \left\{ P_z * PMAX * \tanh\left[\frac{\alpha PAR_z}{PMAX*} \right] \right\}$$

$$\frac{NH_{4z}}{d_{NH4} + NH_{4z}} + \xi H_z \left[\frac{0.4e(P_z - P_o)}{f + P_z - P_o} + \frac{0.4(mH_z)}{h + H_z} \right]$$

$$+ \xi(0.4wD_z) + K_v(NH_{4z-1} + NO_{4z+1} - NH_{4z-1})$$

Some data for choosing appropriate parameters and initial conditions for the model were derived from year-around sample series collected in the 1970s from weather ships occupying Ocean Station "P" (50°N, 145°W) (Figs. 4.8a to 4.8d). Those included temperature, daily irradiance, and mixed-layer depth data. Some starting values were also derived from weather-ship data, chlorophyll and nitrate values in particular. Some parameters were taken from the data (Fig. 4.8e) of an observational program in

the 1980s (Miller 1993). Those include the *P* vs. *E* relation (a very steep initial slope, α, of *P* vs. *E*) and an effect by which the concentration of ammonium, favored for uptake, reduces nitrate uptake (Wheeler & Kokkinakis 1990).

The original model was written in Fortran (six to eight pages), with outputs as time-series (Figs. 4.9a to 4.9d) of chlorophyll, herbivore stock, nitrate, ammonium, detrital nitrogen, and sundry other variables. The calculation not only reproduces a seasonally flat chlorophyll time-series at the right level and with the right annual total of primary production, it also gets the low-level, short-term, inverse oscillations of chlorophyll and ammonium about right in terms of amplitude and period. The Frost (1993) model remains a good representation of the basic production processes in an HNLC system, although Strom *et al.* (2000) pointed out some problems. In particular, the model depends upon a grazing threshold for microzooplankton to make HNLC conditions persist. Unfortunately, a threshold is consistently not found in experimental work. However, Leising *et al.* (2003) have shown that the effect of thresholds can be mimicked by increasing the half-saturation constant for grazing (the variable "*f*" in the $\Delta H/\Delta t$ equation above). It remains to be seen whether that is the case for microherbivores in the subarctic Pacific. That is the value

Table 4.2 Symbols used in the model (simplified from Frost 1993).

Biological and chemical state variables

P, P_z	Phytoplankton carbon concentration in the mixed layer, and at depth Z in the intermediate layer (mgC m^{-3})	
H, H_z	Herbivorous zooplankton carbon concentration (mgC m^{-3})	
D, D_z	Concentration of detritus (mgC m^{-3})	
N, N_z	Nitrogenous nutrient (NO_3 or NH_4) concentration (mmol N m^{-3})	

Physical environment input data (Fig. 4.8)

I_0	Incident solar radiation (ly day^{-1}) (not explicit in equations shown here)	
Z_m	Mixed-layer depth (m)	
T_{Z_m}	Mixed-layer temperature (°C)	

Derived environmental properties

k	Attenuation coefficient of irradiance (m^{-1}) (not explicit in equations shown here)	
PAR_z	Photosynthetically available irradiance (E m^{-2} day^{-1}) at depth z	

Fixed and variable parameters

		Value
K_v	Vertical eddy diffusivity (cm^2 s^{-1})	0.1–1.80*
α	Initial slope of phytoplankton photosynthesis vs. irradiance response (mgC [mg Chl a]$^{-1}$ [E m^{-2}]$^{-1}$)	21.0
P_{max}	Maximum carbon-specific photosynthetic rate (mgC [mgC]$^{-1}$ day^{-1})	0.47–1.38**
ξ	N:C in organic matter (mmol N m^{-3})	0.0126
d_{NH_4}	Half-saturation constant for phytoplankton NH_4 uptake (mmol N m^{-3})	0.1
e	Herbivore maximum specific ingestion rate (mgC [mgC]$^{-1}$ day^{-1})	1.01–1.66[†]
f	Half-saturation constant for herbivore ingestion (mgC m^{-3})	17.0
ρ_0	Herbivore grazing threshold (mgC m^{-3})	10.0
γ	Herbivore growth efficiency	0.3
m	Herbivore maximum specific mortality rate (mgC [mgC]$^{-1}$ day^{-1})	0.30–0.50[†]
h	Half-saturation constant for herbivore mortality (mgC m^{-3})	35.0
w	Detrital degradation rate (fraction day^{-1})	0.03–0.05[†]

*Seasonally variable.
**Dependent on daylength and temperature.
[†]Dependent on temperature.

of models; they suggest logically feasible hypotheses for testing at sea.

The vertical mixing scheme in Frost's (1993) model is simple, but works because the mixing coefficients (diffusivities) were tuned to reproduce vertical profiles of hydrographical properties and to provide the same upper-layer nutrients at the end of a model year as at the beginning. More sophisticated mixing schemes are required if model ecosystem cycles are to respond to some elements of climate change: more-or-less surface irradiance (a difference in cloudiness and thus surface warming) and changes in wind climatology. Denman *et al.* (2006) applied to a model of the subarctic Pacific problem a so-called turbulence closure system (Mellor–Yamada 2.5) in which vertical mixing is subject to surface warming and wind, modeling their effects in layers 2 m thick down to 120 m. Other schemes are in use for this, particularly one called Large–McWilliams–Doney (LMD) or K_v profile parameterization (KPP), not so different from the Frost scheme, but incorporating more explicit processes.

The model from Denman's group also has a more complex set of state variables, i.e. seven (Fig. 4.10), and thus more interactions. There are two size classes of phytoplankton, a pico–nano group and a diatom-like group, phytoplankton growth reduced by iron limitation (more so for the larger cells) and protozoan grazers eating both phytoplankton groups at different rates and themselves

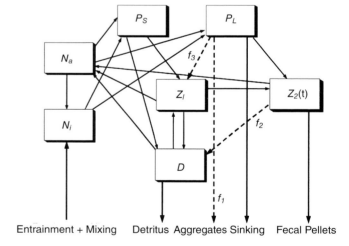

Fig. 4.10 State variables and transfers in the subarctic Pacific model of Denman *et al.* (2006). Schematic representation of the nitrogen version of the ecosystem model. P_S represents nano- and picophytoplankton, P_L diatoms, Z_1 microzooplankton, N_i nitrate, N_a ammonium, D detritus and bacteria, $Z_2(t)$ the imposed annual cycle of mesozooplankton, f_1 formation of sinking diatom aggregates, f_2 loss to detritus of unassimilated food of mesozooplankton that does not end up as ammonium, and f_3 grazing on diatoms by microzooplankton. (After Denman *et al.* 2006.)

being eaten by mesozooplankton with a fixed cycle of abundance based on seasonal data. There is also a silica cycle with three more state variables, added as a secondary control of the diatoms. At this level of complexity, a model has very many parameters, 44 in this case, some of them constrained by data and some chosen by trial-and-error in numerous runs of the program. Whatever uncertainty that leaves about the reality of the actual mechanisms, the model produces some fascinating results. The basically flat total phytoplankton stocks of the region are again reproduced, but the small and large phytoplankton have different seasonal cycles (Plate 4.1).

Small cells pulse at the surface in spring; diatoms become abundant during summer in the deeper layer between the seasonal thermocline and the permanent halocline, and then their stocks expand upward to the surface in fall. October peaks in chlorophyll were noted in time-series sampling done from the weather ships several decades ago. The model also produced blooms of first small and then large cells in response to reducing the iron-limitation parameters, increases on roughly the same time scale as the phytoplankton stock changes that occurred in the SERIES iron-addition experiment in July–August of 2002 (Boyd *et al.* 2005). The vertical separation of different classes of

primary producers constitutes a hypothesis that should be examined in the field.

ERSEM-PELAGOS, a model of pelagic processes in European and global seas

Model ocean ecosystems are also expanded to two and three spatial dimensions, at the expense of more computing. Typical two-dimensional models represent explanations of onshore–offshore, surface-to-seafloor patterns of production and grazer distributions in coastal areas such as the upwelling zones (e.g. Edwards *et al.* 2000). In three-dimensional models, patterns and cycles are constructed for spatially variable ocean sectors and for the entire world ocean (Zahariev *et al.* 2008, and many more).

Pelagic ecosystem simulations are frequently taken to extremes, attempting to include virtually every significantly distinctive component known to be part of the system. The ERSEM-PELAGOS model is an example. It has 44 state variables (Fig. 4.11), more than appear in the diagram

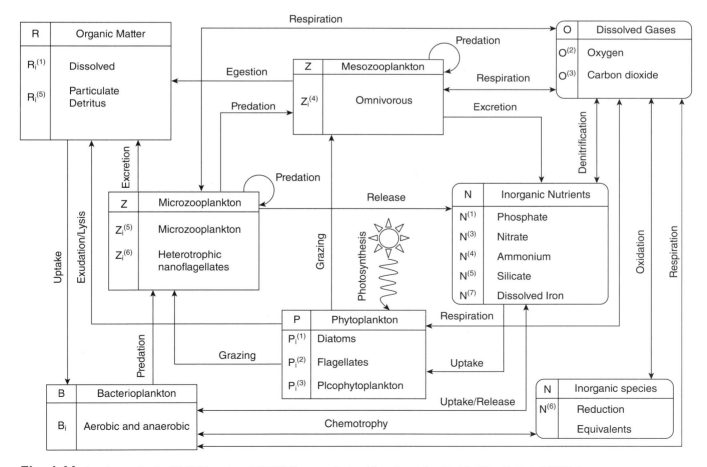

Fig. 4.11 Box diagram for the PELAGOS version of ERSEM (European Regional Seas Ecosystem Model). (After Vichi *et al.* 2007a.)

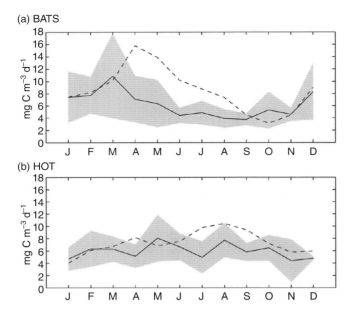

Fig. 4.12 Seasonal cycles of observed (continuous line) and the ERSEM-PELAGOS simulation (dashed line) of gross primary-production rates (monthly means averaged in the mixed layer) at (a) the Bermuda Atlantic Time-series and (b) the Hawaii Ocean Time-series (HOT). Gray shading is the standard deviation of the 10-year data sample. (From Vichi *et al.* 2007.)

because C, N, P, and in some cases Si and Fe, are independently tracked within living or detrital components. There are about 300 parameters (the authors did not count them for us). Because many of the parameters are tuned to make the model match data, the model appears to generate reasonably realistic profiles, seasonal cycling (Fig. 4.12) and maps (see Vichi *et al.* 2007a & b). However, while the patterns appear realistic, actual quantities are often off by two-fold. For some purposes it may not matter whether the incorporated processes and exchanges are close representations of real ocean physics and ecology, if the output patterns fit the maps and cycles. On the other hand, if the mechanics are just a fit forced in some way, then a model will not correctly predict effects of changed environmental forcing (climate change, overfishing, eutrophication, . . .). To an increasing extent, human activity constitutes management of the oceans' ecology, as it does of all Earth habitats. Wisdom in that management depends increasingly on these complex models. If their mechanics are not realistic, then they will provide misleading predictions and misleading advice for the design of our interactions with nature.

The life and times of individual animals

This section is placed here because it is about modeling. The reader may want to postpone study of it until after reading more about the zoology of zooplankton in Chapters 6 through 8. To show that models can have a completely different form and address completely different issues, we will consider some population-dynamical problems using individual-based (IBM) models. These are also called agent-based models.

A simple example

Several lines of evidence suggest that some, possibly many, species of copepods have environmental sex determination (ESD; see the copepod section of Chapter 6). In brief, instead of genes clustered on one chromosome (for example, "Y" in *Homo sapiens*) from one parent controlling development as male or female, sex is determined at a receptive stage in development, sometimes very late, by some aspect of the habitat. Sex in many reptiles is set by the temperature at which the embryos develop. In the copepod genus *Calanus*, clutches of eggs can mature entirely as females, which usually occurs in laboratory rearing, or as 90% males, which is sometimes observed in the field. The controlling factors are not reliably identified, and may differ among genera or species.

With only that much background, and without fully proving that ESD operates, one can ask what selective advantage could accrue to populations from having flexible sex ratios? Hypotheses can be proposed, and since relative numbers of distinctly different types of individuals are involved, the feasibility (not the reality!) of those hypotheses can be tested by numerical modeling. The strong prevalence of males maturing in some field observations, combined with the certainty that males have much higher field mortality rates than females (males also mature earlier), suggests that an advantage should accrue to stocks with many more males at maturation than females. That can be tested with an IBM.

Let $\female$ = No. of young maturing as females, $\male$ = No. as males and $T = \female + \male$. From simple encounter theory, the expected number (or rate) of male–female encounters (E) for mating will be proportional to the product of their abundances: $E = C\male\female$, in which C is an "encounter kernel" (see Chapter 1). Then, since $\female = T - \male$, $E = C\male(T - \male)$ and $E = CT\male - C\male^2$. Thus, the number of encounters should have a parabolic relationship to the relative numbers of males and females. Also, at least with no differential mortality, since setting $dE/d\male = CT - 2C\male = 0$, we find the number of males (and females) for maximum E is $\male = T/2$ (i.e. equal abundance of the sexes). However, those are not necessarily the relative proportions generating the most fertilized eggs (a function of fertilized females), especially if there is much greater post-maturation mortality for males. There is doubtless an analytical approach, but some answers are readily found with an IBM, also recognizable as a Monte Carlo test. The process can be grasped by study of the program in Box 4.5. It calculates the number

Box 4.5 Program for an individual-based model of variation in copepod mate-encounter rates as a function of adult sex ratio

This takes a while to run. Take a walk in the sunshine, then check.

```
%Copepod-mating encounter-rate model

%This plots total matings as a function of proportion of males.
%It also counts matings/male (no limit) & matings/female
%(the latter limited to 2).
T=20000; %Total adults in population
C=0.0000045; %c=small daily probability a specific male will mate
  %with a specific female
%Prepare some storage:
A=zeros(9,1); Propmales=A; Storem1=A; Storem2=A; Storef1=A; Storef2=A;
MpM=zeros(9,8);
for h=1:9 %Loop over proportions of males from 0.1 to 0.9
 m=2000*h; f=T-m; Propmales(h)=m/20000;
%Set matrices of individual vectors for males and females,
%If males(x,1)=1, that male is alive; =0 means dead.
 males=zeros(m,3); males(:,1)=1; fems=zeros(f,3); fems(:,1)=1;
%Zero some summing registers
 em1=0; em2=0; ef1=0; ef2=0;
 for i=1:45 %Loop over 45 number days of mating
%To apply differential mortality to males & females, loop over both:
% remove all % (comment signs) down to storage process part.
% PdeathM = 0.15; PdeathF=0.015 (try changing these)
   %for j=1:m
   %g=rand; g is compared to probability of death, for males 15%
   % if males(j,1)==1 && g<=PdeathM; males(j,1)=0; end
   %end
   %for k=1:f;
     % g=rand; for females
     % if fems(k,1)==1 && g<=PdeathF; fems(k,1)=0; end
   for j=1:m %loop over all males
   %if males(j,1)==1 %this skips dead males
    for k=1:f ; %loop over all females
     %if fems(k,1)==1 %skip dead females
      Ctest=rand; %ctest
      if Ctest<C
        if fems(k,2)==0
        fems(k,2)=1; males(j,2)=males(j,2)+1;
        elseif fems(k,2)==1 ;
        fems(k,3)=1; fems(k,1)=0; males(j,3)=males(j,3)+1;
         end
        end
      %end
     end
    %end
   end
  end %Storage processes:
 for j=1:m %Get frequency distribution of first femle matings per male
  for n=0:6
```

(Continued)

```
      if males(j,2)==n; MpM(h,n+1)=MpM(h,n+1)+1; end
   end
   if males (j,1)>=8; MpM(h,8)=MpM(h,8)+1; end
   end
     for k=1:f %Total up matings per female:
       ef1=ef1+fems(k,2); ef2=ef2+fems(k,3);
     end
       Storem1(h)=em1; Storem2(h)=em2; Storef1(h)=ef1; Storef2(h)=ef2;
       h %To show program progress in the Command Window
       MpM(h,1:8) %List matings per male for current proportion of males, h
end
MpM
figure
plot(Propmales(:),Storef1(:),'r');
xlabel('Proportion of Males')
ylabel('Number of Females Mated, 1x=red, 2x=blue')
hold on
plot(Propmales(:),Storef2(:),'b');
```

of mating encounters among 20,000 adults (increase that to get greater precision; reduce it to make the program finish sooner) over 45 "days" (arbitrary time steps) with an arbitrary value of C and sex ratios varying from 10% to 90% males. The likelihood of encounter at each time step for each male with each female (a pair-by-pair equivalent of C) is set very low, which would be correct for almost any number of adults in a sizeable search volume. When a female is mated twice she is removed from the female matrix, likely a realistic feature of mating in *Calanus*.

For the value of $C = 4.5 \times 10^{-6}$ per day for every possible male–female pair and no mortality, running the model showed that the sex ratio producing the most females fertilized once (probably enough to maximize egg output) is about 33% males (Fig. 4.13), but about 18% of females went unfertilized. It took relatively more males to maximize the number of females mated twice (filling both sperm receptacles, a feature of copepod reproduction, possibly the limit allowed by females). Thus, a more demanding requirement for successful mating requires a greater proportion of males. If C is greater, both more females can be fertilized and the proportion of males required for doing it is less (try that "experiment"). If C were considerably smaller, such that it was extremely difficult for females to attract males or males to find females (a function of population density, search volumes, and much else), then the maximum number of fertilized females would be highest with a strong majority of males (try that, too).

There is an evolutionary aspect to this problem. If males are relatively few, and search problems or female behavior limit the number of matings, then those males are likely to participate as gene providers to more zygotes than females

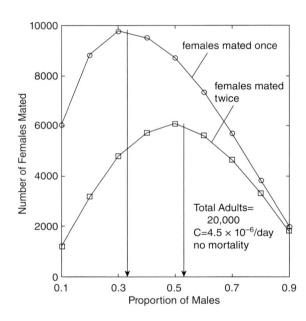

Fig. 4.13 Output from an individual-based model of copepod mating encounter rates at varying sex ratios represented as proportion of males.

(each zygote has only one father and one mother). Thus, in the experiment of Fig. 4.13 at 10% males, a majority were partners in the egg output of three or more females (Fig. 4.14).

That would lend selective advantage for any genetics producing males over genetics producing females (Fisher 1930). That effect should operate up to the point at which an overabundance of males left most of them to die without

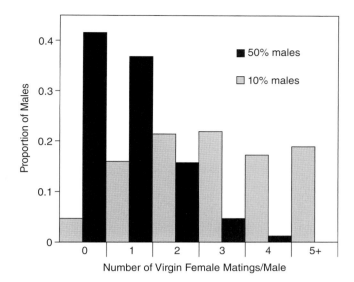

Fig. 4.14 Proportions of males failing to mate, mating once, twice, etc. for the encounter conditions of Fig. 4.13 when there are 10% and 50% males at maturation.

issue (as the saying goes). The balance usually lies at about 50% males, which evolution has fixed in many species by rigid, chromosomal sex determination. For copepods to sustain variable sex ratios and ESD, there must be some mechanism by which a variable aspect of the habitat informs an individual before it matures about its reproductive odds as male or female, directing the choice. This, too, could be modeled with a multigenerational IBM, which would be more complex to program but equally simple in conception. "Adaptive" models are indeed applied in testing at least the logic, though not the reality, of evolutionary theories. For an example, see the model by Fiksen and Giske (1995) of the adaptive advantage of diel vertical migration (it is not an IBM, *per se*).

Developing individuals embedded in flow regimes

Individual-based models are good tools for simulating the effects on individual organisms of growth conditions and advection. An early representation of habitat effects on reproductive and development timing in plankton populations is Batchelder and Miller's (1989) life-history model of *Metridia pacifica* (a copepod). The basic idea is that a stock of individual animals can be represented in a computation by a large number of vectors (up to $>10^6$) whose elements represent their developmental and reproductive status. Elements can include alive-vs.-dead (1 or 0), age, stage, age-within-stage, nutrition, readiness of the ovary for spawning, and anything else necessary to represent the likely contribution of the individual to the status and dynamics of the population. Vector elements are changed

at successive time steps according to functions based on biological information describing developmental progress and mortality under varying habitat conditions (temperature, food availability, time of day, . . .). An initial set of vector animals is best established from field data about a particular ocean place and season. When vectors reproduce, the "eggs" are assigned new vectors.

In a population-dynamical IBM for *Calanus finmarchicus* moving in the flow of the Gulf of Maine (Miller *et al.* 1998), some model elements were added to each individual's development vector representing its spatial address in a site-specific circulation model. The model was initiated at the winter solstice, December 21, by defining a set of 1000 vectors, each representing a single, resting fifth copepodite (C5), the G_0 generation. At suitable time steps, the vector animals emerge from rest, move to the surface layer and, as they mature, are advected along by applying current velocities specified at each one's spatial address by the Quoddy model of regional flow (Lynch *et al.* 1996). They are also given random moves representing mixing dispersion. (For nekton, changes in position could also be applied to represent swimming.)

As the G_0 C5s mature, they are assigned new vectors appropriate for spawning females, and as eggs are produced they are assigned to larval vectors. Each vector has six elements (Table 4.3). The probability that any given C5 will emerge from diapause and mature at a time step starts from zero at the winter solstice (Day 1), then rises. Progressively, all individuals mature and are assigned to a female vector. Clutch-readiness fraction (CR) at maturation is set to a random number, 0 to 1.0, to force spreading of spawning time around the day, which may not be realistic but smooths the shifting of stage abundances. After activation (vector assignment), 7 days (age as ♀) are allowed to pass before reproduction begins, then CR is incremented at each hourly time step by 1/24th of the inverse of the clutch-to-clutch interval in days predicted by a Bělehrádek function (Fig. 4.15a) fitted to some data for clutch interval as a function of temperature. This inverse is the fraction of clutch development occurring in the one-hour time step. When CR reaches 1.0, a clutch of 50 eggs is produced and each is assigned to an egg-C4 vector, and CR is zeroed. Temperature for this and other purposes in the model is drawn from a seasonal function developed from a long set of field data. Each day, each female and larval vector is subjected to a random chance of death (e.g. female daily survivorship = 0.975). One of the main problems with such modeling is that we have only weak clues about the stage-wise mortality rates (see Chapter 8).

Durations of each larval stage at the temperature of the date are calculated from Bělehrádek temperature functions (Fig.4.15b) fitted to full-nutrition stage-duration data (Campbell *et al.* 2001). At each hourly time step, 1/24th of the inverse of the stage duration in days is added to the molt cycle fraction (MC). When MC reaches 1, it is zeroed

Table 4.3 Vector values and meanings for three distinctive life stages as modeled by Miller *et al.* (1998). Pixel number refers to a spatial element in the physical model. CR is a female's clutch readiness variable that increases as oocytes mature, leading to spawning when CR=1.0. MCF is a larva's molt-cycle fraction that increases as it grows; it molts (and "stage" is incremented) when MCF=1.0. When "stage" reaches 12, larval vectors are converted to C5 vectors.

ELEMENT NO.	RESTING C5 VECTORS		FEMALE VECTORS		EGG-C4 VECTORS	
	VALUE	MEANING	VALUE	MEANING	VALUE	MEANING
1	1 or 0	Alive/dead	1 or 0	Alive/dead	1 or 0	Alive/dead
2	0–10	Molt readiness	Variable	Age as ♀	1 to 11	Stage (E–C4)
3	Degrees	Latitude	0 to 1	CR	0 to 1	MCF
4	Degrees	Longitude	Degrees	Latitude	Degrees	Latitude
5	Meters	Depth	Degrees	Longitude	Degrees	Longitude
6			Variable	Map Pixel no.	Variable	Map Pixel no.

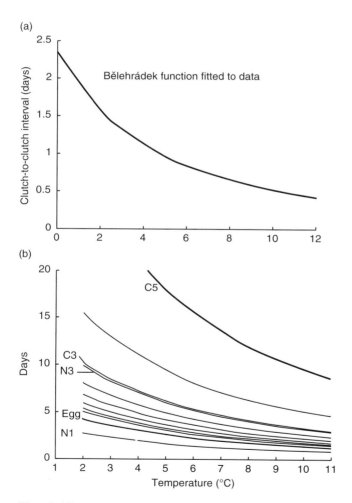

(a)

(b)

Fig. 4.15 (a) Bělehrádek function fitted to inter-clutch interval of *Calanus finmarchicus* females based on data from field-collected females. (b) Bělehrádek functions describing stage duration variation as a function of temperature in *C. finmarchicus*, based on data of Campbell *et al.* (2001).

and stage is incremented by one. Finally, the physical sub-model moves each individual at each hourly time step by changing the location elements of its vector. All females and developing young are consistently advected at the mean velocity for a 25 m mixed layer taken from the Quoddy model. Numbers of vectors in G_1 approach 200,000 per 100 G_0 females.

When a winter- or spring-generation individual completes its C5 development, it either matures or enters diapause by a random process. The fractional rates are 50:50::mature:diapause for winter and 10:90 for spring, values based on rough estimates from a seasonal jaw development study (Crain & Miller 2001). Half of maturing individuals are assigned as male (random process) and simply counted; half are female and are stored with maturation date and location. Resting C5s are stored with rest initiation dates and locations for mapping. The mature females of the G_1 generation are assigned to female vectors, and the model continues.

Population dynamics

Biological aspects of the model output (Fig. 4.16), without advection, show substantial similarity to the GLOBEC Broad Scale Survey data on *C. finmarchicus* development in the southern Gulf of Maine and over Georges Bank.

Trajectories

The most effective presentation of the interaction of development with flow is by animation, showing movement over a map of individuals initially spread across an interesting subregion, such as Wilkinson Basin in the Gulf of Maine. Life stage is shown by changing the color of dots locating each individual in the cohort. Such animations are available on the worldwide web at: http://www-nml.dartmouth.edu/

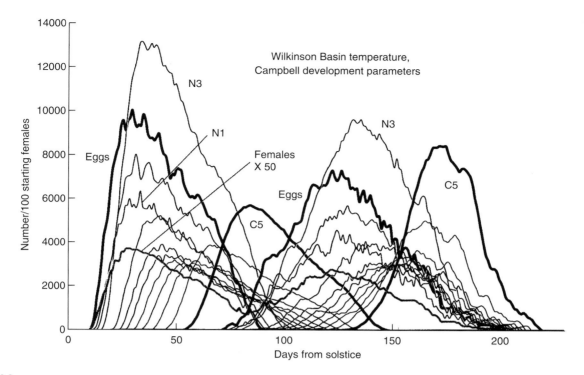

Fig. 4.16 Output of individual-based model of *C. finmarchicus* life-stage abundances in the Gulf of Maine and Georges Bank area. Fifth copepodites moving to diapause are not shown; equal numbers of the first generation matured and entered diapause. Female abundances were quite low and represented by a line at 50 times their numbers.

Publications/internal_reports/NML-98-7/. Models of this type have been constructed to study the life history and spatial dynamics of species from clam larvae to euphausiids in a variety of flow regimes. For example, there are informative models of the effects of diel vertical migrations on advective population transfers (e.g. Batchelder *et al.* 2002).

Overview

Many observationalists among oceanographers have long doubted the value of modeling, and they often remain uncomfortable around modelers. Modelers tend to rip into papers about observations, tear out just the numbers useful to them – perhaps the mean of wildly varying estimates – and then rush on to the next paper. On the other hand, modeling enforces a reading style that cuts through extraneous detail, exposing the tiny useful nuggets that are all many papers really offer. Modeling creates a powerful intellectual sorting mechanism to separate what is known from what needs to be learned. Knowledge of modeling techniques and their limits is essential to all oceanographic practitioners – observers as much as specialist modelers.

Chapter 5

A sea of microbes: archaea, bacteria, protists, and viruses in the marine pelagial

New organic matter generated by phytoplankton in the sea is partly added to the structural and storage components of the producing cells, but a substantial fraction is released directly into the adjacent water. Some of this secretion, particularly that of diatoms, is transparent exopolymer (TEP), which exists as amorphous flakes and strings among the phytoplankton cells. Some is small organic molecules. In addition, both protist and mesozooplankton grazers release organic matter when undigested remains are emptied from food vacuoles, when spine- or tooth-punctured food is squeezed into throats, and when fecal pellets leach soluble matter. The portion of this material that passes through a filter is termed "dissolved organic matter", and the acronym "DOM" is standard terminology (as are "DOC" and "DON" for the carbon and nitrogen contents of the dissolved organic matter, respectively).

DOM is a resource for heterotrophic, pelagic bacteria. Bacteria are fed upon by protists, mostly nanoflagellates, and infected by viruses. Nanoflagellates are preyed upon by larger protozoans (heterotrophic protists) and to some extent by either young mesozooplankton (e.g. crustacean nauplii) or mucoid filter-feeders (e.g. appendicularians). Herbivorous protists feed upon phytoplankton, release some DOM in the process, and serve as a food source for mesozooplankton. These food-chain links "return" the DOM to the progression from unicells toward whales, fish, and fishermen. Thus, the sequence from DOM to bacteria to protists to mesozooplankton is another pathway in the marine food web parallel to the direct consumption of phytoplankton by mesozooplankton. The *microbial loop* (Azam *et al.* 1983) refers specifically to the loss of DOC from all organisms and its recovery in the food web by heterotrophic bacteria (Fig. 5.1). Viral infection causes bacterial lysis, which releases cell contents back into the DOM pool, in a process termed the "viral shunt", which lowers the overall efficiency of the microbial food web. Viral infection of phytoplankton also releases organic material to the DOM pool.

Realization that heterotrophic bacteria are numerous in ocean waters came late in the development of biological oceanography (Pomeroy 1974; Hobbie *et al.* 1977). This was because the dominant means up to that time for estimating their abundance was plate culture. Seawater agar, usually with substantial nutrient enrichment ("peptone", beef extract, or other amendment), would be sterilized, gelled, inoculated with a small volume of seawater collected with a sterile sampler, and allowed to incubate for a few days. Typically, the plates would show a few hundred colonies per milliliter of near-surface seawater, each presumed to have started with a single bacterium. Compared to marsh water, sediment or soil, these were minuscule numbers, so bacteria were taken to have a very limited role in the marine pelagial. Some of the cultured forms were shown to have specific roles, including mediation of specific steps in the marine nitrogen cycle. Others were studied for the dependence of their growth on cold (*psychrophiles*) or pressure (*barophiles*). Lots of interesting work was accomplished with these bacteria, but, given the vanishingly small populations, it did not appear to have much relevance to trophic processes in the ocean.

Biological Oceanography, Second Edition. Charles B. Miller, Patricia A. Wheeler.
© 2012 John Wiley & Sons, Ltd. Published 2012 by John Wiley & Sons, Ltd.

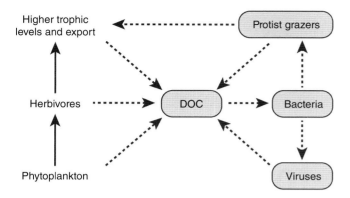

Fig. 5.1 All components of the microbial loop are shown in gray. "Phytoplankton" includes cyanobacteria, which are major components of the phytoplankton community in most ocean regions. Protists are single-celled eukaryotes. The "loop" refers to losses of dissolved organic carbon (DOC) from all organisms, and its recovery into the food web by heterotrophic bacteria. (After Kirchman *et al.* 2009.)

Then, in the 1970s, perhaps suggested by reading of Russian work (Sorokin 1964), but with hints from work in the West (Jannasch & Jones 1959), the alternative technique of direct microscopic counts was tried. Pelagic bacteria are small; cell diameters of 0.3 to 1.0 μm are typical, the mean about 0.6 μm. An object of 0.3 μm is near the resolution limit of light microscopy. So, special techniques are needed to make reliable counts that distinguish bacterial cells from tiny bits of organic or inorganic detritus. A moderately priced scheme introduced by Francisco *et al.* (1973) is to strain bacteria from a water sample with a flat, black plastic filter, then to stain the cells retained on the filter with fluorescent dyes, such as acridine orange, that bind specifically to nucleic acids. Hobbie *et al.* (1977) developed the now standard procedure utilizing Nuclepore™ filters. Cells are counted with an epifluorescence microscope (see Box 2.2), producing what is called an "acridine orange direct count". Following the biological acronym habit, this became AODC. Typical numbers are 0.5 to 2×10^6 cells ml^{-1} – huge values compared to plate counts. This method continues in use with very little change, except that other stains, including DAPI, YoPro and SYBR, are often substituted for acridine orange (Plate 5.1).

Within a very short time, microbiologists were making direct counts at every depth in every part of the world ocean. They were also faced with an array of new questions for which some answers are in, but many are still under study. What is the biological character and activity of these numerous bacteria that don't grow readily on seawater agar? Why do they not respond to the stimulus of food enrichment and form plate colonies? Are they "oligotrophs", essentially poisoned by concentrated molecular food? Are all of the cells alive, or, if alive, are they meta-

bolically active? If they are metabolically active, how fast are they growing and respiring? What different kinds of bacteria are present in the mixture? What controls their abundance? What is their role in the elemental cycles of the sea and the biosphere generally? All of these questions were immediately obvious when AODCs first became available. Some of them were also evident in basic respirometry studies of coastal waters. There was more respiration in filtrates than could be accounted for by plate-count bacteria. Pomeroy (1974) listed the questions in a paper considered a founding manifesto of modern marine microbiology. So, after some biological preliminaries, we will review the status of those questions.

Prokaryotes

These are the cellular organisms having no nuclear membrane surrounding a specialized cell organelle (nucleus or "karyon") housing DNA and processing its information. Rather, the strands of genetic material in prokaryotes are suspended within the general cytoplasm of the cell, although there may be differentiation of a subregion of the cell, a *nucleoid*, specialized for genetic processes. Based on biochemical and genetic sequence differences, two strongly distinct prokaryote groups, often now termed "Domains", are recognized, the Archaea and the Bacteria. Controversy exists over whether the term "prokaryote" is still valid. Pace (2006) argues that since Archaea and Bacteria are not monophyletic, the term "prokaryote" is obsolete. However, others find the term useful for grouping organisms without nuclei (Whitman 2009). We choose to continue using the prokaryote/eukaryote distinction here, but remind the reader that this distinction does not involve an evolutionary relationship.

Archaea

Microbiologists have only recognized this group as strongly distinct since the mid-1970s, initially from the sequence difference of their ribosomal RNA (see below) compared to that of "true" Bacteria. The lists of their biochemical differences from both Bacteria and Eukarya are long enough that Archaea must have taken a separate path very early in the evolution of cellular life. Most strikingly distinct are archaeal cell membranes. All cell membranes are constructed (Fig. 5.2) of molecules with one hydrophilic and two hydrophobic groups bonded to glycerol. In Archaea, these attach by *ether* bonds, while in other organisms attachment is by *ester* bonds. Archaeal hydrophobic groups are poly-isoprenes, while in bacteria and eukaryotes they are fatty acids. Finally, the glycerol in archaeal membranes is consistently a different stereoisomer from that in other life forms. Membranes are an essential feature of cellular

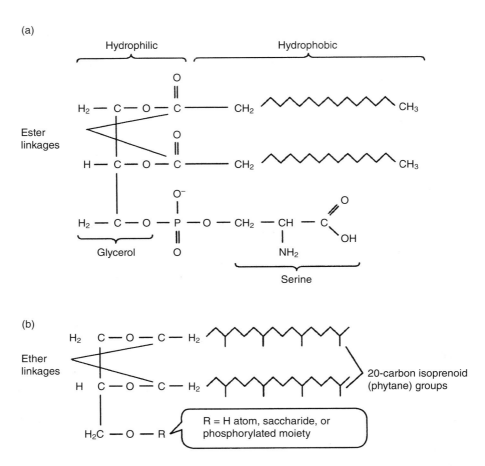

Fig. 5.2 (a) Chemical structure of the phospholipids that are the main constituents of bacterial and eukaryote cell membranes. A glycerol backbone is linked by ester bonds to two long-chain fatty acids and a phosphatidyl group (several moieties can substitute for the phosphatidyl serine shown). (b) Structure of archaeal membrane lipids. The hydrophobic moieties are polyisoprenes linked to the glycerol by ether bonds. One terminal glycerol carbon may carry a hydroxyl group, a sugar or a phosphatidyl group. In all organisms, the hydrophilic ends form both inner and outer membrane surfaces; the hydrophobic long-chain fatty acids project into and form the membrane's central core layer.

life. Strong divergence in membrane chemistry between Archaea and other groups implies a very deep, certainly ancient division. Archaeal membranes are a single lipid layer, while bacterial membranes are lipid bilayers. Cell membranes also are the biosynthetic sources of outer-cell coverings such as cell walls. Archaeal cell walls are never constructed of proper murein (see below) like those of true Bacteria, although some subgroups are covered with a "pseudo-murein" composed of cross-linked, ammoniated polycarbonates. More typically, archaea have a protein coat outside the cell membrane. Some distinctions in archaeal biochemistry, for example RNA-polymerase and DNA-polymerase forms and functions, imply closer relationships to Eukarya than to Bacteria.

Archaea were initially thought to be dominant only in habitats extreme with respect to heat, osmotic stress, or reducing capacity. They were often characterized as dividing into three main ecological groups: thermophiles, halo-philes, and methanogens. Most of the prokaryotes growing in very hot (>80°C and to >100°C at depths with sufficient pressure to prevent boiling) or very salty environments are archaea. All of the methanogens that gain energy by reducing carbon dioxide with hydrogen, yielding methane as a product, are archaea. They do fix carbon dioxide and can also assimilate some small-molecular-weight organic compounds such as acetic acid. The biochemistry of this chemosynthesis is distinct from those modes involving RuBisCO, including photosynthesis. The metabolic portfolio of Archaea as a group also includes the oxidation of ammonium, sulfur, and metals. While Archaea were initially considered to be mostly "extremophiles", they are now known to exist also in colder waters below the euphotic zone (DeLong 1992; Fuhrman *et al.* 1992). They account for 10–20% of the cells in the deep ocean (Varela *et al.* 2008) and play a major part in carbon and nitrogen cycles (Fuhrman & Steele 2008).

Bacteria

Bacteria have a moderately stout cell wall constructed of a polymer of alternating acetyl glucosamine and acetyl muramic acid. The chains are strongly cross-linked into a three-dimensional web by short chains of amino acids. This wall structure is termed peptidoglycan or *murein*, a molecular structure diagnostic of Bacteria (see Mathews *et al.* 2000). Bacteria tend to dominate the microbial communities of less-extreme habitats, although forms capable of metabolizing inorganic substrates – chemolithotrophs and thermophiles – are found among them as well as in the Archaea, although Archaea survive to greater extremes. All Bacteria (and Archaea) are specialized in respect to habitat conditions supporting their growth, the substrates they can consume, and the products generated by their metabolism. Many support their growth and division using organic matter with energy derived from glycolysis. A few are more versatile. Some use energy from "inorganic" reactions to drive synthesis of organic matter.

In a long period after their discovery in the mid-19th century, the tools available for classifying bacteria were limited. Light microscopy revealed variation in shape and in presence of fine flagellae. Thus, the classification began with shape: round (coccoid, e.g. *Pneumococcus*), tubular (rods, e.g. *Bacillus*), bent rods (e.g. *Vibrio*), and helices (spirochaetes, e.g. *Spirillum*). Rods are most likely to have flagellae and modest motility. Most of the rest of the characteristics of bacteria relate to their infectivity and the biochemistry of both their constituents and the metabolic reactions in which they participate. Infectious bacteria produce rather specific syndromes and tend to specialize in respect to hosts, which is a huge and well-studied subject. Bacteria (and archaea) mediating chemical transformations in natural habitats to derive sustenance and energy are classified by those reactions. The tendency to specialization makes this an effective tool. Hans Christian Gram introduced a broad division related to cell-wall biochemistry by application of iodine and the stain crystal violet to bacteria stuck to microscope slides. Cells retaining the stain after an alcohol rinse are Gram-positive; those rinsing clear of it are Gram-negative. Gram-positive bacteria have relatively simple cell walls in which the outer layer is murein. Gram-negative cell walls incorporate large fractions of lipoglycans, molecules of mixed lipid and carbohydrate character, which coat the murein and prevent dye bonding.

The vast majority of free-living pelagic bacteria are coccoid or slightly elongate and Gram-negative. Some rods are found as well, often with one flagellum. Rods are more common among bacteria attached to particulate aggregations such as "marine snow". The refined stages of classifying bacteria occasionally depend upon morphology (flagellae present, flagellae absent) and always depend upon determinations of metabolic activity (oxidase present, oxidase absent). Sochard *et al.* (1979) provide a good protocol for this mode of identification, with examples of a marine application. More recently, bacterial classification is primarily based on nucleic acid structure.

Molecular systematics of planktonic prokaryotes

Much insight has come from classifications of life forms based on differences in their DNA sequences (Box 2.4), particularly those for the genes coding the RNA sequences in ribosomes (rRNA) – the organelles that assemble proteins and are components of every cellular organism. The rRNA is separable into a smaller (SSU rRNA) and a larger subunit. Results from thousands of SSU rRNA sequence comparisons show a deep division between the Archaea and Bacteria (Fig. 5.3).

Archaea branch into two groups, Crenarchaeota and Euryarchaeota, initially distinguished by differences in their SSU rRNA and protein trees. Further comparative genomics support the profound divergence between Crenarchaeota and Euryarchaeota, since a number of genes present in euryarchaeal genomes are missing altogether in crenarchaeal ones and vice versa. These differences are not trivial, and suggest different strategies for cellular processes such as maintenance of chromosome structure, replication, and division. These differences are more fundamental than the ones usually observed at the phylum level, and it may be more appropriate to consider Crenarchaeota and Euryarchaeota as subdomains. The marine bacteria are themselves deeply divided in the rRNA phylogeny (Fig. 5.4), showing a suite of 11 well-separated groups. One of those, the Proteobacteria, divides into five major subgroups, designated alpha-, beta-, gamma-, delta-, and epsilon-proteobacteria. Bacteria are everywhere in natural systems, many of them not requiring extremes of any sort. They make up the bulk of water-column prokaryotic cells in surface water.

The SSU rRNA classification gives a phylogenetic picture of the prokaryotes (and every other organism, too). The value for pelagic marine bacteria, which are very difficult to culture, is to give them identity, to place them in the phylogenetic scheme, and to examine their metabolic and ecological functions. A cultivation-independent approach based on earlier work on bacteria from ponds (Olsen *et al.* 1986) was adapted to PCR techniques and applied by Giovannoni and coworkers (Britschgi & Giovannoni 1991; Mullins *et al.* 1995) to marine bacteria. Rather than grow bacteria in quantities sufficient for identification, the approach (detailed in Box 2.4) is to sample the DNA of the entire bacterial assemblage, and then randomly amplify dozens to thousands of SSU rRNA genes to see where their original owners fit in the phylogeny, a "shotgun" approach

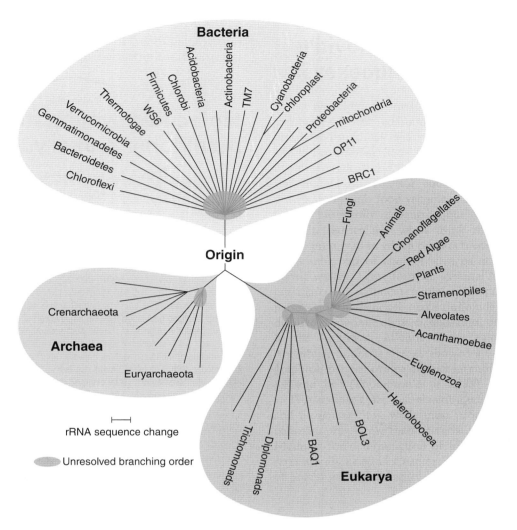

Fig. 5.3 A molecular tree of life based on rRNA sequence comparisons. The diagram compiles the results of many SSU rRNA sequence comparisons. Only a few of the known lines of descent are shown. The gene sequences define three broad groups, two of them prokaryotes, the third with a membrane demarcated cell nucleus (karyon, hence the name Eukarya). Organisms represented by codes (e.g. BRC1) have not yet been completely identified. (After Pace 2009.)

to the collective microbial gene pool in the habitat. If a few gene types predominate, then those are presumably the most abundant types in the bacterioplankton.

A gene clone survey of this sort was first done in the Sargasso Sea near Bermuda, and it has been repeated in all of the world's oceans. Similar dominant groups are found in all places. Groups are designated by their initial Sargasso Sea clone numbers, as for example SAR11 or SAR324. In the upper water column, the dominant group is alphaproteobacteria (Fig. 5.5). The most abundant of these, SAR11, is about 25% of pelagic bacteria. For about a decade, the SAR11-clade defied attempts at being cultured, but this was eventually achieved (Rappé *et al.* 2002). Seawater samples with native bacteria were diluted with ultrafiltered (0.2 μm), autoclaved seawater to densities of ~22 bacterial cells ml[−1]

and enriched with 1 μM ammonium and 0.1 μM phosphate. Some samples were also given 0.001% (w/v) of simple organic molecules (e.g. sugars and amino acids). After 12 days at sea-surface temperatures, the cells appeared above the detection limit in DAPI counts, 3000 cells ml[−1], and then increased at rates from 0.40 to 0.58 d[−1], reaching apparently resource-limited levels of 350,000 cells ml[−1] after 27–30 days. Cultured cells were comma-shaped and very small, $0.2 \times <0.9\,\mu m$. They could be enumerated with a SAR11 fluorescent probe, and SSU rRNA sequence variations were no greater than usually associated with SAR11 rRNA from the field.

Having cultured SAR11, Rappé *et al.* (2002) suggested the "candidate" name *Pelagibacter ubique*. Its growth was inhibited altogether by a very dilute addition of peptone.

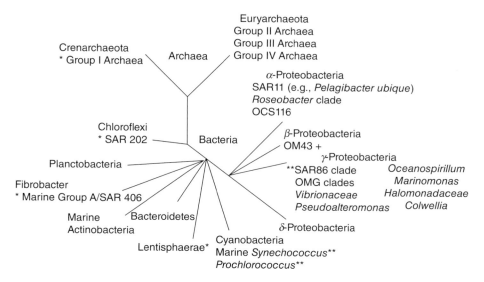

Fig. 5.4 A more detailed phylogenetic tree, also based on SSU rRNA, for the domains Archaea and Bacteria, showing only the major marine groups. Groups with a single asterisk are mostly found in the mesopelagic zone and in polar surface waters during winter; those with two asterisks live mostly in the euphotic zone; those with a + are mostly coastal. Others seem ubiquitous in seawater. (After Giovannoni & Stingl 2005.)

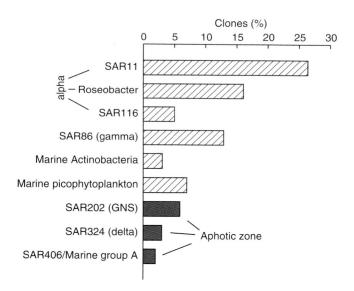

Fig. 5.5 Frequencies of bacterial types identified among 578 DNA clones of SSU rRNA. SAR stands for Sargasso Sea, where Mullins et al. (1995) first began defining types from sequences. Each type is narrowly defined, not a broad class of bacteria. Thus, SAR11 is the most common "species" of bacteria in the marine pelagial. (After Giovannoni & Rappé 2000.)

Thus, the characterization that abundant, hard-to-culture bacterioplankton are oligotrophs, inhibited by any but very dilute organic substrates, appears to be confirmed. *Pelagibacter ubique* has one of the smallest genomes of any free-living organism, and appears to be adapted for efficient growth under limited nutrient resources (Giovannoni *et al.*

2005). Comparison of proteins expressed during exponential and stationary growth, shows that *Pelagibacter ubique* adapts to growth limitation by increased synthesis of a small suite of proteins which allow quick response to fluctuations in nutrient supply (Sowell *et al.* 2008).

Many culturable pelagic bacteria are gammaproteobacteria, a group that also includes *E. coli* and *Vibrio*, and with which SAR86 associates in the phylogeny. However, SAR86 is distinct from the culturable forms in DNA sequence, and it has not yet been cultured. Its SSU rRNA sequence resembles that of some bacterial methanotrophs, and there are hints that SAR86 subgroups are related to hydrothermal vent chemolithotrophs (Giovannoni & Rappé 2000). Some progress is now being made on identifying aspects of function in near-surface variants of SAR86. Béjà *et al.* (2000) sequenced SAR86 DNA contiguous with the SSU rRNA gene and found a gene similar to those for archaeal rhodopsins, which they called proteorhodopsin. Rhodopsins are photopigments; in archaea they are membrane-associated parts of a mechanism generating ATP from the action of sunlight. The cells do not fix CO_2 and thus are not autotrophic, but the availability of "free" ATP for energetic purposes supplements their primarily heterotrophic metabolism. They generate NADPH for other redox-reactions by organic carbon oxidation. Apparently SAR86 has similar physiology. A study of membrane-bound proteins in open-ocean bacterioplankton (Morris *et al.* 2010) indicates association of transporter molecules with rhodopsin proton pumps. Thus, these photoheterotrophs may use light energy to facilitate transport of substrates for growth.

Kolber *et al.* (2000) found evidence of another type of photoheterotroph. They report that fluorescence responses to strobe flashes resembling those of phototrophic alpha-proteobacteria (*Rhodobacter* and *Rhodospirillum*, examples of "purple photosynthetic" bacteria, again actually phototrophs, not autotrophs) are a consistent part of the bacterioplankton in the eastern tropical Pacific. Their fluorescence output was >5% that of associated phytoplankton in very oligotrophic waters. These bacteria are now known to be aerobic anoxygenic phototrophic (AAP) bacteria: they require oxygen (they are aerobic), but they do not evolve oxygen (anoxygenic) by their simple, one photosystem reaction center. It remains to be seen how important these two photoheterotrophic bacteria are in the sea. The gene for proteorhodopsin is found in about half of all bacteria in the ocean, while AAP bacteria typically make up 1–7% of the bacterial community in oligotrophic areas and up to 30% of total prokaryotes in more productive environments (Koblizek 2011). So, photoheterotrophic bacteria are likely to be important in the ocean, but their exact roles remain to be determined.

Not surprisingly, fully photosynthetic cyanobacteria, such as *Synechococcus*, show up in the SSU rRNA clone libraries, but they are not as dominant as groups that remain uncultured. Some gram-positive forms, the marine Actinobacteria, are present (~7% of clones). Their relatives in richer environments grow on such a wide variety of substrates that few clues are provided about their activity in the marine pelagial. Tracking down the ecological roles of these many uncultured, but abundant, bacteria remains an active line of research in marine microbiology.

Distribution and molecular systematics of planktonic archaea

In addition to planktonic bacteria, there are planktonic archaea, and archaeal SSU rRNA clones appear in shotgun DNA libraries derived from prokaryotic plankton. However, better quantitative evaluation of their relative importance has come from studies (e.g. DeLong *et al.* 1999) with molecular probes specific to archaea. Water samples are suitably preserved, and fluorescent polyribonucleotide probes are added that leach through prokaryote cell walls. Cells filtered on to black backgrounds only appear in epifluorescence microscopy when the probe binds to the highly specific target RNA sequence, a technique called fluorescent *in situ* hybridization (FISH, of course). See Plate 5.1. Karner *et al.* (2001) found that archaea are only a few percent of prokaryotes above 100 m depth, but they increase below that to maximal densities of ~10^5 ml^{-1} and ~20% of prokaryotes below the euphotic zone.

Knowledge of the distributions and activity of archaea in coastal and open oceans has advanced rapidly. The use of probes and biomarkers has now documented the distributions of archaea in all of the oceans and peripheral seas. The main findings of these studies are that: (i) some of the microbes previously counted as bacteria were actually archaea; (ii) crenarchaea are generally 10–20% of prokaryotic biomass in the water column and are much more abundant than euryarchaea; and (iii) relative contributions to prokaryotic biomass of crenarchaea increase with depth. Euryarchaea include many methanogenic forms, and they appear to be more abundant in sediments than in the pelagial.

Clearly archaea are widely distributed in the ocean. What are their metabolic capabilities, and how do those compare with bacterial metabolism? Ouverney & Fuhrman (1999) showed that, like some bacteria, crenarchaea can assimilate free amino acids, suggesting a heterotrophic role. Evidence for a quite different role has been found by geochemists. Pearson *et al.* (2001) looked at the compound-specific distributions of stable isotopes and bomb ^{14}C from nuclear tests during the fall of 1961 and 1962 to distinguish the sources of carbon to the sediments. Different lipid biomarkers allowed identification of the relative contributions of phytoplanktonic, zooplanktonic, bacterial, archaeal, and terrestrial sources of carbon reaching the sediments in the Santa Barbara and Santa Monica Basins. Most of the lipid biomarkers were derived from marine euphotic zone primary production or heterotrophic consumption of that biomass. In contrast, the abundance of ^{14}C in isoprenoids (ether-linked lipids found only in archaea) showed no change over time (i.e. between pre- and post-bomb levels) indicating that the carbon source for these organisms remained isolated from the atmosphere. This finding suggests that archaea are acting as chemoautotrophs in the dark waters below the euphotic zone. Venter *et al.* (2004) first detected archaeal ammonium oxidizing genes in the Sargasso Sea, and Francis *et al.* (2005) used PCR techniques to demonstrate the widespread presence of these archaeal genes in the water column and sediments of the ocean.

Wuchter *et al.* (2003) confirmed that crenarchaea are autotrophic organisms capable of light-independent bicarbonate uptake by measuring ^{13}C incorporation into the unique ether-bonded membrane lipids of natural populations in the North Sea. A cultured species of crenarchaea was found to be autotrophic and could also oxidize ammonium to nitrite. Positive correlations of the abundance of crenarchaea with nitrite and the detection of putative archaeal genes for an enzyme required for ammonia oxidation in seawater hint that crenarchaea may also play a role in the marine nitrogen cycle as nitrifiers. Until this study by Wuchter *et al.*, marine nitrification had been attributed to two groups of bacteria belonging to the beta- and gamma-proteobacteria. Wuchter *et al.* (2006) conducted a time-series experiment in the North Sea and showed that the abundance of archaeal genes for ammonium oxidation is correlated with a decline in ammonium and an increase in the abundance of crenarchaea. Bacterial genes for ammonium oxidation were one to three orders of magnitude less abundant than the archaeal gene. Their results and genomic studies (Walker *et al.* 2010) suggest a major role for archaea in oceanic nitrification.

Ingalls *et al.* (2006) compared the natural radiocarbon levels in DIC and in crenarchaeal ether-bonded membrane lipids in surface and mesopelagic waters to identify the source of carbon used in each part of the water column. Values of $\Delta^{14}C$ in DIC (+71‰, relative to an oxalic acid standard, see Box 9.1) and archaeal lipids (+82‰) in surface water were close, indicating that archaeal lipids were produced from DIC or freshly produced DOC. Values of $\Delta^{14}C$ in DIC and archaeal lipids in mesopelagic waters were (−151‰ and −77‰, respectively) indicating a much older (deeper) source of DIC for the mesopelagic archaea. Ingalls *et al.* (2006) calculated that 83% of archaeal metabolism is autotrophic rather than heterotrophic. This suggests that the archaeal community either includes heterotrophs and autotrophs or is a single population of mixotrophs. The abundances of archaeal genes for the enzyme required for ammonia oxidation and for the dark fixation of carbon decrease dramatically with depth below the mesopelagic zone (Agogue *et al.* 2008). Although archaea are present in both zones, they appear to play different metabolic and ecological roles. In the mesopelagic zone, archaea are predominantly autotrophic, with bicarbonate as the carbon source and ammonia as the energy source. This chemoautotrophic fixation of carbon in the mesopelagic zone is about 1% of annual primary production in the euphotic zone (Ingalls *et al.* 2006), and it makes a significant contribution to the carbon budget of deeper waters. Moreover, if archaeal autotrophy is fueled by the oxidation of ammonia, these organisms generate >1.2 Gt yr^{-1} of N as NO_2^-. This estimate is enough to account for all of the first step of nitrification below the photic zone.

In the bathypelagic zone (below 1000 m) where ammonia is essentially undetectable, archaea most likely utilize organic matter and live heterotrophically. Bathypelagic archaea have a larger genome size than those in shallower waters (suggesting an opportunistic lifestyle), and they also have a gene repertoire indicative of a surface-attached life, e.g. genes for pili (hair-like appendages for attaching to surfaces) and genes for exoenzymes. Baltar *et al.* (2009) found a strong correlation between the distribution of suspended particulate material and electron-transport activity in waters below 1000 m, further suggesting that deep-water prokaryotic activity is predominantly located on suspended particles. The relative contributions of bacteria and archaea to carbon and nitrogen metabolism in the bathypelagic zone remain to be shown.

Bacterial abundance and production in the euphotic zone

Given that the most abundant marine bacteria are not readily cultured, it is understandable that microbiologists have turned to bulk measures of activity to evaluate what the typically 10^6 bacteria per milliliter are doing metabolically and in generating new bacterial biomass. Most of the techniques involve uptake or metabolism of radio-labeled substrates, but oxygen consumption is also determined as a measure of microbial respiration. The most widely applied measurements of bacterial production are based on DNA (Fuhrman & Azam 1982) and protein (Kirchman *et al.* 1985) synthesis, with protein now generally preferred (Box 5.1).

An interesting "field experiment" confirms a proportionality between tritiated thymidine (TdR)-uptake (a measure of DNA synthesis) and bacterial abundance. In May 1988, an expedition to the Gulf of Alaska arrived after the passage of a vigorous storm, which deeply mixed the water column, reducing surface bacterial stocks to less than half their usual level. Kirchman (1992) initiated a time-series of bacterial counts and TdR-incorporation measures lasting 21 days (Fig. 5.6). The change in bacterial abundance was not exponential, presumably because growth was partly balanced by grazing. However, TdR-uptake was proportional to stock numbers (and thus, to increase) for fully 15 days. This implies that at least the new bacteria added to the post-storm residual were all actively dividing and that during this time the bacterial production rate exceeded the grazing rate of bacterivores.

Single observations of TdR incorporation rate from the field are converted to cell increase by application of a mean TdR conversion factor (TCF) like the 2×10^{18} cells (mol TdR)$^{-1}$ mentioned in Box 5.1. It is usually desired to compare these rates of cell increase to bacterial standing stock (biomass), which can be approximated by multiplying direct counts by a typical value of the amount of organic carbon per cell. Lee & Fuhrman (1987) showed that there is a somewhat surprising negative relationship between organic-matter density and cell volume in bacteria (Fig. 5.7) such that the carbon per cell is roughly constant at about 20 femtograms C per cell, (20×10^{-15} g C per cell). For oceanic regions, this is thought by many to be somewhat too high. Nevertheless, most estimates range from ~10 to 30 femtograms C per cell. Note that the typical cell abundance, 10^9 bacteria liter^{-1} only represents 10^{-6} g carbon, or 1 μg C liter^{-1}.

After converting direct counts to bacterial carbon and TdR- or leucine incorporation to cells produced per time, a rough relationship emerges between bacterial standing

Box 5.1 Determination of bacterial growth rates

Thymidine, a constituent needed for replication of DNA, is added to seawater samples as tritiated thymidine (abbreviated TdR, <u>T</u>hymine <u>d</u>eoxy<u>R</u>ibose). The choice of [H³]-thymidine (TdR) is dictated by the fact that it is not utilized in RNA production as are the other three DNA nucleotide bases. There is also evidence that thymidine is not taken up by eukaryotes or cyanobacteria using nM concentrations in short incubations. Thus, it is strictly indicative of heterotrophic, bacterial activity. Thus, samples need not be filtered with a large pore filter to remove non-bacterial organisms before the measurement, avoiding disruptive effects of filtration. It is assumed that tritium in thymidine taken up and not metabolized is returned to the medium. After a short incubation under simulated natural conditions (temperature, light), the bacteria are filtered out (using very small pores), then radioassayed for their tritium activity. Incorporated TdR is taken to be a measure of cell reproduction by the bacteria. The results are compared to calibrations done as follows. A culture is established by removing grazers from a seawater sample by relatively coarse filtration (~1.0 μm) and then by diluting the filtrate with bacteria-free, filtered (~0.2 μm) seawater, typically at a ratio of 9 : 1. Next, the increase of bacterial numbers is traced by direct counts under natural habitat conditions of temperature and light over a period of several days to a week. Uptake of TdR is measured repeatedly in small samples of the culture throughout this incubation. Assuming exponential growth, the bacterial increase, which can be determined by direct counts, provides a measure of growth rate, $dN/dt = \mu N$. It is expected that the change over time in TdR-uptake, $v(t)\,mol\,h^{-1}$, will be proportional to dN/dt:

$$v(t) = \frac{dN/dt}{C}$$

in which C = cells mol^{-1} TdR. A relatively clean example (Kirchman *et al.* 1982; Box Fig. 5.1.1) of this relationship comes from a study of bacteria in salt-marsh water. Several other more empirical formulations are also in use. A general grand mean value of marine estimates for C is 2×10^{18} cells mol^{-1} TdR (Ducklow & Carlson 1992), with a range of 1 to 4×10^{18}.

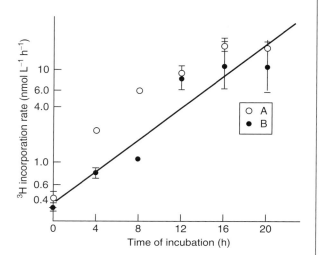

Box Fig. 5.1.1 Progressively increasing rates of tritiated-thymidine incorporation by bacteria in a seawater sample (actually from Great Sippewissett Marsh, Massachusetts) after dilution. Each point on the graph represents a relatively short-term uptake determination. Points A and B are from replicate experiments. Over the same interval, bacterial abundance increased from 1.6×10^5 to 30×10^5 ml^{-1}. (After Kirchman *et al.* 1982.)

stock and production rate for all pelagic habitats taken together. For example, Billen *et al.* (1990) and Ducklow (1992) converted cell counts to biomass (20 fg C/bacterial cell), then plotted the result vs. production rates (Fig. 5.8). As put by Thingstad (2000) and others, the overall production rate is roughly set by the trophic richness of the system. Thus, TdR-uptake increases steadily from the most oligotrophic, oceanic waters in subtropical gyres to the most eutrophic estuarine waters. There is no necessary relationship between the number of bacteria and this measure of new-cell production rate. In a suitably arranged system, stocks could be the same everywhere, with huge variation in turnover rates. That does not occur; much of the varia-

tion in production is reflected in variation of bacterial stocks (Fig. 5.8). Clearly, richer habitats support larger standing stocks. Elaborate arguments (e.g. Thingstad 2000) have been developed from the fact that biomass covers two to three orders of magnitude, while production covers five. The relationship actually curves upward to the right: the ratio of biomass to production is higher in the more eutrophic habitats, suggesting a weaker coupling of production and grazing there as compared to more oligotrophic habitats.

In addition to study of cell-number increase by TdR, other labeled substrates are used to study biomass increase. The most common incubation substrate for this purpose is

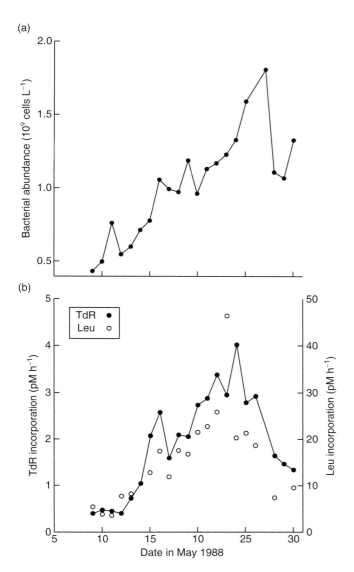

Fig. 5.6 Time-series of (a) near-surface bacterial counts, and (b) two bacterial production-rate estimates from an oceanic station (50°N, 145°W) in the Gulf of Alaska. Bacteria were increasing after a strong storm mixed the water column, diluting the surface layer and reducing bacterial numbers. A close correspondence is implied between bacterial abundance and bacterial metabolic activity, indicating a constant growth rate. (After Kirchman 1992.)

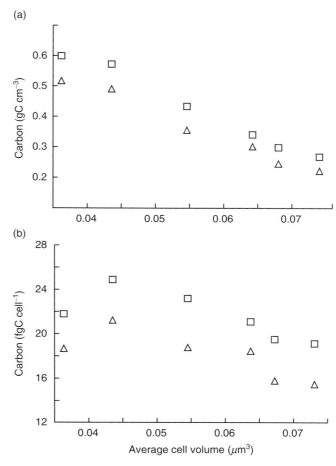

Fig. 5.7 (a) Relation in marine bacterioplankton between carbon density ($gC\,cm^{-3}$) and cell volume. The negative relation causes carbon per cell to be *roughly* constant at ~20 fg, as shown in (b). (After Lee & Fuhrman 1987.)

[^{3}H]-leucine, an amino acid. After the incubation, bacteria are washed in TCA and ethanol to extract radiolabeled substrate that may have been taken up by cells but not yet into macromolecules. Only leucine incorporated in protein, a measure of growth, will appear in the scintillation counts. Since they measure different processes, leucine- and thymidine-uptake rates need not be related. Sometimes the two rates are well correlated (Fig. 5.9), sometimes with outliers. A leucine:TdR incorporation ratio of 10 implies balanced growth, i.e. bacteria are synthesizing protein and DNA at approximately the same rates (Ducklow 2000).

Both the Antarctic data (Fig. 5.9) and the subarctic Pacific data from Kirchman (1992; Fig. 5.6 here) have a typical Leu:TdR ratio around 10, although there is seasonal and regional variability. Wide variations in the ratio of Leucine:TdR have been found in field studies, and these need to be interpreted with caution (Sherr *et al.* 2001).

Bacteria can utilize a wide suite of dissolved organic compounds and, by combining microautoradiography and fluorescence *in situ* hybridization (MICRO-FISH) the relative contribution of different phylogenetic bacterial groups can be determined. Cottrell and Kirchman (2000) found that, in the Delaware Bay estuary, no one group dominated the consumption of all DOM, but that a diverse assemblage of bacteria appears to be necessary for the degradation of the composite mixture of DOM present in the water. Cottrell and Kirchman (2003) also used MICRO-FISH to compare the TDR and leucine incorporation by the major phylogenetic groups of bacteria present in Delaware Bay, and found that alphaproteobacteria were the dominant

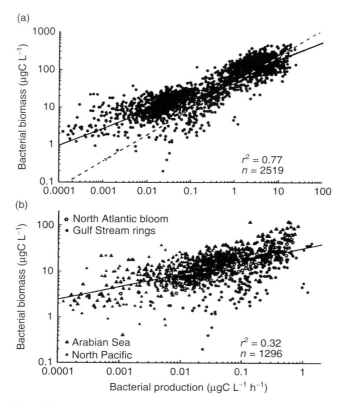

Fig. 5.8 Comparisons of bacterial biomass and bacterial production rates. (a) Open-ocean sites and Chesapeake Bay sites – most of the latter with production rates greater than 1 µg C liter^{-1} h^{-1}. (b) Just open-ocean sites (also shown in A). Lines are least-squares fits. (After Ducklow 1992.)

substrate-active group at salinities > 9 psu, whereas beta-proteobacteria were more important in fresh water. A review of studies of the substrates commonly used by the most abundant groups of bacteria (del Giorgio & Gasol 2008) showed that in general between 10% and 60% of each phylogenetic group is actively incorporating a given substrate. Thus, at any given time only a small fraction of all bacteria are actively utilizing a specific substrate.

Bacterial respiration and growth efficiency

Early and recent studies with size-fractionated samples showed that the majority of total marine community respiration occurs in the <1 µm size fraction. Most of this respiration is due to the activity of heterotrophic bacteria. How much of autotrophic primary production is consumed by heterotrophic bacteria? An obvious comparison is between estimates of bacterial production from TdR-uptake or leucine uptake and primary production estimated by

carbon uptake. If both measures are made at the same station using the same water samples, the importance of bacterial growth can be scaled to the autotrophic activity of the ecosystem. Ducklow (2000) assembled a number of open ocean euphotic zone comparisons (most of them his; Table 5.1) meeting this standard and that provided multiple estimates to allow some averaging. The conclusion is that the ratio of bacterial production to primary production is 10 to 25% in a wide range of habitats. Bacterial growth efficiency (BGE) is the growth yield or the amount of biomass produced (BP) relative to the total carbon required for growth which is calculated as the sum of BP and bacterial respiration (BR). Thus, BGE = BP/(BP + BR). The values of BGE efficiency can be determined by simultaneous measurements of bacterial respiration and bacterial net production in relatively short (<36 h) incubations or in dilution cultures (see Box 7.5). Values for BGE in natural aquatic systems generally range from 0.05 to 0.5. There is a strong positive correlation between BGE and phytoplankton production (del Giorgio & Cole 2000). In oligotrophic waters, BGE < 0.15, while in productive environments BGE approaches 0.5.

Food-chain transfer via dissolved organic matter (DOM)

If bacterial production is 10–25% of primary production, and even if bacterial growth efficiency is high enough that bacteria utilize 30–50% of primary production in growth or metabolism, then a very large fraction of photosynthate must be released (directly or indirectly, e.g. via grazing or viral lysis) from phytoplankton cells into the ocean. This is simply examined by an extension of the ^{14}C-uptake technique for quantifying photosynthesis. Seawater is sampled from the euphotic zone, and ^{14}C-labeled bicarbonate is added, followed by incubation under natural or simulated-natural temperature and illumination. Phytoplankton are filtered out and evaluated for ^{14}C content, producing the primary productivity estimate. The filtrate is saved and acidified to remove carbon dioxide, including the labeled tracer. The sample is mixed in scintillation fluor, and then the ^{14}C which comes from labeled, dissolved organic matter (DOM) not removed by the acidification is measured. In natural waters, DOC as a percentage of assimilated, particulate carbon varies from a few percent to 80%. High values occur toward the peak or termination of blooms under nutrient-limited conditions (Wetz & Wheeler 2007). This is believed to result from continued production of carbohydrates, for which nitrogen and phosphorus supplies need not be limiting. Indeed, when release of DOM is a high fraction of primary production (e.g. Biddanda &

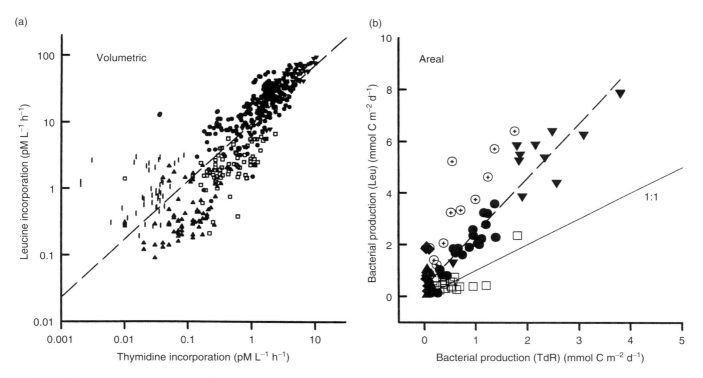

Fig. 5.9 Correlations from Antarctic waters of thymidine uptake (a measure of cell-division rates) and leucine uptake (a measure of protein synthesis) in bacterioplankton. Both single sample results (a) and vertical integrals through the euphotic zone (b) are compared. Dashed line in (a) is 10:1. Bacterial production rates (mmol C m^{-2} d^{-1}) for vertical integrals are derived using empirically derived conversion factors. Symbols represent cruise results from different seasons. (After Ducklow *et al.* 2001.)

Benner 1997), more of it is carbohydrate, both small molecules and polymers. An overall average is toward the lower end of the range, about 13% (Baines & Pace 1991). Polymeric exudates have been termed transparent extracellular polymers or TEP (Alldredge *et al.* 1993). Some DOM release may be the inevitable result of an organic-matter concentration gradient of roughly a million-fold across the cell membrane.

Thirteen percent of primary production (on average) is not enough DOM generation to cover bacterial production and respiration, so there must be additional sources (Williams, P.J.L. 1981). The bulk of the remaining requirement probably comes from losses to the water during feeding by grazing protists and larger zooplankton. Strom *et al.* (1997) showed that 16 to 37% of carbon from ingested phytoplankton quite rapidly shows up as DOC. Protist digestion is not complete, with the remains in food vacuoles returned to the water. Copepods, euphausiids, and other mesozooplankton break cells during ingestion and digestion, with some immediate loss near the mouth. Additional DOC leaches from zooplankton fecal pellets and some is present in excretory products. Since phytoplankton growth is nearly balanced most of the time by

grazing, the transfers to DOM during grazing are probably sufficient to fill out the budget of DOM production. Balance implies that ~100% of primary production is eaten daily. Thus, grazer generation of DOC is 16–37% of primary production. The sum with direct release of 13% is 29–40%. Additional DOC release can come from viral lysis of both bacteria ("recycled DOC") and phytoplankton. Still-further production will come from losses when bacteria are eaten by protists, a recycling of DOC (from DOC to bacteria, back to DOC). Given the uncertainties of all constituent numbers in this calculation, a rough balance is implied. No substantial amounts need be supplied by rivers, for example, to sustain the balance. In coastal systems, primary production often exceeds the bacterial carbon demand, but, in oligotrophic systems, bacterial carbon demand equals or exceeds the local rate of primary production (Duarte & Regaudie-de-Gioux 2009).

Approximately 50% of bacterial production (using DOC) takes place in the epipelagic zone of the water column. The other half of heterotrophic production and most archaeal production occurs in the mesopelagic and bathypelagic zones (the dark ocean). In the dark ocean, most bacterial production is supported by sinking or

Table 5.1 Bacterioplankton and phytoplankton properties in the open sea. All stock estimates are based on 20 fgC per cell. (From Ducklow 2000.)

PROPERTY	NORTH ATLANTIC	EQUATORIAL PACIFIC – SPRING	EQUATORIAL PACIFIC – FALL	SUBARCTIC N. PACIFIC	ARABIAN SEA	NEAR HAWAII	NEAR BERMUDA	ROSS SEA
Euphotic zone (m)	50	120	120	80	74	175	140	45
Biomass (mgC m^{-2})								
Bacteria	1,000	1,200	1,467	1,142	1,448	1,500	1,317	217
Phytoplankton	4,500	1,700	1,940	1,274	1,248	447	573	11450
Bacteria : Phytoplankton	0.2	0.7	0.75	0.9	1.2	3.6	2.7	0.02
Production (mgC m^{-2} day^{-1})								
Bacteria	275	285	176	56	257	nd	70	5.5
Phytoplankton	1,083	1,083	1,548	629	1,165	486	465	1248
Bacteria : Phytoplankton	0.25	0.26	0.11	0.09	0.22	nd	0.18	0.04
Growth rates (day^{-1})								
Bacteria	0.3	0.13	0.12	0.05	0.18	nd	0.05	0.25
Phytoplankton	0.3	0.64	0.8	0.50	0.93	1.1	0.81	0.11
Bacteria : Phytoplankton	1	0.2	0.15	0.1	0.19	nd	0.06	2.3

nd, no data.

suspended particulate organic carbon (POC) rather than direct export of DOC from the epipelagic zone (Aristegui *et al.* 2009). In these environments, a large percentage of the total number of bacteria can be found attached to particles. Utilization of POC is primarily via extracellular enzymatic activity that releases bioavailable DOM (the primary substrate for heterotrophic prokaryotes) from particles. Meso- and bathypelagic bacteria have a larger genome size and higher per-cell respiratory activity than those in the euphotic zone and, like bathypelagic archaea, they have a gene repertoire indicative of a predominantly surface-attached mode of life (Robinson *et al.* 2010). Exohydrolases release dissolved material faster than it can be absorbed by the attached bacteria, thereby also supplying DOM for free-living prokaryotes (Aristegui *et al.* 2009).

Chemical characteristics of DOM and POM

DOM and DOC must be defined operationally, which turns out to be simply the organic matter (or organic carbon) that passes the finest filters available. Work on quantifying DOM has used filters from 0.2 to 1 µm (Benner *et al.* 1993), with 0.7 µm glass-fiber filters often used because filters with this nominal porosity are cheap and readily cleaned. Recent developments provide filters down to 0.01 µm (Poretics®), but not much experience is available with pores smaller than 0.2 µm. It is ferociously difficult to get much water through the finest holes now available; these filters retain particles previously said to be in colloidal suspension. While even living bacteria, for example SAR11, can be smaller than, say, 0.7 µm pores, the amount of organic matter they contain is small, despite their large numbers, compared to truly dissolved organic matter. Thus, there is little difference in DOC determinations between 0.7 and 0.02 µm filters (Williams, P.M. *et al.*, 1993). The DOC in the global oceans overall amounts to about 6×10^{17} g (Hedges 1992). This is a relatively small amount compared to the total of inorganic carbon (entirely as oxidized, equilibrated CO_2, HCO_3^- and CO_3^{2-}), about 38×10^{18} g C, which is the largest reservoir of labile carbon. Nonetheless, marine DOC is approximately equal to CO_2 in the atmosphere. Most DOC is not particularly labile, which means that it is not immediately available to bacteria for nutrition. Bacterial activity is dependent upon the roughly 1% that is amenable to assimilation and metabolism and thus subject to rapid turnover. There is also a seasonal cycle in concentration of DOM in surface waters (Williams, P.J.L. 1995), which implies that some components (termed semi-labile) have intermediate turnover times, weeks to seasons, not hours to days. The DOM in the deep Pacific at the downstream end of the deep-sea, density-driven circulation has radiocarbon ages as great as

6000 years. So, it is very stable (recalcitrant to metabolism), with long turnover times. The sink processes for this deep-sea DOC are poorly known.

While DOC is a relatively big global organic-carbon reservoir, it is still very dilute in seawater, about 30–150 µmol C liter^{-1} (~1 mg C liter^{-1}), the upper range near the surface and inshore, the lower range at depth (Benner 2002). This has made measurement of amounts very difficult, plagued by contamination, high blanks (water suitably free of organic carbon is extremely difficult to make), adsorption to container walls, and sample preservation problems. The now standard technique is high-temperature oxidation (HTCO) of samples after acidification and removal of carbonates, with quantification by infrared spectroscopy of the CO_2 produced. It is also very difficult to accumulate significant amounts of material for compositional analysis, and we lack suitable standards equivalent to refractory marine organics. However, work on this progresses. Aluwihare *et al.* (1997) have concentrated the high-molecular-weight fraction by dialysis and tangential flow ultrafiltration (0.1 µm pores), showing that it accounts for about 30% of DOC and is about 80% highly polymerized carbohydrate, including a wide range of simple sugars. Dissolved organic nitrogen (DON) ranges from 3.5 to 7 µmol liter^{-1}. Identified components include amino acids and amino sugars; however, the bulk of DON is not characterized (Benner 2002). "Bacterial biomarkers" that appear in DOC and DON are methylated sugars, amino sugars, D-amino acids and muramic acid (Benner 2002). Using D-amino acids and muramic acid as markers, Kaiser and Benner (2008) estimated that 25% of POC and DOC and 50% of PON and DON are derived from bacterial residues. There is no sign that much of the refractory DOC in seawater derives from terrestrial sources such as lignin and humic acids (Hedges *et al.* 1997).

Nutrient regeneration in microbial food webs

An important process in microbial food webs is nutrient regeneration, especially mineralization of organic nitrogen to ammonium and organic phosphorus to phosphate. These originally bound nutrients can be released directly by bacteria, released by grazers during the feeding process (sloppy feeding), or by viral lysis of bacteria and protists. Release rates have been measured in culture with controlled food sources and predicted by comparing elemental ratios of food source and cell composition. Bacteria release ammonium when the C:N of the DOM utilized is <4.5, but take up ammonium when the C:N of DOM utilized is >6.6 (Goldman *et al.* 1987; Kirchman 2000). There remains some controversy over the relative roles of bacteria and protists for ammonium regeneration. Both can release ammonium, but bacteria collectively appear to take up as much ammonium as they release (Kirchman 2000). Therefore, the availability of ammonium for other organisms (phytoplankton in particular) appears to depend on the activity of heterotrophic protists.

Bacterivores, protist consumers of bacteria

Ocean waters contain a diverse bacterial assemblage. It is not clear what all of them are, what they are doing, or even if all of them are metabolically active, but progress is being made on those problems. Since the growth implied by the TdR- and leucine-uptake rates does not fill the water solid with bacteria, the bacteria must consistently be removed at very close to the rates at which they grow. There are three candidate explanations for their removal. First, about half of the new cells produced at division just die. There is no evidence for that. Second, bacteria are eaten at rates about the same as their growth rates. Third, there is substantial and recurring cell loss due to viral lysis. Both of these latter possibilities are part of the explanation.

Ocean water harbors an elaborate array of bacterivorous protists (Sherr & Sherr 2000). These can be classified by either a phylogenetic or a functional scheme. In practice, the two schemes overlap. The most numerous group, especially well out to sea, are the heterotrophic nanoflagellates ("HNAN"), ranging from 2 to 20 μm in cell size. Most are "heterokont", that is they have two functionally and often anatomically different flagellae. Bacterivorous protists come from a wide variety of phylogenetic groups and vary in shape, flagella arrangement, and feeding mode. Sherr and Sherr (2000) list the groups of more common forms as chrysomonads, bicosoecids, pedinellids, choanoflagellates, bodonids, and small ciliates. All of these eat bacteria, including autotrophic bacteria (cyanobacteria), and many also eat small phytoplankton up to sizes approaching their own. In addition, some flagellated phytoplankton, thought of as primarily autotrophic, may eat bacteria. The prymnesiophytes (e.g. *Coccolithus*), prasinophytes (e.g. *Micromonas*) and dinoflagellates include such "mixotrophs".

Bacterivorous protists appear to face a fairly severe problem finding bacteria in typical ocean water. Although 10^6 bacteria ml^{-1} is a large number, at ~0.6 μm diameter they are only 0.1 parts per million by volume, so lots of water must be processed by relatively small sensing and ingesting structures to generate significant nutrition from captured bacteria. Protists do that either by swimming and shifting water in the oral area, achieving clearance of bacteria from up to 10^5 body volumes per hour (Hansen *et al.* 1997) or by actively searching for food particles following chemical cues. The exact mechanics of particle capture and chemical sensing are still conjectural (Strom 2000).

Box 5.2 Grazer consumption rates

Single grazer consumption rates can be determined by providing fluorescently labeled bacteria (FLB), usually dead or rendered non-dividing (Sherr *et al.* 1987). After a short incubation, single protists are examined by epifluorescence microscopy, and the labeled bacteria within them counted. Averages for large numbers of grazer cells give I = bacteria ingested per grazer per time. Supplied FLB concentration, [FLB] = FLB ml^{-1}, can be used to estimate clearance rates: $C = I/[FLB]$, ml per grazer per time. If it can be assumed that natural bacteria are cleared at the same rate, then the result can be applied to the community. However, it has been shown that grazers do select for live motile particles over heat-killed non-motile particles (Gonzalez *et al.* 1993). Live bacteria can be engineered to express a green fluorescent protein and then used as a cultured tracer to measure ingestion rates (Fu *et al.* 2003).

Several methods are available for estimating rates of bacterivory by single cells and by the whole protist assemblage. The most widespread technique is the use of fluorescently labeled bacteria. (See Box 5.2). Total bacterivory can also be estimated by serial dilution experiments. Seawater samples are diluted at several different fractions with bacteria-free water. Net per-bacterium increase rates (growth − grazing) rise as dilution increases, since diluted grazers must search more water to get prey. Grazing rates can be calculated from the relative increase at greater dilutions (at lower bacterial and grazer concentrations). Bulk grazing on the order of 3–5% of water volume per hour is commonly measured (Vaqué *et al.* 1994). This amounts to 25–100% of the daily bacterial production.

Protist control of bacterial stocks is implied by time-series data from a fjord (Fig. 5.10) (Anderson & Sorensen 1986). Counts of both heterotrophic flagellates and bacteria showed complementary oscillations, strongly implying classical predator–prey population cycling and, thus, a predator–prey relationship. A further demonstration of the predator–prey relations among very small organisms, in this case photosynthetic protists, is provided by an experiment with filters of different sizes performed by Calbet and Landry (1999, 2004). Net growth rates (μ = increase − grazing) of autotrophic *Prochlorococcus* with cells of ~1 μm diameter and of heterotrophic bacteria mostly <1 μm were determined in incubations of whole seawater, and of filtrates from 1,2, 5, 8, and 20 μm filters. Results were as follows:

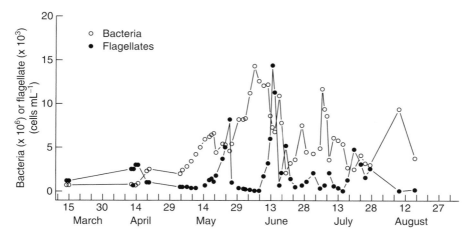

Fig. 5.10 Time-series of bacterioplankton and bactivorous nanoflagellates in a shallow-marine fjord (Limfjorden, Denmark). Points are means of counts from 1 and 2 m depths. This appears to be a classic predator–prey oscillation. (After Anderson & Sorensen 1986.)

REMAINING SIZES	NET GROWTH RATE DAY⁻¹ *PROCHLOROCOCCUS*	NET GROWTH RATE DAY⁻¹ HETEROTROPHIC BACTERIA
Control	+0.20	−0.04
<20 μm	+0.15	−0.06
<8	−0.11	−0.10
<5	−0.40	−0.18
<2	+0.04	−0.06
<1	+0.09	−0.03

Keep in mind that only the growth rates of the smallest components, i.e. those passing the 1 μm filter, were measured. The 20 μm and 8 μm filters apparently removed predators on the grazers of smallest cells, so the burgeoning grazers increased, reducing the net increase rate. The 2 μm and 5 μm filters removed those predators, but also grazers, so apparent growth increases again, and that effect is even stronger with the 1 μm filter. Thus, there must be a "trophic cascade" of at least three trophic levels among organisms less than 20 μm, even those less than 8 μm. Calbet *et al.* (2001) have pursued such evaluations further. A trophic cascade involves progressive interactions among trophic levels. For example, one way to remove planktonic algae from a lake is to introduce piscivorous fish to eat planktivorous fish, which then do not eat zooplankton, which increase and graze down the algae. Sometimes that works.

A summary (Fig. 5.11; Strom 2000) of studies determining the relative rates of bacterivory and bacterial production shows that at low production rates, that is, in oligotrophic or oceanic habitats, there is near balance between bacterial growth and protist grazing. In more eutrophic, essentially nearshore, habitats, grazing lags somewhat at almost all sites; bacteria are not eaten as fast as they grow. The methods are imperfect, and sometimes imbalance up to a factor of two or so can be simply an artifact. Note that viruses are present in dilution experiments, further complicating interpretations. However, the result is consistent enough that another bacterial control mechanism must usually operate in inshore waters.

Viruses, viral lysis of bacteria, and the viral shunt

The ocean not only contains ~10⁶ bacteria per ml, it also contains ~10⁷ viruses per ml. Counts are made by centrifuging particles from filtered (pore size, say, 0.45 μm) seawater on to electron microscope grids (copper-wire grids covered

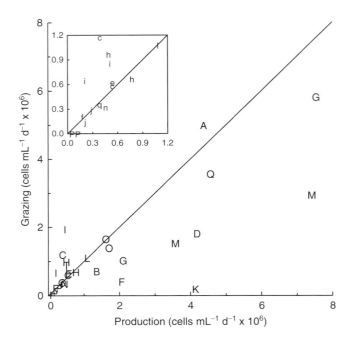

Fig. 5.11 Comparison of bacterial production and grazing rates on bacteria based on a literature survey. Grazing balances production in oligotrophic, mostly oceanic waters (inset), while production, when it is high, usually exceeds grazing somewhat. (After Strom 2000.)

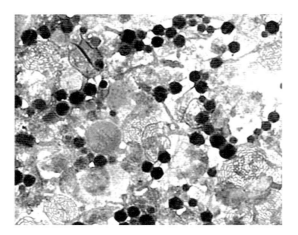

Fig. 5.12 Transmission electron micrograph of viral plankton, probably mostly bacteriophage, centrifuged from seawater on to an EM grid. Dark, roughly hexagonal spots are virus "heads". Some clearly show their attachment ducts. The web-like, nearly circular objects are probably the collapsed protein aprons which surround the attachment ducts of some bacteriophage. (Courtesy of K. Eric Wommack.)

with thin, plastic film) and examining them by transmission electron microscopy (TEM; Fig. 5.12) or by epifluorescent microscopy (Weinbauer & Suttle 1997; Chen *et al.* 2001). All viruses live by parasitizing the cellular machinery of larger organisms. They consist of a modestly complex protein coat covering a small strand of DNA or RNA as its genome. Upon contact with a suitable host cell, say a

marine bacterium, the protein coat attaches to the cell surface and the viral genome is inserted through the cell wall, often through a specific transport protein. The genome then redirects the cellular machinery to generate multiple copies of the virus, which are eventually released into the habitat by lysis of the cell to repeat the cycle. Most marine bacteriophages sustain populations by such cycles of infection, multiplication in the host, then lysis.

Fuhrman and Noble (1995) studied bacteriophage activity in cultures of bacteria collected at the end of Santa Monica pier, a decidedly neritic site. They examined viral TdR incorporation by a method analogous to that for bacteria and determined viral infection levels of bacteria (viruses per cell). They estimated protist bacterivory by the FLB method (see above). They also examined the rates of disappearance of labeled bacterial DNA with protists present and with protists removed, attributing the difference to viral lysis. Both protist feeding and viral infection balanced major fractions of bacterial growth.

The field of aquatic viral ecology is moving rapidly. Early reports of numbers and activity show variable results that may in part be due to the specific methods used. Viral abundances are typically determined by TEM (transmission electron microscopy), EFM (epifluorescent microscopy) or FCM (flow cytometry). The most widely applied method for measuring rates of virus production is "virus reduction and grow-out", which is similar to the dilution technique for measuring rates of grazing by microzooplankton. New infections are decreased by reducing total virus abundance (and virus–host contacts). This allows for enumeration of viruses released from already infected cells. Water is ultrafiltered (0.2 µm) to reduce the viral population, and added to water with naturally occurring bacteria and viruses to reduce the populations of both to about 10% of their initial values. Frequent sampling (every 4–6 h) over a period of 20–36 h allows determination of short-term increases of viral infections of natural communities of bacteria. Winter *et al.* (2004) found that viral lysis generally occurs around noon, and that infections generally occur at night (Fig. 5.13). The two peaks observed in these incubation studies could result from two different virus–host systems or two different infection events prior to collecting the samples.

Microscopic analyses (TEM, Weinbauer *et al.* 2002) suggest that 2–24% of bacterial cells are infected by viruses and that viral lysis removes 20–40% of the prokaryotic standing stock per day (Suttle 2007). Rates of production are 12×10^9 to 230×10^9 viruses per ml per day in coastal waters (Weinbauer 2004). Viruses are subject to mortality from non-biotic factors such as solar radiation, and temperature and biotic factors like inhibitory compounds released by bacteria (Weinbauer 2004). Decay rates (loss of infectivity) can be measured experimentally, and rates range from -0.05 to $-0.11\,h^{-1}$. The turnover times for viral populations are 1.6 days in coastal water and 6.1 days in oceanic waters.

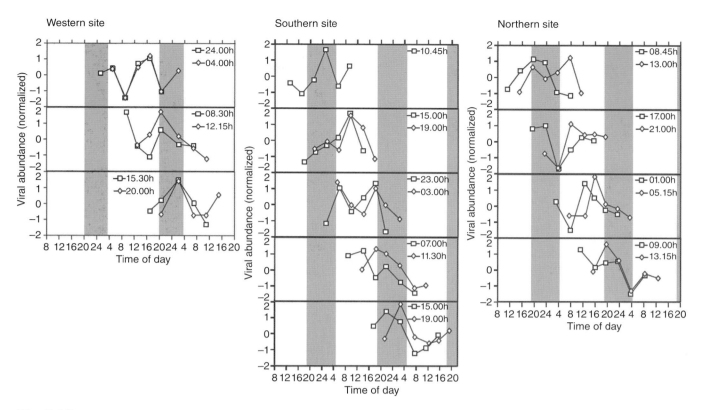

Fig. 5.13 Viral abundance over time in the virus dilution approach at three sites in the North Sea. Abundances are normalized to a mean of zero to allow comparison of relative changes among the three stations. Difference between average values of duplicate incubations plotted against sampling time and the start of the incubations corresponds to time needed to process the sample. Shaded areas represent night; error bars are not shown for the sake of clarity. (After Winter *et al.* 2004.)

Most, or all, viruses are host-specific, so the effect of viruses on bacterial populations depends on abundance and activity of the host population and on host-virus contact rates. Infection rates depend on contact rates, so the more abundant components of bacterial communities are more likely to be controlled by viruses. This leads to a "kill the winner" hypothesis: viruses infect and lyse the fastest-growing bacteria, and then the less-abundant bacteria become dominant. Thingstad (2000) developed a model showing how non-selective grazing by heterotrophic protists and host-selective viral lysis of fast-growing bacteria could account for the co-existence of many bacterial species and also for the 1:10 ratio of bacterial to viral abundance. Possibly, abundances of host bacterial species and their viruses could oscillate in a predator–prey relationship, yet maintain relatively constant total levels of bacteria and viruses (Fig. 5.14). Experimental and field data provide some support for this concept; however, the oscillations of specific host bacteria and their viruses have not been shown experimentally, just predicted from model results.

Viral lysis results in release of new virus particles (a range of 20–50 per cell) plus other cellular contents consisting mostly of dissolved organic material, and as much as 25% of primary production may flow through the viral shunt. The cellular components are rich in nitrogen and phosphorus, and may enhance rates of cycling of carbon and nutrients. Wilhelm and Suttle (1999) estimated that viral lysis could supply 80–95% of the bacterial carbon demand in waters of the Strait of Georgia (Western Canada), and that on a global scale virally mediated DOM release is approximately 3–20 GtC per year. Middleboe and Jorgensen (2006) conducted experiments with a marine bacterium (*Cellulophaga* sp.) and a virus specific to it, and quantified the amounts of dissolved free and combined amino acids released (DFAA and DCAA). The DCAA constituted 51–86% of the total DOC released. Glucosamine and DFAA each accounted for 2–3% of the total. Most of the released material (83%) was reassimilated by the remaining live bacteria. Given the high N content of this released material, it is reasonable to assume that a large fraction of the amino N could be mineralized to ammonium. Poorvin *et al.* (2004) have shown that virally induced mortality in coastal waters results in enough release of iron to support as much as 90% of primary production. Viral lysis constitutes an internal recycling of bacterial carbon and other constituent elements within the microbial food web, while zooplankton feeding on bacterivorous protists moves carbon and other elements toward higher trophic levels.

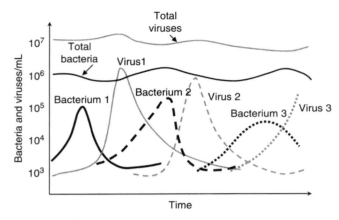

Fig. 5.14 Diagram illustrating "kill the winner" dynamics. Although the total concentration of bacteria and viruses remains relatively stable over time, the concentration of individual bacterial and viral strains changes dramatically. As a bacterial species becomes abundant (Bacterium 1), viruses that can infect those bacteria (Virus 1) will specifically lyse that host. This will lead to a decrease in the population size of Bacterium 1 and an increase in the population size of Virus 1. However, as the host population declines, there will no longer be hosts to produce more Virus 1, leading to a subsequent decrease in its population size. The virally mediated decline of the host population creates an open niche, allowing another bacterium to become abundant (Bacterium 2). A virus that can infect Bacterium 2 then becomes abundant, and the cycle continues. Selective virally induced mortality is a driving force for fluctuations in the structure of bacterial communities and contributes to maintaining high bacterial diversity. (After Breitbart *et al.* 2008.)

Herbivorous microzooplankton

Larger planktonic protists, 20–200 μm in diameter, are primarily herbivores (consuming phytoplankton). These are mostly spirotrichous ciliates (choreotrichs and oligotrichs) and larger dinoflagellates (see Plate 5.2). Choreotrichs are ovoid or conical, and have a full or partial circlet of cilia surrounding the anterior end of the cell. Ingestion by drawing prey into food vacuoles occurs within this circlet. Genera are readily recognized, but species distinctions are difficult, particularly when using the standard Lugol's iodine preservation which renders them opaque. Tintinnids are a subgroup of choreotrichs, particularly common in neritic waters. These form an external lorica, often covering that with sediment particles. Shapes and decoration patterns of their loricae make tintinnids amenable to very detailed taxonomic division, and, thus, distributional study. In the southwest Atlantic Ocean, along the coast of Argentina, tintinnid species showed a very strong biogeographical pattern of five distinct latitudinal bands between 34° and 58°S (Thompson *et al.* 1999).

Protists that graze picoplankton consume cyanobacteria and picoeukaryotes, while most protists larger than 5 μm consume prey larger than bacteria, mainly phytoplankton. Methods for determination of microzooplankton grazing rates are discussed in Chapter 7. We cover the importance of protist grazers on larger phytoplankton in Chapter 9.

Chapter 6

The zoology of zooplankton

For some practititioners of biological oceanography, zooplankton are simply "Z", the grazers on phytoplankton in pelagic ecosystem models. This quantity Z is varied as necessary to supply a loss term to reduce the spring bloom or balance the growth of phytoplankton. Real zooplankton do the same things, but we are severely limited in determining their rates of activity both individually and on average. On the other hand, they are structurally, developmentally, and behaviorally elaborate, and we have a wealth of detailed information about them. Some of that knowledge is sketched here and in Chapters 7 to 10. The zoological range of the studies is wide, since marine zooplankton include a number of protozoan groups and members of most of the phyla, from Cnidaria and Platyhelminthes to Chordata.

Zooplankton are free-swimming animals that live in oceans and lakes. Most are small, a few centimeters or less, but some jellyfish and pyrosomes are a meter across and several meters long. Again, the Greek word πλαγκτος, which is the source of the term "plankton" introduced by Viktor Hensen, means "wandering or drifting". Thus, it implies relative passivity. In using the term, we separate the weaker swimmers from more active forms, nekton (νεκον), that swim with sufficient strength to maintain their geographical positions or to travel at will despite ocean currents. Fish, porpoises, and squid are examples of nekton. This definition of zooplankton is not, however, an operational definition. That is, it contains no rule for deciding whether a given animal captured at sea is planktonic or nektonic. An operational definition is simply that animals caught in plankton nets are zooplankton. Planktologists distinguish between *holoplankton* that live in the water column through their entire life cycle, and *meroplankton* that are planktonic larvae. Meroplankton either grow into swimming capability qualifying them as nekton or settle to the seafloor becoming benthos.

Collection

The basic method of capturing marine (lacustrine, too) zooplankton is necessarily filtration (Box 6.1) (Wiebe & Benfield 2003). There are two usual approaches: nets (cones of mesh towed through the water), and pumps that deliver water to the deck of a ship for filtration aboard. Generally we use industrial filter-cloth woven from monofilament nylon melted together at the crossings of warp and woof. The holes are square, measured by the length along their sides. For larger zooplankton (euphausiids, adult copepods) 200 or 333 µm mesh is typically used. To catch copepod nauplii, you need 50 µm mesh. Finest meshes available are 10 µm or less, but the fabric is delicate and hard to push much water through. Plankton nets in common use (Fig. 6.1) have mouth areas from 0.2 (a circular, half-meter net ring) to 10 m² ("MOCNESS-10"; a 1 m² "MOCNESS" is shown). A 0.4 m² net (a 70 cm "bongo" net, for example) towed at 1 knot will filter 12 m³ per minute. A 15-minute tow will filter a volume equivalent to a small swimming pool. A variety of mechanisms have been developed for opening and closing nets at depth so as to catch only the plankton in vertically restricted layers. MOCNESS systems are widely used for that at present (Frost & McCrone, 1974; Wiebe *et al.* 1985). They are sets of nine nets attached by their tops and bottoms to bars held by cables at the top of a frame. The cables (and bars) are dropped one after another by a computer-controlled motor, closing and opening nets in sequence as the system is raised through the water. Sensors on the frame report depth, temperature, distance traveled, and other variables, via the conducting tow cable to a computer and its operator on deck. Smaller zooplankton are often filtered from water collected by pumps or in bottles closed at depth. Pumps of reasonable size for deployment aboard ships cannot filter

Biological Oceanography, Second Edition. Charles B. Miller, Patricia A. Wheeler.
© 2012 John Wiley & Sons, Ltd. Published 2012 by John Wiley & Sons, Ltd.

(a)   (b)   (c)

Fig. 6.1 (a) A simple ring net rigged with a bridle, and in this case weighted at the cod end for vertical hauling. (b) Retrieving 70 cm bongo nets after an oblique haul, that is, the net was towed forward by a ship, while wire was paid out and retrieved with a winch. (c) A 1 m² MOCNESS multiple net system ready for lowering. Nine nets can be released sequentially by a stepping motor that releases restraining cables attached to bars holding the upper edge of the closing net and the lower edge of the opening net.

as much water as nets, but they allow for precise control of depth of sampling and use of fine mesh for small, resilient forms like copepod eggs and nauplii.

Captured plankton can be examined alive or preserved for various purposes (chemical analysis, identification, and counting; determination of biomass as displacement volume or weight; and so forth). Animals for experimentation should be uninjured, and, for those only readily captured by nets, more healthy specimens can be caught with fine mesh nets (<100 µm causes less abrasion at the cost of more clogging with phytoplankton), towing very slowly and using large cod-end containers with fine mesh ports for gentle draining. Even with all precautions taken, specimens must be subjectively evaluated for good condition after capture.

In recent decades, it has been popular to distinguish gelatinous zooplankton from the rest. Many gelatinous forms are delicate, and much new information has come from gentle collection by divers and submersible- or ROV-borne enclosure devices. They are species that have very large ratios of water to organic matter; often 98% of their wet mass is water. This allows them to grow large bodies with minimal acquisitions of food. Gelatinous composition allows also for rapid population growth, since large bodies can search or process more water to find food, while not

much food is required to make new bodies. Large bodies are also protected from smaller predators just because they are hard to ingest. Gelatinous zooplankton now have a separate group of enthusiasts represented by websites with URLs like David Wrobel's "www.jellieszone.com". Diving techniques have been widely applied by these enthusiasts since William Hamner (1974) and colleagues initiated studies by blue-water diving in the 1970s. There is no commonly accepted name for the other plankton, which are typically ~70% water (as are land animals). "Hard-bodied" zooplankton isn't quite right, and non-gelatinous defines them by what they are not. Groups will be characterized here as gelatinous when they are.

An introductory description of the forms of planktonic animals

There are planktonic representatives of every major (but not every "minor") phylum, and there are important phyla largely restricted to the plankton. We depart now on a Cook's Tour, visiting each phylum briefly. If no picture of

Nets for capture of plankton are usually cones of loosely woven fabric. Originally made from silk, modern netting is precisely woven of nylon melted together at the thread crossings. The holes are square and mesh size is specified as the length along the sides of the holes (Box Fig. 6.1.1). Smallest holes are about 5 μm; the largest in common use are 1 mm. Since fabric can twist and stretch, mesh sizes should be selected so that the diagonal measure is shorter than the narrowest axis of the target organism. Phytoplankton nets are usually 20–60 μm; zooplankton nets are commonly 50–1000 μm.

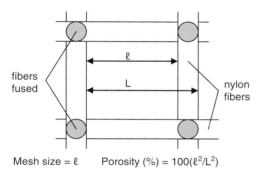

Mesh size = ℓ   Porosity (%) = $100(\ell^2/L^2)$

Box Fig. 6.1.1 Diagram of fibers surrounding a mesh hole in a plankton net.

an organism is provided, it is because space and budget for figures are limited in textbooks. Good pictures can be found by typing almost any genus name into a suitable worldwide-web search-engine (e.g. Google). Try for the "heteropod" snail *Carinaria*, for example, and you will find lovely pictures by Roger Seapy, Steven Haddock, and others.

Protista

These are organisms complete in a single, eukaryotic cell. "Protozoa" are now not considered "animals", and the once formal term "animal" is lately supposed to be used only for multicellular beasts. Protists are not one, monophyletic "phylum", but an agglomeration of forms variously related. Many marine phytoplankton share animal-like characteristics: they are motile, have sensory organelles like those of heterotrophic protists, and respond actively to

environmental stimuli. Several groups of flagellated phytoplankton, many prymnesiophytes for example, are "mixotrophs" that both photosynthesize and graze on particles. Quite likely such mixotrophic nanophytoplankton, and similar but unpigmented flagellates, account for much of the grazing on pelagic bacteria, including cyanobacteria. The most obviously animal-like algae are the dinoflagellates, many of which are strictly heterotrophic. Some dinoflagellates (Chapter 2) capture prey, often phytoplankton, by encapsulating them in a mucous sac (a pallium). Immobilized prey are ingested, forming food vacuoles. Other dinoflagellates insert a tubular peduncle into prey, harvesting nutrition without engulfing them.

Diverse heterotrophic protozoa reside in the plankton, and they have received attention in Chapter 5. Those of the sorts common in freshwater are present in the sea as well: ciliates, flagellates and amoebae. Ciliates have a major role in pelagic food webs, which remains under evaluation. Naked ciliates are mostly members of the subgroup Choreotricha, which are abundant in all ocean areas. They belong to a small number of genera such as *Lohmanniella*, *Strombidium*, and *Laboea*. Some choreotrichs are facultative "mixotrophs", using chloroplasts from ingested phytoplankton to provide them with photosynthate. The complexity this imposes on pelagic food webs will be considered in Chapter 9. Not all ciliates are naked; the tintinnids found more abundantly in coastal regions have leathery, cone-shaped tests that are often covered with cemented mineral particles. Naked flagellates are also abundant but delicate, which together with their small size (many <10 μm) delayed recognition of their importance, particularly as bacterivores. They are a complex assemblage, most of which are "heterokont", that is, having two flagellae of different types. Flagellate size and form are highly varied (Fig. 6.2). All of these "microheterotrophs", together with mixotrophic phyto-flagellates, are the grazing component in the microbial food web. It is typical for microheterotrophs to be the preferred foods of mesozooplankton such as copepods (Gifford 1993), which were long considered to be the principal pelagic herbivores.

Naked amoebae (Rhizopoda, protozoa with pseudopods) are quite rare in pelagic habitats. However, five shell-bearing groups of Rhizopoda are common and reach modest abundances of several per liter: Heliozoa, Acantharia, Phaeodaria, Radiolaria, and Foraminifera. The relationships of these groups (and all eukaryotes) are under review, based on DNA sequences and possession (or not) of particular proteins. There has been a tendency to create new names for possible evolutionary lines (new kingdoms, supergroups, phyla) with every shift in the phylogenetic reconstructions. Agreement regarding phylogeny will eventually emerge when enough sequences from enough genes (and amino-acid sequences in proteins, readable from DNA, to reduce the effects of code synonymies), and from

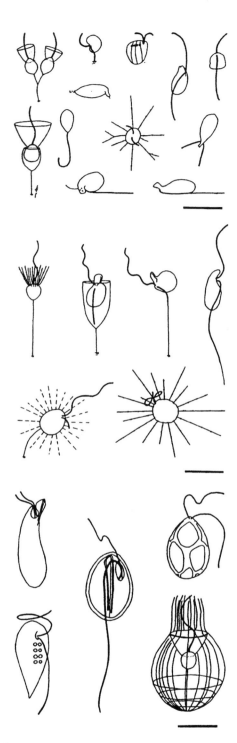

Fig. 6.2 A range of cell types among heterotrophic flagellates, grouped as small, medium, and large. Those with attachment filaments are indicated by a little "holdfast" at the bottom. Those are commonly seen on mesozooplankton surfaces, particularly crustacean exoskeletons. All scale bars are 10 μm. (After Sherr & Sherr 2000; sketches by Naja Vors.)

enough distinct groups, are included in the analysis. At present, Radiolaria and Foraminifera appear to be one clade (i.e. related on an evolutionary branch) recently named "Retaria" (almost certain to change); the others are more distantly related to them and to each other (Moreira *et al.* 2007). It is fair to suggest that pseudopods evolved very early, possibly more than once (there are several distinct types), and are retained by numerous, now distantly related, groups.

Cells of the Acantharia are constructed around radiating spicules of crystalline strontium sulfate (celestite), which dissolves rapidly after cell death. They are probably the most abundant protozoans with hard skeletons. Heliozoa and Radiolaria ("rads"), although not necessarily closely related, have a siliceous central capsule within the cytoplasm surrounding the nucleus and several other organelles; the outer cytoplasm is partly occupied by food vacuoles. Radiolarian skeletons are quite durable, and although small they sink with particulate detritus to form sediments called radiolarian ooze. Polycystine rads have a billion-year fossil record; their position in phylogenetic analyses suggests that most of the rhizopod groups were distinct long before the Cambrian period. Most Phaeodarians have proteinaceous or chitinous capsules, in some cases incorporating amorphous silica, bearing elaborately branched spines. A few, like *Phaeodina*, are naked. A brownish mass near the nucleus is termed a phaeodium and believed to be accumulated food waste. Ernst Haeckel produced some wondrously elaborate pictures of Phaeodaria in the *Challenger Reports*, together with a complex taxonomic system. He was accused of taking excessive artistic license with his drawings, but eventually his observations have been confirmed. "Forams" have larger, calcareous capsules surrounding their nuclei that sediment as foraminiferan ooze. There are not only planktonic forams, but morphologically distinct benthic forms. Together with radiolarian deposits, these sediments provide us with valuable stratigraphic records. In many cases, distinctions in isotopic composition between benthic and planktonic forams in the same buried sediments are informative about past hydrographic differences between deep and shallow ocean layers.

In all of the shelled Rhizopoda, the pseudopods slide in and out along a spray of fine spines extending from the cell center. They capture particulate food with those pseudopods, ingesting it by phagocytosis. There is a second layer of cytoplasm around the shell and base of the spines. In forams and rads this layer is highly vacuolated (Plate 6.1) and may provide buoyancy by favoring lighter ions internally (e.g. ammonium replacing sodium ion). Many acantharians, rads, and forams (not Phaeodaria) contain zooxanthellae, autotrophic dinoflagellate cells as symbionts in their cortical cytoplasm. Thus, they acquire nutrition by a combination of farming the zooxanthellae internally and contact capture of various animals ranging upward in size to small copepods like *Oncaea*. Among

Foraminifera, there are also deep-living pelagic species without symbionts and a distinctive group of benthic forms. Many planktonic foraminifera range up to 1 cm in overall cell diameter (some benthic forms are much larger). Gowing (e.g. 1989) has shown that Phaeodaria are generalist particle feeders, finding bacteria, diatoms, dinoflagellates and both protozoan and crustacean remains in their food vacuoles. They can be collected from the surface down to mesopelagic depths, tending to be most abundant just below ~100 m.

Cnidaria

These are animals bearing nematocysts or stinging cells on their outer surfaces, particularly along their tentacles; all the planktonic forms are gelatinous. Cnidaria is usually considered one of the "lower" phyla of metazoa, or multicellular animals, because their level of organization is relatively simple, with no centralized nervous or circulatory systems. There are two basic tissue layers: an outer covering and a gut lining. Between those is firm, jelly-like matter called mesoglea. There are moderately sophisticated sensory organs in many forms, particularly in medusae, including eyes and statocysts (gravity and acceleration sensors). Three cnidarian groups are represented in the plankton: Hydrozoa, Scyphozoa, and Cubozoa.

Hydrozoa

The hydromedusa stages of this group can be an important component of the zooplankton. Medusae (Plate 6.2) are bell-shaped, with the mouth at the end of a stalk, the manubrium, hanging from the top inner surface like a clapper. They are predators, capturing prey with tentacles on the bell edge and then transferring them to the mouth. This stage alternates with a sessile, usually colonial, polyp stage. Hydromedusae are mostly small jellyfish, the gamete-producing life-phase of these dimorphic animals. Zygotes develop into small, ciliated larvae called planulae that settle to become polyps, then colonies. The bell has a flexible band of tissue, a velum, extending into the bell opening from its edge, which serves to narrow the aperture and concentrate the propulsion jet when the bell contracts. Some of the oceanic genera, particularly among the narcomedusae, do not have alternating generations, and their zygotes develop directly into medusae.

An abundant tropical–subtropical "neustonic" (living right on the sea surface) hydrozoan group are the Porpitidae (e.g. *Porpita* and *Vellela*), the hydroids of which form colonies suspended downward from a stout, gas-filled float 3 to 10 cm across. In *Vellela*, this is topped by a sail. The hydroids on one float are male or female that release very small male and female medusae with gonads (Larson 1980). Zygotes develop into new hydroid colonies. *Vellela* often appears in massive wash-ups on west coast beaches of North America.

Studies in the 1990s of the plankton over Georges Bank offshore from New England (Madin *et al.* 1996) rediscovered that the enclosed anticyclonic circulation over and around the bank provides a spring–summer habitat in which the colonial polyp phase of *Clytia gracilis*, commonly found attached to shallow surfaces and sea-grasses, can be free-floating and reach high density by recurring colony expansion and breaking. They can color the sea surface yellowish green. They do not, however, have algal symbionts, and they feed by predation on zooplankton. Similar populations have been reported from the northeast Atlantic.

Siphonophora are holoplanktonic (Plate 6.3), hydrozoan cnidaria with complex body forms termed "colonies". "Individuals" in the colony (Plate 6.4) are specialized for propulsion, feeding, and reproduction. Typically, there is a long, tubular stolon emerging from a large swimming bell, carrying a curtain of tentacles extending from feeding polyps arrayed along the stolon. Nematocysts on the tentacles sting and capture prey, which are then moved to the polyp opening for ingestion. Siphonophore structure can be dramatically complex. In one abundant group (Physonectidae) a vertically oriented body mass of many swimming bells is topped by a terminal individual that secretes a carbon monoxide bubble internally to regulate colony buoyancy, sometimes with resulting vertical migrations. Feeding curtains of Calycophoran siphonophores can be many meters in length, making them very important predators of small zooplankton, including larval fish. A well-known neustonic siphonophore, the Portuguese Man o'War, *Physalia physalis*, has a gas-filled float as large as 40 cm. Its long beard of tentacles can produce very painful stripes of welts on unwary swimmers.

Scyphozoa

This is another group of medusae (Plate 6.4). Most of the common genera grow to large body sizes; in some forms the bell can grow to 1 m diameter with several meters of layered tissue folds hanging from the manubrium. Some genera (e.g. *Chrysaora*, Plate 6.3, *Aurelia*, *Nemopilema*) have a small asexual, benthic polyp phase that is never colonial like the benthic phase of the Hydrozoa; other genera (e.g. *Pelagia*) do not. Larval medusae (scyphistomae) are produced from the polyps by repeated, transverse constrictions just below the oral opening. Scyphozoan medusae lack a velum, but like hydromedusae they are the sexually reproducing life-phase. In recent years, massive population outbursts of scyphozoans have become a problem for trawl fisheries along many coasts, the most prominent example being summer–fall infestations of the 2 m diameter (and ½ ton!) medusae of *Nemopilema nomurai* in the Japan Sea and beyond.

Cubozoa – box jellies

The name was given to these jellyfish because the very transparent medusae of the group are square in transverse section, with one to three tentacles depending from each corner. The bell has a hydrozoan-like velum. Cubozoan planula larvae settle and become polyps that eventually metamorphose into medusae, detach, and swim away, a pattern different from those of both Hydrozoa and Scyphozoa. The medusae have a sensory complex on the bell edge at each corner, including several types of eyes, some with a cornea, lens, and retina, and a statocyst (the other groups with medusae also have distinctive sensory systems). Data from this complex apparently direct cubomedusans' rapid (for medusae) swimming, to ~20 cm s^{-1}, around obstacles and away from predators. The nemato-cysts are in very large batteries and the venoms are more toxic than those of most other cnidarians, occasionally killing swimmers in tropical or subtropical waters. One small species (1 cm bell) abundant in reef areas, *Carybdea sivickisi*, is active in the water column at night but attaches with sticky pads on the bell to the undersides of rocks during the day (Hartwick 1991).

Ctenophora

This is a pelagic phylum, apart from a group of extremely derived benthic forms found primarily in the tropics. Planktonic ctenophores are commonly called comb jellies (indeed, they are gelatinous). While still retaining a struc-ture of two tissue layers with intervening mesoglea, they have added an anus, and thus have spatially separated feeding from waste elimination. The simpler forms are spherical (e.g. *Pleurobrachia*) with diameters to about 2 cm. Eight "comb rows", series of plates or ctenes of fused cilia, run from the upper (aboral) to the lower (oral) end. Ciliary lineations on the ctenes act as diffraction gratings, casting moving rainbows as propulsive waves sweep in sequence down the rows and across the light field. The resulting locomotion is slow and smooth, usually with graceful turns. Internally, the gut has two side pouches containing retract-able tentacles that extend through tubes adjacent to the mouth. The tentacles bear cells called colloblasts, which discharge much like nematocysts, but that entangle prey with sticky pads rather than stinging them. Tentacles arc through the water as the ctenes row the animal ahead. One common genus (*Beröe*) has dispensed with tentacles alto-gether, and has a powerfully extensile mouth. They feed entirely on smaller ctenophores, such as *Pleurobrachia*. The body in some forms (e.g. *Mnemiopsis*, *Cestum*) stretches during late development into a strongly bilateral pattern, forming large flaps which propagate a wave-like motion that results in oscillatory swimming. *Mnemiopsis* in coastal and estuarine areas of the western Atlantic, and *Pleurobrachia*

off California and Oregon, exhibit population outbursts which nearly eradicate their predominantly copepod prey and then subside. Many ctenophores, but not *Pleurobrachia*, are bioluminescent, producing a faint blue light along the digestive canals under the comb rows. Some deep-sea ctenophores can release bioluminescent mucus to distract approaching predators. Observations and collections from deep submersibles are revealing a previously unrealized diversity in this phylum, including some very delicate species as large as a meter in size. New genera, families, and orders have been added based on these deep collections (e.g. Madin & Harbison 1978).

Platyhelminthes

Flatworms are not important members of the plankton, but there are planktonic species. Specimens are often found swimming in the vicinity of tropical reefs. These probably come and go from the bottom.

Lophophorate phyla

Brachiopods, bryozoa, entoprocts, and phoronids do not contribute to the holoplankton, but some species in these groups release abundant planktonic larvae in coastal areas, larvae of bizarre form compared to the adults. Resemblances to the space cruisers of science fiction are unmistakable.

Nemertea

Around 100 species of pelagic nemertean worms have been described, although the systematics are far from complete (Roe & Norenburg 1999). Nemerteans are unsegmented, predatory worms, the planktonic forms as large as 20 cm in length, which capture prey with an eversible proboscis armed with hooks. Shallow-living forms live closely associ-ated with the bottom, but swim up off it. A number of holoplanktonic species are found in mesopelagic and bathy-pelagic depths.

Annelida

There are holoplanktonic polychaetes, fewer than 100 species, but in six distinct families, suggesting that a pelagic lifestyle has evolved repeatedly. They usually have trans-parent bodies, large flap-like parapodia, and very large, probably image-forming eyes. All of these are predators, capturing prey either with strong jaws or protrusible oral stylets. The most common genus is *Tomopteris* (Plate 6.5), worms bearing very long setae on the ends of its elongate tubular parapodia. *Tomopteris* are never abundant, but occur in modest numbers in most regions. According to Latz *et al.* (1988), *T. nisseni* can spray yellow, luminescent particles from its parapodia when disturbed. Most benthic annelids have planktonic larvae with biconical shape and ciliary locomotion, called trochophores, a larval type shared with mollusks and other "Trochozoan" phyla.

Mollusca

Mollusks are important as both holoplankton and as meroplankton. The larvae of all major classes contribute to the meroplankton. The snails (gastropods) are the primary con-tributors of holoplanktonic forms, which are well described and well illustrated by Lalli and Gilmer (1989). There are planktonic representatives of two major gastropod groups, the opisthobranchs and the prosobranchs.

Opisthobranch groups
Euthecosomata

These are shelled *pteropods*, an informal term meaning "wing foot". The foot of these snails modifies during

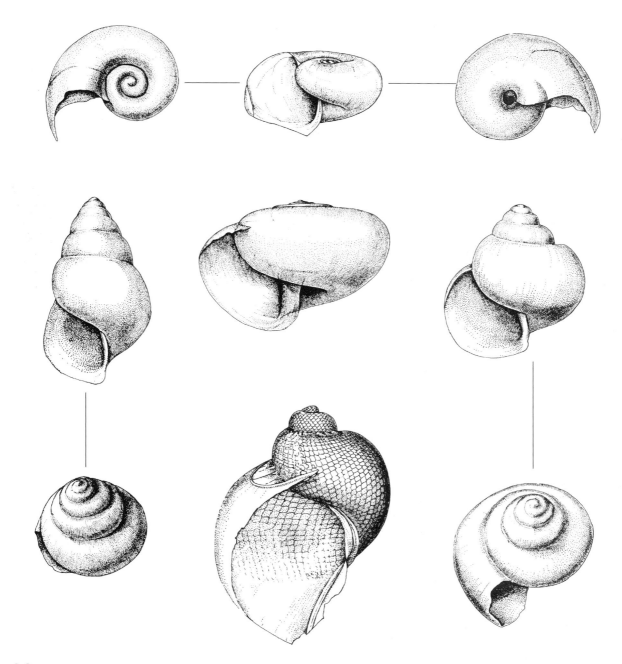

Fig. 6.3 Shell variations in the common euthecosome genus *Limacina*. The lines connect different views of the same species. (After McGowan 1968.)

development into paired wings, which act as swimming organs. In the most common genus, *Limacina* (=*Spiratella*, Fig. 6.3), the wings extend from the mouth of a fairly typical but left-coiled snail shell, usually about 3–5 mm diameter in older adults. Left-coiled or sinistral means that when the shell is held with the top of the whorl up and the opening facing you, the opening will be to your left. A vast majority of benthic snails are right-coiled or dextral. In other common Euthecosome genera, the shells are not coiled but are variations on a cone, e.g. *Cavolinia* (Plate 6.6), *Clio*, *Creseis*, *Cuvierina*, and *Diacria*. Head and wings emerge from the opening of the cone, the tip of which usually points down. Euthecosome shells are formed of aragonite (Fig. 6.4), a mineral form of calcium carbonate shared with true corals and distinct from the mineral calcite of benthic snails, clams, foraminifera, and the red algal family Corallinaceae. In a few relatively shallow parts of the Atlantic, sediments termed pteropod ooze are largely composed of pteropod aragonite.

Euthecosome feeding apparently occurs in two modes. Internal filtration takes place on a ciliated mucous gland inside the mantle cavity. Gilmer and Harbison (1986) have observed an external mode of "filtration": many euthecosomes actually hang suspended from a mucous "bubble", or float (Plate 6.7), which almost certainly also serves as a food-gathering structure and is reeled in periodically. The floats explain how pteropods can afford the energetic costs of keeping their weighty shells in the upper water column. However, Gallager and Alatalo (pers. comm.) have reared *Limacina retroversa* in tanks, and report that they survive and grow for long periods without ever forming mucous floats, apparently feeding during active swimming with the wings. Euthecomes are sequential hermaphrodites, the gonad switching from testis to ovary when growth is at about half the final body size (which is 5–10 mm in most species).

Pseudothecosomata

These are pteropods with a gelatinous pseudoconch or false shell. In this group, the mineral shell has been lost altogether, and the visceral organs are surrounded by the slipper-shaped pseudoconch, which is internal to a very thin skin. These are also filter-feeders. Elaborate balloons of mucus set out into the water are used for feeding and suspension. Body size can be as much as 10 cm. All pseudothecosomes are subtropical or tropical. The most common genera are *Corolla* and *Gleba* (Plate 6.8).

Gymnosomata

These carnivorous, unshelled pteropods are little cones of muscle, typically about 1 cm long with two small wings. Most specialize in eating shelled pteropods. A common

(a)

(b)

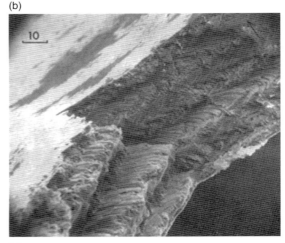

(c)

Fig. 6.4 Shell of *Cuvierina columella*, a euthecosomatous pteropod. (a) Whole shell. (b) SEM of cracked shell surface, scale bar = 10 μm. (c) Enlargement of aragonite ($CaCO_3$) crystals, scale bar = 1 μm. (After Bé *et al.* 1972.)

high-latitude genus is *Clione*, a predator of *Limacina*. It pulls the soft tissue from the shell using its extrusible radula, then drops the usually undamaged shell, which falls to the ocean floor. Other genera (e.g. *Paedoclione*, *Crucibranchaea*) have tentaculate structures around the head and extrusible spines for grabbing prey. Apart from *Clione*, very little is known of gymnosome life and times.

A prosobranch group

The Prosobranchia are a large group of mostly benthic snails is represented in the plankton by the Heteropoda. Many of the heteropods retain a calcite shell. In *Atlanta*, which are the smallest representatives (to about 5 mm), the shell is coiled all the way around to the opening, and the animal can withdraw completely into it. In *Carinaria*, typical of the larger heteropods, the body has enlarged and no longer fits in the shell, which takes the form of a conical hat covering the gonad and digestive gland. The rest of the body is a transparent tube with a large feeding structure anteriorly, an undulatory tail, and a ventral fin opposite the shell. They have eyes located on the trunk and directed to look beyond the mouth. Black retinal elements and spherical lenses imply that these eyes are image forming. *Carinaria* swim with the ventral fin and mouth directed upward, presumably taking most prey from below by grasping them with the radular tooth rows. *Carinaria japonica* can be as large as 30 cm long. Variations on this body plan are found in *Pterotrachea* and *Cardiopoda*.

Arthropoda

Class *Crustacea* represents this diverse phylum in the plankton. Crustaceans contribute by far the largest number of zooplankton species, and they are the largest component of the total biomass at most times, places, and depths. Members of the following subclasses are important in the marine holoplankton.

Branchiopoda

The Cladocera are the only order in the subclass Branchiopoda, and are familiar to freshwater biologists as water fleas, such as *Daphnia*. The cladoceran body of 1–2 mm is dominated by the head and wrapped in a posteriorly pointed, bivalve carapace resembling a clam shell. However, the first few segments, which bear large, hemispherical, compound eyes, are outside this shell. Swimming is by rowing movements of the antennae (the second limbs) that extend outside the carapace behind the eyes. Feeding is predatory in marine Cladocera, which grab prey with limbs surrounding the mouth. Those limbs are exposed, the valves being almost incorporated into the posterior body. There are 10 marine species, with most individuals belonging to the genera *Evadne* and *Podon*. While these genera are found all across the oceans, they are never abundant or dominant. In estuaries, however, they can be numerous in some seasons. Like the daphnids, *Evadne* and *Podon* can reproduce parthenogenetically, occasionally alternating a sexual generation including males. The genus *Bosmina* is often an abundant component of the plankton in the brackish water of upper estuaries.

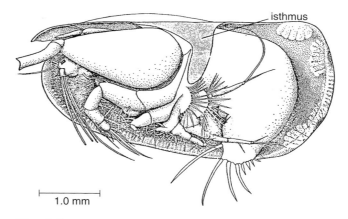

Fig. 6.5 *Conchoecia borealis*, sketch of the body layout and limbs. The right valve is in the background, the left valve is removed. The adductor muscle passes through the isthmus. The teardrop structure on the left is the antennal basis housing the propulsive muscles. (After Iles 1961.)

Ostracoda

Ostracods have a bivalve carapace over the whole body, including the head. This closely resembles a clam shell, with a distinct dorsal hinge and closure by a transverse muscle. The body is strongly dominated by the head, with the trunk reduced to two appendages and the abdomen to a posterior coil. Propulsion is by rowing with the antennae, which bear long sweeps of setae and extend out of the shell through oarlock-like notches. The number of described species is increasing rapidly, due to attention from Martin Angel, Kasia Blachowiak–Samolyk, Vladimir Chavtur, and others, but the total remains a few hundred, a majority in the family Halocypridoidea of the order Myodocopa (many more ostracod species, not all of them Myodocopa, are epibenthic, have calcified shells and, hence, a fossil record). The vast majority of halocyprid individuals used to belong to species lumped in the planktonic genus *Conchoecia* (Fig. 6.5), which is now divided by the small group of experts into 30 or so genera.

Halocyprid feeding is at least partly predatory. Larger *Conchoecia* grab copepods or other prey with the mandibular palps that swing out ventrally through the opening between the valves. Prey are swung in against the blades at the base of the mandibles, then sliced into serial sections and swallowed. The valve margins bear glands, the secretion from which can make a filtering surface across the valve opening. This can be found in the gut in an accordion fold containing food particles. There are other, more prominent glands near the shell edge, some of which in some species eject bioluminescent blobs of mucus as decoys as the ostracod departs to escape from a predator.

Male ostracods deliver sperm to females through a penis. Females of pelagic groups carry eggs internally until hatching, which occurs at a relatively advanced stage compared to copepods. Juveniles ("ostracodites") are small versions of the adult form. In mesopelagic layers, there are larger (to ~2 cm) forms of the genus *Gigantocypris* (Cypridinidae) in which the shell is nearly spherical. Unlike the eyeless halocyprids, these have elaborate visual systems. Holoplanktonic ostracods are oceanic, only occasionally being mixed shoreward over continental shelves and rarely washing into estuaries. Some epibenthic, tropical cyprinids of the genus *Vargula* swim above tropical grass beds and reefs at night, producing luminescent mating displays.

Copepoda

Copepods usually dominate the mesozooplankton in respect to numbers and biomass in all marine waters. The abundant, free-living orders are Calanoida (Plate 6.9), Cyclopoida (family Oithonidae), Poecilostomatoida (families Oncaeidae, Corycaeidae, and Sapphirinidae), and Harpacticoida. Huys and Boxshall (1991) have reviewed copepod anatomy, anatomical variation, and phylogeny, and Boxshall and Halsey (2004) provide a guide to copepod systematics. Claude Razouls and colleagues maintain a website with documentation for each of the 2462 currently accepted planktonic species (http://copepodes.obs-banyuls. fr; *Diversity and Geographic Distribution of Marine Planktonic Copepods*), complete in French and English. With so many species, marine copepods constitute a large fraction of all mesozooplankton diversity, and they are abundant, usually over half the specimens in any net tow and often much of the biomass. In turn, copepodologists constitute a large fraction of all biological oceanographers, but copepodology is more than an oceanographic concern. It is a lively discipline in its own right. There is an active World Association of Copepodologists, and there are several hundred capable systematists around the world. This is in sharp contrast to most other planktonic groups. Ostracods, chaetognaths, euphausiids, pteropods, salps, and appendicularians all have fewer than ten significantly active systematists, and no organizations promote study of their biology and taxonomy. Mauchline (1998) has provided an extensive review of planktonic copepod biology and ecology.

The copepod body has two principal sections: the rice-grain-shaped prosome (six head segments plus, primitively, six thoracic segments) and the urosome that is much narrower, consisting of four to six tubular segments). Slow swimming is by sculling movements of the second antennae. Rapid escape swimming is driven by sequential flapping movements of the thoracic legs on the rear half of the prosome. While the feeding modes and mouth limbs are diverse, the basic body form and operation of the thoracic legs are quite uniform, apparently a key and conserved design feature of the entire group. It makes possible, at least in larger forms, escape velocities greater than 1 m s^{-1} (200–500 + body-lengths s^{-1}). Many copepods (e.g. *Calanus*,

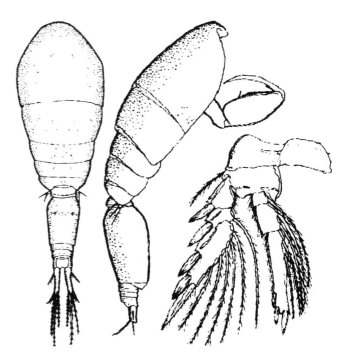

Fig. 6.6 *Oncaea frosti* Heron 2002. Left, 1.0 mm female in dorsal view. Middle, 0.72 mm male in side view, showing maxilliped modified for clipping into soft surfaces. Paddle-like second thoracic leg to right. Stylized, simplified drawings like this (e.g. the thoracic legs are not shown in the lateral views) are typical in the taxonomy of small crustaceans. Böttger-Schnack and Huys (2004) regard *O. frosti* as a *species inquirenda*. (After Heron 2002.)

Acartia, *Oithona*) feed by picking or filtering phytoplankton and protists from the water. The mechanism of particle feeding will be considered in Chapter 7. Predatory copepods (e.g. *Euchaeta*, *Candacia*) impale their prey with large spines on limbs at the back of the head section, then move them to the mouth. Many mid-water forms (*Lophothrix*, *Spinocalanus*, *Tharybis*, . . .) find and ingest sinking detritus, including marine snow and fecal pellets from epipelagic plankton. The mandibles in most families bear teeth for crushing or tearing food. In many species of both particle-feeders and carnivores, the teeth are tipped with hard, opal caps. Several distinctive families, the Oncaeidae (*Oncaea*, Fig. 6.6) and Corycaeidae (*Corycaeus*), have mouthparts reduced to a small cluster of pincers, and their antennae and maxillipeds are modified for gripping on soft surfaces, including gelatinous plankton and marine "snow", apparently sucking in nutrition or nipping small bites from them. *Oncaea* moves over surfaces like discarded appendicularian houses or pteropod feeding-"balloons"; *Corycaeus* probably sucks nutrition from soft-bodied animals (jellyfish, large chaetognaths). *Corycaeus* has paired, anterior cuticular lenses focusing light through light guides that converge to pigmented retinas far back in the prosome. These eyes occupy about half the body, suggesting that *Corycaeus* see their hosts and move to them.

Typical planktonic copepods are mostly 1–3 mm total length as adults, but some are smaller and a few, for example, the deep-living genus *Bathycalanus*, grow to 16 mm. Sexes are separate throughout the Copepoda, and aspects of sexual reproduction, mate finding, and development are covered in Chapter 8. However, it can be noted here that in many copepods the male bears elaborate clamps for catching and then holding the female during copulation. Males have specially adapted limbs for spermatophore transfer, the pair of fifth thoracic legs that copepodologists are famous (among invertebrate zoologists) for studying. The reason for this interest is that characteristics separating species are most often found on that leg. Separation of species is partly ensured by the appropriate fit of the male fifth leg to the female's size and details of her shape.

Females of some genera (e.g. *Euchaeta, Pseudocalanus*) carry their eggs in sacs until hatching; others (e.g. *Calanus, Acartia*) release them free into the water. Hatching of the eggs releases small larvae called nauplii (Plate 6.10), which are rugby-football-shaped and bear three pairs of anterior limbs near a bulky labrum (mouth cover). Nauplii of some genera are supplied with yolk and do not feed; others begin to feed at the earliest stages. Nauplii are a significant part of the smaller heterotrophic plankton, often serving as the diet of larval fish. The six naupliar stages end with a metamorphosis to adult-like forms termed copepodites. There are six copepodite stages, the last being the sexually mature adults.

Copepods are readily studied both preserved and alive, so much is known about them, some of which will be reviewed further on.

Cirripedia

Barnacle larvae are important in coastal plankton, and larvae of oceanic gooseneck barnacles (*Lepas*) are found in all tropical seas. Two distinctive developmental stages are planktonic: shield-shaped nauplii and the cypris stage, which has a bivalve shell reminiscent of the ostracods. Both of these stages are particle-feeders.

Amphipoda

The evolutionarily more advanced crustacea are the *Malacostraca*, and a number of malacostracan groups contribute to the plankton: Amphipoda, Mysidacea, Euphausiacea, and Decapoda. The peracaridan subset of the Malacostraca (those carrying their eggs on plates attached to the thoracic legs (amphipods, isopods, mysids, and six other orders in recent systematics) are more dominant among benthic fauna. However, one family of amphipods, the Hyperiidae, has diverse representatives in the marine plankton. They are not shrimp-like and have no fused carapace; segmentation of the thorax retains its dorsal articulations. Body plans vary widely from sleek, fast swimmers (*Streetsia*) to sprawling tangles of grasping limbs (*Phronima*). Most have very large, compound eyes just behind and surrounding the stubby antennules and antennae. Anterior thoracic legs usually bear pincers for grappling with prey, while rear legs are tipped with hooks for attachment to surfaces. Hyperiids are usually associated with gelatinous plankton, catching rides on salps or medusae. Some associations seem to be obligate, others more general. The relationships are rarely obvious in plankton samples, where the rider and ridden have generally been knocked apart. *Phronima*, however, is often caught with its "host", the hollowed-out gelatinous (but resilient) barrel of a salp or doliolid. It resides inside the tube, propelling it forward with a jet stream generated with its abdominal legs.

Reproduction is sexual. Like marine benthic and freshwater peracarids (most of which are benthic or epibenthic), the female hyperiids carry their eggs under the thorax attached to oöstegites, plates forming a brood pouch at the base of the legs. Development in the egg produces juveniles resembling the adult in body plan and activity. Much remains to be learned about every aspect of the biology and ecology of this diverse group.

Mysidacea

Mysids are the most shrimp-like of the peracaridan crustacea. Older stages have stalked eyes, and the basic body form is close to that typical of shrimp and euphausiids. However, the mysid "carapace" is only a posterior expansion of the first thoracic segment, not a dorsal fusion of thoracic segments. Members of the family Mysidae, typically 1.5 cm long as adults, are transparent but spotted with chromatophores. They have large statocysts on the uropods (tail fan), a certain identifier of shrimp-like swimmers as mysids. Mysidae (*Mysis, Hemimysis*, and others, sometimes called opossum shrimp) are bentho-planktonic in nearshore habitats, often abundant in swarms along the bottom in or near the surf zone. Mysids are very important food for fish in this high-energy habitat, and swarms over rocky reefs are the food resource allowing year-round residence of a small number of gray whales off Oregon, USA (Newell & Cowles 2006). Some mysids related closely enough to be called by the same species name are found in both marine and freshwater habitats and all along the salinity gradients between them. There are populations in many lakes, and they are a major part of plankton biomass in some very large lakes. *Mysis relicta*, abundant in Lake Michigan, is also found in coastal seas and estuaries. Other mysid families, for example Lophogastridae (*Lophogaster*) and Gnathophausiidae (e.g. *Gnathophausia ingens*) are mesopelagic. They are larger, some over 10 cm, and bright reddish-orange. They have reduced eyes, no statocysts, very long antennular flagellae, and feed by scavenging and predation. *Gnathophausia* is one of the few mesopelagic animals that has been maintained in captivity (reviewed in Chapter 12).

Euphausiacea

Commonly called krill, euphausiids are important as food for whales, commercially important fish like hake and salmon, even for very large squid that would seem to have no means of catching them. They are shrimp-like (Plate 6.11), with seven abdominal segments and the dorsal exoskeleton of the thoracic segments smoothly fused into a carapace that attaches to the body along its whole length (distinct from mysids). Euphausiid gills are lateral branches from the thoracic legs that extend outside the carapace, and euphausiids have no maxillipeds, meaning that none of the anterior thoracic legs are modified as additional mouthparts. Shrimp have maxillipeds, and their carapace covers their gills. The size of euphausiid juveniles and adults ranges from about 1 to 10 cm. Some euphausiids are strictly carnivorous, feeding mostly on other crustacea. Several genera (*Nematocelis*, *Nematobrachion*, *Thysanoessa*, *Stylocheiron*) have one or two pairs of greatly elongated thoracic legs equipped with a claw or bundle of spear tips for prey capture. Other genera are predominantly herbivorous, particularly the more-abundant species in coastal habitats, but all forms supplement their diet with animal prey, particularly copepods. Filter-feeding mechanics are described in Chapter 7.

Sexes are separate in the euphausiids and, since mating is apparently required for each spawning (or at least quite often), males and females are found together in roughly equal numbers. Sperm transfer is by spermatophore, with the two anterior pairs of pleopods (abdominal limbs) modified in the male for manipulating these sperm sacs. Some genera carry the eggs in masses on the thoracic legs (e.g. *Nematocelis*), others release them into the water (e.g. *Euphausia*). Hatching produces a nauplius or "metanauplius", which shortly molts to a calyptopis stage, a limpet-shaped larva with a stick-like abdomen. Later larvae with stalked eyes outside the carapace and one or more abdominal legs are called furcilia. Furcilia grow progressively larger and add limbs through an indeterminate number of molts, such that more stages can be added or stages can be skipped, to produce juveniles (e.g. Knight 1984). Juveniles and adults molt on a regular, temperature-determined schedule, regardless of either food availability or achieved growth. Without food, they continue to molt but to progressively smaller sizes. However, growth adds elements to the compound eye, and they are not lost during size reduction (Sun *et al.* 1995). Trace metals have been shown to accumulate in the exoskeleton and are discarded at molt, so perhaps regular molting is a form of excretion or toxin elimination. It may also be adaptive during intervals of food shortage to repackage the body in a smaller shell, thus sustaining a compact, robust body format despite weight loss.

Euphausiids have photophores along the midline on the ventral side of the body. These produce countershading illumination to fill in the silhouette from below. Many species are strong diel vertical migrators. Species of at least *Euphausia* and *Thysanoessa* can gather into swarms or schools. The prime example is *Euphausia superba*, the Antarctic krill, which lives in massive shoals at older ages. These are true "schools", with individuals maintaining close discipline with respect to spacing and common orientation (more about *E. superba* in Chapter 10). In *Thysanoessa*, the massing is more properly called swarming, with individuals moving alternately toward and away from the swarm center. Swarms in that genus, and perhaps in *Euphausia pacifica*, seem to be an adaptation for mating.

Systematic work on krill includes a CD-ROM expert system for identification (including larval stages) by Brinton *et al.* (1999/2000). It includes references to virtually the entire literature to that date on Euphausiacea, and a wealth of systematic and biological information. Sampling has been globally extensive. With the exception of cryptic species that may be identified by molecular genetics and possibly a few more bathypelagic forms (like *Bentheuphausia amblyops*), the systematics are nearly complete. There are 86 species in 11 genera.

Decapoda

In the plankton, these advanced crustaceans are represented by several families of "true" shrimp, which are distinguished by a carapace attached along the entire back and covering the gills. At least some thoracic legs are modified as maxillipeds. Some shrimp are holopelagic and qualify as zooplankton or small nekton. The most widespread of these in the upper ocean belong to the genus *Sergestes*, which are "half-red" shrimp. They have red pigment covering the food crop in the thorax, so that luminescent meals don't give away their position to other predators. The rest of the body is transparent. At mesopelagic depths there are a number of dark-red shrimp belonging to the caridean and penaeoid groups. Most of these are about 10 cm long, but have antennal flagellae reaching as much as a meter away from the body. Thus the shrimp's vibration detectors extend through a very large volume, giving it a chance to find prey where very few are passing through. In addition, most benthic shrimp and crabs contribute larvae to the plankton.

Urochordata

There are two important chordate groups in the oceanic plankton, the urochordates and the vertebrates. Urochordates, familiar to tide-pool zoologists as tunicates or sea squirts, are represented in marine plankton by the Thaliacea (salps, pyrosomes, and doliolids) and Appendicularia.

Salpidae

Salps

Salps (e.g. *Salpa*, Plate 6.12, and *Thalia*) are tube-shaped, gelatinous zooplankters with flap valves at either end. Fully grown, they are a few centimeters to about 20 cm long, typically with an aspect ratio of about three (length/width). The tube, or test, is a stiff, gelatinous structure supported by cellulose fibers. Muscles around the anterior, incurrent opening will close it, and then muscles in the test wall will compress the cylinder and produce a propulsive jet from the excurrent opening. Re-expansion is driven by the resilience of the tube, which fills through the incurrent opening. A cone of mucus is spun inside the cylinder, which acts as a water filter for obtaining particulate food. Particles are progressively wadded together along a ciliated band, the hyperpharyngeal lamina, and moved along for digestion to the posterior gut mass near the excurrent opening. Some selection is exercised; filtration can stop based on olfactory signals, and sufficiently malodorous water or particles will cause the expulsion of the mucous cone.

Salps have a moderately elaborate alternation of generations. There is asexual reproduction from an individual called a *solitary*, by pinching off of sections of a long tubular extension of the body, a stolon. Many of these sections (10–50) remain attached to each other side to side, and develop into chains (or a closed circle in *Cyclosalpa*) of new, fully formed, filtering cylinders – new salps. This *aggregate* stage reproduces sexually. The gonad, which is wrapped into the posterior mass with the gut, operates first as an ovary, producing a single egg (thus, the sexual phase reproduces but does not multiply the stock), which is fertilized by sperm pumped in from the surrounding water. The egg is retained in a chamber in the test wall and nurtured through an umbilical connection to the aggregate individual. The gonad then proceeds to produce and release sperm. Eventually, the embryo breaks out through the body wall, becoming the solitary phase, and the cycle repeats. The cycle is rapid, a few days, with the production of many chains of many aggregate individuals leading to very strong population outbursts – salp blooms. Towing a plankton net through an unsuspected salp bloom can produce a mass of watery tissue too heavy to swing aboard. We do not have clear understanding of the conditions stimulating, or conversely preventing, these blooms.

Doliolidae

These are very like salps, but differ in specific points of anatomy and details of the life cycle. In doliolids, the test muscles completely circle the body, whereas in salps there is a space along the ventral surface with no muscles. The doliolid solitary produces young by budding, which migrate along and then attach to a tube of tissue (stolon) trailing behind their "mother". The solitary stage tows this column of its young aggregates as they feed and grow. Eventually, the solitary loses most of its structure, becoming simply a muscular towing engine for the train of aggregates that feed it through the stolon. Some of the individuals in the aggregate train separate and mature sexually, with gonads that produce both eggs and sperm. Their embryos hatch as "tadpoles" (tailed larvae) that grow into new solitaries. The typical genus is *Doliolum*.

Pyrosomida

Pyrosomes are colonial in a different format from salps or doliolids, with the barrel-shaped pumping and filtering individuals embedded across the wall of a gelatinous tube that is closed at one end. Their incurrent openings are on the outside of the tube, with the excurrent ones on the inside. Thus, the colonial tube is propelled by the water jet from its open end. Typical colonies are 2 cm diameter by 8–20 cm long, but giants occasionally develop that are >50 cm diameter and several meters in length. The individuals are dramatically bioluminescent, hence the name meaning "fire body". Individuals in the tube wall respond with prolonged flashes to rubbing or poking on the colony sides. The individuals in the colony reproduce sexually, as hermaphrodites, and the larvae produced grow into filtering barrels with budding stolons that generate new colonies. Both salps and pyrosomes, at least some of them sometimes, execute extensive diel vertical migrations.

Appendicularia

Appendicularia (Plate 6.13) resemble the larvae of the benthic sea squirts (tunicates), and thus have also been called Larvacea. The body proper is a small pharynx below a bulb of digestive gland and gonad. This basic layout resembles a sea squirt adult. The resemblance to tunicate larvae is the long tail with its central notochord and adjacent muscles, which in appendicularia persists through adulthood. From a patch of secretory cells, the oikoplast, located near the mouth, the animal forms an elaborate "house" of mucus (Plate 6.14), typically the size of a walnut but in a few species much larger, even 1 m in diameter, which acts as a feeding filter. Water is pumped through the house by oscillations of the flat, muscular tail. The house is an elaborate apparatus with prefilters, filters, channels for excurrent flows, and more, the function of which is described in Chapter 7. Appendicularian houses, both occupied and abandoned, are at times abundant in the ocean. Their mucous surfaces are important accumulators of particulate matter, providing a habitat for small crustaceans, most prominently the copepod genus *Oncea*, that crawl over them and harvest delectable bits. The houses make up part of the flocculent detrital matter called "submarine snow". The most frequently encountered appendicularian genus is *Oikopleura*. Like the other planktonic urochordates, they are sequential hermaphrodites.

Vertebrata

This phylum includes the fish. Fish larvae, or ichthyoplankton, are meroplankton. Virtually all fish larvae look like little fish; most are a few millimeters long or a little larger. Some hatch before fully absorbing their yolk sac, which remains as a large ventral bulge. Such larvae do not need to feed immediately, but survival chances are often better if feeding can begin before the yolk is fully absorbed. Other species are more precocious and hatch ready to swim and hunt. Most fish larvae have relatively large eyes. There are few adult fish that could be classed as plankton, except perhaps the deep-sea family Gonostomatidae ("bristle-mouths").

Chaetognatha

This is an exclusively marine, epibenthic (the Spadellidae) and planktonic phylum commonly called arrow worms (Fig. 6.7).

Chaetognaths are frequent and abundant predators in the marine zooplankton, including at present 98 accepted, holoplanktonic species. They are elongate, usually transparent cylinders divided by transverse septa into head, trunk, and tail sections. Adult lengths are 2 to 12 cm. Tail and trunk sections bear thin fins. The head is armed with long, curved, chitinous, opal-tipped fangs that jam captured prey (mostly copepods) into the jawless, ventral mouth. The chaetognath hunting strategy is to hang motionless, waiting for actively swimming animals to pass within striking distance, a few to 20 cm. Alerted by vibrations transmitted through the water and sensed by hair-like receptors on the body surface, the worm flicks forward, driven by longitudinal trunk muscles, and impales the prey with its fangs. Predators with this strategy make swimming very dangerous for prey; the less swimming, the better. The copepod genus *Metridia* has not evolved in accord with this wisdom, individuals keeping constantly on the move. They are, therefore, among the most common prey of all pelagic "ambush" predators, particularly arrow worms (Sullivan 1980). Other copepods are also important. Thuesen *et al.* (1988) showed that at least some chaetognaths "still" their prey at capture with tetrodotoxin, a powerful neural poison. The likely importance of this is to stop vibrations that could give away the predator's position to its own predators.

Chaetognath reproduction is hermaphroditic, generally protandrous. The ovaries are in the trunk, the testes in the

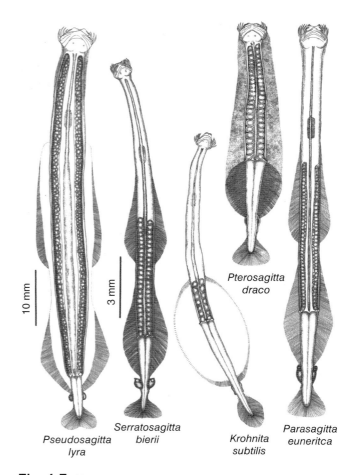

Fig. 6.7 Chaetognatha: several genera and several body-form variants. The 10 mm scale bar applies to *Pseudosagitta lyra*, the 3 mm bar to all the others (Drawings from Alvariño 1965.)

tail section. Sperm transfer is by spermatophore, but the transfer, while not well described, is mutual. Spermatophores constructed inside seminal vesicles on the lateral sides of the tail section fill with sperm through an aperture in the tail-section wall. They are placed next to the female gonopores of both partners at the posterior end of the trunk section. Most experts believe self-fertilization is also possible. Sperm migrate into the tubular ovaries that extend forward along the trunk wall from the gonopores just anterior to the trunk-tail segment. Ova are generally in single lines on each side. The sperm, which have a worm-like structure (Shinn 1997), must burrow through "guard" cells in the oviduct to reach and fertilize the ova. Species of the Sagittidae (previously the genus *Sagitta*, now divided into 12 genera: *Parasagitta*, *Serratosagitta*, . . .) release eggs freely, but at least some species of *Eukrohnia* carry them in marsupial sacs. Hatchlings resemble simplified adults.

The phylum has received intense recent interest because it appears to have separated very early from the common ancestor of the Bilateria (i.e. all metazoans except the Cnidaria and Ctenophora). A consensus of genetic sequences (e.g. Marlétaz *et al.* 2006; Helmkampf *et al.* 2008) implies a relation to the protostomes, the major bilaterian branch that includes arthropods, mollusks and annelids. That is a change from the long-standing assignment of arrow worms to the other major branch, the deuterostomes, which includes chordate-like groups, chordates, and echinoderms. Kapp (2000) has re-evaluated chaetognath development, showing that it differs in many respects from that of either major group, and Ball and Miller (2006) provide a catalogue of respects in which chaetognaths differ from all other metazoans. Their body plan is shown by recently reported and convincing fossils from the Maotianshan and Burgess Shale deposits to have remained essentially stable since the early Cambrian (Chen & Huang 2002; Conway Morris 2009). Molecular genetics (Papillon *et al.* 2006) have provided a reasonable characterization of the relationships among chaetognath family groups. A peculiarity of their genetics is a radically reduced mitochondrial genome compared to nearly all other phyla (Helfenbein *et al.* 2004).

Aesthetics

All that dry terminology does no justice to the elegance and beauty of planktonic animals. They are a joy to the eye, and even preserved specimens can fire the imagination with questions. Why are there fancy curved spines and patches of hairs on copepod legs apparently used only as paddles? Of what use is an eye on the edge of a jellyfish's swimming bell? How can a flat patch of gland cells, the appendicularian oikoplast, produce a gob of mucus that unfolds into a multistage filter with coarse and fine pores, trapdoors, and flow tunnels? How can useable information about up and down come from a statocyst in the middle of a mysid's tail fan that must constantly flip up and down, back and forth? Or, is gravity not the datum sought by that sensor? Is the sea that we experience at its surface as so wild with waves and churning, really quiet enough just a little way down that delicately linked salp chains can grow to lengths of several meters without breaking? Apparently so. Some such questions we can answer; some answers are too well hidden behind a watery curtain ever to be found. A dose of lingering mystery is a good thing.

Chapter 7

Production ecology of marine zooplankton

"Zooplanktologists" are interested in the animals themselves, their systematics (taxonomy), their adaptive schemes for survival, and their places in the marine food web. Particular studies examine distributional patterns; feeding mechanics and rates; food selectivity; growth rates and patterns with age; secondary productivity; reproductive biology and fecundity; mortality (rates, causes, and age distribution); life-history variations; and vertical migrations. This chapter will consider issues of feeding and growth, or the trophic aspect of zooplankton, which is the topic of greatest interest to ecosystem modelers. A very large fraction of the work has been done on copepods, making other groups seem somewhat neglected by comparison. That is mostly because copepods are so reliably present in samples. Therefore, they are always available to study, which is less true of other groups. However, other groups have had attention, and we know a great deal about them. Distribution patterns and adaptive strategies are treated in separate chapters.

Feeding mechanics

Mesozooplankton capture, sort, and ingest food by diverse mechanisms. Particle-feeders are those animals feeding on prey much smaller than themselves, such as phytoplankton and protozoans. They essentially filter the water in one way or another, although filtering the water from around a particle is often a final step after it is located by a searching procedure.

Phytoplankton and protozoans are small ($\sim 1\,\mu m$ to hundreds of μm) compared to most mesozooplankton ($\sim 200\,\mu m$

Box 7.1 Oceans are a trophically dilute medium

Relatively rich water has chlorophyll concentrations from 1 to $20\,\mu g\,liter^{-1}$. Take, for example, $2.0\,\mu g\,Chl\,liter^{-1}$, a typical coastal or oceanic spring-bloom level. At carbon:chlorophyll = 60–200 (use 100), carbon:dry weight = 0.4, and dry weight:wet weight = 0.3, we have:

$$2\left[\frac{\mu g\,Chl}{liter}\right] \times 100\left[\frac{\mu g\,C}{\mu g\,Chl}\right] \times \frac{1}{0.4}\left[\frac{\mu g\,DW}{\mu g\,C}\right] \times \frac{1}{0.3}\left[\frac{\mu g\,WW}{\mu g\,DW}\right]$$

$= 1667\,\mu g$ wet weight of food $liter^{-1} = 1.7\,mg/1025\,g$, which is <2 parts per million in coastal water or oceanic blooms. In the vast oligotrophic ocean stretches, chlorophyll is less than $0.3\,\mu g\,Chl\,liter^{-1}$ and food <0.3 ppm. These quantities of "food" are two-thirds seawater (at $1025\,g\,liter^{-1}$).

---

to $\sim 20\,mm$) that feed upon them, and they are dilute relative to the water that suspends them (Box 7.1). A few parts per million is a rich soup. Therefore, lots of water must be processed to acquire meals. The animals most correctly described as filtering are the salps and doliolids, which strain water through a literal mesh constructed of mucus. By injecting particles of various sizes (usually size-graded plastic beads) at the incurrent opening, then collecting at the excurrent opening those particles not captured on the filter, it is possible to determine the effective pore size of the filter cone (Fig. 7.1; Plate 6.12). The results show that

Biological Oceanography, Second Edition. Charles B. Miller, Patricia A. Wheeler.
© 2012 John Wiley & Sons, Ltd. Published 2012 by John Wiley & Sons, Ltd.

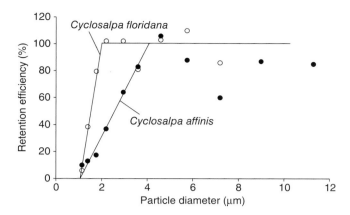

Fig. 7.1 Retention efficiency of the mucous filtering screens of two salp species as a function of particle size. (Data from Harbison & McAlister 1979.)

particles larger than 2–4 µm are captured with high efficiency. Since the polysaccharides in mucus are usually charged molecules, it is also possible that some organisms smaller than the mesh can be attracted to it, then stick.

Appendicularians also feed by pumping water through a mucous mesh filter (Plate 6.14) that is part of their house structure, but the mechanics are subtler than in salps. Observations of transparent mucous structures and the movement of tiny particles within them are extremely difficult, but Flood (1991) provided a convincing description of the feeding process in *Oikopleura labradoriensis*. The house itself has walls about 1 mm thick, a rather random meshwork of hydrated mucopolysaccharide fibrils carrying at some points clusters of bioluminescent granules. Entering water passes through outer screens in which the mucous strands are arranged in a rectangular grid that keeps particles larger than ~13 µm out of the house. Funnel passages carry the flow from the outer screens toward a chamber containing the trunk, and from which the tail chamber extends to the back of the house, just wider than the tail and just taller than the wave-like tail oscillations. It narrows at the back to an exit spout. Passages along the house sides and extending upward carry water from the back of the house toward entrance slots in an "organ-pipe" filtering apparatus. This is a pair of porous membranes kept from flying apart under flow pressure by an internal cross-lacing of mucous fiber connections (see Flood's original diagram), which create the curving, organ-pipe shape of the filter. Flow is mostly parallel to the planes of the filter, but, under sufficient pressure, water passes through all along their surfaces, depositing particles on them. Pores are 0.24×1.43 µm in the lower surface, 0.18×0.69 µm in the upper. Along the midline, the two curving mucous surfaces attach to the sides of a central duct, communicating with it by valved pores. The central duct attaches to the edges of the mouth.

This sort of apparatus has also been devised by filtering engineers and is called a tangential flow filter; particles are only held lightly against the mesh pores by the outward component of flow, since the overall transfer is spread over such a wide area and is locally very slow. When flow stops, the particles mostly drop back off the filter into the central space, and slow flow from another direction will move them to an accumulation point. The appendicularian does exactly that, the tail pumping stops and the house deflates somewhat, shaking particles off the outer screen, cleaning it, and freeing particles from the surfaces of the inner filter. A new, slower, flow moves water sideways through the filter to the central duct and along to the mouth. Force for this new flow comes from the cilia surrounding the spiracles (holes) in the sides of the pharynx behind the mouth. Particles are filtered again by mucous sheeting across the inside of the pharynx. Flood's description of this complex mechanism is good reading.

In another species, *Oikopleura vanhoeffeni*, the pharyngeal filter is much coarser than the house filter, with typical pore openings of 3–5 µm (Deibel & Powell 1987). Thus, the most numerous smaller particles of ~1 µm may not be very efficiently retained, and the mean size of prey actually reaching digestion may be >3 µm. Apparently no actual determinations exist of retention efficiency as a function of particle size. The bacteria-sized particles that are retained may be captured by direct impact with filter fibers, possibly becoming embedded in the fluid boundary layers of the fibers. The boundary layers should be thick at flow velocities measured by Acuña *et al.* (1996), suggesting fiber Reynolds numbers $\sim 10^{-5}$.

Euthecosomes and pseudothecosomes (the two groups of particle-filtering pteropods) are also mucous filter feeders, but they do not push water through a mesh; rather they are dependent upon particle attraction to a mucous surface. According to Gilmer and Harbison (1986), they expand a bubble of mucus (Plate 6.7) generated in the pallial cavity (inside the shell in euthecosomes), hang from it for some minutes, then ingest it with the particles it has attracted. It remains possible that ciliary–mucoid filtering within the pallial cavity, described by Morton (1954), is used while the animal actively swims. Cilia in the pallial cavity drive water and particles over the pallial gland, which continually secretes mucus and gathers it with accumulated particles into a string that is moved along ciliary tracts to the mouth for ingestion. In either mode, hard, meshing teeth in a gizzard between esophagus and stomach break captured particles.

Extraction of particles from their watery background by crustaceans occurs by several qualitatively different mechanisms. *Euphausia*, typical of filter-feeding krill, comes closest to the simple screening of the salps. The anterior surfaces of the long inner branches (endopods) of the paired thoracic legs bear long, anteriorly directed setae which close the spaces between successive legs (Fig. 7.2),

(a)

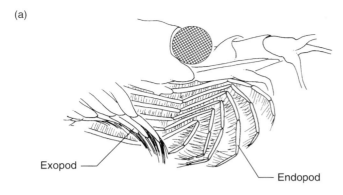

Exopod ⎯⎯⎯⎯   ⎯⎯⎯ Endopod

(b)

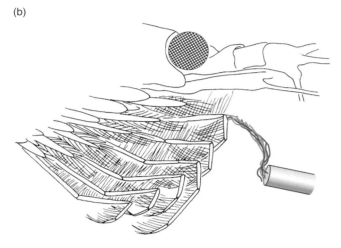

Fig. 7.2 Thoracic legs of particle-feeding euphausiids bear long anteriorly directed setae, forming a filter basket (a). This fills from the front, beneath the head, when the legs are opened (b) (flow is shown by the dye stream moving from a pipette tip at the right). During opening, the filter surface is covered by the exopods (outer legs) to keep water from moving in through the screen. (After Hamner 1988.)

and the setae bear a series of setules, which in turn bear secondary setules, creating tightly woven filter screens beneath the body on either side. In addition, each leg bears short combing setae on the medial surfaces, and these comb through the filter setae on the leg behind, from leg surface to seta tip. According to Hamner (1988), water is drawn between the thoracic leg screens at an opening below the head as the animal thrusts the legs to the sides. The short, flap-like outer branches of the legs (exopods) lie over the screened spaces between the endopods, preventing the basket from filling through the filter screens. Then the leg tips close the filter at the bottom, pushing water vortices (and particles) upward into the basket, which then closes from the sides, squeezing water out through the screens that retain particles. As the legs are pulled back again, the combing setae move particles toward the seta tips, which are cleaned by other combs on the mouthparts. Food is shoved into the mouth under the labrum (the bulky upper

lip), ground by the mandibular blades, and ground again by meshing teeth in the crop.

Copepods probably have several mechanisms for particle feeding. That of *Eucalanus pileatus* is well documented. Rudi Strickler and colleagues used high-frequency movies (to 2000 frames s^{-1}) to capture the movement patterns of feeding limbs cycling at 50 Hz and higher. Elaborate measures were needed to obtain such movies, including laser illumination of animals held in position before a microscope objective by hairs glued to their backs and special mechanisms to move film and shutters at such high frame frequencies. Strickler developed and applied all that. One problem was that only a few seconds of filming at 2000 frames s^{-1} required whole reels of movie film, which then were evaluated frame-by-frame to determine the sequence of limb movements. Work along these lines was popular in the 1980s, and lately more is being done with similarly high-frequency video. Video has also been used for studies of swimming mechanics. In the form of particle-image velocimetry (PIV), it has allowed definition of the flow field around feeding copepods, both tethered and free swimming.

The work on *Eucalanus* is likely the least affected by tethering (see below) before a microscope objective, since feeding movements of its mouthparts generate only very slow motion of the whole body. Koehl and Strickler (1981) described the basic feeding mechanism. They emphasized that all water movement at the scale of the copepod mouthparts (Box 7.2) is in the viscous regime of flow, that is, at low Reynolds numbers. The aspect of this most obvious in their movies is that there is no inertial "carry" of the flow after a given propulsive limb motion is complete. Water moves while a limb applies force, but it stops almost instantaneously when the force is off. If a particle advances toward the body, and the animal stops its limb motions, the particle stops also. It does not swirl off in residual eddies. Thus, the effect of recurring limb movements is a sequence of step-like advances of the surrounding water and contained particles up to and around the searching animal.

The cycle of limb movements is as follows (Fig. 7.3): the second antennae move forward and the maxillipeds move backward, advancing water one "step" toward the ventral side from the surrounding volume. A ventral-to-dorsal sweep of the mandibular palp then advances it laterally and dorsally, just past the end of the maxilla. The sequence repeats, moving water toward and around the body in a series of definite, individual steps. The limb motion abruptly stops when a particle of food (or in some cases any particle) approaches the animal. Then the maxillae swing laterally with a spreading of the setae. The flow replacing the water moved aside carries the particle between the extended setae of the maxillae, which then close over it. The setae are pulled closer to the body wall and closer together, squeezing the "viscous" water out through the spaces between the setules. Water does not move anteriorly parallel to the setae

Box 7.2 Feeding limbs of *Calanus*

To understand Koehl and Strickler (1981), you the must know the head limbs of copepods (shown in Box Fig. 7.2.1).

The most anterior pair are the antennules, unbranched (uniramous), tubular limbs that stick out to the sides of the body during feeding. In Box Figure 7.2.1 they are lopped off and appear as circles. The biramous antennae are major generators of the feeding "current" which moves water past the body to be searched for food. The mandibular palps are also involved in generating the feeding current. An "endite" (a side branch off one segment), termed the mandibular gnathobase, at the base of each mandible bears the copepod's teeth and extends into the mouth under the labrum. The palps operate almost independently of the gnathobases. Next are the maxillules (or first maxillae), which are large flaps looking a little like elephant ears. A basal endite on each maxillule bears a comb of very stout setae. The uniramous maxillae (or second maxillae) follow, bearing a palisade of slender, long setae that are the final food-capturing screen. These setae extend anteriorly from the limb base, pointing to the posterior opening of the mouth. They can be raked by the endite comb of the maxillules. Last come the uniramous maxillipeds, which are involved in feeding-current generation and also grasp large prey and lift them toward the mouth. Now you are ready.

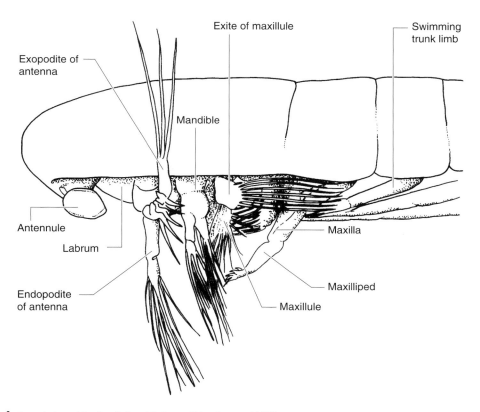

Box Fig. 7.2.1 Lateral view of feeding limbs of *Calanus*. (After Cannon 1928.)

of the maxillae because the maxillules move posteriorly at the same time, closing the anterior gap. Koehl and Strickler term maxillary particle capture a "fling and clap". There is also a sequence of reverse moves that can push water out of the space between the maxillae when a particle is to be rejected. In some movie sequences, it appears that particles are simply sucked into the mouth at this point. The esophagus has striated muscles attaching its walls to internal skeletal elements, and those muscles could suddenly pull it open, sucking in water and particles close to the mouth.

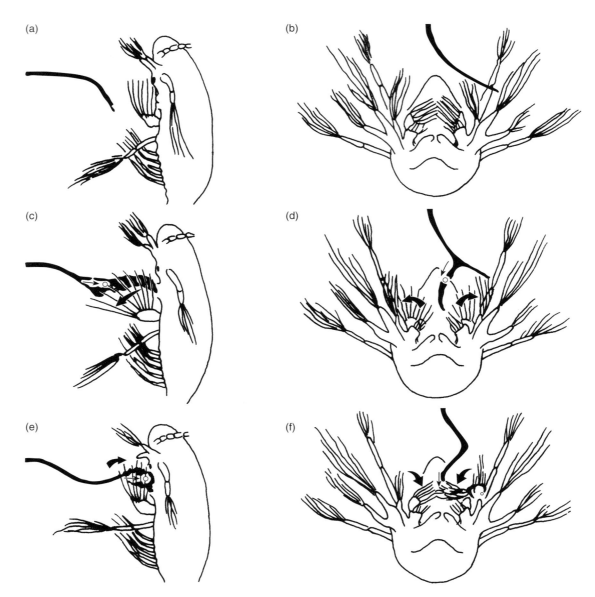

Fig. 7.3 Particle capture by *Eucalanus pileatus*, side views left, end views right; dark line is marker dye stream. In (a) and (b), antennae have moved apart, drawing water toward the mouth area, then around to the sides. In (c) and (d), a particle (circle) is sensed and the maxillae swing out to surround it and draw it between them. In (e) and (f), the maxillae close and squeeze water out from around the particle. (After Koehl & Strickler 1981.)

Larger particles, such as small animal prey, are probably combed from the maxilla by the stiff bristles on the medial lobe of the maxillules just below the body. Those then push the meal through the medial space between the labial palps, which close the posterior side of the mouth, and between the toothed mandibular gnathobases (the jaws) that meet under the labrum below the mouth.

Particles are detected during this sequence at a distance, before they touch any part of the copepod. Detection likely is primarily olfactory. The principal evidence is that receptors all over the mouth area are of a type shown to be

olfactory in other crustaceans. An important part of a convincing proof would be proper comparison of the diffusivity of substances that might be coming from food particles with the time and distance requirements set by the movies. Paffenhöfer and Lewis (1990) observed that initial perception distance is greater as particle concentration decreases, which they interpreted as expected for chemoreception. Phytoplankton cells are thought to have a *phycosphere* about them, a boundary layer in which cell products are diffusing away from them at the slow rates of molecular diffusion. Price and Paffenhöfer (1984) showed an intriguing aspect of this detection mechanism. They took repeated films of *E. pileatus* after various durations of experience with a given phytoplankton food. Inexperienced feeders responded to cells at a mean distance of 273 μm from the maxilla, while those that had had the food available for 24 or more hours responded to them at 345 μm. The experienced copepods also responded to 31% of cells appearing in the vicinity of the maxilla, as opposed to only 12% for the inexperienced. Price and Paffenhöfer took this to be evidence that olfaction is indeed the sense by which phytoplankton are detected. At least it shows that experience tells the animal that weaker signals of the effective kind are associated with food. Bundy *et al.* (1998) have shown that copepods also will move to and capture polystyrene beads that are not likely to have much odor. They speculate that the presence of a bead distorts the streamlines of the feeding current as the flow approaches the animal's boundary layer, and flow-sensing organs (innervated setae) on the limb surfaces, particularly the antennule, detect the distortions.

While the mechanism described by Koehl and Strickler is definitely filter feeding, it is not that of a strainer cleaning plants out of a thin soup to get together a thick porridge for swallowing. Rather, each food particle is found, then moved closer, separated from surrounding water, evaluated, and eaten or rejected *individually*. We know from large differences in feeding mouthparts that not all filter-feeding copepods collect particles by this exact mechanism. One variant mechanism, that of *Diaptomus sicilis*, a freshwater copepod, is described by Vanderploeg and Paffenhöfer (1985), again based on films of tethered individuals. In *D. sicilis*, the limbs generate a flow from anterior to posterior. Particles coming quite close to the maxillar setae will invoke a fling and clap response comparable to that of *Eucalanus*. However, part of the flow is deflected along the inner surface of the maxillar setules which seem to funnel it. Some particles appear to be ingested from this directed flow without any fling and clap, a passive capture mode. However, Vanderploeg and Paffenhöfer insist that the setal web does not act as a filter.

Video studies at 1600–2200 Hz (Kiørboe *et al.* 2009) of some small copepods, the largely neritic, calanoid genus *Acartia* and the widespread and abundant cyclopoid genus *Oithona*, describe feeding without generating a feeding current (ambush feeding) on motile cells approaching the antennules. When a cell approaches within ~200 μm, the thoracic legs flick back in a posterior to anterior sequence, similar to escape swimming (Chapter 8) but less sustained, driving the body forward over the cell. The lunge of the copepod does not significantly push the food particle away, and the tail fan may be used to rotate the mouth area next to the particle. This is followed by opening of the mouth limbs, similarly to the *Eucalanus* feeding-limb sequence, drawing the cell close for capture by the maxillae. Time from initiation of the lunge to final capture varies, several

flings of the mouth limbs can be required, but it is short, typically 3–30 ms in *Oithona*. This leaves very little time for any prey escape response.

Some diatoms encased in hard opal are far too large and stiff to swallow. However, Jansen (2008) has observed by simple microscopy that *Temora* can grasp a hatbox-shaped cell of *Coscinodiscus* with diameter a third of its length and chew a hole in one side. This causes the cytoplasm to withdraw to the opposite side, likely due to release of turgor pressure. So, the copepod rotates the cell, bites another hole, sucks out the cytoplasm, and drops the frustule.

Reports can be expected soon from video studies of copepods generating feeding currents (as opposed to ambush feeding) without being tethered. Being held in place changes the flow patterns. In the meantime, flow around an untethered copepod while it feeds has been examined by PIV. A thin (1 mm) plane of laser light is projected through an aquarium containing small particles that reflect the light. Video cameras record the fields of particles in the plane at frequencies on the order of 60 Hz, and the field of fluid velocities is determined (by analyses at several levels of sophistication) from the shifts in position of the same particles in successive frames. Catton *et al.* (2007) provide images (Plate 7.1) of the flow field around copepodites of *Euchaeta antarctica* cruising in the illuminated plane at 1–2 cm s^{-1}, views looking down on the dorsal side and at the side. Locomotion at this low speed is driven by a sculling motion of the antenna. Flow is accelerated only a very short distance in front of the animal, about half the body length, minimizing predator-alerting disturbance in the direction of travel. However, some flow carries water in from well to the side and over the antennules, likely for olfactory evaluation. Much of the water necessarily pushed from in front of the animal passes over it dorsally. Water is drawn up from below the body into the area of the feeding limbs, presumably checked for food scent, then accelerated under the center rear of the body, and dispersed aft and down. Comparative pictures for tethered specimens are in the paper by Catton *et al.*, indeed showing that flow in their vicinity is strongly modified, particularly with zones of acceleration extending much farther in all directions. Plankters sensing predators or prey in fluid adjacent to their bodies respond to shear (strain rates) on the order of 0.5 s^{-1} (that is, Δcm s^{-1} along stream per cm across stream). The gradients in velocity shown in Plate 7.1 when converted to shear by Catton *et al.* show 0.5 s^{-1} extending over about two body lengths and three body diameters. To move at all, some enhanced risk of predation has to be accepted; evidently, nothing ventured, nothing gained. The lesser volume of water accelerated near the free-swimming copepod also implies that less work is required for cruise feeding than would be calculated (e.g. van Duren *et al.* 2003) for a tethered animal. The work involved in cruise feeding appears to be a small component of the animal's overall energy budget.

Feeding rates and factors affecting them

From a trophic–dynamic viewpoint, the interesting aspect of zooplankton feeding is not how it is done, but the rate of consumption. Feeding rates are determined in several ways presently. The before-and-after method is to bottle some plankters with a known concentration of algal cells for a period of several hours or a day, then remeasure the cell concentration. The measurements can be done by direct microscopic counting on hemocytometer slides or by electronic particle counter (Box 7.3).

The results are usually expressed as a filtering or clearance rate, F, with units of volume per time per animal. Since filtering as such is not necessarily how an animal feeds, *clearance* rate is the better term. An animal feeding on suspended particles is not like you straining peas from a pan. You would pour the water and peas through a sieve, each lot of water removed from the pan being discarded and separated from the peas. The rate of change of peas remaining in the pot at successive times, N_t, would be given by the solution of:

$$dN/dt = -FN_{initial}/V \text{ (i.e. } dN/dt = \text{a constant)},$$

where F is the volume sieved/time and V is the original volume of cooking water. A plot of N_t against time would be a straight line with negative slope, intercepting the time axis ($N_t = 0$) when $t = V/F$. In filter feeding, the water filtered is not removed from the experimental bottle (or ocean) after the animal removes particles from it; the filtered water returns to the suspending volume, diluting the remaining particles. Therefore, reduction in concentration follows an exponential decay law. That is, the animals must filter progressively more water for each unit of food obtained. This is expressed by:

$$dN/dt = (-FC/V)N,$$

where C is the number of, say, copepods in the bottle volume V, and F is the volume cleared by each copepod per unit time at an assumed 100% filtering efficiency. Harvey (1937) apparently first applied this equation to filtering-rate determinations. Note that the model here is a strainer, not an encounter predator. However, since the animal is very small relative to the volume of water it must "search" by the Koehl–Strickler or other mechanism, the model is still the right one. The per capita clearance rate is an appropriate measure of the search effort that the animal applies in feeding. Separating variables and integrating, we have:

$$\ln(N_t/N_0) = -FCt/V, \text{ or } N_t = N_0 \exp(-FCt/V),$$

where N_0 and N_t are the cell concentrations at the beginning and end of the observational interval t.

If the cells are growing significantly, a control container without the grazers must be used to evaluate that rate, μ, and the associated equations must take account of that growth:

$$\ln(N_t/N_0) = (\mu - FC/V)t$$

This, or a mathematical equivalent, is often termed the Frost equation (Frost 1972). Frost also showed that the average concentration, N_{mean}/V, of cells over the interval of grazing in the experiment is calculated as follows: let $FC/V = g$ (for "grazing"), then

$$N_{mean}/V = N_0[e^{(\mu-g)t} - 1]/Vt(\mu - g).$$

This number can be used in calculating the average individual ingestion rate: $I = F N_{mean}/V$, or from the units: (ml cleared per hour) $\times N_{mean}/$ml = cells ingested h^{-1}.

Copepod studies

The effects of a variety of factors on F and I have been studied. These include: food density; container volume; particle size of food; grazer size and life-cycle stage; mixtures of several types of food; previous feeding history; and temperature. More work remains to be done in each area. Frost (1972) gives the basic result for copepods of the

---

Box 7.3 Automated particle counting

Electronic particle counters made experimental analysis of filter feeding fairly simple, probably deceptively simple. These devices, deriving from an original design by Wallace Coulter, count microscopic particles moving with a stream of electrolyte (blood or seawater) through the space between two electrodes embedded in the wall of a glass tube. Counts are electronically cumulated of the changes that the particles cause in the resistance to electric current flow between the electrodes. It is possible to gather changes in resistance in separate counts according to their magnitude, partly a function of particle size, and thus to obtain particle counts in a number of size-related "channels". Resistance change is roughly proportional to particle volume, so feeding experiments are often characterized in terms of total particle volume in a feeding chamber before and after an animal has fed there for some known time.

effect of food density, the so-called "functional response". The gently stirred experimental vessel was 3.5 liters, well above the threshold of any described volume effects (small containers reduce feeding rates, a very early result). Each vessel contained centric diatoms as food and 10 to 30 adult female *Calanus pacificus* that had not been starved prior to the experiment but were feeding steadily. The functional response (Fig. 7.4) has the following basic features:

1 Filtering rates are variable, but do not change significantly with food concentration at low concentrations.

2 When a sufficient food density is reached, filtering rates begin to decrease so as to keep the ingestion rate constant. Exactly this sort of functional response represents zooplankton grazing in practically all pelagic ecosystem models.

3 Copepods do not (at least in this data-set) eat superfluously, despite high food availability; that is, they do not eat more than growth and metabolism require.

4 The maximum ration is obtained at lower concentrations of large cells than of small cells.

Calanus pacificus females are about $170\,\mu g$ dry weight, which at 40% carbon includes $68\,\mu g\,C$. Thus, the ingestion of $1.1\,\mu g\,C$ per copepod per hour ($26.4\,\mu g\,C$ per day) is 39% of the body carbon per day.

The maximum filtering rate exercised by the animals at low food density is about $8\,ml$ per copepod per hour, or $192\,ml$ per day. This turns out to be less than they can do when previously starved. Runge (1980) showed that F values for this species can be $49\,ml$ per copepod per hour (over 1 liter per day) feeding on large cells after a period of starvation, and when collected in particular seasons. Debate about the adequacy of these rates to provide enough food at natural phytoplankton concentrations appears to be endless. Many copepods are found where less than $1\,\mu g$ of particulate carbon is available in $200\,ml$; they should be on short rations. Older (e.g. Mullin & Brooks 1976; Derenbach *et al.* 1979) and newer (Cowles *et al.* 1998; Benoit-Bird 2009) observations of thin horizontal layers with phytoplankton concentrations two to five times above background suggest that, at least in very nearshore areas, copepods may solve the problem by sojourns in such layers. Acoustic profiles do show strong associations of zooplankton reflecting $120\,kHz$ sound associated with nearshore thin layers (Benoit-Bird *et al.* 2009). The fact that hungry copepods can suddenly filter very fast when finally given a meal implies that they are equipped to take advantage of happening into such strata. The acoustic results also suggest that they can stay in the layers and open wide spaces in their own layering around predatory fish.

Perhaps a hundred worthy papers could be cited here to illustrate various refinements and extensions of these observations on copepod feeding. We will consider only a few of the important results:

1 The functional response curve is not very fixed. Copepod ingestion measured soon after collection at a range of food concentrations will tend to become asymptotic at close to the field concentration of phytoplankton (Mayzaud & Poulet 1978). More food will not stimulate more eating immediately. However, copepods collected from high food situations will have higher asymptotes than those from more dilute situations. Thus, the functional response is flexible over some interval longer than the term of the usual feeding experiment (Donaghay & Small 1979).

2 There are feeding thresholds: at low food concentrations, at least of some foods, filtration rates are reduced (*Calanus* – Frost 1975; *Acartia* – Besiktepe & Dam 2002). Thresholds are potentially important because they give the phytoplankton a refuge from annihilation, although that is not likely to be the reason that grazers have thresholds.

3 There are a variety of selection processes. Copepods can eat different sorts of particles at different rates. Many factors affect which available foods will be eaten or at least preferred. Particle size is not the primary determinant of this. Richman *et al.* (1977) demonstrated selection by *Acartia tonsa*, from Chesapeake Bay, feeding on natural particle assemblages. They found that any peak in the particle-size spectrum could be consumed at significant rates, while particles just larger and just smaller were ignored completely (Fig. 7.5). This cannot be done with a feeding mechanism that is essentially a flour sifter, unless particles can be identified and chucked back out after sifting. The feeding mechanism proposed by Koehl and Strickler allows selection, and particle rejections are obvious in the high-speed films.

4 Animal size affects feeding as you would expect: bigger individuals filter faster than smaller ones. Paffenhöfer (1971, 1984) has provided some good comparisons (Fig. 7.6). Saiz and Calbet (2007) reviewed size vs. feeding rate data for copepods in laboratory and field studies. Rates were found to be approximately equalized across a range of temperature, which is an expected adaptive scheme. Maximum ingestion in the laboratory scaled with body weight to the 0.74 power (Fig. 7.7a), a recurring relation of physiological rates to size (but see below). Field feeding rates showed the same maxima relative to size and scattered down to low values (Fig. 7.7b), probably as a function of food availability (lower ingestion when less food is available).

5 The importance of olfactory cues from the phytoplankton is suggested by a variety of experiments, although none of them is very definitive. Poulet (e.g. Poulet & Oullet 1983: "Copepods are French, they prefer foods that taste good.") showed that Sephadex® beads with algal flavor attached are ingested more readily by copepods than those without it.

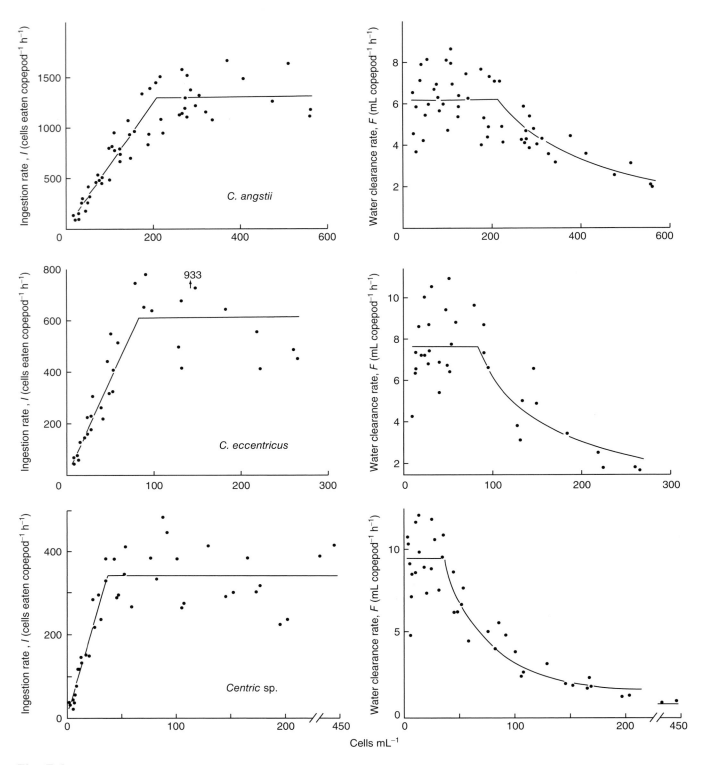

Fig. 7.4 Effect of cell concentration on ingestion rate (*I*, cells eaten copepod⁻¹ h⁻¹, left) and water-clearance rate, (*F*, ml copepod⁻¹ h⁻¹, right) in *Calanus pacificus* females feeding on small-, medium- and large-sized, centric diatoms. All three ingestion asymptotes are at ∼1.1 µg cell carbon h⁻¹. (After Frost 1972.)

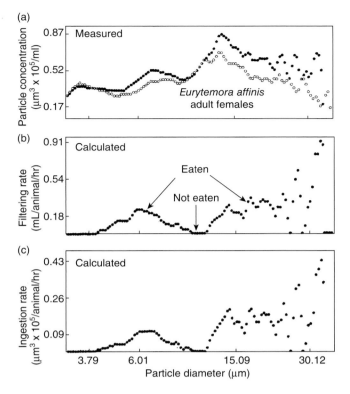

Fig. 7.5 An example of selection by a particle-feeding copepod, female *Eurytemora affinis*, from Chesapeake Bay. (a) Particle-size abundance spectra before (dark circles) and after (open circles) a day of feeding. No particles of about 10 μm equivalent spherical diameter were removed, and the (b) calculated filtering rate and (c) ingestion rate spectra reflect that. (After Richman *et al.* 1977.)

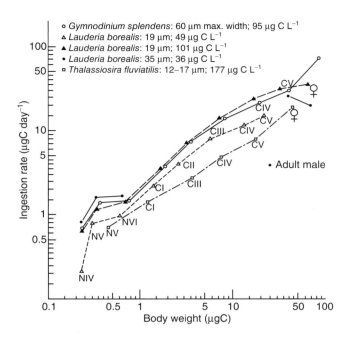

Fig. 7.6 Ingestion rate of *Calanus pacificus* as a function of body weight, all at 15 °C. Filtering rates increase similarly. Naupliar and copepodite stages, indicated by Roman numerals, varied somewhat in weights attained and ingestion rates among four species of food, represented by the separate lines. Ingestion increases roughly as weight to the 0.87 power. (After Paffenhöfer 1971.)

6 In the 1990s, experimentalists pursued a suggestion of Rothschild and Osborn (1988) that small-scale turbulence in the sea should enhance encounter rates between grazers and prey, enhancing ingestion rates. The basic idea is that a grazer and the array of prey around it will have their relative velocities increased by small-scale shear, thus passing more prey within the detection radius of the grazer. Peters and Marrasé (2000) reviewed the literature to that date, finding the results inconclusive. However, in at least some cases there is a dome-shaped relation with increase of ingestion rate at intermediate turbulence levels. Caparroy *et al.* (1998) found a level of turbulent energy dissipation, $\epsilon \sim 0.3\,\mathrm{cm^2\,s^{-3}}$, that enhanced feeding capability compared to calmer conditions in experiments with *Centropages typicus*. That much turbulence enhanced encounters (or capture efficiency, it is impossible to say which) such that ingestion was readily saturated, and clearance rate fell off rapidly with increasing prey density. Higher turbulence seemed to interfere with prey capture. Intense or recurring shear perhaps eroded the feeding current or disrupted transfer of prey location information. It is complex (if possible)

to duplicate ocean turbulence convincingly in laboratory containers at the size scales of plankton animals. For example, in the Caparroy *et al.* experiments, a grid was moved up and down through the feeding chamber. Mixing was high just as the grid passed any point, then decayed. Different mean energy dissipation (estimated ϵ), is achieved by different speeds (and frequency) of grid pulsing. Thus, the effects could have been due to the frequency of extreme disturbance rather than the general level of turbulence.

Nevertheless, turbulence certainly affects some processes. Yen *et al.* (2008) using a quasi-spherical, completely enclosed aquarium of 40 cm diameter and eight pulsing actuators at symmetrical positions have approximately duplicated typical upper-ocean turbulence in terms of energy dissipation and eddy dimensions. Indeed, copepods in this apparatus move more or less independently from water motion up to dissipation rates of $0.1\,\mathrm{cm^2\,s^{-3}}$, above which their motion begins to follow the eddies. The order-of-magnitude agreement with the results of Caparroy *et al.* and others is striking. It can be argued that most turbulence occurs at length scales (basically eddy diameters) that are large relative to the body sizes of the mesozooplankton. This argument comes from estimates of the Kolmogorov length scale, the length at which viscosity effectively damps transfer of

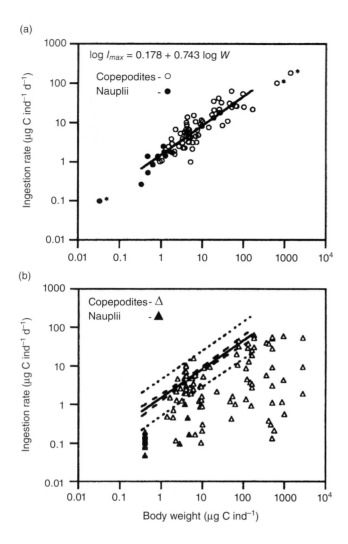

Fig. 7.7 (a) Maximum laboratory feeding rates as a function of body biomass (*W*, carbon) for calanoid copepodites (○) and nauplii (•). The equation is for the fitted line. (b) Ingestion rates measured and compared to the line from (a) for calanoid adults (△) and nauplii (▲) collected in the field eating their field rations. (After Saiz & Calbet 2007.)

momentum into progressively smaller eddies. However, a plankter of whatever length within an eddy will be turned over by it, and many of them, particularly copepods, have preferred orientations but small righting moment (vertical stability). Thus, turbulence will require recurrent righting moves regardless of scale, and sufficient eddy motion will make it difficult to stay oriented.

Other mesozooplankton groups

Filtering-rate determinations for euphausiids, also important in the planktonic economy, are fewer. One by

McClatchie (1985) shows a different approach to the measurement. He used a spherical 250-liter flow-through chamber with a magnetic stirring bar at the bottom. Diatom culture was added to the inflow in pulses, mixed through the chamber quickly, and then diluted away at a rate measured from chlorophyll fluorescence. The increase in apparent rate of dilution with *Thysanoessa raschii* present was the basis for determining their ingestion rate, and thus water-clearance rate, *F*. The euphausiid ingestion rate closely tracked (Fig. 7.8a) the diatom concentration through four-fold pulses, which implied a constant $F = 15.4$ ml krill^{-1} h^{-1}. The functional response curve (Fig. 7.8b), then, increases linearly over the range of food concentration tested. While there must be an upper limit to ingestion rate for this, or any, animal, it wasn't reached at the very high food levels represented by 7 µg chlorophyll liter^{-1}. Results from clearance-rate measures for other euphausiids vary with animal size. McClatchie's *T. raschii* were 17 mg dry weight. The very large Antarctic krill, *Euphausia superba* at ~250 mg dry weight, filters diatom suspension at 25 to 300 ml krill^{-1} h^{-1} (Antezana *et al.* 1982). The lower rates occur at <3 µg chlorophyll liter^{-1}. At 4 to 12 µg liter^{-1} they average 210 ml krill^{-1} h^{-1}. This appears to be a threshold effect, although some feeding occurs even at 0.6 µg liter^{-1}.

Hernández-Léon *et al.* (2001) applied gut-content pigment analysis (Box 7.4) to compare diel feeding cycles of the copepod *Metridia gerlachei* and juvenile (8–21 mm) *Euphausia superba* in Antarctic waters in summer. The copepod ascended to surface layers at night from below 400 m and filled with pigmented cells, carrying some down near dawn. The euphausiids ate pigmented phytoplankton near the surface during daylight, and then dispersed through the water column to at least 600 m at night, switching almost entirely to predation on crustaceans. Intermittent feeding, night vs. day or shorter term, makes application of gut pigments for quantitative ingestion rate estimates perhaps too complex, but gut-content pigments are excellent for demonstration of ingestion cycling between day and night. Karaköylü *et al.* (2009) have developed a laser technique for evaluating plankter gut fluorescence that could be adapted to provide very large numbers of estimates from field samples very rapidly.

Some clearance rate estimates and functional response curves are available for other particle-feeding plankton. Bochdansky *et al.* (1998) determined clearance rates for a large appendicularian, *O. vanhoeffeni,* derived from a modified pigment replacement method. In this animal, the defecation rate is essentially constant, one pellet every ~13 minutes regardless of either small temperature changes or food level. It was observed that the digestive tract consistently contained three fecal pellet volumes of food, and that the pigment in the gut was only 21% of that required to replace it. Thus, the turnover of gut content occurs each 39 minutes, and replacing the pigment requires 4.76

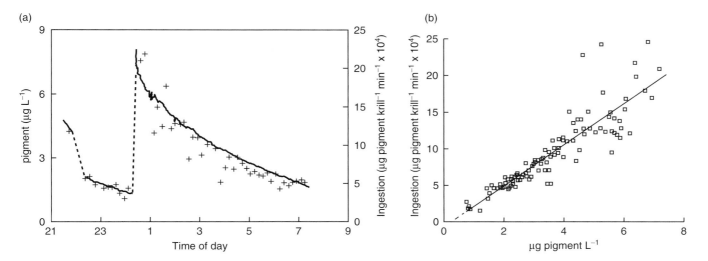

Fig. 7.8 (a) Time course of pigment in a feeding chamber containing 25 specimens of *Thysanoessa raschii*, an euphausiid. A pulse of phytoplankton culture (5.4-fold increase) was provided, then traced as it was both washed out of the flow-through container and eaten by the animals. Ingestion rate is calculated from the implied filtration rate and concentration. (b) Ingestion rate varies directly with food concentration over the range provided. (After McClatchie 1985.)

Box 7.4 Clearance rates from pigments in gut contents

Mackas and Bohrer (1976) developed an ingestion-rate measurement technique that has mostly been evaluated for use with copepods (a similar method exists for fish). An animal eating algae containing chlorophyll will have some in its gut. If that food is suddenly taken away, then the gut chlorophyll, or (chlorophyll + breakdown pigments) will decline at a rate equal to the sum of internal pigment degradation and pigment defecation rates. Typically, this decline with time is a steep exponential curve, but the pattern depends upon the type of animal and whether individual gut clearance or reduction of total gut content of a group is examined. In an animal feeding steadily, the input rate might equal this output rate, sustaining its gut pigment content close to constant. Gut chlorophyll or total fluorescent pigments are readily measured by fluorometry. A captured animal is immediately(!) ground in acetone, extracting the pigments, which have a calibratable fluorescence signal. Standard laboratory instruments give strong signals from the amounts of pigments in one zooplankter, and the fluorescence blanks of animals with empty guts are very small. Considerable averaging among animals should be done, since at any moment a fraction will not be feeding. Sets of animals are also immediately sorted live into filtered seawater, and then sacrificed every few minutes for an hour or more to determine the rate (say, min^{-1} or h^{-1}) of decrease of contained pigment (Box Figure 7.4.1).

Emphasis on sets of animals comes from the fairly wild individual variability (Mobley 1987). The time course of decrease is generally well fitted by an exponential curve. Then it is assumed that pigment in algae must be grazed from the habitat at the same rate in order to maintain the initial chlorophyll content, and the clearance rate is calculated as the volume of water at the ambient chlorophyll concentration that has to be cleared per time to replace the internal pool at the decrease rate.

A great deal of discussion has been occasioned by the fact that some, or often most, of the pigment is actually digested to non-fluorescing breakdown products, with components perhaps assimilated. Some have supposed that this makes the estimates biased. It would matter if dietary pigment input were evaluated by fecal pigment output. But that is not the measurement. Both egestion and destruction are included in the rates based on time-series of whole-body measures. There is no need to multiply the rate by the ratio of chlorophyll (or pigment) ingested to that defecated, thus supposedly accounting for the chlorophyll "lost" internally in crustaceans. However, such a correction can be needed, as in the study of appendicularian feeding by Bochdansky and Deibel, a study considered in the main text.

(Continued)

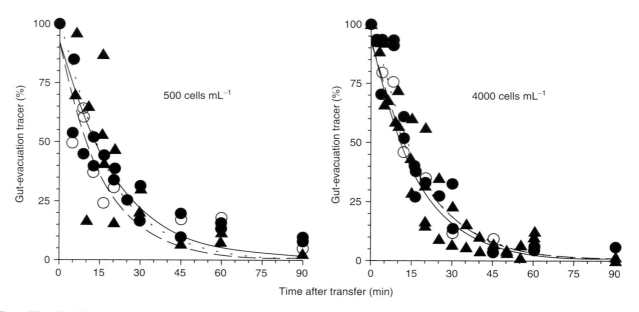

Box Fig. 7.4.1 Time course of decline in gut pigment content (triangles) in *Calanus marshallae* originally feeding on different cell concentrations. Open circles are in filtered seawater. Closed circles are decline in gut pigment content of germanium-68-labeled food after transfer to filtered seawater (open circles) or to unlabeled food suspension (filled circles). (After Ellis & Small 1989.)

There are assumptions to test. Principally, food may be retained in the gut longer when the animal is in filtered water, than while food is available. Thor and Wendt (2010) have shown that assimilation efficiency is greater with low rations, likely associated with longer gut retention. Thus, the reliability of the pigment-loss rate depends upon how fast the animal recognizes that food has been withdrawn. One test (Ellis & Small 1989) showed that the change to longer retention is slow enough not to matter. Rates of decline were checked by feeding diatoms labeled with radioactive germanium in the frustule, then switching to unlabelled food just before the loss-rate determination. This allowed a comparison based on defecation of germanium-labeled frustules. Pigment-loss rates were greater than label-loss rates, which is not a problem. Rates of label evacuation were the same for fed animals as for those switched to filtered seawater, which verifies the assumption that withdrawal of food doesn't change behavior instantly. This result is not consistent; Penry and Frost (1991)

found that the shift to filtered water accelerated defecation, and others have seen that also. Some caution is required, but not correction for digestion of pigment.

A similar approach to ingestion of specific organisms, rather than simply all bearers of chlorophyll, is under development in several laboratories based on quantitative PCR (qPCR) of species-specific DNA (Durbin *et al.* 2007; Nejstgaard *et al.* 2008). Comparisons are made between the copy number of a gene, say mitochondrial cytochrome oxidase I or SSU rRNA, in a potential prey and the copy number of the same sequence in a predator's gut content (the predator can be ground up whole; it will not itself have the sequence used as a primer for prey DNA). Digestion of DNA is proving to occur very rapidly, but qPCR is very sensitive, so the method will at least give qualitative indications that a particular prey species of interest to an investigator (who has primers for a species-specific stretch of DNA) has been eaten. The method has yet to be moved to studies in the field.

(= 1.0/0.21) times that content for each turnover. The clearance rate in field-captured *O. vanhoeffeni* (ml h^{-1}) was determined as the volume of water containing that scaled-up content each 0.65 h (39 min). Rates depend upon body size, measured as trunk length: 40 ml h^{-1} at 3 mm, increasing linearly to 175 ml h^{-1} at 5 mm.

Rates for large specimens are about 10-fold higher than those of copepods. Bochdansky and Deibel (1999) show successive estimates of cell abundance in bottles containing one feeding *O. vanhoeffeni*. Because of the logarithmic scale, the slopes of the lines (Fig. 7.9a) are proportional to the filtering rates. Food levels were allowed to drop, falling

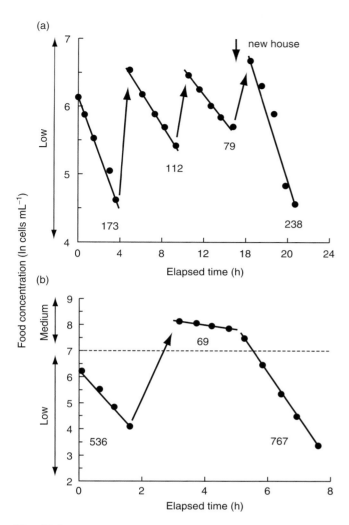

Fig. 7.9 Time course of cell abundance in containers with feeding *O. vanhoeffeni*. Clearance rates in ml h^{-1} are given beneath each downslope. Upward arrows represent additions of concentrated food particles. (a) Specimen feeding at around 250 ml h^{-1}. Its feeding rate declined as its house aged, then recovered when it made a new house. (b) Specimen with high clearance rates at low food concentrations (<1100 cells ml^{-1}), lower rates at higher concentrations, indicating that ingestion becomes saturated. (From Bochdansky & Deibel 1999.)

exponentially in a clearance rate relation (straight line on the semi-log plot), then replenished. At low food levels, filtering is faster, over 500 ml h^{-1} for one tested animal (Fig. 7.9b), than at medium or high levels, at which it cleared at 69 ml h^{-1} (Fig. 7.9b). Evidently, appendicularians have a saturable functional response, saturating for this animal at diatom concentration greater than 10^7 cells ml^{-1}. The time series for one animal feeding at low food concentrations showed progressive slowing as the filters of its house clogged and deteriorated (Fig. 7.9a). Upon replacing its house, a matter of a few minutes, it returned to a high rate. Apparently a house can deteriorate substantially before

replacing it is worth the cost. Actual rates in these experiments varied greatly between individuals – more than was explained by their size.

Protistan feeding and feeding rates

Since about 1985 – surprisingly late – we have been aware of the importance of heterotrophic protists (a term now generally preferred over protozoa) in marine pelagic habitats. They have been shown to consume a larger fraction of algal production than do mesozooplankton, so they constitute a major pathway in the marine food web. Protists are relatively more important in oceanic than in neritic habitats, although abundant and active in both. The delay was occasioned by (i) the intense interest generated by larger grazers, and (ii) the fragility and transparency of oceanic protists. The most-abundant planktonic microheterotrophs are small flagellates. The swimming behavior and feeding mechanisms of these animals are dominated by the properties of viscous flow and by boundary-layer processes. Protists search for food particles using clues from water motion (in the viscous regime of the very small) and from chemicals dispersing from prey. Sensed chemicals can be quite simple, like those amino acids with small side-chains (e.g. proline). Strom *et al.* (2007) have shown that the simpler amino acids can be strong inhibitors of feeding in the tintinnid ciliate *Favella* at nanomolar concentrations. The adaptive value of this is unclear, but it indicates that complex systems for interpreting extremely dilute chemical information are used by protists, their prey, and their predators.

Protists grapple with food particles by several mechanisms. At the smallest scales (microns), e.g. nanoflagellates feeding on picoplankton (both auto- and heterotrophic), hydrophobic forces associated with cell surfaces of both predator and prey can pull prey out of the streamlines around a moving predator, creating contact leading to ingestion (Monger *et al.* 1999). Some flagellates, like the colorless cryptophyte *Katablepharis* (Lee *et al.* 1991) have modestly elaborate "mouths", which can manipulate and engulf food using microtubular organelles that extend into the cell as a digestive tract. Others (choanoflagellates) have feeding collars coated with sticky material to which nanoparticulate food adheres, followed by eventual ingestion at a localized site. Ciliates move to particulate food items, ingesting them with mouth-like organelles specialized for phagocytosis. Ciliate feeding mechanisms are reviewed in Hausmann *et al.* (1996).

Three techniques are often applied to determine the feeding rates of marine protists, often referred to as "microheterotrophs".

1 As for mesozooplankton, prey in a suitable container can be counted before and after a timed interval of feeding

by counted protists. Prey are assumed to be decrease exponentially, allowing calculation of both water-clearance rate (volume per grazer per time) and grazer ingestion rates.

2 The FLB (fluorescently labeled bacteria) technique (Box 5.2; Sherr *et al.* 1987) can be applied to algae (as "FLA"). Fluorescently labeled cells are visible, glowing with fluorescence, in food vacuoles inside the protists. The rate of water clearance, F, by a cell is given by:

$$F = [\text{count of FLB in a cell}]/([\text{FLB concentration}] \times [\text{incubation duration}]).$$

Given suitable averaging and a density estimate for protists from the same slide, an overall rate (volume cleared time^{-1}) of microheterotrophy can be obtained. This is most useful for grazing rates on *Synechococcus* and *Prochlorococcus*.

3 The dilution-series technique (Landry & Hassett 1982; Box 7.5) has been more widely applied than FLB, perhaps because some versions do not require extensive microscopy.

Strom *et al.* (2001) used dilution experiments to show that very large fractions of phytoplankton production are consumed by protists, even in coastal waters and even for relatively large phytoplankton including diatoms. Over a large series of experiments, the ratio g/μ = grazing rate/growth rate was 80% for cells smaller than 8 mm and 42% for cells larger than 8 mm, with an overall average of 64%. They found that large ciliates and heterotrophic dinoflagellates tend to increase as stocks of large phytoplankton increase, consuming larger relative fractions of production at high phytoplankton standing stocks. This is possible in part because many heterotrophic protists, particularly dinoflagellates, can digest phytoplankton externally, inserting a feeding tube into prey cells or surrounding them with a pallial sac.

While all protists are small, relative to adult copepods or appendicularians, they vary over a wide size range and can occupy multiple successive trophic levels. Very large ciliates like *Strombidinopsis* (to 150 μm) can capture and eat heterotrophic dinoflagellates like *Oxyrrhis* or *Oblea* (20–25 μm) that are in turn eating small to large algal cells (Jeong *et al.* 2004). Feeding by individuals at each level follows functional response patterns to varying food availability like those of mesozooplankters (Fig. 7.10).

Ingestion and growth rates rise gradually with increasing prey abundance, whereas clearance rates drop. There are no signs of a threshold of food availability below which clearance rates are low (Strom *et al.* 2000). Protists apparently expend maximum effort seeking food when none is available. As for copepods, small-scale shear (turbulence) can enhance encounter rates between protists and potential prey, as shown for the helioflagellate *Ciliophora* by Shimeta *et al.* (1995). Food-web processes among the very small denizens of the sea can also double back on themselves, with heterotrophs extracting chloroplasts from their prey and maintaining their function for long periods (Stoecker *et al.* 2009).

Some of the importance of microheterotrophy arises because of the rapid potential growth of protists with division rates similar to those of phytoplankton. Thus, the overall rate of grazing can more or less keep pace with phytoplankton stocks. Estimates of division rates for the ciliates *Strombilidium* and *Strombidium* (Montagnes 1996) are mostly in the range 0.6 to 1.0 d^{-1} (recall that one doubling each day equals an exponential rate of 0.69 d^{-1}), about the same as typical phytoplankton growth rates, and can reach 2.2 d^{-1} with saturating food at the highest habitat temperatures. Strom and Morello (1998) found that coastal ciliates were growing at 0.77 to 1.01 d^{-1}. Heterotrophic dinoflagellates increased more slowly, at 0.41 to 0.48 d^{-1}, which is still fast enough to make them significant contributors to grazing. These are maximum rates from laboratory batch cultures, not gross rates in the field and not net rates after predation.

Biological oceanographers are still assimilating the importance of protistan grazing. It is clear that well over half of the metabolic processing of pelagic primary production occurs in the microbial food web. That is true both in oligotrophic regimes far out to sea and in richer coastal zones. Mesozooplankton like appendicularians, copepods, euphausiids, and pteropods take a large fraction of their nutrition from microheterotrophs, acting as "giant" carnivores, and planktivorous fish have stepped out perhaps two trophic levels compared to the place they were believed 40 years ago to occupy.

Evaluation of mesozooplankton production

Next, zooplankton must be fitted into the scheme of production analysis that we started on with phytoplankton. Zooplankton are "secondary" producers, so we want to determine the rates of their production, the amount of new zooplankton tissue elaborated each day or year. However, at least mesozooplankton are not finely dispersed particles like phytoplankton, and they do not take up a well-defined and easily labeled substrate (CO_2), such that they can simply be filtered and their productivity calculated from the content of an incorporated tracer, although that has been tried and may be quite appropriate to protistan zooplankton. Estimation is difficult, and we have no complete sets of values for any defined portion of the community. However, work on the problem generates

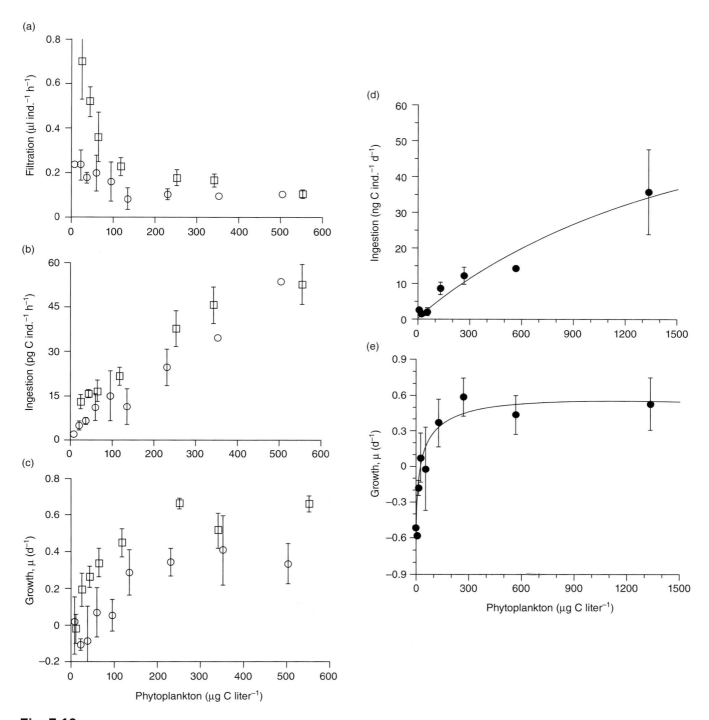

Fig. 7.10 (a–c) Water-clearance ("filtration") rates, ingestion rates and cell growth rates of the pallium-feeding dinoflagellate *Oblea rotunda* as a function of food availability. Foods were the diatom *Ditylum* (□) or the chlorophyte *Dunaliella* (○). (After Strom & Buskey 1993.) (d & e) Ingestion and growth rates of the 149 × 70 μm ciliate *Strombidinopsis jeokjo* feeding on the dinoflagellate *Gyrodinium dominans*. Note carefully the different ingestion-rate units (After Jeong *et al.* 2004.)

Box 7.5 Microzooplankton grazing rates by dilution series (Landry & Hassett 1982)

Dilution series produce bulk estimates of the rate of microheterotroph ingestion, not feeding rates of individual protists. Collect seawater from a station and depth of interest using a clean and gentle (no slamming closures) sampling bottle. Filter a suitable quantity ("F") with membrane filters (0.45 or 0.2 mm pores) to remove all particulate organisms. Remove mesozooplankton from another portion ("–M") of the original sample. Establish a series of incubations of –M diluted in different proportions with F, e.g. 1.0, 0.75, 0.5, 0.35, 0.2, and 0.1 of –M. Determine the per capita rates of increase during an incubation period, say 24 hours, for phytoplankton (usually, but heterotrophic bacteria can also be evaluated) by increase of cell counts, chlorophyll content, or ^{14}C-uptake. The concept is that dilution will not change the per capita growth rates, but it will decrease per capita mortality rates from grazing. These rates are then regressed (see Box Figure 7.5.1) against fraction of –M; the expectation is a negative slope, with higher net (growth – grazing) rates at greater dilutions. The regression intercept is the "true" phytoplankton growth rate (μ, d^{-1}); the slope is the bulk fractional grazing rate (g, d^{-1}).

Problem: derive that relationship

As the figure shows, this often works. There are, however, cases of excess scatter or positive slope (apparent "negative grazing"). The usual practice is to discard such results as bad runs, even if no reason for the problem is evident.

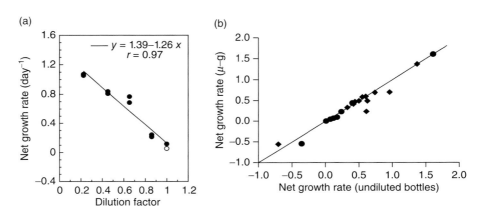

Box Fig. 7.5.1 (a) A dilution-series regression for an experiment from the eastern tropical Pacific. Net growth was determined from chlorophyll concentrations before and after 24 h in shipboard incubators. This is an extreme, so very clear, example, in which net growth in undiluted seawater (dilution factor = 1.0) was nearly zero, and net growth at dilution factor = 0.0 was very large, 1.4 d^{-1}. (After Landry *et al.* 2000.) (b) Comparison for dilution runs in coastal waters (Puget Sound and northern Gulf of Alaska) of net growth in undiluted incubations with the estimated difference between dilution experiment intercepts (μ) and slopes (g). The line is the 1:1 line, showing excellent agreement. (After Strom *et al.* 2001.)

insights about zooplankton themselves. The hope has been offered at times that, if we knew the rates of secondary production in pelagic ecosystems, we could estimate the ecological efficiency of the lower trophic levels and develop an expectation for amounts of production (of fish, squid, shrimp, . . .) that might be harvested from higher levels.

The place of particle grazers in the trophic–dynamic scheme can be presented as follows:

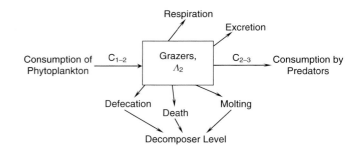

Secondary productivity is the rate of change in the biomass of grazers, $\Delta \Lambda_2/\Delta t$, *plus* the portion of increase of biomass that is balanced by predation, C_{2-3}. Some would add the non-predatory death and molting amounts. This rate is appropriate for a system in steady state, or it is a suitable representation of the production rate averaged over a seasonal cycle. The problem is to measure this rate for all grazers or even for one abundant species. Distinct approaches are sometimes termed growth methods and physiological methods.

I Growth methods

The basic notion of a production measurement for a defined component of an ecosystem, say one growth stage of one species, is to determine its growth rate per unit biomass (called a *specific* growth rate, g, i.e. {mass added} mass^{-1} time^{-1}, say $g\,C\,(g\,C)^{-1}\,d^{-1}$) and to multiply that by the standing stock or biomass (B): $P = gB$. With such measures for all stages of that species, we could determine the production rate for its whole population, using the Boyson (1919) equation:

$$2^\circ P = \sum_{\text{Youngest stage}}^{\text{Oldest stage}} \text{stage growth rate } X = \text{stage biomass}$$

where $2^\circ P$ *is* the secondary production *rate* of the population.

One simple (if labor-intensive) approach is to let this summation happen in an incubation, much like a primary-production estimate with isotopes, the tracer being simply change in biomass measured as dry weight or organic carbon (it is possible to use isotope tracers, but added mass itself will serve). In principle, one removes the predators from an enclosed sample of the habitat, and then observes over a suitable incubation period the immediate increase in herbivore biomass that results. This direct technique has been applied, for example, by Kimmerer (1983), working with the tiny tropical copepod *Acrocalanus inermis* abundant in Kaneohe Bay (Hawaii). He collected water, animals, and their particulate food by dipping with buckets; filtered the water through 333 µm mesh to remove larger animals; picked all stages of *A. inermis* from replicate sets of both initial samples and samples incubated for 20–40 hours; and measured their carbon content with a combustion analyzer. He argued that the ratio ($P:B$) of carbon added in the incubation interval (production rate, P) to carbon biomass (B) is the same as the *specific* rate of increase in B (i.e. $\Delta B/[B\Delta t]$), or equivalently the slope of $\ln(B)$ vs. time. In several of Kimmerer's experiments, the relation of $\ln(B)$ to time was indeed reasonably linear, and almost all the $P:B$ results ranged from 0.2 to 0.32 mg C (mg C d)$^{-1}$. Most of the production was growth of the copepodite stages, since nauplii grow as fast or faster but are relatively much smaller.

Production totals for volumes or areas can be obtained by estimation of the biomass density, say milligrams of *A. inermis* below each square meter of Kaneohe Bay, and multiplying by the $P:B$ ratio. Kimmerer (1983) did not report such a result.

Artificial cohorts

There have been very few studies that precisely followed that "Kimmerer protocol", but a large number have used various approximations to arrive at a guess about the weight increment during incubation (reviewed by Kimmerer *et al.* 2007). Most popular are variations of an "artificial cohort" technique applied to copepods. A gently collected net sample is poured through a fine screen to remove young stages of one or more target species. The animals on the screen are then resuspended in filtered water and refiltered through a coarse screen to remove old stages and predators. Animals passing the coarse screen are an "artificial cohort" (AC). They are given a substantial volume of the fine screen filtrate as both temporary habitat and food supply, and then incubated, sometimes in containers suspended at depth in the ocean (or estuary). With the youngest stages removed, there will be no recruitment to the younger stages in the AC. Stages older than those in the initial AC and appearing during incubation will all have been produced by growth. In one approach, the stage composition is evaluated both before and after the incubation. If incubated for a substantial fraction of the stage development time, some will have molted, generating a difference. The before-and-after stage compositions are assigned biomass estimates by multiplying numbers of each stage by measures of its stage weight. Mean stage weights from field samples were applied by Kimmerer and McKinnon (1987). Using:

$$P = (B_{\text{after}} - B_{\text{before}})/[B_{\text{mean}} \times \text{incubation time}],$$

they found P:B estimates of 0.025 to 0.25 d^{-1} (mean = 0.11 d^{-1}) for *Acartia fancetti* in a tropical coastal embayment, which were multiplied with a time-series of biomass estimates giving 130 mg C m^{-3} yr^{-1}, about 1% of primary production.

A substantial number of AC studies have been done instead by measuring samples of copepodite lengths (which change substantially only at molting) before and after incubations, then used length–weight regressions based on field samples to determine both the biomass and the change in biomass during incubation. This is fraught with issues that are reviewed by Hirst *et al.* (2005). However, for smaller tropical copepods that grow and molt rapidly, and with incubations lasting through, say, all of the copepodite stages (~5 days), these issues are less important. There will be strong changes in length, changes that can be checked at several time intervals. Hopcroft *et al.* (1998) made multiple such estimates in Jamaican coastal waters at 28°C for five calanoid and three cyclopoid species:

SPECIES	$g \pm$ SE (N) (d^{-1}) AT KINGSTON HARBOR (MORE FOOD)	$g \pm$ SE (N) (d^{-1}) AT LIME CAY (LESS FOOD)
Acartia spp.	0.81 ± 0.17 (6)	0.59 ± 0.13 (4)
Centropages velificatus	—	0.85 ± 0.15 (2)
Parvocalanus crassirostris	0.73 ± 0.06 (13)	0.69 ± 0.12 (6)
Paracalanus aculeatus	—	0.78 ± 0.06 (15)
Temora turbinata	0.93 ± 0.10 (5)	0.72 ± 0.23 (8)
Corycaeus spp.	—	0.23 ± 0.04 (6)
Oithona nana	0.65 ± 0.05 (8)	0.46 ± 0.03 (4)
Oithona simplex	0.35 ± 0.04 (8)	0.35 ± 0.03 (7)

The last three are cyclopoids, which grow relatively slowly. The others (calanoids) all have rates at or greater than one doubling in mass per day. That is realistic for small, tropical copepods, since they can go from egg to adult in less than 2 weeks. In the slightly larger species like *Temora*, for which stage-by-stage growth can be estimated, there can be slowing of both relative weight gain in successive stages (from 2-fold for C2/C1 vs. 1.5-fold for C5/C4) and, thus, growth rate varies despite relatively constant stage durations. Larger copepods of more temperate areas like *Calanus finmarchicus* in the North Atlantic grow at overall average rates (Campbell *et al.* 2001) that vary with temperature and rations:

TEMPERATURE	FOOD	GROWTH RATE (d^{-1})
12°C	High	0.28
8°C	High	0.21
8°C	Medium	0.13
8°C	Low	0.09
4°C	High	0.13

Those are laboratory rearing data, but some field data show similar rates and relationships: both food and temperature are determining factors for growth, which is secondary production.

There are other variants of the AC technique; for a worked example, see Renz *et al.* (2008). However, most applications have severe statistical issues and suffer from uncontrolled effects of sample sieving and unnatural maintenance of collected animals.

Boyson's (1919) equation, given above, also referred to as Ricker's (1946) equation, is usually converted to symbols:

$$2^\circ P = \sum_{i=1}^{\text{Adult}} G_i B_i,$$

where $2^\circ P$ is the daily, secondary production of the population, i signifies the life-cycle stage (or any marker with respect to age or size), G_i is the weight-specific growth rate of the ith stage (weight added per weight per day), and B_i is the mean biomass of the life-cycle stage in the habitat. A rather different approach to secondary production is to evaluate growth directly from field observations of the increase in size of individuals in a cohort (see below), sometimes with help from lab rearing to determine stage duration. Total biomass of stages is evaluated from the same field sampling data as numbers × stage biomass. If samples are taken often enough, then these short-term estimates of $2^\circ P$ can be added to give the cumulative production of the population during some suitable period, such as a year or a growing season.

This is very difficult to do convincingly in the ocean. The water and its contained plankton are on the move, and the fauna present on any given date usually are different from those present the day before or the day after. Species and stage composition are usually (not always) too erratic for reliable estimation of growth. Individuals captured today may be younger and smaller on average than those captured yesterday, which probably isn't a biological change in a consistently sampled population. It happens because populations shift with the flow. Observations of B_i are afflicted by large variability, variability so great that estimates must be treated as having multiplicative confidence intervals (½ to two-fold, or worse).

For confined populations in lakes and estuaries, this method has been used with some success, as for example in Landry's (1978) study of *Acartia hudsonica* production in Jakle's Lagoon, a small, enclosed, marine pond of 3.5 m depth adjacent to Puget Sound. Over 2 years (we will only look at one), Landry sampled this small copepod in the lagoon; his collections included the whole water column so that abundance could be estimated on a per unit area basis (Fig. 7.11).

First nauplii (N1) were obviously undersampled, possibly hatching and staying too close to the bottom to catch. Recruitment to N2, mostly determined by egg input, clearly had strong variability in time. This is usually due in part to variability in female abundance and fecundity, but intense and variable mortality of eggs and N1 hatchlings is also important. Whatever the source of variability, its pattern both persists and evolves as the animals proceed through their developmental stages. Landry picked out somewhat

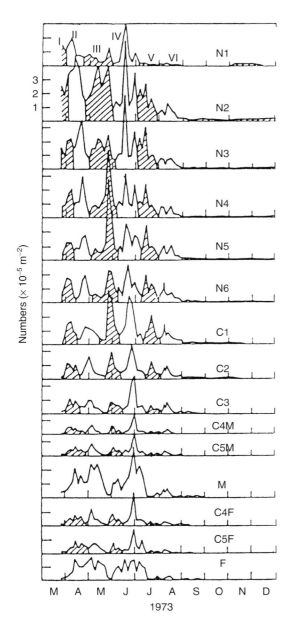

Fig. 7.11 Stacked time-series of stage-abundance estimates for *Acartia hudsonica* in Jakle's Lagoon, Washington State, USA, in 1973. Groups (I through VI) identified as "cohorts" are alternately shaded or white. (After Landry 1978.)

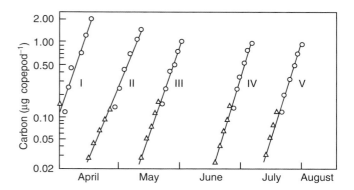

Fig. 7.12 Growth progression (as carbon content vs. time) for *A. hudsonica* cohorts in Jakle's Lagoon. Slopes of these semi-log plots are the cohorts' weight-specific growth rates. Nauplii – triangles; copepodites – circles. (After Landry 1978.)

arbitrary "cohorts" and shaded them alternately in his time-series representation (Fig. 7.11). Carbon contents of each stage in each cohort were determined by CHN (carbon, hydrogen, nitrogen) analyzer. The date at which a cohort was said to be in a given stage was determined as the balance point of the cohort block in each stage time-series, that is at the abundance-weighted mean date. Plotted together, these produced growth curves (Fig. 7.12), the slopes of which are the weight-specific growth rates. Given

sufficient nutrition (which seems to have been available in Jakle's Lagoon), these animals grow exponentially all through development, so the same rate can be used for all stages.

Applying the production equation, the field estimate of biomass in each stage on each date was multiplied by the cohort growth rate and summed. Estimates of egg production by females, often an important part of secondary production by a planktonic species, were added. This gives a time-series of production by the copepod stock (Fig. 7.13) in units of carbon incorporated in organic matter and assimilated into new copepod tissue and eggs. Naupliar growth contributed 15%, copepodite growth 47%, and egg production 38%. Production had clear seasonality, deriving mostly from variation in stock size. Total annual production was $\sim 8\,\mathrm{g\,C\,m^{-2}}$. *Acartia hudsonica* is the only abundant mesozooplankter in Jakle's Lagoon, so its production must represent most of the "secondary" production. At $\sim 70\,\mathrm{mg\,C\,m^{-2}\,d^{-1}}$ during summer, that production was probably a significant fraction of the primary production. That wasn't measured but was likely on the order of $200\,\mathrm{mg\,C\,m^{-2}\,d^{-1}}$. Clearly, much of the primary production is consumed by something else. In a shallow lagoon, the benthos would be important, partly eating settled phytoplankton, partly just filtering the water column above.

There are few similarly detailed studies of the Boyson type. A critical issue is the lack in the ocean itself of stationary populations. Students interested in the approach should examine Uye's (1982) estimation of the production of *Acartia clausi* in Onagawa Bay, Japan, a somewhat more open system than Jakle's Lagoon.

More about growth

Vidal (1980) provided a thorough study of the responses of copepod growth to temperature, food availability, and

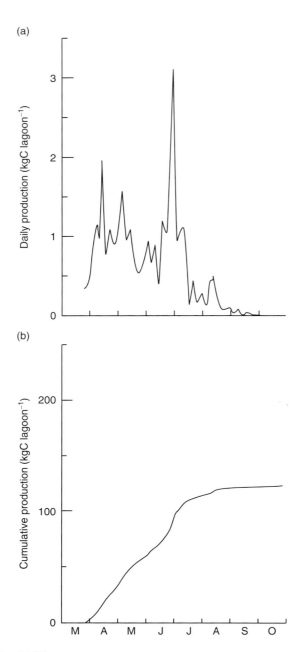

(a)

(b)

Fig. 7.13 Daily (above) and cumulative production of *A. hudsonica* in Jakle's Lagoon. Oscillations in daily production mostly reflect variations in abundance of biomass, since growth rates of all cohorts were quite close (Fig. 7.14). (After Landry 1976.)

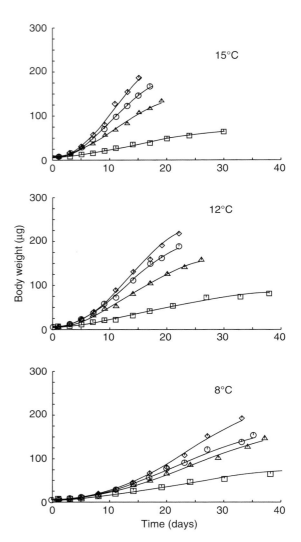

Fig. 7.14 Body weight of *Calanus pacificus* at different ages as a function of temperature and of available food as parts per million by volume (ppm): 0.67, □; 2.28, △; 4.70, ○; and 9.39, ◇. (After Vidal 1980.)

individual size. The animal was *Calanus pacificus* collected from Puget Sound, where it is the dominant, herbivorous zooplankter. Gravid females were collected in the field, maintained at the temperature of collection, and left to lay eggs. After some acclimation, eggs of ages even as young as 12 hours were collected. These were reared with heavy feeding at 12°C through the naupliar stages. Then they were transferred to jars held at 8, 12, and 15.5°C and fed

at various food (*Thalassiosira* spp.) densities, which were reset to the desired levels twice each day. Along with general aquarium keeping, samples were removed every 1 to 6 days, depending upon size and stage, for evaluation of stage composition and growth. Vidal fitted the growth data by the Chapman–Richards equation representing the time course of weight increase:

$$W_t = W_{max}(1 + Be^{-kt})^{-m},$$

where W_t is the dry body weight of a copepod at time t; W_{max} is the maximum weight attained at maturity; and B, k, and m define the initial weight, slope, and inflection point of the sigmoid curve of weight vs. time. Size does increase in sigmoid fashion with time as described by the equation (Fig. 7.14).

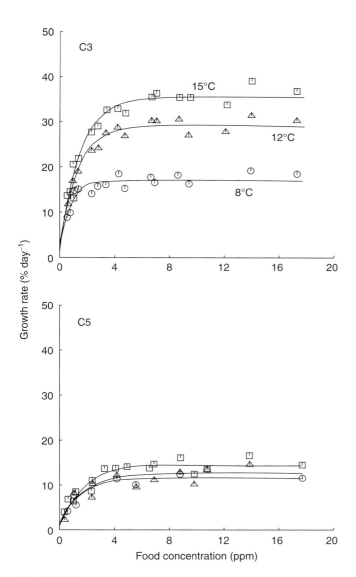

Fig. 7.15 Growth rates of two copepodite stages of *C. pacificus* at three temperatures (8°, ○; 12°, △; and 15°, □) plotted vs. food concentration. (After Vidal 1980.)

160 μg C liter⁻¹), there is a very strong effect of temperature on growth rate: it is faster in warmer water. The effect of temperature on size at a given stage is progressively less for older stages, but colder water mostly produces somewhat larger adult sizes. That is commonly but not always observed for plankton in the field. Higher temperatures allow a considerably faster start, especially for the change from 8° to 12°C. However, the effect of higher temperature on progress through the stages (development rate) is even greater, so that final adult sizes are smaller at warmer temperatures.

Vidal combined his results into a complex predictive equation for growth (secondary production) as a function of body size, temperature, and food availability. In principle, the equation could be used, given measures of *T*, food concentration, size distribution and abundance of *Calanus pacificus*, as well as phytoplankton food availability, to determine secondary productivity of this copepod in Puget Sound. However, an unsolved problem is how to estimate food availability in the field in the same way that the copepods see it. We cannot do that. As a result, secondary-production studies, no matter how rich in careful laboratory rearing results, often end without reaching a satisfying *field* result.

II Physiological methods

The secondary-production diagram is an input–output relation, so one approach is to estimate all of the input and output rates and calculate from them the secondary productivity. It is also necessary to measure the population biomass of the animal or animals of interest, a measure usually afflicted with variability on the order of half to double or greater. This is called the *physiological method*. It turns out to be rather theoretical, since practically all applications to the field lack one or more variables and some *ad hoc* fix is applied. For a given species, we need estimates of ingestion rate (*I*), defecation rate (*D*) and thus, absorption efficiency (*A* = [*I* − *D*]/*I*), respiration (*R*), molting (*E*, for ecdysis), and mortality (*M*). With estimates in hand, production, 2°*P*, would be calculated by:

$$2°P = IA - R - E - M.$$

Units can be dry weight or mass of carbon added (increase of *Λ₂*) per day or per year. As for direct estimates of growth, production estimates are often expressed as a ratio of increase to standing stock, say, g C g C⁻¹ d⁻¹, known as the *P:B* ratio. At steady state or on average,

$$2°P = B[P:B].$$

All of these rate variables will change with animal size and age, so we cannot take a sample of zooplankton, weigh it,

Food, temperature, and body size all affect the growth rates. Individuals do not grow so rapidly, or so large at a given development stage, with less food as they do with more (which is not a surprise!). On low rations, the slope of the curve (dW_t/d*t*, the growth rate at a given size) is more affected by food availability than by temperature only (Fig 7.15).

The effect of food on size at a given stage is hyperbolic in form (Fig. 7.15), becoming asymptotic at about 4 to 6 ppm of phytoplankton by volume, equivalent to 160 μg C liter⁻¹. Clearly, growth does have a required food level, a nutritional threshold. This minimum doesn't change much with temperature but is surprisingly low, perhaps 50 μg C liter⁻¹. With full nutrition (above about and get secondary production from relationships of I, A, R, E, and M to habitat variables like food availability and temperature, although recent attempts to make that work are reviewed below. All processes must be considered in detail with respect to size and species composition. The most difficult process rate to estimate is mortality, M, especially M as distinct from $C_{2–3}$. However, in principle, if we could get rates for enough components (species, stages, times of year) then we could add everything up and estimate secondary productivity.

Probably many zooplankton workers will disagree, but there seems to be no physiological estimate of secondary production that is good enough to present as a well-worked example. Part of the difficulty is uncertainty about what and how much animals actually eat in the field, i.e. which of the available particles they actually ingest. Thus, the "IA" term is unreliable. For example, an ingestion rate obtained from the phytoplankton pigment method will neglect nutrition obtained from colorless herbivorous protists, which are major components in the diets of many mesozooplankton. Another problem is the necessity to rely upon laboratory data, considered here mainly for respiration, which in this context is a general term covering catabolism of organic matter and excretion.

Respiration

The respiration term is generally measured as oxygen consumption rates in sealed containers over intervals of hours or a day. Change in oxygen concentration is determined by Winkler titrations, with electrodes or (lately) with optodes. There are complications in selecting and interpreting oxygen consumption data. Immediately after capture there is generally a strong spike in oxygen consumption. This may be due to capture stress, or it may be close to the real metabolic rate. In laboratory containers of any size small enough for a reasonably low density of animals to provide measureable oxygen uptake, their ambits are restricted and no predators can give chase. These effects may possibly explain the drop in rates, rather than recovery from capture. There is no certain way to tell which rates are applicable to the field. Fed and unfed animals have different rates, a well understood effect. Tsutomu Ikeda, who generated a great part of the available data, has generally preferred to ignore (or not measure) the high initial values, assuming they represent stress, and starts measurements 12 hours after capture. Subsequent oxygen utilization rates are reasonably steady in fed animals, slowly declining in starved animals. Nitrogen and phosphorus excretion rates are also high initially, but they continue to drop for several days, even weeks, after oxygen use stabilizes. The difficulty of assuring normal measurements makes the physiological approach to secondary productivity problematical. However, useful approximate measures can be produced.

Many studies have been done of zooplankton respiration rates, which vary with temperature and body size. Ikeda (1974, 1985) has shown these relations over wide ranges of plankton type (seven phyla) and size in both the tropics and polar zones (Fig. 7.16).

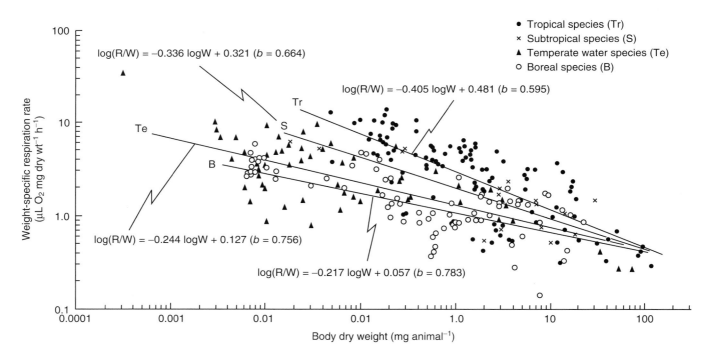

Fig. 7.16 Weight-specific oxygen consumption of zooplankton individuals from seven phyla as a function of body weight, grouped by region. Tropical plankton respire faster at a given weight and respond more strongly to body size than boreal plankton. (After Ikeda 1974.)

Oxygen consumption *per unit weight*, measured by before-and-after Winkler titrations, decreases with body size. That is true of animals in general, of course. Hummingbirds have higher metabolism per gram than eagles, and babies have higher metabolism per gram than parents. For zooplankton the decrease is more pronounced in the tropics than in high latitudes. In the figure the scatter is fit by regression lines equivalent to:

$$\text{Respiration/Weight} = a(\text{Weight})^{(b-1)}$$

or:

$$\log(R/W) = (b^{-1})\log W + \log a$$

That derives (divide by Weight) from Rate $= a(\text{Weight})^{b}$, an allometric relation applicable to many biological rate-to-size comparisons. At least for the data as Ikeda grouped them, b (slope) decreases and a (position) increases with habitat temperature (boreal $\sim 8°C$, tropical $\sim 28°C$). The latter effect is obvious: warmer conditions drive most processes faster, including biological rates. However, respiration differences are not as large between regions as they would be for a single species experiencing such large temperature changes. That is because species of cold and warm habitats adapt to produce rates of movement and activity as closely similar as changes in biochemistry allow. For example, speed capabilities for predator attack and prey escape will be pushed by selection toward hydrodynamic limits not set by temperature. Change in slopes between regions is not altogether understood. For some reason, log[respiration] compared among species in the tropics and subtropics is almost proportional to body surface area ($b \sim 2/3$), while in higher latitudes it is more affected by body mass (b closer to three-quarters). The change in slopes represents an interaction of size with habitat temperature such that smaller organisms are more affected than large ones. Predicted shift in respiration rate between tropical and boreal habitats is three-fold for 1 mg animals but only 1.1-fold for 100 mg animals. Huntley and Boyd (1984) provided major-axis regressions for these data, and the relative slopes and positions were unchanged. A substantial literature and polite (if heated) controversy on the explanation (e.g. West *et al.* 1997) and variation (e.g. Glazier 2005) of allometric scaling of metabolism have appeared in recent years.

A very large part of the available data are for copepods (as in everything else about mesozooplankton). Ikeda *et al.* (2007) summarized the effects of temperature, body mass, and oxygen availability in three graphs (Fig. 7.17). For epipelagic copepods, the temperature range from -2 to $28°C$ corresponds to an increase in mean, weight-specific respiration a little greater than three-fold – not really very great. Again, this reflects the adaptation of metabolism to provide maximum potential for movement and growth at all temperatures. Total respiration "corrected" (by a Q_{10} function) to $10°C$ varies with the three-quarter power of the dry mass. Mesopelagic and deeper copepods respire less, which appears to be partly a function of relatively low oxygen availability (but, compare Childress *et al.* 2008). Similar data and conclusions exist for euphausiids, and a few other groups.

Thus, in a general way, respiration is well described. On the other hand, in order to use the fitted equations in a production study, you have to tolerate errors of about five-fold, as indicated by the scatter (Fig. 7.17b). With an

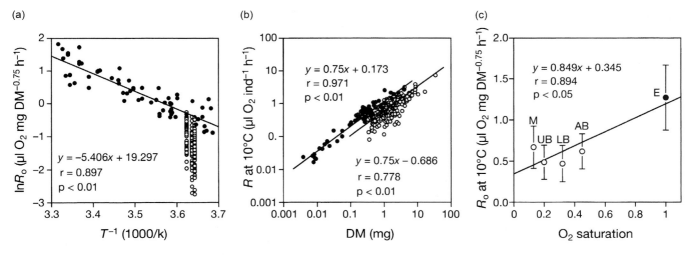

Fig. 7.17 Results for diverse (•) epipelagic and (○) mesopelagic or deeper copepods collected from the ocean and incubated at the temperature of collection. (a) Oxygen consumption rates vs. the Arrhenius function of temperature (translated also to Centigrade). (b) The same data standardized with a Q_{10} estimate to $10°C$ plotted vs. body dry weight. Slopes, 0.75 of the log–log regressions, are typical of metabolic rates relative to individual biomass. (c) Grouped means relative to habitat oxygen saturation (both greater in epipelagic habitats (•) than in meso- [M], upper [UB] and lower [LB] bathypelagic and abyssopelagic [AB] zones). (From Ikeda *et al.* 2007.)

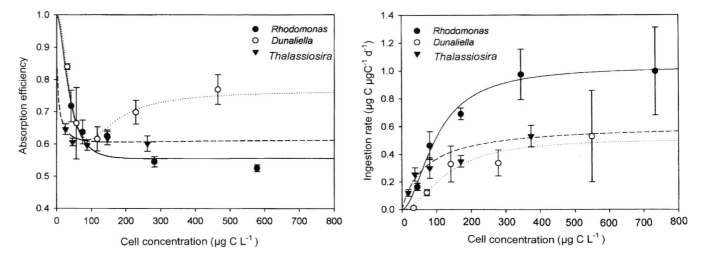

Fig. 7.18 Ingestion rates and absorption efficiency of *Acartia tonsa* on three different algae. (After Thor & Wendt 2010.)

oxygen-to-carbon conversion factor and very broad-scale averaging of biomass estimates by region, Hernández-León and Ikeda (2005) have used a sweeping compilation of such respiration data to evaluate the role of mesozooplankton globally in the world ocean carbon cycle. They came up with ~13 Gt C yr^{-1}, 25–30% of likely global marine primary production. Three-quarters of the grand total respiration occurs above 200 m. This is a high proportion, given recent estimates of bacterial and protistan respiration. However, it is unlikely that we will ever exactly determine the roles of different heterotrophic categories in the bulk metabolism that must roughly balance primary production at the annual and global scales.

Absorption efficiency

Measures of ingestion rates were discussed above. Food inside the gut is not, however, yet inside the animal, the topology of which is really toroidal: the lumen is outside the body like the hole in a donut. To get food inside, the animal breaks it up with grinding and enzymes into small molecules for absorption. Components that do not break up, or are grabbed by gut-dwelling microbes, are defecated, D. The ratio of absorbed to ingested food is the absorption efficiency, AE = $(I - D)/I$. "Assimilation" efficiency is a term often used for this ratio, but full assimilation involves more steps than just getting nutrient through the gut wall. Both I and D can be amounts or rates; AE is unitless, a fraction. There are only a few studies of AE in mesozooplankton. Thor and Wendt (2010) showed with the coastal copepod *Acartia tonsa* that AE varies with the type and amount of food. To measure AE, they used phytoplankton cells labeled with two radioisotopes, ^{14}C and ^{51}Cr. The carbon labels the organic matter and is partly absorbed, the chromium adheres to the outside of the cells and mostly is not

absorbed, with the $AE = (1 - \Phi_{\text{feces}}/\Phi_{\text{prey}})$, in which $\Phi = {}^{14}\text{C}/{}^{51}\text{Cr}$ ratios in feces and food (algae). Derivation is left as an exercise. They also measured ingestion rates by cell counts before and after some grazing. On *Rhodomonas* (Fig. 7.18) the AE declined with food availability; on *Thalassiosira* it was roughly constant; and on *Dunaliella* it varied in an odd but definite pattern. For some foods, at least, digestion is more thorough if less food is available. The range from ~50 to 85% is broad enough to distort general estimates of secondary production based on single AE values, like those generally applied in NPZ models. Besiktepe and Dam (2002) have shown that fecal pellet size and production rates vary in parallel with the shifts in AE.

Is mesozooplankton secondary production controlled by temperature or food availability?

Reviewing the approaches to secondary production makes clear how difficult it is actually to measure all of secondary productivity, or even tissue production by one species population. Obtaining field estimates can depend upon measurements that we have no means to make (as for applying Vidal's results) or upon assumptions difficult to evaluate. Still, we would like a measure of how much production there is of, say, mesozooplankton, or even just the copepods. To repeat, we would like to know how much food the ocean produces for animals that we harvest like herring, anchovy, cod, and pollock. In order to make predictions, perhaps about the effects of climate change, we need to know what controls the rates. The likely candidates are temperature and food availability. Several short-cuts to evaluating control have been suggested and rather extensively studied. We will examine two.

The Huntley–Lopez model

A paper by Huntley and Lopez (1992) caused a minor storm in the community of zooplanktologists. They suggested, based on recalculations of sets of old data, that copepods, and possibly other zooplankton, grow everywhere at all times at rates controlled primarily by habitat temperature. First, they made the broad assumption that copepods grow exponentially at a nearly constant rate from hatching to maturation. This is almost true for some copepods; see, for example, data from Lee *et al.* (2003) for *Pseudocalanus* at full nutrition. Huntley and Lopez surveyed the literature, finding data from many species at all latitudes for the mass of an egg, for adult mass (carbon or dry weight), and for development time, *D*, from egg to adult. In applying their assumption, they used the equation:

$$\text{Adult Mass} = \text{Egg Mass} \times e^{gD}$$

to obtain a growth-rate estimate, *g*.

The results (Fig. 7.19) showed two seemingly odd things: (i) both *D* and *g* were independent of body size, and (ii) both appeared to depend very strongly on temperature. Both of these seemed odd because "everybody knows" that physiological rates depend upon body size, being approximately proportional to $W^{0.7}$ (where W is weight), or some similar rule, and because everyone thought of growth as just as much dependent upon food, which varies in time, as upon temperature. Thus, Huntley and Lopez were suggesting basically that food variation is much less important than temperature variation.

They fitted a function (shown in Fig. 7.19b) to their growth-rate curve:

$$g = 0.0445\, e^{0.111T},$$

where *T* is the temperature in degrees Celsius. As in the Boyson equation, *g* is the rate of growth *per unit biomass*, that is, mass added per unit mass per day. Next, leaping tall obstacles at a single bound, Huntley and Lopez suggested that these rates could be applied to general net-tow biomass estimates. Go to the field, estimate the carbon weight of mesozooplankton caught with nets (gC m^{-3}), measure the water temperature, then apply the equation:

Rate of 2° production (g C m^{-3} d^{-1})

$= \text{Biomass (g C m}^{-3}) \times 0.0445 e^{0.111T} \text{(g C(g C)}^{-1}\text{ d}^{-1}).$

The approach remains the subject of review and argument. Huntley and Lopez in their development chose some data rather cannily to show that attempts at "physiological" estimation of growth rates were wildly variable. For example, from some studies examining the effects of food, they included data in their evaluation ranging from full nutrition down to none. This gave a much wider range of

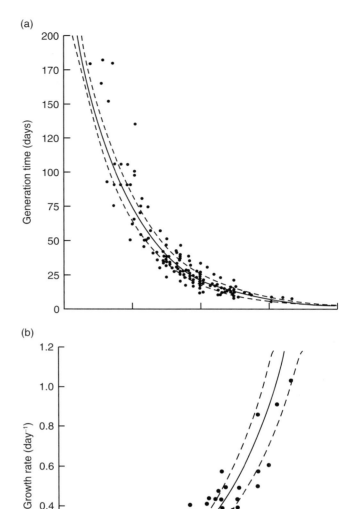

Fig. 7.19 (a) Time from egg to adult for copepods of many species and adult sizes from a wide range of latitudes as a function of temperature. (b) Relationship, for the same copepods, of body growth rate (estimated from a simplified exponential growth relation) to temperature. (After Huntley & Lopez 1992.)

g estimates than could fairly be expected from field results. In a response to critics, Huntley (1996) made an interesting point: that the goal of the 1992 paper had been to move the issue from laboratory growth studies to getting good estimates of biomass. We cannot do the latter, yet, since net-tow, acoustical and other estimates of zooplankton abundance are strongly affected by patchy distributions. Growth-rate estimates are unlikely to be off much more than two-fold, while biomass estimates are generally uncertain by a factor greater than two-fold, sometimes 10-fold.

"Hirst" models

Were Huntley and Lopez right? Is temperature the dominant control on a global scale? Several papers by Andrew Hirst and colleagues have looked at the literature from a different perspective. These authors began with a critique of the Huntley–Lopez analysis, saying that it assumes that growth is always food saturated. In fact, Huntley and Lopez only claimed that, over the full range of latitude, the data show strong dominance of variation in g by temperature. If you examine the Huntley–Lopez graph, there is still more than two-fold variability on the g scale at any temperature, providing lots of room for nutritional and body-size effects. Hirst and Bunker (2003) extracted from the literature all of the stage-specific growth rates that they could find. The usual technique in the papers that they reviewed was related to the artificial cohort method discussed above: sample the zooplankton, count out a large number of living individuals of a given stage from one species (say, copepodite 3 of *Calanus finmarchicus*), dry and weigh a few (W_{start}), let the rest feed and grow for a day or two (t), then dry and weigh those (W_{end}). Alternatively, carbon content was measured by combustion. Usually, growth, g, in copepods is roughly exponential over intervals ranging from the whole of development (*Acartia*) down to a stage or so (e.g. results for *Calanus* in Vidal 1980), so they calculated exponential growth rates as $g = 1/t \ln(W_{end}/W_{start})$. The assembled results show the expected temperature dependence (Fig 7.20a).

Development rate (1/[egg to adult duration]) varied from ~0.01 d^{-1} at 2°C to 0.1 d^{-1} (and 0.2 d^{-1} for some very small species) at 28°C (Fig. 7.20a). The modest scatter agrees with Huntley and Lopez. Mean weight-specific growth rates of both freely spawning (Fig 7.20b) and egg-carrying copepods increase from ~0.1 to ~0.5 d^{-1} in the range 7 to 28°C, again in agreement, but with very large scatter, especially for free-spawners. Rates below 0.1 d^{-1} were likely to be food limited. Rates above the regression line approach 1 d^{-1} above 10°C and would be primarily those for small copepods, in modest disagreement with Huntley and Lopez. Significantly, growth was nearly independent of body mass up to ~20 µg carbon, falling off somewhat for older stages of large species, most of which in the data were free-spawners. Finally, both free- (Fig. 7.20c) and sac-spawners showed strong effects of food availability, approximated as chlorophyll concentration near the depth of capture. However, even at very low chlorophyll values, many species showed growth near the asymptote of Hirst and Bunker's hyperbolic function. All in all, the Huntley and Lopez analysis comes out reasonably well.

Having found definite effects of several factors, Hirst and Bunker offered multiple regression formulas for free

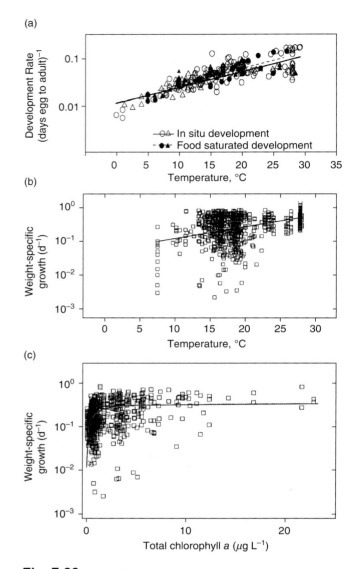

Fig. 7.20 Relationships of development rates and growth of pelagic copepods to habitat factors and body mass. (a) Inverse of time from spawning to final molt to adulthood ("development rate") vs. temperature". Circles, broadcast spawners; triangles, sac spawners; open symbols, field data; filled symbols, maximum rations (laboratory). (b) Weight-specific growth rate ($\Delta W/W\Delta t$) of broadcast spawners vs. temperature. (c) Weight-specific growth rate of broadcast spawners vs. chlorophyll concentration at the depth of collection. (After Hirst & Bunker 2003.)

and sac spawners: $\log_{10}$(growth rate) as functions of temperature, body weight, and, for free-spawners, chlorophyll. To get at secondary production, one would measure T and chlorophyll concentration, appropriately analyze a plankton sample into classes by spawning mode and body weight, and then apply them in those equations. However, the equations "explain" (R^2 values), respectively 39% and 29% of the variances, so there is considerable room for variation among species, seasons, and locations. Finally, one would sum the products, g_iW_i, over all body weights, obtaining a secondary production estimate for copepods.

Such an analysis has been done by Roman *et al.* (2002) using a similar equation (not including chlorophyll) from Hirst and Lampitt (1998). They used a series of sieves to partition 200 µm mesh plankton samples (0–200 m) from time-series stations (HOT near Hawaii and BATS near Bermuda) into five size categories, determined their carbon biomass, and ran the calculation. There was seasonal variability driven by both T and BW. Averages for samples over four years were 13 and 6 mg C m^{-2} d^{-1}, respectively, at HOT and BATS. These were ~2.7% and ~1.3% of mean measured primary production (1°P) available for transfer to predators. Perhaps these are reasonable estimates for the very large subtropical oceanic areas. The amount of food ingested would have been on the order of three-fold greater, 9% and 4% of annual 1°P. Compare those estimates to the global estimate, mentioned above, of 25–30% of 1°P of Hernández-León and Ikeda (2005). Additional copepod production as egg output was not included. Similarly approximate estimates for higher latitudes are scattered in the literature. Zhou (2006) shows an alternate approach in which community biomass is partitioned by size (using automated, optical, size-determining counters) into a spectrum (numerous small animals grading progressively to few large ones) and production is calculated from sundry approximations.

As can be seen by examining the scatter in Fig. 7.20, we have not solved the problem of quickly and reliably estimating mesozooplankton productivity in the sea. Perhaps some averages are of the right order of magnitude, but estimates from different perspectives come up with very different fractions of measured primary production.

Secondary production by microheterotrophs

There is, as yet, no direct observational technique for determining the rate of organic-matter assimilation (growth) in microzooplankton, primarily protists. Landry and Calbet (2004) provide a scheme for broadly averaged estimation of the rates, which we discuss in Chapter 9 on pelagic food webs.

Chapter 8

Population biology of zooplankton

Population dynamics of protists, including phytoplankton, are relatively simple. A cell divides into two daughter cells; each runs risks of being eaten or fatally infected, then divides again. If the risks are small, then the population grows exponentially at a rate dependent upon the intervals between divisions and upon the risks. If the risks are large, then the population declines. There can be complications introduced by occasional mating, auxospore formation in diatoms, resting spore formation in dinoflagellates, and other life-cycle variants. In comparison, the life cycles of metazoans are consistently complex. So, their population dynamics involve longer delays between reproductive periods, chancy mate-finding and mate-selection, in many cases repetitive and multiple reproduction by single females, changing risks of death as development proceeds and elaborate strategies for surviving periods of bad conditions. This applies to zooplankton, benthos (annelids, isopods, clams, . . .), nekton (tuna, squid, porpoises, . . .) and seabirds. We cannot cover the details for all of these; each is a specialized study. To provide examples of the sorts of facts, measurements, and calculations involved, we present some zooplankton studies that are familiar to us. In every animal group, reproductive intervals, fecundity, longevity, and age-specific mortality all vary at the levels of species, regional populations, and generations, because it is those aspects of adaptation that are most strongly and immediately tuned by natural selection, and most strongly affected by short-term conditions.

A good deal of mathematics has been developed describing population dynamics. Unfortunately, for most organisms, including zooplankton, the assumptions do not fit well enough to make that mathematics useful. For example,

plankton animals in middle and high latitudes do not tend toward unchanging ("stable") age- or stage-frequency distributions. Thus, the classical calculations of population-increase rates, which depend upon that stability (e.g. calculation of "little r", the intrinsic rate of natural increase, from $1 = \Sigma r^{-x}l_x m_x$, the sum over all ages, x, of survivorship to x times reproduction at x) will not work. In tropical areas, where stable age distributions are more likely, if not adequately proved, r is usually close to zero and oscillates around it. So, all of the great math for evaluation of population-dynamical parameters is useless. Sorry. Our information about abundance cycles is better for many species than our evaluation of the component birth and death rates that produce the cycles. Copepods, with their fixed sequence of stages and their reliable presence in samples from all marine waters, offer advantages for studies of population dynamics. So, much of the work focuses on them.

Most mid- and high-latitude plankton have a population cycle involving a resting stage or period of quiescence. During this phase of the life cycle, all individuals stop growing in one or a few adjacent life stages. In the copepod genus *Calanus*, it is the fifth copepodite (C5) stage, and to lesser extents C4- and C6-adults, which accumulate copious lipids and then swim down out of the near-surface growth habitat and rest for months, a phenomenon called *diapause*. When unfavorable conditions approach, estuarine and nearshore copepods like *Acartia* produce diapause eggs that accumulate in sediments. The eggs hatch in a burst when good conditions return with the cycle of the seasons. Recurrent stopping of population processes with all individuals in just one life stage creates persistent cycling of

Biological Oceanography, Second Edition. Charles B. Miller, Patricia A. Wheeler.
© 2012 John Wiley & Sons, Ltd. Published 2012 by John Wiley & Sons, Ltd.

stage and age composition. Zooplankton population dynamics must be studied in the face of that cycling, but can also take advantage of it. It involves specialized estimation procedures and lately modeling of reproduction, development, and mortality.

Mortality rates are the most difficult parameters to estimate, and with exceptions they become a row of free dials on numerical models, one dial for each distinctive stage. We simply turn them until the model output looks like our time series of stage-abundance data. Possibly, this will prove to be the most effective scheme for obtaining mortality estimates. Development timing is obtained from rearing experiments, or by following cohorts as they develop in the field. Those have been adequately covered under secondary production (Chapter 7). The experiments are flawed because we cannot exactly mimic the nutritional situation in the field. Cohort following is flawed by sampling statistics with staggering variance; by advection moving variations in population timing past a sampling site (thus, development can sometimes appear to back up, the age distribution shifts toward youth); and by the fact that the progress of cohort age- or stage-structure is confounded in field data with effects of mortality. Reproduction, however, is readily measured, and it has become very popular to measure it.

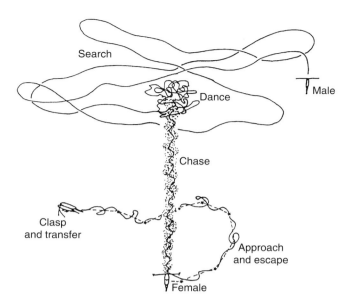

Fig. 8.1 Mate-finding behavior in *Calanus marshallae*. The female slowly sinks, leaving a vertical trail of pheromone. The male searches horizontally for trails. When he finds one, he does a lively dance in a space about 12 cm across, then follows the trail down to the female. He bumps her, she moves, he follows, eventually there is an embrace and spermatophore transfer. (After Tsuda & Miller 1998.)

Reproductive biology and fecundity

For most kinds of zooplankton, even the various hermaphroditic forms, reproduction begins with mate finding and copulation. In some groups, females advertise their readiness for mating with chemical signals (pheromones) dispersed in the water. Males follow the signals, find the source, and initiate mating. Among copepods, *Calanus* (Fig. 8.1), *Temora*, *Pseudocalanus*, *Centropages*, and *Oithona* attract and find mates in that fashion (Weissburg *et al.* 1998; Kiørboe & Bagøien 2005). Many other species must also, given the impressive array of chemosensory setae (aesthetascs) on the antennules of adult males. Indications for a few genera are that pheromones are not particularly species-specific. For example, Goetze (2008) has shown that males of both *Temora* and *Centropages* will track pheromones from females of congeneric but definitely distinct species, at least in laboratory containers. Some of this "heterospecific" trail-following in these partially sympatric pairs proceeds to mate capture.

The search process enforces much higher mortality rates for males, which, by being constantly on the move, encounter not only females but the attack perimeters of ambush predators like chaetognaths. To reduce this mortality (and save energy) *Calanus* males search (Fig. 8.1) only when low levels of female attractant pheromone are present, indicat-

ing females someplace nearby. At least some species of *Acartia* do not detect pheromones of potential mates, but follow their shorter trails of hydrodynamic disturbance from swimming (Kiørboe & Bagøien 2005). Kiørboe and Bagøien have modeled mating rates by modification of the encounter rate (ER) equation: ER = β($\female$ density)($\male$ density). The rate constant, β, is partly a function of male swimming speed and accounts for both meeting frequency and fractional success of matings per encounter. Since hydrodynamic disturbances from swimming are shorter lived, and thus less extensive than pheromone tracks, the *Acartia* strategy requires much higher population density. It may also force mate-seeking females to move as much as males, leading to more equal predatory mortality rates for the sexes. Indeed, *Acartia* species generally have more equal adult sex ratios than, say, *Pseudocalanus*. Females of many copepod genera (e.g. *Calanus*) and likely chaetognaths, store sperm from one or a few matings to fertilize all the eggs that they will produce over a long spawning interval. Females of other groups (e.g. *Acartia* and *Temora* among copepods) require recurring mating. The latter strategy, common in free-spawning euphausiids, requires greater proportions of males and a persistent association of the sexes that may promote long-term swarming. A few groups of zooplankters release sperm into the water, and those must make their way to free ova or to females bearing unfertilized ova. Close grouping during release is often involved in raising the fertilization success rate. For example, a subarctic Pacific salp, *Cyclosalpa bakeri*, migrates from considerable daytime depths shortly after sundown, right to the sea surface. This is not a feeding migration, as no mucous feeding webs are in place, but a trip to a readily identified location for mating. Sperm are released from older male individuals and pumped through the tests of younger individuals with ripe ova (Purcell & Madin 1991).

As for several other topics, fecundity has been most extensively studied in crustaceans, although there are sketchy data for other groups. Very gently collect some adult females from the ocean, ladle ,or pipette them into containers of the water (and food) they came from, hold the temperature steady, wait a day, then count the eggs in the container. It's easy, but immediately there are problems. Cooped up close to their eggs, the females can eat them. So, since most eggs sink, a mesh screen is placed under the females so the eggs can sink out of reach. Or, the females are held in flat petri dishes. The eggs rest on the bottom, and the females cannot move easily to get them. This seems to give the highest fecundity estimates. These methods have many variants with investigator-specific and species-specific details. First, of course, we will examine some copepod data.

Copepod reproductive rates

Jeffrey Runge started a continuing wave of interest in egg production rates (EPR) with papers (e.g. Runge 1984, 1985) on the fecundity of *Calanus pacificus* in Puget Sound, Washington. The key thing sparking interest was his demonstration that egg production depends upon relatively recent feeding, such that a correlation exists between measures of available food (the ever-accessible chlorophyll concentration) and egg output. Runge has also studied egg production in *C. finmarchicus*, the dominant large copepod of the subarctic Atlantic. The general conclusions are:

1 Egg production depends upon available food (Fig. 8.2), in a hyperbolic fashion.
2 Clutch size is greater for larger females: in *C. finmarchicus* 45 eggs per clutch at 2.5 mm prosome length, increasing to 120 eggs at 3.3 mm.
3 Egg production is mostly late at night, sometimes extending into morning.
4 Inter-clutch interval is set by interaction of food availability and temperature, which thus determines egg production rate.
5 Egg production rate for the population as a whole can be determined for a stock by estimating the fraction of females approaching spawning readiness ("ripe and semi-ripe") and multiplying by the mean clutch size. This can be done with preserved samples.

The apparent functional relation with chlorophyll doesn't always hold up. Plourde and Runge (1993) showed a long

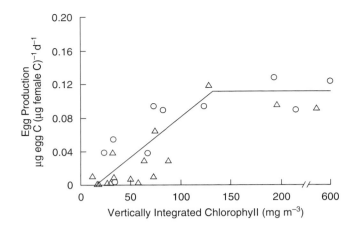

Fig. 8.2 Egg production by *Calanus pacificus* (egg carbon as a fraction of female carbon) from Puget Sound measured during the day after capture, plotted against vertically integrated water-column chlorophyll. (After Runge 1985.)

delay of any reproduction in the St Lawrence estuary until abundant phytoplankton showed up in late March–early April. However, the subsequent reproduction of this stock did not seem to depend very tightly upon chlorophyll abundance; probably the diet switched to microzooplankton. Runge often expresses egg production as a fraction of female body mass (carbon, usually) per day (Fig. 8.2). This measure tends to increase almost smoothly with temperature, about a doubling in *Calanus* sp. between 5° and 10°C, mostly due to shorter inter-clutch intervals. Ambler (1986) demonstrated egg production rates in well-fed *Acartia tonsa* as great as 1.6 body masses per day at 28°C. A great deal of mesozooplankton production goes into eggs.

Niehoff (2000), working with *C. finmarchicus* from a Norwegian fjord, confirmed that, in this genus, feeding is requisite for continued egg production (Fig. 8.3). In contrast, several *Neocalanus* species of the northern Pacific do not support egg production with current nutrition. In fact, the adults have no functional feeding limbs. They mature from their diapause stage spent below 500 m, remaining there to mate and then spawn. Nutrition for egg formation is drawn from stored lipids and other tissue components. The nauplii are provided with abundant oil droplets and rise as they develop (Saito & Tsuda 2000).

Spawning rate data of high quality have been gathered for many copepod species in many habitat situations. Significant and separate issues are how many eggs hatch and how many of the first nauplii (N1) are healthy. Poulet *et al.* (1994) discovered that, in fact, there can be substantial rates of hatching failure that seem to correlate with the prevalence of diatoms in the diet. They issued a challenge to everyone on Earth to prove or disprove that diatoms eaten by copepod mothers are toxic to their eggs. Quite a few folks took up the challenge. One of the better

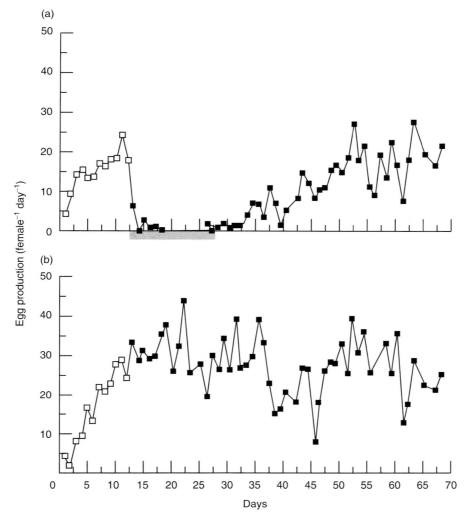

Fig. 8.3 Mean daily egg production of two groups of *C. finmarchicus* females that immediately after capture were first fed copiously on diatoms. (a) After 12 days of increasing per capita egg output, one group (*N* = 10) was switched to filtered water (shaded bar on the abscissa). (b) The other group (*N* = 25) was continued on a rich diet. Egg production dropped quickly to zero in the starved group. After feeding of starved individuals, their spawning slowly resumed but did not reach the level mostly sustained by the continuously fed group. (After Niehoff 2000.)

experiments was by Uye (1996), working with *C. pacificus* collected off the Oregon coast. He alternately fed females with several different diatoms and with several dinoflagellates. The results are striking (Fig. 8.4). Females eating diatoms soon spawn eggs that don't develop. When switched to dinoflagellates, the viability of their eggs recovers. Nauplii that do hatch from diatom-fed females are often teratogenized, with missing limbs or just bulbous stumps. They are a sad sight. Uye showed the active principle is a chemical; he soaked eggs produced by females eating a strictly dinoflagellate diet in diatom extracts, producing high rates of teratogenesis.

Nauplii with deformed limbs are not found in field collections, where diatoms are certainly sometimes the food of female copepods. Normal nauplii are capable of remark-able escape jumps, but those with abnormal limbs are likely to be removed very rapidly from the stock. However, the phenomenon does occur in the field; Miralto *et al.* (1999) found a case in which spawn from *Calanus helgolandicus* females that were definitely eating diatoms had very low hatching rates. The effect is toxic, not nutritional. Miralto *et al.* demonstrated that the toxic compounds are a small class of polyunsaturated aldehydes with molecular weight around 152 daltons. Not all diatoms produce these toxins, but when those that do, for example *Thalassiosira pacifica*, are abundant and eaten they can cause sharply reduced egg viability in *Calanus* and, particularly, *Pseudocalanus* (Halsband-Lenk *et al.* 2005). The toxic mechanism involves disruption of mitotic spindle formation at cell division. This suggests that in addition to poisoning of embryos by

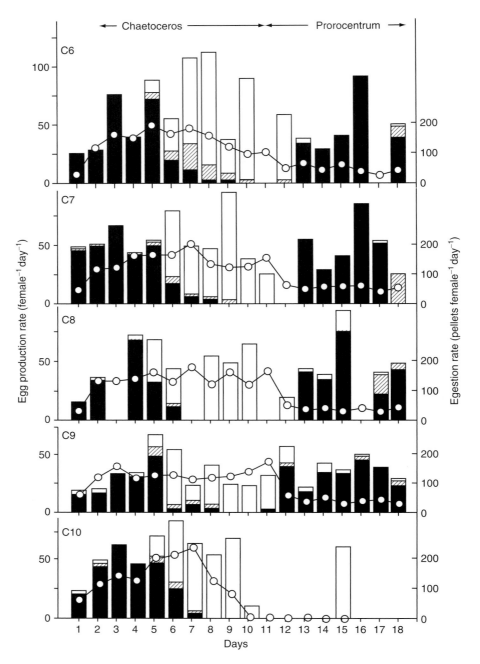

Fig. 8.4 Egg production rate (bar height) and egg viability (black = healthy nauplii, hatched = deformed nauplii, white = inviable, no hatching) of *Calanus pacificus*. Females were initially fed *Prorocentrum minimum*, a dinoflagellate, then shifted to the diatom *Chaetoceros difficilis* on day 2, then switched back to *Prorocentrum* on day 11 (as indicated by the bars along the top). Fecal pellet production (connected circles) indicates feeding level. (After Uye 1996.)

a maternal diatom diet, malformations of segmentation of the body and legs can be induced in copepodite stages that eat toxic diatoms (C. Miller, unpublished). Newly forming articulations are the principal sites of mitosis in later stages. This may produce more immediate protection of diatoms from herbivores than poisoning of embryos. Leising *et al.* (2005) showed that species-specific grazing rates of female *Calanus pacificus* were significantly lower on toxic diatoms

than on others, even when the toxic species were dominant. Possibly, toxic aldehydes or their chemical precursors can be detected and appropriately evaluated by grazers, providing both the copepod and the diatom with a direct selective benefit.

From egg production experiments, we have some idea of the input of new young to plankton populations. Egg mortality rates for copepods that spawn freely into the water (as opposed to those carrying their eggs to hatching) are high, which is likely true for all freely spawning plankton. Evaluation of the relative abundance of *Calanus* eggs at different embryonic stages shows an excess of early division stages, with implied mortality before hatching on the order of 60% (Miller *et al.* in prep.). Probably the bulk of that mortality is predation. Whatever the loss, healthy young must often hatch in good numbers (40% may be very good), because freely spawning forms are consistent and abundant components of the plankton community.

There are also many kinds of copepod (*Pseudocalanus*, *Euchaeta*, *Euchirella*, *Oithona*, and others) in which the females carry eggs in pouches until they hatch. Pouches are formed by pushing eggs at spawning into secretions from glands adjacent to the genital opening. The female then tows these pouches until the eggs hatch. Survival to hatching is dramatically increased because the eggs benefit from the mother's capability to escape from predators (Kiørboe & Sabatini 1994). Therefore, development can be slower and more complete. In such species, first nauplii are relatively larger than those of free-spawners, more completely developed, and can begin to feed at hatching, instead of at N3 as is typical of free-spawners. Greater survival to hatching allows much smaller numbers of eggs and longer inter-clutch intervals. Trade-offs between many young each at high risk given low parental care and fewer young each given greater parental investment are seen all through the animal kingdom.

Euphausiid fecundity

Euphausiids, too, divide into free-spawning (e.g. *Euphausia*, *Thysanoessa*) and egg-carrying (e.g. *Nematoscelis*, *Nyctiphanes*) groups. Available fecundity data are mostly from free-spawners, for example the species of *Euphausia* (Ross *et al.* 1982; Gómez-Gutiérrez *et al.* 2007). Euphausiids close to spawning have a large, colored ovary loaded with so-called Stage IV oocytes. In *E. pacifica*, of the subarctic Pacific, spawning occurs within a day (at most two) after the egg mass appears darkly blue or purple and is bulky enough to spread the carapace laterally. Similar criteria apply (a more grayish color) to the antarctic *E. superba* (Ross & Quetin 1984). After spawning, *E. superba* exhibits a clear space in the ovary, which also helps to evaluate field spawning activity. Ross *et al.* (1982) distinguished "ripe" from Stage-IV females of *E. pacifica* as those that spawn within 24 hours after capture. In their Puget Sound study site, the fraction of females with Stage-IV oocytes rose to 100% in late April, stayed high through May, then tapered off. Ripe females were about half of those in Stage-IV. Ross estimated the inter-brood interval as the inverse of the fraction of females spawning during the day after capture, obtaining a 2–3 day interval. Egg number per spawning episode is highly variable, with greater maxima at greater body size (Fig. 8.5a). Some, but only some, of the variability

results from availability of food (Fig. 8.5b), with clutch size usually greater at high chlorophyll concentrations. However, low chlorophyll levels only sometimes correspond with low clutch size, suggesting that other foods can replace phytoplankton; indeed all filter-feeding euphausiids are omnivores. *Euphausia pacifica* spawns from April through September in Oregon (USA) waters, and a female might produce a total of 45 clutches in that time (Gómez-Gutiérrez *et al.* 2007). If she grows across the size-range of spawning females (13 to 25 mm length) and produces the average egg outputs for her increasing sizes, she could produce over 5800 eggs.

Ross and Quetin's (1984) observations for *E. superba* in the Antarctic are similar in kind. Mature females have a red thelycum (gonopore area), and thus are readily distinguished from juveniles, regardless of ovary status. Ross and Quetin placed net-caught females singly into 4 liter jars in a water bath on deck, and then watched. First-day spawning was fairly limited, and a large fraction had clear spaces in the ovary, an indication of recent spawning, possibly during capture. On days 2, 3, and 4 of captivity, frequency of spawning rose, reaching about 16% per day, implying an inter-clutch interval of 6 days (if specimens in a collection are not in phase). Indeed, at collection some females are in each phase of the ovarian cycle, in proportions for a 6 day spawning interval. This seems fast for the prevailing temperatures of −1 to 2°C, but cold adaptation is expected in antarctic krill. Maturity comes as early as 36 mm total length, with initial clutches numbering ~450 eggs. Clutches increase with body growth, and females of 56 mm produce ~5000 eggs per clutch on average. Maintenance of live antarctic krill suggests 9 to 10 spawning episodes over two summer months, a mean of ~2500 eggs per spawning and total output of about 22,000 eggs per female per season. Ross and Quetin point out that these results are in contrast to earlier work in which it was assumed that females produced one brood per year; they warn of the danger in guessing what animals do in the field from characteristics of preserved specimens. The parallel with the chaetognath case below is exact.

The material and energetic outputs involved in spawning are substantial, amounting to ~40% of body mass in each clutch (Nicol *et al.* 1995), substantial growth to accumulate every six days. Quetin and Ross (2001) quantified the interannual variability of Antarctic krill fecundity near Palmer Peninsula. The reproducing fraction of the female stock varied from 10 to 98% over 7 years, corresponding to variations of food availability that depended, in turn, on timing of sea-ice retreat in spring. A collation of development-time data from free-spawning euphausiids (Pinchuk & Hopcroft 2006) shows (Fig. 8.6) a remarkably consistent relation to temperature. The prolonged development at antarctic temperatures allows sinking of *Euphausia cystallophorias* and *E. superba* eggs to great depths before hatching.

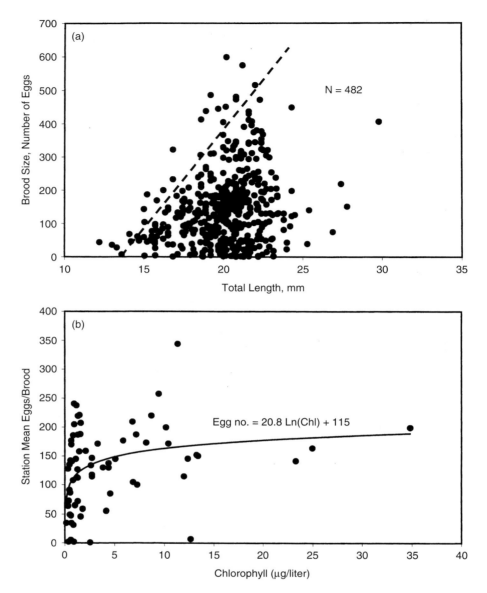

Fig. 8.5 (a) Initial after-collection brood sizes of *Euphausia pacifica* from Oregon (USA) coastal waters. The line indicates maximum egg numbers for 95% of the observations; 5% of females had larger broods. (b) Average brood sizes for variable numbers of females from different collection sites plotted vs. near-surface chlorophyll concentration. Data from J. Gómez-Gutiérrez (see Gómez-Gutiérrez *et al.* 2007.)

European workers often refer to *Meganyctiphanes norvegica* as northern krill, which are abundant in coastal waters from Norway to the Ligurian Sea and are found off North America from the Gulf of Maine northward. Cuzin-Roudy and Bucholz (1999) have shown that *M. norvegica* has a strong relationship between its egg production and molt cycles. A group of eggs matures for spawning just before cells of the epidermis detach from the old exoskeleton. Spawning itself occurs in two roughly equal bursts (and sometimes a smaller third) during the premolt phase in which new skin and setae form. Once the ovary is clear of large eggs, yolk formation (vitellogenesis) begins to enlarge a new lot, and that continues past the shedding of the old skin and on through an entire second molt cycle. Thus, the female alternates between a spawning molt cycle and a vitellogenic molt cycle. New small oocytes are produced continuously, restocking a layer from which those undergoing vitellogenesis are drawn. Molt cycles at a given temperature are shorter (with variation) for animals during the Mediterranean spawning season from January to May than during the summer–fall period of reproductive quiescence. Not only are spawning and molting cycles coordinated, most euphausiids molt at night, and *M. norvegica*, which is often a diel vertical migrator, spawns shortly before dawn. Thus, all the strong periodicities in its biology are coupled: spawning, molting, and vertical migration. Similar coupling will probably be found for more euphausiids.

Sac-spawning euphausiids extrude their eggs into paired sacs that are attached to and surround the posterior thoracic legs. In the small, mostly subtropical species *Nyctiphanes simplex*, the egg count increases with female size from ~32 at 9.5 mm to ~70 at 12.2 mm (Lavaniegos 1995). Development to hatching takes 5 days at 16°C, and the mother's ovary is engaged in oogenesis as development of an attached brood proceeds. Females molt soon after a brood hatches and then fill a new pair of egg sacs, the total

cycle time being 7 to 12 days for most individuals (Gómez-Gutiérrez *et al.* 2010). The commitment of growth to egg production is considerably less than for free-spawning species, but certainly with some energetic cost for carrying the eggs and moving water past them for oxygen and waste exchange.

Chaetognath egg production

Chaetognaths usually get abraded in net collections, so reliable fecundity data for oceanic forms are elusive. Nagasawa (1984) worked with *Sagitta crassa*, a small, robust, coastal species that she collected in a salt pond connected to Tokyo Bay by a sluice gate. She kept them in dishes and fed them *Acartia clausi* from the same source. Only three specimens from a collection of 100 held until death were significantly reproductive. Spawning by those three was cyclic, increasing then decreasing over periods of ~10 days, but not corresponding to cycles in feeding activity. Lifetime production for the longest-lived specimen of *S. crassa* was 952 eggs over 33 days, a mean of 29 eggs per day, cycling from zero to almost 90 (Fig. 8.7).

Nagasawa was too polite to contradict an old and widely accepted idea put forward by Kuhl in 1938, that chaetognaths only produce one "brood", the lifetime egg output being the oocyte counts in larger ovaries at collection. Rather she redefined Kuhl's term "brood" as lifetime egg output, so that spawning only occurs once in that sense. Clearly, spawning recurs over many cycles of oocyte production, maturation, and release. The Sagittidae are all free spawning, but the deep-living *Eukrohnia bathypelagica* and *Eukrohnia bathyantarctica* carry their eggs (up to >60 and 6 per sac, respectively) and young in "marsupial" sacs very similar to those of copepods or euphausiids. Eggs are extruded into a film-like substance secreted near the gonopores, forming the sacs. Embryos develop and hatch there

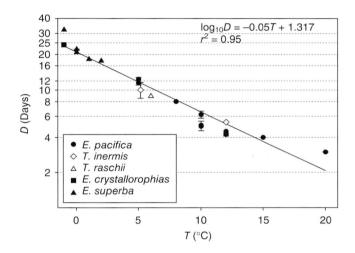

Fig. 8.6 Inverse exponential relationship of embryonic development time with temperature for various species of polar and subpolar euphausiids. At least this subset of euphausiids share a common relationship of development rate to temperature. *E.* = *Euphausia* spp. *T.* = *Thysanoessa*. (From Pinchuk & Hopcroft 2006.)

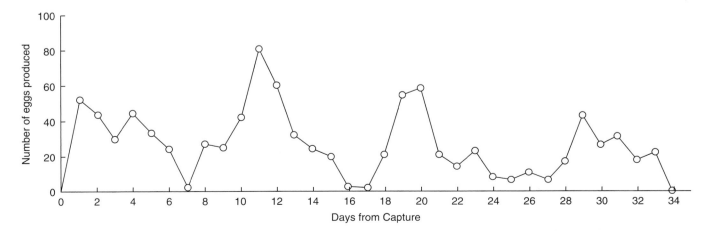

Fig. 8.7 Cycling of daily egg production rate in *Sagitta crassa* collected from Tokyo Bay and fed to repletion on *Acartia* copepodites. (After Nagasawa 1984.)

and are retained until yolk absorption is complete (e.g. Kruse 2009). Evolution in mesoplankton groups has repeatedly favored both free-spawning and egg protection strategies.

Appendicularians are semelparous

Because they can be maintained in culture, it is possible to measure appendicularian egg production. A common species in temperate waters is *Oikopleuroa dioica* that, unlike most members of the group, has male and female function in separate individuals. After a growth period of about 10 days at 14°C, ova zip through vitellogenesis in under one day. Then the eggs burst through the anterior end of the trunk, killing the mother. Thus, at least one group of zooplankton is semelparous (one clutch of eggs, then death). Fertilization is external. Fenaux (1976) found a correlation between egg number and length of the trunk (pharynx, digestive gland, and gonad) in four Mediterranean specimens: 21 eggs at 800 μm, 94 at 920 μm, 118 at 1000 μm, and 187 at 1100 μm. Greater fecundity occurs in colder waters, to >500 eggs. Eggs of *O. dioica* and many other species are about 100 μm in diameter, an ideal food size for copepods. Experiments by Sommer *et al.* (2003), with different concentrations of copepods in mesocosms suspended in a Norwegian fjord, suggest that appendicularian abundance may be controlled by copepod predation on their eggs, a notion supported by inverse population oscillations of copepods and *Oikopleura dioica* in the Baltic Sea.

Mortality rates and age distribution of mortality

A key issue is how mortality is apportioned among the life stages of plankton animals: what does the *survivorship curve* (l_x, proportion surviving, vs. x, age) look like from birth to the oldest ages attained? The shape of this curve affects every aspect of a population's productivity and reproductivity. For example, if only a few individuals ever enter the larger size classes whose growth can contribute significant amounts of tissue production per capita, then stock production can only be small. Moreover, if most potential reproduction is lost to early mortality, then high individual egg output is required for survival over multiple generations, and secondary production comes to focus in the adult female. Survivorship curves can be expected to vary among generations of the same species growing up at different seasons or in different locations.

There have been relatively few attempts to draw survivorship curves for plankton, because advection and patchiness of sampled stocks make the estimation process difficult and the results uncertain. Attempts fall into classic categories defined by ecologists long ago, the *horizontal* and *vertical* life-table approaches. A horizontal life table is obtained by enumerating a cohort of very young individuals born at about the same time, and then following their numbers until eventually they are all dead. A classic exercise of this kind in intertidal ecology is to scrape a square of rock and count all of the barnacles that arrive there during one pulse of larval settlement from the plankton. A map is made showing each individual. Then this square is visited at low tide for several years as those barnacles grow, push each other off, get bored into by snails, reproduce, and finally are gone entirely. A vertical life table is made by visiting a stock of animals once at some suitable season and determining the age structure. The relative numbers at different ages are tabulated and survivorship estimated from the decline of numbers with age. Abundance in horizontal tables cannot increase with age, because cohort entry is closed and the only source of change is death. Vertical tables take statistical rambles, because production of young may have been unusually high (or low) at the time when animals now middle-aged were born. A reproductive peak occurred in the American human population after World War II; giving birth became stylish, generating the baby boom. That massive pulse remains in the population, and it produced its own, lesser pulse of children – the "echo" of the baby boom. Thus, the age structure does not exactly represent the mortality schedule, and vertical life tables aren't reliable as predictors of future population survivorship. In the same vein, of course, a horizontal life table can trace the fate of a cohort that is unusual and can be of no general application to populations of the species. Nobody who knows about it will tell you that ecology is easy. Estimation of mortality rates has seldom been attempted for plankton other than crustacea, so our examples will come from that group.

The vertical method can be used in a number of ways to obtain stage-specific mortalities. One is to sample repeatedly and do lots of averaging between generations, years, etc. Brinton (1976a & b) used the massive sample set of the CalCOFI investigations in the Southern California Bight area to study the age structure (actually the *size* structure) of *Euphausia pacifica*. If you simply add up all of the size–frequency distributions over hundreds of samples from many years and seasons, you get a sort of average vertical survivorship curve (Fig. 8.8). There are three or four stanzas of roughly constant mortality per millimeter of body length: larval, early juvenile, adult, and senescent phases. This survivorship pattern is plotted on a semi-logarithmic scale and appears roughly linear in successive growth stanzas, suggesting that exponential mortality with age is an appropriate model. Rates ($-m$, mm^{-1}) have been calculated and are shown along the graph. Brinton (1976) dissected the dataset into cohort patterns (figures 9 and 10 in his paper) and derived more information about mortality patterns, achiev-

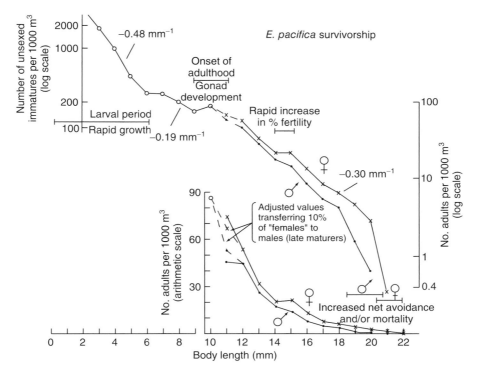

Fig. 8.8 Long-term averages for numbers of *Euphausia pacifica* in the Southern California Bight by length category. Abundances (1000 m³)⁻¹ are plotted for larvae and immature specimens on a log scale above and for adults on a linear scale below. This generates a survivorship curve according to size. Mortality has three stanzas with relatively constant rates: a larval rate from 3 to 6 mm, a lower juvenile rate to 12 mm, and an adult rate slightly higher again from 12 to 20 mm. Mortality rates certainly change, but the slope changes in the progressive decline of abundance with size are also affected by changes in growth rate and net avoidance capability. (After Brinton 1976.)

ing a horizontal approach for some of the better-defined age groups that moved through the population from 1953 to 1956. For example, there are differences in reproduction timing and growth rates among seasons that recur from year to year. Close study of this landmark paper will show you why such studies are so very rare: the labor is enormous.

The horizontal method has been applied also to copepod populations, with stage distinctions used as an approximation of age. Johnson (1981) estimated stage abundances twice per week through the growing season for the *Acartia californiensis* population active in summer in warm upstream waters of Oregon's Yaquina Bay (USA). Nauplii, unfortunately, could not be distinguished from those of the co-occurring *Acartia clausi*, so they were not counted (also saving labor). The data are shown for 1973 (Fig. 8.9). From fecundity estimates, females present in the plankton on any given day were assumed to spawn 20 eggs each, giving a very rough egg number to start each cohort. Stage-to-stage survivorships (Fig. 8.10) were then calculated as reductions in the number of "copepodite-days" on the graph for each of the year's cohorts (distinguished similarly to Landry's work in Jakle's Lagoon, Chapter 7). Clearly, there must be distinct mortality stanzas in the life history of copepods as

well as euphausiids. Very large numbers of eggs and nauplii die at high rates, leaving a few copepodites, which then have relatively high survivorship to adulthood. Earlier cohorts with low abundance have better naupliar survival than later, more abundant cohorts. Similar estimates were made by Landry (1978) for *Acartia clausi*, Twombly for diaptomids in lakes, and others. Eventually, it was realized (Hairston & Twombly 1985) that changes in copepodite-days occupied by a cohort are not dependent upon mortality alone, but on development rates. If development is nearly isochronal, that is each stage passes in equal time, then the procedure isn't too bad. That is true of well-fed *Acartia*, so the results of Johnson and Landry are probably useful. But, when stage durations vary, a more complex model is required.

One such model was developed by Simon Wood and applied to a *Pseudocalanus newmani* population in Dabob Bay, Washington, by Ohman and Wood (1996). The method needs data on the progression of stage abundances. In order to avoid obtaining both stage duration and mortality from the field data, the user must apply laboratory estimates of development rate under the food, temperature, and other conditions pertaining in the field. Unfortunately, there is no guarantee that development rate in the field is the same as

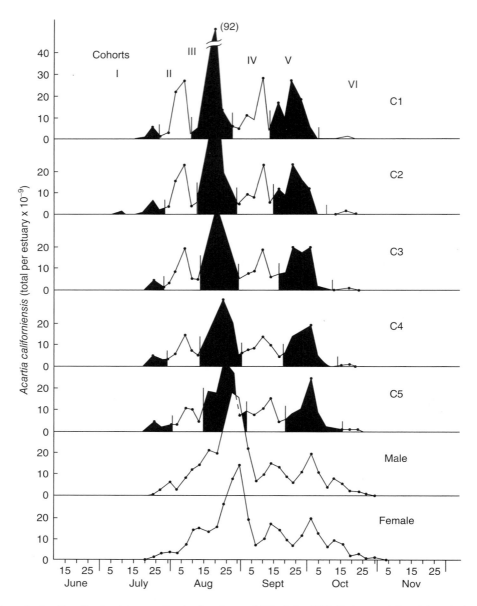

Fig. 8.9 Cohort patterns for *Acartia californiensis* in upper Yaquina Bay, Oregon (USA), during 1972. Each group followed through development is designated by a Roman numeral. The abundance pattern does roughly repeat in successive stages. (After Johnson 1980.)

that measured in experiments. The progression of stage abundances is then fitted by a smoothing function with several assumptions. One is that the population is closed, which is, of course, the basic assumption of any horizontal life-table method. For zooplankton, this restricts the method to a few, rather odd sites with very little net advection to carry animals either in or out. Ages at advance to each stage are assumed to be fixed: ages in successive stages do not overlap. Unfortunately age-within-stage at molt can vary by a factor greater than two for copepods raised in conditions as nearly identical as can be arranged or imagined to occur in the field (Carlotti & Nival 1991). Even

from the same clutch of eggs, some fourth copepodites will be younger than some second copepodites! Wood's method assumes that can't happen and may or may not be robust against it. A main feature of the method is fitting of a surface represented by cubic splines to the abundance data. That provides mathematical functions that can be integrated explicitly, as well as getting rid of *some* of the troubling chaos of real data. The splines are then modeled (the method is very complex) by equations for each stage reproducing the function in terms of throughput (stages advance at rates specified from rearing experiments) and stage mortality (the final output).

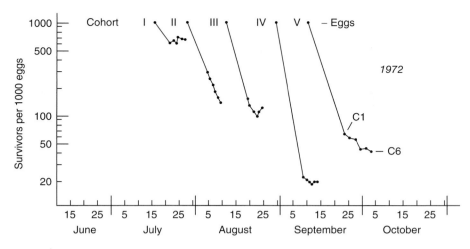

Fig. 8.10 Survivorship curves for *A. californiensis* calculated from stage-abundance time-series in Fig. 8.9. (After Johnson 1980.)

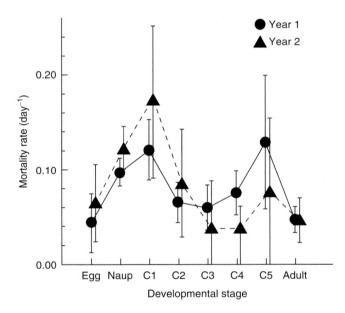

Fig. 8.11 Stage-by-stage instantaneous mortality rates of *Pseudocalanus newmani* in Dabob Bay, Washington. Rates estimated from weekly data by the statistical procedure of Wood (1994). The symbols ● and ▲ represent data from two years. (After Ohman & Wood 1996.)

The results (Fig. 8.11) are exponential stage-specific mortality rates (d⁻¹). For *P. newmani* in Dabob Bay, the variations among stages were reasonably consistent between two years, suggesting that mortality varies strongly from stage to stage. As in Johnson's study, the nauplii were not distinguishable from those of sympatric congeners, so their mortality was approximated from likely input estimated from female abundance. Egg mortality is relatively low and about the same as for females, presumably because eggs are carried until hatching by the females. Male mortality, not

estimated, must be very high, since males in *Pseudocalanus* are very few compared to females. This could also result from disproportionate maturation of C5s as females, not males. There are strong hints that adult sex in many copepods is determined by environmental conditions experienced by late copepodites. Elevated mortality rate of fifth copepodites is attributed to growing into the size range of interest to abundant predators, while reduced female mortality is attributed to their adoption in summer of reverse vertical migration (down at dusk, up at dawn; Ohman *et al.* 1983) to avoid predators which migrate normally (see below) and respond to vibrational stimuli from prey. Every feature of such results can be explained somehow; testing the explanations is another matter.

A simple vertical method devised by Mullin and Brooks (1970) has also been applied to the Dabob Bay data for *P. newmani* by Aksnes and Ohman (1996). Assume that mortality rates will be very close for adjacent stages, the younger with abundance X and the older with abundance Y. If the duration of both is a days, and input and output are roughly steady, then $Y < X$, and:

$$Y/X = S^a,$$

where S is the fractional survival per day. The exponential mortality rate is $m = -\ln(S)$. That's simple, but Mullin and Brooks went on to consider the common case in which older stages take longer to complete than younger. This made the equations more interesting (see their paper, or Aksnes & Ohman 1996), but not conceptually different. Results for *P. newmani* (Fig. 8.12) compare quite well to those from the elaborate horizontal method. The problem which recurs with this technique is that sampling bias is often not constant with stage; older stages for various reasons are often more susceptible to capture than younger

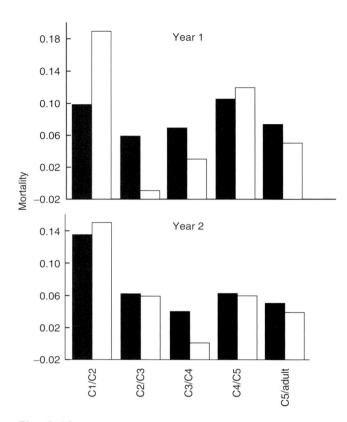

Fig. 8.12 Results of the Mullin–Brooks vertical mortality-rate estimates for the same data as were used to generate Fig. 8.11. Mortality rates from the Wood (1994) technique are solid bars (averaged between stage values in Fig. 8.11); those from the Mullin–Brooks calculation are open bars. (After Aksnes & Ohman 1996.)

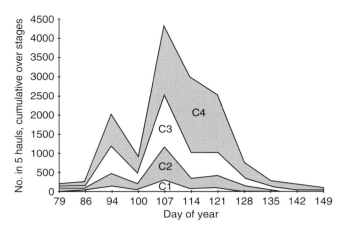

Fig. 8.13 Plot of data from Marshall and Orr (1934): time-series estimates of stage abundance (C1, C2, etc.) of *Calanus finmarchicus* in Loch Striven. Abundance integrated through time increases in older stages much more rapidly than expected from their somewhat longer stage durations. Apparently they are more susceptible to capture, which is not attributable to mesh size.

ones (Fig. 8.13), leading to apparently negative mortality, which is against our biological belief system.

An additional problem with application of the Mullin–Brooks calculation is that most copepod populations never have steady-state population composition, as required. Whether averaging stage abundances, stage ratios, or estimated mortality rates over months of data, for example as done by Ohman *et al.* (2004) with the 5-year January to June GLOBEC sampling program on Georges Bank, can compensate for the generality of short-term violations of the assumptions is unclear. The two-weeks-on, two-weeks-off sampling of Georges Bank must also have introduced gaps in the series of stage-ratio estimates, gaps with at least some impact on the estimates, despite the very large number (thousands) of samples evaluated. More positively, mortality-rate estimates for newly spawned eggs to N3 of *C. finmarchicus* on Georges Bank by a modified Mullin–Brooks method (Ohman *et al.* 2008) are likely of the right general magnitude, about 20% per day in March, about 50% per day in May. Those authors attribute a substantial share of that mortality to cannibalism by *Calanus* females. Experiments in several laboratories (e.g. Basedow & Tande

2006) confirm such feeding, at least by females cooped up with their eggs and nauplii in one-liter containers.

An alternative route to mortality-rate estimation is to model population stage composition and abundance by fitting mortality rates to time series of field data for stage abundances. The greater the number of population-dynamical variables that can be approximated from field data, the more accurate the fitted mortality rates will be. Egg production per female can be measured, and multiplied by female abundance to obtain egg input. Stage duration is more difficult. Inverses of stage molting rates (fraction molting per day among incubated samples from the ocean) would equal stage durations at steady state, but *again* steady state does not develop. The oldest individuals in a cohort that is just reaching a stage will have zero molting for a while. Even the youngest individuals in a cohort just finishing a stage will have fractions near 1.0 that molt in a day. Laboratory rearing data for effects of temperature and nutrition can be applied, although characterizing available nutrition in the field is an unsolved problem, and even the actually effective temperature can be difficult to determine for a population vertically dispersed in a stratified water column.

Despite the difficulties, there are well-worked examples. Li *et al.* (2006) combined a model of the circling water flow (anticyclonic, driven by tidal rectification) over Georges Bank with a model of the progress of the advected *C. finmarchicus* population through its generation cycles. They attempted to estimate not only birth and stage-by-stage mortality, but also the entrainment of stock over, and dispersive losses from, the bank. Stage durations, strongly affected by seasonal temperature variation, were accounted for based on a laboratory rearing study by Campbell *et al.*

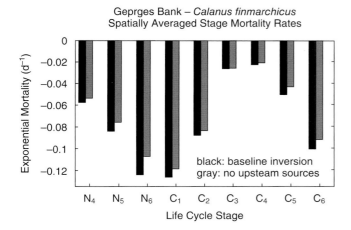

Fig. 8.14 Mortality rate (d^{-1}) estimates for *C. finmarchicus* from a model combining advection and horizontal diffusivity over Georges Bank with the developmental population dynamics of the resident stock of that copepod. Reproduction and mortality were tuned to fit spatially variable plankton abundance and stage composition data. Black bars represent best estimates of mortality if some stock can recruit from outside (based on data from outside), while gray bars represent mortality with no recruiting. Very small increments to mortality are required to account for in-bound advection of stock. Clearly on-bank, reproductive stock recruitment is dominant over recruitment by advection. (After Li *et al.* 2006.)

(2001). Li *et al.* included some slowing of development from food limitation by using an averaged time-series of chlorophyll data. A model copepod stock was dispersed into a model of the average hydrodynamics, and its mortality rates were estimated by so-called data assimilation (the "adjoint method") of a climatology (an average spatial and temporal pattern) of stage abundances developed from the GLOBEC Georges Bank sampling program. The mathematics of such modeling are complex and are rarely fully presented in papers applying them, so we leap here to the conclusions. Additions to stock were primarily from on-bank reproduction (Fig. 8.14). Biological sources of mortality (predation, parasitoids, . . .) were strongly dominant compared to dispersion off the bank. Mortality rates were 20–40% less than those found by the Mullin–Brooks method, leaving room for development of an adequate number of C5 to supply the large diapause population that develops off the bank by July.

Models based on full representation of population-dynamical processes are developing rapidly, and estimates of mortality rates can be expected to shift up and down for some years to come. For example, Neuheimer *et al.* (2009) have reworked the *C. finmarchicus* over Georges Bank problem again, noting that the numbers of nauplii estimated by the model of Li *et al.* (2006) for May (recruited from the last major reproductive cycle) fall one to 15 standard deviations outside the data. They propose varying the naupliar mortality rates more radically in both time and space (much greater in late April and May, greatest over

the bank crest). Seasonal and spatial variations of mortality rates are not a particularly surprising conclusion. Neuheimer *et al.* (2010) also propose an individual-based modeling approach with a mechanism for tuning mortality rates to match time series of stage-specific abundance estimates. Application to more studies will show whether such methods produce reliable mortality rates.

Apart from euphausiids and copepods, explicit estimates of mortality rates in zooplankton have not been attempted.

Causes of mortality

Death results from predation, starvation, disease, and old age. We don't have good aging methods for any full-grown or adult plankton, so we don't know how many reach the ages to which we can hold them in the laboratory. Very few female copepods or euphausiids that appear to be reproductively spent are collected in the upper water column. It is likely that the daily risk of predation is so great that few reach such an age. There are masses and masses of gut-content data showing that zooplankton are eaten by other zooplankton, by larval fish, adult fish, and whales. Being eaten is the fate of practically all individual zooplankters. Since the risk is continuous and high, virtually every species in every group has a repertoire of adaptations to foil predators. Pelagic habitats lack hillocks, trees, shrubs, and rocks to hide behind, soil to burrow in, and other elements of *cover*. If you walk through a wood, it is usual to see very few animals. Birds and some insects are moderately conspicuous, depending upon their capability for flight to move them quickly from advancing predators. Mammals, reptiles, and most invertebrates, however, are rarely seen unless special provisions are made. They are there, but they are in cover, hiding behind, inside, and underneath the obscuring elements of the landscape.

In pelagic habitats, the lack of cover requires different adaptations. One is that many plankton animals are very nearly transparent and difficult to see at all. Some that are mostly transparent have slightly variegated color patterns that "break up" against flickering light from above, leaving no clear image of the animal against its background. Pigmented or luminescent gut contents are impossible to make transparent, so feeding often stops during daylight. Animals in several groups, for example some species of ostracods and several copepod genera (prominently *Metridia* and *Pleuromamma*) leave behind blobs of luminescing mucus as they dart away from approaching predators. Diel vertical migration to evade visual predators is considered separately below.

Life in the open requires a constant state of alertness, coupled with the capability to dart away. Escape darting in copepods is a response to shearing motion in the water around the antennule, which is sensed by setae extending in multiple directions. Shear rates on the order of $1.5 \, s^{-1}$ (i.e. $\Delta[cm \, s^{-1}]cm^{-1}$] evoke escape responses in the copepod *Acartia* (Fields & Yen 1997). For ordinary current shear, that would be a large number, but, for detecting an eddy announcing the approach of a predator, it represents refined sensitivity. Fields and Yen have demonstrated that these shear thresholds vary such that species in habitats with high background shear are less sensitive than those living in quiet water. Time to initiation of escape response is of order of 1.5 ms in the most anxious copepod species (Lenz *et al.* 2000), about 100 times quicker than human reflexes. Copepods with the briefest response times have myelinated axons (Lenz *et al.* 2000), while those in other families do not and can only initiate jumps after about 10 ms (still relatively fast). Myelin, a multilayered lipid wrapping of axons, is found in very few invertebrate groups and is arranged in different ultrastructural patterns than in vertebrates. It has evolved independently in copepods. Mechano-sensory dendrites in the bases of setae extending in three or more directions and at varying distances along the antennules, provide directional information about the approach of predators (or an experimenter's vibrating needles), allowing rapid rotation onto an optimum escape trajectory. Reflex arcs are short and motor impulses to the swimming legs are carried by giant axons in the ventral nerves, a feature common in many invertebrates. The muscles of the thoracic legs driving the acceleration have a fine structure adapted for exceedingly rapid ATP recharge (Fahrenbach 1963).

Velocity during escape swimming pulses dramatically between the sequence of power strokes by the four or five thoracic legs followed by a tail flap, moving the animal into the range of inertial drag, and coasting during return of the legs anteriorly, decelerating it back toward the viscous drag of low Reynolds numbers. In *C. finmarchicus* C5, the cycling of escape velocities is zero to $750 \, cm \, s^{-1}$ in 8 ms, dropping to ~$380 \, cm \, s^{-1}$ in the next 4 ms, with acceleration pulsing greatly with the power stroke of each successive leg (Fig. 8.15).

If this suite of tricks did not work some of the time, then extinction would follow. However, it does not always work, since predators must evolve matching quickness and low shear approaches or face extinction in turn. While the jumps are rapidly initiated and reach high velocity relative to body size, they do not necessarily move the animal very far from any sizeable predator, only 4 to 6 cm for *Acartia* spp. of ~1.3 mm length (Buskey *et al.* 2002). However, for some larger copepods, sequences of thoracic leg rowing can last on the order of a second, moving them a half meter or more from a frightening stimulus. In all cases, rapid movement also accelerates the surrounding fluid into eddies and wakes that can alert predators possessing motion sensors to passing meals. Thus, moving may eliminate one danger and create another. Most of the available results are for adults and late-stage copepodites, but work is in progress on

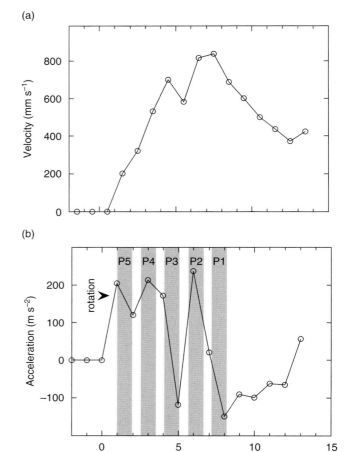

Fig. 8.15 Dynamics of escape swimming in *C. finmarchicus*. (a) Velocity build-up and fall-off as a free-swimming individual rows (*cope-pod* is from Greek for "oar-foot") with sequential rearward sweeps of its five thoracic legs (P5 at the rear first, P4, . . . , P1), but then must be slowed by drag as the legs return anteriorly, feathered. Setae open flat on the power stroke, close and trail on the return. (b) The acceleration sequence showing successive responses to the serial leg sweeps. An escape jump involves multiple power and recovery cycles. (From Lenz *et al.* 2004.)

escape jumps by nauplii. They can move very quickly away from approaching predators, but not particularly far, <1 cm.

Other zooplankton also have predator avoidance behaviors. Appendicularia burst out and away from their mucous house-filters when touched. At least solitary salps accelerate their jet propulsion when touched, blasting out their mucous feeding webs. Shelled pteropods release their mucous webs when attacked, pull their "wings" into their shells, and sink rapidly. Chaetognaths, which attack small sources of shear, dart away from larger ones. Euphausiids stuck to ctenophore tentacles will tail-flip backward and often escape. Antarctic krill, *Euphausia superba*, swim evenly spaced in schools many meters in horizontal dimensions at densities up to more than 25,000 m⁻³, making *en echelon* turns and dives. The schools open meters-wide spaces around the attacking predators and plankton nets. Individuals induced to tail flip ("lobstering") can disrupt this organization, something used to advantage by penguins and other large krill predators (Hamner & Hamner 2000). Some other juvenile and adult euphausiids (*Euphausia, Thysanoessa, Nyctiphanes*) have been observed to school or swarm, sometimes apparently for breeding, but more generally for avoidance of visual predators. The advantages of schooling are that most of the water is left empty, requiring more searching by predators, and that the prey gain by combining their abilities to detect predator approach.

Starvation can be induced in the laboratory. Some species can suspend growth, reduce metabolism, and wait it out. Others can't and die quite quickly (Dagg 1977). Debate regarding how often starvation actually occurs in the field remains unresolved, but it probably is not very frequent. There are diseases and parasites of zooplankton. One group of ciliated protozoa, the Apostomatidae (no mouths), have an abundant "phoront" stage which encysts on the setae of euphausiids and copepods. The smallest wound anywhere on the host's body is detected, certainly from the scent of leaking tissue fluids, and stimulates the phoronts to leave their cysts. They swim to the wound, enter the body, rapidly consume the tissue by "osmotrophy", divide repeatedly and finally exit to seek suitable new hosts. The dead host is just a wisp of exoskeleton. Gómez-Guttiérez *et al.* (2006) report that an apostome ciliate parasitoid (animals live with their parasites, die from their parasitoids), *Collinia*, can infect several species of euphausiids, at least occasionally causing mass mortality.

Life-history variations

Zooplankters, like many other animals, vary their activities and reproduction according to the season; the pattern of a particular species or population is termed its *phenology*. Many phenologies include a prolonged resting phase or *diapause* in each generation or in the generations of some seasons, others do not. Diapause is not found only at high latitudes, but also in monsoonal areas like the Arabian Sea. Some zooplankton at high latitudes may take more than one growing season to mature and thus live several years, requiring two or even three diapauses. Like most aspects of the biology of zooplankton, life-history patterns are best known for the reliably collected crustaceans, particularly copepods. Their abundance changes are great enough to provide a strong signal against the severe sampling noise that afflicts abundance estimation. Finally, we can rear or at least maintain copepods in the laboratory. Copepods exhibit three basic life-history patterns. First, some tropical species simply hatch, pass through successive naupliar and copepodite stages, mature, mate, and spawn. That cycle is then repeated by the offspring. Individuals can be found in all stages of the cycle at any given time. Similarly continuous patterns have been demonstrated in chaetognaths.

Second, copepod species for which the range is inshore over shallow bottoms are in many cases absent from the water column during some part of the year. Those inhospitable seasons are avoided by producing eggs adapted for a period of rest in the sediment before they hatch. Workers puzzled surprisingly long about the periods of total absence before these *resting eggs* were discovered in several Black Sea species by Sazhina (1968). At the beginning of more favorable seasons, the resting eggs develop, hatch, and the nauplii re-enter the water column, a substantial stock developing very quickly from the sediment "egg bank". Females ready to spawn in periods of favorable conditions produce eggs which hatch immediately, termed *subitaneous* eggs. Common genera exhibiting this pattern are *Acartia, Centropages,* and *Labidocera*. In *Acartia*, individual females can switch from spawning subitaneous eggs to spawning resting eggs, and they can switch back. The cues are changes of water temperature, which cycles more evenly in water bodies than in the atmosphere. *Acartia hudsonica* produces resting eggs when the temperature rises above a threshold (Sullivan & McManus 1986); *A. tonsa* and *A. californiensis* produce them when the temperature drops below a threshold (e.g. Johnson 1980). Development and hatching occur when conditions again cool, in *A. hudsonica*, or warm, in *A. californiensis*. Development of resting eggs of the latter species, in which the active population is restricted to the upper reaches of estuaries, remains suppressed until salinity rises in spring. At least in *A. californiensis* there is a short "refractory" period. That is, the eggs must experience a certain interval of cold conditions before they will respond to warming by developing. This isn't very strict, as is the refractory phase of diapause in some insects, but it ensures that transient warmings won't initiate development of all the eggs in the sediment, only to be chilled to death almost immediately. In *Labidocera aestiva*, production of resting eggs also depends upon temperature, but the dependence varies according to the daylength (Marcus 1982).

Third, most species of the abundant oceanic copepod family Calanidae can rest in one or more of the older copepodite stages, termed a copepodite diapause. Most typically, if a generation is going to include a diapause, the fifth copepodite (C5) stage feeds near the surface and accumulates a large mass of oil in a thin-walled sac (Plate 8.1) as a reserve of organic matter and energy. Then it descends to a considerable depth in the ocean and remains there during a diapause on the order of a half year. During this time there usually is no feeding, although *Calanus finmarchicus* has been caught taking "midnight snacks" in the Gulf of Maine during November. Respiration is strongly suppressed, and activity is nil. Recent work shows that many genes are down-regulated (no mRNA generated), while others are up-regulated during the onset and maintenance of diapause (Tarrant *et al.* 2008). Apparently, only one gene for a small heat-shock protein (an HSP or a "molecular chaperone", a protein that protects enzymes from unfolding and degrading), HSP22, is up-regulated during diapause (Aruda *et al.* 2011). Large HSPs (e.g. HSP70 and HSP90) are typical of diapauses in many other animals. Fifth copepodites seem to approach neutral buoyancy closely and just hang suspended. They do remain alert and dart away when poked or when the water eddies near them. Because their metabolism is shut down, resting individuals do not recover rapidly from such an escape effort, and they essentially get only one or two tries at escape from a predator. In most Calanidae, diapause is initiated in late spring, with the warmest, less-productive summer–fall period spent at depth. One tropical species, *Calanoides carinatus*, spawns and grows actively near the surface during periods of large-scale coastal upwelling. It is found most abundantly in the Gulf of Guinea and the Arabian Sea, sites where upwelling occurs for about half the year. The C5s from the last of about six generations (Binet & Suisse de Sainte-Claire 1975), and perhaps some C5s from all generations, enter diapause at mesopelagic depths far seaward of the upwelling centers occupied during the growing phase of the phenology.

Diapause phases apparently evolve readily. There are about as many variant patterns as there are species, and within species the patterns vary across the oceans. This suggests that since habitats suitable for feeding and growth are only available part of the year, it is an urgent task to make a close match of life-cycle events to those periods. Emergence from copepodite diapause is coupled to the maturation molt, C5 advancing to C6 males and females. In *Calanus finmarchicus* of the subarctic Atlantic, newly emerged and mated females (termed the G_0 generation) move near to the surface to feed and spawn. Food is required to make eggs (G_1 generation), although spawning can begin well before the spring bloom. Timing of G_0 maturation from diapause by *C. finmarchicus* varies strongly by region (Planque & Batten 2000). In the Gulf of Maine it starts in late December; off Scotland it begins in February; west of Iceland and in the southern Labrador Sea it is even later. Late winter and spring see completion of one generation, which at least partly matures in many subregions, but not off northern Norway, where all but a few percent of G_1 enter diapause. The G_1 females produce G_2, which completes growth by a time varying from May to August and mostly enters diapause as C5, sometimes as C4. In some areas part of G_2 also matures and produces G_3, which may or may not survive significantly to reach diapause. Possibly, G_3 is the source of most of the C4 individuals included in the final, November diapause stock.

Variation for Canadian Atlantic waters has been nicely summarized by Johnson *et al.* (2008) from multi-year time-series of samples (Plate 8.2). The consistency within subregions is striking, as are differences in timing among subregions. Adults emerge from diapause in December off Newfoundland and on the shelf south of Nova Scotia, similarly to the Gulf of Maine, in some years allowing a strong G_2 generation. In the Lower St Lawrence (west of Nova Scotia), diapause breaks much later, although that time-series has long winter sampling intervals. The St Lawrence series usually shows a greater proportion of C4 in the diapause stock.

We would like to know which changes in the habitat inform an individual that it should enter diapause or mature immediately. Unfortunately, experiments have not answered this, since *C. finmarchicus* will not enter diapause in containers, even in very large, so-called mesocosms. Arousal from diapause appears to depend upon internal time-keeping, although exactly how that works remains unclear.

Not only diapause timing, but depths occupied during rest, are variable. *Calanus finmarchicus* spends its diapause at ~450 m depth south of Georges Bank, but well above the ~300 m bottom in the Gulf of Maine north of Georges Bank. In basins on the Nova Scotian Shelf, it is found just above the bottom. In deep areas (>2000 m) of the eastern North Atlantic, the diapause stock is widely dispersed, mostly between 400 and 1200 to 1600 m (Heath *et al.* 2000). In Norway, the depth of diapause varies strongly from one fjord to another. Dale *et al.* (1999) have suggested that this variation is a response to vertical changes in predation pressure from place to place. Either layered predators remove diapause stock everywhere that we do not find it, or the copepods avoid the layers with predators. We don't have a test to distinguish these hypotheses.

Other species show other patterns. In the Gulf of Alaska and west across the subarctic Pacific, the endemic copepod *Neocalanus plumchrus* enters diapause as C5, mostly in June. The resting individuals scatter downward from 500 to 2000 m and begin slowly maturing in August, sustaining a roughly constant number of adults from September through January. As stated above, they spawn in place without feeding. Individuals probably do not spawn for longer than a month, but population reproduction is continuous over about 5 months. In years observed to date, development is only successful toward the end of that reproductive period. The timing of peak abundance of copepodites feeding near the surface in spring can vary by 6 weeks or more, so presumably the metering out of reproduction makes some young available to grow whenever growth is favored. This seems like a population "strategy", but there are individual-based selection mechanisms that could make it work: sometimes early spawners succeed, sometimes late ones do. Another species, *Neocalanus flemingeri*, is sympatric with *N. plumchrus*, but less abundant. It grows from February through May, then descends to depth and matures to adulthood. The males stop part way down and are joined by the females. Mating occurs and the males die off. The females carry a large mass of oil, and they are the diapause stage. Ovary ripening and spawning are narrowly focused in February, and the cycle repeats. Every species examined has a somewhat different mode in the diapause and reproductive phases. *Calanus hyperboreus* in the Arctic Ocean appears to diapause twice, placing parts of its life cycle in each of three summer production seasons. There are five large, particle-feeding copepods in Antarctic waters. *Calanoides acutus* and likely *Rhincalanus gigas* have winter resting stages with features of diapause similar to those of subarctic species: abundant wax accumulation, reduced activity and metabolism, and ontogenetic down-migration. On the other hand, *Calanus propinquus*, *Calanus simillimus*, and *Metridia gerlachi* appear only to slow their usual activities in winter, which may be feasible because temperature actually varies rather little seasonally. More numerous small species, including *Microcalanus pygmaeus*, *Ctenocalanus citer*, and *Oithona similis*, also remain active.

We do know something about life-cycle timing and resting periods in euphausiids and chaetognaths. There are strong pauses in growth during winter periods, so that peaks in size–frequency distributions do not move. The characterizations of rest phases are not as sharp as for copepods, however. It looks as if in some seasons they simply stand by, waiting for things to get better, but that may be because we do not know how to observe them, and other groups (ostracods, pteropods, appendicularia, . . .) are very little studied in this regard. Most life-history studies, even of copepods, consider species to be epipelagic during growth. The few studies for mesopelagic species indicate continuous population processes, with reproducing individuals present all year. However, in seasonal regions, some mesopelagic zooplankton, such as the copepod *Gaidius variabilis* (Yamaguchi & Ikeda 2000), accelerate reproduction after organic matter from the spring production pulse sinks through their habitat.

Diel vertical migration

Nineteenth-century studies of zooplankton in both oceans and lakes showed greater abundance estimates near the surface at night than during daylight. For a time, two competing hypotheses could have explained this observation (Franz 1912). Either plankton moved upward at night and returned to depth during the day, or they avoided nets better in daylight than in the dark. A clear test requires vertically stratified sampling, that is, abundance estimates for animals filtered from restricted layers in the water column. If animals depart the lower layers at night, as well as departing the upper layers during the day, then an actual population movement occurs. Since many kinds of animals are involved, it is not very surprising that each explanation applies to some species. Daytime net avoidance is restricted to animals with advanced visual systems such as euphausiids. It is not seen in those with simpler eyes such as copepods and chaetognaths. A statistically strong data-set (Fig. 8.16) was provided by Brinton (1967a & b), who examined the migration and net avoidance of euphausiids in the California Current using opening–closing ring-nets. Species like *Nyctiphanes simplex* are harder to catch during daylight, so the vertical integral during the day was much less than that at night, but they showed no sign that deeper-living individuals moved up at night. Presumably, those individuals near the surface see the nets coming and move out of the way, while those in dim light at depth are less successful at dodging. Species like *Euphausia hemigibba* had nearly constant vertical integrals; apparently they did not interpret the oncoming net as a threat to avoid, but their modal depth shifted strongly upward from day to night. For many years, such migratory behavior was called *diurnal* vertical migration, but in the 1970s that changed to *diel* vertical migration, since diurnal specifically refers to activity in daytime, while the migratory behavior is characterized over the whole day–night cycle. Under either terminology the behavior is often referred to by the acronym DVM.

The basic vertical migration pattern is simple: down at dawn and up at dusk. A colorful demonstration of this has been provided by Wade and Heywood (2001) using an acoustic Doppler current profiler (ADCP), a 153 kHz echo-sounder mounted on a ship's hull. It records the shift in frequency of sound reflected from particles in a series of successive time intervals, and thus distances, after a "ping". Shifts are interpreted as velocity relative to the ship. In most applications, ADCPs are used to provide vertical profiles of horizontal current velocity, but Wade and Heywood used it to resolve the vertical velocity of the particles. Echo amplitude (backscattering strength from a volume of suitable size) can also be interpreted as particle abundance. Most of the particles with a substantial backscattering "target strength" at this frequency would be larger zooplankton such as euphausiids, or small animals with an especially strong target strength, like shelled pteropods. Fish with swim bladders likely are also important. A day-long record of depth-distribution of backscattering strength (Plate 8.3a), shows a strong layer at ~350 m (just where the ADCP results begin to break up) during daytime, which moves up slowly at first, then accelerates, hitting the surface just after sunset. The same animals returned to depth well before sunrise. Vertical velocity (both up and down) of "particles" (Plate 8.3b) is maximal just when the layer is moving fastest, as expected. Sustained swimming at greater than ~10 body lengths s^{-1} is unlikely, so ~6 cm s^{-1} implies animals on the order of nearly a centimeter or longer, such as *Euphausia* or myctophid fish. Enright (1977a) showed that the strongly migrating copepod *Metridia* can move vertically at sustained speeds up to 2.5 cm s^{-1}, about 10 body lengths s^{-1}, so while they often move 350 m or more

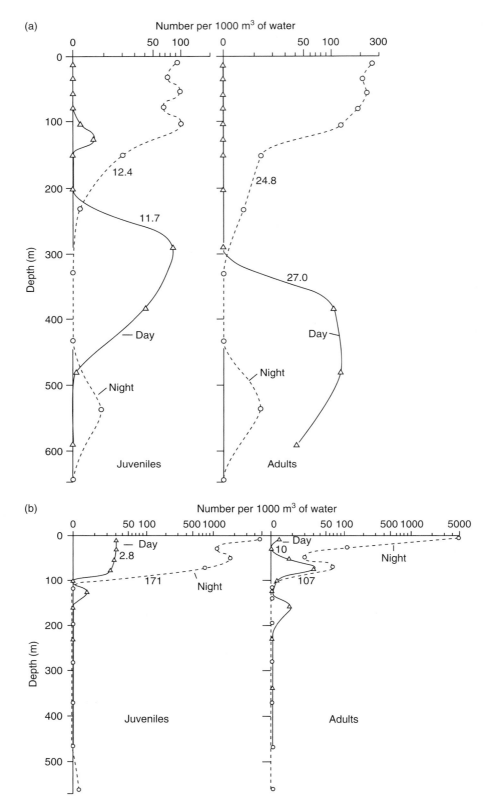

Fig. 8.16 Comparisons at stations in the California Current of day and night net-tow profiles for (a) *Euphausia hemigibba*, a vertically migrating euphausiid, and (b) *Nyctiphanes simplex*, a non-migrating euphausiid that avoids nets during daylight. Numbers beside the curves are vertical integrals (number m⁻²) (After Brinton 1967b.)

in their migrations, they probably cannot precisely follow an isolume. They must start up earlier than a specific isolume, arrive in the surface later, or both. All available data (e.g. Plate 8.3) indicate active swimming both up and down. Even a dead fish can't sink that fast. Only physonect siphonophores with gas-filled floats rise and sink with just buoyancy changes – probably quite slowly, since the float is small relative to the body.

In habitats where some species populations migrate strongly, there are usually others that don't migrate. The common supposition that DVM is a typical or nearly invariant behavior in zooplankton is not correct. Marlowe and Miller (1975) compared day and night distributions for the entire mesozooplankton community in the oceanic sector of the Gulf of Alaska, finding that only 10% of species showed statistically reliable indications of vertical migration. Some groups of zooplankton migrate more regularly and farther than others, for example the copepod family Metridiidae (*Metridia*, *Pleuromamma*). Others migrate sometimes, sometimes not, and still others show no variation in vertical distribution coordinated with illumination.

Reasons for diel vertical migration are commonly divided into two sets: (i) the *proximate* responses to habitat changes, i.e. the cues which say it is time to move up or down; and (ii) the *ultimate* or adaptive value of moving, the reason that selection has repeatedly installed this behavior in many populations. The proximate response is generally upward or downward swimming as illumination changes in the upper water column. The response to light can be modified by other variables, for example temperature structure, food availability, and the presence or absence of visual predators. Many studies demonstrate the response to light, and some of the nicest are studies of the movements of the "deep-scattering layers" of animals that reflect downward-directed sound back up to a listening ship, as in the ADCP study considered above. Kampa and Boden (1954) used a 12 kHz echosounder, which means that most of the sound returned from midwater was reflected by swim bladders of small fish. They identified a sound-scattering layer well before sunset, and lowered an upward-looking irradiance meter to its top surface. Then, as the sun set, they raised the meter so as to keep the irradiance constant at the meter, and recorded its depth continuously. Thus, they obtained a time-series of the depth of the rising isolume, which eventually passed out through the sea surface. The depth of the scattering layer, recorded independently after the first data point, tracked the isolume upward exactly (Fig. 8.17).

Clearly, the reflecting fish moved to maintain the illumination around them at a preferred level of irradiance. Moreover, when this same irradiance re-entered the water column at dawn, the scattering layer followed it back down. Observations of such 12 kHz scattering show steep rises during solar eclipses, too, although just as the moon fully

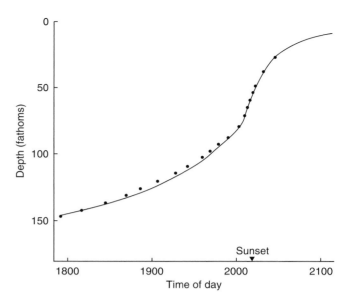

Fig. 8.17 Comparison of acoustic-scattering-layer movement from an echo-gram (continuous line) with the vertical position of a single isolume (6.6 × 10⁻⁴ foot-candles, dots) determined by irradiance metering. (After Kampa & Boden 1954.)

covers the sun the isolumes rise too fast for the swimmers to keep pace. This is a clear illustration of the simplest response pattern.

Clean experimental demonstrations of the response to light can be produced with freshwater cladocerans. Harris and Wolfe (1955) held *Daphnia* in a glass cylinder filled with an India-ink suspension to raise the light extinction coefficient, thus collapsing into a short length the gradient from surface irradiance to darkness. The cylinder was periodically photographed from the side with red light as a programmable rheostat varied the surface light intensity through a cycle simulating a day in a few hours. *Daphnia* were counted in short increments down the photos of the column (Fig. 8.18).

In complete darkness the animals sank to the bottom. As the illumination increased, they swam up to near the surface, a phenomenon known as a "dawn rise". Then as illumination increased more they swam back down, finding a "mid-day" depth. The mid-day depth showed signs of "accommodation" to the light intensity; that is, the animals moved up somewhat during the period of strongest illumination. Then, as the illumination faded, they accelerated toward the surface. Once their preferred level of illumination was gone, they sank back to the bottom. All the features of vertical migration in the field are reproduced and can be manipulated. A full migratory cycle can be compressed into an hour or several, since intensity variation is such a dominant control on upward and downward

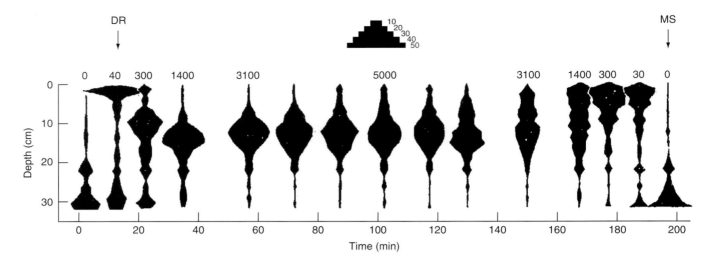

Fig. 8.18 Artificially induced, accelerated, and vertically compressed migration cycle in *Daphnia* held in a 32 cm cylinder of India-ink suspension. Positions of 50 *Daphnia* are represented for 10-minute intervals by widths of the "kite diagrams", as shown by the scale at the top. Surface illumination in lux was progressively changed as indicated by the numbers above the diagrams. DR indicates behavior equivalent to a dawn rise; MS indicates behavior equivalent to midnight sinking. (After Harris & Wolfe 1955.)

movement. Close coupling of the movement of the population mode to the variable cycle length shows that, for this animal, staying close to a preferred light intensity is a proximate cause for vertical movement. The scattering layer experiment, particularly the response to an eclipse, shows that at least some marine plankton or micronekton also shift vertical position in response primarily to changes in irradiance. Elaborate extensions of such experimental work have been carried out. They show many variants, including species which respond to the rate of change of light intensity, not to intensity *per se*.

The obvious ultimate value of moving down into dark layers during daytime is to avoid visual predators, and that is certainly the principal selective advantage that has driven the recurring evolution of DVM in many distinctive animal groups, including every phylum represented in the plankton. As stated above, the darkness at depth is one of the few forms of cover available in pelagic habitats. This adaptive value of DVM is simpler to propose than to prove, but clever observational designs and some luck have provided proof that predator avoidance is primary.

The strongest indication that predation avoidance is the adaptive value of DVM comes from species in which it seems to be optional. Bollens and Frost (1989a) compared the night and day vertical distributions of *Calanus pacificus* females in a fjord between the spring periods of two years. In 1985 the day and night distributions were identical, while in 1986 there was a strong shift between deeper day and shallower night distributions. In both years, strong migrations set in eventually and were marked in August (Fig. 8.19). This difference was accompanied by absence from net samples of sand lance larvae early in 1985 (result-

ing from failed spawning), while trawl tows in May 1986 showed ~13 juvenile sand lance (and a few other planktivorous fish) per 10,000 m³ in the upper layer of the fjord. By summer of both years, other planktivorous fish had moved into the fjord, producing threats from visual predation equivalent to that from sand lance in spring 1986. Presence of predators seemed to correspond to adoption of DVM by the copepod, and more importantly, the copepod did not migrate vertically *unless* the predators were present. At least in this instance, the cost of migrating is clear, the migrating *Calanus* had to forgo surface feeding during daylight. There is no phytoplankton food in this fjord at the depths occupied in daytime when migrating, and copepod guts are empty at depth from dawn to dusk (Dagg *et al.* 1989).

Field data are always somewhat indirect, often incomplete in one way or another. For example, in the study by Bollens and Frost (1989a) there were no actual fish trawls in April 1985. Seeking more assurance, Bollens and Frost (1989b) ran an experiment on the migrations of the copepod *Acartia hudsonica* in mesocosm bags hung from a raft in a coastal lagoon. Some bags had just copepods, some had copepods plus fish, and some had copepods plus fish confined in mesh containers suspended near the surface. Fish free in the container induced copepod vertical migration, while copepods alone or with caged fish did not migrate. The caged fish show that chemicals from the fish do not mediate induction of migration; the fish are detected from the effects of their motion. While this was true in that one marine study, the results of many freshwater studies have shown that fish odor can be a strong migration cue. For example, Loose (1993) pumped water from an

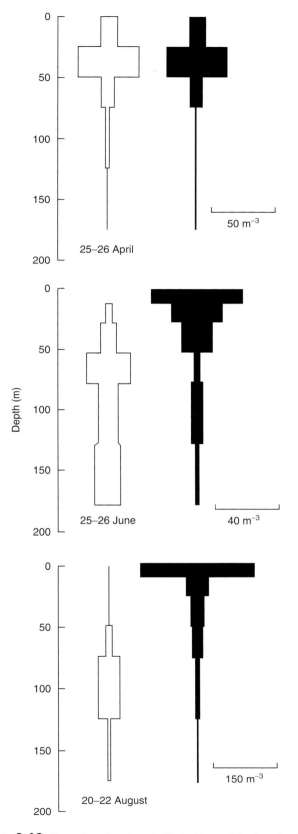

Fig. 8.19 Comparisons based on stratified net tows of day (open bars) and night (dark bars) vertical distributions of *Calanus pacificus* females in Dabob Bay, Washington State. Diel vertical migration was absent in April, obvious in June and August. (After Bollens & Frost 1989a.)

aquarium with no fish, then one or more fish, into the top of a tower tank holding a stock of the cladoceran *Daphnia* and a food stock of algae. The tank was 11 meters deep with algae mostly above a thermocline at 3 meters. In the absence of fish in the aquarium, the *Daphnia* spread out through the layer above the thermocline. When fish were placed in the aquarium, *Daphnia* spread through the upper layer at night, but sank to the thermocline during daylight. With more fish, the effect was stronger. Since the fish were not in the tank with the cladocerans, the transfer of information about predation risk must have been relayed by something exuded by the fish and carried in the water. Such substances are called *kairomones*, chemicals transferring information that the fish would be better served to keep secret, as opposed to pheromones that an animal secretes in order to transfer information that it is useful to make known to other animals.

While predation avoidance is reasonably accepted as the dominant adaptive value of DVM, other adaptive purposes have been suggested. Before it was realized that many species actually stop migrating when predators are absent, these alternative explanations held considerable interest. It remains possible that they play a part for some animals and circumstances. Alister Hardy pointed out that a vertical migrator that finds its food in one layer of the ocean can shift position substantially (many kilometers) by spending a period in a different layer going in another direction or at a different speed. Ocean currents do vary directionally with depth, providing a zooplankter with a means to search for "greener pastures". Maximum benefit would require that migration stop once a really good "grocery" had been located. No strong observations have been reported of such cessation of migrations. It has been hinted that the phytoplankton patches produced by iron fertilization experiments retain some of the regularly migrating copepods like *Metrida* and *Pleuromamma* during daytime. However, net-tow results (Zeldis 2001) from "SOIREE", an iron fertilization experiment in the Southern Ocean, showed little if any such effect. Ian McLaren suggested that migrants might benefit from DVM because they switch between ingestion at the surface, where food is abundant but it is warm, and assimilation at depth where it is cold. Thus, they could gain in fecundity from the larger terminal sizes that can be achieved in the cold. This may be important to some species, but it always comes at the expense of longer generation times, which can offset much of the gain in fecundity. Some species, particularly in high-altitude lakes, can avoid light- and ultraviolet-light damage by leaving surface layers while the sun beats directly down on the water surface. Detrimental effects of UV are readily demonstrated experimentally, and perhaps migrations to a few meters from the surface help to avoid those. Such slight displacements are hard to demonstrate in field data. There are also arguments (e.g. Enright 1977b) that phytoplankton are most nutritious at the end of the day and should be eaten then, after photosynthesis has had maximum time to increase cell organic content and quality.

Study of hundreds of species in hundreds of locations and situations has revealed many variants on the basic down-at-dawn, up-at-dusk theme. An early observation (Michael 1911) of DVM in chaetognaths seemed to show that they swim up at dusk into a fairly restricted near-surface layer, presumably having followed a rising isolume until it passed out through the surface, then slowly scatter downward through the middle of the night, a phenomenon called "midnight scattering". In the following dawn, when there is again visible irradiance, but at less than the preferred intensity, the animals rise toward the source, the surface, possibly seeking the ideal isolume. This is called the "dawn rise", and is shown explicitly by the *Daphnia* experiments of Harris and Wolfe (1955) discussed above. While they can be modeled in the laboratory, the importance of neither midnight sinking nor dawn rise is well established in the ocean. Even though the data suggesting the phenomena and their mechanism came from the ocean, Michael's original results were not strongly convincing. Some of Wade and Heywood's (2001) ADCP data show something like midnight sinking, but only well after midnight, long after any preferred isolume was absent from the upper ocean. This suggests operation of a food saturation effect.

Pearre (1973) suggested that the subarctic chaetognath *Sagitta elegans* exhibits something akin to midnight sinking as a result of satiation. The arrow worms rise at dusk, then begin to hunt in the well-stocked surface layers. Once they have ingested a copepod or two, the food quota for the day is essentially met, so they can return to depth early, gaining the greater safety of the depths sooner. That this happens is indicated by the presence of surface-dwelling copepods in the guts of chaetognaths captured at depth in the middle of the night or later, but long before the slowest individuals, or at least those unlucky in the hunt, descend at dawn. Similar early departures for deep layers have been claimed for *Calanus* (Simard *et al.* 1985; Durbin *et al.* 1995) and sundry other plankton and fish. Durbin *et al.* showed data for fourth copepodites of *C. finmarchicus* at one southern Gulf of Maine station in May at which phytoplankton suitable for ingestion, i.e. cells >7 μm, were only available in the upper 20 m. However, fourth copepodites with substantial gut contents were present at all depths from near the surface to below 80 meters from at least 03:00 h. To make this descent of ~85 m while pigment was eliminated from the gut at a rate of $-0.022\,\text{min}^{-1}$ and still retain 3 of the original 5.7 ng pigment copepod^{-1}, requires the distance to have been covered in 29 min, nearly 5 cm s^{-1} or 25 body lengths s^{-1}. That is fast, perhaps a quarter of the predator

escape velocity, rather than a typical steady swimming velocity less than 10 body lengths s^{-1}. More work is needed, but if descent on satiation actually occurs, it appears from the data of Durbin *et al.* to have started well before midnight.

These observations of surface-acquired food in gut content at depth require that some individuals are *en route* up or down throughout the night, and perhaps in daytime as well. Some daring individuals may leave the security of the dim depths to grab a noon lunch; some safety-conscious night diners may leave for deeper layers out of the moonlight, even starlight, as soon as they have eaten. Trapping at night in a Washington State (USA) fjord of just upward-bound and just downward-bound individuals (Pierson *et al.* 2009) shows that many more copepods of several grazing species that are swimming down have full guts than do those swimming up. It is of related interest that non-migrating zooplankton, particularly species that are primarily herbivorous, often cease feeding during the day, even though they remain around the clock in the stratum with the most phytoplankton. Durbin *et al.* (1990) demonstrated with a time-series of gut-content pigment estimates for the copepod *Acartia tonsa* from Narragansett Bay that it ceases feeding altogether near dawn and does not resume feeding until dusk. This cycling of behavior also is certainly a predation avoidance adaptation. Casual observation of copepods in glass containers shows that those with their guts full of dark phytoplankton are far more visible than those that are empty.

In addition to its importance in understanding the individual lives and population dynamics of zooplankton (and much nekton), diel vertical migration has been suggested to be the largest mass movement of biomass on Earth. It moves significant amounts of carbon from surface layers deeper into the ocean, where all migrators respire and some die. Longhurst *et al.* (1990) suggested that the carbon transport is on the order globally of $2.7 \times 10^{14}\,\text{g C yr}^{-1}$, in their estimates about 20% of flux to sediment traps at ~150 m. Ontogenetic downward migrations of animals to their diapause depths add to this, particularly at high latitudes. Bollens *et al.* (2011) found a range of migratory transport estimates in the literature from 10 to 50% of particulate flux, and Hernández-León *et al.* (2010) suggest that nocturnal feeding on surface mesozooplankton by migratory midwater fish, particularly when aided by moonlight, generates substantial transfers to depth. Variation and uncertainty in these estimates remain large. Of course, some of what goes down each dawn comes back up at dusk, confounding the difficulty of evaluating net transfers via vertical migration. Like particle flux, most of the gut contents and tissue that remain at depth are metabolized in the water column, not on the seafloor.

Chapter 9

Pelagic food webs

Food chains were given an equivalent name by an Arab scholar as early as the eighth century CE (Egerton 2002), and the idea is obvious from sequences like grass eaten by goats that are eaten by people. Simple, direct food chains of that sort do exist in the sea, and they support some of the charismatic megafauna. For example, massive and persistent diatom production during summer in parts of the Southern Ocean is consumed by *Euphausia superba*, the antarctic krill. Juvenile and adult krill live in voluminous, relatively dense schools, making them suitable prey for rorquals: blue, fin, sei, and minke whales. These whales ingest great gulps of water and schools of krill, squeeze the water out through baleen, then swallow. They are well defended from predators (other than explosive harpoons) by size and have great longevity, but they can be attacked and eaten by killer whales. That is a food chain. However, even this system, with its obvious direct links, is more complex. Diatoms are also eaten by dinoflagellates, salps, copepods, and pteropods, and part of the diatom stock mixes or sinks to the seafloor and is filtered out by clams and worms. Krill are also the main diet of penguins, crab-eater seals (the teeth of which form a filtering lacework) and the cod-like notothenid fishes of the antarctic shelves and slopes. As commercial harvests devastated the rorqual stocks of the Southern Oceans in the decades before and after the 1985 moratorium on commercial whaling, populations of crab-eater seals (*Lobodon carcinophagus*) and penguins increased dramatically (Laws 1984), as did the leopard seals (*Hydrurga leptonyx*) that feed on both crab-eater seals and krill. Thus, even the diatom–krill–whale food chain is just one sequence of transfers in a complex *food web*. Marine food webs reside out of sight and underwater, so there are powerful challenges to characterizing them and quantifying the rates of trophic transfers. To a degree, those challenges have been met, and we have many insights.

An early illustration (Fig. 9.1) of the complexity of pelagic food webs is Hardy's 1924 diagram of North Sea trophic transfers from microplanktonic algae (diatoms, dinoflagellates, small flagellates) to herring. Hardy drew the arrows from predator to prey, implying the effects of predatory population control. More recent diagrams often run the arrows in the other direction, that of organic-matter transfer. There are so many links that not all of them could be drawn, even though Hardy knew that they exist. The euphausiid *Nyctiphanes*, for example, has no arrows, but it feeds on smaller zooplankton, mostly copepods, and phytoplankton. *Tomopteris*, a predatory planktonic polychaete, could have many more arrows, eating virtually any smaller plankter passing by. Adult herring almost certainly eat larval herring when they encounter them, just as they eat sand lance (*Ammodytes*) juveniles. Moreover, herring are not the top of the food web with which they interact, but are eaten by cod, hake, small sharks, tuna, harbor porpoises, seals, and seabirds, among a longer list.

We are not done, yet, with even the complexity well known by Hardy in 1924. Each of the metazoan elements of the food web has a development sequence from eggs to ~1000-fold larger adults, with shifts in diet and predators as they grow. A copepod nauplius is much the same size, eats the same diet, and dies in the mouths of the same array of predators as a tintinnid protist. The entire dynamics of such skeins of connections cannot be fully evaluated, but we have learned since Hardy's work that there are multiple food-web levels involving smaller organisms: autotrophic bacteria, nanoplanktonic grazers, abundant herbivorous, and predatory protists. Organisms release organic matter into solution directly and during feeding. It is taken up by heterotrophic bacteria and returned to the particulate food web. Some of these features, as well as pelagic primary production, have been discussed in earlier chapters and need no review here.

Biological Oceanography, Second Edition. Charles B. Miller, Patricia A. Wheeler.
© 2012 John Wiley & Sons, Ltd. Published 2012 by John Wiley & Sons, Ltd.

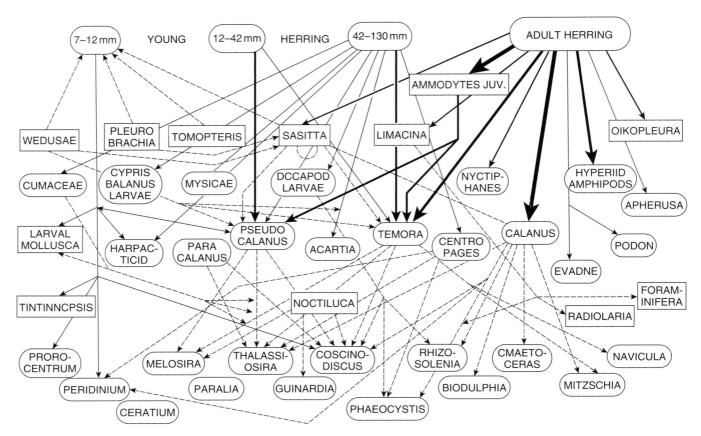

Fig. 9.1 Alister Hardy's (1924) classic pelagic food-web diagram from large phytoplankton to herring. All of these links are accurate. Many of the predators of herring (and trophic levels above those) were also known in 1924. Recent insights introduce much more complexity at the primary production and herbivory levels. (After Hardy 1924.)

Approaches to what animals eat

Gut-content analysis

What specifically does a particular class or species of animal eat? The obvious way to find out is to collect some of them when they have been eating, open their stomachs and see what is there – an approach termed gut-content analysis. Second best is to poke through feces for identifiable residues from prey. These are obvious approaches, but they involve modest difficulties. First, meals taken in the course of capture, rather than before, must be identified somehow. "Net feeding" is very common in the capture of zooplankton and fish with nets, and is likely the best explanation for observations like copepods holding chaetognaths in their mouthparts (Davis 1977). For many predators, such meals can be taken out of consideration by their forward position in the gut, lack of digestion, or simply improbability.

Second, regurgitation or defecation of partly digested food under the stress of capture is demonstrated for many animals, reducing feeding rates estimated from frequency of prey in the guts of individuals. And third, a substantial part of gut contents is likely to be unrecognizable due to chewing and enzymatic breakdown, leading to food categories like "green matter" or "gray stuff". To date, specific identification with DNA sequence probes and PCR of prey that have been converted to gray stuff has been hampered by the very rapid digestion of DNA.

Despite its shortcomings, gut-content analysis provides useful information. Sullivan (1980) examined the meals in collections of chaetognaths from the Gulf of Alaska. These tubular animals impale their prey with grasping hooks, manipulate them through the jawless mouth, then rapidly transfer them along the gut to just ahead of the anus (at the trunk–tail septum), where they are digested. An undigested residue is eliminated after several hours as a compact fecal mass. Thus, meals eaten well before capture can be examined by cutting across the body anterior to the digestion site, then pressing out the meal. Virtually all meals of *Sagitta*

(*Parasagitta*) *elegans* and *Eukrohnia hamata* were copepods. So, the next questions were: which copepods and do the predators eat selectively from the smörgåsbord of copepods available? To answer those, it is necessary to identify copepod remains after nearly complete digestion, which turns out often to be possible. One anatomical part, the tooth-bearing jaw, is resistant to digestion, and copepod jaws are distinctive among genera and often among species (Sullivan *et al.* 1975). Their size gives the developmental stage.

Sullivan found for juvenile *Sagitta* (4–14 mm long, living mostly in the top 25 m) that the predominant meals were *Oithona similis*, a ~1 mm copepod found in 37% of worms (14 to 73% in seven samples of 100). The sampling was done with paired bongo nets of 183 and 333 μm mesh, and the chaetognaths were taken from the coarse mesh sample that included almost no *Oithona*, so net feeding on at least that species was likely nil (it was up to 50% of meals for chaetognaths from the 183 μm mesh samples). From the fine-mesh sample came an estimate of the relative abundance of *Oithona* and other species, allowing an evaluation of how selective the chaetognaths had been in taking their

dominant prey over other candidates. An electivity index ranging from −1 (rejection) to +1 (exclusive diet),

$$E = (\% \text{ in guts} - \% \text{ in prey mix})$$
$$\div (\% \text{ in guts} + \% \text{ in prey mix}),$$

averaged −0.05, not significantly different from zero. A few other prey were taken, with *Metridia* copepodites having positive average electivity. *Eukrohnia* juveniles, living mostly below the seasonal thermocline, ate lots of *Oithona* but selected against them ($E = -0.33$), favoring *Oncaea* ($E = +0.73$), a copepod of similar size that clips on to bits of mucus, jellyfish surfaces, and the like. *Sagitta* and *Eukrohnia* larger than 14 mm switched progressively to the copepodites of *Calanus* and *Neocalanus*, swallowing animals as wide and slightly wider than their heads, with electivity for *Oithona* becoming negative. Like snakes, arrow worms can stretch their mouths and bodies around prey somewhat wider than they are. Prey size increases with body size, perhaps most importantly mouth size (Fig. 9.2; Pearre 1980).

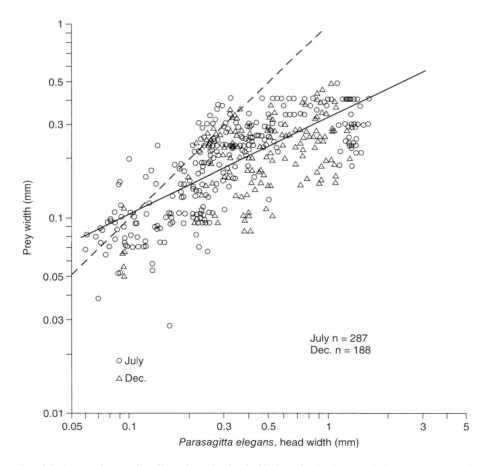

Fig. 9.2 A demonstration of the increase in prey size with predator size, head width here, for the chaetognath *Parasagitta elegans*. The spread of prey-grasping spines increases with head width. Both the trend for increase and the scatter are typical of many (not all) predators. The dashed line is 1 : 1, and the best-fit exponential function shows that head width increases faster than prey size. (After Pearre 1980.)

Since 1980, there have been perhaps 30 significant studies of chaetognath gut contents along the same lines. Baier and Purcell (1997) tested the importance of immediate defecation (or vomiting) of recently eaten prey on capture for *Sagitta* (*Flaccisagitta*) *enflata*, which eliminates meals that should have been counted. They compared 2-minute hauls (immediately preserved) to 5-minute and longer hauls, finding a drop from more than half of worms having gut content to only a quarter. This would shift gut-content results dramatically in respect to both chaetognath nutrition and the quantification of their impact on prey stocks. That conclusion can be generalized: gut-content studies are of great qualitative value and are consistently problematic for quantitative purposes. While calculations can be made of organic-matter intake rates (for chaetognaths see Pearre 1981) by multiplying (meals per predator) × (meal mass) × (a gut-residence time of meals from experiments, often a function of temperature), such estimates suffer from multiple biases. On the other hand, gut contents do identify what animals eat, the times of day that eating occurs, and approximately the degree to which prey are eaten as encountered or are chosen selectively.

There are studies like those for chaetognaths for a wide range of animals, especially fish. Hardy's (1924) herring study was based on gut-content analysis. Tuna are a convenient subject because they eat relatively big prey (but also small zooplankton like hyperiid amphipods, those perhaps as riders on swallowed jellyfish) that remain identifiable in the stomach. Different species differ somewhat in diet, but all are visual predators feeding primarily in daylight and crepuscular periods. They swim at high speeds (bursts to five body lengths per second), usually attacking relatively large prey and sometimes working in groups to corral them. Olson and Boggs (1986) provide a gut-content analysis (Table 9.1) for yellowfin tuna (*Thunnus albacares*) caught in purse seines in the eastern tropical Pacific (ETP). At least in this region and on across to Samoa (Buckley & Miller 1994), fish are the dominant prey, particularly mackerel and smaller tuna. Squid are important, especially for younger tuna. Galatheid crabs eaten in the ETP are the small epibenthic *Pleuroncodes planipes* that occasionally swarm at the surface. Over island banks near Samoa, crustacean prey are stomatopods (mantis shrimp). Vertically migrating mid-water fishes like *Vinciguerria* are probably eaten during dusk as they arrive near the surface. Recent work with pressure-recording internal tags shows that tuna also make occasional dives into mesopelagic layers, where they likely feed on bioluminescent prey (squid, gonostomatids, myctophids). It appears that some prey, squid for example, decrease in the diet as yellowfin tuna grow larger, while other prey like frigate tuna increase. Data in Table 9.1 are percentages of the overall diet, and the shifts with age are relative. In fact, at 13.4% of 10 g of food in its stomach, an "average" age-1 tuna has eaten 1.3 grams of squid, while at 1.5% of 310 g of food, an age-4 tuna has eaten 4.7 g of squid. Some foods, like pompano, a fish, do drop from the diet as the tuna grows, but most are still on the menu, and the population impacts of growing tuna increase for most of their prey.

Gut contents (and other data) show that tuna may frequent areas where specific prey are aggregating for feeding or mating. Around the mid-Atlantic São Pedro and São Paulo Islands just north of the equator there is an aggregation of flying fish (*Cypselurus cyanopterus*) from November to January. Yellowfin move inshore to feed on them, and flying fish become the dominant meal in tuna guts (Vaske *et al.* 2003). The observations are certainly biased toward flying fish as prey, because the tuna are captured by handline fishing under night lights deployed to attract the flying fish. However, Vaske *et al.* found they were only 42% of observed meals in 210 fish with gut contents (of 395 total), and an ommastrephid squid, *Stenoteuthis pteropus*, was also common (27%).

Another example occurs in the Gulf of Guinea several degrees north of the equator. Zooplankton are relatively rare along that latitude, and *Vinciguerria nimbaria*, usually diel vertical migrators, apparently cannot gain enough food at night to support their metabolism. So, they remain at the surface, feeding throughout the day. From sonar and trawl results, Ménard and Marchal (2003) showed that they form schools about 15 m thick and 30 m diameter (some schools larger) at between 60 and 75 m depth, and the schools cluster in non-random patterns. The fish, mostly adults 38–48 mm long and ~0.6 g, are fed upon by skipjack and juveniles of both bigeye and yellowfin tuna, all ~46 cm long and weighing 1.9 kg, that are in turn preyed upon by a purse-seine fishery. Indeed, the gut contents of tuna from the fishery are *V. nimbaria*; neglecting a few empty stomachs there were 1 to 150 (mean 45) of these small fish recently eaten by each tuna. Using approximate digestion times, Ménard and Marchal estimated the daily ration at 66 to 133 g d^{-1}, or 3.5 to 7% of tuna body mass.

Similar examples and similar variations can be found for a wide variety of pelagic animals from all ocean areas and depths – checking gut contents is an almost reflexive observation for biologists. It generates good and often true stories. Some results for zooplankton were discussed in Chapter 7. The tuna results are quite usual in that, like tuna, most animals eat anything within suitable size limits that presents itself and moves slowly enough to overtake and ingest. As both the chaetognath (Fig. 9.2) and tuna data (Table 9.1) show, the size of prey tends to increase as a predator grows, but food less than the maximum ingestible size continues to be part of the diet.

Trophic-level assignment by stable isotope proportions

Above we mentioned "food-web level", for which the standard jargon is "trophic level". Much attention has been

Table 9.1 Food of yellowfin tuna, *Thunnus albacares*, in the eastern tropical Pacific (5° to 15°N from Central America out to 140°W). Data are gut-content analyses by Olson and Boggs (1986). Data are percentage of the total mass of prey in the guts.

Class	Family	Description	AGE 1	AGE 2	AGE 3	AGE 4
Number of specimens			53	637	1897	994
Crustacea	Galatheidae	Benthic crabs that swim to the surface	3.1	3.5	1.8	1.2
	Portunidae	Benthic crabs that swim to the surface	0	0.5	2.1	3.6
Cephalopoda		Squid	13.4	8.8	4.1	1.5
Osteichthyes	Gonostomatidae	Small mid-water fish	14.8	11.8	4.3	3.4
	Excoetidae	Flying fish	10.4	12.9	7.7	3.7
	Bramidae	Pomfret, fanfish	12.7	3.2	2.1	0.1
	Carangidae	Pompanos, jacks	0	0.5	0.9	2.2
	Scombridae	Tunas, *Auxis* ("frigate tuna")	39.7	42.4	58.2	55.5
	Nomeidae	Driftfish (perch-like)	0	11.1	12	21.7
	Balistidae	Triggerfish	1.4	0.1	0.4	1.3
	Tetraodontidae	Pufferfish		1.1	3.9	3.2
	Other		4.5	4.3	2.5	2.6
Mean stomach content (grams)			10	50	116	310

given to determining the trophic levels of marine organisms: from gut-content analysis and in recent decades from shifts in the relative abundances of the stable isotopes of carbon and nitrogen (e.g. Fry & Sherr 1984). However, trophic level is an abstract concept; trophic levels do not swim about in the sea, available for collection and study as such. Primary producers might seem a possible exception, being identified by their photosynthetic pigments, but many of them (dinoflagellates, other flagellates, ciliates with borrowed chloroplasts, . . .) are actually "mixotrophs", getting nutrition both auto- and heterotrophically. Most pelagic consumers feed at more than one trophic level. For example, ciliated protists can ingest *Synechococcus* or a prymnesiophyte in one hour and flagellated microheterotrophs in the next, and a tuna can eat a mackerel in one swallow and a chain of salps in the next. Thus, few animals "belong to" a single, integral trophic level. Nevertheless, like some other abstractions that do not exist (an ideal gas, an infinitely dilute solution, a stable age distribution), trophic levels are a useful notion for evaluating how much primary production is metabolized by an ecosystem in producing its higher-order carnivores, i.e. for evaluating the transfer efficiency of the food web.

As carbon and nitrogen bound in organic matter are passed progressively up a food web, the tissue that omnivores and carnivores construct from them becomes progressively enriched in the less-abundant stable isotopes ^{13}C and ^{15}N relative to the dominant ^{12}C and ^{14}N. That happens primarily because compounds of the lighter isotopes fit somewhat more readily into the active sites of metabolic enzymes, and are preferentially removed from the tissues. The rarer isotopes amount to only several atoms per thousand, and "per mille" (‰) units are the favored expression of their abundance. Generally, those ratios are presented as comparisons to the ratios in standard substances, symbolized $\delta^{13}C$ and $\delta^{15}N$ (Box 9.1). Use of the standards simplifies the mass spectrometry required to determine the isotopic ratios.

Changes in both $\delta^{13}C$ and $\delta^{15}N$ occurring at each trophic-level step are quite strongly variable (Post 2002), which is most often handled by using overall averages from a mixture of laboratory studies and food-web studies for which trophic position is relatively obvious. Changes in $\delta^{13}C$ are small from diet to consumer, ~+0.39‰, but $\delta^{13}C$ varies for other reasons, and we will return to it. According to Vanderklift and Ponsard (2003) and Caut *et al.* (2009), $\delta^{15}N$ changes by ~+2.4‰ at each trophic transfer, termed the trophic enrichment factor (TEF). Because this is much larger than for ^{13}C, nitrogen-isotope ratios are favored for trophic-level estimation. A well-worked example has been

Box 9.1 Determination and expression of ratios of carbon and nitrogen stable isotopes

Measurements of $^{15}N/^{14}N$ in tissue are made by drying and grinding it; burning the powder with pure oxygen at 1000°C in a carbon–nitrogen analyzer; drying water from the gases produced; passing the carbon and nitrogen oxides through a reduction column (shredded copper) to obtain N_2; separating N_2 from CO_2 in a gas-chromatographic column; and finally determining relative isotope abundances (^{14}N, ^{15}N, ^{16}N) in a mass spectrometer. Measurements of $^{13}C/^{12}C$ are made in the same general way: the CO_2 from the burn is directly analyzed by mass spectrometry, with $C^{18}O^{16}O$ separated from $^{13}C^{16}O_2$ by mass difference. Multiple heavy-isotope combinations are rare, but also have distinctive masses.

Both $^{15}N/^{14}N$ and $^{13}C/^{12}C$ ratios are conventionally characterized by comparison to standards. For nitrogen, it is atmospheric nitrogen, for which $^{15}N/^{14}N = 0.003660$ (3.66‰). Since these ratios amount to parts per thousand, ‰, the comparison of a sample to the standard is made as a ratio of ratios in that unit and symbolized as $\delta^{15}N$:

$$\delta^{15}N = 1000[(^{15}N/^{14}N_{sample}/^{15}N/^{14}N_{standard}) - 1]‰$$
$$= 1000[(^{15}N/^{14}N_{sample}/0.003663) - 1]‰.$$

This ratio of ratios equation is usually used to present $\delta^{at.wt}X$ values. However, if you multiply the quantity in brackets by $(^{15}N/^{14}N_{standard})/(^{15}N/^{14}N_{standard}) = 1$, you will see that it is really a *difference* of the sample and standard ratios.

One standard for carbon in early work was $^{13}C/^{12}C$ from fossil cephalopod shells, *Belemnitella americana*, found in the Pee Dee Cretaceous limestone formation in South Carolina, USA, and thus called Pee Dee Belemnite (PDB). The ratio is very high for a natural substance, 0.01111 (11.1‰), such that the $\delta^{13}C$ values for natural organic matter are mostly negative:

$$\delta^{13}C = 1000[(^{13}C/^{12}C_{sample} \div ^{13}C/^{12}C_{standard}) - 1]‰$$
$$= 1000[(^{13}C/^{12}C_{sample} \div 0.01111) - 1]‰.$$

The advantage of limestone is obvious: add a little acid and you get CO_2 instantly. It is too late to consider whether it is advantageous for almost all relative ^{13}C contents to be reported as negative numbers. The standards must be repeatedly analyzed to check the mass spectrometry so, when stable isotope studies became very popular, the original PDB standard ran out. A laboratory in Vienna generated a cross-calibrated substitute limestone standard now commonly used, termed Vienna-Pee Dee Belemnite (V-PDB). There are several other standards that are precisely calibrated to PDB. Unless a report specifies otherwise, $\delta^{13}C$ values are made equivalent to PDB as standard.

---

provided by Olson *et al.* (2010), an estimate of yellowfin tuna trophic level in the northern portion of its eastern tropical Pacific population. They compared the $\delta^{15}N$ of tuna muscle sampled from commercial catches to $\delta^{15}N$ of "omnivorous" copepods sorted from net tows taken across the same area as the tuna. Copepods were chosen as a "proxy for the base of the food web", to simplify the skein of the initial pathways from primary production to tuna, and Olson *et al.* assigned them an approximate trophic level of 2.5, which may be as good a guess as any. Thus, the trophic level (TL) approximation from $\delta^{15}N$ values of yellowfin tuna (YFT) and copepods (COP) is:

$$TL_{YFT} = TL_{COP} + (\delta^{15}N_{YFT} - \delta^{15}N_{COP})/TEF,$$

in which $TL_{COP} = 2.5$ and $TEF = 2.4‰$.

In the comparison of tuna to omnivorous copepods, the $\delta^{15}N$ values of copepods turned out to vary in relatively smooth fashion across the sampling region (Fig. 9.3a), a pattern explained by the $\delta^{15}N$ distribution of upper-water-column nitrate in this area. That in turn is a product of the regional distribution of $\delta^{15}N$ in nitrate, which is increased by bacterial denitrification in the shallow suboxic layers, a process that favors $^{14}NO_3^-$ and elevates ^{15}N abundance in residual nitrogenous nutrients (Sigman *et al.* 2005). Moreover, the $\delta^{15}N$ of tuna muscle had a similar geography, implying that the tuna move about over only modest distances within the larger distribution pattern, at least relative to the tissue turnover time of tuna, which has been measured in laboratory studies as ~37 days. Olson *et al.* fitted the statistical surface shown in Fig. 9.3a to the copepod $\delta^{15}N$ at the station points, and then plotted the tuna $\delta^{15}N$ against the copepod $\delta^{15}N$ predicted by the surface (Fig. 9.3b). Dividing the differences by 2.4‰ and adding $TL_{COP} = 2.5$, the estimated TL_{YFT} varied from 4.1 to 5.7, averaging 4.7.

Olson *et al.* (2010) also compared $\delta^{15}N$ and gut-content TL_{YFT} results to those from a more recent $\delta^{15}N$ technique:

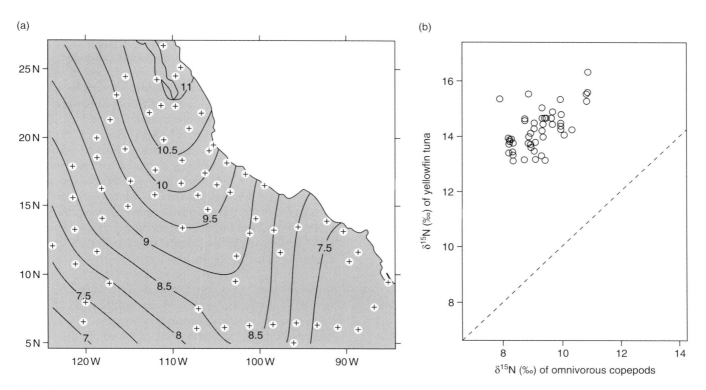

Fig. 9.3 (a) Contour plot of $\delta^{15}N$ values (‰) of omnivorous copepods fitted by a smoothing routine to 68 sampling stations in the Pacific west of Mexico. (b) Measured $\delta^{15}N$ values of yellowfin tuna versus $\delta^{15}N$ values of omnivorous copepods from the fitted contours in (a). The dashed line is the 1:1 relationship. (After Olson *et al.* 2010.)

compound-specific isotope analysis (CSIA) of constituent amino acids (McClelland & Montoya 2002; Popp *et al.* 2007; Hannides *et al.* 2009). Stable nitrogen isotopes in different amino acids (AA) have different enrichment responses to trophic transfers because they have strongly different metabolic pathways. Some AAs (most strongly and consistently glutamic acid, alanine, aspartamine) progressively accumulate ^{15}N at each trophic transfer and are termed "trophic AAs". These are those most frequently deaminated and then metabolized, with strong fractionation. Other "source AAs" (mostly consistently glycine and phenylalanine) are incorporated from food without significant fractionation, retaining the ^{15}N signature of the base of the food chain. The distinction does not parallel the difference between essential amino acids (synthesized only by autotrophs and some bacteria, and thus a dietary requirement for metazoans) and those synthesized by metazoans. It is definite that some AAs, including glycine and phenylalanine, retain the $\delta^{15}N$ of regional primary producers through multiple trophic transfers. Glutamic acid accumulates $\delta^{15}N$ at a recurring rate at each trophic transfer $\sim$7‰ per level (McClelland & Montoya 2002).

The analysis for CSIA involves hydrolysis of protein, conversion of the amino acids to heavier derivatives, gas-

chromatographic separation of individual amino acid derivatives, and finally mass spectrometry to determine the specific $\delta^{15}N$ for each. Using glutamic acid and glycine, the trophic level of tuna muscle was estimated from:

$$TL_{YFT(Glu-Gly)} = 1 + (\delta^{15}N_{Glutamic\ acid} - \delta^{15}N_{Glycine})/7‰.$$

The difference divided by 7‰ gives the number of trophic steps to reach tuna, and 1 is added for the tuna themselves. For six stations running northeast to southwest across the region of study (Fig. 9.3), the results (Table 9.2) for $TL_{YFT(Glu-Gly)}$ averaged 5.2, higher than the bulk muscle $\delta^{15}N$, but not significantly so.

Hannides *et al.* (2009) applied CSIA to four copepod species and one euphausiid from the North Pacific subtropical gyre (NPSG), using glutamic acid as the trophic AA, and compared it to glycine and phenylalanine as source AAs. The differences among the species (Table 9.3) were as expected from knowledge of these copepods and their likely diets (*Oithona* and *Neocalanus* TL = 2.1 to 2.2; *Euchaeta* and *Pleuromamma* TL ≈ 2.8 to 3.3), and all trophic-level estimates were surprisingly low, given the

Table 9.2 Estimated $\delta^{15}N$ ratios (‰, means with standard errors in parentheses) in glutamic acid and glycine from white dorsal muscle of yellowfin tuna from the eastern tropical Pacific, compared to $\delta^{15}N$ of bulk muscle. Trophic level, TL, was calculated with the McClelland and Montoya equation. (From Olson *et al.* 2010.)

SAMPLE NUMBER	BULK WHITE MUSCLE	GLUTAMIC ACID	GLYCINE	TL_GLU–GLY
10	14.3	27.0 (0.01)	−0.1 (0.01)	4.9 (0.02)
13	13.1	27.0 (0.28)	−2.1 (0.00)	5.2 (0.08)
16	14.0	25.0 (0.04)	−5.4 (0.13)	5.3 (0.06)
31	14.6	28.7 (0.03)	−3.9 (0.34)	5.7 (0.09)
33	14.8	26.7 (0.05)	−1.5 (0.14)	5.0 (0.06)
34	14.0	27.1 (0.83)	0.4 (0.99)	4.8 (0.19)

Table 9.3 Trophic levels (TL ± SE) estimated from $\delta^{15}N$ differences (glutamic acid – phenylalanine) and (glutamic acid – glycine), for four copepods and one euphausiid (*Thysanopoda*), (After Hannides *et al.* 2009.)

SPECIES	USUAL DIET	TL(GLU – PHE)	TL(GLU – GLY)
Euchaeta rimana	Small zooplankters	2.9 ± 0.05	2.9 ± 0.09
Pleuromamma xiphias	Small zooplankters	2.8 ± 0.08	2.7 ± 0.1
Neocalanus robustior	Omnivorous: protists and phytoplankton	2.2 ± 0.07	2.0 ± 0.08
Oithona spp.	Small moving particles, including flagellates	2.1 ± 0.1	
Thysanopoda spp.	Phytoplankton, protists, microplankton	2.3 ± 0.2	

known complexity of the picoplankton to microplankton food web in these oligotrophic waters. We review that complexity below. There were substantial seasonal differences in the $\delta^{15}N$ of both the source and trophic AAs, but the variations of both were in the same direction and amounts for all of the species, so that the TL estimates were nearly constant. Seasonal variation seems to derive from the isotopic variation of the available nitrogenous nutrients. Because the life cycles of at least the copepods are on the order of <1 to ∼2 months, and given the almost daily turnover of organic carbon at lower trophic levels, the $\delta^{15}N$ of their AAs should, indeed, follow seasonal changes in nutrient isotope composition. The CSIA-estimated TL should be roughly constant despite the shifts in nutrient $\delta^{15}N$.

We have found no satisfactory field-based studies of stable-isotope amplification in the marine pelagic microbial food web; possibly we missed them. There are many studies of $\delta^{15}N$ in particulate organic matter (e.g. Bode *et al.* 2007) and some culture studies. Hannides *et al.* (2009)

suggest that their low trophic-level estimates for NPSG copepods, using CSIA-$\delta^{15}N$, likely result from low amplification due to intense recycling. Virtually all organic nitrogen can be recycled to the inorganic form at high frequencies, likely every few days. Thus, amplification of stable isotopes may only take a permanent hold on nitrogenous organic compounds after they are transferred to moderately long-lived animals. Bode *et al.*, working with "plankton" collected with 20 μm mesh and then partitioned with nested sieves, found no mean or median differences among three size classes from 20 to 500 μm, mean $\delta^{15}N$ = 5.6‰. Plankton (actually particles) <20 μm were only 0.52‰ less, while plankton 500–1000 μm were 0.80‰ greater. Plankton retained by 1 mm, but passing 2 mm screens, had $\delta^{15}N$ = 6.8‰, very close to values for three copepod species. These small shifts among small but distinct size categories are in agreement with the proposal of Hannides *et al.* that rapid recycling of nutrients from small organisms reduces amplification of heavy isotope fractions. Bode *et al.* found $\delta^{15}N$ of 9.3 to 11.3‰ in muscle of planktivorous fish and 13.1‰ in the piscivorous common dolphin. Apparently, amplification occurs mainly in longer-lived, more-stable tissues. However, the results from Hoch *et al.* (1996) for a flagellate (*Pseudobodo*) and ciliate (*Uronema*) feeding on bacterial cultures, suggest that enrichment on the order of 3‰ can occur. Possibly, if organisms like heterotrophic flagellates, ciliates, and dino-flagellates can be separated in sufficient quantity from bulk POM, a single-step enrichment will be discernible. Such separations are a task for flow cytometry without the initial use of filters.

Estimates by Olson *et al.* (2010) of yellowfin tuna trophic level from both whole-muscle $\delta^{15}N$ and CSIA-$\delta^{15}N$ were reasonably close to an analysis based on mass-weighted averages of approximate TL values from gut contents of tuna from the same catches: $TL_{YFT(gut\ content)} \approx 4.6$. You are entitled to ask: Why would one bother with all the extraction of tissue nitrogen or with CSIA chemistry, and then with the mass spectrometry in both methods, when gut-content values give the same results? For tuna, perhaps, isotope evaluations can be taken as mainly an expensive check. However, the level of expertise required to identify and evaluate gut contents with results resembling quantification should not be underestimated. Because diets of generalist feeders can vary wildly (they eat everything of suitable size that they can overtake), very large numbers of full stomachs should be studied. Observers must be able to recognize at least general categories of prey from torn and digested remains, perhaps just a jaw or some fin rays. In some instances, suitable experts could be harder to find or fund than a mass-spectrometry laboratory. Also, some predators regurgitate gut contents at capture, so trophic placement requires other methods. When $\delta^{15}N$ is the method of choice, replication at both the field and laboratory levels should be applied to overcome the substantial inter-individual variability of the isotopic ratios (Post 2002). In any case, $\delta^{15}N$-based estimates have been made for marine biota ranging from zooplankton to albatross, including polar bears. The results are of uneven quality and should be used bearing that in mind.

Primary source identification for organic matter from $^{13}C/^{12}C$ ratios

The change in $\delta^{13}C$ at each trophic transfer is also an enrichment on average, but much less than for $\delta^{15}N$. Mean $(\delta^{13}C_{consumer} - \delta^{13}C_{diet}) \approx +0.39‰$, with a standard deviation of 1.3, based on a summary of 107 studies reported by Post (2002). Thus, the shift can be either positive (a small majority of cases) or negative, ranging from -3 to $+4$, providing minimal (not useful) trophic-level informa-

tion. However, in nearshore and estuarine waters, $\delta^{13}C$ has some value for determining the photosynthetic source of organic matter. There are two distinct sets of photosynthetic carboxylation reactions: the Calvin–Benson or C3 cycle typical of algae and some higher plants, and the Hatch–Slack or C4 cycle of grasses (including reeds and many crops – sugarcane, maize, wheat). The $\delta^{13}C$ values of C3 photosynthate are in the range from -24 to $-34‰$, whereas C4 values range from $-6‰$ to $-13‰$. Cacti and other xeric plants have an alternate C3 system, CAM, with $\delta^{13}C$ from -10 to $-22‰$, but their carbon mostly does not reach estuaries or oceans. For seagrasses, $\delta^{13}C$ ranges from -3 to $-24‰$, likely because isotope ratios of seawater carbonate available for photosynthesis are different from and more variable than those of CO_2 in air (Lin *et al.* 1991). For nearshore animals, snails or sea urchins for example, the difference of $\delta^{13}C$ from $-24‰$ can indicate the importance of algal vs. seagrass carbon. Also, despite the seagrass values, $\delta^{13}C$ greater than $\sim -24‰$ can be indicative of a partially terrestrial source for organic matter in the marine food web, while most marine organic matter will have considerably more negative $\delta^{13}C$. Possibly more useful, marine organic matter in riparian zones, sometimes delivered by salmon or other anadromous fish and distributed by large animals like bears and river otters, can be detected by unusually negative $\delta^{13}C$. There are examples of both applications in the literature, sometimes supplemented with other tracer isotopes such as sulfur-34 (Ailing *et al.* 2008). They consider a topic that we leave readers to pursue if they are interested.

Lower-level trophic transfers in the sea

Dilution experiments

Two methods have been widely used to examine trophic-level processes among organisms smaller than the larger metazoan zooplankton, often defined as those less than 200 μm: the micro-, nano- and picoplankton. Both methods were initiated by Michael Landry and colleagues: dilution experiments and size-fraction outgrowth experiments. Dilution experiments were explained in Chapter 3, and size-fraction outgrowth was introduced by Calbet and Landry (1999, 2004). Not surprisingly, dilution and fractionated outgrowth experiments have lately been combined in dilution experiments using just biota small enough to pass some small filter pores, say, 10 μm.

Of course, dilution experiments have been subjected to modeling (e.g. First *et al.* 2009), which shows some complexities that should be reviewed by those planning further applications. Those include the effects of grazer growth during incubations, which make the m vs. μ relationship non-linear, an effect observed in actual experiments by First *et al.* For present purposes we ignore that possibility.

Calbet and Landry (2004) reviewed the dilution results (see Box 7.5) of the many (788) experiments providing estimates of autotroph growth and mortality rates during the two decades after the method was introduced (Landry & Hassett 1982). These were not, of course, of uniform quality, but Calbet and Landry exercised some quality control, and the results are informative. Recall that the regression of apparent growth rate, call it μ_a, versus the fraction of sampled seawater, reduced in steps by adding filtered seawater (0.45 μm or smaller pores), provides estimates of autotroph growth rate, μ, unaffected by grazing as its intercept and the grazing mortality rate, m, as its slope. With greater dilution, two things progressively allow μ_a to increase: (i) relatively fewer grazers (G) in the incubation container, and (ii) fewer autotrophs (A) and so greater volume that each grazer must search to find one. The method is, then, an application of modified encounter theory, $m = k[G][A]$; the modification being that $[A]$ changes during the incubation (see Landry *et al.* 2000 for an approximation). To get realistic evaluations of the interactions, it is necessary to run the incubations for 24 hours, because photosynthesis occurs only in daylight, but grazing is continuous and phytoplankton cell division can also occur in, and often favors, the night. It can be argued that the best procedure is to obtain m from nutrient-enriched dilutions and μ from separate, unamended dilutions, and the data-set included 392 such pairs of m and μ that are plotted in Fig. 9.4.

The implication of the fitted line is that, on a global and all-season average, 57% of phytoplankton growth is consumed by microzooplankton on the day that it occurs.

While the regression in Fig. 9.4 is statistically significant, the scatter represents variability occurring and expected in the pico- to microphytoplankton and the community of grazers feeding on them. A preponderance of μ values near or even below $1\,d^{-1}$ still represents reasonably rapid growth, especially in cold regions. Values $>1.4\,d^{-1}$ (>2 doublings d^{-1}) are very high, and the modest number exceeding that are likely to represent the rapid cell cycles of very small phytoplankton. A substantial number of points near the 1:1 line are typical of oligotrophic habitats where pico- and nano-sized autotrophs and heterotrophic flagellates sustain a close balance between autotroph increase and mortality. The points both well above and well below the 1:1 line mostly (apart from unrealistic experimental outcomes) represent short-term measures from longer-term shifts: ups and downs in small autotroph abundance over periods of many days like those illustrated in Figs. 11.5a and 11.5b. Those are likely to occur in all oceanic systems, with lower amplitude where conditions are more oligotrophic.

Calbet and Landry (2004) also summarized dilution results for different ocean habitats (Table 9.4) in terms of proportions of primary production grazed daily by microzooplankton. With the exception of estuaries, that fraction is on the order of two-thirds. That may imply a greater role for mesozooplankton grazing in estuaries, where a greater fraction of phytoplankton are usually of larger sizes and somewhat less accessible to microherbivores (but see discussion of heterotrophic dinoflagellates below). However, much of the consumption of estuarine phytoplankton is by benthic filter-feeders, and a good portion is mixed away

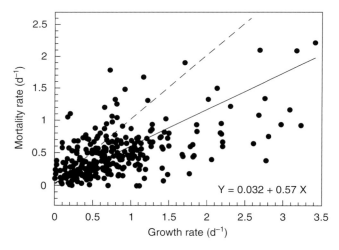

Fig. 9.4 Scatter plot of paired rate estimates from dilution experiments for phytoplankton grazing mortality, m, and phytoplankton growth rate, μ. Data are all from experiments in which mortality was from nutrient-enriched dilutions and growth from unamended dilutions. The solid line is a Model II linear regression ($r = 0.6$). The dashed line represents a 1:1 relationship (After Calbet & Landry 2004.)

Table 9.4 Percentage of primary production grazed daily by microzooplankton, as determined from dilution experiments. Values are mean percentages and standard errors for subsets of the data in Fig. 9.4. From Calbet and Landry (2004).

HABITAT	PERCENT OF PRIMARY PRODUCTION
Oceanic	78.0 ± 1.8
Coastal	56.6 ± 2.9
Estuarine	38.6 ± 2.5
Tropical/subtropical	71.3 ± 2.3
Temperate/subpolar	68.8 ± 2.3
Polar	65.2 ± 3.7

when flow carries it out to sea. The remaining one-third of primary production (a very broad average) in oceanic areas can be eaten by mesozooplankton, is lysed by viruses, and contributes to multi-day increases in phytoplankton stock. It can also sink to depth, to be eaten on the way down to or on the seafloor. Recheck the variability in Fig. 9.4, which shows that grazing often exceeds growth; those events are crushed by the standard-error calculation in Table 9.4, which should only be taken as measure of the validity of the grand means as such, not a measure of how much the autotroph–micrograzer interaction varies.

In a companion paper, Landry and Calbet (2004) discuss the implication of these estimates, basically values of $m/\mu \times 100$, for the rates of secondary production by grazing protists. Based on Straile (1997, discussed below) they assign an approximate gross growth efficiency (GGE) of 30%. Thus, if the global average percentage of primary production grazed by protists is ~70%, their production should be on the order of 21% of primary production. If GGE is somewhat higher (lower), then the proportion is higher (lower). When there are multiple trophic levels among the microheterotrophs, the total production is somewhat greater. For three levels, total production would be:

$$21\% + 0.3 \times 21\% + 0.3^2 \times 21\% = 29\%$$

of primary production. More levels would add trivially more. If global primary production is approximately $44\,\mathrm{Gt\,C\,yr^{-1}}$, then production of grazing protists (and tiny metazoans; all heterotrophs left in dilution incubations) is likely about $10\,\mathrm{Gt\,C\,yr^{-1}}$.

Induced trophic cascades

Meta-analysis of dilution experiments implies intense interactions among ocean microheterotrophs and phytoplankton, and intense interactions can be revealed by short-term manipulations that induce trophic cascades. That is the idea of size-fraction grow-out experiments. Ocean water and the organisms that it contains are poured gently through filters of several pore sizes to initiate incubations of progressively smaller size fractions. Abundance of pico-autotrophs, and often heterotrophic bacteria (Hbact), are determined before and after a 24-hour incubation in each filtrate and in the original whole seawater (sometimes filtered through a $200\,\mu\mathrm{m}$ screen) and growth rates (assumed exponential) are calculated. In the original Calbet and Landry (1999) experiment in the North Pacific subtropical gyre, the pore sizes were 1, 2, 5, 8, and $20\,\mu\mathrm{m}$. Each filter from $20\,\mu\mathrm{m}$ down to $5\,\mu\mathrm{m}$ decreased the apparent growth rates of both *Prochlorococcus* and Hbact (text table, Chapter 5 page 111; Fig. 9.5). The interpretation is that the coarser filters remove progressively more of the predators on the grazers of bacteria-sized organisms, while the finer filters (2 and

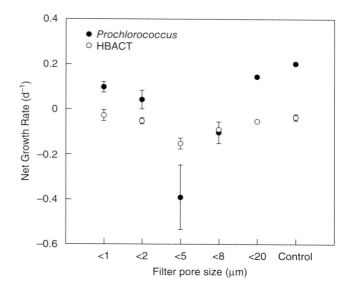

Fig. 9.5 Net growth rates of *Prochlorococcus* and heterotrophic bacteria in the NPSG after removal of organisms larger than the filter pore sizes along the abscissa. The control rates are those in unfiltered water. Decreasing growth in water retaining progressively smaller organisms, those >2 μm, and of increasing growth after removal of organisms >2 μm implies a multi-step food-chain among organisms smaller than 20 μm. Vertical bars are standard errors of four replicates. (After Calbet & Landry 1999.)

1 μm) remove both the predators and the grazers. Both the predators and grazers are primarily heterotrophic flagellates (lately termed Hflag (Fig. 6.2) that move to and ingest even smaller organisms.

Since the dominant primary producers of vast oligotrophic ocean regions, and during long seasons in temperate zones, are *Synechococcus* and *Prochlorococcus* and an array of miniscule autotrophic eukaryotic phytoplankton, Hflag are responsible for a very large part of the initial transfer of primary production into pelagic food webs. That first grow-out experiment was very clean. Later work (e.g. Calbet *et al.* 2001) shows that the outcome can have many patterns. However, in most cases with near-zero apparent growth in unfractionated water, filtrates of very fine filters generally show elevated growth rates. Microflagellates (Hflag) are generally present and feeding. Evaluation by Calbet *et al.* (2001) of Hflag with epifluorescence microscopy using proflavin (distinguishes Hflag from autotrophs) and nuclear stains showed shared numerical dominance of cells <2 and 2–3 μm, and an inverse dominance of biomass by relatively rare (a few percent by numbers) forms >10 μm (Fig. 9.6), both in the mixed layer and at the deep chlorophyll maximum.

Fonda Umani and Beran (2003) and Calbet *et al.* (2008) have combined a dilution protocol with size fractionation to evaluate grazing on pico- and nano-autotrophs by the <10 μm community. Studies were quarterly in the Adriatic

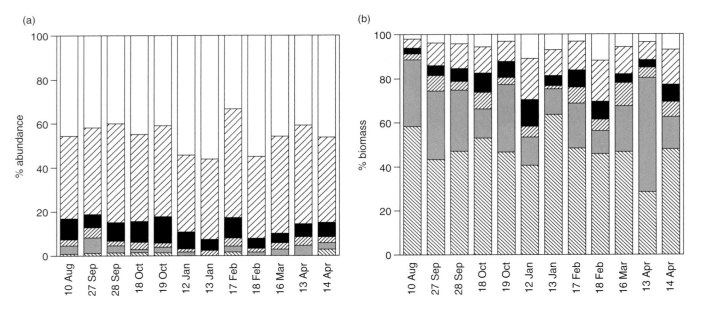

Fig. 9.6 (a) Proportions of numerical abundance and (b) proportions of biomass of size fractions of heterotrophic flagellates in the North Pacific subtropical gyre. Bars from top to bottom are size fractions: <2, 2–3, 3–4, 4–5, 5–10, and >10 µm. (After Calbet *et al.* 2001.)

and monthly off Barcelona, respectively. This combined approach is certain to be applied more widely to evaluate the feeding rates of specific size classes of microzooplankton. For the moment the results need replication. One of the problems of both dilution and cascade experiments has been the long intervals between trials. Sometimes experiments have been monthly – as off Barcelona, or quarterly – as in the Adriatic, or just once, such that the likely grazer–autotroph and micropredator–grazer oscillations are not revealed and can be at any stage when the experiments are performed. The labor involved in just one measurement is substantial, so time-series at near-daily frequency will require some further innovation. Another issue is that the quantity measured in many experiments is change in chlorophyll concentration, which is relatively simple but subject to changes not related to growth.

Some comprehensive regional comparisons of phytoplankton growth with grazing

Many attempts have been made to partition comprehensively both phytoplankton production and the grazing of it according to the sizes of the responsible agents. Studies on a quarterly basis in coastal Mediterranean waters by Fonda Umani and colleagues (discussed in the next section) are an example. Landry *et al.* (2011) assigned rates to size and taxonomic categories sampled and tested in the HNLC

eastern equatorial Pacific during cruises in December 2004 and September 2005. The 32-station series both crossed the equator at several longitudes (110° and 140°W) and ran east–west at 0° and 0.5°N. Simplified dilution experiments (just full seawater and 64% seawater) were conducted at all stations with samples from eight light levels (100% to 0.1% of E_0) down the water column. After 24-hour incubation at ocean temperature in irradiance simulated for the depths of collection, the change in chlorophyll content was evaluated for all bottles. Regressions of apparent growth vs. dilution were very good, and the intercepts (growth rate, $\mu\,d^{-1}$, for phytoplankton) and slopes ($m\,d^{-1}$, mortality from grazers; essentially all being microzooplankton and mostly protists, given the sample volumes) were compared (Table 9.5). Almost all contributions to the vertical integrals of growth rate, μ, came between 100% and 1% of E_0, which agrees with a very old approximation of the lower boundary of the euphotic zone. There was some photoinhibition (25%) above 50% of E_0. Mortality (grazing by heterotrophic protists) averaged 72% of increase, a fraction typical for oligotrophic habitats in all oceans.

Landry and colleagues extended the calculation, approximating the additional grazing likely from mesozooplankton by using gut-fluorescence measures (with the usual assumptions), and found an exact average balance over all stations between phytoplankton growth and total grazing mortality. In any case, the residual daily increase in phytoplankton growth after protist grazing is only about 30%, which must go to mesozooplankton, sinking and mixing

Table 9.5 Vertical integrals from eight depth profiles of two-point dilution experiment results: phytoplankton growth rate, μ, and microzooplankton grazing rate, m, for 32 eastern tropical Pacific stations. Bold values of >100% are likely to represent declining stocks; grazing can exceed growth. (After Landry et al. 2011.)

DATE	LATITUDE	LONGITUDE (°W)	μ (d^{-1})	m (d^{-1})	μ/m (%)
11 Dec 04	4°N	110	0.15	0.09	60
12 Dec 04	3°N	110	0.12	0.14	**117**
13 Dec 04	2°N	110	0.33	0.22	67
14 Dec 04	1°N	110	0.34	0.19	56
15 Dec 04	0	110	0.45	0.20	44
16 Dec 04	1°S	110	0.39	0.17	44
17 Dec 04	2°S	110	0.37	0.23	62
18 Dec 04	3°S	110	0.53	0.43	81
19 Dec 04	4°S	110	0.37	0.23	62
22 Dec 04	0	116.7	0.54	0.29	54
23 Dec 04	0	120	0.49	0.30	61
24 Dec 04	0	122.8	0.55	0.36	65
25 Dec 04	0	125.5	0.66	0.42	64
26 Dec 04	0	128.2	0.65	0.45	69
27 Dec 04	0	131.6	0.63	0.47	75
28 Dec 04	0	135.2	0.64	0.49	77
29 Dec 04	0	138.7	0.45	0.37	82
30 Dec 04	0	140	0.53	0.41	77
10 Sep 05	4°N	140	0.32	0.18	56
11 Sep 05	2.5°N	140	0.45	0.27	60
12 Sep 05	1°N	140	0.40	0.21	53
13 Sep 05	0.5°N	140	0.25	0.12	48
14 Sep 05	0	140	0.32	0.31	97
15 Sep 05	0.5°S	140	0.30	0.32	**107**
16 Sep 05	1°S	140	0.34	0.46	**135**
17 Sep 05	2.5°S	140	0.51	0.40	78
20 Sep 05	0.5°N	132.5	0.56	0.29	52
21 Sep 05	0.5°N	130.2	0.49	0.43	88
22 Sep 05	0.5°N	128	0.40	0.32	80
23 Sep 05	0.5°N	128.7	0.27	0.28	**104**
24 Sep 05	0.5°N	123.5	0.54	0.43	80
25 Sep 05	1.7°N	125	0.57	0.39	68
		Mean	0.43	0.31	72.56

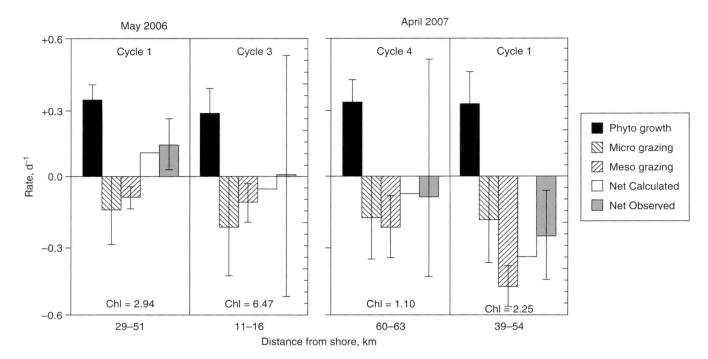

Fig. 9.7 Weighted (by chlorophyll concentration of each depth stratum) average rates of change of chlorophyll concentration due to separable processes in dilution experiments suspended from drifters down through the euphotic zone of the upwelling system offshore of Southern California. The averages include three to five repeated trials with each drifter (work cycles). Also shown are estimates of change due to mesozooplankton grazing and of net changes both observed in the water column and from differences of the average rate estimates. Results are shown for four selected cycles at different positions offshore and different chlorophyll standing stocks (surface concentrations shown in $\mu g\,liter^{-1}$. (After Landry *et al.* 2009.)

export. Landry *et al.* claim that viral lysis of phytoplankton is "invisible" to dilution experiments, although, if substantial, it is a transfer directly to heterotrophic prokaryotes. The grazing of 70–100% of each day's phytoplankton production does not eliminate the availability of produced organic matter for export to depth. Grazers defecate some; some is transferred up several trophic steps to migratory zooplankton and nekton that carry it down; and some eventually reaches whales that die and sink to the seafloor.

Landry *et al.* (2009) performed an essentially similar study in the more eutrophic California Current and close inshore off Point Conception during the upwelling season. A difference was attaching the incubation containers for eight, two-point dilution experiments to a weighted line suspended to 140 m from a Langrangian float with a drogue at 15 m. There were two float deployments in waters with chlorophyll concentration greater than $1\,\mu g\,liter^{-1}$ in each of May 2006 and April 2007. New incubations were attached to the suspended line three to five times in each deployment, producing a good deal of averaging. Overall, phytoplankton growth and microzooplankton grazing rates, again measured from changes in chlorophyll abundance, were integrated after "weighting the mean rates in

each depth stratum to the proportion of the total water-column Chl-*a* that it represented". These rates, averaged among days and integrated vertically (Fig. 9.7) show that microzooplankton grazed 43% and 80% of phytoplankton growth in the two 2006 observation cycles and 55% and 58% in the two 2007 cycles. The mesozooplankton grazing rates also shown (Fig. 9.7) are approximations from measured gut-content chlorophyll of bulk zooplankton coupled with approximate turnover times from a temperature-dependence relationship from Dam and Peterson (1988: fractional gut Chl reduction ≈0.0124 exp[0.077 $T(°C)$] min^{-1}). Landry *et al.* also listed mixed-layer phytoplankton growth rates that were two- to three-fold greater than the rates when integrated to 140 m (growth at depth is slow, while chlorophyll concentration can remain substantial).

The memorable message from both of these elaborate dilution studies is that microzooplankton grazing is a major fraction of phytoplankton growth, especially in oligotrophic but even in nutrient-rich, substantially eutrophic pelagic ecosystems. Similar studies worldwide have similar results: protists primarily, plus some metazoans <200 μm, usually consume half or more of primary production. Relatively greater consumption by macrozooplankton than

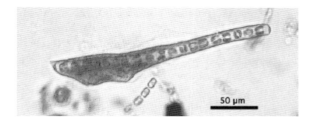

Fig. 9.8 Micrograph of a heterotrophic dinoflagellate, *Gyrodinium*, that has ingested a diatom chain that is much longer, 165 μm, than its original body dimensions. (Photo by Evelyn and Barry Sherr, published by Calbet 2008.)

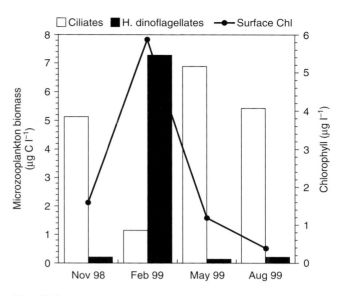

Fig. 9.9 Abundances of chlorophyll, ciliated protists, and heterotrophic dinoflagellates (*Gymnodinium* sp.) during four seasons in nearshore waters of the Gulf of Trieste. The intense phytoplankton bloom was a strong increase of diatoms (*Lauderia* spp.) Data are from Fonda Umani, S. & Beran, A. (2003). (Figure after Calbet 2008.)

microzooplankton sometimes occurs in eutrophic ecosystems, because large phytoplankton are often a dominant part of autotrophy there.

The importance of protist grazers on larger phytoplankton

Sherr and Sherr (2007) have provided a literature review (including some new observations) establishing that microplanktonic (20 to 200 μm), heterotrophic and mixotrophic dinoflagellates, most prominently *Gymnodinium* and related forms, are significant and at times the dominant grazers of microplanktonic phytoplankton, particularly including diatoms. They credit Evelyn Lessard (1991) for early recognition of that. Some of that feeding is accomplished using peduncles (Fig. 2.14) inserted through the cell membranes of prey. However, other dinoflagellates can ingest remarkably large cells, even diatom chains, using a pallium (Fig. 2.13) or simply by stretching around them to greater dimensions than their unfed size (Fig. 9.8). Heterotrophic dinoflagellates (Hdinos) increase in parallel with diatom blooms (Fig. 9.9), although their growth does not accelerate until diatom increase progresses to provide enough food. Thus, diatom stocks do, under quite consistent conditions, escape control by microzooplankton grazing and bloom (reviewed by Sherr & Sherr 2009).

The proportional level of Hdino grazing as blooms decline is likely more significant than recognized until lately, but is also strongly variable. Figure 9.9 was drawn by Calbet (2008) from data in Fonda Umani and Beran (2003), which specifies that the peak in chlorophyll was attributable to a massive and unusual bloom of *Lauderia* species. Fonda Umani confirmed for us from her data sheets that the "Protozoa non-Ciliophora" listed in the 2003 paper were heterotrophic *Gymnodinium* spp. So, very likely the increase of dinoflagellates (to >7 μg C liter⁻¹) was supported by the *Lauderia* as food. However, the *Lauderia* stock came to exceed 1000 μg C liter⁻¹(!). Fonda Umani *et al.* (2005) extended the Gulf of Trieste data-set to four

years; the second greatest diatom bloom was 300 μg C liter⁻¹, not *Lauderia* , with no substantial increases in Hdinos. Regarding the February 1999 *Lauderia* bloom, they state, "surprisingly, this high biomass was not eaten at all [take that to mean "significantly eaten"] by either mesozooplankton or by microzooplankton". Their sediment traps showed that virtually all of the bloom was exported to the bottom uneaten. Other observations are available (e.g. antarctic data in Archer *et al.* 1996) documenting modest effects from Hdino grazing on diatoms during the decline of blooms. The numerous but scattered data suggest Hdino grazing is sometimes significant, sometimes not. Access to prey nearly as large as the predator apparently does not extend to particle-feeding ciliate protists, which are largely limited to prey less than a tenth of their body dimensions.

Micrometazoans

In addition to larger protozoans, smaller metazoans are important as predators of protists. Small copepods (*Oithona, Paracalanus, Parvocalanus, Clausocalanus, Microcalanus, Ctenocalanus,* . . .) are by far the most numerous small mesozooplankton, much more abundant than usually shown with the classic nets of 200 and 333 μm mesh (Turner 2004). The adults are ~0.6 to 1.2 mm long, but much narrower and readily pass through the mesh. As a group they have relatively high fecundity, for example, up to >80 eggs ♀⁻¹ d⁻¹ for *Paracalanus parvus* in California coastal waters when food is abundant (Checkley 1980). Thus, very large numbers of nauplii <200 μm are part of the microplankton community, feeding partly as mid-level predators on Hflag and other nanoplankton. Their importance relative to that of ciliates and dinoflagellates remains to be adequately quantified. Copepods strongly favor ciliates as food (Calbet & Saiz 2005), possibly a main transfer link from the microbial to the mesozooplankton trophic levels.

Certainly the feeding impacts of the various components of the very small grazing community (2 to 200 μm) will vary greatly, so that a *spectrum* of possible combinations of sizes and taxonomic relationships must be considered and characterized. It turns out that experts on different groups are often also champions for their significance, which may or may not help with moving this task toward completion. At least they are out to sea gathering data for their teams.

Top-down cascades

As will be discussed in the chapter on fisheries, stocks of large, predatory fish have been dramatically reduced worldwide by fishing. For example, long-line catch rates of yellowfin tuna in northeastern Brazil's EEZ declined from 9.6 fish per 100 hooks in 1956 to 0.77 in 1971, stayed low for a decade, then recovered some to ~1.5 fish per 100 hooks from 1988 to 2003 (Vaske *et al.* 2003). In reviewing fisheries, we learn there are many cases of ~10–40% residuals of the once unfished or lightly fished stocks of large pelagic predators in the sea. It is expected from lake and terrestrial ecology that these reductions must have had and continue to have, trophic cascade effects. When hunters kill nearly all the coyotes, wolves, and carnivorous bears in a temperate ecosystem, populations of large herbivores such as deer expand, overgraze the vegetation, starve, and eventually die back. This has been well and repeatedly observed. Stocking of large piscivorous fish in a lake without them can sharply reduce populations of planktivorous fish. Zooplankton then increase and graze down the phytoplankton, "cleaning" the water to improve the view (and sometimes the air quality) for nearby luxury homes. Presumably, starvation then moves up the food web, but eventually a new equilibrium becomes established. Such an induced cascade is an anti-alga strategy that sometimes works.

Exactly what effects the dramatic reductions of apex predators are having on oceanic ecosystems is not particularly clear. The dramatic population shift of krill predators in the Southern Ocean from over-exploited rorquals to seals and penguins, mentioned above, is a top-down effect. Tuna fisheries, to continue that example, mostly take larger, older age classes that eat a preponderance of smaller tuna-like fishes (Scombridae) and squid. Those in turn eat small fish that primarily eat zooplankton. This oceanic food web has at least one additional step in these levels compared to adding or removing fish predators in a lake. In areas with lots of nutrients the zooplankton reduction might increase populations of large phytoplankton and lead to more rapid exhaustion of nutrients by them. By and large, however, oceanographers did not gather detailed data about ocean biota during the draw-downs of large predatory fish that would allow us to see the resulting cascades at lower levels. There must have been and continue to be such effects.

A widely cited (e.g. Baum & Worm 2009; Perry *et al.* 2010) case study by Shiomoto *et al.* (1997) is a possible interaction of pink salmon (*Oncorhynchus gorbuscha*) with zooplankton in the subarctic Pacific and Bering Sea. Pink salmon have a two-year life cycle, and all life stages eat zooplankton. Stocks of pinks spawning in coastal rivers of northeastern Asia alternate year-on-year between strong and weak year classes, and the data seem to show that zooplankton from 1989 to 1994 (three cycles) were low when pink salmon catches (CPUE) were high. Unfortunately, the salmon data came from the central Bering Sea, while the zooplankton estimates were from south of the Aleutians in the Alaska stream. Shiomoto *et al.* show data (also from the Alaska stream) that perhaps show the cascade extended to chlorophyll concentration. At least the oceanic migration patterns of pink stocks from Kamchatka to Anadyr Bay include both areas.

Strong evidence for top-down trophic effects comes from two studies by Frank *et al.* (2005, 2006) of fisheries on eastern Canadian shelves. These ecosystems are not strictly pelagic, but involve the interaction of demersal (near bottom, partly bottom-feeding) fish species (cod, haddock, flatfish) and small pelagic fish (herring and others), pink shrimp and crab. In one statistical subdivision, "area 6", as throughout the region from the Grand Banks to New England, a radical collapse of demersal fish (Fig. 9.10), particularly cod, but also haddock, hake, flatfish, and others, started in the mid-1980s, led to fishery closures in the early 1990s, without strong recovery through the years to 2010. There was a parallel increase in many of their prey, particularly well documented in area 6 for epibenthic shrimp (*Pandalus borealis*) and snow crab (*Chionocetes opilio*), other benthic invertebrates, and small pelagic fish. A food-chain effect is a reasonable explanation. There may have been a modest decrease of larger zooplankton, like *Calanus* and krill, and a substantial increase in phytoplankton was documented by continuous plankton recorder (CPR) surveys between the 1960s and 1990s (Frank *et al.* 2005).

Comparisons (Frank *et al.* 2006) of fisheries agency experimental trawling from 1972 to 1994 for sectors of the northwest Atlantic shelf from eastern Nova Scotia to the

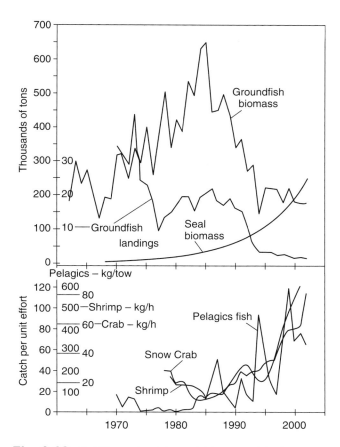

Fig. 9.10 (Top) Time-series of demersal fish ("groundfish") biomass estimates and fisheries landings on the eastern continental shelf of Nova Scotia in statistical area 6 (Fig. 9.11a). (Bottom) Time-series of abundance estimates in the same region for small pelagic fish (clupeids primarily), shrimp (*Pandalus borealis*), and snow crab. (After Frank *et al.* 2005.)

reasonably attributable to top-down effects may instead (or also!) have had bottom-up causes. Ocean ecology is not an experimental science with excellent control of all conditions except those under test.

Shrimp-for-cod substitutions similar to those adjacent to maritime Canada and New England have been observed in coastal Alaskan fisheries. See Baum and Worm (2009) for a catalogue of other, possible examples of top-down cascade. There are multiple-step trophic cascades in marine ecosystems. We just have a very slim file of convincing demonstrations. Despite that, understanding of stock-size interactions among predators and prey has come to hold a prominent place in the ecological understanding and lately the management of ocean ecosystems, both oceanic and coastal. Food-web models now hold a prominent place in codifying this understanding and making it useful. We will examine just one such approach, Ecopath models.

Marine food-web modeling: Ecopath and Ecosim

An important issue in modeling food webs is assignment of correct trophic efficiencies. The ratio between tissue formed and food eaten is an animal's *gross growth efficiency* (GGE). Values of GGE must be greater than *ecological efficiencies*, which are defined in terms of populations (or trophic levels) and incorporate more losses than metabolism and defecation. Specifically, any non-predatory mortality (due to genetic diseases, parasites, and infections) are additional transfers to the decomposer trophic compartment of the ecosystem. Thus, ecological efficiency refers to transfers "up" the food web. However, keep in mind that there are returns from the decomposer level to the "main" prey-to-predator trophic path. For example, copepods feeding at depth on falling fecal pellets can be eaten by myctophid fish that are later eaten by squid. Gross growth efficiencies set an upper limit on ecological efficiency, and are somewhat helpful with problems like approximating how much fish might be harvested from a region based on its primary production (see Chapter 17). The GGE of an animal is not fixed, but changes with size, age, and habitat conditions. This is obvious for people, who grow very rapidly immediately after birth, their weight doubling in a few months, but normally (obesity problems excepted) stop growing altogether before 20 years of age. We do not stop eating, but our growth efficiency drops to zero. Many fish continue to grow throughout life, but growth slows strongly. Given a replete diet, growth efficiency of larvae and early juveniles is typically ~30%, then drops off. For example, tank-reared, juvenile bluefish, *Pomatomus saltatrix*, initially

Labrador Sea (Fig. 9.11a) showed strong negative correlations (the opposing time-series trends shown in Fig. 9.11b) between benthic fish stocks (again, cod and company) and small pelagics (herring, etc.). The implied food-chain effect apparently extended down the food web to phytoplankton estimated from CPR records. It was low when fish stocks were high. Application of the trophic-cascade hypothesis seems straightforward, but ocean ecology consistently provides alternative hypotheses. Greene and Pershing (2007) have pointed out that, simultaneous with the cod collapse, Arctic Ocean and North Atlantic circulation shifted, reducing temperature and salinity downstream all the way to the mid-Atlantic Bight. They suggest that enhanced stratification due to salinity reduction allowed stronger phytoplankton blooms and increased abundance of small copepods (like *Pseudocalanus* and *Centropages*), for which some data show strong shifts from about 1991 in the Gulf of Maine and over Georges Bank. Thus, lower trophic-level impacts

(a)

(b)

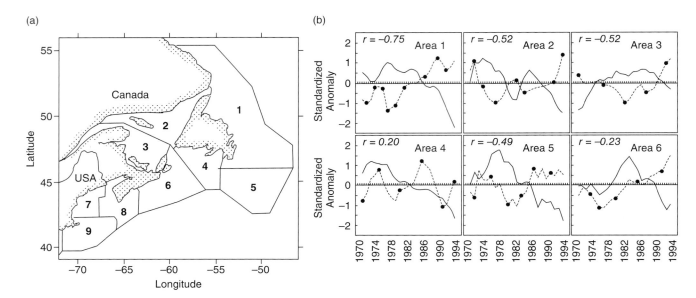

Fig. 9.11 (a) Subdivision into fisheries statistical areas (1 through 9) of the northwest Atlantic continental shelf. (b) Comparison from 1975 to 1994 from fishery-agency census trawling of groundfish (solid lines) and small pelagic fish abundances (dotted lines with some points as •). The plots are values of annual mean numbers/trawl haul plotted as standardized anomalies from the long-term means. The data included a total of 55,043 trawls that collected 26,286,369 fish of 412 species. (After Frank *et al.* 2005.)

weighing 1–2 g, grow at GGE = 29%. Over 90 days they increase to 86 g, with GGE dropping to 15% (Buckel *et al.* 1995). At ages of several years, GGE drops to a few percent at most. Both growth rates and GGE depend upon temperature, generally higher in the warmer half of the habitable range, and food abundance and quality. Restricted rations require larger proportions expended on metabolism, with a lower limit on GGE of zero, although some increases in assimilation efficiency are possible.

Growth efficiencies of zooplankton, both protists and metazoans, are more uniform through life, in some cases with egg production of adults sustaining output efficiencies nearly equivalent to early growth. Straile (1997) reviewed data to that date and stated that "all taxa (i.e. nano/microflagellates, dinoflagellates, ciliates, rotifers, cladocerans, and copepods) were found to have mean and median GGE of ~20–30%". The most important factor influencing variability was food availability, with lesser but positive effects of warmer temperatures within the usual habitat limits. The comparison of zooplankton with fish, much of the food consumption of which is by older, slow-growing individuals, suggests there is a major reduction of ecological efficiency as nutrient ascends the food web, simply due to progressive reduction of GGE. Estimating the overall ecological efficiency between any given trophic levels, even of a species population, remains largely a matter of guessing well.

Ecopath models are complex and can only be sketched here in a general fashion. They were initially developed by Jeffery Polovina in the 1980s, and then extended by Daniel Pauly, Villy Christensen, Carl Walters, and many others (Christensen & Walters 2004). They are steady-state models with populations represented by their biomass (or energy) and connected to each other by trophic relations. The basic assumption is a conservation of biomass (energy) such that input of food equals a sum of outputs for each population in a food web:

Food eaten = tissue formed + metabolism + defecation

Tissue formed (production) is then partitioned according its ultimate fate:

Production = predation mortality + catch mortality
+ net migration + biomass increase
+ other mortality

In the models, each population (stock) in a food web is assigned one of a linked set of equations based on those two. Generally, it is necessary to define at least some food-web "populations" in a general way, say as ecological guilds: "phytoplankton" and "zooplankton", for example, or "small squid" and "small mesopelagic fish". Other populations, those of central interest, can be particular species, perhaps chinook salmon or yellowfin tuna. A given species can be both prey and predator. As prey species i, it is subject to predation by a series of predators given subscripts j. The input–output (or mass-balance) equation (Heymans *et al.* 2007) becomes:

$$B_i \cdot [P/B]_i \cdot \mathrm{EE}_i = C_i + \sum_j [B_j(Q/B)_j \cdot DC_{ij}] + E_i + BA_i.$$

The product on the left is: prey stock biomass (B_i) times prey production to biomass ratio ($[P/B]_i$) times "ecotrophic efficiency" (EE_i = fraction of prey production consumed by predators *or* emigrating from the area of interest). This is the total of all growth by the prey stock over some interval (implied by the production rate P). It may seem odd to write $B_i \bullet (P/B)_i$ instead of just P. The reason is that B_i can be approximately measured as stock estimates, while the ratio $(P/B)_i$ can be a sum across age classes of year-on-year growth of individuals: weigh samples and age specimens from otolith growth rings or from guesses. This product represents all the growth "output" of the prey.

On the right are fisheries catch (C_i), plus net biomass change (E_i) due to prey emigrating from and immigrating into the region modeled (perhaps an unfished reserve, perhaps the whole Mediterranean Sea), plus biomass accumulation in the region (stock increase, BA_i), plus the sum over all predators, j, of their biomass times their consumption per unit biomass ($[Q/B]_j$) times the proportion of Q taken from prey species i (DC_{ij}). Again, B_j and $[Q/B]_j$ are more readily estimated than Q directly. (Beware: many papers on this subject are printed with subscripting errors: i for j, j for i.) Generally, such models are simplified by assuming, at least for some period of interest, that emigration is zero or balanced by immigration ($E_i = 0$) and that the prey stock does not change ($BA_i = 0$). Given N-linked equations of this type representing major stocks (or guilds) in a food web, with estimates or guessed values for most quantities, up to N missing quantities can be estimated. That generates a static picture of the transfer of organic matter (and energy) through that food web.

Almost all enthusiasm for Ecopath models comes from fishery scientists. They have estimates of sundry stock biomasses from fishery returns. Most species they work with can be aged (to give P/B from size vs. age data), have identifiable gut contents, and often remain identifiable in the diets of their predators. For example, in the northern Benguela current off Namibia, sardine, anchovy, hake (two species), horse mackerel, snoek (*Thyrsites atun*), several tunas, several sharks, and more, are all fished and fur seals are regularly culled. These harvests produce catch records of variable quality for most of these species that are engaged with each other and in the wider pelagic community as predators and prey. A static Ecopath model of this trophic

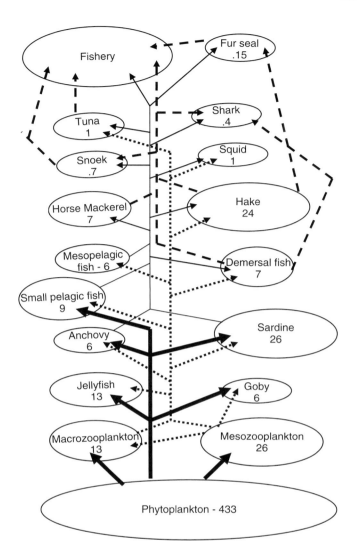

Fig. 9.12 Generalized diagram of trophic connections among major ecosystem components in the northern Benguela ecosystem off Namibia. Diet characterizations are from discussions in Heyman and Sumaila (2007). Phytoplankton grazing, heavy dark lines. Zooplanktivory, dotted lines. Predation on clupeids and other small pelagic fish, thin solid lines. Upper-level predation, dashed lines. Arrows point in the direction of nutrition transfer. Vertical lines in the center column imply potential prey switching and/or competition. Numbers in ovals are approximate annual production tonnages (millions of metric tons).

interaction scheme with 32 trophic groups (equations) was created by Heymans and Sumaila (2007). The main stocks and their interactions can be sketched as an Ecopath-like food web (Fig. 9.12).

Next, concern for sustaining both fisheries production (to generate food, money, and jobs) and ecosystem health creates the need to go beyond static models, like Ecopath, to evaluate the effects of changes in catches, in C_i values, often for stocks of many species at once. Effects of changing habitat conditions must also be evaluated. Several steps toward this are possible. Ecopath models have been converted by Carl Walters and colleagues to dynamic equations ($dB_i/dt = \ldots$) that accept various and varying values of Q_j, DC_{ij}, C_i, (P/B_i) etc. and can be implicitly integrated in a programming system called Ecosim (Christensen & Walters 2004). Heymans *et al.* (2009) used the parameter-fitting routines in Ecosim to generate a trophic-web history based on fisheries data for 1956 to 2003. They started from their Ecopath model for the system in 1956. The Ecosim model was fitted to time-series for the Benguela of (i) estimated fish-stock biomass (based on catch per unit effort and virtual population assessments.VPA) (Fig. 9.13a), and of (ii) fisheries catches off Namibia. The fitting was to

year-by-year biomass and catch estimates modified by a water-temperature function (Fig. 9.13b). The key variables, of course, were the C_i, for which the fisheries provided the data. Fits for catches are better than those for biomass, primarily because there are better data. The complexity of such models may in fact be sufficient to capture a substantial (if uncertain) portion of the significant interactions among major stocks of nekton (i.e. predator–prey interactions), major habitat factors (reasonably represented by temperature in an upwelling ecosystem), and fisheries. An obvious problem, as Ecosim was applied by Heymans *et al.* (2009) is the absence of prey-switching. The Ecosim program does allow for that, supposedly provided that

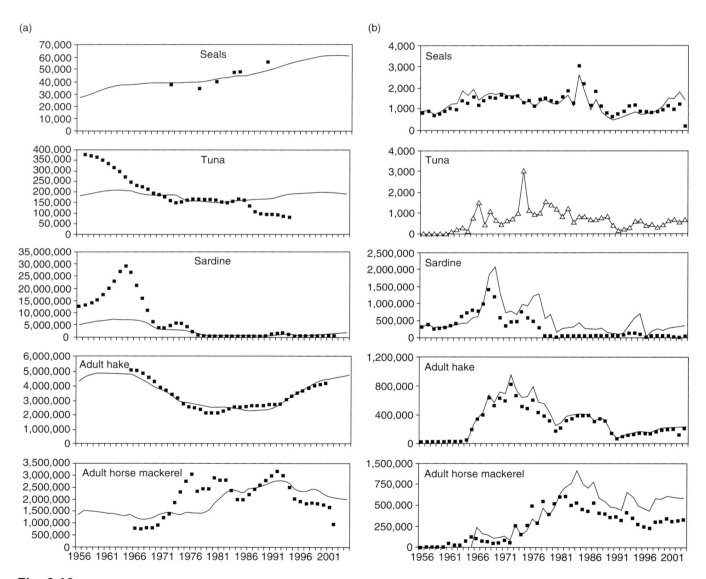

Fig. 9.13 Comparisons of data and Ecosim model results for populations of the northern Benguela upwelling ecosystem: (a) biomass (tonnes) and (b) fisheries catches (tonnes) from 1956 (baseline Ecopath community model) to 2001. Ecosim results are lines and annual data are squares, except that the tuna-catch model (and some other groups not shown) was forced to exactly fit catch data (triangles) to improve overall model results. The model is an overall best fit of Ecosim to data for all 32 community categories. (After Heymans *et al.* 2009, their Figs. 1 and 2).

actual data show what the switches are. Surely snoek, for example, are not particularly fussy about eating sardines when anchovy are unavailable.

This model and others like it are being used as a basis for "ecosystem-based management" of regional fisheries. Catches, C_i, can be regulated on the basis of the partial insight embodied in the models, and to an extent results from those directed C_i values can be used to improve the models. However successful this might be, interaction of such management with political forces, especially the interest of fishermen and fish processors in taking fish immediately, can be rocky.

Chapter 10

Biogeography of pelagic habitats

Throughout the history of biology, ideas of broad scope have been derived from studies of the distribution patterns of organisms upon the Earth. Darwin and Wallace were biogeographers, and many evolutionary mechanisms have been discovered through comparison of the distributions of closely related species and of subspecies. Because of this record of success for biogeography, biological oceanographers have determined distributional patterns for a modest fraction of the oceanic biota, particularly epipelagic zooplankton. As expected, the patterns suggest hypotheses about the history and ecology of the sea. Some hypotheses have been tested by examination of ancient patterns of distribution preserved in ocean sediments.

What is a "species"?

The basic unit in the study of distribution is the *species*. The notion of species has a long – and sometimes contentious – history that cannot be reviewed here. At present, several species concepts are in use. There is the "*biological species*" concept: a species is an interbreeding, or potentially interbreeding (if individuals could be moved over some barrier to mate), population of organisms. They are sufficiently similar to be inter-fertile. In the cases of *Homo sapiens* and other domestic animals, we have a great deal of direct experimentation with the question of inter-fertility. We have tried most of the possible crosses, and we know which work and which do not. However, for the typical marine animal or alga this experience is missing. We don't know which deep-sea fishes would be successful at mating, if they tried it. We have no operational way to decide which variation is *intraspecific* and which *interspecific* according

to the biological species definition. The typical plankton or benthos sample is a jar of dead bodies. It contains no information about inter-fertility. In this case, a species, sometimes called a "morpheme", is a group of organisms that share a great many characteristics, usually aspects of body form. Morphemes are separated by gaps in a continuum of degree of overall similarity. The degree of alikeness can be evaluated by sophisticated numerical techniques or by the common sense of experienced systematists. It is actually "species" described in the latter, somewhat subjective, mode that make up the vast majority of those we recognize. As an old saw goes, "a species is whatever a competent, recognized systematist says it is".

When that systematist assigns a name to a set of animals newly recognized as very similar and also distinct from obviously related sets, she or he provides a detailed description and (when preservation makes it possible) stores "type" specimens in some accessible, curated facility, generally a museum. Those practices allow new workers to determine the identity (or not) of new specimens with named species. Identification, also subjective to a degree, is central to biogeography. Species designated and identified in this fashion are termed "*typological species*".

In recent decades, most systematists have adopted the philosophy that classification of organisms should reflect their phylogeny (evolutionary relationships) as closely as possible. The revolution in molecular genetics promoted this adoption by providing semi-quantitative estimates of degrees of relationship: similarity (or difference) in DNA sequences from selected genes. These are eminently suitable for explicit systems (computable algorithms) that construct phylogenetic trees. At the level where species names are assigned to related sets of sequenced individuals at the tips

Biological Oceanography, Second Edition. Charles B. Miller, Patricia A. Wheeler.
© 2012 John Wiley & Sons, Ltd. Published 2012 by John Wiley & Sons, Ltd.

of the branches, this involves deciding what proportion of DNA base pairs (among hundreds to thousands) might vary among individuals likely to be exchanging genetic material (something like the biological species definition) versus larger proportions. Once this decision is made, usually based on within-group versus among-group variation in previously defined typological species (circularity seems inevitable), it can be applied to new cases within suitable limits, say just among diatoms or snail fish. Species defined in this way are termed "*phylogenetic species*". They play an increasingly significant role in marine biogeography, because many morphologically defined species are being found on the basis of gene differences to be clusters of closely related, "cryptic" species. Included in this trend has been development of DNA "bar codes" based on mitochondrial DNA (mtDNA): partly with the goal of identifying individuals from their "codes", partly to identify new, particularly cryptic, species (reviewed by Goetze 2010). Bar coding with mtDNA was the basis of a large-scale census of marine life (COML), recently completed. It would be premature to evaluate the results here. Species are not static population entities. On sufficient time scales they are adapting to changing circumstances, splitting, sometimes hybridizing across previous mating barriers.

There are not very many species of either algae or animals in the ocean water column compared to the number on land, and this is particularly true for epipelagic habitats. Among crustacea for example, extensive cataloguing by Razouls *et al.* (2005–2011) found 2454 described species of marine pelagic copepods, the most diverse group, Vinogradov *et al.* (1996) listed 233 hyperiid amphipods; Blachowiak-Samolyk and Angel (2008) listed 217 ostracods, and a catalogue updated to 2011 lists 305 ostracods, some recognized but still to be described (Martin Angel, pers. comm.), and Baker *et al.* (1990) listed 86 euphausiids. Some planktonic groups, for example, copepods and ostracods, remain understudied, with species numbers increasing, particularly from mesopelagic and deeper levels, while others, including euphausiids and chaetognaths, appear to be almost completely catalogued and named. The total for all mesozooplankton is <5000, and microplanktonic heterotrophs are not notably diverse. We have done a great deal of sampling and describing of species, and while there may yet be much lumping and splitting and some completely new finds, for the upper ocean we know most of the species. The Census of Marine Life program of recent years is adding somewhat to the species counts of deep-sea groups, but it is too soon to say what the new totals will be. There are only about 2000 species of all pelagic vertebrates (fish, mammals, etc.) and other nekton (squid, large shrimp, etc.). There are certainly fewer than 6000 described species of phytoplankton. Thus, in pelagic habitats there are only a few more than 10^4 specifically distinct organisms (apart from microbes, for which species definitions are different and a count would still be premature). The numbers

for the land are surely greater than 2×10^6; there are ~300,000 species of beetles alone, and described vascular plants exceed 300,000. Recently, claims have been made that there may be $>10^7$ species. While we are frequently regaled with the notion of the marvelous diversity of marine life, and it is diverse at high systematic levels (phyla, classes), there are actually relatively few kinds of specifically distinct organisms, especially in pelagic habitats. Keep that in mind and wonder why.

Global patterns

We will examine patterns for epipelagic zooplankton because they are the best studied and good maps are available. All other pelagic groups, from phytoplankton to cetaceans, exhibit very similar patterns. A few species, like the blue whale, are more wide-ranging than the zooplankton patterns, migrating seasonally between feeding areas in antarctic or subarctic waters to calving and mating areas in the tropics. Mesopelagic patterns remain obscure, although few species are cosmopolitan despite similar deep-water conditions globally. Most of the zooplankton samples were collected by oblique tows with ring-nets from about 300 m to the surface and filtering 500 to 1000 m³. Examples of most of the basic distribution types are provided by Bruce Frost's (1969) worldwide study of *Clausocalanus*, a temperate–tropical genus of small, epipelagic copepods. The 13 species of this genus fall into three morphologically defined groups. Frost analyzed over 800 globally distributed samples to determine distributions. His results are presented in seven maps (Figs. 10.1 & 10.2).

The first shows the distribution of samples. Contours on the other maps show the zones in which each of the 13 species were found. A later study by Frost (1989) of distributions of the seven species of *Pseudocalanus*, another genus of small copepods, gives us patterns not seen in *Clausocalanus* (Fig. 10.3).

Six basic patterns occur in these genera:

Circumglobal, temperate–tropical – all three oceans, excluded only from north of 50°N and south of 45°S:

C. parapergens
C. furcatus
C. paululus
C. mastigophorus

Circumglobal, bi-antitropical – all three oceans, excluded from the equatorial zone, in some cases excluded from the equatorial zone of the Pacific only (the Pacific Ocean's tropics are colder than its subtropics, which is not so in the Indian Ocean). The term "bi-antitropical" refers to living on either side of the equatorial tropics but not in them.

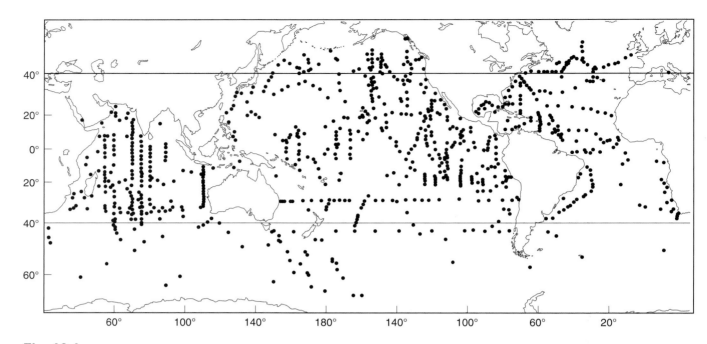

Fig. 10.1 Distribution of plankton net tows used by Frost (1969) in determining the distributions of the 13 species of the copepod genus *Clausocalanus*. (Kindly provided by B.W. Frost.)

C. lividus
C. pergens
C. arcuicornis

Indo-Pacific, temperate tropical – Indian and Pacific Oceans, absent from Atlantic Ocean.

C. farrani
C. minor

Circumglobal, Southern Ocean – these can be latitudinally restricted to various degrees:

C. ingens (subantarctic)
C. brevipes (subantarctic/Antarctic)
C. laticeps (subantarctic/Antarctic)

Broadly neritic – patches of suitable habitat scattered near continents:

C. jobei

Arctic Ocean and boreal: patterns of varying latitudinal extent in and circling the Arctic Ocean:

P. major – narrowly restricted to Arctic waters
P. minutus and *P. acuspes* – ranging from ~50°N Pacific and Atlantic to 90°N

P. elongatus – strictly Atlantic
P. mimus – strictly Pacific
P. newmani and *P. moultoni* – both Atlantic and Pacific but not in Arctic Ocean.

All of these patterns have parallels in other groups. Already, however, we can draw some generalizations for oceanic biogeography:

1 Temperate, tropical, antarctic, and boreal species tend to be distributed in broad, latitudinal bands across the oceans.
2 Species vary in the width of the latitudinal belt they inhabit. Some are broadly tolerant; others require a very specific hydrographic regime. Frequently, if that regime is found in several places around the globe, the species will be found in many or all of them. Bi-antitropicality and patterns like that of *C. jobei* are the basis for this statement.
3 The three major oceans are not the same, and they share some but not all species. In many cases, there are clear habitat differences explaining the zones of exclusion. For example, the surface of the equatorial Indian Ocean is warmer than the equatorial Pacific because the westward phase of the monsoonal circulation produces equator-ward flow, and thus a stable lens of heated surface water. Probably that explains the extension of *C. arcuicornis*

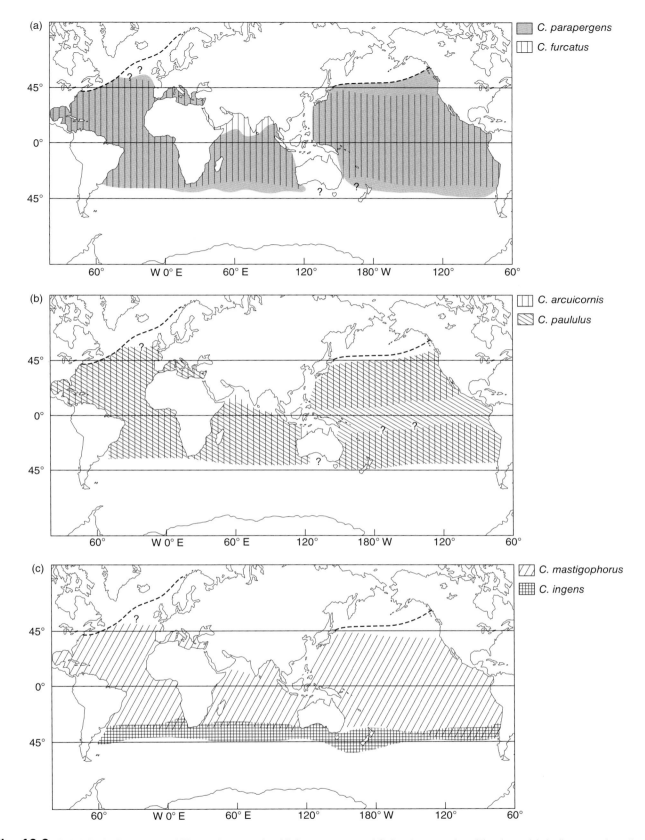

Fig. 10.2 Global distribution patterns of *Clausocalanus* species: (a) *C. parapergens* and *C. furcatus*, examples of the circumglobal, all-warm-water pattern with typical differences between them in the exact northern and southern limits; (b) *C. arcuicornis*, another circumglobal, all-warm-water pattern, and *C. paululus*, a circumglobal, all-warm-water pattern with tropical Pacific hiatus; (c) circumglobal, all-warm-water species *C. mastigophorus* and the circumglobal subantarctic species *C. ingens*; (d) *C. lividus*, a circumglobal temperate (or central gyre) species, and *C. laticeps*, a circumglobal Antarctic–subantarctic species; (e) *C. farrani* and *C. minor*, two Indo-Pacific temperate–tropical species, and *C. jobei*, a species occurring in patches around the world's oceans; and (f) *C. pergens*, a transition-zone and cool–tropical species (eastern tropical Pacific, north equatorial Atlantic) and *C. brevipes*, a circumglobal, subantarctic species. (All after Frost 1969.)

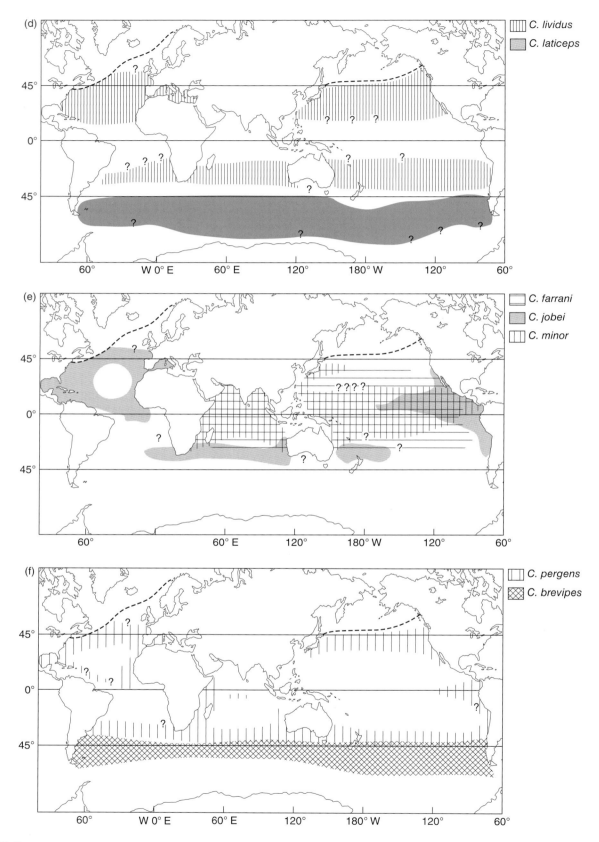

Fig. 10.2 (*Continued*)

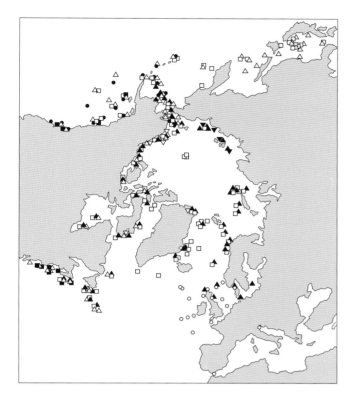

Fig. 10.3 Geographical distribution on a polar projection of the species of the arctic–subarctic copepod genus *Pseudocalanus*: □, *P. minutes*; ○, *P. elongates*; ▲, *P. acuspes*; ▼, *P. major*; ■, *P. moultoni*; △, *P. newmani*; •, *P. mimus*. (After Frost 1989.)

across the equator in the Indian Ocean, but not the Pacific Ocean. Habitat differences explaining exclusion of Indo-Pacific species from the Atlantic remain obscure. There are massive intrusions of water at times around the Cape of Good Hope from east to west. Possibly, the warm-water species cannot transit the full extent of the cold Benguela Current.

Pacific patterns

Work at Scripps Institution of Oceanography, led by Martin Johnson in the 1950s, provided distributions for species from a variety of groups in just the Pacific Ocean as far south as the subantarctic convergence. The British Discovery Expeditions extended the analysis to Antarctica. Other work, particularly Japanese studies (see Nishida 1985), has added more species distribution results, but remarkably few new patterns. Species patterns shown were selected as typical.

Pacific subarctic
There are two variants:

All across the subarctic gyre north of 40°N, not carried south in the California Current:

Sagitta elegans: Chaetognath (Fig. 10.4a)

(a) *Sagitta elegans*

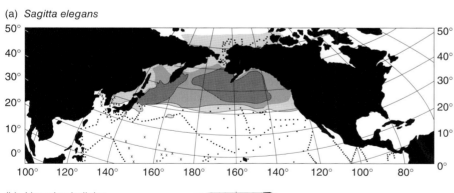

(b) *Limacina helicina*

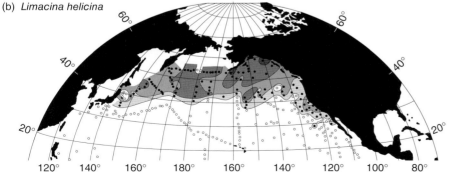

Fig. 10.4 Subarctic Pacific distribution patterns of (a) the chaetognath *Sagitta elegans* (after Bieri 1959), and (b) the euthecosome pteropod *Limacina helicina* (after McGowan 1963). Dots are sampling sites. Variations in shading represent factors of 10 in population, darker for higher abundance. The pteropod pattern illustrates extension of the pattern southward in the California Current – a feature of some subarctic species.

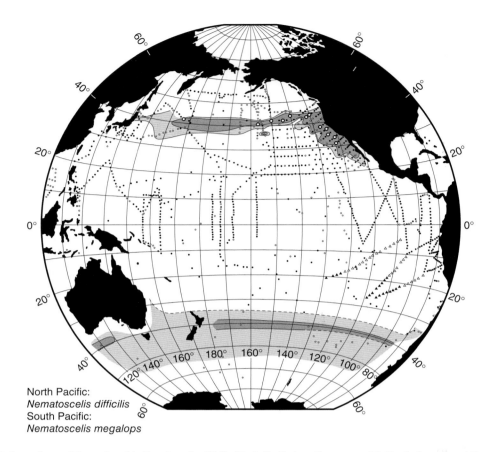

North Pacific:
Nematoscelis difficilis
South Pacific:
Nematoscelis megalops

Fig. 10.5 Distribution patterns of the euphausiids *Nematoscelis difficilis* (North Pacific transition zone and California Current) and *Nematoscelis megalops* (subantarctic). (After Brinton 1962.)

It is the same with the extension in the California Current, often to central Baja California:

Limacina helicina: Pteropod (Fig. 10.4b)

Subarctic–central transition zone

Species that inhabit all or part of the belt from 35 to 45°N, always with the California Current extension, usually not extending west to Asia. Most have another population or close relative in the north edge of the subantarctic.

Nematoscelis difficilis: Euphausiid (Fig. 10.5)

All warm water – some circumglobal, some Indo-Pacific

Clausocalanus parapergens: Copepod (Fig. 10.2a)
Globigerina sacculifera: Foraminiferan

Central water or subtropical

Such species are usually bi-antitropical.

Euphausia brevis: Euphausiid (Fig. 10.6)

Equatorial forms

These exhibit various degrees of restriction to the equatorial zone. Two examples (Fig. 10.7) show mild and extreme restriction.

Euphausia diomediae: Euphausiid
Eucalanus subcrassus: Copepod

Eastern tropical pacific endemics

Euphausia distinguenda: Euphausiid (Fig. 10.8a)

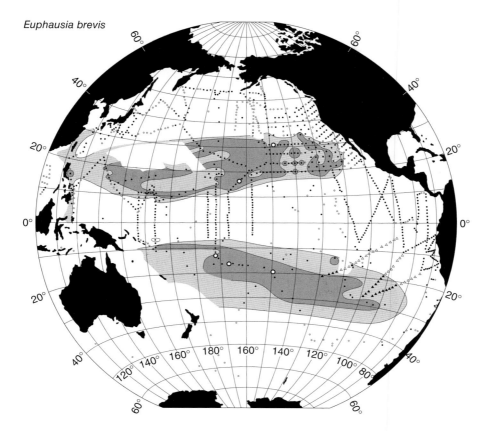

Fig. 10.6 Pacific distribution pattern of the euphausiid *Euphausia brevis*, a subtropical or central-gyre species. (After Brinton 1962.)

Warm water with eastern tropical Pacific hiatus

Nematoscelis microps: Euphausiid (Fig. 10.8b)

Subantarctic

Clausocalanus brevipes: Copepod (Fig. 10.2f)
Euphausia longirostris: Euphausiid

Antarctic

Calanoides acutus: Copepod (Fig. 10.9)
Euphausia superba: Euphausiid

There are some distributions that do not fit exactly. *Euphausia gibboides* is an example; it is distributed in the southern sector of the transition zone and in the eastern tropical Pacific, with a definite and wide gap between them. Like most odd patterns, it is a combination of two usual patterns. Nishida (1985) has given us an example (Fig. 10.10) of a distribution with extreme latitudinal extent, the pattern for *Oithona similis*, a small copepod. Its pattern includes the subarctic Pacific, the California Current

(our observations), the eastern tropical Pacific, the Peru Current, and the subantarctic. This pattern makes sense over most of its extent in terms of water temperatures available fairly close to the sea surface, <~300 m.

Pattern maintenance

The resemblance of the Pacific patterns to the distribution of water masses (Fig. 10.11), defined on the basis of temperature vs. salinity (TS) diagrams, is striking. A TS diagram is simply temperature plotted vs. salinity for a vertical profile of these variables in the water column. Over wide ocean expanses the TS curves fall into narrow bands, or envelopes, on the plots, bands identifiable with regional water types or *water masses*. These are in a sense the oceanic equivalent of specific habitats. At least their co-extension with species ranges implies that they are. This is confirmed by comparing the TS plots for stations at which given species are captured with TS envelopes for given water masses. There is a general correspondence, usually with the array of plots for positive sites somewhat larger than the standard envelope (Fig. 10.11). This is because

(a) *Eucalanus subcrassus*

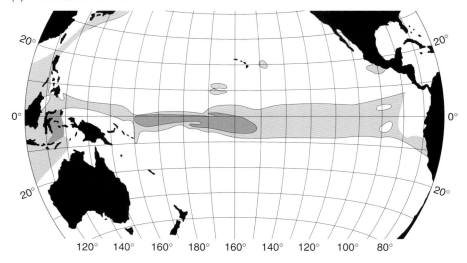

(b) *Euphausia diomediae*

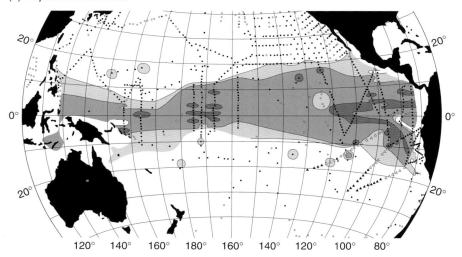

Fig. 10.7 Pacific distributions of (a) *Eucalanus subcrassus*, an equatorial belt copepod (after Lang 1965), and (b) *Euphausia diomediae*, a tropical euphausiid (after Brinton 1962).

(a) *Euphausia distinguendo*

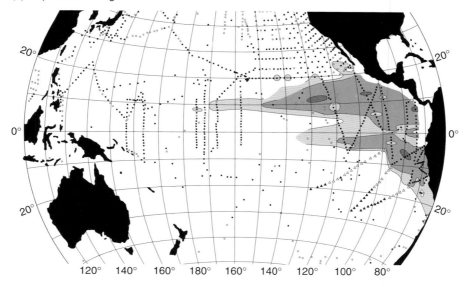

(b) *Nematoscelis microps*

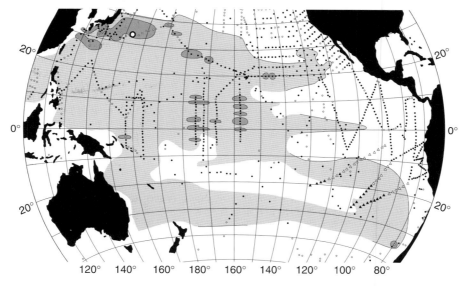

Fig. 10.8 Complementary distribution patterns of (a) the euphausiids *Euphausia distinguenda*, an eastern tropical Pacific endemic, and (b) of *Nematoscelis microps*, a tropical species with a distributional hiatus in the eastern tropical Pacific. (After Brinton 1962.)

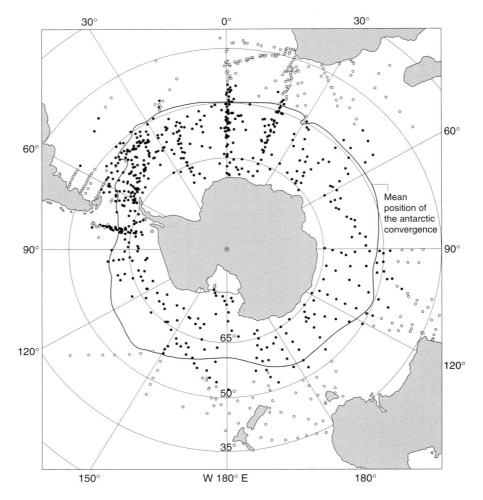

Fig. 10.9 The distribution pattern (filled circles) of *Calanoides acutus* at stations (all circles) sampled by the Discovery Expeditions. This is a copepod mostly confined to south of the Antarctic convergence. (After Andrews 1966.)

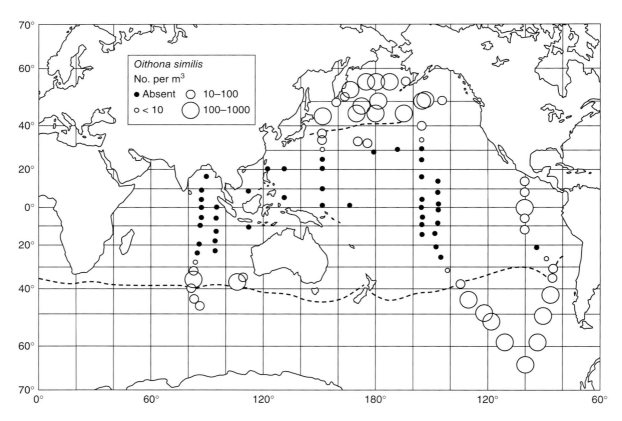

Fig. 10.10 The distribution pattern of the small copepod *Oithona similis* (open circles) is continuous from the subarctic Pacific, along the California Current (our observations), across the equator and south to the subantarctic and Antarctic waters. (After Nishida 1985.)

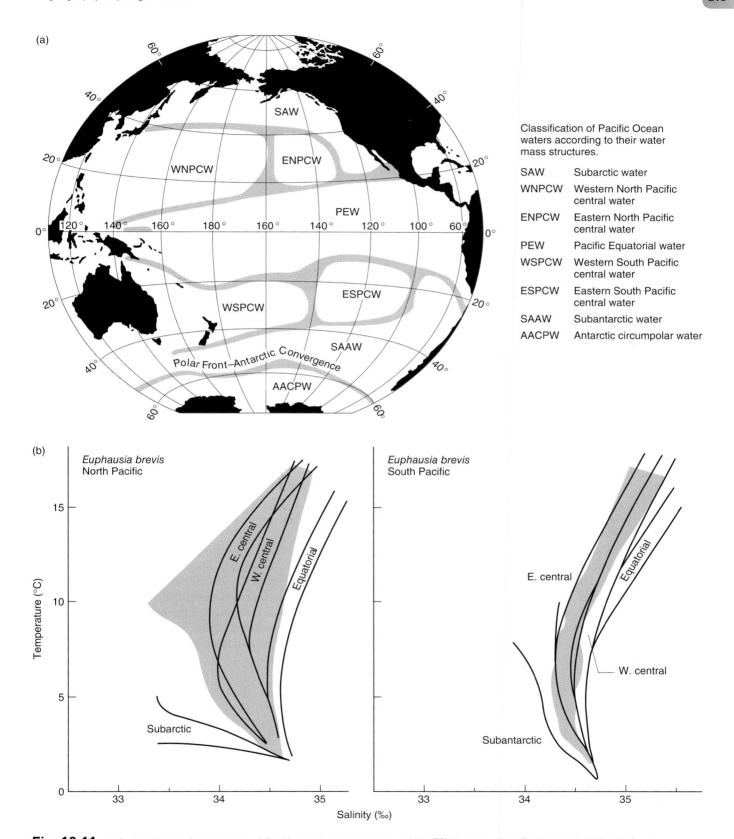

Fig. 10.11 (a) Regional extent of water masses defined by consistent temperature–salinity (TS) diagrams. (After Sverdrup *et al.* 1942.) (b) Envelopes (stippled) of TS diagrams from stations where *Euphausia brevis* was collected by Scripps expeditions (Fig. 10.6) in the 1950s, compared to general envelopes (lines) of TS diagrams associated with North Pacific and South Pacific central waters. (After Johnson & Brinton 1963.)

individuals are frequently expatriated from their usual habitat by mixing processes, and they can survive quite some time outside their usual range. Of course, many species occupy two or three water masses, particularly those with distributions divided by the tropical belt. Apparently, no subtropical species distinguishes only east- or west-central water masses as an unsuitable habitat. If an animal is found in one, then it is found in the other.

Why should distribution boundaries be generally consistent with those of water masses? If a planktonic species is to persist, it needs a semi-closed circulation within a suitable habitat, or at least intermittent return flows. Therefore, species tend to evolve ranges of tolerance that encompass all of the conditions encountered within a region of closed flow. Natural selection for sufficiently wide tolerance of environmental variation to live throughout an entire circuit must be strong. For sound physical reasons, the regions with semi-closed circulation are largely the same as water masses defined by TS diagrams. Moreover, because water masses are internally mixing, they tend to be more homogeneous than larger regions. Thus, the expansion of tolerance to encompass the whole range of conditions for a water mass is a reasonably obvious direction for evolution.

In confirmation of this notion, there are many patterns of distribution along the continental coasts, for example, the California Current distribution of *Nyctiphanes simplex* (Eu) (Fig. 10.12). Actually, the *N. simplex* pattern extends south along the American coasts to south of the equator, and it is abundant around the Galapagos Islands. Planktonic populations occur in places where coastal effects on the circulation create small gyres or intermittent return flows.

The transition zone pattern requires more analysis to be fit into this hypothetical scheme. It stretches east–west across the North Pacific between about 34° and 43°N, right along the main axis of the Oyashio–Kuroshio confluence and west wind drift, which might be expected to push its contents east against North America. This space between the cool subarctic and warm subtropical gyres is bounded in upper layers by fronts in salinity and seasonal temperature, both increasing southward. It is also the range of latitudes (an oscillating band of 6° to 8° latitude centered around 40°N) in which there is no strong density stratification for part of the year. Subarctic waters to the north have a persistent halocline at about 110 m that survives winter mixing, and the subtropical gyre to the south has a near-surface thermocline right through the year and a deep, permanent thermocline. There is formation of subtropical-mode water to the western side of the subtropical gyre with mixing to 500–600 m, but it does not cool the surface below ~16°C. Just in the transition zone there is winter destratification with potential for deep mixing and stronger cooling. This latitudinal band is not narrow, except relative to the whole expanse of the Pacific. It occupies a major swath of ocean. A modest list of zooplankton species live in this band, including at least *Nematoscelis difficilis*,

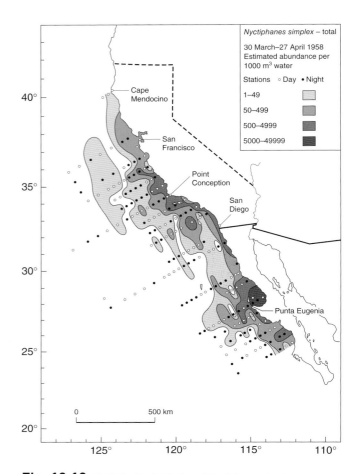

Fig. 10.12 North Pacific distribution of *Nyctiphanes simplex*, a euphausiid. Another population is found in the vicinity of the Galapagos Islands. (After Brinton 1967a.)

Thysanoessa gregaria, Euphausia gibboides, Clausocalanus pergens, Eucalanus hyalinus, a subspecies ("type B") of *Limacina helicina,* and *Pseudosagitta lyra.* Several species of small nekton have similar patterns. Distributions for the euphausiid species, and most of the others, do extend all the way to Japan, but numbers consistently fall off in the west end compared to longitudes east from 165°E.

The puzzle of transition-zone patterns is solved in principle (Olson & Hood 1994; Olson 2001), although a necessary quantification of the population budgets has not been undertaken. East of the Emperor Seamounts at about 170°E the flow is sluggish. The mean flow at 158°W is eastward with a mean speed of only a few cm s^{-1}. Moreover, the mesoscale eddy field has local velocities about the same, i.e. very slow. Drifters drogued at 15 m show the water moving mostly east at a very stately pace. This is true on both sides of the transition as well, with eddy velocities much higher just south of the transition than in it or to the north. The mesoscale eddies extend way down, to approximately 600 m, and thus include the entire vertical range of any epipelagic zooplankton. Probably in this range of longitudes, the critical net loss is due to eddy diffusion past the northern and southern boundaries of habitat conditions to which the animals are adapted, and the drifter data suggest that those losses are small. The net flow coming in from the west replaces losses to net flow out to the east. So, sufficient births plus good survival readily sustain transition endemics along their meridional band, despite the lateral advective losses.

West of the Emperor Seamounts, the relations are different. Out to about 150°E, the net flows in the Oyashio (c.41°40′N) and Kuroshio (34°N) push water through in nearly coherent and unidirectional fashion, and quite fast, especially in the Kuroshio extension (Plate 10.1). In the Oyashio, the coherence is harder to see. The strong, coherent flow probably explains the strong drop-off in abundance of transition zone species at this western extreme. But just east of 150°E, lots of back-eddying sets in; the Kuroshio breaks up. In fact, to the north of 36°N the eddy activity is high almost to the coast of Japan. All the way from 150°E to the Emperor Seamounts, there is very strong eddy activity to return plankton upstream. Mesoscale features reaching 1000 m down the water column and roughly of 250 km scale alternate all across north–south hydrographic sections in this vicinity. Net throughput is on the order of 15–20 cm s^{-1}, while eddy velocities to move animals upstream are of the same order, in agreement with the drifter tracks. Since there is no stock input in the inflow, it must be return of stock via recurring back-eddies that allows population maintenance. Transition-zone endemics peter out at the farthest upstream end, exactly as one would expect. Enhancement of production by deep seasonal mixing, which does not occur in either subarctic or subtropical waters to the north and south, is certainly important in providing transition endemics with the high rates of population increase needed to compensate for the advective losses.

The distribution of *Calanus finmarchicus* (copepod; Plate 10.2), *Meganyctiphanes norwegica* (euphausiid) and some other species in the North Atlantic Current (NAC) from New England to Iceland to the Norwegian Sea likely has a similar explanation. The through-flow of the NAC, that sustains the warmth of northern Europe, has upper-ocean return flows along the northern limbs of three large cyclonic gyres (Norwegian Sea, Irminger Sea, and the slope water from New York to the Grand Banks). Exchanges between adjacent gyres are thought to connect planktonic populations across the whole region with subarctic conditions (Bucklin *et al.* 2000). Bucklin and colleagues applied gene analysis (methods for reading of gene sequences are summarized in Box 2.4) and found a suggestion of genetic differentiation in *C. finmarchicus* based on proportions of small differences in a 72 base-pair sequence from a mitochondrial pseudogene. A pseudogene is a copy in nuclear DNA of the mtDNA. However, the dominant haplotype is the same across the whole range (save for one sample of 10 specimens off Iceland). Recently, Provan *et al.* (2010) found no evidence for geographical differentiation of *C. finmarchicus* in either nuclear microsatellite DNA or mitochondrial cytochrome B genes. The degree of population-genetic separation appears to be somewhat greater in *M. norvegica* (Papetti *et al.* 2005), which has a much wider range, including the eastern boundary current and even the Mediterranean. On the other hand, the mtDNA sequence studied by Papetti *et al.* has two major haplotypes, two minor ones, and 31 rare haplotypes. All the populations have substantial proportions of the major and some of the minor forms, except for those to the far southeast. Intergyre exchange among the northern stocks could either be substantial or the major haplotypes represent coalitions of previously separate populations and have no current selective significance.

Mesoscale and larger eddies also expatriate organisms from zones of suitable habitat. Dramatic examples of eddy effects are the cold- and warm-core rings formed along the Gulf Stream southeast of New England, in the Kuroshio–Oyashio confluence and along western boundary currents in the southern hemisphere. As the current jets turn seaward, they meander at the boundaries between cold water poleward and warm water equatorward, forming loops to both north and south. A loop to the south of the Gulf Stream can bend enough for the downstream end to join the upstream end, forming a cyclonic current ring with cold northwest Atlantic slope water in its core. As the ring moves south into the Sargasso Sea, the warmer, less-dense ring water spreads toward the center, submerging the cold core. Cold rings can persist under the Sargasso surface layer for about 1.5 years. At least the larger zooplankton, such as euphausiids (Wiebe & Boyd 1978; Endo & Wiebe 2007), with home ranges in the colder waters to the north and east, die out very slowly. *Nematoscelis megalops* distributed in slope water from the surface to 600 m, peaking at ~300 m, slowly moves down as the rings submerge and dissipate (Fig. 10.13), showing progressive decline in body condition (carbon content; Boyd *et al.* 1978), eventually going extinct in the oldest rings. *Euphausia krohni*, another cold-water species, is a strong diel migrator and continues that habit when captured by a cold core ring, exposing it eventually to the 25°C summer warmth of the Sargasso Sea. It dies out much more quickly (Endo & Wiebe 2007). It is soon replaced in surface layers by a suite of subtropical euphausiids (Fig. 10.14). Surely similar displacements, mortality and replacements occur for the whole plankton community.

Similarly, loops to the north can join into anticyclonic rings with warm Sargasso water cores. The warm cores of these rings are hundreds of meters deep, providing a temporary habitat for subtropical plankton in slope waters as they move west and south along the North American coast before eroding away over a few months (Wiebe *et al.* 1985).

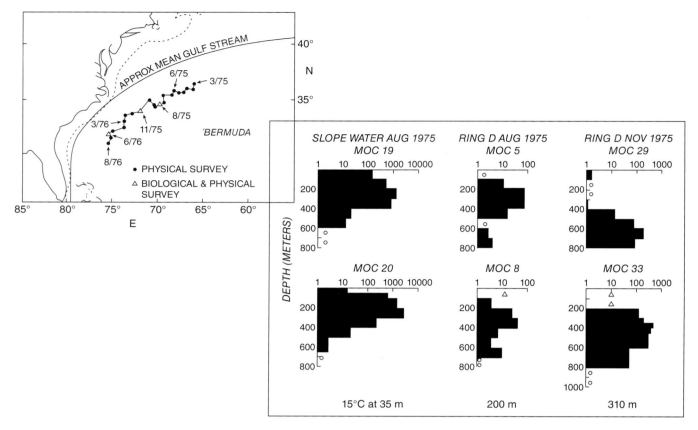

Fig. 10.13 Trajectory of Gulf Stream cold-core ring "D" through the Sargasso Sea in 1975, shown with the vertical distributions in the ring at night from multiple-closing-net tows of the cold-water euphausiid *Nematoscelis megalops*. Tows were in Northwest Atlantic slope water (left) and in the ring center at 6 months (August 1975) and 9 months (November 1975) from ring capture. Columns show two replicate tows from each date. Most of the population in the ring stayed below the 15°C isotherm which progressively deepened. No *N. megalops* were captured when that isotherm was very deep in June 1976. (After Wiebe & Boyd 1978.)

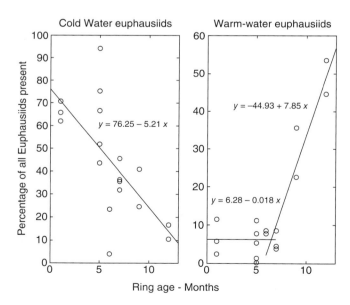

Fig. 10.14 Proportions of cold- and warm-water euphausiids in the water column above and in cold core rings of increasing ages. Cold-water species, dominated by *Euphausia krohnii* and *Nematoscelis megalops*, are displaced downward and replaced by a complex suite of warm-water species. (After Endo & Wiebe 2007.)

They can "rub" along the southern flanks of Georges Bank, an advective heat wave in a usually subarctic habitat. Eddies and current meanders transfer expatriate individuals between distinct but adjacent habitats along all oceanic distributional boundaries, sometimes generating uncertainties about the extent of the patterns and the tolerance limits of species, unless the physical circumstances of sampling stations are fully recognized. Coastal plankters are carried seaward from their home waters by topographically steered upwelling jets and features like the recurring Haida eddies off British Columbia.

Pattern maintenance in the equatorial zone depends similarly upon current circuits. Patterns like that of *Euphausia distinguenda* (Fig. 10.8a) can be sustained by the circuits of the north equatorial current headed west and north equatorial countercurrent headed east and the comparable current pair to the south of the equator. Southern subtropical gyres are closed circuits, much like those of the northern hemisphere. The Southern Ocean is a suite of concentric circumglobal currents, loops closed around the sphere of the Earth matching numerous species distributions (e.g. Fig 10.2d – *Clausocalanus laticeps* and Fig. 10.2f – *Clausocalanus brevipes*).

Again, while we have examined mesozooplankton distributions, the basic patterns demonstrated by them recur in all high-seas pelagic organisms: phytoplankton; squid; epipelagic and mid-water fish; and cetaceans. On the other hand, nekton that swim substantially faster than currents can move between distinctive habitats, and many of them regularly migrate to exploit the seasonal advantages of different regions.

Diversity and community structure

The patterns that recur for numerous species do not have equal numbers of subscribing species. Diversity, particularly the length of the species lists, is much higher in tropical and subtropical seas than in subpolar and polar ones. For example, according to Razouls *et al.* (2005–2011) the North Pacific central gyre is home to more than 520 species of copepod, while the subarctic region has around 300. The North Atlantic appears to have more in both regions, but that is likely to be due to the concentration there of taxonomic effort. Many species on these lists in all regions are deep-living, many rare. There is no certain answer to why the tropics are more diverse, but it applies to most terrestrial groups as well – tropical rainforests have more plant, bird, insect, and other species than mid- and high-latitude rainforests. Likely a major part of the story is that over long geological times warm habitats have been more stable, i.e. less subject to the strong temperature and productivity shifts imposed on high-latitude systems by glacial–interglacial cycling. Plankton community structure in higher latitudes also tends to be more strongly dominated by only a few species, although the dominant forms may shift with season. Aspects of regional distinctions in community processes are covered in the next chapter.

Speciation in pelagic habitats

Charles Darwin's seminal book, *The Origin of Species*, considered how a species would evolve into a substantially different form (a new species) over time, with little attention to how one species evolves into two or more. Later work and thinking have shown that the majority of subdivisions into new species occur by an *allopatric* process. The term derives from Latin for "different country", and the process is also referred to as geographical speciation. There are a number of *sympatric* ("same country") speciation mechanisms, but they are harder to demonstrate and certainly operate less often. Allopatric speciation has three steps:

1 The geographical range of a species, throughout which there is consistent exchange of mating individuals and thus "gene flow", becomes divided into two or more smaller ranges by a new, strong barrier to transit by the organism. For example, a distribution on two mountain ridges and contiguous across the valley between can divide onto the two ridges by climatic change making the valley uninhabitably warm or dry.

2 Both genetic drift and differences among the new ranges in selective advantages cause the genetics of the populations to diverge. This must reach sufficient levels of difference that matings between individuals from different ranges are disadvantaged, producing fewer or less fit offspring.

3 The barrier breaks down, the ranges again become contiguous, and opportunities for breeding with individuals from other stocks become frequent. However, because these outcrosses produce fewer or less fit offspring, there will be a *new* selective advantage to any trait that allows mate selection according to origin, causing like to favor like. Such traits are often differences in mating ritual or in form of the copulatory apparatus, particularly the latter in arthropods.

The last fact explains why crustacean (and insect) taxonomists distinguishing species spend concentrated effort examining and describing sex organs. An alternative outcome of rejoining previously disjunct distributions is that matings between members of the different populations work well or even benefit from a version of "hybrid vigor", the young having high proportions of heterozygous genes. After recombination and selection over multiple generations, a markedly adaptive suite of genes may spread through both original ranges. In these cases, effects from the period of allopatry vanish. Step 3 is critical and is often forgotten in discussion of allopatric speciation. Division of a species into several not only requires genetic divergence but subsequent selection for mating barriers in renewed sympatry.

Edward Brinton (1962) suggested several mechanisms by which allopatric speciation could have operated in the sea. These were theoretical ideas in 1962, but subsequent developments have made possible some very direct tests, genetic comparisons of closely related species. The problem of accurate calibration of molecular clocks may eventually be sufficiently solved to show the timings of divergences. The mechanisms all depend upon the warming and cooling effects of global glaciation upon ocean temperature patterns during the Pleistocene. Before that, the dominant mechanisms must have been different, and they were probably fewer. Brinton's patterns were reconstructed from the array of different patterns observed among modern pelagic species, much as Darwin explained the formation of coral

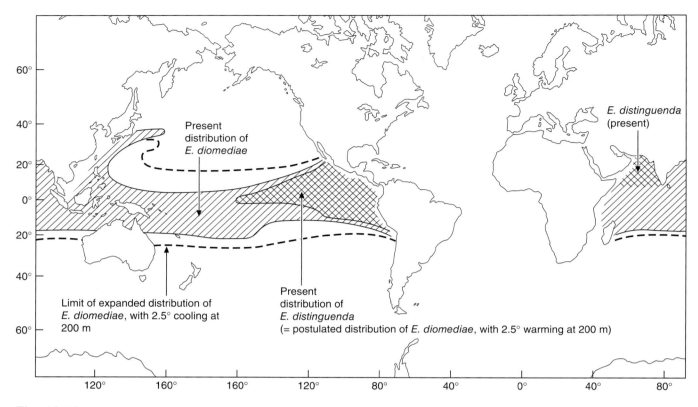

Fig. 10.15 Present distribution of eastern tropical Pacific and Arabian Sea populations of *Euphausia distinguenda* compared to the real distribution of *Euphausia diomediae*, which Brinton proposes as the hypothetical distribution of *E. distinguenda* in the event of global cooling. (After Brinton 1962.)

atolls from the array of variations he observed on the *Beagle* expedition. Brinton used euphausiid patterns based on his own work and global records.

Thysanoessa gregaria occupies the subpolar–subtropical transition zones in all of the world's oceans (it is bi-antitropical), and it would have been able to survive at much lower latitudes during a cooler interval, particularly across the eastern boundary of the oceans. This former continuity of the pattern is an explanation for the bi-antitropicality of the species at present. The present situation is conducive to division, and speciation is presumably in process at present. The *Nematoscelis difficilis–N. megalops* pair (Fig. 10.5) is a further example. In 1962 there were no species known to have a subpolar and eastern-boundary-current distribution continuous across the equator, but as we have seen, *Oithona similis* (Fig. 10.10) provides an example.

Central species possibly would be oppositely affected by general cooling. At present, the warmest waters, except for those at the very surface, are in the central gyres, not along the equator (excepting the Indian Ocean). For a species like *Euphausia brevis* (Fig. 10.6) with a vertical range of hundreds of meters, warming could cause trans-equatorial coalescence of ranges, while cooling could intensify their

separation. *Euphausia brevis* is presumably undergoing speciation at present. The range of an all-warm-water form like *Clausocalanus parapergens* (Fig. 10.2a) might be expected to divide during a colder interval and speciation could occur.

Global warming could break up the present Indo-Pacific or circumglobal distributions of equatorial species like *Euphausia diomediae*, leaving isolated populations in the cooler equatorial regions like the present distribution of *Euphausia distinguenda* in the eastern tropical Pacific and the Arabian Sea (Fig. 10.15). Sebastian (1966) has shown that the Arabian Sea population of *E. distinguenda* has small but consistent morphological differences, naming it *Euphausia sibogae*, which implies that the development of two separate species is already well advanced.

Abraham Fleminger (e.g. Fleminger & Hulsemann 1974) pursued this last class of mechanisms, proposing that for circumglobal tropical and temperate–tropical species the continents must serve as barriers to dispersal between the Atlantic and the Indo-Pacific, and sometimes between the Indian and Pacific Oceans. During an interglacial warming, tropical–subtropical species like *Clausocalanus arcuicornis* (Fig. 10.2b) can spread around the Cape of Good Hope into the Atlantic. Subsequent cooling then squeezes the distribution until the Atlantic and Indo-Pacific populations are not in communication, and the stocks could speciate. When warm conditions again prevail, the process is completed by formation of biological mating barriers, and in some cases both forms spread to circum-global distributions. Withdrawal of water from the oceans during major glacial episodes would make the landmass along the arc from Malaysia to Australia more complete, a second barrier, allowing the development of three species from one original form in a single glacial cycle by separating stocks in each of the Atlantic, Indian, and Pacific Oceans. There are many epiplanktonic copepod genera with three or four groups of three or four similar species. Thus the separate-ocean mechanism (in combination with Brinton's equatorial–subtropical one) could have had two to four chances to operate, in line with the number of major Pleistocene glaciations, four.

Fleminger and Hulsemann (1974) showed some of the possible ramifications of the continental barrier notion by a study of the copepod genus *Pontellina*. Prior to 1974, all of the reports of this tropical group were attributed to one species: *P. plumata*. Fleminger and Hulsemann described three new forms: *P. morii*, *P. sobrina*, and *P. platychela*. They retained *P. plumata* for a fourth form, the one most widely distributed. Since the new species were not recognized in a very large body of work before this study, it is clear that the genus is made up of very similar or "sibling"

species. We cannot represent the sorts of differences that distinguish these species without going into extreme detail, so in order to learn those distinctions please refer to the original paper. All of the morphological differences are on anatomical parts that are important in mating. Fleminger and Hulsemann's tabulation of differences implies that the overall relationships are as follows:

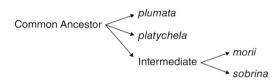

Pontellina plumata, which ranges from 40°N to approximately 40°S (Fig. 10.16), has a nearly constant body form worldwide. The species more restricted latitudinally are those that also are restricted longitudinally. As an exercise, try to come up with a sequence of range extensions and restrictions that could have produced the speciation sequence suggested by the relationships.

To discuss tests of these speciation hypotheses, some different background is required: paleontological stratigraphy and molecular genetics. For the latter, please refer back to Box 2.4.

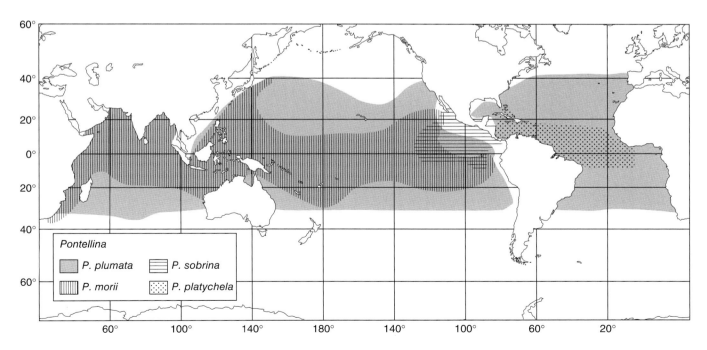

Fig. 10.16 Current distributions of four closely related species of the copepod genus *Pontellina*: *plumata*, *platychela* (dotted in Atlantic), *sobrina* (horizontal lines in eastern Pacific) and *morii* (vertical lines across tropical Indo-Pacific). (After Fleminger & Hulsemann 1974.)

Paleontological stratigraphy of the ocean basins in relation to planktonic biogeography

Basic notions

There is a fundamental interdependence of studies of oceanic stratigraphy and studies of the biogeography of plankton. Stratigraphy is the description and interpretation of the content of layers of sediment that have accumulated on the seafloor. These sediments may be lithified and uplifted, thus occurring on land; or they may remain unconsolidated and submerged. The latter will be our main concern. Several devices for collecting sediment sequences from the ocean bottom are in use, including free-fall and piston corers and drilling devices. All of them amount to driving a pipe into the sediment, then pulling it out to retrieve a vertical cylinder of sediment, a core. In general, the farther down along a core that a sample is taken, the longer the interval since its deposition. There are exceptions caused by slumping of sediments down submarine slopes or winnowing by currents with subsequent redeposition. These usually can be detected at a small price in evidential circularity. Thus, any process in the water column that affects the character of the materials deposited will have its history fuzzily recorded in the sediments. Fuzziness is caused by the extremely slow sedimentation rates over much of the ocean, as low as $1\,cm\,1000\,yr^{-1}$, coupled with mixing ("bioturbation") of the upper 2 to 10 cm by benthic animals. Thus, annual and even centennial events and cycles are almost always obscured. There are useful exceptions, especially deposits in anoxic basins with no benthic animals.

Careful sorting of sediments from most locations will yield a substantial number of microfossils. These are the siliceous shells of diatoms and radiolaria; the calcareous shells of prymnesiophytes (coccoliths) and foraminifera; and aragonitic pteropod shells. Microfossils present in a given sedimentary sequence vary in two ways. On long time scales individual species evolve. On shorter time scales the relative abundances of different fossil types vary markedly with depth in the sediments. Evolutionary changes cannot be expected to tell us much about environmental conditions in the past until the adaptive significances of the observed changes in microfossil morphology are known, significances that are extremely difficult to determine. Shorter-term abundance variations, however, are powerful indicators of environmental history, particularly during the Pleistocene. We will use a single, old example to illustrate the relation between short-term variations and zoogeography.

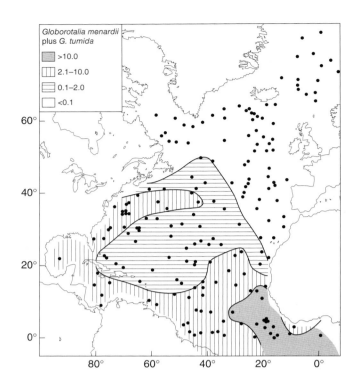

Fig. 10.17 Distribution pattern of relative abundance (%) of *Globorotalia menardii* plus *G. tumida* (which is similarly distributed) in sediments from the tops of cores collected at the scattered stations represented by dots. (After Kipp 1976.)

Globorotalia menardii, a foraminiferan, is a circumglobal, temperate–tropical species (Bé & Tolderlund 1971). It constitutes a substantial fraction of the planktonic foraminifera in many warm-water areas. Phleger *et al.* (1953) showed that the spatial distributions of many kinds of foraminifera from the tops of sediment samples are coincident with the main hydrographic provinces of the oceans, just as we have seen to be the case with living animals. Thus, sediment distributions for *G. menardii* (Fig. 10.17) from core tops, that is, from the surface of the sediment, resemble the pelagic pattern closely. Not only are the overall ranges the same, but the zones of maximum relative abundance around the circumference of the North Atlantic central gyre were also observed by Bé and Tolderlund (1971) in the plankton. The similarity of the patterns implies that the shells do not drift very far from the near-surface sites at which they originate before sinking to the bottom. The appearance of *G. menardii* in the sediment is, therefore, an indication of its presence in the overlying water.

Ericson and Wollin's (1968) plots of the number of *G. menardii* per gram of sediment vs. distance downward in cores from the eastern Caribbean Sea (Fig. 10.18) show that it drops to very low relative abundance for long sections of the core. The "events" represented by variations in species composition can be dated by radiochemical age

Scale is number of *G. menardii* mg⁻¹ foraminifera in sample

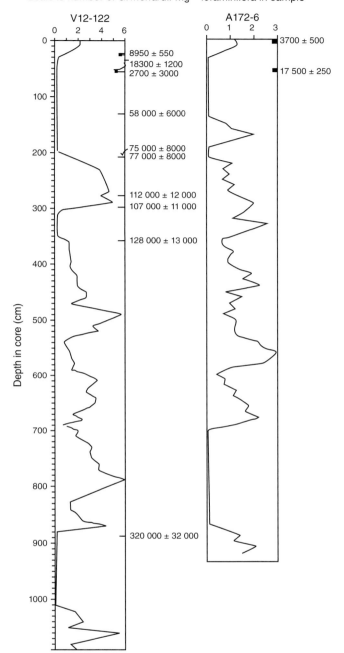

Fig. 10.18 Several examples of core profiles of the relative abundance of *G. menardii* in core samples from the Caribbean Sea. Radiocarbon dates are shown on the right-hand edges of the profiles. Abundance scales are varied between cores to make the plots look more similar. (After Ericson & Wollin 1968.)

determinations. The increase in *G. menardii* at about 10,000 years ago (YBP) coincided with the recession of the Wisconsin ice sheet as dated in terrestrial deposits. It seems likely, then, that global cooling during the last ice age caused the paucity of *G. menardii* from 75,000 to

10,000 YBP. So, we can use the rest of the record to date prior glacial and interglacial epochs, provided that the core has no gaps. Correlation of large numbers of cores has provided the general pattern, so that gaps in particular cores can be detected.

What happened to the stock of *G. menardii* during the glacial epochs, when it appears to have nearly vanished from the Caribbean area? Either, the range contracted into the remaining warm zone as cold isotherms moved toward the tropics, or abundance dropped off everywhere as the species reproductive success or survival declined throughout its range. CLIMAP, an elaborate study of ice-age distribution patterns has provided the answer.

Mapping the zoogeography of the past

During the 1970s, the CLIMAP program studied the pelagic zoogeography during a glacial epoch by mapping the distributions of relative abundance of forms like *G. menardii*. The questions of importance to those doing the study did not concern the fate of a given species; they concerned global climate patterns. Therefore, while distributional maps for species were presumably studied, they were not published, except for the comparison study of surface sediments (Kipp 1976). Maps published by McIntyre *et al.* (1976) show the distribution of faunal assemblages identified by a factor analysis procedure. This classification of species into assemblages was performed on faunal data from the tops of cores (Kipp 1976). For example, *G. menardii* belongs to a set constituted principally of five species called the "gyre margin" assemblage. All these foraminifera are most abundant, or are only found, around the margin of the North Atlantic central gyre. *Globoquadrina deutertrei* is another abundant member of this assemblage. Next, the degree of occurrence of an assemblage in samples from the 18,000 YBP level, the peak of the last glaciation, was determined and plotted. At this point, interpretation takes off like a great pterodactyl.

Note the circularity in the argument before we proceed. CLIMAP used the stratigraphy of $(g\,CaCO_3\,(g\,sediment)^{-1})$ to determine the location in each core of the 18,000 YBP level, a variable known to parallel the calcite oxygen-18/ oxygen-16 ratio, a measure mostly of glacial withdrawal of water from the seas. However, sediment concentration of $CaCO_3$ is mostly due to foraminiferan shell production, the same biological process used in the next steps to characterize conditions. Moreover, ". . . floral and faunal curves and stratigraphic boundaries proved useful for corroborative time control within the period 0 to 130,000 BP" (McIntyre *et al.* 1976). Nevertheless, the results are extremely interesting. A sausage should be judged by its final flavor and texture, not on what goes into it!

The 18,000 YBP distribution of the gyre margin assemblage (Fig. 10.19) shows that these forms retreated during the glacial into the equatorial zone of the eastern Atlantic, off Africa. During the ice ages, tropical and sub-tropical pelagic species probably are not reduced every-where to low abundance, rather their ranges contract to small refugia where conditions remain suitable for their survival.

CLIMAP (Kipp 1976; McIntyre *et al.* 1976) carried the argument in another direction. First, they fitted a multiple, second-degree (squares of variables are entered as variables) regression equation for the relation between some present-era oceanographic variable (e.g. mean winter temperature) and sediment-surface faunal data (actually "factor loadings" from the original analysis) as dependent variables. Second, they used this equation to predict temperatures at 18,000 YBP from faunal data for that time. Comparison of today's sea-surface temperature map with the glacial era paleotemperature map (Fig. 10.20) shows the strong equatorward collapse of isotherms resulting from glaciation.

Speciation, again

The most instructive CLIMAP comparison for those inter-ested in pelagic speciation processes is that of the North

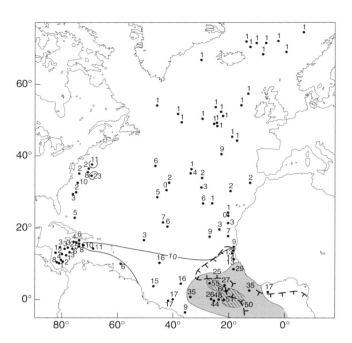

Fig. 10.19 Distributional map of percentage contribution (numbers by station dots) of the central gyre margin assemblage (mostly *G. menardii*) at the 18,000 YBP level in selected cores. The very low percentages in the North Atlantic and Caribbean indicate either recurring expatriation from the eastern tropical Pacific refugium of the glacial period, or mixing by bioturbation of some *G. menardii* group fossils from interglacial layers (After McIntyre *et al.* 1976.)

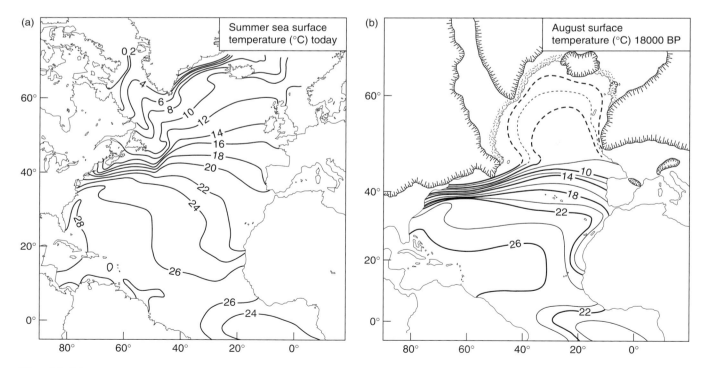

Fig. 10.20 (a) Modern August surface-isotherm map for the North Atlantic Ocean. (b) August surface isotherm map for 18,000 YBP estimated from transfer functions based on foraminiferan faunal composition. (After McIntyre *et al.* 1976.)

Atlantic subpolar assemblage (Fig. 10.21). In the core tops it extends south to about 15°N in the east, while at 18,000 YBP it crosses the equator. Thus, the patterns of former sympatry suggested by Brinton (1962) for species like *Thysanoessa gregaria* and the *Nematoscelis difficilis–N. megalops* pair clearly did exist for species preserved in the fossil record, like *Globorotalia truncatulinoides* dominant in the subpolar assemblage. Brinton's theoretical suggestion has been borne out: allopatry of bi-antitropical species has developed in the post-glacial era exactly as he proposed.

Many more and very interesting applications of this approach to questions of climatic history and climate process have been made by the CLIMAP group and others. Zoogeographical shifts of pelagic organisms underlie all of them. Of greatest importance is the test of an explanation of glacial–interglacial cycling based upon "Milankovitch cycles" in the rotation and solar orbit of the Earth. We will consider this relationship further in Chapter 16.

Does speciation really occur in this way?

Do disjunct populations, or possibly even the distant ends of very extensive conjunct populations, of animals that obviously are closely related on the basis of morphology show genetic divergence? If so, how great is the divergence? Some of the cases that would provide the best answers have yet to be examined by molecular geneticists. For example, sequences for a suite of genes from the genus *Pontellina*, discussed above (Fig. 10.16), and particularly for *P. plumata* collected in the three oceans, would show the range of genetic divergence installed by recent evolution in distant, possibly isolated stocks. However, other cases providing indications of amounts of divergence have been examined. The study of *M. norvegica* by Papetti *et al.* (2005), mentioned above, is one.

A sequence of studies by Erica Goetze shows that strong divergence occurs. She began by sequencing several genes from copepods identified initially as *Eucalanus hyalinus* (Claus 1866), which is bi-antitropical in the Pacific, subtropical in the southern Indian Ocean, and lives in all warm-water areas of the Atlantic. She found that gene differences indicated two quite distinct subsets and initiated a close morphological inspection of many specimens from across the range. She found two groups that were separable by a set of subtle morphological distinctions: consistently different head shape, different asymmetries between the two antennules, different lengths of the terminal tail segments, and a size difference in adults. The smaller form was described from the Gulf of Guinea by T. Scott (1894), who gave it the name *Eucalanus spinifer*. However, *E. spinifer*

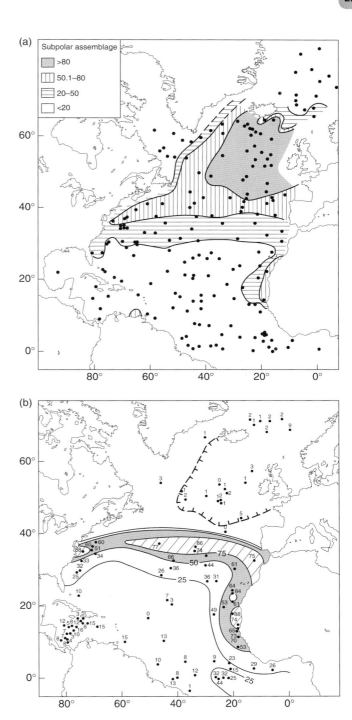

Fig. 10.21 Maps comparing the extent of sediment deposits dominated by a subpolar foraminiferan faunal group between the present (a, after Kipp 1976) and at 18,000 YBP (b, after McIntyre *et al.* 1976), when these cold-water forms constituted 25% of the foraminiferan fauna right across the equator. The ticked dashed line in (b) shows the core of distribution from (a).

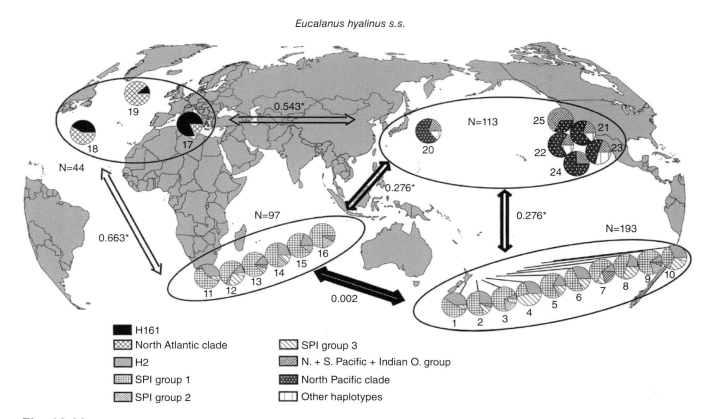

Eucalanus hyalinus s.s.

H161
North Atlantic clade
H2
SPI group 1
SPI group 2
SPI group 3
N. + S. Pacific + Indian O. group
North Pacific clade
Other haplotypes

Fig. 10.22 Frequencies of haplotypes of COI in *Eucalanus hyalinus*. Overall regional differences are represented by the Φ_{st} values (significant values with *) alongside the interchange arrows, which are darkened in proportion to the likely gene flow. (After Goetze 2005.)

had long since been called a synonym of *E. hyalinus*. More extended genetic study confirmed that the two species differed consistently in the bases of two mitochondrial genes, genes for 16S rRNA and cytochrome oxidase I (COI), and of nuclear ITS2 (a spacer in the gene for ribosome structure that does not code rRNA). For COI, the worldwide average within-species differences were 2.7 of 348 base pairs (bp) for *E. hyalinus* (N = 450 individuals) and 0.8 of 348 bp for *E.spinifer* (N = 383), whereas the between-species difference was 36.1 of 348 bp. That level of within vs. between differences is typical, or relatively large, compared to other pairs of closely related but distinct species. On the basis of both the morphological and genetic differences, Goetze and Bradford-Grieve (2005) removed *E. spinifer* from the limbo of synonymy, declaring it to be a real population entity living in the ocean.

Goetze (2005) followed that discovery with identification and sequencing studies of both species from sites across their global range. She found that while they live in general sympatry, they do have distinct habitat preferences, with *E. spinifer* more abundant in the centers of oligotrophic gyres, and *E. hyalinus* dominant at the gyre edges and in adjacent upwelling zones. Among 450 sequences for a 349 base-pair

region of COI from individuals of *E. hyalinus* collected worldwide, she found 239 haplotypes (recall that those differ on average by only 2.7 bases). Those fell into seven distinct haplotype groups, two of which were strongly dominated by single haplotypes (coded H161 and H2). The proportions of individuals in each group differed dramatically among sampling areas, and differed with strong statistical significance (large values of a proportional distinction measure, Φ_{st}, among three regional clusters (Fig. 10.22): North Pacific, Atlantic, and South Indian–South Pacific. A sample of 337 *E. spinifer* sequences showed similarly distinct groups, except that (i) all sampled areas had substantial numbers of the globally dominant haplotype (H1), and (ii) haplotype proportions in Atlantic and South Indian stocks were similar, while North and South Pacific stocks were distinct from each other and from the Atlantic–Indian group. All of that leaves no doubt that continental and ocean-circulation barriers restrict gene flow in planktonic populations, and that genetic distinction does emerge in isolated populations. It can be taken as likely that the speciation process for both *E. hyalinus* and *E. spinifer* is in progress through the genetic drift and selective differences between ranges – a key step in speciation.

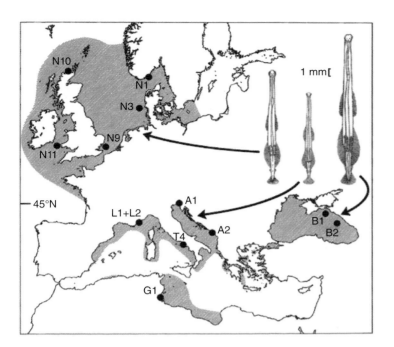
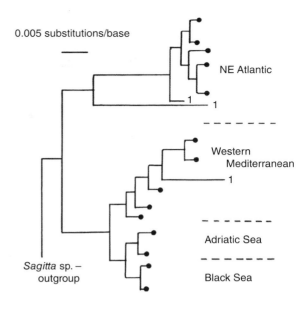

Fig. 10.23 Species distribution map and COII (mtDNA) cladogram for *Sagitta setosa*, an eastern North Atlantic and Mediterranean/Black Sea chaetognath. Map dots are sampling sites: A , Adriatic Sea; B, Black Sea; G, Gulf of Gabès; L, Ligurian Sea; N, Northeast Atlantic; T, Tyrrhenian Sea. Dots in cladogram represent refined branching to many specimens; 1 represents a distinctive individual. (After Peijnenburg *et al.* 2004.)

Do more contiguous populations spread across wide ocean basins also diverge? The general within-gyre similarity of Goetze's data for subtropical *Eucalanus* suggest not, that circulation generates enough genetic mixing across up to 12,000 km to sustain homogeneity. While a number of studies have addressed this issue, none yet proves the *generality* of this conclusion. Among more coastally bound species, it is common to find statistically significant differences in proportions of haplotypes or alleles in different parts of the distribution. These occur within the sequence-variation range usual for morphologically recognized species, within ~1–4% in mtDNA. An example is the variation in cytochrome oxidase II (mtDNA) among stocks of the chaetognath *Sagitta setosa* around the British Isles and in several of the basins of the Mediterranean (Peijnenburg *et al.* 2004, 2006). Haplotypes among 551 base pairs for 148 specimens from scattered sites (Fig. 10.23) showed strong divergence between North Atlantic, Mediterranean, and Black Sea sites, and there were significant regional differences within the Mediterranean. At first glance, the NE Atlantic vs. Mediterranean difference is not surprising, since *S. setosa* has a distribution gap from 45°N in the Atlantic to the Spanish coast off Barcelona, apparently not at present passing the Straits of Gibraltar. This study has an added advantage: the authors tell us that, while almost every single specimen had an at least slightly unique haplotype, only seven specimens showed a *coding* difference for the COII protein and in those cases the same difference.

Oddly, those specimens were from both the Black Sea and the NE Atlantic, at the maximum separation in distance and minimum likelihood of transfers. An attempt by Peijnenburg *et al.* to assign separation times to even the most obvious divisions among subsets of haplotypes fell short of providing close dating, primarily because the evolutionary rates for genes (the "molecular clock"), even for presumably selectively neutral, synonymous mutations, remain radically uncertain. The strongest arguments suggest that a NE Atlantic–Mediterranean split occurred in *S. setosa* in early to mid-Pleistocene, forced by northern glaciation. The ancestral stock transferred south and into the Mediterranean due to cold, then a separation was forced by drying of the straits. On glacial retreat, the southern stock moved back toward Britain, and the Mediterranean stock remained east of the straits, likely subject to subsequent isolation.

Apart from distribution maps for foraminiferan and other microfossils sedimented in glacial eras, is there evidence for glacial impacts on pelagic populations? Bucklin and Wiebe (1998) made the claim that the low diversity of mtDNA in subpolar zooplankters is in fact an indicator of glacial impact. Among 216 sequences of mitochondrial SSU ribosomal DNA from *Calanus finmarchicus*, they found that 79% were just one haplotype, much lower genetic diversity than expected for a reproductive population on the order of 10^{15} females. Diversity was much lower than that of subtropical species of the same family with stocks of ~10^8 females. They propose that during at least the most recent glacial era the subpolar habitat was both modified and reduced in area by 75% (based on CLIMAP results), forcing the *C. finmarchicus* stock through a population-genetic "bottleneck" Of course, this glacial-era bottleneck effect could have operated repeatedly through the Pleistocene. Provan *et al.* (2010), in contrast found no DNA-sequence evidence for genetically significant glacial population restriction in *Calanus finmarchicus*. All of these issues will remain unsettled; we cannot go back to sample the past. The likelihoods will become clearer with larger samples, more sequences, and more secure methods in bioinformatics.

It might seem that divergence in haploid mitochondrial genes would not imply general divergence, since mitochondria are inherited from mothers only and are strictly clonal organelles. Perhaps having any functional mitochondrion would serve for oxidative metabolism. That is likely true for the substantial majority of haplotype differences that do not code for different amino acids; those are simply an indicator of genetic drift and possibly time of stock separation (evidently substantial between oceans and northern vs. southern hemispheres). However, mtDNA does acquire non-synonymous mutations, and likely more rapidly than nuclear DNA. Mitochondria (and chloroplasts) are not constructed solely from proteins coded in mtDNA, but include proteins (and protein parts) coded in the nucleus. Such changes in mtDNA must be matched by changes in nuclear genes. Mismatches are a significant cause of mating failures in experimental crosses between distant populations (e.g. Burton *et al.* 2006). Thus, rapidity of mitochondrial evolution is likely a major component in genetic divergence of allopatric populations. Better phylogeographical insight will come with study of more genes, particularly developmental and other nuclear genes. Eventually, molecular sequences are going to give us a blazing spotlight on timing issues. It's just not fully powered up, yet.

Coastal distributions and the indicator species concept

The strong El Niño events of recent years along the west coast of the Americas include extensive poleward flow along the Americas that affects the distributions of plankton. El Niño events are lapses in the trade winds, allowing the east to west upslope of the tropical Pacific to collapse eastward, moving the boundaries of distributions rapidly, and changing the composition of the plankton at a given point along the route. El Niño provides an excellent application of the notion of "indicator species."

The idea that planktonic species composition can be informative about water movements has a long, honorable history. In global perspective, distribution patterns appear to correspond with the spatial extent of the major gyral circulations of the oceans. That is established. However, at most coastal sites there is considerable fluctuation in the pelagic fauna between sampling times, and these fluctuations often have a rhythm or regular sequence. Unlike a site such as 35°N, 150°W, where a standard central Pacific gyre fauna is always present, the fauna alongshore varies with season and with changes in the flow. This was first studied in detail in the vicinity of the British Isles, for which Russell (1939) suggested the notion that particular forms were indicative of the sources of the water that happened at any given time to be lapping the shore.

Russell dealt primarily with species of chaetognath, and he named different water types after the dominant species of *Sagitta* (*sensu lato*) they contained: "*serratodentata* water", "*setosa* water", and "*elegans* water". These chaetognaths appear in waters surrounding the British Isles under different conditions of wind and flow, so he called them "indicator species". The indicator-species notion must, to be fully effective, be calibrated by a survey of the distributions on a very large scale. That has been done, and the following sources have been identified:

1 *S. serratodentata*: a form of the waters well offshore to the west of Britain, waters of the North Atlantic West Wind Drift, or north branch of the Gulf Stream;

2 *S. setosa*: a shallow-seas form, usually the dominant in most of the nearshore or shelf waters of Western Europe;

3 *S. elegans*: occurs in the zone of mixing between the two. It is closely related (the same or called by the same name) to a species of the oceanic sector of the Atlantic (well north of the West Wind Drift) of the Arctic Ocean, the Bering Sea, and subarctic Pacific.

At any given site, say Plymouth, UK, the species present allow identification of the advective sources of water resident at any given time. Oceanographic conditions correlate with appearances of a given indicator species. Appearance of *S. serratodentata* off the British Isles usually corresponds to warmer than usual coastal temperatures, low phytoplankton nutrients, low primary productivity and poor fishing.

This idea can be applied to any coastal region, and it has been applied along the US West Coast. The California Current is not a simple, one-way flow from north to south. Rather it has a complex and varying structure involving a northward undercurrent and seasonal oscillations of the direction and speed of nearshore flow. In spring and summer there is southward and seaward surface flow, the offshore component producing nearshore upwelling. The intensity and timing vary with latitude. Satellite pictures show large onshore-to-offshore jets as a feature of upwelling. Winter flow is northward and onshore. During El Niño events the winter pattern is more extreme and can persist into summer,

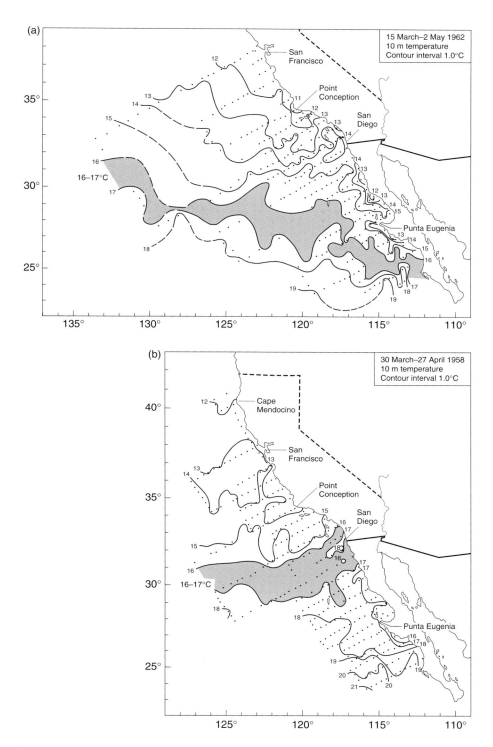

Fig. 10.24 Comparison of 10 m isotherm patterns in the California Current area between (a) a typical year, 1962, and (b) an El Niño or warm year, 1958. Offshore, at 130°W, say, the positions of the 14° to 18° isotherms are not very different. But inshore they are strongly changed between these two situations. (After Anonymous 1963; Wyllie & Lynn 1971.)

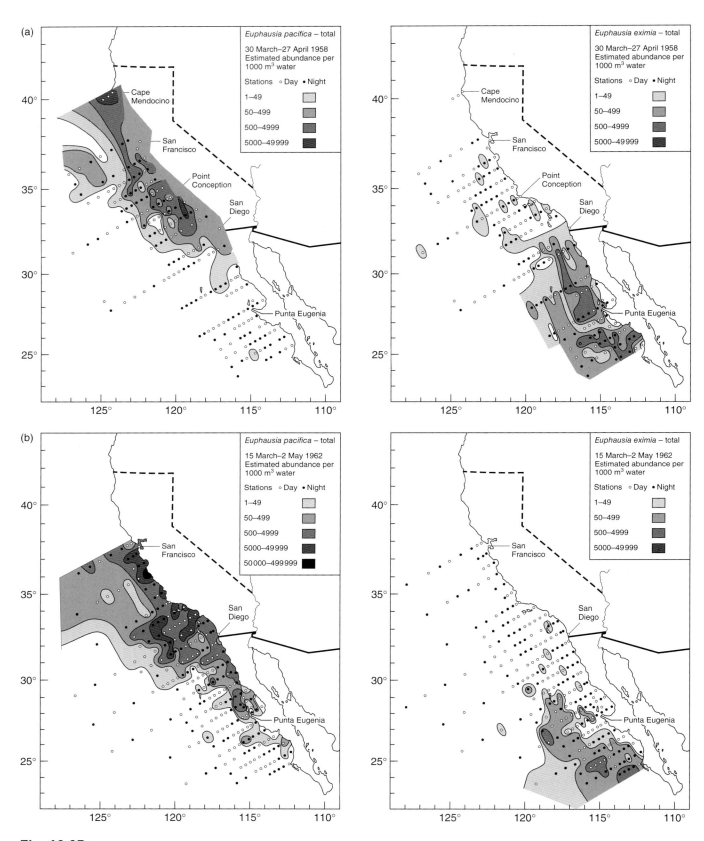

Fig. 10.25 Comparisons of distribution patterns of a subarctic species, *Euphausia pacifica* (left), and an eastern tropical Pacific species, *Euphausia eximia* (right), between (a) 1958, a warm year, and (b) 1962, a more typical year. The northern species was not carried so far south and the tropical species was carried much farther north in the warm year. (After Brinton 1967b.)

often characterized as a difference between warm years and cold years (Fig. 10.24).

The shifts in isotherms could result either from a change in advection of cold water to the south and warm water to the north by the current complex, or they could come from changes in the regional budget of heating and cooling. Which? West Coast indicator species provide the answer. For example, *Euphausia pacifica* is endemic in the subarctic Pacific and northern California Current, while *Euphausia eximia* is an eastern tropical Pacific endemic with a distribution usually just extending into the southern end of the California Current. The distributions of these two species change in complementary fashion between years of warm and cold conditions in the sea (Fig. 10.24). The euphausiid distributions (Fig. 10.25) show that advection is the main factor. The data shown are for March–April, usually the period of maximum extent of northward intrusion of *E. eximia* and, conversely, the minimum extent of southward extension of *E. pacifica*.

The notion of species populations remaining established in source regions, then being transported elsewhere and giving information about source and flow, has been applied (Peterson & Miller 1977) to the standard cycle of zooplankton species off Oregon (USA) corresponding to the seasons. During summer, the very nearshore zone usually has a copepod fauna dominated by an array of northern species: *Pseudocalanus mimus*, *Calanus marshallae*, *Acartia hudsonica*, *Centropages abdominalis*, *Acartia longiremis*. Those species are present all year round. However, in winter they are mostly replaced by an array of copepods with southern affinities: *Paracalanus parvus*, *Ctenocalanus vanus*, *Clausocalanus* spp., *Corycaeus anglicus*, and others. These are forms that have their main ranges of distribution to the south and in some cases offshore in central waters. They are carried into the region during the late fall and winter months. The seasonal switching is in agreement with the direction of the wind, current meter observations and other faunal changes. Summer winds from the north bring in species of northern affinity, and their populations increase during the cooling and strong production induced by near-shore upwelling. In the fall, the winds reverse and inshore flow switches to northward and onshore, so southern forms are carried into the region and are present through the winter, usually until April. They do not prosper, specifically reproduction is low or zero and abundance remains low.

During the strong El Niños of 1983 and 1997, southern plankton were carried into the region in winter, just as in normal years, but were not carried away again in spring. During the abnormally warm sea conditions of those summers, southern forms remained along the shore and were able to spawn and grow. They remained dominant or subdominant components of the fauna right through the entire summer of 1983 and 1997 (Keister & Peterson 2003). For example, the dominant copepod on many dates in the summer of 1983 was *Acartia tonsa*. This is a circumglobal species (or "species group") of nearshore waters in warm, temperate zones. Along this coast they are a dominant summer form south of Point Conception. It is not usual for them to get as far north as Oregon in most winters. In 1982–1983 they did, and they were able to increase and stay all summer.

In general, the El Niño results (in agreement with others) suggest that because they "go with the flow" planktonic life forms have an advective flexibility. They are carried to newly suitable habitats immediately as those develop, primarily because habitats are usually made "newly suitable" by advection. Once arrived, new forms replace the usual inhabitants immediately. However, those usual inhabitants are themselves displaced farther downstream, and somewhere along the line they may be doing fine. There is no radical reduction in the global total of the population and no danger of extinction. When conditions reverse, stocks are carried back to their waters of origin and re-establish quickly. An important aspect of this flexibility, of course, is that most plankton are fast growing with relatively short generation times. Thus, they are resilient to large changes in "ocean climate". El Niño is no exception. We present the Oregon example because it is closely familiar to us. Similar applications of the indicator species concept have been or can be made for virtually every coastal zone.

Chapter 11

Biome and province analysis of the oceans

Longhurst's analysis

Alan Longhurst was among the group (Abbott, Banse, Brown, Esaias, Longhurst, McClain, McGowan, Pelez, Platt, Sathyendranath, and others) who took an early interest in the information from images made with the first visible-wavelength satellite radiometer, the CZCS, which was active from 1978 to 1986. In 1998, he published a book, *The Ecological Geography of the Sea* (2nd edition, 2006), attempting to define a geography of ecosystem types in the world's oceans based on chlorophyll and temperature images. He concluded that basically there are four "biomes" in the world ocean. *Biome* is a term from terrestrial ecology, where we recognize rainforest, desert, savannah, and other habitats as general biome types, regardless of the specific organisms living in any given example. Longhurst suggested a division of oceans into polar, westerlies, trades, and coastal zone biomes. Immediately after defining them, however, he divided them into a substantial number of "provinces", zones that are more homogeneous physically and are "recognized" by large sets of species as suitable habitats (Chapter 10). For example, at least the two polar and all the westerlies biomes vary enough from each other that they must be considered separately. It is also clear that the trades biome, in which Longhurst included central gyre and equatorial areas, covers some fairly heterogeneous physics and biology. He did obtain a division comparable to that indicated by the distributional patterns of organisms (a source of information that he rather disparages) by the "provincial" subdivision, ultimately arriving at a necessarily complex scheme (Fig. 11.1). The general biogeographical

observations reported in the previous chapter match remarkably well with Longhurst's provinces, a set of ocean sectors which the satellite observations distinguish reasonably well. We will treat those as distinctive habitats requiring individual analysis. We agree with Longhurst's distinction of coastal biomes, which again divide into many types, such as zones of upwelling or of large river influx.

Longhurst's province analysis has become the subject of recurring review by the International Ocean-Colour Coordinating Group (IOCCG 2009), which posts its reports on the worldwide web in the publications-and-reports section at http://www.ioccg.org/. The 2009 report is an extensive discussion of the modes of province identification from satellite data. One rationale for this work is to define zones in which the parameters affecting primary production will be roughly the same at least within seasons; that is, the subsurface profile of phytoplankton abundance and the P vs. E relations (specifically P_{max} and α) can be expected to be relatively constant. In addition it is hoped that the full suite of food-web relations from nutrients to whales will be consistent and, thus, somewhat predictable from satellite color observations.

Recent research in oceanic biological oceanography has been partly organized as programs targeting specific biomes or ecosystems, or sometimes repeatedly crossing several of them. As examples, the subarctic Pacific was approached by the SUPER, SERIES and SEEDS programs (Miller 1993; Boyd *et al.* 2004; Uematsu 2009); the North Pacific subtropical gyre by the Hawaii ocean time-series (HOT); the North Atlantic subtropical gyre by the Bermuda Atlantic time-series (BATS) studies (e.g. Siegel *et al.* 2001); the Antarctic by the US Southern Ocean JGOFS–AESOPS

Biological Oceanography, Second Edition. Charles B. Miller, Patricia A. Wheeler.
© 2012 John Wiley & Sons, Ltd. Published 2012 by John Wiley & Sons, Ltd.

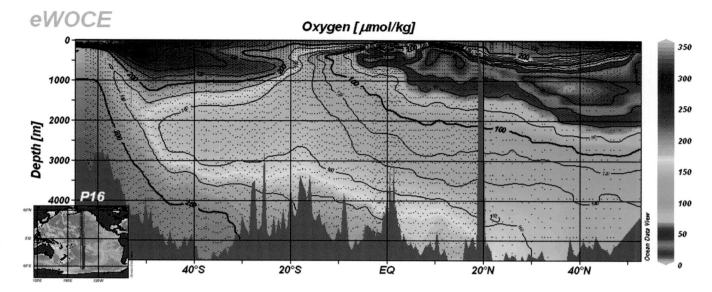

Plate 1.1 An "oceanographic section" of oxygen concentration from cruises by research ships along the line shown in the inset: along 150°W in the Pacific Ocean from Antarctica to the Aleutian Islands. Water was collected with closing samplers at each depth indicated by a dot and dissolved oxygen concentration (µmole kg⁻¹) was estimated on board the ships. The most intense oxygen minimum (violet shading) is in the subarctic and northern subtropical gyre, beneath the intermediate water subducted from the subarctic. Low-oxygen water outcrops near the equator. The minimum rides above the influx of water from the deep global thermohaline circulation. Maximum concentrations (red) are in Antarctic intermediate water. Data are from WOCE (World Ocean Circulation Experiment) profile *p16*.

Primary endosymbiosis

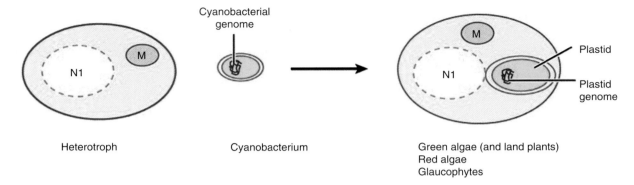

Heterotroph   Cyanobacterium   Green algae (and land plants)
Red algae
Glaucophytes

Secondary endosymbiosis

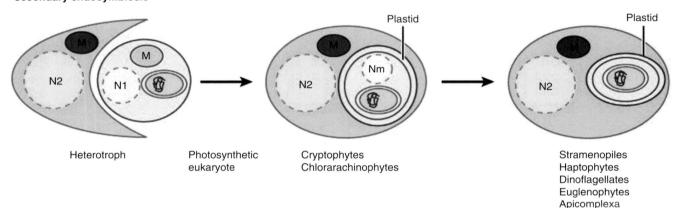

Heterotroph   Photosynthetic   Cryptophytes   Stramenopiles
eukaryote   Chlorarachinophytes   Haptophytes
Dinoflagellates
Euglenophytes
Apicomplexa

Tertiary endosymbiosis

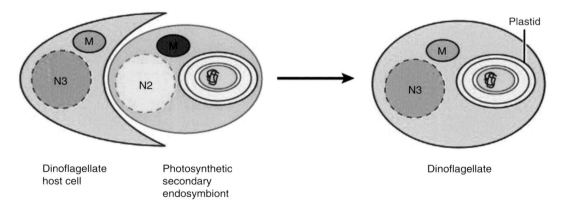

Dinoflagellate   Photosynthetic   Dinoflagellate
host cell   secondary
endosymbiont

Plate 2.1 Schematic depiction of endosymbiotic events establishing plastids as organelles, and the consequences for the genomes of the symbionts. In primary symbiosis, a cyanobacterium was engulfed by (or invaded) a heterotrophic eukaryote. The cyanobacterial genome was reduced over time, but, in all known photosynthetic eukaryotes, at least a few genes are still retained as a plastid genome. In secondary endosymbiosis, a heterotrophic eukaryote acquires a photosynthetic eukaryote. The nucleus of the endosymbiont (**N1**) is severely reduced to a nucleomorph (**Nm**) or lost altogether in the following transfer of many genes to the host nucleus. "Stramenopiles" is an alternative name for the Heterokontophyta, including the diatoms. Tertiary endosymbiosis followed when dinoflagellate host cells engulfed stramenopiles, haptophytes, or cyptophytes. (After Parker *et al.* 2008.)

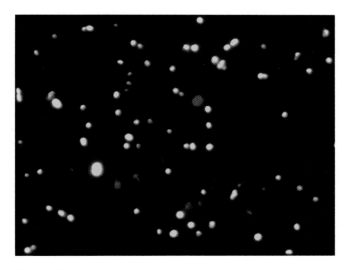

Plate 2.2 An epifluorescence micrograph of unstained cells showing *Synechococcus* (orange) and eukaryotic picoplankton (red). (Photograph by E. and B. Sherr.)

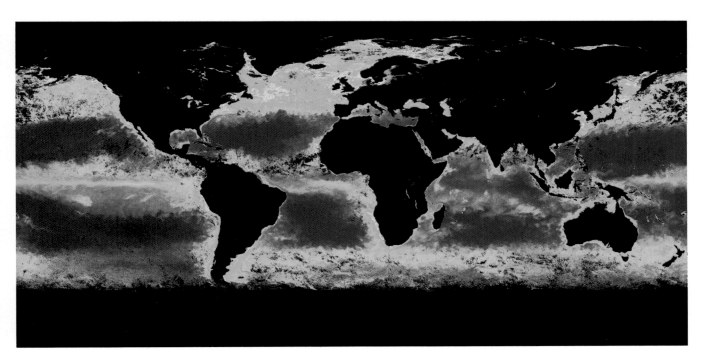

Plate 2.3 Global SeaWIFS image generated by NASA: monthly mean for May 2000. In ocean areas, purple represents very low chlorophyll and the progression to blue, green, yellow, and red represents progressively more chlorophyll. Highest values are in subpolar areas, lowest values in subtropical gyres. Equatorial areas have intermediate, but still low, values. (© GeoEye, with permission.)

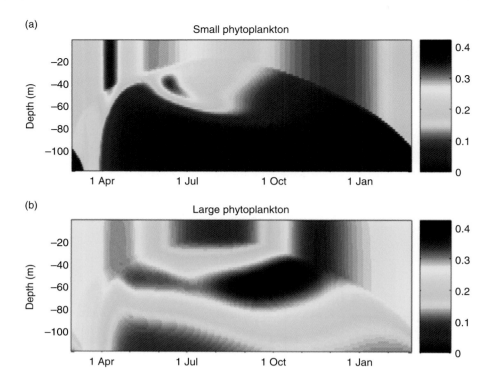

Plate 4.1 Annual cycles and abundance variation with depth of (a) nano- and picophytoplankton and (b) large phytoplankton (diatoms) in the Denman *et al.* (2006) model of production processes in the iron-limited (HNLC) ecosystem of the Gulf of Alaska. (After Denman *et al*. 2006.)

Plate 5.1 Marine bacteria and Archaea made visible by using polyribonucleotide probes and fluorescent *in situ* hybridization (FISH). Seawater sampled at 80 m depth at 177 miles offshore of Moss Landing, California, USA. Bacteria are labeled with a fluorescein (green) stain and Archaea are labeled with a CY-3 (red) stain. Scale bar, 5 μm. Separate images were overlaid in Adobe Photoshop. (After DeLong 2001.)

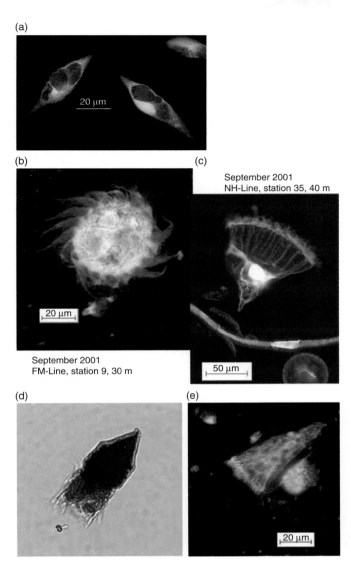

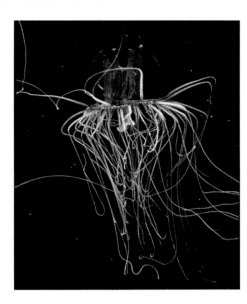

Plate 6.2 *Polyorchis penicillatus*, a hydromedusan. Typical bell height about 1.5 cm. (© Monterey Bay Aquarium.)

(a)

20 μm

(b)

20 μm

September 2001
FM-Line, station 9, 30 m

(c)

September 2001
NH-Line, station 35, 40 m

50 μm

(d)

(e)

20 μm

Plate 5.2 Heterotrophic protists. (a) Heterotrophic dinoflagellate with ingested coccoid cyanobacteria (red particles). (b) *Leegaardiella* sp. and (c) an unidentified ciliate formalin fixed and stained with DAPI. (d) *Strombidium* sp., a ciliate, and (e) *Tintinnid* sp. epifluorescence. (Photos courtesy of E. Sherr.)

Plate 6.3 A calycophoran siphonophore. The nectophores or swimming bells are at the left, a long chain of tentacle-bearing gastrophores extending behind. (© Monterey Bay Aquarium.)

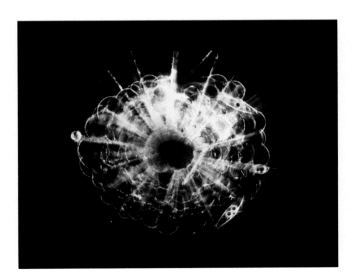

Plate 6.1 *Hastigerina pelagica*, a planktonic foraminiferan, about 4 mm in diameter. A small copepod, an *Oncaea* with purple pigment, has been engulfed by the cytoplasm. (Photograph by Allan Bé.)

Plate 6.4 *Chrysaora fuscescens*, a scyphomedusan. Size of this animal is up to about 30 cm. (Photograph by James M. King, courtesy of Alice Alldredge.)

Plate 6.5 *Tomopteris*, a pelagic polychaetous annelid. This specimen was approximately 1.5 cm long. (Photograph by Jaime Gomez Gutierrez.)

Plate 6.8 *Gleba*, a pseudothecosome with extended proboscis and "foot" attached to a large mucous float. The "wingspan" of the animal is about 10 cm. (Photograph by James M. King, courtesy of Alice Alldredge.)

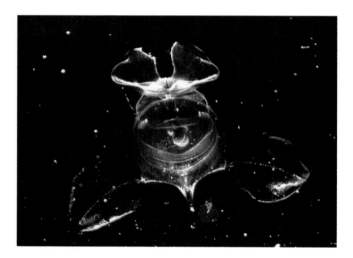

Plate 6.6 A *Cavolinia*, an euthecosome (shelled pteropod) photographed by a diver. The "wings" are at the top, mantle expansions merging from small ports in the shell are at either side. (Photograph by Ron Gilmer, published previously by Hanmer 1974.)

Plate 6.9 *Euchaeta norvegica*, a large (about 3.5 mm), predatory copepod. The antennules extend to the side and the urosome is down in this typical resting posture. Eggs in this genus are held in a sac attached to the anterior urosomal segment. (Photograph by Jeanette Yen.)

Plate 6.7 *Cavolinia inflexa* suspended from a mucous float which has been stained with carmine particles and photographed by a diver. (Photograph by R. Gilmer; Gilmer & Harbison 1986, with permission.)

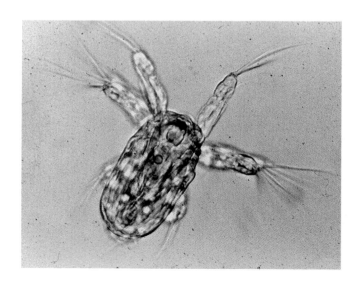

Plate 6.10 Second naupliar stage of the estuarine copepod *Acartia californiensis*. Length about 60 μm. (Photograph by J. Kenneth Johnson.)

Plate 6.11 *Euphausia pacifica*, a typical euphausiid. (Photograph by Lisa Dilling, courtesy of Alice Alldredge.)

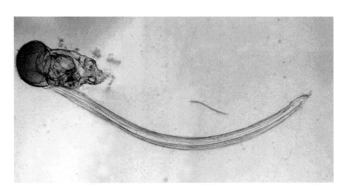

Plate 6.13 Appendicularian body (Photograph by J. Cavanihac, amateur de biologie marine).

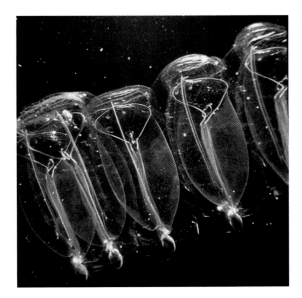

Plate 6.12 A salp in the aggregate phase. The mucous feeding cones have been stained with carmine particles before the photograph was taken by a diver. The incurrent openings are up, the digestive gland and gonad in the mass at the bottom. (Photograph by Larry Madin; published previously by Hanmer 1974.)

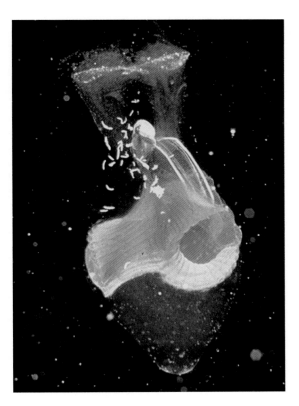

Plate 6.14 Appendicularian in its mucous house photographed by divers after staining with extremely fine carmine particle suspension. The coarse entry filters are at the top; the tangential flow filter is below the animal and its cluster of fecal pellets. (Photograph by James M. King, courtesy of Alice Alldredge.)

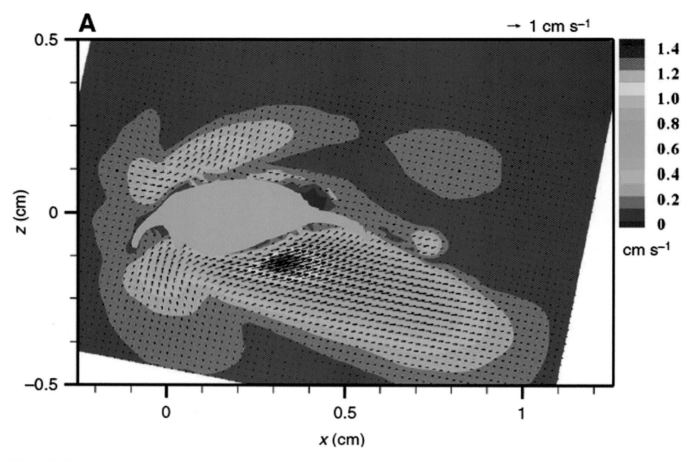

Plate 7.1 Streamlines of flow around an untethered specimen of *Euchaeta antarctica* (fifth copepodite) swimming slowly forward while searching for food items. Seen from above (top) and the side. Velocity vectors are contoured in color. (After Catton *et al*. 2007.)

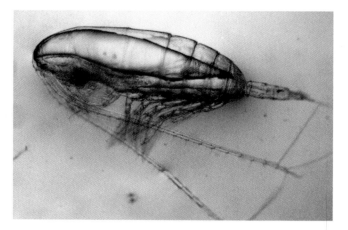

Plate 8.1 Photograph (by Jaime Gómez-Gutiérrez) of *Calanus marshallae* with a large, median oil sac (the posterior end of the sac has red pigmentation).

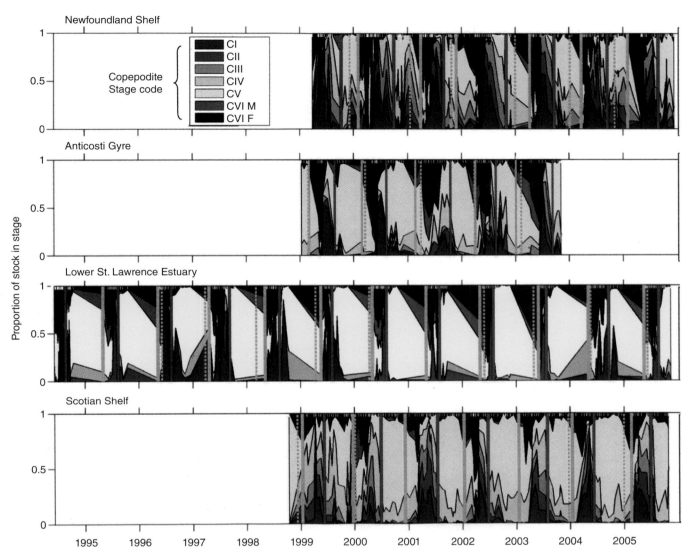

Plate 8.2 Time-series demography of *C. finmarchicus* at western Atlantic time-series stations. Proportions of copepodid stages for years with data. Red lines: estimated times of onset of diapause. Solid green lines: estimated times of exit from dormancy based on proportion of females present. Dashed green lines: estimated time of exit based on back-calculation from presence of early copepodid stages (assuming saturating food). (After Johnson *et al.* 2008.)

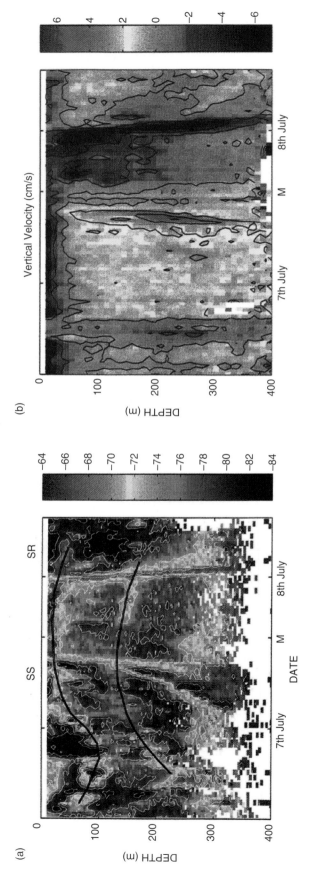

Plate 8.3 (a) Volume backscattering (color scale in decibels – red is more scattering, blue is less) analysis at 48°N for particles reflecting sound from an acoustic Doppler current profiler. Note the several combinations (one obvious, the others emphasized by lines) of layering and layer movement with time of day. (b) Vertical particle velocity from sound-frequency shift. Layers move rapidly near sunset (SS) and sunrise (SR). (After Wade & Heywood 2001.)

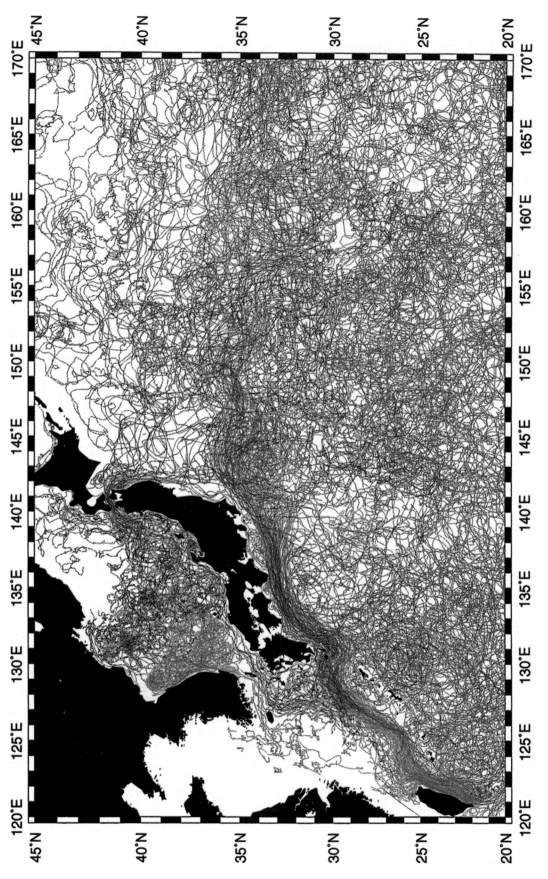

Plate 10.1 Accumulated tracks of satellite-reporting drifters (a "spaghetti diagram") for the Kuroshio–Oyashio confluence area out to the Emperor Seamounts. (Diagram by Peter Niiler, with permission.)

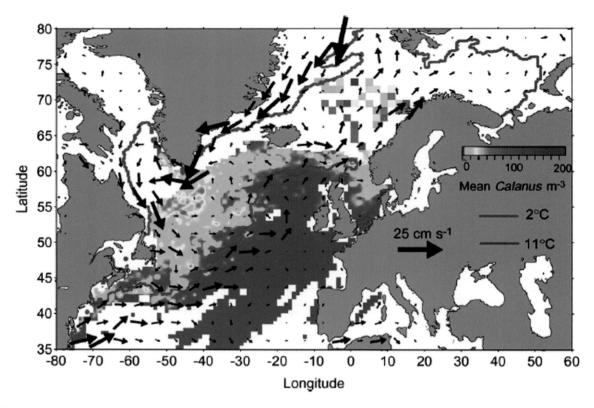

Plate 10.2 Time-averaged *C. finmarchicus* C5–C6 abundance (m^{-3}) at 7–10 m depth from a combination of CPR samples collected between 1958 and 2000 and plankton sampling in the Norwegian Sea during 1997. Arrows showing annual average velocity at 10 m depth and the 2°C and 11°C annual average isotherms are from a global circulation model for 1997 conditions. (After Heath *et al.* 2008.)

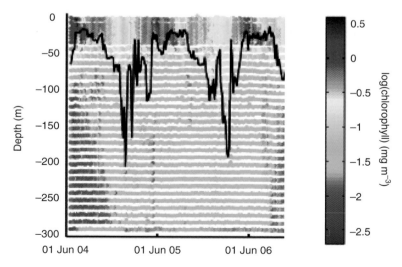

Plate 11.1 Two-year time-series of midnight fluorometer profiles interpreted as chlorophyll concentration (color scale) collected every 5 days by ascending ARGOS drifters in the vicinity of 50°N, 47°W. Dark line is a time-series of mixing-layer depths defined as the shoalest level with σ_t greater by 0.125 from the surface. Phytoplankton are consistently distributed evenly down to that boundary. (After Boss *et al.* 2008.)

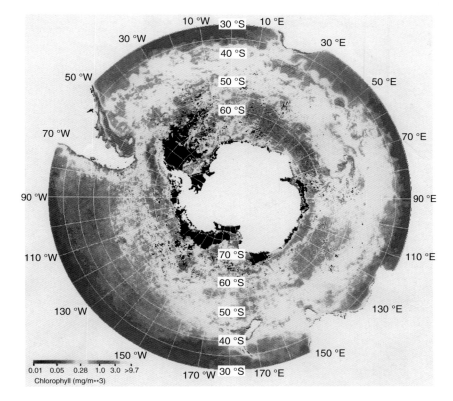

Plate 11.2 Analysis based on SeaWIFS data of chlorophyll distribution in the Southern Ocean. Yellow represents approximately 0.3 mg Chl m^{-3}, a typical value in subpolar HNLC regions. (After Moore & Abbott, 2000.)

(a)

(b)

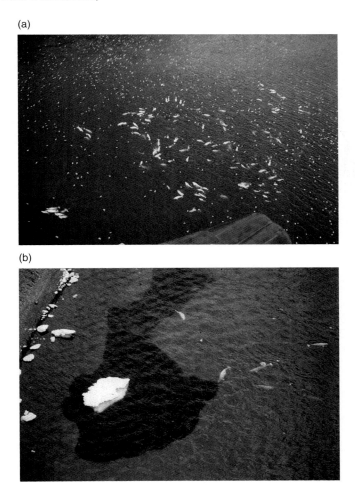

Plate 11.3 Photographs of Arctic cod, seabirds, and beluga whales near Patrol Point, Radstock Bay, in July 1991. (a) Aerial view of beluga and flying birds 100–200 m offshore seaward of a cod school. There are >130 whales in the picture. Remnants of a cod school in shallow water are visible in the top left corner. (b) The forward end of a large school of Arctic cod. The school is 500 m long, of which about 50 m is shown. Predation pressure is light (note the absence of birds) but a few whales are present. (After Welch *et al.* 1993.)

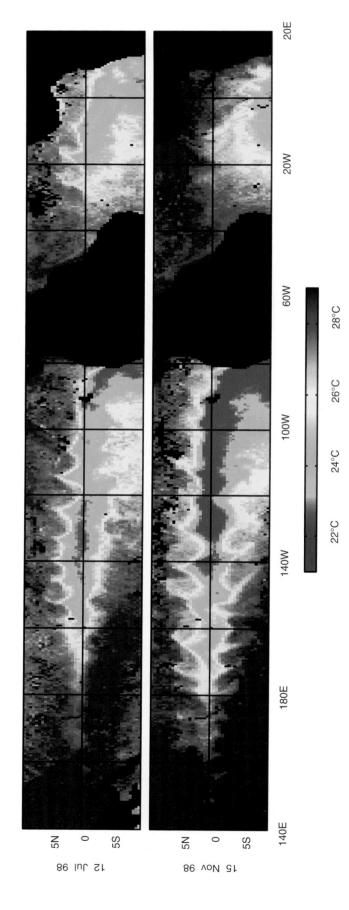

Plate 11.4 Satellite microwave sea-surface-temperature observations of tropical instability waves in the Pacific and Atlantic Oceans. These are 3-day composite-average maps for July 11–13, 1998 (upper) and November 14–16, 1998 (lower). Black areas represent land or rain contamination. (After Chelton *et al.* 2000.)

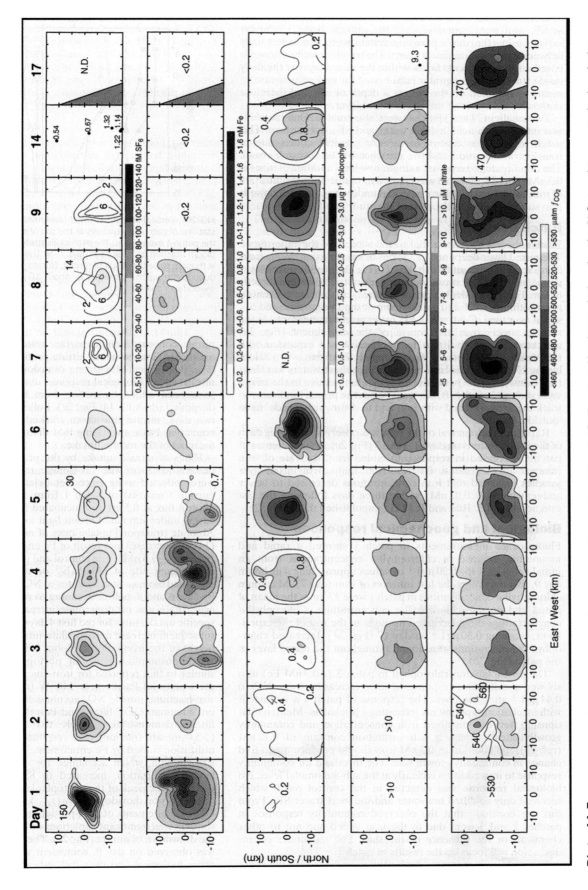

Plate 11.5 (a) Temperature during iron-enrichment experiment at a site in the oceanic, eastern tropical Pacific contoured in color on a depth *vs.* time plot. (b) Contour plots on successive days after iron addition. Rows of plots are sulfur hexafluoride (SF6) tracer, iron concentration, chlorophyll concentration, nitrate (contoured as difference from 10 μM), and fugacity (% partial pressure) of CO_2. (After Coale *et al.* 1996.)

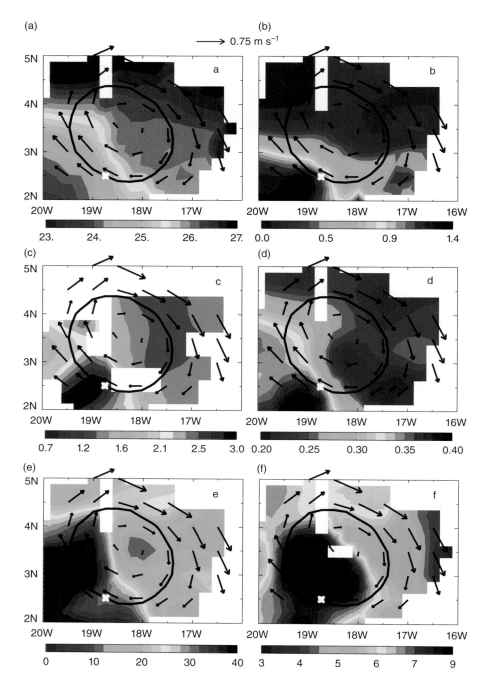

Plate 11.6 Circulation, nitrate concentrations, and biological parameters in the field of an anticyclonic vortex of a tropical instability wave in the Equatorial Atlantic. (a) Temperature and velocities averaged to the pycnocline depth. (b) Nitrate (μM) averaged down to the pycnocline depth. (c) Gross primary production (measured by natural fluorescence) integrated down to the 0.1% light level in $g\,C\,m^{-2}\,d^{-1}$. (d) *In situ* chlorophyll averaged down to the pycnocline depth in $mg\,Chl\,m^{-3}$. (e) Zooplankton biomass (acoustic measurements calibrated with data from net tows) averaged over 150 m as dry weight m^{-3}. (f) Micronekton biomass sampled acoustically and averaged over 500 m. (After Menkes *et al*. 2002.)

Plate 12.1 Photo of the ventrum of *Abraliopsis pacificus*, showing photophores in several colors, blue with white centers and red. Red is achieved by a color filter. The large white spots are the ocular photophores located well inside the translucent skin. (After Young & Mencher 1980).

Plate 13.1 Photograph of the ocean bottom on an abyssal plain at 2°N, 140°W. The curving track was made by a sea urchin seen in the foreground. Older tracks and sundry animal mounds are also visible. (Photograph by C.R. Smith, from Smith 1996.)

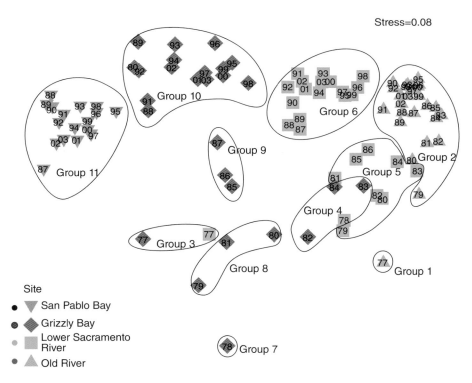

Plate 14.1 Plot of time-series samples from four stations in the upper San Francisco Bay estuary versus the first two principal components of the benthic macrofauna composition. PCA-1 is the "virtual" abscissa, PCA-2 the ordinate. (After Peterson & Vayssières 2010.)

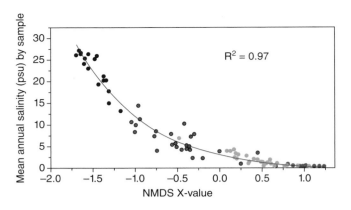

Plate 14.2 Position along PCA-1 (Plate 14.1) versus salinity (lower upstream). (After Peterson & Vayssières 2010.)

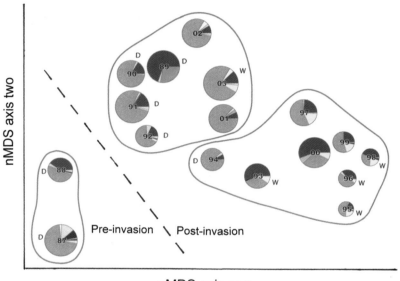

Plate 14.3 Principal-component plot for annual mean composition of samples from Grizzly Bay (central San Francisco Bay estuary). Years (numbers inside each pie diagram) cluster in before and after the *Corbula amurensis* invasion groups. There is a partial tendency for samples from wet (W) years to cluster apart from samples from dry (D) years. Some years (no label) had average precipitation and thus freshwater flow in the estuary. Species composition in each year shown as pie diagrams. Pie-sector colors are by phylum: blues for clams, reds for crustaceans, etc. (After Peterson & Vayssières 2010.)

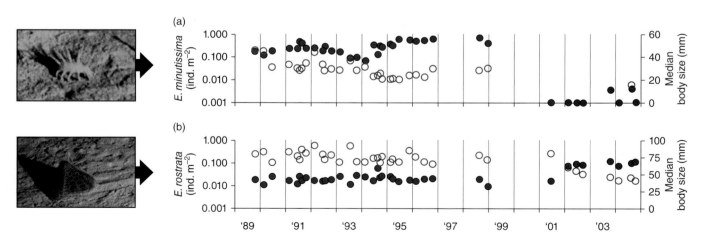

Plate 14.4 Time-series from transects of sediment-surface photographs of abundances (filled blue dots) of the holothurian *Elpidia minutissima* and the echinoid *Echinocrepis rostrata* at Station M seaward from southern California, USA. Time-series of median individual sizes (open red circles) of the two species were measured for samples from a trawl attached behind the camera sled. (After Smith *et al.* 2009.)

Plate 15.1 A montage of animals from deep-sea hydrothermal vents. (a) Vestimentiferan worms, *Riftia pachyptila*, at the Galapagos Ridge in 1977, © the late J. Edmond, courtesy of Woods Hole Oceanographic Institution. (b) Vent mussels, *Bathymodiolus thermophilus*, on the East Pacific Rise at 21°N. (c) Vent clams, *Calyptogena magnifica*. (d) A black smoker vent with sundry animals nearby. (Frames b, c and d © J. Baross, courtesy of the Woods Hole Oceanographic Institution).

Plate 15.2 Photograph of the "scaly footed snail" from the Central Indian Ocean Ridge. (After Warén *et al*. 2003.)

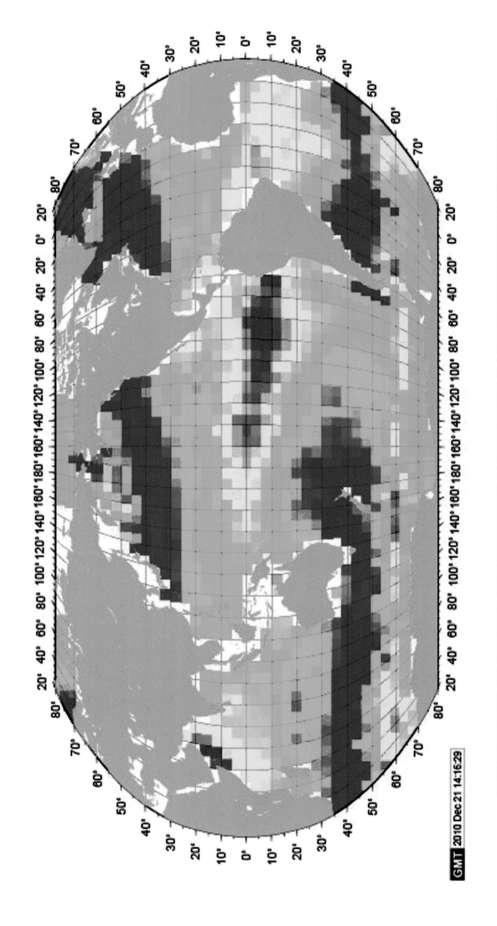

Net Flux (grams C m^{-2} year^{-1})

-108 -96 -84 -72 -60 -48 -36 -24 -12   0   12   24   36   48   60   72   84   96 108

GMT 2010 Dec 21 14:16:29

Plate 16.1 World map of ocean CO$_2$ influx (negative, cold colors) and efflux (positive, warm colors) calculated from ~3 million surface pCO$_2$ values from 1980–2000 projected forward with geographically distributed trends to represent the year 2000. (Figure based on a recalculation of December 2010 from the worldwide web with permission of T. Takahashi; methods in Takahashi *et al.* 2009.)

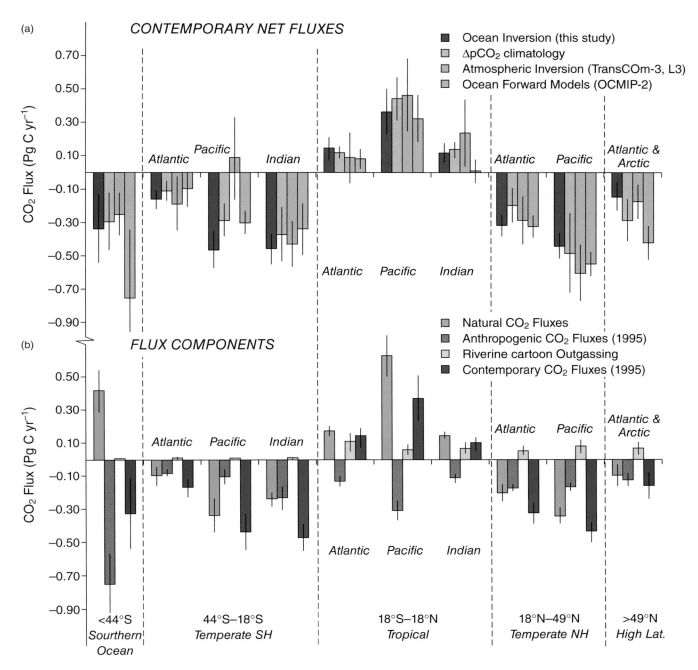

Plate 16.2 Air–sea CO_2 fluxes for 10 regions (positive: out of the ocean). Approximate standard deviations ("uncertainties") are shown by vertical whiskers. (a) Estimates for 1995 by four modes of calculation shown with bars of different colors. (b) Approximate preindustrial ("natural" – blue green), anthropogenic (pink) and estimated 1995 exchanges by the inversion method of Gruber *et al.* (2009) (the blue bars in both (a) and (b)). (After Gruber *et al.* 2009.)

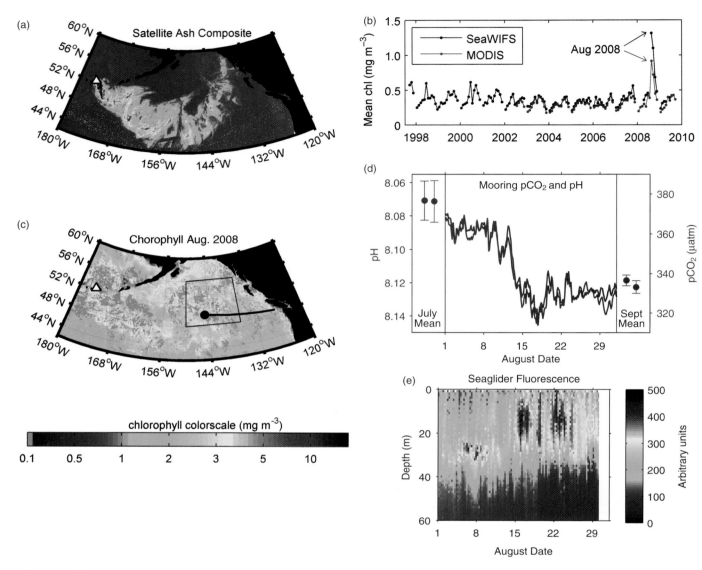

Plate 16.3 Aspects of the phytoplankton bloom induced by the Mt. Kasatochi volcanic dust plume over the Gulf of Alaska in August 2008. (a) The initial pattern of the plume from dust reflections recorded from a satellite. (b) Time-series of monthly SeaWIFS and MODIS satellite chlorophyll averages for the box in part frame c, showing the strong and unusual increase of August 2008. (c) MODIS chlorophyll distribution pattern August 2008; see more typical distribution patterns in the paper by Hamme *et al.* (2010). (d) Rise in pH and drop in pCO$_2$ at Station P (dot in part c) after plume arrival about 12 August. (e) Raw seaglider fluorescence data from the vicinity of Station P before and after plume arrival. (After Hamme *et al.* 2010.)

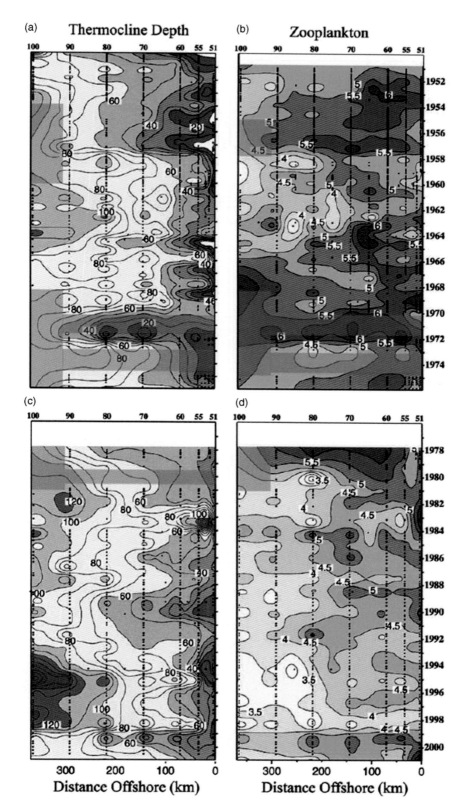

Plate 16.4 Time vs. offshore distance at CalCOFI Line 80 (Fig. 16.22a) contoured for (a & c) depth (m) of the 12°C isotherm as a proxy for thermocline and nutricline depths and for (b & d) log$_e$[macrozooplankton displacement volume, ml (1000 m)$^{-3}$]. Above (a & b) 1950–1975 and below (c & d) 1976–2000. Dots (1392 for thermocline depth and 1750 for plankton volume) represent stations. Gray shading covers times with long gaps between cruises. (After McGowan *et al.* 2003.)

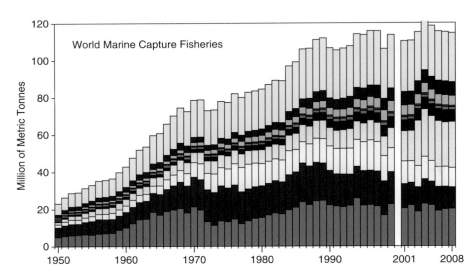

Plate 17.1 FAO estimates of annual world capture-fisheries yields since 1950. Colors of bar segments represent, from the bottom: blue, clupeids (herrings, anchovy, . . .); maroon, demersal fish (cod, hake, flatfish); cream, jacks and other pelagic; pale blue, miscellaneous; purple, tuna, bonito; pink, anadromous (salmon, shad), followed by sharks, crustacea, mollusks (about half squid); and yellow, approximation of at-sea discards. Because of category selections, these bars do not add exactly to the FAO world totals. Categories changed in 2000–2001, mostly by moving some demersal and pelagic catch to "unidentified marine" (combined from 2001 here in the pale blue). However, some demersal catches declined at the same time, partly through closures.

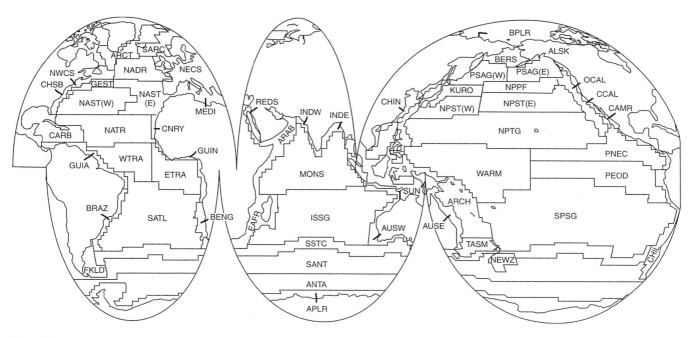

Fig. 11.1 The layout of pelagic provinces identified by Longhurst (2006) based primarily on satellite chlorophyll mapping. Readers should decode the province names using general knowledge of ocean and global geography. The oddly rectilinear boundaries are an artifact introduced by Longhurst's boundary selection algorithm; real ocean boundaries are often extensive gradients, are generally smoothly curved, and move about substantially. (After Longhurst 2006.)

Program (Smith & Anderson 2000a,b) and the Australian BROKE–West study (Nicol 2010); and the Atlantic from subarctic to subantarctic by the AMT project (Robinson *et al.* 2009). There have been numerous other regional projects in the last three decades, mounted by national and international teams. Output from these programs is sometimes (but often not) in the form of single volumes of *Deep-Sea Research II* or *Progress in Oceanography*, stacks of papers by participating research groups, in most cases including a synthesis of the overall significance of the disparate components. Full summaries of what is now known of each ecosystem type would take a pile of books larger than this one. The following are some rudimentary basics for some of these biomes.

Westerlies biomes

Subarctic Pacific

The westerlies zones are those also termed "subpolar": the subarctic Atlantic, subarctic Pacific and subantarctic. Much of our understanding of the subarctic Pacific came about because of its contrast to the subarctic North Atlantic. The subarctic Pacific is one of the cold versions of a high-nitrate–low-chlorophyll, or HNLC region. The other cold HNLC provinces are the antarctic and subantarctic areas

of the Southern Ocean. These have been studied to determine why phytoplankton do not exhaust the surface supplies of nitrate, phosphate and silicate during the warmer, illuminated summer season when the water column is stratified. At least in part, the mechanism is that iron limitation constrains the phytoplankton to small size (<10 μm). This was initially demonstrated by Martin and Fitzwater (1988), who compared phytoplankton growth in large water samples taken at 50°N, 145°W (Station P) augmented and not augmented with iron. After a lag in which enough large phytoplankton, primarily diatoms, accumulated to contribute significantly to chlorophyll, there was a nitrate-depleting bloom in augmented containers, but not in the others (Fig. 11.2). An explanation has been well tested in the field: small algae with large relative surface area are not iron-limited, while large algae (diatoms, dinoflagellates) are. The small size of the prevalent primary producers allows protozoans to be the principal grazers in the ecosystem, and protozoa can increase at least as rapidly, likely up to two doublings per day, as their phytoplankton food, when that food is abundant. So, they crop down incipient blooms, recycling nutrients in the upper layers, preventing exhaustion of nitrate, phosphate, and silicate.

The subarctic Pacific region is a cyclonic gyre, with the North Pacific west wind drift moving along the southern side in a broad band north of ~40°N. This flow turns offshore of North America into the Alaska Stream, carrying water north then west around the coast of the Gulf of

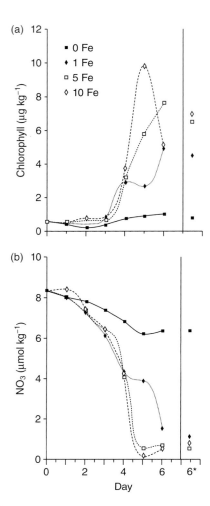

Fig. 11.2 Time-series of (a) chlorophyll and (b) nitrate concentrations in incubation bottles of subarctic Pacific (Ocean Station P, 50°N, 145°W) surface water with and without iron enrichment. Three levels of enrichment were tested. Cells breaking out of background chlorophyll levels on Day 4 were large diatoms. Symbols over Day 6* are for separate incubations not opened for sampling until Day 6. (After Martin & Fitzwater 1988.)

surface. The low surface salinity results from a combination of strong precipitation over the coastal rim of mountains draining to the sea, abundant glacial meltwater and reduced wind stress curl (relative to the North Atlantic subarctic) resulting from frictional erosion of westerly winds over Asian mountains (Warren 1983). Also, the whole circuit of flow is constrained by land at the north, a major difference from the Atlantic.

The halocline acts as a strong mixing barrier, preventing full ventilation to the surface layer from the water beneath that contains abundant nutrients (40 µM nitrate at 300 m). In the central Gulf of Alaska, gradual fall (autumn) and winter erosion of the seasonal thermocline raises maximum surface-layer nitrate to only ~17 µM in March. By April or May, reduced winds and solar heating establish a seasonal thermocline at about 35 m, after which surface nitrate drops by about 8 to 10 µM from March to August, but it does not drop below ~6 µM (Fig. 4.8d), a level unlikely ever to limit phytoplankton growth. Phosphate and silicate are not reduced to limiting levels, either. Another important part of this is that limitation of vertical mixing by the halocline keeps the microplanktonic food-web in place right round the year. It is never disrupted by winter mixing to depths as great as occur in the subarctic North Atlantic, 300–400 m.

As stated, the dominant phytoplankton are in the nano- and pico- size range. A significant but never dominant part of the biomass is *Synechococcus*, particularly in late summer and fall beneath the thermocline and above the halocline. *Prochlorococcus*, common in even more oligotrophic waters, are nearly absent. There are generally eukaryotic phytoplankton as major constituents, accounting for more carbon than the picophytoplankton, but exactly which are abundant varies with every visit to the oceanic sector. Booth *et al.* (1993) observed coccoid Chlorophyceae (*Chlorella* and *Nannochloris*), Prasinophytes (*Mantoniella* spp.), Chrysophyceae, small diatoms (*Nitzschia cylindroformis*), and Prymnesiophyceae (*Phaeocystis* spp.) as dominant forms on different dates, and dozens of other species of secondary importance. Dominance of any particular species or group has not yet been connected to specific conditions. All are species expected to be readily accessible to protozoan grazers, and observations show abundant heterotrophic and mixotrophic dinoflagellates, choanoflagellates, flagellates (e.g. *Bodo* cf. *parvulus*) and ciliates (Strombidiidae, including *Laboea* and *Strombidium*).

The mesozooplankton in the subarctic Pacific are dominated, especially from March to July, by five species of copepod: *Neocalanus plumchrus*, *Neocalanus flemingeri*, *Neocalanus cristatus*, *Eucalanus bungii*, and *Metridia pacifica*. Copepodite stages of the first two are abundant in the layer above the seasonal thermocline; those of the latter two are abundant below it and probably primarily scavenge fecal matter sinking down from the surface layer. *Metridia* is a strong diel migrator (>200 m in older stages); the others

Alaska and Aleutian Island chain. Return flow into the west wind drift occurs along Kamchatka and Hokkaido as the Oyashio Current meets and turns eastward as it encounters and mixes in a strong eddy field with the Kuroshio flowing north along Honshu. There is also exchange near the dateline between the westbound northern limb and west wind drift, amounting to a separation of western and eastern subarctic gyres. The cyclonic circulation creates doming of density isopleths in the gyre centers that should correspond to upward Ekman transport. However, the entire region is also characterized by a surface layer of reduced salinity, $S \approx 33$ down to a halocline (Fig. 11.3) at 80 to 120 m (variation from internal waves), where it increases to $S \approx 34$. The southern limit of the region is often defined as the line running west to east where the $S = 34$ isohaline rises to the

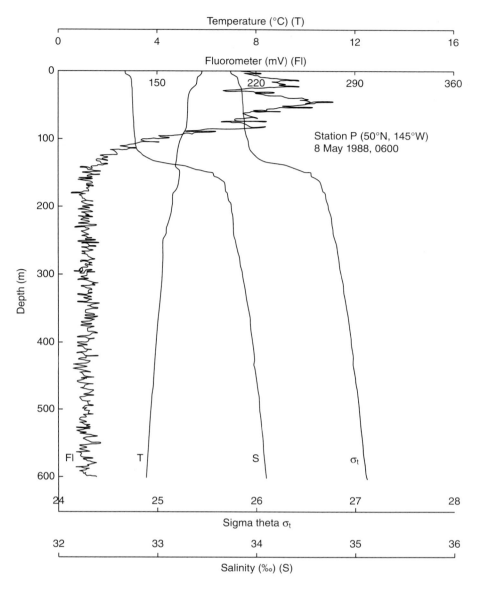

Fig. 11.3 Spring water-column stratification in the oceanic subarctic Pacific at Ocean Station P. The "Fl" profile is chlorophyll fluorescence. The halocline, which obviously determines the density (σ_t) profile, is close to the deepest descent made on recurring internal waves. More surface warming develops as the seasons progress through summer and fall. (After Miller *et al.* 1991.)

are not. All but *Metridia* have a resting stage as fifth copepodites (C5). This stage loads with large amounts of nutriment stored as liquid wax, then descends to depths below 400 m (to >2000 m) as a refuge from late summer warming and predation. The different species have somewhat different reproductive schedules, as discussed in Chapter 8. The dominant species, *N. plumchrus*, distributes its spawning at depth from late summer through to early spring, apparently "spreading its bets" across the entire season. The winning nauplii that get food to grow are those arriving at the surface as first-feeding stages when production increase is

strong due to rising insolation and surface thermal stratification. All of these copepods feed primarily on protozoa, and they constitute the trophic level that shows strong seasonal cycling (Fig. 11.4). Thus, seasonality of primary production is first exhibited as biomass accumulation two to three trophic levels away from the primary producers.

The interactions of these lower trophic levels result in a still unexplained variation in phytoplankton stocks (Fig. 11.5). While the variation is confined within narrower limits than in coastal temperate or North Atlantic waters, it slowly but recurrently wanders between about 0.15 and

0.6 µg liter⁻¹. There are also daily cycles, partly day–night fluorescence variation, and partly stock increase in the light, and decrease due to grazing in the dark (Bishop *et al.* 1999). The longer oscillations result from the small daily differentials, which can be either positive or negative. Those are smaller than the 10–40-day variations, suggesting that the latter correspond to real variation in phytoplankton standing stock. Moreover, at the >10 day periods, there are inverse variations of upper-water-column ammonium

concentration: ammonium is taken up as stocks increase and is regenerated as stocks decrease (Miller *et al.* 1991a & b). Presumably, protist grazers increase during the regeneration phase and their grazing overtakes and reverses phytoplankton increase, returning their fixed nitrogen to the water column.

The overall pattern (Fig. 11.5) could be:

1 A predator–prey oscillation: the protozoa eat down the phytoplankton and excrete ammonium, then decrease themselves because they become food-limited, releasing phytoplankton to grow, which eventually raises protozoan stocks that then limit the cycle at its top, and the oscillation repeats.

2 Variation could be driven by intermittence in the iron supply; soluble iron inputs, as dust (probably falling in rain squalls) or from vertical mixing, would enhance phytoplankton growth rates, pulling their stocks up, later to be grazed down as the grazers increase on their improved diet. Light-transmission profiles gathered by ARGO floats rising twice daily in the vicinity of Station P (Bishop *et al.* 2002) indicated a strong increase in particulate carbon during a two-week period after passage overhead of a dust storm originating in the Gobi Desert and observed by satellite. During the ∼12 days after the dust cloud had passed, grazing fell behind, leaving ∼10% of the previous day's net gain at each dawn to generate the overall increase. While the coincidence with dust delivery was impressive, there was also mixed-layer shoaling during the stock increase and deep-

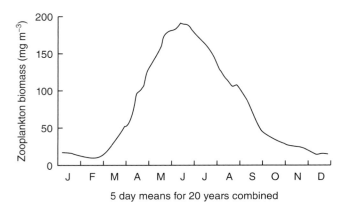

Fig. 11.4 Annual cycle of zooplankton biomass in the upper 150 m in the oceanic subarctic Pacific. The build-up in spring is from growth, mostly of large copepods. Decline in early summer is from descent of these copepods to depths below 500 m for diapause. (After Fulton 1983.)

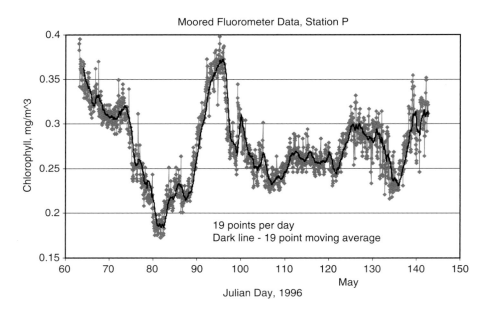

Fig. 11.5 Short-term cycling of phytoplankton biomass (measured as chlorophyll standing stock) near 50°N, 145°W. (Plotted from moored-fluorometer data graciously provided by Phillip Boyd.)

ening during its decrease, suggesting the next hypothesis. Another example of a dust-driven phytoplankton outburst in the subarctic Pacific is detailed by Hamme *et al.* (2010; see Chapter 16).

3 Mixed-layer shoaling would allow more consistent illumination of phytoplankton, increasing their growth, while storms would dilute phytoplankton in a deeper mixing layer. Thus alternation of calmer and stormier weather could drive the cycling.

4 Similarly, a rare period of open skies could enhance phytoplankton growth and deepen the euphotic zone (but without swirling phytoplankton down to darker layers), allowing an increase in excess of daily grazing.

5 Oscillations could occur as a manifestation of a trophic cascade (Dagg *et al.* 2009). An increase or decrease in mesozooplankton (copepods, euphausiids, salps) would decrease or increase protozoans, releasing or cropping down phytoplankton, initiating a phytoplankton (and fluorescence) oscillation that would progressively dampen, much as in the data shown (Fig. 11.5). That just pushes the uncertainty along the causal chain to a new uncertainty, but there are many mechanisms to initiate stock increases and decreases among the predators of microherbivores.

Of course, more than one, or even all, of these mechanisms may operate in different intervals, but grazers almost always catch up to phytoplankton stock increases within a few weeks, limiting stock measured as chlorophyll well below $1.0 \mu g \, liter^{-1}$. Strom *et al.* (2000) pointed out the problem that there is a relatively high lower limit to these oscillations. In more oligotrophic subtropical waters, where major nutrients as well as required trace metals are driven to near-vanishing levels, chlorophyll stays below $0.1 \mu g \, liter^{-1}$ for very long intervals. Why do subarctic Pacific levels stay above that seemingly false bottom? Possibly the complexity of the microbial food-web includes a self-limiting activity: e.g. larger microheterotrophs eating more of the smaller ones as the pico- and nanophytoplankton are reduced. We do not know in any fundamental way how microherbivore grazing is regulated.

The reality and importance of iron limitation in the subarctic Pacific and other HNLC areas have been conclusively demonstrated by enrichment with iron of modestly large, typically $10 \times 10 \, km$, patches of open ocean. Martin and Fitzwater's (1988) initial demonstration of iron limitation of larger phytoplankton types was based on adding iron in nanomolar amounts to incubation containers maintained on a ship's deck. This left open the possibility that something about the containment interfered with alga–grazer or alga–nutrient interactions. Field iron-enrichment studies in the subarctic Pacific were carried out primarily by Canadian ("SERIES", Harrison *et al.* 2006) and Japanese ("SEEDS I & II", e.g. Tsuda 2005) oceanographers, morally supported by the North Pacific Marine Sciences Organization

("PICES"). SERIES involved stirring acidic ferric chloride solution (in seawater with some SF_6 as a patch tracer) with a ship's wake into an $8.5 \times 8.5 \, km$ patch of ocean centered on a drifter buoy starting at 50.14°N, 144.75°W. A second iron addition was made 7 days after the first, covering the then N–S elongate patch shape. By day 26 the patch was $\sim 35 \times 10 \, km$ with some side lobes. Iron was raised at least initially after each addition to $\sim 2.4 \, nM$, and the response of phytoplankton was dramatic, if somewhat slow and prolonged. Chlorophyll rose to $\sim 5 \mu g \, liter^{-1}$ in the patch center by day 18, with dominance of relatively large, pennate diatoms (Marchetti *et al.* 2006). Over the same interval, silicate was drawn down from 15 to $1 \mu M$, and nitrate was reduced from 11 to $4 \mu M$. The SEEDS I experiment was similar, but in the western gyre (48.5°N, 165°E) in July–August with a mixed layer of only $\sim 10 \, m$, which kept the added iron concentrated and generated a bloom that reached $20 \mu g \, Chl \, liter^{-1}$, almost entirely accounted for by one centric diatom, *Chaetoceros debilis*.

The entire inshore rim of the subarctic Pacific gyre has a substantially different production regime. The eddy-rich British Columbia and SE Alaska coast; the close-in coastal waters of the northern Gulf of Alaska; the inshore Alaska Stream along the Aleutians; and the Oyashio from Kamchatka to its offshore turn off Hokkaido, all have strong, diatom-dominated spring blooms. These do draw down NO_3^-, PO_4^{3-} and $Si(OH)_4$ to low levels, and evidently there is at most intermittent iron limitation, likely because both sediment resuspension and river influx supply iron to the euphotic zone. The boundary between bloom-supporting and HNLC zones occurs at various and fluctuating distances from shore. The mesozooplankton community of these more coastal waters is constituted of the same species as that of the oceanic area. The *Neocalanus* species have faster developmental rates in coastal and fjord waters, more-focused periods of active feeding in the surface layers, and more-focused times of spawning, particularly in the Sea of Japan and in fjord areas like Georgia Strait and Prince William Sound. *Calanus pacificus* and several species of *Pseudocalanus* are more abundant than offshore, and the coastal copepod *Calanus marshallae* is added to the mix. Several euphausiid species, for example *Thysanoessa spinifera* and *Thysanoessa raschii*, tend to be much more abundant in or restricted to the coastal zones.

The subarctic Pacific is habitat to an array of endemic squid, fish, and marine mammals, both in oceanic reaches and along the coast. Among schooling fish, the Pacific saury (*Collolabis sauri*), an important food fish for salmon, is both oceanic and coastal, migrating north and south across the region seasonally. The pomfret (*Brahma brahma*) migrates north in summer from more subtropical waters, crossing far into the subarctic. Offshore coastal waters are home to several species of anchovy and sardine. Of course the totemic fish of the region are the five species of *Onychorhynchus*, the Pacific salmon, anadromous fish that home to their natal streams around the whole arc of the subarctic coastline. Some species are more coast-bound than others, especially as the young enter the ocean, but sockeye, pink, and older chinook salmon range across the entire oceanic reach, south in winter and north in summer. Massive volumes of research results are available on the biology of salmon, partly because of commercial importance, partly because their migrations and physiology are so amazing.

There are both large oceanic squid, *Onykia robusta*, and more coastal squid, including the Japanese flying squid, *Todarodes pacificus*. The latter grows during about a year of migrations all around the islands of Japan, then gathers to spawn at sites in the East China Sea, near Kyushu and in the northern Sea of Japan. The nearshore zone off Hokkaido is the site of a fishery based on attracting *T. pacificus* to lights hung from boats over the water. Dahl's porpoise, *Phocoenoides dalli,* is endemic to the region, living mostly well offshore, and both killer whales (*Orca*) and several smaller porpoise live around the perimeter, as do Steller sea lions (*Eumetopias jubatus*), harbor seals (*Phoca phoca*), and sea otter (*Enhydra lutris*). Humpback whales (*Megaptera novaeangliae*) that overwinter in subtropical waters, many of them around the Hawaiian Islands, migrate north in spring, then swing east or west on reaching subarctic latitudes, feeding through summer months in Alaskan and Russian fjords. Gray whales (*Eschrichtius robustus*) migrate along the coast from Mexican wintering areas, basically bypassing the subarctic to feed in the northern Bering and Chukchi Seas. Thus, the region has a rich nektonic fauna, richer than listed here, and many of these larger animals have whole populations of experts devoted almost exclusively to studying them.

Subarctic Atlantic

The contrast of the subarctic Pacific with the North Atlantic north of the Gulf Stream and south of the polar front has four main parts. In the North Atlantic, (i) winter mixing is deep, to ~200–400 m (Fig. 11.6), and phytoplankton stocks are reduced then to very low levels (<0.2 µg Chl liter^{-1}). (ii) Spring blooms are typical, reaching 2–4 mg Chl m^{-3} from New England to Norway. Large diatoms dominate Atlantic blooms, whereas diatoms are only occasionally a major fraction of phytoplankton biomass in the oceanic Pacific and always small (<7 µm) species. (iii) Proximity to continents apparently provides enough iron to fuel nearly complete exhaustion of major nutrients from surface layers, particularly nitrate. It is important, also, that the concentration of nutrients at depth is about half that in the Pacific, so depletion is more readily reached before trace metal limitation is severe. (iv) The large grazing copepods (the dominant mesozooplankton) are from the genus *Calanus*,

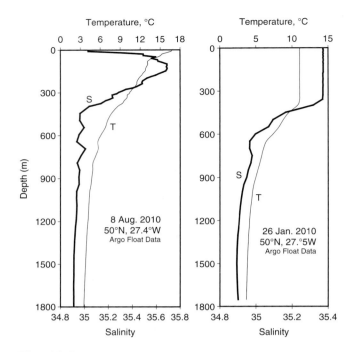

Fig. 11.6 Profiles comparing summer (August) and winter (January) temperature (T) and salinity (S) profiles from the subarctic Atlantic. The summer water column is strongly stratified. The winter profile shows deep mixing (to 360 m). These are Argo float data from the World Ocean Database maintained at the US National Ocean Data Center (NODC).

species that ascend to the surface and generate egg biomass from active feeding, unlike the subarctic Pacific *Neocalanus* species.

The mechanisms controlling the timing and intensity of spring blooms in inshore, temperate waters, across the North Atlantic and elsewhere, are an unending source of fascination to oceanographers. Models of seasonal cycles of phytoplankton stock, as considered in Chapter 4, are a sub-industry in our profession. Measurements from boats and ships, moorings, autonomous floats, and satellites are all enlisted to improve our insights, and we are far from finished. Phytoplankton stock size depends upon the past history of productivity and also upon the history of cell mortality. "Death" (or at least loss to the surface layer stock) comes from grazing, mixing out of the euphotic zone, sinking, or disease. The latter was largely ignored until recently; marine waters contain viruses that are capable of lysing phytoplankton cells (Chapter 5).

Seasonal cycles vary markedly among subregions, but there is a classic cycle that is observed in coastal Atlantic waters, such as the North Sea or off Cape Cod, and all across the high-temperate Atlantic. Characteristically, textbooks show this cycle in a schematic way (Fig. 11.7). That is symptomatic of a problem of the discipline. Because getting out to sea on a regular or sustained basis is difficult,

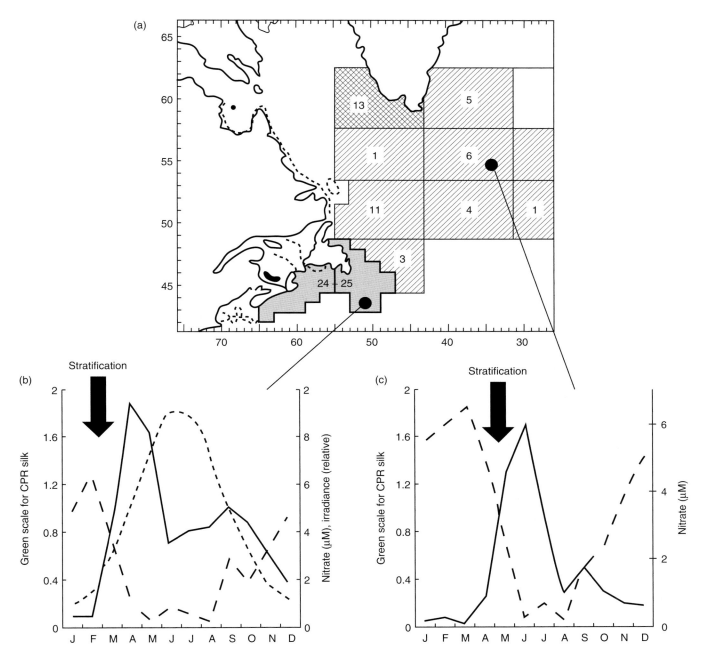

Fig. 11.7 Average annual cycling of phytoplankton biomass (solid line) at two North Atlantic sites. Cycles include a spring bloom, fall bloom, and extreme winter low. Estimates are from color of 240 mm continuous plankton recorder "silks" after towing through the regions neighboring the points indicated in all seasons of many different years (based on Robinson 1970). Peak stock at the more northern site occurs later. Relative irradiance (short dashes) and nitrate (long dashes) curves are hypothetical.

we still have no single time-series quantifying the progression of a spring bloom with good (daily, or near-daily) resolution of required variables, although some data series from moorings, and lately ARGO floats, supply part of what is required. Measurements needed are phytoplankton stock and species composition, irradiance, water-column density structure, nutrients, and grazer community data, all from well before the stock increase until it has subsided again. Evaluation of exactly which species make up the phytoplankton stock during the progress of the bloom would be particularly useful, because the overall bloom can be constituted of several, sequential blooms of different species. Estimates of grazing and phytoplankton sinking would also be good, but harder to obtain.

The standard explanation of the schematic (Fig. 11.7) goes as follows: in winter, mixing increases surface-layer nutrients to their seasonal high, setting the stage for phytoplankton growth in spring. However, persistence of low winter light, averaged over a deeply mixing water column, prevents rapid growth, and the mixing keeps loss rates high. In spring, illumination increases and warms the surface so that mixing is inhibited by stratification of the water column. It also raises the phytoplankton growth rate. Thus, stock can accumulate, producing a "spring bloom". By late spring or early summer, nutrients become exhausted in the surface layer, growth slows, and loss to increasing grazer stocks reduces the phytoplankton stock to a low but varying level. Summer variations come from intermittent injections of nutrients from depth (storms). In the fall, illumination is still good, many of the grazers have gone into resting phases in anticipation of winter (or because late summer–fall water temperatures are the highest of the year), and nutrients begin to be supplied to the surface by strengthening winds. The result often is a fall bloom. The onset of winter winds mixes this away and returns the system to low winter stocks and low winter activity rates. That explanation is basically right. For sites that exhibit such blooms, it holds up under quantitative investigation. There are many details to consider.

Critical depth theory

It is not uncommon for the onset of the spring bloom to be explained with only passing reference to the grazing process. This produced the *critical depth* theory of Gran and Braarud (1935) and Sverdrup (1953). As originally formulated, grazing was incorporated in a vague way. The notion is that the relative rate of phytoplankton growth, i.e. stock increase (dP/dt) per unit stock ($1/P$), equals:

[gross photosynthesis – respiration]:
$$1/P(dP/dt) = PS - R.$$

Sverdrup suggested that photosynthesis, PS, decreases exponentially with depth, following the exponential decrease of irradiance, while respiration, R, might be *roughly*(!) constant with depth. He took respiration to be "community metabolism", that is, all removals including both phytoplankton respiration and stock reductions by grazers (as emphasized in a comment by Smetacek and Passow 1990). Several different vertical levels are defined by the interactions of PS and R (Fig. 11.8). The community *photosynthetic compensation depth* is the vertical level at which the local value of ($PS - R$) = 0. Again, this definition applies to the metabolic activity of the whole community, not just of phytoplankton. A physiologist would define a photosynthetic compensation depth as the level where net primary production (photosynthesis less cell respiration) = 0. That would be somewhat below the level intended

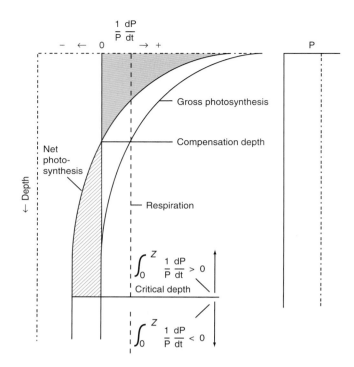

Fig. 11.8 Features of Sverdrup's (1953) critical depth model for spring bloom initiation. Gross photosynthesis decays exponentially downward in response to progressive diminution of irradiance according to Beer's Law ($I_z = I_0\,e^{-kz}$), where k is the coefficient of extinction due to absorbance and scattering, and z is the depth. Net photosynthesis is shifted left by the assumed depth-invariant community respiration. Shaded areas show zones of positive (above the compensation depth) and negative below) net photosynthesis down to the critical depth where the integrals are equal. (After Sverdrup 1953.)

by Sverdrup, because grazing increases the loss term. Well below either version of compensation depth is the "*critical depth*", the level at which the *vertical integral* of ($PS - R$) = 0. The only losses involved in this original definition of critical depth are from community metabolism of photosynthate. In addition, however, there will be losses from the upper, lighted layer through vertical mixing. In general, mixing *above* the first significant thermal step in the water column is rapid, for some practical purposes instantaneous. Mixing *through* this thermal step is slow. Sverdrup predicted that spring blooms would get under way at the time during heating of the water column when the thermocline rises above the critical depth. Before that time, mixing would make $1/P(dP/dt)$ negative for the euphotic zone overall. Afterward, mixing losses would be small and stock would accumulate above the thermocline.

Critical-depth theory works, in a general way. Sverdrup (1953) showed data (Fig. 11.9) from a weather ship stationed in the high-temperate North Atlantic (66°N, 2°E), where phytoplankton, evaluated with nets and thus larger algae, did increase shortly after water-column stratification. In general, blooms, where they are important, begin shortly

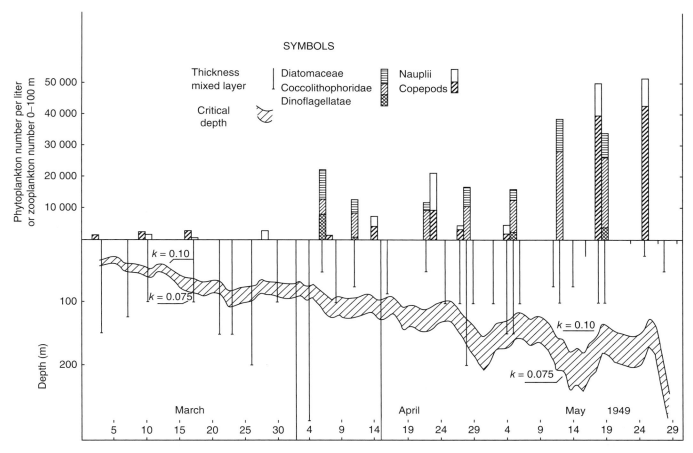

Fig. 11.9 Data for 1949 from the weathership at Station "M" (66°N, 28°E), showing the relationship between the approximate critical depth (shading between approximate k values of 0.075 and 0.10) and mixing depth. Phytoplankton counts increased in April–May, when critical depth exceeded the mixing depth. (After Sverdrup 1953.)

after the first, non-transient establishment of the seasonal thermocline. Bloom timing varies between years according to variation in the surface heating. Where grazer stocks are very low in winter and early spring, the theory works well because grazing then is a minor part of "community" respiration. This is true in the mid-latitude North Atlantic and elsewhere close to coasts, so the theory works well for the "classical" seasonal cycle problem. In other regions, the mixing depth remains important, but it sometimes works in a somewhat different fashion (e.g. see Nelson & Smith 1991). Seasonal mixing effects on phytoplankton stocks are different over shallow shelves, since the bottom can act as a lower limit to vertical mixing, a limit always above the critical depth. A study by van Haren *et al.* (1998) provides a high-resolution time-series of chlorophyll and other relevant data from a site at 45 m depth in the central North Sea. Chlorophyll increased from the mid-winter level of 0.5 to 3–6 mg m^{-3} starting in mid-February, as soon as photosynthetically active radiation (PAR) exceeded 6.5 µmol photons m^{-2}s^{-1}. That is approximately the physi-

ological compensation (PS − R ≥ 0) intensity in those waters (Tett 1990). Water-column stratification was only established long after the bloom became nutrient limited and diminished. Such early-season blooms are reported for many shallow, coastal sites, including Georges Bank and Narragansett Bay.

The JGOFS North Atlantic bloom experiment

In the oceanic Atlantic north of 45°N, particularly on the eastern side (Glover & Brewer 1988) winter mixing to >200 m replenishes major nutrients in surface layers, with nitrate exceeding 6 µM, and flushes most of the phytoplankton (Chl to 0.05 mg m^{-3}) and the microheterotrophs out of the illuminated upper strata. Stratification re-establishes in late March through May, progressing northward, and a bloom ensues, lasting about 50 days. This was studied by a cooperative international program, the Joint Global Ocean Flux Study (JGOFS) in 1989 and the

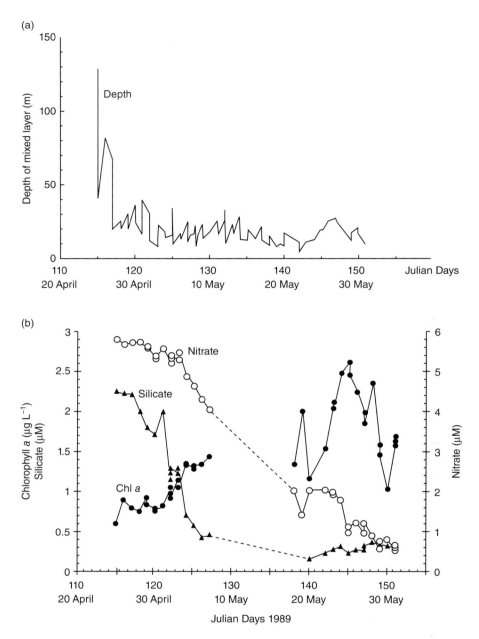

Fig. 11.10 (a) Time-series of mixed layer depth at 47°N, 20°W, April–May 1989. Observations began just before sharp shoaling of the seasonal thermocline. (After Lochte *et al.* 1993.) (b)Time-series of chlorophyll, nitrate, and silicate at 46°N, 18°W in April–May 1989. (After Sieracki *et al.* 1993.)

Biogeochemical Ocean Flux Study (BOFS) in 1990. One of the most intensely studied areas encompassed three stations ranging from 46° to 49°N along about 18°W, well out to sea due west of Cornwall. By the time that all the ships and scientists arrived on April 25, 1989, stratification was just setting in (Fig. 11.10) and phytoplankton stocks were still low (0.5 mg Chl m^{-3}). A bloom ensued, raising chlorophyll to 2.6 mg m^{-3}, and lowering major nutrients (Fig. 11.10). Similarly, at 49°N, 18°W in 1990, stocks increased until mid-May, when there was a peak at about 2.8 mg Chl m^{-3},

a typical peak for the spring bloom in this region. Most of the enhanced stock was very close to the surface, only extending below about 25 m at the very peak of the bloom (Savidge *et al.* 1992). In both studies, nitrate came down as phytoplankton went up, generating a strong inverse correlation (Fig. 11.11). The connection is certainly causal: the growing phytoplankton reduce the fixed nitrogen, incorporating it in their organic constituents. When the nitrate, phosphate, silicate, and other nutrients are reduced, phytoplankton growth slows.

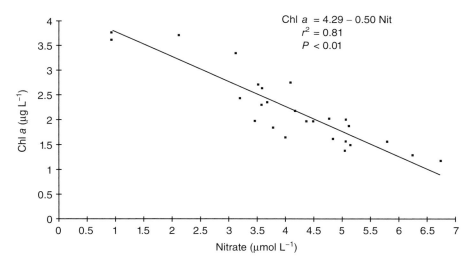

Fig. 11.11 Scatter diagram showing the relation between chlorophyll-*a* and nitrate concentrations during the BOFS observations at 47°N, 20°W in 1990. Effectively, this is a time-series from lower right to upper left during the increase phase of the bloom. Subsequent points would drop vertically to the abscissa. (After Barlow *et al.* 1993.)

Over this entire region on any given date, the bloom appears to be extremely patchy, to occur in *mesoscale* blobs (scales of a few tens to hundreds of km). The JGOFS North Atlantic bloom experiment showed (Robinson *et al.* 1993) that this is an effect of mesoscale eddies, which are always scattered over this region (Fig. 2.4). There were three persistent cyclonic (anticlockwise in the northern hemisphere) eddies (Fig. 11.12) evident from satellite altimetry in the region of the observational study during April–May 1989. The sea surface of a cyclonic eddy slopes up from the middle to the rim due to the Coriolis effect. This slope can be estimated by satellite radar and the rotary velocity approximated from geostrophy (Fig. 11.12, right inset). Cyclonic eddies are regions of greater vertical stabilization of the water column, and the spring bloom tends to advance in them earlier or faster so that they become high-chlorophyll patches (Fig. 11.12, top left inset).

Similar patterns also occur in the phytoplankton distributions after the main spring bloom, as seen in a June satellite image of waters south of Iceland (Fig. 2.4), a picture derived from reflections from the calcite plates of coccolithophores. The importance of the mesoscale eddy field to oceanic phytoplankton is evident from this picture. Coastal chlorophyll concentrations estimated from satellite data collected over the western north Atlantic (cover of this book) show that phytoplankton stocks there also vary in swirling, active patterns.

Species successions in the spring bloom

No floristic analysis based on microscopy was included in the JGOFS North Atlantic bloom study. However, Barlow *et al.* (1993) did examine this question indirectly

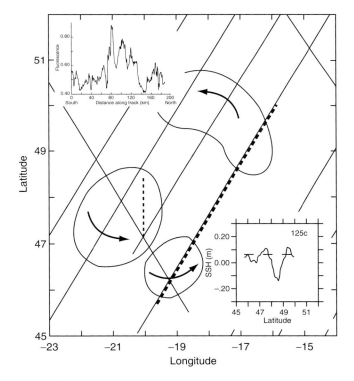

Fig. 11.12 Outlines of three cyclonic eddies identified by satellite altimetry in the JGOFS North Atlantic bloom observation area. Thin, oblique lines are the tracks of the satellite. The heavy dashes indicate the portion of satellite track for which sea-surface height (SSH) is shown in the right inset. (After Robinson *et al.* 1993.) The dashed line in the middle eddy is the track of a LIDAR-carrying aircraft. The left inset shows along-track chlorophyll estimated by flash fluorometry. A downward-directed, blue laser (LIDAR) is pulsed downward from an airplane. A detector quantifies the red fluorescent return, which is converted with calibration data to estimates of near-surface chlorophyll-a. (After Yoder *et al.* 1993.)

in the JGOFS project, applying a chromatographic analysis to identify the main algal groups at different times from differences in their accessory chloroplast pigments. At a so-called Langrangian station following a drifter with a drogue at 20 m from 49°N, 19°W along a southeasterly track, the relative pigment composition changed (Fig. 2.20), from predominantly fucoxanthin to predominantly 19′-butanoyloxyfucoxanthin, implying a shift in phytoplankton from mainly diatoms to a preponderance of prymnesiophytes. This happened in less than 10 days, after the peak of chlorophyll-a but while it remained relatively high (>1.5 mg m^{-3}). Thus, spring blooms are not driven by just the production needed to raise stock levels once. The phytoplankton are turning over rapidly and an initially dominant species can be replaced by another. Shifts of this sort, from diatoms to flagellates, can be caused by slowing of diatom growth when silicate depletion occurs before nitrate or phosphate depletion limits growth rates for all phytoplankton (Sieracki *et al.* 1993).

A nearshore, weekly sample series from Long Island Sound examined microscopically by Conover (1956) shows another feature of spring blooms. In shallow water the bloom can be superimposed over chlorophyll levels that are greater than 2 μg liter^{-1} (mostly 5 μg liter^{-1}) throughout the year, levels that would be bloom maxima in oceanic areas. The cyclic pattern is the same, occurring against a higher background. As in the North Sea case discussed above, the bottom is above the critical depth, so the "spring" bloom can begin with only a slight increase in day length and sun angle. Nutrients, particularly nitrate, were used very rapidly during the diatom bloom, and nitrate stayed low until September. During the bloom peak, diatoms, mostly *Skeletonema costatum*, made up the majority of cells counted in formalin-preserved samples. They were replaced by much lower numbers of dinoflagellates, mostly *Ceratium*, in summer. Diatoms-then-dinoflagellates is the typical sequence among larger cells. Conover was aware that microflagellates were also present – phytoplankton smaller than 5 μm diameter from several, distinct algal divisions. They carry most of the chlorophyll from May through December, but they were not preserved by Conover's technique. The original data ran to 95 weekly samples, a standard that we need in some oceanic studies.

Automated phytoplankton identification is coming into use to characterize seasonal changes in community composition of cellular plankton. These include flow cytometry systems suitable for mooring that store cell counts for general classes of pico- and nanoplankton (*Synechococcus*, *Prochlorococcus*, chrysophytes, other microeukaryotes, etc.) and recording devices called "image-in-flow" cameras (Sieracki *et al.* 1998) that store vast numbers of pictures of individual microplankton such as diatoms and dinoflagellates. Computer image analysis, currently operating mostly at >95% accuracy for generic identification of diatoms, allows automated conversion of these images to abundance

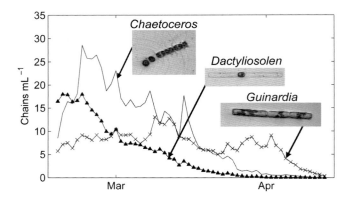

Fig. 11.13 Abundances of chain-forming diatoms during a 2005, 2-month deployment in Woods Hole Harbor of a video instrument, the "Imaging Flow Cytobot". Automated image analysis separated proportions of *Chaetoceros* spp., *Dactyliosolen* spp., and *Guinardia* spp., as indicated by arrows. (Figure by Dr. Heidi Sosik.)

time-series. For example, Sosik and Olson (2007) moored a recording camera of this type in Woods Hole Harbor, collecting picture sets every two hours from mid-February to mid-April (Fig. 11.13). The dominant larger cells were diatoms of three morphologically distinctive genera that replaced each other in a remarkably smooth temporal sequence, eventually all diminishing in abundance to very low levels. While these levels of cell abundance were not particularly great, they show again that chlorophyll concentration and bulk cell counts are only sums that can mask dynamic shifts in community composition. Short-term oscillations removed from the time series (Fig. 11.13; see Sosik & Olson 2007) were due to tidal flow in and out of the harbor, moving cell-abundance gradients past the moored camera. We look forward to oceanic deployment of these automata and expect clarification of bloom dynamics at the genus level.

The fate of bloom phytoplankton

A key aspect of spring bloom dynamics is the fate of phytoplankton once they have grown. Most often the phytoplankton increasing dramatically in spring blooms are diatoms of intermediate to large size (>10 to 70 μm or more in diameter) with opal cell walls (Chapter 2). They have a large central, water-filled vacuole with electrolytes modified relative to seawater so as to maintain nearly neutral buoyancy. This capability fails under nutrient stress, the cells become "senescent" and they begin to sink. In both coastal waters and the oceanic North Atlantic, the spring bloom most usually terminates by sinking-out of these relatively large cells (Smetacek 1985). They progressively flocculate on the way down and reach the bottom in a few days or weeks as a major pulse of organic particle flux. Throughout the course of the bloom, however, substantial amounts of phytoplankton are eaten by mesozooplankton.

Table 11.1 Critical depths as a function of date and latitude. (From Platt *et al.* 1991.)

DATE	LATITUDE (°N)	CRITICAL DEPTH (m)	
		WITH JUST PHYTOPLANKTON RESPIRATION	WITH ALL LOSSES INCLUDED
1 February	40	361	131
	50	274	97
1 March	40	447	164
	50	385	141
1 April	40	551	193
	50	521	238
1 May	40	635	237
	50	639	238
1 June	40	691	258
	50	723	270

The exact grazing rates during the bloom are a continuing issue, but grazing creates considerable phytoplankton stock turnover well before nutrients are depleted. The grazers return some of the elements in the phytoplankton they eat to the water as excretory products, and a significant fraction of the organic matter sinks in fecal pellets.

Critical depth theory, again

If critical depth theory works anywhere, it must work in the oceanic North Atlantic (Smetacek & Passow 1990). The North Atlantic is a strongly seasonal ocean that mixes deeply enough in winter for vertical exchange to keep phytoplankton stocks low until stratification sets in during spring. Platt *et al.* (1991) reformulated Sverdrup's theory to take account of modern data describing rates of photosynthesis as a function of available light (the *P* vs. *E* relation). The reformulation was mathematically abstruse, and they showed that it differed by a maximum of about 10% from Sverdrup's simple linear *P* vs. *E* relation (which gives a simple exponential decay in photosynthesis vs. depth). Next, Platt *et al.* guessed at the loss terms:

- phytoplankton respiration (4% biomass day^{-1} plus a fraction varying with photosynthesis);
- excretion of unrespired organic matter (set at 5% of photosynthesis);
- grazing by mesozooplankton (4% of biomass day^{-1}) and protozoa (5% of biomass day^{-1}); and
- cell sinking (set at 1 m day^{-1} at all depths).

They found no information upon which to estimate the variation of these losses with depth, so they stuck with the constant vertical profile adopted by Sverdrup. Fractional losses of biomass were "assumed independent of depth". That is radically unsatisfactory, but Platt *et al.* were right that we do not have the data to do much better. They then proceeded to calculate the critical depths for specific dates and latitudes. Their table is reduced here (Table 11.1) to show the trends.

The result approximately predicts bloom dates in the North Atlantic, for example in the venerable data from the weathership at Station "M" (Fig. 11.9). The key point is that the greater the daily irradiance (that is the lower the latitude or the later in the spring), the deeper the critical depth. Increasing algal-growth rates near the surface increase the vertical integral of production, driving the critical depth down. Spring blooms do not occur in this ocean until significant stratification sets in above the levels calculated. Roughly, the theory works.

Alternative scenarios

Boss *et al.* (2008) deployed an ARGO-type CTD float, augmented with a light-scattering sensor (particle density) and a fluorometer, which remained in the vicinity of 50°N, 47°W (subarctic waters east of Newfoundland and south of Greenland) for over two years. Most of the time it resided at 1000 m, making profiles to the surface every five days and recording variables each 50 m up to 400 m, then at closer intervals on upward. This provided a good record of the relation between mixed-layer depth (defined as the first depth with density – σ_t – greater by 0.125 kg m^{-3} than the sea surface) and both chlorophyll and beam scattering. All profiling was near midnight to get the diel maxima of

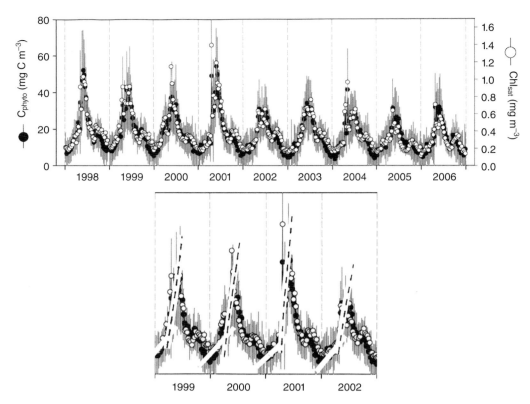

Fig. 11.14 Above: SeaWiFS OC4v4 chlorophyll estimates averaged weekly for a box bounded by 45°N and 50°N and by 25°W and 35°W. Below: the same data enlarged for 1999 to 2002 to emphasize the sharp changes in slope between slow increase from about the solstice to rapid increase in early spring. (After Behrenfeld 2010.)

chlorophyll estimates. Comparisons of chlorophyll levels (calibrated before launch in a laboratory) to pixels recorded in nearby satellite passes were within expected uncertainties. Spring blooms indicated by both particle scattering and chlorophyll (Plate 11.1) peaked close to June 1 each year, and levels above 1 µg liter^{-1} lasted ~40 days in 2005, ~30 days in 2006. The actual increases from the winter low <0.2 µg liter^{-1} were gradual and started very early: from 11 December of 2004 for the 2005 bloom, and from ~21 February in 2006. The decrease after the bloom was approximately the mirror image of the long increase.

Throughout the two years, chlorophyll was very evenly mixed down to the point defined by $\Delta\sigma_t = 0.125$ and almost absent below that level (Plate 11.1). Clearly, mixing depth is critical, but it was never great enough to eliminate all phytoplankton even in the middle of winter. Moreover, the winter–spring increase developed gradually at first, following quite exactly the slow shoaling of mixing. Then, as the chlorophyll concentration approached 1 µg liter^{-1}, there was a month-long doubling or tripling when the mixing depth leveled off above 50–75 m. A shadow of particle scattering below $\Delta\sigma_t = 0.125$ (Fig. 6 in Boss *et al.* 2008) occurred during the spring bloom. A likely interpretation is that light and nutrients in winter are always sufficient for some algae, almost certainly pico- and nanoplankton, to

sustain and increase their stocks. Dilution of protozoan grazers and slowing of their activity by cooling are likely also important. Then the dramatic spring increase occurs when illumination is at its annual maximum and mixing is constrained to a shallow enough layer to allow rapid diatom growth. The diatoms then rain out as they become nutrient limited, and there are greater losses to grazing by mesozooplankton and possibly also by dinoflagellates.

Behrenfeld (2010) has confirmed this pattern for the central subarctic Atlantic from 25° to 35°W using nine years of SeaWiFS data. Specifically for 45° to 50°N (Fig. 11.14), the surface chlorophyll is minimal (~0.2 µg liter^{-1}) close to January 1, well before the maximum mixed-layer deepening, then immediately begins a gradual increase amounting to a doubling. There follows a sharp acceleration of stock increase on a date ranging from mid-March to late May (Fig. 11.14), the classical spring bloom. This strong acceleration occurs during the steepest ascent of the mixing-layer depth (approximated by Behrenfeld from a wind-data assimilating model of mixing) and rapid increase of illumination (PAR, estimated by SeaWiFS). These are exactly the conditions proposed by critical depth theory for promotion of diatom blooms. The interesting feature revealed from both the ARGO floats and satellite analysis is the slower increase of algal stocks from the winter solstice to about the spring equinox. Very likely the modification of critical depth theory required by this will be found to be a very large winter reduction in the community "respiration", primarily grazing on pico- and nanoplankton. The euphotic-zone protozoan community must be greatly diluted by vertical mixing and its feeding and reproductive potential must be reduced by cold winter temperatures. This interpretation is akin to what Behrenfeld refers to as a "dilution-recoupling hypothesis". However, he has a substantially different interpretation, for which we refer you to his paper. The approximation that community catabolism of algal production is constant seasonally and with depth was forced on Gran and Braarud and Sverdrup by lack of information. It was never satisfying, and these new observations call for its adequate quantification.

The critical depth mechanism probably operates in the most accelerated phase of North Atlantic (and coastal) spring blooms. However, those light-mixing-growth rate relationships are not the only possibility. Townsend *et al.* (1994) have suggested that the key aspect of vertical water-column structure is not stratification but actual mixing. In the absence of recurring winds, mixing can slow to the modest rates of diel convection. Thus, a very calm period could lead to a near-surface bloom without stratification, simply because the phytoplankton growth rate is maximal near the surface. That opens the possibility that an established phytoplankton stock will intercept more light in the upper water column, thus enhancing upper-layer warmth and establishing thermal stratification. In other words, the order of events (and causation) can be calm → bloom → stratification, rather than calm → stratification → bloom. Townsend *et al.* (1994) also argued that early blooms may be enabled by the greater inhibition by cold temperature of grazing than of photosynthesis. The data that they adduce in this regard are suspect, but the idea may have some validity. For example Stramska and Dickey (1993) showed chlorophyll and temperature data from fluorometers and thermistors moored in deep water south of Iceland during April–May 1989. A bloom, which eventually reached $4\,mg\,Chl\text{-}a\,m^{-3}$, was under way, if still incipient, a week prior to measurable stratification above $100\,m$ (which is, however, well above the critical depth; Table 11.1). Stratification then set in immediately. This seems to fit the alternate causal order. However, Stramska and Dickey's convincing model of the interactions among available irradiance, enhancement of absorbance by phytoplankton pigments, and mixing (a function of measured wind speed) suggests that the effect of pigment on stratification is at most a one-day acceleration of the bloom.

Mesozooplankton

Like the subarctic Pacific, the dominant mesozooplankton are copepods, particularly species of *Calanus* and *Pseudocalanus*. Among larger copepods, *Calanus helgolan-*

dicus is dominant to the east and just north of the Gulf Stream, *Calanus finmarchicus* in a northeasterly expanding stripe from Cape Cod crossing in the vicinity of Iceland then into the Norwegian and Barents Seas, and *Calanus glacialis* more abundant approaching and under the arctic ice. All of those share the family trait of resting in late copepodite stages (C4 and C5). *Calanus finmarchicus* matures at different dates according to subregion (Plate 8.2): January in the Gulf of Maine, February in the Norwegian Sea and March in the Irminger Sea (Planque *et al.* 1997). In each case, the timing anticipates the spring bloom by weeks to months, so that females are matured and actively spawning when the spring bloom gets started.

Unlike the subarctic Pacific *Neocalanus* species, *Calanus* females must eat to spawn. Niehoff *et al.* (1999) showed for the Norwegian Sea that pre-bloom spawning produces copepodites that use the bloom to grow – the main stock production of the year. Females are actually much less numerous during the bloom than before, although spawning at maximal rates, and copepodites from eggs produced during the bloom must depend after the bloom upon micro-heterotrophs that constitute suitably sized food in that interval. In all regions except the Irminger Sea, diapause of *C. finmarchicus* at depth resumes in June or July. Irminger Sea stocks only rest after about October. Over most of the range, except for the northern Norwegian Sea, there are two generations – the first maturing immediately without rest (at least without prolonged rest) to produce the second. Many of the smaller copepods are sac spawners, *Oithona* and *Pseudocalanus* in particular. The relative importance of large and small copepods in grazing is not well quantified. The abundant epipelagic North Atlantic euphausiids belong to only a few species: *Euphausia krohni* (southern subarctic and farther south), *Meganyctiphanes norvegica* (particularly abundant in shelf waters), *Thysanoessa inermis* and *Thysanoessa longicaudata* (both abundant in more northern reaches above Iceland). The dominant chaetognath, *Parasagitta elegans*, is also found in the North Pacific.

Upper trophic levels

Small schooling fish of the subarctic Atlantic include several distinct populations of Atlantic herring (*Clupea harrengus*), which migrate to nearshore spawning areas in varying seasons, then redistribute seaward, and capelin (*Mallotus villosus*) in northern reaches around Iceland and Norway. Both of those stocks support significant fisheries, while most of the larger fish of commercial importance (cod, hake, haddock, and several flatfish) are demersal and are fished primarily over shelves and banks. The North Atlantic supported very great stocks of the *Calanus*-eating northern right whale (*Eubalaena glacialis*) centuries ago. These whales were hunted to commercial extinction before whalers had motorized ships; stocks currently number a few hundred. Fin whales and killer whales remain moderately numerous, as do several species of porpoise, dolphin, and seals.

The circumglobal subantarctic band is certainly a "westerlies biome", but it is convenient to consider it with the rest of the Southern Ocean.

Polar biomes

By polar ocean biomes, Longhurst meant the areas covered permanently or seasonally by sea ice and consistently cold, with surface temperatures remaining below about 5°C. Included are the Arctic Ocean, part of the Bering Sea, and the Antarctic out to nearly the antarctic circumpolar convergence (below called the *subantarctic front*). Ice has a high albedo and shades the water column below. Of course, where the sea is frozen in winter the sun is low or below the horizon, so photosynthetic production is slight. When the sun does not rise significantly above the horizon, the satellite sensors do not see ocean color for estimation of pigments, so they provide no data. There are substantial variations in seasonal cycles, on a regional and subregional basis. The Arctic Ocean actually is a medium-sized marginal sea, but an extremely important one with distinctive ecology, which we will treat separately.

Southern Ocean provinces

Most of the Southern Ocean from the antarctic continent out to the subtropical convergence (STC) at around 40°S is in a sense also a set of "westerlies" provinces. The persistent winds are westerlies, and water flow, the Antarctic Circumpolar Current (ACC), is from west to east around the entire ring and essentially all the way down the water column. North-to-south and south-to-north velocity components are added at different depths by density-driven flows. South of about 65°S, the winds reverse to easterly, forcing westward flow alongshore of Antarctica, augmented by buoyancy from ice melt, and generating cyclonic (clockwise in the southern hemisphere) gyres at the boundary with the ACC, particularly in the Ross Sea south of the Pacific and Weddell Sea south of the Atlantic. Westward flow inshore is not continuous; the southern edge of the east-bound ACC sweeps along the coast from 120°W to the tip of Palmer Peninsula, and is quite close in to the continent across the Indian Ocean and Australian sectors.

The density-driven meridional flows are associated with ascent of temperature and salinity isopleths toward the south, their rise occurring in steps. Bands of horizontal density gradients alternate with zones (fronts) with steep slopes of density isopleths, which are thus the cores of strongest current. These fronts divide the region into zones distinguished by temperature, nutrient availability, and biota. The most important are the polar front (PF) at

~50°S, and the subantarctic front (SAF) at ~46°S (for both at the Greenwich meridian east of the Weddell Sea; Orsi *et al.* 1995). Both fronts are associated with sea-surface temperature rises south to north. However, the vertical rise of density isopleths extends quite generally from the SAF to 60°S, associated with the very strong, very deep flow of the ACC. Density isopleths are flatter and temperature is more uniform from the SAF out to the subtropical convergence (STC) at ~40°S. The STC stays close to 40°S, except for swinging south around Tasmania and New Zealand, then north to almost 30°S as it crosses the Pacific toward Chile.

As the concentric rings of flow approach Drake Passage (Mar de Hoces) between South America and the Palmer Peninsula, part of the more northern, subantarctic flow peels off to the north, becoming the Humboldt Current. The rest of the flow and the frontal sequence compress through the gap. At many other points, the flow and thus the fronts are subject to topographic steering over submarine ridges and plateaus, causing the concentric rings to oscillate across latitudes. There are topographically steered excursions of the SAF toward the north to the east of New Zealand, at the Crozet Islands and Kerguelen Plateau. In winter, sea ice forms and solidifies progressively northward, reaching ~60°S (varying ~5° longitudinally, also interannually), then it melts back toward the south in summer, with at least open leads reaching the outer shelf waters around much of the continent: in the Bellingshausen Sea, along the west side of Palmer Peninsula, and the northern reaches of the permanently (at least up to now) frozen Ross and Weddell Seas.

Antarctic waters to the edge of seasonal ice

In the Antarctic, in the seasonal ice zone, the upwelling of nutrient-rich deep water under the receding ice provides copious major nutrients. There is a phase in which epontic algae living in frazil (loose needle-shaped crystals) on the undersurface of the ice are abundant, followed by a strong diatom bloom in the open meltwater as the melt proceeds southward. Dust accumulated on the ice over winter provides iron. Meltwater provides stability and critical depths are well below the mixing depth, enabling blooms of >10 mg Chl m^{-3}). Diatom production is sufficiently prodigious that formation of opal sediments in this circumglobal band constitutes the principal removal term in the dissolved silicate budget of the world ocean (Nelson *et al.* 1995). However, Nelson *et al.* also maintain that the input to sediment of diatom silica in this region is not particularly unusual, and that high rates of opal preservation in sediment are the main factor.

Northward from the winter ice to the PF and from there to the SAF, there is also upwelling from global-scale thermohaline circulation that provides abundant major nutrients. In the vicinity of the PF there are significant, but not huge, spring (November–December) blooms (~1.5 mg Chl m⁻³, Fig. 11.15) with increase above the annual background (~0.3 mg Chl m⁻³) lasting about a month. It is thought that, as this bloom moves southward with seasonally increasing irradiance, depletion of trace-metal nutrients, most importantly iron, terminates the increase of phytoplankton stock well before major nutrients are consumed. Nitrate and phosphate remain in excess throughout the year (e.g. >7 μM nitrate at 76°S in the Ross Sea, Gordon *et al.* 2000), sustaining modest levels of phytoplankton stock (~0.3 mg Chl m⁻³) until darkness and ice return. Silicate, however, can be exhausted as far poleward

as 64°S (Fig. 11.16), which is likely important in termination of diatom production all across the northern extent of the ACC to the PF.

In the eastward outflow from Drake Passage, particularly in the vicinity of South Georgia, which is probably a local iron source, the diatom production continues right through the summer, sustaining high secondary productivity, particularly of krill and large stocks of whales, seals, fish (e.g. the cod-like notothenioids of the southern hemisphere), and seabirds (e.g. penguins). Diatoms are the principal food of antarctic krill, predominantly *Euphausia superba*, particularly over the outer continental shelves (Bellingshausen Sea, the outflow from Drake Passage around South Georgia Island, Ross Sea), but there are also stocks in the cyclonic gyres between the ACC and coastal currents all around the shelf (Nicol 2006). Lesser numbers scatter out to the PF. Krill group in massive schools (up to kilometers in breadth), which are as organized, in terms of even spacing and unitary movement, as fish schools. Paradoxically, this both protects them, at least on average, from predation, and suits them to the gulp-and-strain feeding modes of baleen whales and crab-eater seals. Penguins and leopard seals also depend heavily upon krill for nutrition. The whales move in winter to the tropics to breed and calve, returning to the Antarctic to feed in summer. Seals and penguins come and go from the ice on the continent.

Krill spawn in summer (their fecundity is discussed in Chapter 8), and it is thought that females migrate out over deep water to release eggs, possibly to avoid predation on eggs by their own dense schools that are capable of clearing virtually all particulate matter from the water. Possibly for

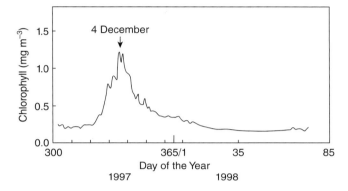

Fig. 11.15 A time-series of mean polar front-chlorophyll concentration in 1997–1998, estimated from nine fluorescence sensors moored in the surface mixing layer from 60° to 61°S near 170°W. A phytoplankton bloom was initiated in late November and ended (Chl < 0.3 mg m⁻³) by 15 December. (After Abbott *et al.* 2000.)

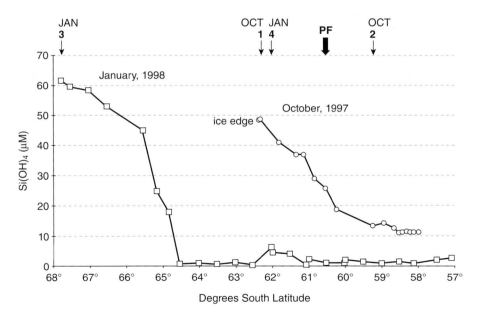

Fig. 11.16 Silicic acid concentrations on spring (October) and summer (January) transects along 170°W from well north of the polar front to the seasonal ice edge. Depletion from 60° to 65°S results from the diatom bloom that follows the receding ice edge. (After Franck *et al.* 2000.)

the same reason, eggs sink to mesopelagic depths before hatching, so that hatched nauplii must swim back up to layers with particulate food by the molt to first-feeding calyptopis larvae. Krill overwinter in the marginal ice zone (Daly 1990), along and under the ice. Juveniles and adults survive the season of low food supply by some carnivory, by feeding on ice-associated algae, and by progressively repackaging their tissue biomass as smaller bodies through consistently timed molts (Ikeda & Dixon 1982). As light returns in spring during the melt-back of the ice, *E. superba*, particularly the larval stages and also larvae of *Euphausia crystallophorias*, feed on the bloom established by meltwater stabilization, and on the increasing *epontic* algae. As in the Arctic, ice surfaces, particularly frazil, support a modestly complex community adjusted to cold and strongly varying salinity, including forms other than krill and krill larvae. For example, *Stephos longipes*, a copepod belonging to a typically epibenthic family, lives in antarctic frazil, in melt ponds above the ice and melt layers within the ice (Schnack-Schiel *et al.* 2001).

Adult *E. superba* reach a length of 10 cm, with sufficient tail muscle to make them a desirable fishery product. Moreover, its schools constitute the largest, slightly exploited fishery in the oceans, with an average stock estimated by Atkinson *et al.* (2009) at ~379 million metric tons (Mt). Fishing was tried in the 1970s and early 1980s, but tailed off. Special processing requirements and the great expense and difficulty of fishing in the Antarctic have kept exploitation to a minimum.

The oceanic copepods (principally larger biomass dominants *Calanus propinquus*, *Calanoides acutus*, *Rhincalanus gigas* and *Metridia gerlachi*; small numerical dominants *Microcalanus pygmaeus*, *Ctenocalanus citer* and *Oithona similis*) are the principal grazing mesozooplankton seaward of the marginal ice zone in most of the region south of the SAF (Atkinson 1998; Schnack-Schiel 2001). The larger species resemble those of the Arctic (and subarctic) in that they load with lipid as a nutriment store to see them through the winter hiatus in primary production. However, only *C. acutus* and *R. gigas* move to mesopelagic depths during the dark season for a prolonged copepodite diapause. In some years or for some part of the stocks of these two, the life cycle may take two years and involve two diapause intervals. The other species continue through the winter near the surface, living and feeding, albeit at reduced rates. Through much of the year, the copepods feed a trophic level or two removed from phytoplankton, since pico- and nanoplankton are the main primary producers after the spring bloom becomes iron limited, although *Phaeocystis* colonies are particularly important in this ecosystem and are significantly avoided by both microheterotrophs and mesozooplankton grazers. Dilution experiments (Pearce *et al.* 2010) show control of reasonably fast phytoplankton growth by balancing grazing by heterotrophic nanoflagellates, ciliates, and dinoflagellates.

Antarctic particle-feeding copepods are preyed upon by fish, fish larvae, predatory copepods of the family Euchaetidae, and chaetognaths, particularly *Solidosagitta marri*, *Parasagitta gazellae*, and *Eukrohnia hamata*. Population outbursts of *Salpa thompsoni* are also common but not related in an obvious way to habitat conditions. Mesopelagic waters south of the SAF support mesopelagic communities similar to those worldwide, but with many endemic species. Antarctic pelagic ecology is, of course, much more complex than our presentation of it, and recent research has generated massive amounts of data, thousands of papers and almost equal numbers of new questions. For reasons that are largely political and coupled to interest in the common international "ownership" of the Antarctic continent, of whale stocks and the krill resource, it has been studied more thoroughly than the subantarctic that we consider next.

Subantarctic

The subantarctic is a roughly circular band of ocean between the SAF front and the STC, varying (and oscillating) in width from ~14° of latitude (38° to 52°S) at 135°W (eastern Pacific) to very narrow in the Cape Horn Current where all the circumpolar currents squeeze through Drake Passage. It stretches across the "roaring forties" with persistently high winds (recorded on some of the islands in excess of 50 km hr^{-1} for >150 days per year). That ensures persistently deep mixing. Its pelagic ecosystem functions somewhat similarly to the subarctic Pacific. Surface temperatures vary with latitude from 4 to 10°C in winter, rising to ~14°C above a weak seasonal thermocline in summer, but deeper than that in the subarctic Pacific. Surface salinity is reduced mostly by abundant rainfall. There is a halocline, but again deeper, ~200 m, than in the subarctic Pacific. In terms of production ecology, the subantarctic is a cold, high-nitrogen low-chlorophyll (HNLC) region (Moore & Abbott 2000). Major nutrients, with the exception of silicate, are never exhausted, chlorophyll varies around 0.3–0.4 mg m^{-3} (Plate 11.2), and isolation from land results in iron limitation of growth rates of larger phytoplankton. While there are chlorophyll "hot spots" during spring–summer blooms around islands in antarctic waters south of the SAF (South Georgia, Kergulen and Crozet islands), the main hot spots in the subantarctic are downstream of New Zealand and South America. Even near the Crozet Plateau, well into usual subantarctic latitudes at 46°S, the subantarctic front steers north well to the west, putting the archipelago into fully antarctic waters. However, iron from the New Zealand Plateau and from South America generates eastward-billowing flags of elevated chlorophyll in satellite images of the subantarctic during spring and summer. Deeper mixing and winter darkness limit production and stocks in fall and winter.

Unlike other HNLC areas, silicate has very low availability in the subantarctic, <15 µM at the seasonal maximum and <1–3 µM after mid-summer. The general explanation (Zentara & Kamykowski 1981) is that the siliceous shells of diatoms sink to much greater depth – even to the seafloor – before significant dissolution, than do organic nitrogen and phosphate (including the organic portions of diatoms). Also, near-surface N and P regeneration allows those elements multiple cycles of incorporation, each cycle exporting a larger fraction of Si than of N or P. Thus, silicon becomes much less accessible to upward seasonal mixing: the silicocline is much deeper than the nitricline. South of the SAF, the strong vertical velocity over most of the water column compensates for this process, so that levels after the winter darkness exceed 50 µM south of ~62°S.

Several subantarctic mesoscale iron-enrichment experiments (SoFEX and LOHAFEX) have been carried out late in summer, an appropriate season if iron were to be used to enhance deep-sea carbon sequestration. Both failed to generate significant blooms of large diatoms. The SoFEX "north" addition (Coale et al. 2004) in the eastern Pacific subantarctic (~54°S, with ~3 µM Si(OH)$_4$) did stimulate increase of flagellated phytoplankton and some *Pseudonitzschia* (small diatoms), while LOHAFEX (Smetacek 2009) weakly enhanced stocks of flagellated phytoplankton, but also of grazing pteropods and copepods, particularly the regionally endemic *Clausocalanus laticeps*. Predatory amphipods also strongly increased, presumably eating the grazers.

At least one abundant, circumglobal, subantarctic copepod, *Neocalanus tonsus*, has a lipid-loading and diapause scheme resembling that of the subarctic Pacific dominants. Like its subarctic congeners, it matures, mates, and females initiate spawning at the resting depth. However, unlike the Pacific species, females retain feeding mouthparts and ascend to the surface to continue spawning while feeding (Ohman et al. 1989). *Calanus simillimus* shares dominance among larger copepods in the subantarctic, its distribution extending across the PF into northern antarctic waters. It apparently is distantly related to other *Calanus* species (Hill et al. 2001), but it retains their life-history pattern of deep winter diapause, the females feeding to produce eggs. It runs through several generational cycles in the productive season (Ward et al. 1996). A suite of small copepods, *Limacina*, salps, and euphausiids are also present. Clupeids or other small, schooling fish are apparently few, apart from coastal regions off South America. Vertically migrating midwater fish and ommastrephid squid are, however, important and support stocks of island-breeding fur seals, sea lions, elephant seals, and endemic seabirds, e.g. wandering albatross (*Diomedia exulans*), several species of giant petrels (*Macronectes*), and of small petrels (*Procellaria*). Endemic epibenthic fish are (or were – they have been overfished) abundant on shelves and over seamounts, including sundry notothenids tolerant of extreme cold and occurring also farther south. Commercially important species have been Patagonian toothfish (*Dissos-*

tichus eleginoides), orange roughy (*Hoplostethus atlanticus*), hoki (*Macruronus novaezelandiae*), and a southern hake (*Merluccius australis*).

The Arctic Ocean

The Arctic Ocean is the smallest and most land-locked of the named oceans. Pacific Ocean water flows in through the Bering Strait and contributes about 24% of the seawater exchange in the Arctic Ocean. Atlantic water flows in through the Fram Strait and Barents Sea and comprises the remaining 76% of exchange. There is a cyclonic gyre over the Eurasian Basin and an anti-cyclonic gyre over the Canadian Basin. Those central basins are surrounded by six shelf seas that comprise over 50% of Arctic Ocean area (Fig. 11.17). Sea ice is the dominant physical factor nearly year round in the central basins and seasonally in the surrounding shelf seas. The overall extent of summer melting and open water has been increasing markedly in recent decades. Formation of sea ice excludes salt, increasing salinity and density in the water just below the surface. Melting of ice releases fresh water.

The Arctic biome is structured by freshwater entrainment, sea-ice cover and resulting stratification of the water column. Fresh water is entrained from increasing seasonal ice-melt (~800–1100 km^3 yr^{-1}), inflowing rivers (3559 km^3 yr^{-1}), and inflowing Pacific water which is less saline than Atlantic water (~2500 km^3 yr^{-1}). These freshwater inputs lead to a shallow mixed layer over a pronounced halocline. Arctic sea ice is a ~17,300 km^3 reservoir of freshwater. As the Arctic continues warming, freshwater inputs will continue to increase.

When the polar atmospheric high-pressure cell is well developed, the Beaufort gyre extends over most of the Canadian Basin, and Pacific-derived water is found in the Makarov Basin all the way to the Lomonosov Ridge. When the high-pressure cell is weak, the Beaufort gyre is weak and displaced toward the American continent, reducing the Pacific inflow and its extension toward the Atlantic. The front between the Atlantic- and Pacific-derived waters has shifted over the last several decades from the Lomonosov Ridge to the Mendeleyev Ridge (McLaughlin et al. 1996)

Production controls

The Canadian Basin, and the Beaufort, Chuckchi, and East Siberian Seas have low-salinity surface water overlying the warmer but saltier Atlantic water. Nutrient concentrations vary with the proportions of Pacific versus Atlantic waters, with Pacific water having three-fold more nitrate, phosphate, and silicate at a given salinity (Fig. 11.18). Nutrients can be depleted seasonally in the surface water, but tend to range from 2 to 4 µM nitrate throughout the year in the Eurasian Basin. Elevated nitrate, phosphate, and silicate are apparent in shelf water that is advected into the basins

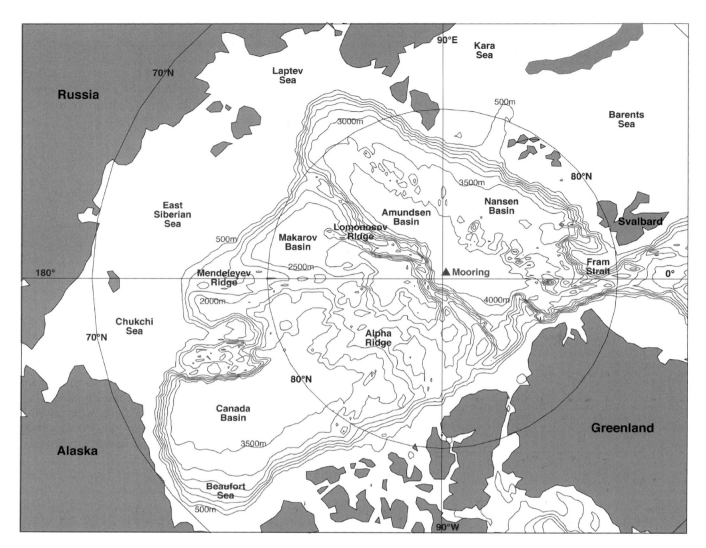

Fig. 11.17 Map of the Arctic Ocean showing the shelf seas and central basins. The Nansen and Amundsen basins together form the Eurasian Basin, and the Makarov and Canada basins form the Canadian Basin. The Lomonosov Ridge separates the Eurasian and Canadian Basins. (After Aagaard *et al.* 2008.)

forming a subsurface nutrient maximum associated with the halocline. There are distinct nutrient and salinity signatures for Pacific- and Atlantic-derived waters (Wheeler *et al.* 1997).

For the shelf seas, the mixed-layer depth and nutrient levels are set by winter convection and circulation. As the ice edge melts and light levels increase in spring, ice algae and then phytoplankton start growing. River water and ice melt lead to enhanced stratification, and the duration and the size of the bloom are regulated by the initial nutrient level. As nutrients are drawn down, a subsurface chlorophyll maximum develops (Fig. 11.19).

Primary production in the Arctic Ocean has been measured with the standard ^{14}C technique, but the spatial and temporal extents of such measurements are limited. During the polar winter, phytoplankton production is obviously light-limited. Seasonal production begins as soon as the snow melts (Sherr *et al.* 2003), with ice algae blooming first, followed by phytoplankton. During the spring and summer seasons, nutrient supply is the more important controlling factor, and by mid-summer the shelf water and basin water often become depleted in nitrate and phosphate. Annual rates of primary production are only about half of those measured in the Antarctic and vary with hydrographic conditions in the marginal seas and the central basins (Table 11.2).

Microbial food-webs are net heterotrophic

Bacterial biomass and growth rates are lower in the polar regions than in other oceans. Growth of heterotrophic bacteria has been thought to be limited by cold polar tem-

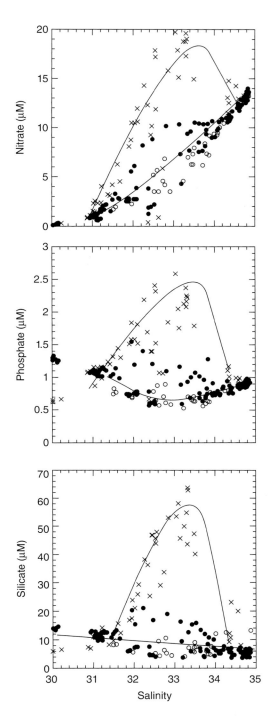

Fig. 11.18 Nutrients versus salinity plots for a 1994 Arctic Ocean section. Stations with Pacific-derived characteristics (×); stations with intermediate chemical characteritics (filled circles); stations with Atlantic-derived characteristics (open circles). At a given salinity, Pacific-derived water has much higher nutrient concentrations than Atlantic-derived water. (After Wheeler *et al.* 1997.)

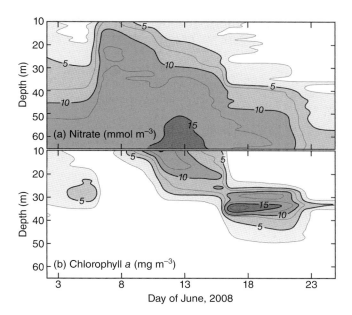

Fig. 11.19 Time-series of (a) nitrate (mmol m^{-3}) and (b) chlorophyll (mg m^{-3}) during an ice-edge upwelling phytoplankton bloom in the Canadian Beaufort Sea. Data from 47 hydrocasts were interpolated and plotted. (After Mundy *et al.* 2009.)

peratures, but a recent comparison of bacterial production in the polar oceans (Kirchman *et al.* 2009) suggests that the major factor limiting bacterial growth is the relatively low amount of labile DOC. In both polar regions, bacterial growth rates increase with increasing levels of semi-labile DOC. The levels of DOC vary seasonally and reflect the seasonal cycle of plankton production and consumption.

The balance between autotrophy and heterotrophy can be determined by measuring net changes in oxygen in incubated seawater samples. Cottrell *et al.* (2006) showed both positive and negative net changes in oxygen. Repeated measurements across a latitudinal gradient in the Chukchi Sea suggest that primary production on the shelf can be transported to the basin, where respiration exceeds the local levels of photosynthesis.

Heterotrophic protists are an important component of the arctic plankton. Dilution experiments in parallel with copepod grazing experiments in the Chukchi and Beaufort Seas showed that ciliates and heterotrophic dino-flagellates consumed $22 \pm 26\%$ of phytoplankton daily growth (Sherr *et al.* 2009), while copepods consumed 13–28% of primary production (Campbell *et al.* 2009). In these shallow seas, meso- and microzooplankton consume about 44% of water-column production, the rest being available for export to depth.

In comparison to the shelf seas, stocks of plankton are lower in the central arctic basins, and there are differences in food-web interactions. During a year-long sampling in the central Arctic, stocks of bacteria and protists doubled

Table 11.2 Primary production in the Arctic Ocean (subregion values from Sakshaug 2004).

	gCm⁻²yr⁻¹	TgCyr⁻¹
Central Arctic		
Canadian Basin		
Eurasian Basin	>11	>50
Inflow seas		
Bering Strait/Chukchi Sea	>230	42
Barents Sea	<20–200	136
Interior shelves		
Beaufort Sea	30–70	8
Kara/Laptev/Siberian Sea	25–50	83
Out-flow		
East Greenland Shelf	70	42
Canadian Archipelago	20–40	5
Total Arctic primary production		
Sakshaug (2004)	>26	>329
Pabi *et al.* (2008)	44	419
Total Southern Ocean primary production		
Arrigo *et al.* (2008)	57	1949

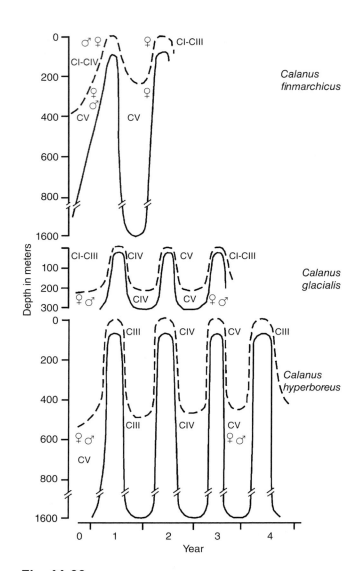

Fig. 11.20 Generalized seasonal migration and stage development of *Calanus finmarchicus*, *Calanus glacialis*, and *Calanus hyperboreus*. Upper and lower lines bound the general depth distributions of the populations. (After Falk-Peterson *et al.* 2009.)

during the growing season (Sherr *et al.* 2003). The central arctic mesozooplankton biomass is dominated by four species of copepods: *Calanus finmarchicus*, *C. hyperboreus*, *C. glacialis*, and *Metridia longa*. *Calanus finmarchicus* enters with incoming Atlantic water and survives but does not reproduce in the Arctic Ocean proper. Patterns of seasonal migration and life stages (see Fig. 11.20) are reviewed by Falk-Peterson *et al.* (2009). *Calanus hyperboreus* is the largest arctic copepod and best adapted to short and unpredictable growing seasons, with a 3–5-year life cycle and very large lipid stores. In the central Arctic Ocean, mesozooplankton standing stocks are approximately four-fold greater than phytoplankton standing stocks (Table 11.3). Copepod grazing rates estimated from their rate of fecal pellet production were 3–20% of expected food-saturated rates. Further, estimates of food-saturated carbon demand

exceeded measured primary production rates. Based on those observations, Olli *et al.* (2007) suggested that arctic mesozooplankton are food-limited, and that the abundant copepods prevent the accumulation of phytoplankton stocks in the central Arctic.

Tremblay *et al.* (2006) have suggested an upper food-web structure for the Arctic Ocean (Fig. 11.21). The zooplanktivorous trophic level of the arctic foodweb includes North Atlantic right whales (*Eubalaena glacialis*), the little auk (*Alle alle*), bowhead whales (*Balaena mysticetus*), and Arctic Cod (*Boreogadus saida*) (Plate 11.3), all mostly feeding on copepods. The cod serves as the main food for the ringed seals (*Pusa hispida*), glaucous gulls (*Larus hyperboreus*), beluga whales (*Delphinapterus leucas*), and

Table 11.3 Standing stocks of phytoplankton, bacteria, heterotrophic protists, and copepods in the central Arctic Ocean. Data from Sherr et al. (1997) and Thibault et al. (1999).

GROUP	STANDING STOCK (mg C m^{-2})
Phytoplankton	773 ± 1076
Heterotrophic protists	544 ± 360
Bacteria	506 ± 146
Copepods	3190 ± 1005

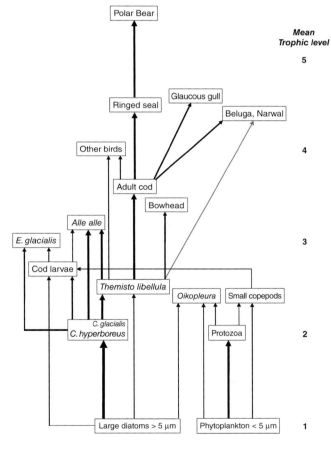

Fig. 11.21 Structure proposed by Tremblay *et al.* (2006) for the pelagic food-web in the North Water Polynya in northern Baffin Bay, based on carbon budgets, stable isotopes, analyses of gut contents, and the literature. Numbers in the margin indicate the trophic level inferred from stable isotopes. The size of the arrows reflects the relative importance of a food item for a given consumer, not the magnitude of the carbon flux. *Themisto* is a planktonic amphipod. *Alle* is a planktivorous bird. Bowhead whales are a threatened species found in shelf areas. *Eubalaena glacialis* is the northern right whale, also a threatened species. (After Tremblay *et al.* 2006.)

narwhals (*Monodon monoceros*), that comprise the next trophic level. Arctic cod form dense schools in the Canadian Arctic and attract large numbers of seabirds and beluga whales (Plate 11.3). The apex predators are the polar bears (*Ursus maritimus*) that consume ringed seals, and the indigenous people who consume seals and whales.

Export of carbon

Export of organic carbon from the upper layer of the Arctic Ocean has been measured with both floating and moored sediment traps. Results from traps at 200 m are highly variable in space and time, ranging from 1.3 to 31 g C m^{-2} yr^{-1} in the central Arctic and from 3.1 to 197 g C m^{-2} yr^{-1} over shelves and within polynyas (open water surrounded by ice) (Wassmann *et al.* 2004). Sedimentation appears to be episodic and patchy, making generalizations and comparisons among subregions difficult. Sinking fluxes are highest in the northern Bering Strait and Chukchi Sea and support an abundant benthos dominated by clams and amphipods. Fluxes are lower in the Fram Strait, and its stocks of benthos are dominated by polychaetes. In other arctic shelf seas, fluxes are moderate and show significant terrestrial inputs.

Significant levels of phytoplankton nutrient (e.g. 3 μM NO$_3^-$) remain in the surface water of the central Arctic throughout the growing season (Wheeler *et al.* 1997; Olli *et al.* 2007). Olli *et al.* argue that in the central Arctic, top-down control of primary production by heavy copepod grazing pressure limits primary production of larger phytoplankton and limits the vertical flux of organic carbon by maintaining low phytoplankton stock. Nutrient levels are lower in the Canadian Basin portion of the central Arctic, and it is possible that phytoplankton growth there is seasonally nutrient limited.

Subtropical gyre biomes

Longhurst placed the boundaries of his *trades* zone at about 30°N and 30°S. Those pass right through the ecologically quite uniform central gyres, although satellite results do show a modest increase of chlorophyll north of 30°N. John McGowan and others have chosen ~28–30°N as the latitude at which to study the fundamentals of subtropical gyre ecology. The Bermuda Atlantic Time Series, BATS, station is at 32°N. BATS is a study of the subtropical Sargasso Sea. There is a basis in biogeography for lumping all of the warmer ocean waters: many species ranges extend from 45 or 40°N to the equivalent southern latitude, stretching right across the equator. However, the satellite images show that the central gyres have lower surface chlorophyll (phytoplankton stocks) and are certainly less productive than the zones ~15°N to 15°S that are affected by equatorial upwelling. Moreover, many species of plankton live in the central gyres only. So, we will treat *subtropical* (*or central*) *gyre* and *trades* (*equatorial*) biomes as separable. Thanks to the CLIMAX studies in the 1960s–1980s (McGowan & Walker 1985; Venrick 1999) and to the Hawaii Ocean Time Series (HOT, sampling at Station "ALOHA", e.g. Siegel *et al.* 2001), work initially part of the JGOFS program, we have substantial data describing the North Pacific subtropical gyre (NPSG). The JGOFS BATS time-series (Steinberg *et al.* 2001) and earlier time-series work near Bermuda provide a Sargasso Sea comparison. The ESTOC time-series in the eastern subtropical North Atlantic (Neuer *et al.* 2007) and many process studies in the Sargasso Sea and elsewhere provide additional insight. HOT (100 km from Oahu) may have been affected somewhat by proximity to the Hawaiian Rise, and ESTOC may have modest effects from proximity to both the Canary Islands and Africa. They are useful nevertheless. Comparable ecosystem features and processes are found in the three southern hemisphere gyres, all of which have been studied on a more occasional basis.

Basis of oligotrophy in the subtropics

The key feature of the subtropical gyres is water-column stability. Examination of north–south sections of temperature or σ_t show flat isopleths in the gyres for thousands of kilometers. The system is said to be "barotropic". This results from the anticyclonic direction of gyral flow, which tends to be convergent, piling water in the center under the tropical sun for stabilizing warming. There is a zone of moderately strong stratification (Fig. 11.22) from approximately 120 m down to about 1200 m, the permanent pycnocline. This is a barrier to mixing which ensures very slow provision of nutrients upward. The surface layer exhibits some temperature cycling (~18°C winter to ~25°C summer) over a seasonal thermocline at about 50–70 m. Because of water clarity, net photosynthesis is possible to about 125 m (Fig. 11.23) and nutrients become depleted down to just above that level. Nitrate and phosphate in surface layers are usually removed to levels undetectable by standard analyses (sensitive to ~0.1 μM). However, preanalysis concentration and ultrasensitive procedures have been invented to measure these macronutrients at nanomolar levels, albeit with some interference from readily hydrolyzed dissolved organic forms. Primary production runs mostly on recycled nutrients (*f*-ratio ≈ 0.05 to 0.1) with some addition from nitrogen fixation. Subtropical phytoplankton have extreme affinity for ammonium and carry surface enzymes for stripping phosphate from organic complexes (e.g. Beversdorf *et al.* 2010; Duhamel *et al.* 2010). Since there are always losses of organic matter to depth, however, the system cannot run entirely on recycled nutrients. At 125 m or just above is the top of a nutricline, which is established at the deepest penetration of light that can support photosynthesis to provide energy for nutrient uptake. The downward increase in available nitrate, phos-

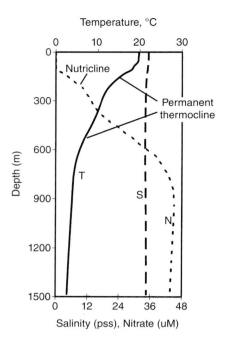

Fig. 11.22 *T, S* and (nitrate + >nitrite) = *N* profiles on 16 March 2006 from the subtropical Pacific at 30°N, 152°W. Salinity variation is minimized by plotting on the nitrogen scale; it varied from 35.26 at the surface to a minimum of 34.00 at ~540 m. The salinity "inversion" is stabilized by the vertical temperature gradient. CLIVAR program profile data from NODC.

phate and silicate is supplied by the slow upward mixing through the permanent pycnocline from high concentrations at depth (16 μM nitrate in the North Atlantic, 40 μM in the North Pacific). Thus, mixing rate ultimately sets the rate of primary production.

The nutricline is also the level of a doubling to tripling of chlorophyll concentration, the deep chlorophyll maximum (DCM, Fig. 3.9, Fig. 11.23a). This is observed in virtually all subtropical and tropical oceanic regions. The significantly increased nutrient availability just above the photosynthetic compensation depth enables phytoplankton to synthesize more chlorophyll (which requires iron and uses much fixed nitrogen) to compensate for the low irradiance: they "spread their antennae" to gather the waning supply of photons. Principally, it is a shift in the amount of chlorophyll per cell, not an increase in the cell counts. On the other hand, Venrick (1982) has shown for the "CLIMAX" station (30°N, 155°W) that at least the diatom component almost completely changes in species composition at a fairly sharp boundary just above the DCM. The flora is not simply shade-acclimated individuals of the phytoplankton above; it is a different, shade-adapted community. There are also increases in the cell sizes of cyanobacteria (DuRand *et al.* 2002). The contribution of the DCM to total production is significant, but the bulk of integrated production occurs only a short distance below the sea surface (Fig. 11.23b).

Primary producers

In all subtropical gyres the dominant phytoplankton in numbers and biomass are picoplankton. DuRand *et al.* (2002) used flow cytometry at BATS (31°40′N, 64°10′W) to characterize their seasonal cycling. Two genera of cyanobacteria, *Prochlorococcus*, and *Synechococcus*, alternate in numerical dominance. Summer to early winter numbers of *Prochlorococcus* integrated to 200 m (most cells above 125 m) are on the order of $10^{13} m^{-2}$ (~0.5 g organic carbon), partly replaced during mid-winter mixing (that often extends to >200 m) by *Synechococcus* that for a few months

increase to $2–3 \times 10^{12} m^{-2}$ (~0.2 g C). Eukaryotic picoplankton are fewer in numbers but about equal in biomass to the combined cyanobacteria, particularly in the late winter "bloom". At this location, however, the periods of greatest cyanobacterial dominance are those of reduced primary productivity, and small eukaryotes account for a majority of annual organic matter production. Worden and Binder (2003) used flow cytometry in March at BATS to examine the numbers of *Prochlorococcus* cells from 50 m depth in different cell-cycle stages. DNA replication (S-phase) occurred from late afternoon through evening. Cells with double DNA complements peaked at midnight, followed by doubling of abundance from 20 to $40 \times 10^3 ml^{-1}$. Grazers returned counts to daytime levels by mid-morning. Similar diel cycling of somewhat lower cell numbers was observed for *Synchococcus*. Growth rates of both species from parallel dilution experiments mostly were just less than one doubling per day.

Campbell *et al.* (1994) performed a comparable flow cytometry study at HOT (22° 45′N, 158° 00′W), finding much stronger and seasonally consistent dominance of *Prochlorococcus* over *Synechococcus*. They also compared the abundances of autotrophic and heterotrophic bacteria (Table 11.4). Liu *et al.* (1997) used flow cytometry to examine the *Prochlorococcus* cell cycle at HOT, using the diel shifts in cycle phasing to estimate growth rates. As in the Sargasso, cell replication was almost entirely in the middle of the night and implied that rates near the surface were less than roughly one doubling per day, less below 80 m. Cell numbers peaked around dawn and were at a minimum around sunset, sustaining approximate daily balance of production and consumption.

Primary productivity in the NPSG at HOT does not have a winter or spring peak, mostly because there is no deep mixing event most winters that would fuel a bloom with nutrients. The 20-year averages of monthly HOT data (Fig. 11.24) show a mid-winter low of about

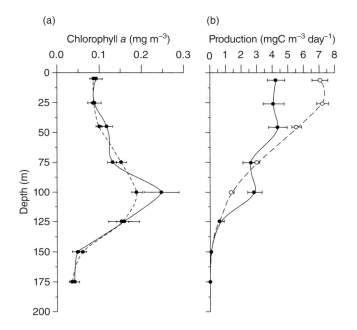

Fig. 11.23 Vertical profiles of (a) chlorophyll and (b) primary productivity from Station ALOHA (22.75°N, 158°W, the Hawaii Ocean Time Series site). Open circles for samples up through January 1991; filled circles for those afterward. (After Letelier *et al.* 1996.)

Table 11.4 Average percentages (Oct, Dec, Jan, Feb, Mar, Apr) of bacteria and algal particles at HOT for the surface mixing layer (0–70 m), the deep chlorophyll maximum (DCM), and integrated carbon biomass. Seasonal variability is modest and 0–70 m abundances are approximately constant, usually tapering to half or less by 120 m, except for March when they taper from 150 m. (After Campbell *et al.* 1994.)

	Z = 0–70 m	DCM	mg C m^{-2} to 200 m
Heterotrophic bacteria	40%	42%	1273
Prochlorococcus	44%	27%	973
Synechococcus	3%	1%	58
Picoeukaryotes (<3 μm)	10%	26%	404
"Large" algae (3–20 μm)	3%	4%	98

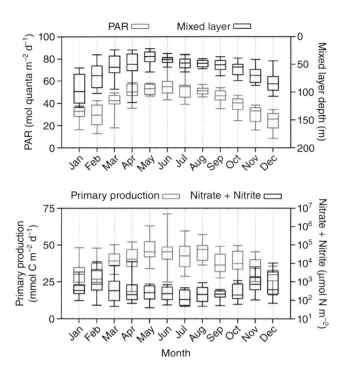

Fig. 11.24 Monthly averages (with range bars and quartile boxes) at Stn. ALOHA of mixed-layer depth, depth-integrated (0–100 m) [nitrate + nitrite], incident flux of photosynthetically available radiation (PAR) and depth-integrated (0–100 m) primary production. (From Church *et al.* 2009.)

30 mmol C m^{-2} day^{-1} (integrated to 100 m), and a May to August high around 54 mmol C m^{-2} day^{-1}. In mass units the annual mean is ~514 mg C m^{-2} day^{-1}; annual total ~190 g C m^{-2}. The effect of the shallow summer mixed layer, maintaining the phytoplankton stock high in the euphotic zone, is likely more important to the summer peak of production than the maximum in irradiance. In addition to production by pico- and nanophytoplankton, there are also in many years subsurface "blooms" of diatoms during which chlorophyll rises to ~0.15 µg liter^{-1} (Venrick 1974; Dore *et al.* 2008). The diatoms belong to the genera *Rhizosolenia*, *Hemiaulus*, and *Mastogloia*, all of which harbor symbiotic, nitrogen-fixing cyanophytes called *Richelia*. As Dore *et al.* document, these subsurface blooms occur principally in cyclonic eddies, which generate upward mixing of nutrients from the vicinity of the nutricline, including the silicic acid to synthesize diatom frustules. Just this brief pulse of diatom production and symbiont nitrogen fixation adds substantially to the annual total of new production in the region and presumably to the draw-down of nutrients.

In the Sargasso Sea (Fig. 11.25) a modest pulse of nitrate usually reaches the upper ocean in late winter, often February or March, when mixing reaches slightly below 200 m. There is a modest increase then in production and phytoplankton standing stocks (Steinberg *et al.* 2001);

much of both coming from the picoeukaryotes and "large algae". Probably the main difference between the Atlantic and Pacific time-series in regard to winter mixing and production occurs because BATS is 10° farther north than HOT, experiencing both greater winter cooling (and thus convective mixing) and greater winter winds. The shift from *Prochlorococcus* to *Synechococcus* during the BATS period of deep mixing corresponds, too, to a general latitudinal shift in their relative dominance northward. It is likely that there are late winter blooms at the poleward edges of all gyres similar to that at BATS and a gradient in seasonality of production. The annual total production average for BATS, 155 g C m^{-2} yr^{-1}, is somewhat less than the HOT estimate of ~190 g C m^{-2} yr^{-1}. Interannual variations in productivity occur in subtropical gyres, but with monthly time-series sampling at a single location, the resolution and spatial coverage are inadequate to estimate variation of annual totals reliably, although an increase of 1.5-fold from 1989 to 2004 is suggested by the HOT data (Corno *et al.* 2008).

There was a large increase in measured primary production rates between the HOT data (1991 on) compared to results from the 1980s and before, that were less than half the later results. Some or all of this (there is no way to be certain) was due a change to very clean carbon-isotope-uptake techniques.

Limiting nutrients

In recent decades there has been consistent interest in determining which nutrients limit phytoplankton production in subtropical gyres, and no study has convinced all of those interested. There are strict "Liebigians" and advocates of limitation by multiple nutrients. Liebig, an early agricultural chemist, suggested that the *one* nutrient compound in least supply *relative to plant requirements* would be the factor limiting plant growth. Complexity is added because different phytoplankton have different requirements, different affinities for any or all nutrients, and are subject to different pressures from grazing, all of which confuse the results of experiments with phytoplankton from the field. Moreover, different gyres have different ratios of major nutrients in the upper euphotic zone and different concentrations available at depth for upward mixing. The main candidates as limiting nutrients are fixed nitrogen (nitrate at depth), phosphate, and iron.

Except after the strongest winter mixing, the upper water columns in both the NPSG and Sargasso Sea are sufficiently depleted in N and P that special techniques are required to determine their concentration. A chemoluminescent method for $NO_3^- + NO_2^-$ can determine levels down to ~2 nM (nanomolar, 10^{-9} mol liter^{-1}). A technique called MAGIC can determine "soluble reactive phosphate" (SRP), including an uncertain part of organic-bound phosphate, down to 3 nM. Some of the organic phos-

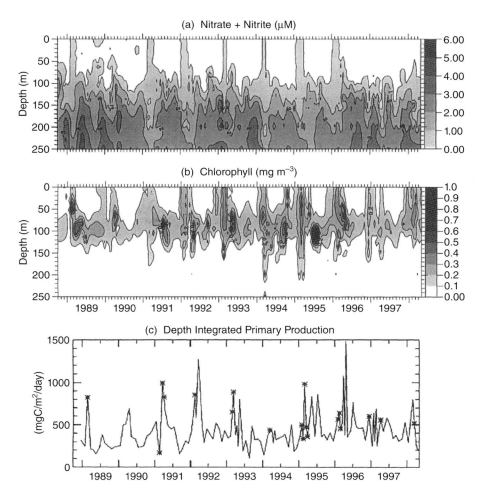

Fig. 11.25 BATS time-series data showing the late winter injections of nitrate into the euphotic zone and the responses of phytoplankton standing stock (chlorophyll-*a* abundance) and primary production rate. The persistent chlorophyll maximum, usually just above 100 m, is typical of all oceanic, tropical waters. (After Steinberg *et al.* 2001.)

phate is available to at least those phytoplankton-bearing phosphatases on their cell surfaces (e.g. *Trichodesmium*, White *et al.* 2010), so SRP may be close to the right measure of sufficiently labile phosphate to support cell growth. At HOT (Karl *et al.* 2001) the bulk euphotic zone values of $NO_3^- + NO_2^-$ (= [N+N]) range from the detection limit to ~8 nM and of SRP from 20 to 100 nM. The ratio of means, [N+N]:SRP is ~3:50. Because ammonium is held extremely low, these are certainly the available forms of the major nutrients, and they are far from the usual Redfield values for phytoplankton composition, N:P≈16. On the other hand, the dissolved organic N and P (Karl *et al.* 2001) have a mean molar ratio DON:DOP = 5 µM N to 0.23 µM P, or ~22. Inorganic N:P = 13.5 (~40 µM NO_3^-: 3 µM PO_4^{3-}) slowly diffusing up from below 800 m. The amounts of DON and DOP available for recycling are uncertain; however, it seems likely, based on the low levels and ratios of inorganic species, that N is consistently more limiting than P.

The upper euphotic zone [N+N] and [SRP] are seasonally variable at BATS, rising from extremely low concentrations most of the year to 0.2–1 µM and 20–100 nM, respectively, during deep winter mixing (N:P ≈ 40). Summer to early winter values are much lower: 2–10 nM and 1–20 nM (N:P ≈ 0.3–7) (Cavender-Bares *et al.* 2001). Thus, nitrogen would appear to be more available relative to phosphate (and to relative phytoplankton requirements) than in the NPSG, although DON and DOP are, again, more abundant and partially available. Explanations mostly depend upon N_2-fixing phytoplankton producing exported organic matter with no requirement for fixed nitrogen, stripping the P from the upper layers.

The obvious tests to determine which nutrients set the proximal limits on phytoplankton growth rates are modest additions of nutrients singly and in combinations that potentially interact, as for example of N and P together, or of Fe and P. Nitrogenase catalyzing N_2-fixation requires a good deal of iron, 15 iron atoms for each active enzyme

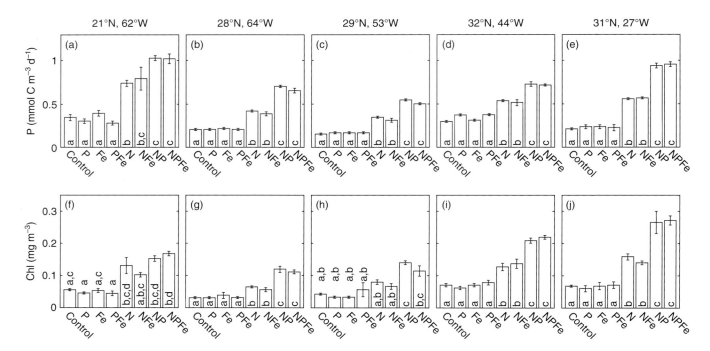

Fig. 11.26 Response of the bulk phytoplankton community abundance and carbon fixation to additions of different nutrients at different locations across the subtropical Atlantic. Upper row: carbon fixation measured from 24 to 48 h after the addition of the indicated nutrients. Lower row: chlorophyll concentration measured at 48 hours. Vertical lines in bar tops are the standard error among replicates. Bars with the different letters were statistically distinguishable (i.e. $p < 0.05$). (After Moore *et al.* 2008.)

complex. It also requires molybdenum, but that apparently is never limiting, despite having the typical surface depletion exhibited by phytoplankton nutrients. Moore *et al.* (2008) in March and April made trace-metal-clean additions of PO_4^{3-} (0.2 μM), $FeCl_3$ (2 nM), P + Fe, N as NH_4^+ (1 μM), N + Fe, N + P and N + P + Fe to samples of near-surface water from five stations across the Sargasso Sea. After 24 hours of incubation they measured ^{14}C uptake, and after 48 hours they measured chlorophyll levels. Results were consistent among stations (Fig. 11.26): no effect from just P, just Fe or P + >Fe. Ammonium supplement, however, gave strong increases, and adding both NH_4^+ and P was even more effective. Iron produced no increment in any combination. Clearly, the immediately limiting nutrient for the resident BATS phytoplankton community, which was typical on the observation dates, that is, dominated by *Prochlorococcus* with modest numbers of *Synechococcus* and picoeukaryotes, was fixed nitrogen. However, given ammonium, the cells benefited further from added phosphate. That result from Moore *et al.* is almost pure Liebig: relief of the most limiting nutrient opens the potential for growth stimulation by the next-scarcest relative to requirements. In the Sargasso, the effect of combined N and P additions is likely due to the low phosphate concentration relative to fixed nitrogen (high N:P) (Fig. 11.27). This experiment is different from determinations of what might limit the rare, often large-cell component of the phytoplankton. Incubations over many days might allow the

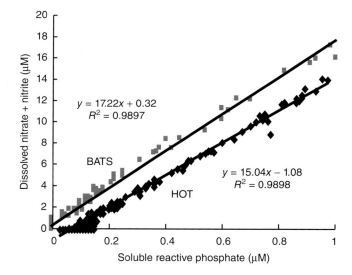

Fig. 11.27 Dissolved [nitrate + >nitrite] vs. soluble reactive phosphate at Station ALOHA (HOTS) and at the BATS station (31°45′N, 64°10′W) in the Sargasso Sea. Probably the difference is attributable to greater relative importance of nitrogen fixation in the subtropical North Atlantic. (After Wu *et al.* 2000.)

"grow-out" of cells with completely different limiting factors.

Nothing exactly comparable appears to have been done in the NPSG, but Van Mooy and Devol (2008) used tracer levels of radioactive phosphate to examine the rates of ribosomal RNA synthesis (PO_4^{3-} required) in *Prochlorococcus*, "eukaryotes", and heterotrophic bacteria at HOT in July of 2003 and 2004. Measurements were made for incubations with added PO_4^{3-}, NH_4^+ and NO_3^-. Somewhat like the Moore *et al.* experiments in the Sargasso, significant effects were only found for ammonium enrichment (Fig. 11.28). The system is nitrogen limited. The absence of stimulation by nitrate may be explained for *Prochlorococcus* by its lack of nitrate reductase (Rocap *et al.* 2003; but see Martiny

et al. 2009). While picoeukaryotes generally have genes for use of nitrate, those may well not be active when [N+N] is at extremely low levels. Clearly, the immediately limiting nutrient is fixed nitrogen. The addition of both ammonium and phosphate might not have had a synergistic effect, as was seen in the Sargasso, given the low N : P ratio in the NPSG. Van Mooy and Devol point out that *Prochlorococcus* is metabolically poised to operate in the face of extremely low phosphorus and iron availability. The phospholipids usual in cell membranes have been replaced by sulfolipids – sulfate is copiously abundant in seawater relative to biological needs. The genome is radically minimized (Dufresne *et al.* 2003), saving on phosphate, and the absence of nitrate reductase saves on iron. Regardless of N : P, the radically low euphotic zone levels of phosphate in subtropical gyres favor this autotroph that needs an absolute minimum of it. Possibly the large South Pacific gyre has completely different nutrient dynamics: while also very oligotrophic, it consistently exhibits 0.2–3 µM inorganic phosphate.

The NPSG ecosystem is likely also iron-limited in a sense, but the long-term rates of eolian and advective iron supply are great enough to overcome the slow rate of fixed nitrogen supply, so that nitrogen becomes fully depleted from the surface. HNLC regions all are much more "baroclinic" and divergent, such that iron supply cannot keep pace with nitrogen supply, so iron becomes, as John Martin liked to put it, the Liebigian nutrient.

Upper-water-column SRP data from the HOT site (http://hahana.soest.hawaii.edu/hot/hot-dogs/interface.html) (Fig. 11.29) show wide and somewhat chaotic variations. The series to 2009 shows that supply and drawdown are intermittent without any obvious periodicity. Phosphate

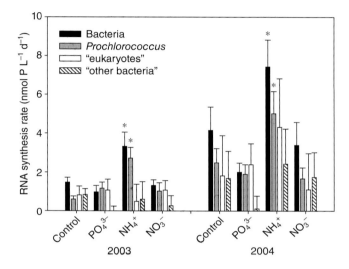

Fig. 11.28 Rates of RNA synthesis by groups of particulate plankton as determined by $^{33}PO_4^{3-}$ labeling. Asterisks indicate statistically significant difference from the control; error bars are standard deviations among three replicates. (After Van Mooy & Devol 2008.)

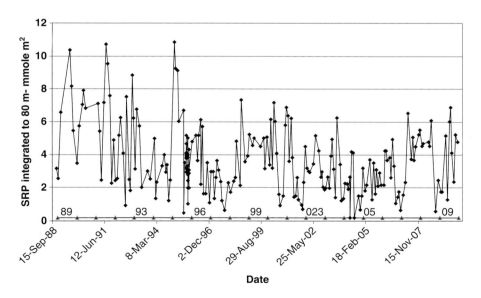

Fig. 11.29 Time-series of soluble reactive phosphate (SRP) at Station ALOHA. From data available as "upper water column SRP" from the HOT web site (http://hahana.soest.hawaii.edu/hot/hot-dogs/interface.html.).

mixes upward with substantial intermittency. Periods of higher levels may result from greater winter mixing, and short-lived up-ticks at HOT may be generated by passing cyclonic eddies. Johnson *et al.* (2010), using an Argo float equipped with oxygen and nitrate sensors, demonstrated that there is such intermittent delivery of nutrients to the euphotic zone. It was launched at HOT, moved mostly east between 22° and 24°N, and profiled from 1000 m to the surface every 5 days for 127 cycles (21 months). There were 12 substantial upward mixing events identified from increased nitrate, particularly in the upper nutricline, and subsequent oxygen increase showed that substantial photosynthesis resulted. Possibly some events were missed between profiles, because of the extreme rapidity of nitrate uptake when it does appear. Intermittency of upward nutrient mixing is consonant with cyclonic eddies being the primary mechanism.

Nitrogen fixation

Subtropical gyres and equatorial zones are the sites of biological nitrogen fixation, the conversion of dissolved gaseous N_2 to ammonium and amine compounds. The general topic is well reviewed by Carpenter and Capone (2008) and covered here in Chapter 3.

The importance of nitrogen fixation remains to be fully determined. The difficulty is that measurement techniques are still developing; diazotrophs are extremely patchy across the oceans and through time; and the geochemical signatures of nitrogen fixed in the past are complex to calculate, subject to errors, and hard both to sum across the globe and to interpret. There are two measurement techniques for bulk diazotrophy in water samples:

1 It turns out that the *nif* enzyme complex also catalyzes a reaction of the carbon–carbon triple bond in acetylene, producing ethylene. Thus, if some acetylene is put in the headspace over a sample to dissolve, the amount of ethylene generated during an incubation is a measure of the N_2-fixation capacity of the phytoplankton, possibly of the gross rate. Gross because, unlike fixed nitrogen in particulates, none of the ethylene is subsequently lost to metabolism or excretion.

2 Nitrogen gas labeled with $^{15}N_2$ can be introduced into the headspace over a sample and dissolves. Organic matter produced subsequently is labeled with ^{15}N, which can be extracted from filtered particulate matter and measured with mass spectrometry. Many of the data to date are likely underestimates because of the slow transfer of $^{15}N_2$ into solution, so that the $^{15}N_2$ available for uptake is less than has been assumed (Mohr *et al.* 2010, whose improved method remains to be widely applied). These $^{15}N_2$ measures are also assumed to be, in a sense, "net", because some of the fixed nitrogen can be metabolized and released to the environment, and NH_4^+ from

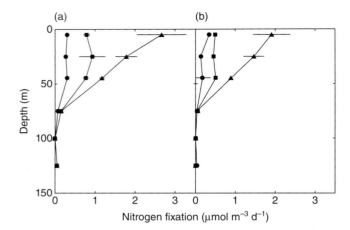

Fig. 11.30 Depth profiles of (a) whole-water and (b) size-fractionated (<10 μm) N_2 fixation rates based on $^{15}N_2$ uptake during November 2004 (●), February 2005 (■), and March 2005 (▲). Error bars are standard errors for means of three replicates. These are likely to be underestimates (Mohr *et al.* 2010), but relative rates are likely to be correct. (After Grabowski *et al.* 2008.)

fixation can simply leach from diazotroph cells. Both of these biases may partly explain why direct rate measurements yield estimated regional rates much smaller than the "geochemical" estimates. However, simply the patchiness of diazotroph abundance and activity would have the same effect.

Rates of diazotrophy peak close to the sea surface (e.g. Grabowski *et al.* 2008; Fig. 11.30) and tail off to zero by 75 m. While fixation depends upon light energy, the timing of the actual reactions varies among species. In *Trichodesmium* it rises during daylight, peaking at noon then decreases again. Some other types of cyanobacterial diazotrophs photosynthesize during daylight and fix nitrogen at night. The factors obviously likely to limit fixation are temperature, irradiance (see Fig. 11.30), phosphate, and iron. Watkins-Brandt *et al.* (2011), contradicting Zehr *et al.* (2007) have shown that small phosphate additions to water samples from the NPSG do increase fixation rates as measured with $^{15}N_2$. The effect would likely be more pronounced in the Sargasso Sea, which has less phosphorus relative to fixed nitrogen. The K_s values for PO_4^{3-} uptake by *Trichodesmium* are low (implying high affinity), as low as 100 nM (Moutin *et al.* 2005) at some ocean sites and for some cultures, but still well above the SRP concentrations widespread in gyre surface layers. A capacity for stripping PO_4^{3-} from organic forms has been repeatedly demonstrated in culture, and alkaline phosphatase serving this function is present in *Trichodesmium* in the field. *Trichodesmium*, particularly in colonies of tangled filaments, has variable density regulated by the interaction of intracellular gas bubbles and carbohydrate loading (e.g. White *et al.* 2006) and can both sink and rise. Possibly, it sinks to the nutricline, takes up phosphate, and returns to better-lighted layers to fix nitrogen and grow. Several studies (e.g. Villareal & Carpenter 2003) have found in some instances that rising colonies have greater phosphorus content than sinking ones. At least *Trichodesmium* can take up ammonium and nitrate, so it might be expected that available fixed nitrogen would suppress fixation. It does, but only to about 30% of maximum rates.

Experiments with field collections (e.g. Reuter 1988) and cultures show that fixation rates in *Trichodesmium* depend on iron concentration, with limitation effects within the range of concentrations observed in the field. Effects on diazotrophy in the field of experimentally increasing iron availability are not well studied, but geochemical arguments suggest that iron availability is key to the global rates and distribution of fixation. The possibility of iron limitation of N_2 fixation leads directly to a partial explanation for the distinction between the Sargasso Sea and NPSG in euphotic zone [N+N]:P ratios: high in the Atlantic, low in the Pacific. The Sargasso is more abundantly supplied with iron from Saharan dust carried westward by trade winds. This is largely depleted by the rains extracted when those winds rise against the Americas, leaving even the eastern Pacific less fertilized. No comparable dust is exported from the Americas, and the eastern Pacific is largely dependent on distant, Asian sources. This enables diazotrophs to utilize phosphate in the Atlantic until it becomes the limiting nutrient, leaving extremely low residual phosphate relative to fixed nitrogen. Intermittent measures of fixation rates may not be particularly high because of the phosphate limitation. In the eastern NPSG, on the other hand, eolian iron is supplied less frequently and in lesser amounts, such that phosphate is not exhausted, while phytoplankton generally reduce fixed nitrogen to very low levels (reducing [N+N]:P).

Several indices of long-term nitrogen fixation effects in different areas have been generated by examination of the discrepancy between the N:P ratio and the canonical Redfield ratio (N:P ≈ 16:1, or a little less). Indices of the differences are:

$$N^* \approx [NO_3^-] - 16[PO_4^{3-}] + 2.79 \ \mu mol \ kg^{-1}$$

(Michaels *et al.* 1996) and

$$P^* \approx [PO_4^{3-}] - [NO_3^-]/16$$

(Deutsch *et al.* 2007).

More sophisticated versions (Gruber and Sarmiento 1997 use an adjusted Redfield ratio; Deutsch *et al.* 2007 do that and subtract the intercept offset from P*) to account for variations of deep ratios from Redfield. Rather oddly, the conclusions about the global patterns of nitrogen fixation, Gruber and Sarmiento using N* and Deutsch using P*, are almost opposite. More strangely, the discrepancy is

not discussed in the later paper, and we leave the whole issue for readers to study in the primary literature. Eventually, these or similar indices may provide a rough evaluation of N_2 fixation in the global nitrogen budget. Both analyses do suggest that fixation by marine diazotrophs is a substantial factor in the global nitrogen cycle, at least of the order of $10^{14} \ g \ N \ yr^{-1}$ – the estimate by Gruber and Sarmiento (1997). Continuing work on evaluating fixation rates using ratios of the stable isotopes of nitrogen in nitrate (e.g. Casciotti *et al.* 2008) may contribute to quantification of global N_2 fixation.

Higher trophic levels

Grazing in the NPSG and similar systems keeps their phytoplankton stocks in almost exact balance between cell division and mortality; effectively, grazers consume all production (not all biomass, all *increase* in biomass) every day. Banse (1995) has made an extended argument for this. He summarized (Fig. 11.31) upper-ocean phytoplankton growth-rate data for oligotrophic mid-ocean areas like the NPSG. Actual phytoplankton growth rates are not strongly affected by nutrient limitation; the very small cells (*Prochlorococcus, Synechococcus,* and picoeukaryotes), are those most capable of obtaining sufficient nutrients from low concentrations. The growth rates range from one to two divisions per day. Banse argues convincingly that loss rates to cell sinking and vertical mixing can amount at most to a few percent of the growth rates, and therefore the apparent steady states which generally pertain must be attributed to cell death, most of it from grazing, although viral lysis is also important. Almost all actual herbivory (eating of phytoplankton) is by protozoans, simply because the dominant phytoplankton are so small. The array of protozoans includes heterotrophic nanoflagellates, larger ciliates, and heterotrophic dinoflagellates, just as in oligotrophic high-latitude habitats. The exact proportions of different groups and species are not well characterized. Copepods, larvaceans, and other mesozooplankton particle feeders are, then, in a strict sense carnivorous, feeding at the third and higher trophic levels, even though most of them are smaller than their high-latitude relatives in the same taxonomic families.

By and large, the mesozooplankton of the NPSG are continuously active, without a seasonal rest phase. Life cycles are not annual or twice-yearly as in subpolar habitats, but are a few weeks or a month in surface layers. Development of reproductive cohorts is not readily followed in these systems. At HOT, collections to 160 m with 200 μm mesh nets demonstrate low amplitude (about twofold) seasonal cycles around an annual mean abundance of approximately $0.7 \ g \ C \ m^{-2}$ in daytime, $1.0 \ g \ C \ m^{-2}$ at night (Sheridan & Landry 2004). In addition, they showed a convincing increase of the overall totals between 1994 and 2002 (Fig. 11.32). Their analysis shows that the change was

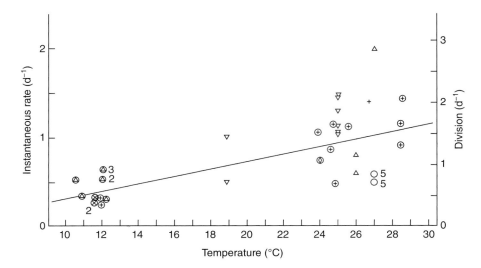

Fig. 11.31 Comparison of phytoplankton growth rates as a function of temperature in several oceanic, oligotrophic systems. Different symbol shapes from different workers. Plain symbols are from nitrate-depleted areas, mostly the NPSG. Circled symbols are from HNLC areas; those below 15°C from the subarctic Pacific. Growth rates obtained in various ways (see original reference); all rates exceed the phytoplankton stock increase rates. Numbers indicate multiple samples. (Simplified from Banse 1995.)

primarily in the non-migrants that remain near the surface during the day, since the night minus day difference in biomass was nearly unchanging. Animals migrating below 160 m are important in carrying carbon and nutrients down out of the euphotic zone by respiring and excreting at depth in daytime. Hannides *et al.* (2009), based on estimates of biomass and oxygen consumption, calculated the amount of migrant-delivered carbon flux at about 15% of the estimates from traps at 150 m, the total being about 2.6 mmol C m^{-2} d^{-1}. The seasonal cycling of both trap flux and zooplankton abundance shows seasonal cycles similar to those of primary production (Fig. 11.24), possibly peaking somewhat earlier in the spring–summer season.

Mesozooplankton in the NPSG are a diverse array despite their low population biomass, with around three-fold more species than are found at higher latitudes. McGowan and Walker (1979), referring primarily to copepods, raised the issue of how so many species can persist together with so few obviously differing niches. The distinctive roles seem to sum up as: (i) eat particles, or (ii) eat each other. As ecologists steeped in the "Law of Competitive Exclusion", our expectation is that the system should simplify to just a few kinds in each role by competitive extinctions. The answer will never be complete, but part of it is that species with similar trophic roles, represented in the data by species in the same genera of copepods, are dispersed vertically (Longhurst 1985 – actually referring to the eastern Pacific equatorial zone; Ambler & Miller 1987 – in the CLIMAX region). Congeners tend to separate into different depth levels.

Like all ocean areas, the subtropical gyres have stocks of nekton in a substantial sequence of trophic levels. Smaller fishes in the epipelagic far from land are dominated by migratory mid-water groups, particularly the Myctophidae. There are some clupeids: sprat, anchovy, and herrings. Those and other epipelagic species, for example the flying fish (Exocoetidae), are much more abundant near islands. Larger, oceanic fish include the so-called dolphin fish or mahi mahi (*Coryphaena hippurus*), blue mackerel (*Scomber australascius*, that reaches 2 kg), sword- and sailfish, several tuna species, and some oceanic sharks. Most of these larger fish are more abundant in the vicinity of islands where topographically driven upwelling increases production; all are subject to active fisheries and most stocks are stressed by those fisheries.

Ommastrephid squids, particularly *Ommastrephes bartrami* in the Pacific at about 70 cm length, are very deep diel vertical migrators that can be attracted at night to lights suspended from ships. A complex assemblage of other squid species is distributed down the water column. Marine mammals are part of the subtropical gyre fauna, although they are much more important in more productive ecosystems, and again greater abundance near islands is typical. Seals, for example the endangered Hawaiian monk seal, pull out to rest on islands. Spinner dolphins (*Stenella longirostris*) are abundant close to shore, and whales – particularly humpback whales – that feed in summer at high latitudes, migrate to subtropical island areas to calve and mate in winter. There are more examples, e.g. beaked whales that are apparently more abundant over seamounts and near the Canaries and Azores than elsewhere, and

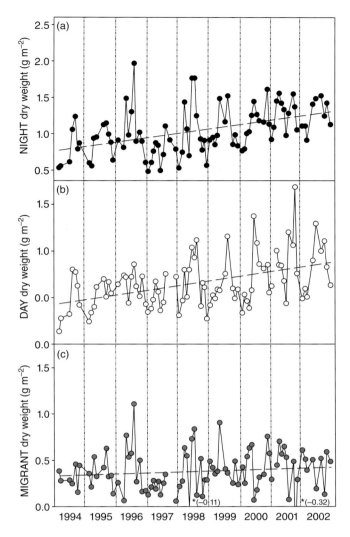

Fig. 11.32 Mesozooplankton dry-weight biomass measured monthly during the HOT program at Station ALOHA from 1994 to 2002. Mean biomass is shown for plankton collected during the (a) night, (b) day, and (c) the biomass of migrating zooplankton (night biomass – day biomass). Evidently, it was primarily the non-migrant biomass that increased. (After Sheridan & Landry 2004.)

Elaborate efforts are under way (e.g. Spitz *et al.* 2001) to model subtropical gyre ecosystem processes. The models do well in representing regional "climatology" (average seasonality) of nutrients and phytoplankton stock, and progress is apparent in predicting general responses of those stocks to variations in mixing, irradiance, temperature, and seasonal timing of events. Models do less well, at least for now, with other components: primary productivity, dissolved and detrital organic matter, and zooplankton. Under development are data-assimilating calculations that seek best-fitting parameters for modestly complex models. The resulting parameters do appear to be better predictors than those selected on a more *ad hoc* basis.

Equatorial biomes

Eastern tropical Pacific

Biomes vary along the equatorial band. We will consider the Pacific region in detail, compare the Pacific and Atlantic regions, and after a section on coastal biomes emphasizing upwelling ecosystems, we will provide a separate section on distinctive features of the Indian Ocean.

Strong trade winds blow east to west from about 20°N to 5°N and from 5°S to 20°S, leaving a band of lighter winds, the doldrums, over the equator. Trade winds are one of the principal manifestations of the global heat-transfer system. They push the ocean westward and the Coriolis effect carries the resulting flow poleward, away from the equator, replaced there by upwelling. The flow then slides under as it butts against the stratified central gyres. This upwelling is rich in major nutrients. In the Pacific, the westward force on the sea surface drives warmed surface water into a "warm pool" (recent terminology) in the west. This generates high evaporation rates and rain, so that a halocline forms above the main thermocline. The warm pool is thus highly stratified by both warming and dilution, and it becomes fully depleted of major nutrients. It closely resembles the central gyres in ecology (Le Borgne *et al.* 2002), and we won't deal with it further.

In the eastern tropical Pacific and extending west past the date line, the elevated nutrients center right on the equator (Fig. 11.33), maximal just west of the Galapagos at 90°W. Typical nitrate concentration in surface water there is $6\,\mu M$, with concentrations decreasing westward ($5\,\mu M$ at 135°W, $3\,\mu M$ at 160°W, and $1\,\mu M$ at 170°E). Fall-off to the north and south is faster, with the southern isopleths more spread than the northern. As the surface current diverges slowly from the upwelling source at the equator, nutrients are assimilated by phytoplankton and moved up the food-chain. There is actually a progression, albeit rather subtle, in the mean trophic level from the equator to higher latitudes. The mesozooplankton species

squid-eating sperm whales that aggregate near northern New Zealand and other islands. All of these mammals do transit long stretches of open ocean, and a few, such as the elephant seals (*Mirounga angoustirostris*) that mate and bear young on the shores of the American West Coast, make extensive migrations seaward to feed, during which they do extremely deep dives. The females make subtropical feeding tours, while the males go toward the Aleutian Islands. Sea turtles, essentially all tropical–subtropical forms, also migrate seaward to great distances from egg-deposition sites on sandy beaches into oceanic waters to feed, primarily on jellyfish. Thus, the system of upper trophic-level links in these waters with the lowest primary-production rates is substantially complex.

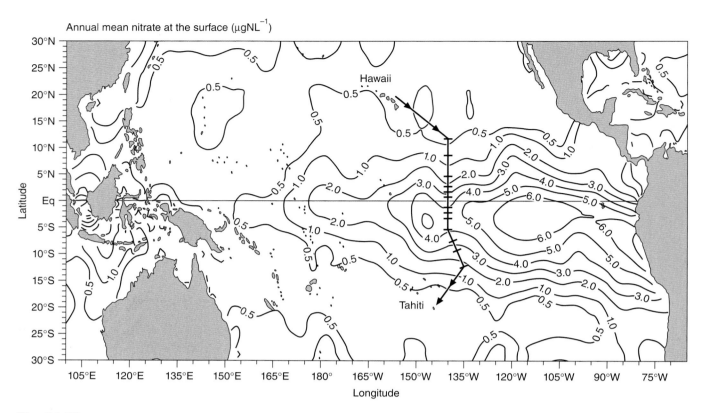

Fig. 11.33 Map of the equatorial Pacific with isopleths of annual mean surface nitrate concentration (μM). Track line shows stations of the US JGOFS equatorial process study. (After Murray *et al.* 1995.)

list closely resembles those of the central gyres, with some deletions and with additions of eastern tropical Pacific endemics. Life-cycle times are short and standing stocks are higher than in central gyres, as are chlorophyll and primary productivity. Primary productivity shows slight surface photoinhibition, reaches a maximum at about 12 m depth, then tails off to zero just below 100 m. At typical surface levels of 0.2 to 0.35 μg Chl liter^{-1}, self-shading is not a major factor. February–March and August–September transects (Fig. 11.34) of primary-production stations across the equator at 140°W (Barber *et al.* 1996) showed a modest difference in vertically integrated primary-production rates, especially from 5°N to 5°S. Barber *et al.* attributed that difference to the presence of El Niño conditions in the northern spring, followed by more normal (colder, richer) conditions in late summer, not to a recurring seasonal difference. The mean measured productivity for that equatorial strip was 1002 mg C m^{-2} day^{-1} (Table 11.5), approximately twice the central gyre average.

This eastern tropical Pacific region is important in the global exchange of carbon between the ocean and atmosphere. Since large volumes of water are brought from depth to the surface and Pacific deep water is highly supersaturated with CO_2, the region is a source of CO_2 to the atmos-phere. Since nutrients also surface, a much larger than average amount of production for the tropics returns a fraction of this carbon to the deep sea. The net flux is nevertheless to the atmosphere. The JGOFS transect work at 140°W included an extended particle trapping study. This gives a near-bottom carbon rain rate of 0.35 mmol C (= 4.2 mg C) m^{-2} day^{-1}, or about 0.5% of surface primary production. The rest is either grazed and respired near the surface, with nutrient recycling contributing to total production, or respired as organic matter sinks through the ocean. A study with transmissometers by Walsh *et al.* (1995) shows that most of the production is consumed every day, as in subtropical areas. The transmissometer gives an estimate of the beam extinction coefficient for a laser diode (laser-pointer) beam through ambient water but shielded from ambient sunlight. From calibration studies with filters, this can be converted to "particle load". Time-series of profile data taken every three hours (Fig. 11.35) show early-morning minima and evening maxima that recur over and over. Primary production and consumption of particles nearly match on a daily basis. While particle load may differ between cruises months apart, the daily cycles almost balance. Differences which could produce a longer-term change in particle standing stocks are invisible in the statis-

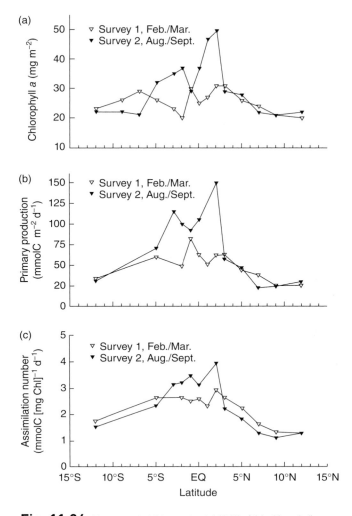

Fig. 11.34 Trans-equatorial transects at 140°W of (a) chlorophyll *a*, (b) primary production integrated to the 0.1% light level, and (c) assimilation number from two cruises in different seasons. Productivity is elevated just along the equator by nutrients supplied by upwelling supporting higher productive biomass. (After Barber *et al.* 1996.)

tical noise of the observations. That is true all over the world's oceans. Except in strong spring blooms, almost all phytoplankton photosynthate is consumed on the day that it is produced.

On a ~3- to 6-year cycle, weakening of the trade winds reduces equatorial upwelling and the warm pool shifts eastward, deepening the thermocline and nutricline in the eastern tropical Pacific. These events, termed El Niños,

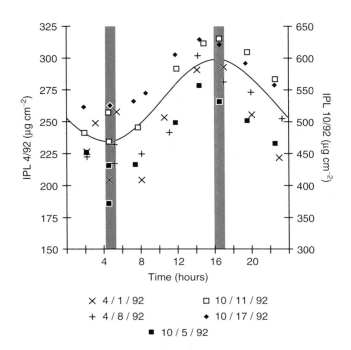

Fig. 11.35 Repeated daily cycles of euphotic-zone particle abundance measured by transmissometry on five different days. (After Walsh *et al.* 1995.)

Table 11.5 Chlorophyll and primary production between 2°N and 2°S integrated to the 0.1% light level in the eastern equatorial Pacific upwelling zone. Data for 1992 from Barber *et al.* (1996) and data for 2004–2005 from Balch *et al.* (2011).

DATE	PRIMARY PRODUCTION (mg C m^{-2} d^{-1})	CHLOROPHYLL (mg m^{-2})	ASSIMILATION NUMBER (mg C [mg Chl]$^{-1}$ d^{-1})
Feb–Mar 1992	720 ± 96	25 ± 1	29 ± 4
Mar–Apr 1992	1,080 ± 36	29 ± 1	37 ± 2
Aug–Sept 1992	1,212 ± 96	32 ± 2	41 ± 2
October 1992	1,548 ± 72	33 ± 2	47 ± 2
Dec 2004	744 ± 108	26 ± 3	29 ± 4
Sept 2005	708 ± 341	32 ± 5	23 ± 10

reduce nutrient supply to the surface from the upwelling that does occur despite the reduced trade-wind activity. Strutton and Chavez (2000) compared nutrient levels, phytoplankton stocks, and productivity between the 1997–1998 El Niño period and normal conditions in late 1998. At the peak of the El Niño, nitrate was less than or equal to 1 μM all along the eastern equatorial zone, a sharp reduction from the 3–6 μM usually present for 3–6° of latitude to either side of the equator out to 155°W. The phytoplankton size distribution shifted sharply to very small cells; chlorophyll concentration was about half of usual, non-El Niño conditions; and daily primary production was equally reduced. This tendency to strongly reduced production extends north and south along the coast of the Americas to the same distance as warming extends in any given El Niño. Thus, coastal plankton and fish populations are exposed to warming, which raises metabolism and, thus, food demand, at the same time that food production drops. Impacts of El Niños on equatorial marine biota are also strong. Those involving plankton, which have short life spans and high population growth potential, come and go rapidly with the events. Effects on larger, longer-lived animals, particularly forage fish and seabirds, involve substantial mortality and long population-recovery times.

Mesozooplankton abundance along the tropical belt drops sharply as well. Dessier and Donguy (1987) developed a 1979–1984 time-series of copepod abundance estimates in the equatorial zone by filtering the nightly filling of a cruise-ship swimming pool along a direct route between Panama City and Tahiti. Serial abundance estimates for the copepods *Clausocalanus* species and *Euchaeta rimana* from the equator to 2°N and for *Calanus minor* from the equator to 2°S (Fig. 11.36) showed both strong (and somewhat different) seasonality and dramatic reduction during the 1982–1983 El Niño. A long list of other species also nearly vanished in that event. At least the stocks of herbivorous species rebounded immediately in 1984; *E. rimana*, a predator, had another reduced seasonal peak in that year. Dessier and Donguy propose that the seasonality is driven by variations in equatorial upwelling strength and production. They note that peaks of herbivorous species (*Clausocalanus* spp., *C. minor*) alternate with peaks of carnivores (*E. rimana*), implying a lag along the food-chain in response to the seasonal variation. Seasonal advective shifts of the strong, trans-equatorial abundance gradients are also a possible explanation.

Zonal gradients of the major phytoplankton nutrients apparent across the cold-tongue of upwelled water are produced by the shoaling of the thermocline in the eastern equatorial Pacific. Iron is supplied from the Equatorial Undercurrent, and concentrations are high in the western Pacific closer to the source water (Slemons *et al.* 2010; Kaupp *et al.* 2011). The shears between the westward-flowing South Equatorial Current, the North and South Countercurrents and the eastward-flowing Equatorial

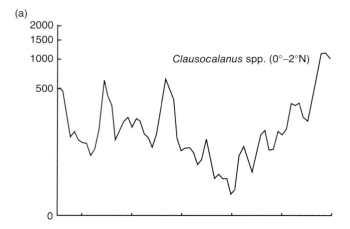

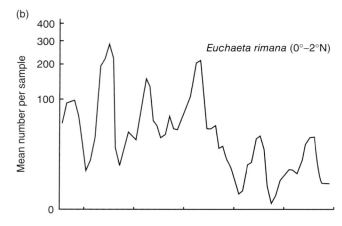

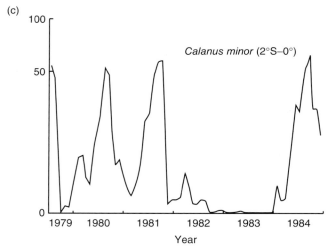

Fig. 11.36 Time-series of abundance estimates for three copepod species sampled along the equator from a passenger liner on a regular schedule. (After Dessier & Donguy 1987.)

Undercurrent produce tropical instability waves (TIWs) that propagate from west to east at 50 km per day. These waves distort the upwelling tongue in a wave-like pattern visible in satellite images of sea-surface temperature (Plate 11.4). Animations of modeled TIWs are posted at: http://

www.atmos.washington.edu/~robwood/images/1999_2000_ct15.avi.

The three-dimensional circulation of TIWs is best visualized as a vortex in the water column (Kennan & Flament 2000). Downwelling and northern transport occur at the west edge, with upwelling and westward transport at the northern edge. Strong TIWs dilute nutrients and chlorophyll by advecting nutrient-depleted water toward the equator, whereas weak TIWs enhance local upwelling and result in higher productivity and biomass accumulation than in the core of upwelled water (Evans et al. 2009).

Waters in the eastern equatorial Pacific resemble the subarctic Pacific in that they are never depleted of nitrate, and phytoplankton are always present in modest, non-bloom quantities. Together with the subarctic Pacific and much of the Southern Ocean, it is an HNLC region. As in the cold HNLC regions, iron limitation constrains the phytoplankton to small size; protozoan grazing regulates the phytoplankton stock to near constancy; and surface-layer recycling of fixed nitrogen as ammonia suppresses nitrate utilization so that it is never completely used up. These ideas about the cascade of iron effects have been tested in two so-called IRONEX studies conducted to the southeast of the Galapagos (on the equator west of Ecuador). The results are described by Martin et al. (1994) and Coale et al. (1996). The second experiment benefited from the experience of the first. Coale et al. added 225 kg of iron (FeIII) as acidic (pH 2.0) iron sulfate solution to the prop wash of their ship as it traced out a grid of 72 km^2 around a central buoy. This raised the iron concentration in the water from less than 0.2 nM to greater than 1.0 nM. The enriched "patch" could be followed by analysis of seawater for minute quantities of the inert chemical tracer SF$_6$ (sulfur hexafluoride) added at the same time. Iron concentration in the patch was sustained for over a week by subsequent 110 kg additions after 3 and 7 days. The result is most clearly shown in the color plots of the patch (Plate 11.5). They rapidly got a big bloom of phytoplankton, eventually reaching 3 µg Chl liter^{-1}, a level never seen under natural conditions in that region. It was large cells, particularly diatoms, which showed the largest increases (85-fold). There was corresponding nitrate and carbonate draw-down.

The IRONEX-I team (Martin et al. 1994) also compared phytoplankton processes upstream (east) and downstream (west) of the Galapagos, showing differences corresponding to a plume of phytoplankton that is usually present downstream of the islands. They claim that this is attributable to iron washing off the Galapagos platform. Probably that's correct. Results of the entire global suite of iron-addition studies are summarized by Boyd et al. (2007). Chlorophyll increases by two- to 25-fold. Greater phytoplankton responses are observed in shallow mixed layers than in deep. More rapid responses are seen in warmer waters. Dominant phytoplankton species change to medium sizes

and more diatoms. Bacteria increase by two- to 15-fold. The duration of experiments is usually too short to see a mesozooplankton response, but in two experiments (IronEx II and SEEDS I) where copepods were abundant, they played a role in controlling the blooms (Boyd et al. 2007). To date, no response of fish to iron additions has been observed. Note that in regard to all of these scenarios that they depend upon more than the phytoplankton and their responses to light, nutrients, and mixing. Protists and animals grazing upon the phytoplankton also play a large part, a point to be repeated shortly.

Biological processes in the eastern equatorial Pacific studied intensively during the EqPAC Program during 1992 (Murray et al. 1995, 1997) showed that the HNLC character of this region results in part from low availability of iron interacting with grazing pressure from micro- and mesozooplankton. Upwelled water has a greater NO$_3^-$ to Si(OH)$_4$ ratio than optimal for diatoms, and Dugdale et al. (2007) suggested that silicate limitation may play a role in restricting levels of primary production. However, it has not prevented diatom blooms after in situ iron additions. The Equatorial Biodiversity (EB) Program (Nelson & Landry 2011) conducted experimental studies in 2004–2005 to clarify the relative roles of grazing, iron-limitation, and silicate-limitation in controlling primary production and the phytoplankton community structure. The phytoplankton community was dominated by small cells that were grazed at the same rate as their growth (Landry et al. 2011). A majority (70%) of the grazing pressure was from microzooplankton. These heterotrophic protists served as a major food source for the mesozooplankton. Prochlorococcus was the only phytoplankton species that showed an in situ increase in abundance with increasing iron concentrations. In 5–7-day shipboard microcosm experiments, additions of iron led to an increase in rare large diatoms and to depletion of NO$_3^-$ and Si(OH)$_4$. Additions of silicate resulted in increased biogenic silica production, but did not result in the depletion of major nutrients (Brzezinski et al. 2011).

Mesozooplankton standing stocks in the eastern tropical Pacific appear to have increased by two-fold between 1992 and 2004–2005; however, different nets likely to capture different subsets of zooplankton in respect to size and activity were used for the two sampling programs (Décima et al. 2011). Rates of primary production and the assimilation index were similar during the two sampling periods (Table 11.5), so it is not clear how a doubling of mesozooplankton stocks could have been supported.

Despite its HNLC character, relatively high productivity in the eastern equatorial Pacific is evident in standing stocks of upper trophic levels and in sinking fluxes of organic material. Planktivorous seabirds (Oceanodroma: Leach's and Galapagos storm petrels) accumulate at the equatorial front (the boundary between the equatorial cold tongue and the warm subtropical water to the north) in densities

Table 11.6 Changes in oceanographic conditions and seabird density (birds per km²) when passing across the equatorial front from the South Equatorial Current (SEC) to the North Equatorial Countercurrent (NECC) on 11 October 1998 at 3°, 34′N, 117°, 37′W. (Table from Ballance *et al.* 2006.)

	SEC	EQUATORIAL FRONT	NECC
Sea surface temperature (°C)	23.7	—	25.8
Sea surface salinity (PSS)	34.20	—	34.00
Thermocline depth (m)	11	—	65
Seabird density	0.18 ± 0.04	8.18 ± 3.40	0.38 ± 0.13
Planktivorous seabird density	0.06 ± 0.04	7.27± 1.58	0.12 ± 0.03
Piscivorous seabird density	0.12 ± 0.05	0.90 ± 0.67	0.26 ± 0.05

more than an order of magnitude greater than seen on either side in the countercurrent flows to the north and south of the equator (Table 11.6). An important feature in the eastern tropical Pacific is the tuna–dolphin–seabird assemblage comprising yellowfin tuna (*Thunnus albacares*), spotted and spinner dolphins (*Stenella attenuata* and *S. longirostris*), and petrels (*Pterodromo* spp.) (Ballance *et al.* 2006). The shallow thermocline there and equatorial upwelling may contribute to the geographical location of this assemblage, which forms the basis of the world's largest yellowfin tuna fisheries.

Buesseler *et al.* (1995) estimated high levels of POC flux (3–5 mmol C m⁻² d⁻¹) to depths of 100 m along the equatorial band, and these were about twice as high as seen north and south of the equator. Honjo *et al.* (1995) measured POC fluxes to the deep ocean (1000–3000 m), and these ranged from 0.2 to 1.0 mmol C m⁻² d⁻¹. Although Honjo *et al.* (1995) concluded that deep fluxes in the equatorial Pacific are less than in the subarctic Pacific, Honjo *et al.* (2008) report similar rates: 158–194 mmol C m⁻² yr⁻¹ in the equatorial Pacific and 163 mmol C m⁻² yr⁻¹ at Station P (50°N, 145°W).

Atlantic equatorial upwelling

The main circulation in the equatorial Atlantic is also driven by the trade winds, and results in divergence of the surface water. Tropical instability waves produce undulations in the tongue of upwelled water visible in remotely sensed temperature measurements (Plate 11.4). The TIWs are most apparent from June to October (five months), as opposed to the nine-month seasonal duration in the Pacific. Integrated studies of the Atlantic equatorial upwelling region were conducted as part of the Atlantic Meridional Transect (AMT) Programme from 1995 to 2005 (Robinson *et al.* 2006). The field sampling was conducted twice per

year and covered 13,500 km between 50°N and 52°S. We compare the physical, chemical, and biological characteristics of the Pacific and Atlantic equatorial systems in Table 11.7. Macronutrients (NO₃⁻, PO₄³⁻, and Si(OH)₄) are higher in the Pacific, while iron is much higher in the Atlantic. Although >1 μM nitrate is sometimes found in the westward-flowing Atlantic equatorial "cold tongue", during the winter (weak upwelling season) nitrate in the surface water typically decreases to <0.1 μM, and nitrogen limitation of phytoplankton growth is likely. Surface chlorophyll is higher in the equatorial Atlantic, but vertically integrated chlorophyll is similar in the two oceans. Primary productivity and the assimilation ratio also appear to be similar, but they both show significant temporal and spatial variability that may not be adequately sampled for precise comparisons.

Phytoplankton in both equatorial Atlantic and Pacific are dominated by picoplankton, and heterotrophic protists are the dominant grazers. Mesozooplankton standing stocks are also similar. Export of particulate organic carbon is approximately the same, but variability is higher in the Atlantic, possibly due to spatial or seasonal differences.

In a series of eight AMT cruises, Tyrrell *et al.* (2003) found high concentrations of *Trichodesmium* spp. between 0° and 15°N at 20°W. The concentrations of these nitrogen-fixing cyanobacteria were correlated with the presence of a shallow mixed layer and high estimated deposition of iron to the surface ocean. "Tricho" abundance was not correlated with temperature, nitrate concentrations, or total dissolved iron in the seawater. The region where *Trichodesmium* is abundant coincides with an area of enhanced general phytoplankton growth. Tyrrell *et al.* suggest that *Trichodesmium* fixes N₂ and releases DON that may enhance growth of other phytoplankton.

Menkes *et al.* (2002) examined the distribution of nutrients, plankton, and nekton in the crest of a tropical in-

Table 11.7 Comparison of eastern equatorial Pacific and equatorial Atlantic upwelling regions.

	PACIFIC	ATLANTIC	(PACIFIC REF., ATLANTIC REF.)
Nitrate (μM)	5–10	>1	(Strutton *et al.* 2011; Pérez *et al.* 2005)
Phosphate (μM)	0.3–3.0	>0.2	(Strutton *et al.* 2011; Pérez *et al.* 2005)
Silicate (μM)	3–8	>1.5	(Strutton *et al.* 2011; Pérez *et al.* 2005)
Iron (nM)	<0.03–0.2	1–2	(Kaupp *et al.* 2011; Tyrrell *et al.* 2003)
Surface Chl (mg m^{-3})	~0.2	>0.5	(Balch *et al.* 2011; Pérez *et al.* 2005)
Integrated Chl (mg m^{-2})	29	32	(see Table 11.5; Marañon *et al.* 2000)
Primary production (mg C m^{-2} d^{-1})	1,002 $\pm$ 341	995 $\pm$ 171	(see Table 11.5; Pérez *et al.* 2005)
Assimilation number (mg C [mg Chl]$^{-1}$ d^{-1})	34 $\pm$ 9	45 $\pm$ 3	(see Table 11.5; Pérez *et al.* 2005)
Bacteria (cells ml^{-1})	8–9 $\times$ 10^5	>10^6	(Taylor *et al.* 2011)
Heterotrophic protists (mg C m^{-2})	200–500		(Taylor *et al.* 2011)
64–200 μm microzooplankton (mg C m^{-2})	4–6	13.9	(Roman *et al.* 1995; Calbet *et al.* 2009)
Mesozooplankton (mg C m^{-2})	780	903	(Décima *et al.* 2011; Calbet *et al.* 2009
Sinking flux (mg C m^{-2} yr^{-1}) (to 2000 m)	1,284 $\pm$ 396	2,256 $\pm$ 1,392	(Honjo *et al.* 1995; Honjo *et al.* 2008)

stability wave in the Atlantic Ocean (Plate 11.6). Maxima within the wave crests were relatively high: chlorophyll 0.8–1 mg m^{-3}, net primary production 1500 g C m^{-2} d^{-1}, zooplankton stocks 40 mg dry wt. m^{-3}, and abundant micronekton consisting of small pelagic fishes (predominantly the deep-migrating *Vinciguerria nimbaria*). The data indicate successively higher trophic levels extending northward downstream away from the equator (Menkes *et al.* 2002). Such mesoscale features appear to be important, in enhancing production at multiple levels, and may support the tuna fishery in the equatorial Atlantic (Lebourges-Dhaussy *et al.* 2000).

Coastal biomes and coastal upwelling ecosystems

Coastal waters generally are more nutrient-rich and more productive than oceanic ecosystems. Inshore over shelves, mixing by tidal currents (of velocity u_s m s^{-1}) stirs the water column to the bottom over depths as great as h m, calculated from $h/u_s^3 > \sim 80$. Maps of that criterion around the British Isles and along the shelf of the northeastern USA correlate with the inshore limit of spring–summer stratification in waters to seaward. The phytoplankton inside that boundary are usually diatom dominated, although silicate can become limiting in summer. Beyond that boundary, diatoms are less important, except during spring blooms.

Coastal seas are less frequently, but sometimes, iron-limited, especially where $h/u_s^3 > 80$, but major nutrients can be removed to very low concentrations both in the tropics and seasonally in high latitudes.

Eastern boundary current systems (EBCS)

Coasts with an equatorward component in the prevailing alongshore winds are located adjacent to eastern boundary currents, and coastal upwelling is generated next to these coasts by seaward acceleration of the surface layer due to the Coriolis effect. After modest reduction in sea-level height, the offshore flow is balanced by shoreward and rising flow in deeper layers. This carries nutrient-laden, relatively cold water into the euphotic zone. After an initial period of incubation, phytoplankton blooms develop and consume the nutrients. Depletion occurs at various distances offshore. The four major EBCS are the California, Peru/Humboldt, Canary, and Benguela Currents (Fig. 11.37). They differ in levels of primary production (as estimated from ocean color) and fish production (Table 11.8). Extensive (but still incomplete) comparisons are reported by Mackas *et al.* (2005), Fréon *et al.* (2009), and Quiñones *et al.* (2010). There are distinct spatial subregions within each of the four EBCS where shelf width, geographical features (e.g. capes), and differences in river inputs have significant effects on nutrient supply, strength of upwelling, retention of plankton, and fish production.

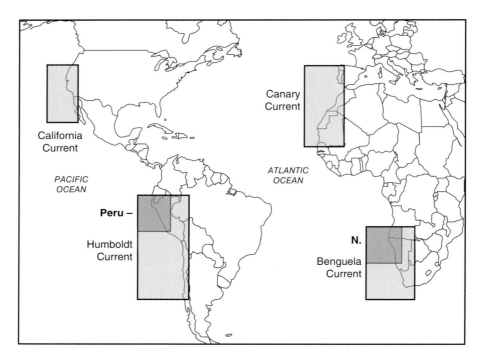

Fig. 11.37 Location (shaded rectangles) of major regional eastern ocean upwelling systems. Darker-shaded rectangles denote the northern subregions of the Peru/Humboldt and Benguela currents. (After Bakun & Weeks 2008.)

Table 11.8 Comparison of coastal upwelling systems.

	CALIFORNIA	PERU/HUMBOLDT	CANARY	BENGUELA	REFERENCE
Nitrate at 60 m (μM)	14.9	16.8	19	16.9	Chavez and Messié (2009)
Chl (mg m^{-3})	1.5	2.4	4.3	3.1	Chavez and Messié (2009)
Primary production (g C m^{-2} yr^{-1})	479	855	1213	976	Chavez and Messié (2009)
Primary production (g C m^{-2} yr^{-1})	361	796	624	909	Carr and Kearns (2003)
Primary production (g C m^{-2} yr^{-1})	345	500	300	450	Jahnke (2010)
Primary production (10^{12} g C yr^{-1})	713	665	816	382	Jahnke (2010)
Zooplankton (g C m^{-2})	2.5	3.34	3.16	2.83	Huggett et al. (2009)
Small pelagic fish (10^3 tons yr^{-1})	479	9210	1292	547	Fréon et al. (2009)
Total fish catch (10^3 tons yr^{-1})	1,278	12,021	2232	1308	Fréon et al. (2009)
Deposition rate (g C m^{-2} yr^{-1})	6.1	7	5.2	6.1	Jahnke (2010)
Total deposition (10^{12} g C yr^{-1})	15.4	12.0	8.3	5.8	Jahnke (2010)

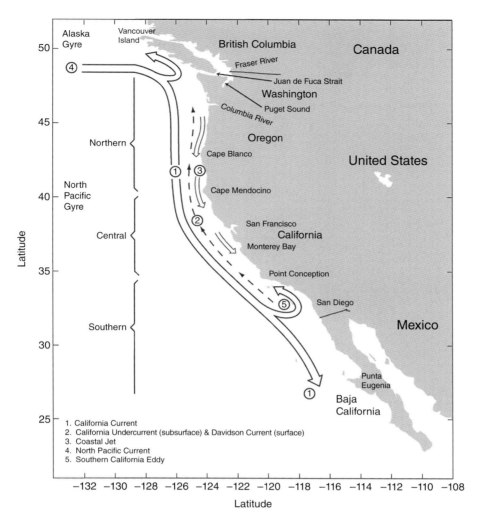

Fig. 11.38 Map of the California Current System. Major regions, currents, and geographical features are shown. The California Current derives from both the North Pacific Current to the north, and the coastal jet to the east. (After Checkley & Barth 2009.)

Ekman transport of surface water offshore in the California Current System (CCS) (Fig. 11.38) leads to a cross-shelf pressure gradient producing a coastal upwelling jet 5–30 km from shore (Fig. 11.39) and leads to rise of the deeper, nutrient-rich, halocline water to the surface inshore of the coastal jet. In addition to this fast vertical coastal upwelling, offshore wind-stress curl leads to slower vertical transport (Ekman pumping) farther offshore (Fig. 11.39). Chavez & Messié (2009) estimated that Ekman transport provides an average of 69–79% of the upwelling across the four EBCS. However, the rate of Ekman pumping is likely to be more significant in some subregions, e.g. wind-stress curl contributes 33% of the upward flux in the central CCS (Dever *et al.* 2006) but 60–80% in its southern subregions (Rykaczewski & Checkley 2008). Turbulent fluxes provide a third mechanism for nutrient input. By combining high-resolution measurements of turbulence and nutrient gradi-

ents, Hales *et al.* (2005) showed that, offshore of the 30 m isobath in the northern CCS, turbulent fluxes lead to cross-isopycnal mixing of nutrients and provide nutrients at 25% of the rate from coastal upwelling. Turbulent fluxes may be even higher inshore. The importance of this nutrient-supply mechanism in other EBCS is not known.

Nitrate appears to be the most limiting nutrient in all four EBCS. In the CCS, surface nitrate ranges from 2–30 µM over the inner and outer shelves (Corwith & Wheeler 2002). Offshore nitrate concentrations are <0.1 µM. Iron can also limit primary production in EBCS subregions. In the CCS iron is replete off the coasts of Washington and Oregon, presumably due to wide shelves and significant river inputs (Chase *et al.* 2007), but is limiting beyond the 200 m isobath in the central region (Kudela *et al.* 2008).

Carr and Kearns (2003) combined a hydrographic and nutrient climatology with primary production modeled

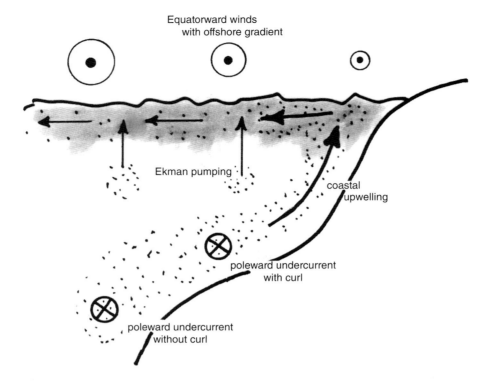

Fig. 11.39 Coastal upwelling and wind-stress curl upwelling mechanisms. Upwelling moves nutrients (dots) into the illuminated sea surface, promoting phytoplankton growth (shading). Equatorward along-shore wind stress generates "Ekman" drift seaward that is replaced by upwelling water from depth. That is "coastal" upwelling. Wind-stress curl, i.e. a gradient in wind velocity (represented by progressively larger arrowheads seaward), generates additional upward Ekman pumping (albeit at lower vertical velocities) offshore. With substantial curl, the poleward undercurrent moves more inshore and shallower, providing relatively more nutrients. (Modified from Albert *et al.* 2010.)

from ocean color to compare patterns of upwelling and biological responses in the four EBCS. The biomass sustained by available nutrients in the Atlantic EBCS was twice as large as that of the Pacific EBCS, presumably due to availability of iron, to biomass retention, or to differences in community structure. Messié *et al.* (2009) expanded this comparison using surface wind data from satellites and vertical distributions of nutrients to estimate nitrate supply from coastal Ekman transport relative to offshore Ekman pumping. Their estimates were that the Peru/Humboldt, Canary, and Benguela EBCS have similar levels of nitrate supply and new production, while California has only 60% of the other three. While estimates of primary production in these areas vary, coastal upwelling systems overall comprise only 0.3% of the ocean's surface area but contribute 2% of global ocean primary production.

Phytoplankton communities can be diverse in EBCS, but blooms are typically dominated by large cells, especially chain-forming diatoms. During the upwelling season, chlorophyll levels range from 1 to 10 mg Chl m^{-3}, and most of the primary production is in the upper 10–20 m of the water column. Small phytoplankton (<10 μm) can be abundant, but at chlorophyll levels above ~2 mg m^{-3}, large cells account for 60–90% of the phytoplankton biomass (Fig. 11.40). Initially, EBCS were thought to have simple and

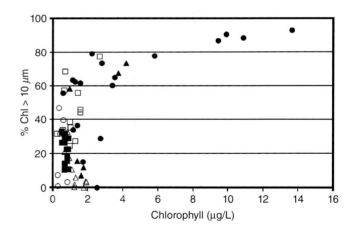

Fig. 11.40 Percent chlorophyll > 10 μm plotted as a function of total chlorophyll. Filled symbols are post-El Niño cruises, open symbols are El Niño cruises. Circles = shelf stations, triangles = slope stations, and squares = off-shelf stations. (After Corwith & Wheeler 2002.)

short food-chains: primarily diatoms grazed by copepods eaten by small pelagic fish. However, microzooplankton (small heterotrophic protists) are abundant and can account for a large portion of the grazing on both small and large phytoplankton (Sherr & Sherr 2007, 2009). Although

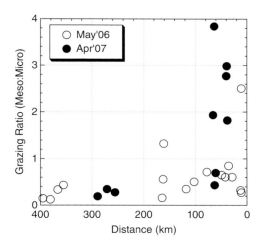

Fig. 11.41 Ratios of depth-integrated grazing rates of mesozooplankton relative to distance offshore for drifter experiments in May 2006 and April 2007. (After Landry *et al.* 2009.)

microzooplankton can account for ~60% of the total grazing pressure, the herbivory of mesozooplankton (especially copepods) is at times more important in nearshore waters (Fig. 11.41). Microzooplankton are an important component of the mesozooplankton diet, just as in oceanic biomes. The recurring emphasis on copepods in upwelling ecosystems may be justified, as both large species (particularly *Calanus* spp.) and small species (of *Acartia*, *Pseudocalanus*, *Paracalanus*, *Oithona*, *Oncaea*, and more) are consistently present and active, often dominating net samples. However, euphausiids can be very abundant, especially in the outer reaches of upwelling ecosystems. In the CCS, there are two dominant euphausiids. *Thysanoessa spinifera* is restricted to shelf waters, while *Euphausia pacifica* resides chiefly in slope waters (Feinberg & Peterson 2003).

Small pelagic fish, especially anchovy and sardines, are abundant in EBCS. Sardines tend to reside offshore and feed on phytoplankton and small zooplankton, whereas anchovy are more abundant inshore and feed on larger plankton. Although the four EBCS show only a two-fold range in primary production, fish production is more than 10-fold greater in the Peru/Humboldt system than in the other three. Most of this is due to the enormous stock of anchoveta (*Engraulis ringens*) off Peru. Bakun and Weeks (2008) attribute that high fish productivity to the strength of nutrient input by upwelling and the extended surface residence time of upwelled water to allow phytoplankton growth. They also hypothesize that periodic El Niño perturbations may prevent the establishment of populations of slow-growing, predatory fish.

Jahnke (2010) compared sedimentary deposition rates for organic carbon in the four EBCS (along with all other continental margins). Deposition rates ($g\,C\,m^{-2}\,yr^{-1}$) were similar for all four regions, but the total deposition ranged from 6 to $15 \times 10^{12}\,g\,C\,yr^{-1}$ (Table 11.8) because of differences in the surface area of each margin (Canary > California > Peru/Humboldt > Benguela). This amount of deposition accounts for 1–2% of EBCS primary production. Overall, the EBCS comprise 16% of total continental margin surface area and account for 22% of global organic carbon deposition on continental margins.

The Indian Ocean

Indian Ocean – the tropics and subtropics

The Indian Ocean is massively blocked on its northern side by the Asian continent, so it has no northern temperate or arctic zones. Moreover, the two very large embayments at the northern side are dramatically affected by continental influence, but in different ways. The Arabian Sea, which is deep with moderately narrow shelves, is surrounded by arid lands to the west and north: Somalia, the southeast face of the Arabian Peninsula, and Pakistan. The eastern side borders western India. The north end has two evaporation basins attached to it by narrow straits, the Red Sea and the Persian Gulf. Like the Mediterranean, these are "negative estuaries" or evaporation basins, with surface water flowing in at the mouths and very saline water flowing out over sills and then sinking into middle depths. The surface over a modest area is diluted by the Indus River outflow at the very north end and by a suite of smaller rivers draining western India. The Bay of Bengal also has narrow shelves and a deep center basin, but has neither desert lands surrounding nor evaporation basins drawing from and draining into it. Two great rivers, the Ganges and the Brahmaputra, freshen the surface out to the open Indian Ocean south of Sri Lanka. Much of the coastal zone is *mangal*, i.e. saline sedimentary areas supporting mangrove forests.

Arabian Sea

Monsoons are the typifying feature of Arabian Sea seasonality, and in fact of the seasonal cycling of the entire Indian Ocean. In winter, moderate winds blow from northeast to southwest and accelerate surface waters in that direction, with the Coriolis effect shifting the flow toward the Arabian coast and sustaining downwelling. Thus, nutrient-depleted waters are carried into the gulf and support only low productivity. In the northern spring, the winds come around to the southwest, blowing both along the Somali–Arabian coast and across the open gulf toward the Himalayas. High evaporation into the initially arid desert air moistens it, and then cooling by uplift over the Indian Peninsula condenses the water, producing a prolonged rainy season on the eastern side. The spring-summer southwest monsoon is very powerful with winds of 40 knots sustained for months that produce waves >10 m and drive the Somali Current along the coast of the Horn of Africa at the world's greatest surface velocity, to 7 knots. It also produces strong coastal upwelling offshore of Somalia, Yemen, and Oman, with prolonged intervals of abundant stocks of large phytoplankton. In addition, the portion of this wind blowing straight across the gulf from the tip of Somalia toward the Himalayas, called the Finlater Jet, has strong wind-stress curl on both sides. To the right there is Ekman pumping, moving nutrients up to support a deep-water bloom 500 km wide and months in duration. To the left, there is Ekman pumping which moves water down, partially separating this "Finlater bloom" from the coastal bloom. The bloom across the northern Arabian Sea can sustain chlorophyll levels in excess of $20 \, \text{mg m}^{-3}$ for months and primary-production rates in excess of $1 \, \text{g C m}^{-2} \text{d}^{-1}$. The combination of a very long seasonal phytoplankton bloom and the countervailing directions of deep and surface flows (toward and away from the continent) creates a level of minimum motion, through which passes copious, oxygen-consuming organic matter exported from the surface. The layer from ~150 to 1000 m is persistently anoxic.

This Arabian Sea anoxic layer does, however, serve as daytime habitat for an abundant fauna of mid-water fish (particularly the myctophids *Benthosema pterotum* and *Diaphus arabica*) preyed upon in the layer (and at night at the surface) by abundant stenoteuthid squid (*Sthenoteuthis oualaniensis*). There are also euphausiids (*Euphausia sibogae*) and shrimp with similar lifestyles. Food is acquired and oxygen debt is paid off at night by migration into surface layers. *Benthosema pterotum* is believed to be extremely abundant, with acoustic studies indicating a stock of 100 Mt (megatonnes) (Gjøsæter 1984).

During the upwelling season, the western Arabian Sea at sites inshore of ~400 km is one of several tropical, Indo-Atlantic areas supporting abundant populations of the large copepod *Calanoides carinatus* (Smith *et al.* 1998). It is also found during the annual, trades-driven upwelling season along the northern Gulf of Guinea coast and occurs less prominently in upwelling intervals off West Australia and Brazil. It is not found in the Pacific. Just as the southwest monsoon starts, it appears in surface layers, runs through multiple and rapid life cycles during the whole season, and then the fifth copepodites store copious lipid and descend to depths offshore below the anoxic layer. This stage is a principal food source for deep-living predators in the Indian Ocean. The rest of the mesozooplankton are a fairly standard set of tropical–subtropical species.

During the winter monsoon, the Arabian Sea becomes a prominent site for *Trichodesmium* blooms and certainly for nitrogen fixation (Wajih & Naqvi 2008). Dust off the deserts of Pakistan provides iron to support this produc-tion, and enhanced denitrification (NO_3^- to N_2) in the relatively shallow oxygen-minimum zone leaves an excess of phosphate in upwelled and upwardly mixed waters. Together that makes ideal conditions for diazotrophy. The blooms become sufficient to color the ocean red in large patches. Similar blooms occur in the Red Sea and are reputed to be the source of its name.

Bay of Bengal

This huge gulf also has a monsoonal circulation, anticyclonic (northward along the Indian coast) from January to October, cyclonic in fall, but much weaker. It is subject to severe cyclones (equivalent to hurricanes) that cause flooding in low-lying sectors of Bangledesh. The whole gulf is strongly stratified by the great freshwater influx (a salinity gradient of several PSS above 20 m from 20° to 16°N, deepening slowly southward), keeping nutrients at depth. The river plumes carry a great deal of terrestrial turbidity, making a very shallow euphotic zone. Together these make for very low primary productivity, mostly about 200 mg $\text{C m}^{-2} \text{d}^{-1}$ year round (PrasannaKumar *et al.* 2006). Higher-trophic-level biomasses, at least mesozooplankton, tend to be surprisingly great, 0.2 to >2 g C m^{-2}, given the low primary production. PrasannaKumar *et al.* attribute that to high bacterial abundance and associated high levels of microheterotrophs. It is likely that the DOC source to support the bacterial stocks is terrestrial.

Indian equatorial zone

Circulation along the Indian Ocean equatorial belt is markedly different from either the Atlantic or Pacific. The trades blow in their usual east-to-west trajectory well to the south of the equator, albeit with strong seasonal variability. Between Sri Lanka (6°N) and the equator there is monsoonal oscillation of winds from both Africa and Australia. These push surface waters both east and west. Headed east they are focused on the equator and accelerated by Kelvin waves. Headed west, they generate Rossby waves. Both wave modes reflect off the land masses at the ends of the long run, creating a sloshing back and forth at remarkable velocities, with alternating periods at some longitudes of a half year or less (Fig. 11.42). This keeps the same water warming under the sun; suppresses most equatorial upwelling; and prevents elevation of water at the western end from driving an undercurrent centered on the equator. The system remains major-nutrient-limited most of the time, with low productivity. Typical surface chlorophyll levels across the equatorial zone are 0.1–0.2 mg m^{-3}, comparable to those of the central equatorial Pacific, somewhat higher than the Pacific warm pool (Antoine *et al.* 2005). There is a deep chlorophyll maximum, about twice surface levels, in the nutricline, that is typically at 50 to 80 m. Primary production runs in the vicinity of 150 to 200 mg C m^{-2} yr^{-1}. There is no eastern tropical tongue of

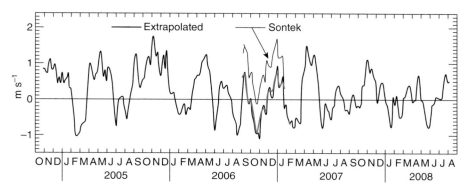

Fig. 11.42 East–west current velocity at 10 m depth at 80°E on the equator (due south of Sri Lanka). The black line is extrapolated from acoustic Doppler current meter data from 35 m using the correlation to 10 m Sontek (a different Doppler system less affected by surface sound reflections) data from August 2006 to January 2007 (shown in gray, overlying the black line and repeated 0.75 m s⁻¹ above to make it more visible). Velocity of 1 m s⁻¹ is 2 knots, relatively rapid oceanic flow. Positive is eastward flow, negative is westward. (After Nagura & McPhaden 2010.)

higher phytoplankton stocks comparable to those occurring seasonally in the Atlantic and more persistently, save for El Niño periods, in the Pacific. Ecological relationships are much like those of a subtropical gyre. Nano- and pico-plankton are the dominant primary producers. Meso-zooplankton and upper trophic-level fauna resemble those of all subtropical gyres, including tropical–subtropical tunas, dolphin fish (*Coryphaena hippurus*) and ommast-rephid squid. There are Arabian Sea endemic species, but few if any species are endemic to the Indian equatorial zone. The southern Indian subtropical gyre is much like those in the Pacific and Atlantic. Phytoplankton stocks are half those at the equator and just to its north.

An invitation

All of this will receive recurring attention from oceanographers in the next few decades. Our understanding of ocean biomes is rapidly improving, and readers of this are invited to contribute to the ongoing effort.

Chapter 12

Adaptive complexes of meso- and bathypelagic organisms

Mid-water habitat, or the mesopelagic zone, ranges from the bottom of the euphotic zone (or somewhat deeper, say 200 m) to 1000 or 1200 m. Carol Robinson *et al.* (2010) have reviewed the ecology and biogeochemistry of this zone, pointing out that much remains to be learned, including full description of both microbial and metazoan diversity. For example, relative abundances of distinct genome types (primarily DNA coding ribosomal RNA) of bacteria and archaea shift progressively downward in the mesopelagic (aristegui *et al.* 2009). About 90% of organic matter exported from the euphotic zone as sinking particles is metabolized above 1200 m, but estimates of organic-matter utilization by picoplankton (bacteria, archaea, protists) and zooplankton both exceed the influx from above as measured with neutrally buoyant traps (Steinberg *et al.* 2008). Partly, that is because quantifying pulsed and spatially erratic input events is difficult. Also, application of uncertain growth-efficiency measures is involved. Steinberg *et al.* propose that much of the supply gap may be filled by vertically migrating zooplankton and nekton feeding at the surface, then both respiring and dying at depth. Precise budgets will continue to elude us. We focus in this chapter on some of the adaptations that permit animal survival below 200 m and that make this zone a showcase of evolutionary problem-solving.

Light in the mesopelagic is insufficient for positive net photosynthesis but remains enough for vision to be important. Since the mesopelagic depends for its trophic support upon organic matter transported from the layers above, the inhabitants must eat detritus as it falls through, participate in a DOC-based food-chain, migrate up to feed in the euphotic zone or eat each other. The last of these possibilities means that modes of predation suitable to the near-dark have evolved, and there are elaborate specializations for "hiding" from predators. The habitat can be described as "trophically dilute": there is very little food, as shown by classic Russian zooplankton biomass estimates from subpolar and subtropical areas in the Pacific (Fig. 12.1). Thus, mid-water animals must have adaptations for surviving on slim rations and for enduring long periods of starvation. Furthermore, the ocean's oxygen-minimum zones center in the mesopelagic (Plate 1.1), with strongly hypoxic levels in the eastern tropical Pacific and northern Indian Ocean. Animals must either avoid these levels or be specially adapted to acquire or live without oxygen: larger, more subdivided gills, respiratory pigments with high oxygen affinity, minimal respiration, dependence on glycolysis, intermittent migrations into oxic layers, etc. Life processes are slow, especially under hypoxia, and there is evidence of long life-cycle times, of the order of years for animals of modest size.

Because food is scarce in mid-water, there are very low standing stocks of the denizens of the deep. That makes it difficult to study the ecology of this zone. Very large volumes of water must be filtered to obtain useful samples of the biota. The usual approach is to use large nets such as the Isaacs–Kidd trawl (10 m^2 mouth area). Much larger nets have also been tried, including Engels trawls with towing warps from two ships, huge otter boards, and a mouth 100 m wide. Huge nets were the rage in the early 1980s, a trend that seems to have died away. They did not catch an *Architeuthis* (giant squid), and drag-forces from the nets tended to pull winches and ships to pieces. On the whole, nothing new was learned from the effort, but upper

Biological Oceanography, Second Edition. Charles B. Miller, Patricia A. Wheeler.
© 2012 John Wiley & Sons, Ltd. Published 2012 by John Wiley & Sons, Ltd.

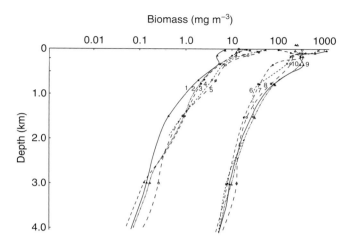

Fig. 12.1 Vertical profiles of zooplankton biomass (mg dry weight m^{-3}) taken from the Soviet oceanographic vessel *Vityaz* with nets hauled vertically over wide depth intervals (e.g. depth = 4 km to 2 km) in subpolar (cluster of curves at right) and subtropical (left cluster) waters. Each curve is for a separate station. Abundance falls by two orders of magnitude between the epipelagic and 3 km. (After Vinogradov 1968.)

size ranges of some species were extended slightly. Studies with submersibles and remotely operated vehicles have been important in recent decades, particularly for observing and capturing gelatinous forms like the diverse and delicate deep-sea siphonophores and ctenophores (e.g. Haddock 2004).

Animal sampling has mostly been done with large trawls, so the animals studied most closely have been fish, squid, and shrimp, the charismatic megafauna of the mesopelagic. Most of our discussion will concern those groups. Spectacular color pictures of those organisms taken in their natural habitat by ROV cameras can be found in *The Deep Sea* by Robison and Connor (1999). However, a larger part of deep-sea biomass is made up of small crustaceans: copepods, euphausiids, amphipods, ostracods, and mysids. There are numerous and remarkable jellyfish and some specialized chaetognaths. Copepods in the deep sea are diverse relative to those in the epipelagic. Common at the tiny end of the size scale are numerous species of *Oncaea*, which crawl about on bits of mucus and submarine "snow", eating attached particles. The larger mesopelagic copepods are diverse calanoid genera, with many species over 5 mm and *Bathycalanus sverdrupi* that grows to 16 mm. Some genera (*Bathycalanus*, *Megacalanus*, *Lophothrix*, *Scottocalanus*) are robust-bodied detritus-feeders that mostly eat fecal pellets sinking through their vicinity from above (Nishida & Ohtsuka 1991). These copepods and some others of less-certain feeding habit (*Gaetanus*, *Gaidus*, and others) are mostly bright-red, opaque, and heavily muscled. In captivity, they alternate zooming about their aquarium with spells of stillness and fairly rapid sinking.

Lurk-and-grab predators are another large group of copepod genera: *Augaptilus*, *Euaugaptilus*, *Haloptilus*, *Disseta*, *Paraeuchaeta*, *Arietellus*. Except for *Paraeuchaeta*, these mostly have very thin muscles and elaborate sprays of long setae to inhibit sinking. In aquaria they hang still in the water, tail down with antennules extended, scarcely sinking or rising. *Paraeuchaeta* behaves similarly in the field, but tends to rest on the bottom in containers. Unlike the red detritivorous copepods (and large red shrimp, for that matter), these predatory copepods have elaborate color variations. *Euaugaptilus* species come in brilliant yellow, lavender, orange, bright green, and other colors. Some have pale tints over the body as a whole, through which is visible a gut brightly colored with a completely different part of the spectrum. Species of *Disseta* come in pale orange and bright white. *Paraeuchaeta* have chromatophores that look like a decorative pattern of tiny brittle-star pictures just under the exoskeleton, and they come in colors and patterns that are species-specific. While dead specimens can be distinguished by experts from very subtle shape differences (following Park 1993), live specimens wear their identities as distinctively colored uniforms. Females of the genus *Euchirella* tow their purple, green, or black eggs along behind them in long zig-zag rows. The adaptive value of this riot of fancy dress in the near dark, with only blue photons to show it off, is hard to imagine. If Annie Dillard is right and we are meant to be the witnesses of creation, then perhaps the show is for us.

Jellyfish and siphonophores are also common in the mesopelagic. They presumably function as tentacular predators, much like their epipelagic cousins. Color is common for them as well; in several groups it is a dark reddish purple, likely to hide bioluminescence from ingested prey. The coronate scyphozoan *Periphylla* has this pigment only in the lower tissue layer beneath a tall conical "lens" of clear mesoglea, an effect that glassblowers have lately been imitating.

Hiding out

Many features of mid-water organisms imply that being very close to invisible is important at all levels where there is some light, say more than 10^{-11} W m^{-2} ($\sim$1100 m at midday in clear tropical water), i.e. very dim illumination indeed. The limit of human vision is of the order of 10^{-9} W m^{-2}. At a maximum of about 800 m in clearest tropical waters, the "sky" is visible as a small, blue circle to a human eye looking through a dorsal window in a submarine. Daytime irradiance just below the sea surface is approximately 10^{3} W m^{-2}. For light at wavelengths near 475 nm, the pure seawater diffuse absorbance is $\sim$0.017 m^{-1}, so illumination is reduced at least 10-fold every 135 m. The absorbance spectrum of very clear ocean waters is quite flat from 400 to 500 nm, but absorbance is considerably greater at longer and shorter wavelengths, so the only light not absorbed above the mesopelagic is blue. Thus, the mid-water lighting is dim and blue only. Sundry observations suggest that deep-sea fish with large eyes, near-perfect tapeta behind the retina (reflective surfaces returning photons not absorbed initially for a second retinal pass); up to five layers of pigmented and long retinal rod cells; and specialized, concentrated retinal pigments, should have useful vision even with about 100-fold less light than required for human vision. Owls, for which behavioral experiments are possible, see effectively in light that dim. In mid-water, the array of visual adaptations for using the residual downwelling and bioluminescent light is complex in single species and also variable and distinctive among families of fish, crustaceans, and squid. Since vision is useful throughout this thick ocean layer, survival for most animals requires means to avoid being seen and then eaten. Adaptations of camouflage and vision to deep-sea illumination have been thoroughly reviewed by Warrant and Locket (2004). We will consider camouflage first.

Color

Most mid-water fishes, especially below ~650 m, are black, and the larger crustacea are dark red. Both black and red appear black where all light is blue. There is a gradient, however, with depth. In levels around 200–400 m, many shrimp are "half-red", that is, red anteriorly to cover luminous food in the foregut, and nearly transparent ventrally and posteriorly. Fish in the upper mesopelagic are dark dorsally, silvered on the sides, and bear photophores ventrally. Blackness or redness over the entire body is typical in the deeper mesopelagic. In the bathypelagic, below about 1200 m, many pelagic fish are gray-brown or chalky.

Mirrors – silvered sides

Silvered sides occur in many fish of near-surface waters: oarfish, sardines, tuna, and others. These mirrors are panels of guanine crystals arranged so that light reflected to the eye of a predator approaching at any angle tends to come from the same direction as if the prey were not there (Fig. 12.2). The silhouette of the opaque fish nearly disappears against the diffuse light of the background (Franz 1907). To make this work, the guanine panels parallel the vertical (dorsal to ventral) axis of the body, regardless of the curvature of the sides (Fig. 12.2). In mid-water fish, such mirrors are only useful on the side planes that are almost exactly vertical, so mirrored scales are reduced to a single row. Deeper yet, they are useless, and fish do not have them. They are not a feature of mesopelagic invertebrates.

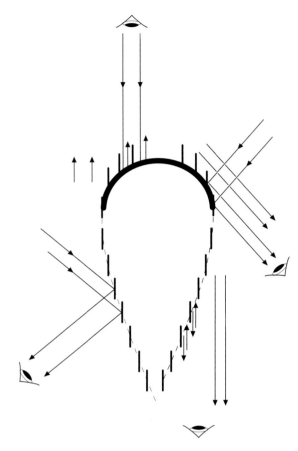

Fig. 12.2 Transverse section of a fish body showing the orientation of reflective guanine plates and the resulting camouflaging effect. The predator eyes at the right and lower left receive reflected light to replace light that it would have seen if the fish wasn't there. Effects of countershading are implied by eyes above and below. (After Denton 1970.)

Countershading

Being dark dorsally and light ventrally is important at all lighted levels of oceans and lakes. You have noticed that fish tend to be dark dorsally, silvery or white ventrally. This "countershading" appears in a range of near-surface organisms: fish, surface-floating snails (*Janthina*), the chambered *Nautilus*. It cannot work perfectly, because there is always less upwelling light than downwelling light. Thus, reflected upwelling light cannot exactly match downwelling light. In mid-water, another possibility arises, and many organisms have evolved this adaptation. At levels where the downwelling light is reduced to an intensity possible to match with bioluminescence, arrays of photophores on the ventral sides of fish, squid, shrimp, and euphausiids match the downwelling light, thus filling in their silhouettes. This reason for ventral photophores was suggested independently several times, starting with Dahlgren (1915–1917).

Young and Roper (1977) and Warner *et al.* (1979) provided experimental demonstrations that this must in fact

Table 12.1 Observed ventral light output for four squid held at different training intensities of downwelling irradiance ("relative light value"). (From Young & Roper 1977.)

RELATIVE LIGHT VALUE	MATCH LEVELS FOR THESE SQUID			
	Abralia trigoneura (1)	*Abralia trigoneura* (2)	*Pterygioteuthis* sp.	*Pyroteuthis* sp.
1	1.0	0.72	1.0	0.12
2	2.0	2.0	2.0	1.0
6.7	4.8	6.7	4.8	3.5
20	20	20	20	20
60	31	43	60	43
120	—	75	4.8	31
200	—	75	1.0	31
300	—	60	—	—

be the function of ventral photophores (which could have other functions, like attracting appropriate mates). They showed that squid, fish, and shrimp closely balance their ventral bioluminescence with the downwelling light; there is an exact match. Young and Roper placed squids or fish in a tank above a mirror at a 45° angle, so that an observer off to the side could watch both the overhead illumination and the ventral side of the animals. A photomultiplier tube (PMT) in line with the observer's eye permitted measures of light levels. Animals were conditioned at a variety of light intensities for 5–10 minutes. Then the observer, who could not see the animal, quickly lowered the intensity until he could. Next, he raised it again until the animal disappeared, indicating a match. The intensity of the match, that is, of the animal's light output, was read from a meter. This required that the matching response be fairly slow, multiple seconds rather than, say, milliseconds. Long response times must be short enough in the slow-moving, mid-water world. The results were a sequence of comparisons of the conditioning intensities and the matches called by the observer (Table 12.1).

Warner *et al.* used more sophisticated equipment for a study of the half-red shrimp *Sergestes similis*. A PMT was positioned below an animal harnessed in a spherical aquarium that could be illuminated from above with a variable intensity of blue light (520 nm). Irradiance could be varied from 0 to $3 \times 10^{-4}\,\mu\mathrm{W}\,\mathrm{cm}^{-2}$. Bioluminescence levels were measured with the phototube in the few milliseconds after the conditioning light was shut off. The result (Fig. 12.3) shows a nearly exact match between the dorsal conditioning intensity and ventral luminescence output. In this

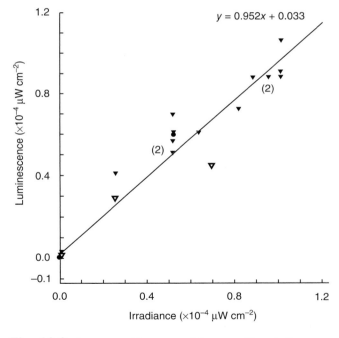

Fig. 12.3 Comparison of light output of the organ of Pesta in *Sergestes similis* to the conditioning irradiance above, showing the close match. (Simplified from Warner *et al.* 1979.)

animal, the organ of Pesta, located below the gut, produces the light. The gut is darkly pigmented and is the only part of the body that would be silhouetted against the downwelling light for a predator below. Warner *et al.* extended the experiment by comparing the matches for shrimp with

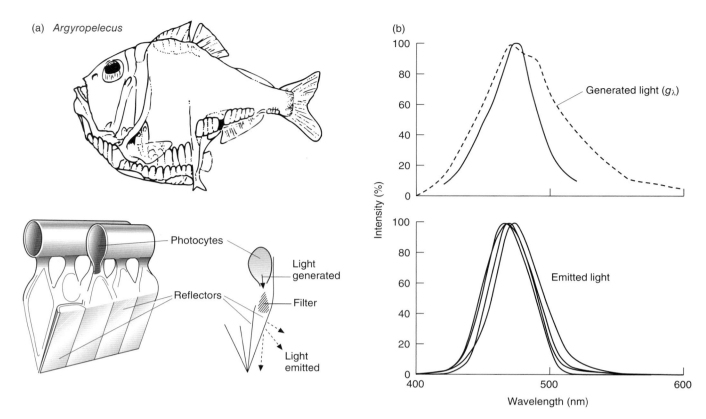

Fig. 12.4 (a) Arrangement of light-producing organs in the hatchet fish, *Argyropelecus*. Light is generated in photocytes in a tubular organ along the ventral side. It is emitted ventrally after passing through a conical filter and is spread below the body by an array of angled mirrors. (b) Above, comparison of the light spectrum generated in the photocytes (g_λ) to the spectrum of downwelling light (solid line); below, spectrum of emitted light (after passing the filter) in four specimens. (After Denton 1991.)

their eyes covered with transparent masks and opaque masks. With opaque masks the organ of Pesta produced no light at all. The eyes must measure the downwelling light to allow the match.

To achieve a countershading match to the downwelling intensity, the animal must (i) measure the downwelling light (eyes); (ii) produce variable levels of light ventrally (photophores); and (iii) be able to determine the ventral output intensity to establish the match. For requirement (ii), the bioluminescence must not only match the intensity of downwelling light, it must match its spectrum. However, the output of luminescing molecules, or luciferins, "discovered" by an animal's biochemical evolution may not provide an exact match. For example, the spectrum (Fig. 12.4) of light produced by the ventral photocytes of the hatchetfish, *Argyropelecus aculeatus*, is broader than the downwelling spectrum, particularly toward longer wavelengths. To perfect the match, the light passes through filters arrayed below the tube of photocytes before it is emitted through a bank of dispersing reflectors. The final match is excellent.

For requirement (iii), most mesopelagic squid, many fish, and most euphausiids have a photophore in close

association with one or both eyes, usually actually built into the eye. It is probable that these produce a neural signal in the visual system proportional to the overall luminescent intensity. The photophore in the eye must be regulated by the same process regulating those producing ventral illumination, thus providing a feedback loop for adjustment to a match. In addition to the demonstration of such a feedback by Warner *et al.*, experiments with squid by Young *et al.* (1979) show that matching is disrupted when light-sensitive vesicles on the dorsal side of the body are covered. Both the principal eyes and dorsal light-sensitive vesicles are involved in determination of downwelling intensity, since matching deteriorates when any of those organs is covered.

Bioluminescent countershading is in some instances even more complex. Young and Mencher (1980) made spectral scans of photophore output at different temperatures (Fig. 12.5) from the ventral skin of *Abraliopsis pacificus*, a small, muscular squid abundant near Hawaii. At the colder (8°C) temperatures of its daytime, mid-water habitat, it produces a narrow spectrum centered at 472 nm, very close to the downwelling spectrum. At the warmer temperature (23°C) of the surface layers, it migrates up into at night, it produces a much broader spectrum, one remarkably close to

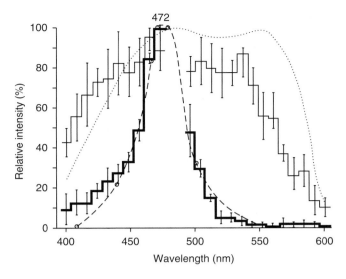

Fig. 12.5 Comparisons of downwelling light spectra to spectra of ventral bioluminescence from the squid *Abraliopsis pacificus*. A stepped spectrum represented by the dark line is produced at 8°C and resembles the deep-sea downwelling spectrum, dashed line; both peak at 472 nm. A stepped spectrum shown by the light line is produced at 23°C, much closer to the spectrum, dotted line, for moonlight at 20 m in the tropical ocean. (After Young & Mencher 1980.)

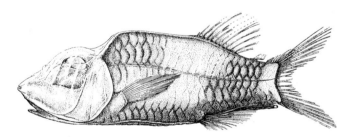

Fig. 12.6 The mid-water fish *Opisthoproctus*. Note the upward-directed eyes with spherical lenses (that are yellow in life). The flat ventral "sole" of this fish is a light organ. (Drawing from Cohen 1964.)

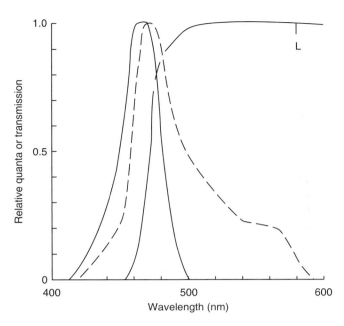

Fig. 12.7 Comparison of the ventral bioluminescence spectrum of a typical myctophid fish, *Myctophum punctatum* (dashed line), which is not filter-corrected as in *Argyropelecus* (Fig. 12.4), to the downwelling spectrum at 500 m (bell-shaped curve) and to the light-transmission characteristics (*L*) of the yellow eye lens of the pearl eye, *Scopelarchus analis*. By eliminating the dominant spectral component (blue), the predator can see the excess of the prey bioluminescence at longer wavelengths (blue-green). This matches the $\lambda_{max} = 503$ nm of one rod pigment of *S. analis* (Fig. 12.9). (After Muntz 1976.)

that of moonlight viewed through 20 m of water. Thus, it not only matches the downwelling intensity, it matches the spectral composition through a dramatic change. To do this, it has three types of ventral photophores, two with blue filters, one with red filters (Plate 12.1). The selection of a matching combination is not made by examination of the spectral quality of downwelling light, as might be expected, it depends only on temperature. Color of the downwelling light has no effect on the output.

These countershading mechanisms are not perfect, and many mesopelagic fish depend upon the imperfection. The eyes of many species look upward (Fig. 12.6). It would be the reduction of visual neural signals caused by shadows from animals overhead that initiates feeding responses. Not only do eyes of many deep-sea fish permanently look up, in many the mouth opens upward for grabbing prey above

them. Probably only those shadows small enough to grab safely are attacked.

The evolutionary back and forth between prey defensive strategies and predator counter-strategies has passed through many cycles in mid-water, as in most habitats. Ventral countershading provides an extended example. While *Argyropelecus* has evolved a very precise spectral match, not all mid-water animals have achieved it; they produce countershading with a substantial spectral "shoulder" above 490 nm. A few predatory fish (*Chlorophthalmus*, *Argyropelecus*, *Scopelarchus*) and squid (*Histioteuthis*) have developed yellow eye lenses which probably enable them to take advantage of this slight mismatch. Yellow pigment in the eye absorbs blue light right at the water-transmission maximum, acting as a sharp cut-off filter that removes the dominant wavelengths. However, it passes light greater than 490 nm (Muntz 1976; Fig. 12.7), which must cause the ventrally luminescent animals above to appear as pale shapes.

Transparency

As in surface layers, some mid-water organisms are transparent, for example the leptocephalus larvae of eels. These have the appearance of a glass tobacco leaf, and achieve near-transparency mostly by virtue of being thin. Their eyes, however, because they require opaque, light-absorbing pigments and tapeta, cannot be transparent. For such animals, seeing raises the risks of being seen. Deep-living jellyfish, amphipods, some squid, and others are likewise nearly transparent. This is effective if it sufficiently reduces the visual *contrast* of the animal against the background. Organic molecules absorb very little more light than water, but both proteins and lipids scatter light differently due to their effects on refractive index. In some transparent tissues, the effects of scattering are reduced by submicroscopic "bumps" in membrane surfaces smaller than the wavelengths of visible light (Johnson 2001). They generate destructive interferences that reduce the overall effects of scattering, a complex subject in physical optics. Contrast, C (negative by convention), for an eye looking at some tissue very close in can be lowered to about $C = -9\%$ (= 91% transparency) by adding water to the tissue and evolving specialized arrangements of lipid and protein structures. However, effective contrast fades exponentially with the distance of a detecting eye from the object. Detectable contrast for deep-sea fish and invertebrates is unknown; it can be small ($C \sim -1\%$) for near-surface fish (Johnson 2001). Nevertheless, nearly transparent animals at mesopelagic depths are effectively invisible.

Vision in near darkness

We emphasize vision because that must be uniquely adapted to mesopelagic conditions, and because much can be learned about eyes from comparative morphology and visual pigments. Of course, other sensory modes remain available. Detection of relative water motion is mediated, as nearer the surface, by lateral lines in fish, innervated setae in crustacea, tango-receptors in chaetognaths, etc. The antennal water motion sensors of deep-living crustacea are particularly impressive, with very thin antennae often extending to over three times the body length. In fish, at least, olfaction appears to be relegated primarily to mate finding: it is highly developed in males, some having elaborate external receptor pads on the head (e.g. *Cyclothone microdon*), but almost absent in females that presumably secrete male-attracting pheromones. Smell appears to be of minimal value for locating food in the mesopelagic zone. Its value re-emerges just above the seafloor, where large smelly meals – such as dead whales – stop falling, and where direction upstream can be distinguished by reference to the bottom.

Mesopelagic vision can be optimized for two quite different sorts of lighting: diffuse background that is brightest from above but illuminates the "scene" from all angles, and bioluminescent point-sources, often flashes, that greatly exceed the diffuse background. Optimizing eyes for low intensities of each of these kinds of light has different requirements (Warrant & Locket 2004), but some designs trade optimization for either to retain some capacity for both types of source. The result is a remarkable array of modifications of both the camera-like eyes of fish and squid and the compound eyes of crustaceans. Images of both scenes, say a squid against background lighting, and points of light, degrade rapidly in water compared to air, due to much more intense scattering. Mid-water animals can only see to ranges of a few tens of meters, but that is far enough to make vision critically useful.

Fish eyes

Most deep-sea fish have only "rod" photoreceptors with a single type of rhodopsin as the visual pigment, and no cones. Pigment absorption maxima, λ_{max}, are mostly between 470 and 490 nm. However, detailed studies of pigments are revealing that some have more than one receptor pigment, in at least one case retinal cones, and thus potential for subtle color differentiation. Because the refractive index of water is $\sim$1.33, about the same as tissue fluids like the vitreous humor (not meaning "funny"!), stronger lenses are required for camera-type eyes than in air, so fish and squid eyes have spherical lenses. Their protein composition varies such that the refractive index shifts from 1.33 at the surface to 1.52 at the center. The focal lengths of such lenses are 2.5 times their radii ("Matthiesen's ratio"), and the gradient of refraction corrects spherical aberration to provide sharp images at the retina. Sensitivity increases with pupil size (more area, more gathered photons). In some mid-water fish, the pupils extend to the full diameter of the lenses, and those protrude through the pupil to admit light from $\sim$180°. Maximum contrast (and thus discernibility) of objects in the visual field is against the brightest light, which is above, and it falls off dramatically to the sides and below. A frequent adaptation to that geometry is tubular eyes looking up, with the retina restricted to the bottom of the tube. Their effective angle is only about 50°, but that provides the sharpest possible images at the nearly flat retina. Focal length sets the overall eye size for a given lens size. Thus, large pupils and lenses require large and distant retinas, dictating that vision in many mid-water fish and squid dominates both space in the head and capacity in the brain. The eyes of the giant squid (*Architeuthis*) are as large as 37 cm in diameter, with a salad-bowl-sized retina. There are many variant eye plans among fish (Collin 1997), but we can only present two examples.

Adults of the circumglobal tropical–subtropical pearl eye, *Scopelarchus analis* are as large as 12.6 cm. They are captured below 500 m in daytime and ascend to as shallow as 275 m at night. Like many mesopelagic fish, their tubular eyes are aligned dorso-ventrally in the head, looking up. The protruding lens (Fig. 12.8) focuses the scene above on

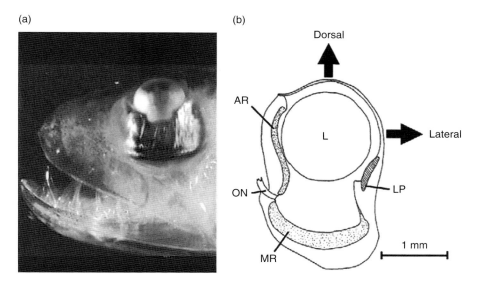

Fig. 12.8 Orientation and anatomy of the eye of *Scopelarchus analis*. (a) Photograph by Shaun P. Collin. (b) Vertical section: AR, accessory retina; L, lens; LP, lens pad; MR, main retina; ON, optic nerve; as drawn by Partridge *et al.* (1992). (After Pointer *et al.* 2007.)

a nearly flat retina at the bottom of the tubular vitreous space. In addition, light entering the lens from the side, some of it passing to the lens through a clear patch of body wall (a "lens pad"), is focused on an auxiliary retina in the medial wall of the tube, providing information about contrasts, or more likely flashes, located laterally. Similarly upward looking, tubular eyes occur in many families, and the addition of a retinal patch for lateral viewing has also evolved repeatedly. In *Bathylychnops exilis* the lateral patch has become a pouch-like subsidiary eye with its own lens (a thickened corneal protrusion) in the body wall looking ventrally (Pearcy *et al.* 1965).

Partridge *et al.* (1992) determined absorption spectra in retinal cells of *S. analis*. They found two pigments, $\lambda_{max} = 405$ nm (violet) and 507 nm (green) in the main retina, the latter only in distal portions of some rods. Both of those were found (Fig. 12.9) in the auxiliary retina, along with a third pigment with $\lambda_{max} = 479$ (blue) found alone just in some cells. Indeed, Pointer *et al.* (2007) have applied molecular genetic techniques to *S. analis*, showing that there are genes for three visual pigments (opsins): two rhodopsins and a third most closely similar to cone pigments of shallow-living fish. The measured absorption maxima for *S. analis* are certainly approximate, but the fact that there are three functionally distinct pigments, proteins coded by substantially different genes, suggests that this fish may discriminate among shades of blue, possibly between sky light and bioluminescence. Whether the neural circuitry for this color distinction is also present remains to be shown.

Eyes (Fig. 12.10) of the bigeye smooth-head (*Bajacalifornia megalops*, lengths to 28 cm) are optimized

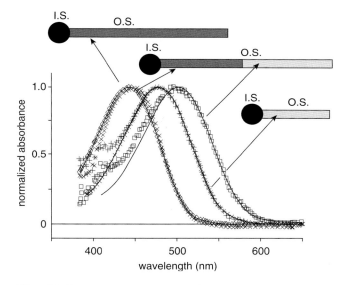

Fig. 12.9 Absorption spectra (from microspectrophotometry) of visual pigments in the accessory retina of *Scopelarchus analis*. A pigment with λ_{max} at 443 nm is found in outer segments (o.s.) of both shorter and longer rods, shown diagrammatically. Part of the o.s. of relatively long outer rods has a pigment with λ_{max} at 503 nm. Those two are also found in rods of the main retina. Compare the lens transmission spectrum (Fig. 12.7). A third pigment, $\lambda_{max} = 479$ is only found in outer segments of short rods of the accessory retina. (After Partridge *et al.* 1992.)

for spotting and localizing point light sources almost directly ahead (Locket 1985). Large specimens of this fish are bathypelagic, collected below 1200 m and as deep as 3000 m, far below levels with any sunlit scene. Juveniles can be caught at depths as shallow as 250 m. Each eye views

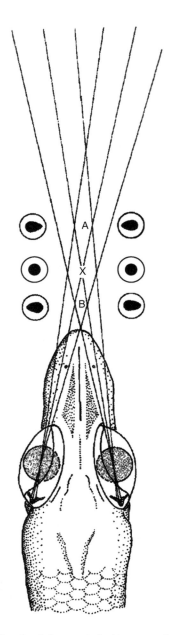

Fig. 12.10 Dorsal–anterior aspect of a bigeye smooth-head (*Rouleina attrita*), showing ranges of light rays focused on the posterior foveae. The circles show hypothesized focal patterns of a point source of bioluminescence on each fovea. Patterns could provide information about source location. Beyond A, only one fovea would receive a focused (and more skewed) image. (After Warrant & Locket 2004.)

Locket hypothesized that the relative shapes of a flash image across the two foveae (Fig. 12.10) could provide information about the location of the source ahead. As in many deep-sea fish, neural impulses from many rod cells are accumulated by ganglion cells in the retina before transmission to the brain, gaining sensitivity at the cost of resolution. Locket observed in sectioned retinas that *B. megalops* has up to 28 layers of rod cells, but that not all of them remain attached to ganglia. He proposed that the layers with lost connections are remnants of a developmental progression during an ontogenetic migration to deeper and deeper levels. However, several layers do remain active, those with the longest pigmented segments and, therefore, the greatest light-absorbing capacity.

A range of distinctive visual designs are described for fish requiring extreme sensitivity to benefit from vision in the near-dark. Enhancements include extremely long pigmented segments in rod cells; retinas backed with reflecting tapeta to return unabsorbed light back through them; wrapping of bundles of rods reporting to a single ganglion cell in reflective tube cells; and extremely long visual integration intervals before ganglion discharge to the optic nerve. Those modifications increase sensitivity but consistently at the cost of spatial or temporal resolution. Possibly, it is more important to see something against the very dim background, than for the scene (or flashes) to be sharply defined.

Crustacean eyes

Some species in the "lower" orders of crustacea have very simple eyes, basically a retina with no associated focusing apparatus. Among copepods, however, some groups have lenses formed from cuticular chitin that do focus light on small retinal elements. Those copepods are not found at substantial depths. The deep-sea ostracod *Gigantocypris* has a parabolic tapeta surrounding its retina and focusing light upon it. This likely provides good sensitivity with almost no image capability, and retrieves minimal information about source location. Amphipods, mysids, euphausiids, and shrimp have complex compound eyes, and deep-sea species have modifications similar to those of nocturnal moths. In amphipods at all depths, vision is often a dominant sense, the ommatidia (facets) forming nearly transparent domes surrounding their heads. The ommatidial light guides narrow down into a cluster of minimized retinal elements (rhabdomes) that must be pigmented to function. Like fish, crustacean λ_{max} values are in the blue portion of the spectrum, mostly from 460 to 500 nm, but in some species another pigment absorbs close to 400 nm (Warrant & Locket 2004), useful for sunlight down to perhaps 250 m, and possibly useful for some bioluminescence. The common offset of the main pigment's λ_{max} toward green wavelengths suggests adaptation to see bioluminescence.

the water ahead and slightly to the opposite side along a deep swale in the snout. Pupils are round at the back but extend anteriorly to allow light from those angles into the lenses. Light focuses onto a fovea (a retinal patch with refined image resolution) at the posterior extremity of the retina. The foveae are pits with curving side walls lined with extreme densities of rod cells with very long outer segments, the part of the cell with photosensitive pigment.

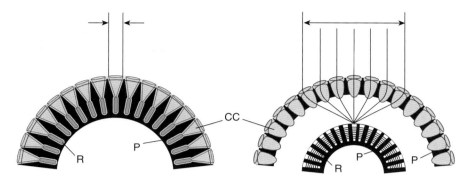

Fig. 12.11 Left: diagrammatic section of an apposition compound eye: short, pigment-cell-shielded crystalline cone cell of each ommatidium in direct contact with the rhabdome; typical of near-surface (and terrestrial) compound eyes. Right: superposition compound eye: crystalline cones more extended; pigment cells absent or clear; light from many lenses accesses each rhabdome. Abbreviations: CC – crystalline cone, P – pigment cell, R – rhabdome. (After Warrant & Locket 2004.)

Specializations for vision in dim light parallel those in fish, sacrificing resolution for sensitivity (Warrant & Locket 2004). In near-surface euphausiids and shrimp, the ommatidia (cuticular lens, light-guiding crystalline cone, and retinal rhabdome) are surrounded by pigmented masking cells that prevent light leakage between them, giving each an image from a narrow angular field (Fig. 12.11 – left). This is an "apposition" eye, meaning there is no unshielded space between the light-directing structures and the rhabdome. The images integrate in the brain as a segmented scene that has extreme sensitivity to motion (including very short integration times), as elements in the scene shift from one facet's rhabdome to the next. Try catching a dragonfly with a butterfly net on a warm day.

In mysid, shrimp, and euphausiid species of the deep sea, the masking cells remain but lose pigment along the crystalline cones, forming a substantial clear space intervening between the crystalline cones and the retinula cells containing the rhabdomes (Fig. 12.11 – right). The crystalline cones act as refracting bodies that guide light from a single direction, across the clear space to the rhabdome centered under that direction. However, gathering of light from lenses of multiple facets enhances sensitivity at the expense of resolution. Eyes with such clear spaces are termed "superposition" eyes, which operate by several variant optical mechanisms (Nilsson 1989) suggesting recurring evolution of the general pattern. Many shrimp and mysids living in dim light have rhabdomes that are star-shaped in cross-section, producing interdigitation of rhadomes and, thus, sharing of scattered light between adjacent cells (Gaten *et al.* 1992), rather than the strict isolation of rhabdomes in near-surface forms. Similarly to fish, signals from many rhabdomes may be gathered by single nerve cells before processing by the brain – making clusters, rather than individual rhabdomes, the units of the visual field. Each of these modifications improves sensitivity at a cost to resolution.

Tapeta at the bases of the rhabdomes are common, raising the proportion of captured incident photons.

Whitehill *et al.* (2009) have demonstrated a developmental sequence from apposition-like to superposition eyes in the mesopelagic mysid *Gnathophausia ingens* as the juveniles (life cycle discussed below) progressively inhabit deeper strata. The first free-living stage, inhabiting depths 175–250 m, has an eye of superposition form, but with almost no clear space and substantial sheathing cells around the crystalline cones. The clear space (actually transparent extensions of the crystalline cone layer that are not surrounded by pigment cells) opens proximally in instar 5 (>400 m), and surrounds the entire rhabdome layer of instar 10 (400–900 m). The eye structure shifts with stage to gain sensitivity, although all stages have interdigitating, stellate rhabdomes, but there is no shift among stages in either visual pigment absorption spectra or electrode-recorded spectra of sensitivity to flashes (Frank *et al.* 2009). Spectral maxima (λ_{max}) were all close to 500 nm. Frank *et al.* identified genes for two rhodopsin pigments in *G. ingens*, but their absorption properties appear to be additive to a smooth, unimodal spectrum ($\lambda_{max} \approx 512$ nm).

Getting fed down deep

Mesopelagic fish that eat zooplankton can have quite regular meals (Moku *et al.* 2000). For example, *Stenobrachius nannochir*, an 11 cm myctophid fish living between 500 and 700 m in the subarctic Pacific, feeds almost exclusively on copepods and, at least in August, eats primarily the large *Neocalanus* spp. that descend in huge numbers for diapause at these depths in mid-summer. Over half the *S. nannochir* collected at all hours of the day and night have some food in the stomach, although it amounts to a small fraction

(0.1% on average) of body mass. Other myctophids that migrate at night to surface layers eat more. For example, stomach content of *Stenobrachius leucopsaurus* is typically 1.5% of body weight: numerous copepods and a greater mass of euphausiids.

However, for many permanent residents, particularly predatory fish in the deep mesopelagic, meals can be few, and waiting times between them long. Tolerating this dietary regime requires special adaptations. Most mid-water fish have very weak bodies. All bones, except the jaw, are weakly ossified, which saves on weight and thus on swimming. Musculature is reduced and has a higher proportion of water than in surface fish. It is not unusual for the body to be reduced to a huge jaw with a wisp of tail appended. Squid are characteristically flaccid, often with body cavities filled with tissue fluid lightened by replacement of sodium with ammonium. In *Chiroteuthis*, for example, there are large coelomic spaces in the ventral arms filled with buoyant fluid.

Dramatic, though infrequent, feeding events are the rule for a number of predatory forms. Some mid-water fish have adaptations like terrestrial snakes: a jaw that unhinges to admit relatively enormous prey, elastic throats, and big folded bags for stomachs. Several species can swallow other fishes substantially larger than themselves. James Childress (pers. comm.) found that *Chiasmodon niger*, the "Black Swallower", adds a new growth layer on its otoliths (ear bones) each time it gets a meal. The evidence is that the outer layer is different when there is a meal in the gut of a captured specimen than when it is empty. The sequence of the two types of layers implies that this species eats about 14 meals in growing from the earliest juvenile stage to reproductive adulthood.

Capture tactics are elaborate, and become more so with depth. As stated above, many fish hang in wait or cruise slowly, their upward-directed eyes finding slight mismatches between downwelling light and counter-illuminating prey – meals to grab with upward-directed mouths. There are anglerfish with bioluminescent lures that dangle before their mouths. Deeper-living mid-water fishes have extraordinarily long, saber-like teeth (Fig. 12.12). The jaws distend and stretch to get prey between the tips of these fangs.

Almost everything is slower

On the basis of a modest array of data, it appears that mid-water organisms are long-lived and grow slowly. At least one tough mid-water animal is amenable to laboratory rearing, the shrimp-like, red mysid *Gnathophausia ingens*. Childress and Price (1978) produced a size–frequency diagram (Fig. 12.13) for the population of the Southern California borderland region (San Clemente Basin). The stock has distinct and non-overlapping size classes, which are the sizes of 11 successive instars after the young leave the brood pouch (there are two instars in the brood pouch).

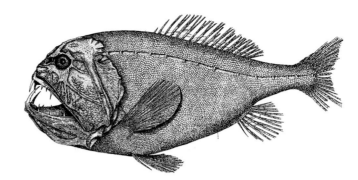

Fig. 12.12 *Anoplogaster*, the fang-tooth or mother-in-law fish. This is one of the instances of the extreme "cephalization" of the body and the extreme lengths of piercing teeth, features which have evolved repeatedly in mid-water fishes. (Drawing from Woods & Sonoda 1973.)

All 13th instar individuals are reproductive females with young in the pouch.

Childress and Price estimated time per instar for *Gnathophausia* in three ways:

1 In the laboratory, newly molted individuals stay soft for 12 days. In a large field collection, 11 of 149 individuals were soft, so the intermolt period approximately equals $(12 \times 149/11) + 6 = 168$ days. [A problem for you: why did Childress and Price add 6 days?]

2 Seventy individuals in a sample molted before dying, and a cumulative distribution indicated that 63 days were required for half to molt. So, the intermolt period approximately equals $(2 \times 63) + 12 = 138$ days (+12 because soft individuals were removed from the sample at capture).

3 Duration of intermolt for individuals that molted twice in the laboratory was a function of size and temperature, with a range from 120 days at 20 mm carapace length and 7.5°C to 200 days at 40 mm and 5.5°C.

Total life span based on all this information is of the order of 6.4 to 8 years. Typical benthic shrimp of comparable size on temperate continental shelves live only 2 years.

Mauchline (1988) has studied size–frequency distributions of various mid-water shrimps collected over a 5-year period from the Rockall Trough in the northeast Atlantic. He grouped collections by month of the year, and showed (Fig. 12.14) that the implied growth pattern is intermittent. Mid-water shrimp, at least at this high-temperate site, grow faster in late summer and fall (some cases) or in spring (other cases) than they do in winter.

Childress *et al.* (1980) provide information about growth and life-cycle timing of mid-water fishes. You can study growth, *if* you can determine age. Age in fish is recorded in the otoliths as depositional layering (growth rings).

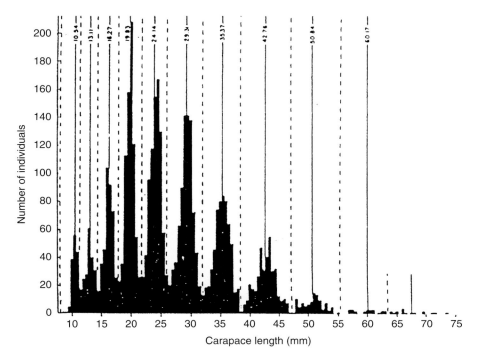

Fig. 12.13 Frequency distribution of carapace length in *Gnathophausia ingens*, a deep-sea mysid, collected off Southern California. Each of 11 size modes represents a life-history stage or instar. Advance from one to the next occurs at molting. (After Childress & Price 1978.)

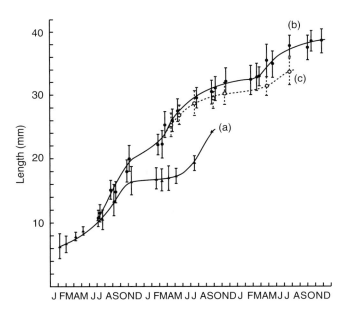

Fig. 12.14 Growth curves for (a) *Boreomysis microps*, a mysid, (b) female and (c) male *Gennadas elegans*, a penaeoid shrimp. Curves were derived from size–frequency distributions based on collections made at all seasons in Rockall Trough. Growth is seasonal, but look at the graph closely. (After Mauchline 1988.)

There are sometimes daily growth rings, and almost always annual growth rings. Rings are made by variations in the crystal structure or in the impurities included as the bone grows. Otoliths are used by the fish as weights suspended on the tips of sensory hairs to determine the direction of gravity and of centripetal effects. Comparison (Fig. 12.15) of length vs. age shows that mesopelagic fish that migrate to the surface at night to feed grow in much the same, decelerating pattern as fish like sardines continuously resident in surface layers. Fish that stay down both day and night add length in nearly equal increments at all ages. At least one bathypelagic fish, *Poromitra crassiceps*, appears to grow in length progressively faster with age, which is not reported for animals from any other habitat. Deep-sea fish must get better and better at finding and capturing food – and thus at growing – as they get bigger.

Mauchline (1988), in commenting on the results of Childress *et al.* (1980), noted that their data were viewed on a year-by-year basis. Roughly annual growth rings (annual annuli) are, however, evidence that growth varies seasonally. Mauchline developed some month-by-month growth estimates based on size–frequency analysis for mid-water fishes in the Rockall Trough, confirming this seasonality in growth for a number of species. As also shown by the shrimp growth data, seasonality telemeters down the water column, not fully damped by the constancy in temperature, salinity, oxygen, and other physical factors.

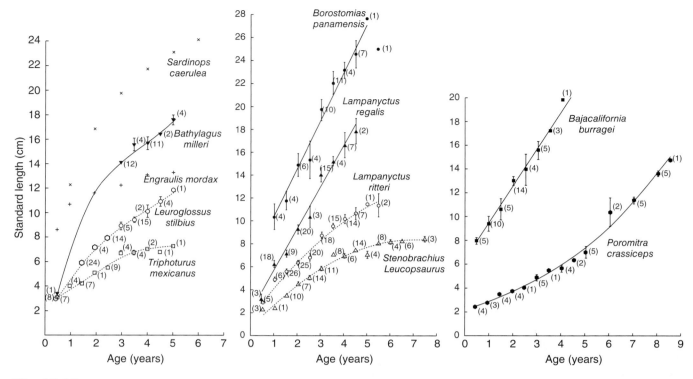

Fig. 12.15 Length vs. age growth curves for epipelagic (×, +), mesopelagic migrators (which come well up in the water column at night, open symbols) and bathypelagic fish (filled symbols). Numbers of otoliths assigned to each age point are in parentheses. (After Childress *et al.* 1980.)

Bathypelagic fish, on the other hand, showed no seasonal changes in growth rates. The vertical telemetering of seasonality is an active area of research.

Energy utilization is extremely efficient

Childress *et al.* (1980) determined the mean masses for fish of known ages, presenting the results as curves of caloric content versus age (Fig. 12.16). Those mesopelagic fish that migrate into surface layers at night grow in body mass in the same pattern as do epipelagic fish, for example sardines. They have the same proportional weight increase with age, although they have a smaller initial size, placing their curves far below that of the sardine. It is hard to find a perfect comparison, since most mid-water fish are taxonomically distant from surface species and are smaller. Growth of bathypelagic fish, species confined to deep layers and short rations, slows less strongly than that of mesopelagic species, and is close to continuously exponential. They grow both faster and larger than mesopelagic fish. Data for *Poromitra crassiceps* are amazing; growth is perfectly exponential for 7 years! A reviewer of this chapter reasonably suggested that minimal movement and flaccid, low-maintenance muscles allow a higher portion of acquired food to be routed to growth.

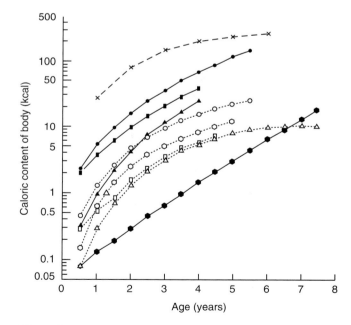

Fig. 12.16 A comparison of growth curves (body mass as caloric content) vs. age for epipelagic fish (sardine, ×), mesopelagic migrators (open symbols), and "bathypelagic" species (solid symbols). The solid hexagons are for *Poromitra crassiceps*, which has a very unusual growth pattern. (After Childress *et al.* 1980.)

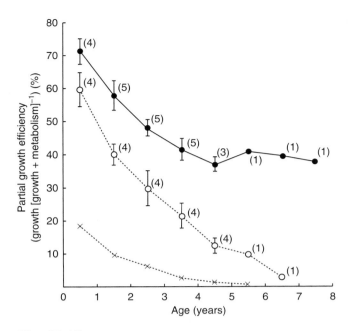

Fig. 12.17 Growth efficiency of fish as a function of age compared for sardine (×), mesopelagic migrators (open circles) and bathypelagic species (•). Numbers in parentheses are sample sizes. The term "partial growth efficiency" is used because food ingested but not assimilated is not accounted for. (After Childress *et al.* 1980.)

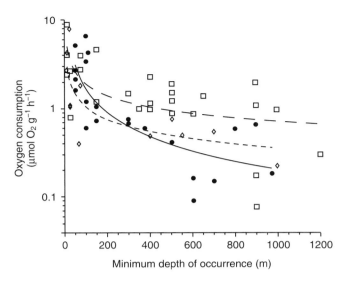

Fig. 12.18 Respiration rates measured as oxygen consumption in fish (diamonds), crustaceans (squares) and cephalopods (filled circles), mostly squid, plotted against the shallowest depth at which the animal is commonly caught, usually the upper range of any vertical migration. Most measurements were for animals captured off Southern California or near Hawaii. (After Seibel *et al.* 1997.)

Oxygen consumption estimates (Torres *et al.* 1979) for fish captured with insulated cod-ends, can be interpreted in terms of metabolic energy expenditure. Metabolism is very strongly reduced in mid-water fish compared to surface fish of comparable sizes at a given age. Mesopelagic migrators use about 25% of the energy used by surface fish, whereas bathypelagic species use less than 10%. Combining growth data with metabolic data, estimates can be made of the relative growth efficiencies of mid-water and epipelagic fish (Fig. 12.17). Since energy consumption for swimming and tissue maintenance declines with depth, growth efficiency of deep-living fish can be and is higher throughout life. The 60–70% growth efficiency of young bathypelagic fish is astoundingly high compared to that of terrestrial or epipelagic animals. Similar growth characteristics probably pertain for all mesopelagic animals, particularly squids and shrimp.

Fish in the general size range examined by Torres *et al.*, with adult length of about 20 cm, all had nearly the same longevity, 4–7 years. However, age at first maturity tends to increase downward (Childress *et al.* 1980), with bathypelagic fish reproducing only at the very end of the life cycle, perhaps only once. Surface fish reproduce in their second or third season, and mesopelagic fish fall in between.

Childress and colleagues have studied metabolism of mesopelagic organisms over several decades. Squid (Seibel *et al.* 1997, 2000), fish (Childress & Somero 1979; Torres *et al.* 1979), and large crustaceans (Childress 1975; Cowles

et al. 1991) show decreasing metabolism with depth of habitat (Fig. 12.18). These rates were either measured at comparable temperatures or corrected for comparability by Q_{10} estimates, so it is nearly certain that rates fall off with depth, due to more than decreasing temperatures. Further indication that downward cooling is not the sole driver of the metabolic slowing with depth comes from a comparison of fish metabolism between the California Current (where there is a strong, vertical temperature gradient), and the Antarctic (where there is virtually no temperature gradient). In both regions, metabolism falls off in the same fashion with depth and reaches an asymptotic value at about 800 m. Childress and Somero (1979) showed that, associated with their decreasing metabolism with depth, squid, fish, and shrimp have decreasing levels of both aerobic and anaerobic enzymes in their bodies (or for some measures just in muscles). It is argued from this that the metabolic-rate reductions are almost certainly real and not attributable to damage by collection procedures and retrieval from depth.

More recent results with chaetognaths (Thuesen & Childress 1993), jellyfish (Thuesen & Childress 1994), and copepods (Thuesen *et al.* 1998) seem to be qualitatively different from results for larger crustaceans, fish, and squid. Childress (1995) has reviewed the significance of these comparisons. In contrast to the larger animals, there is no fall-off of metabolic rate or metabolic-enzyme activities with depth in chaetognaths (arrow worms), jellyfish, or copepods (Fig. 12.19). Apparently, the mesopelagic habitat as seen by chaetognaths or jellyfish does not differ so much

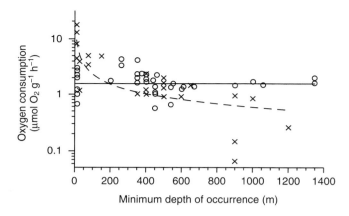

Fig. 12.19 Oxygen consumption per unit weight of copepods (open circles with horizontal regression) compared to that of large crustaceans (×, mysids and shrimp). All measures corrected to 5°C and equivalent weight. (After Thuesen *et al.* 1998.)

from the surface as it does for fish, squid, and shrimp. Thuesen and Childress (1993; also Childress 1995) attribute the difference to mode of prey finding in the two groups. The fish–squid–crustacean group mostly finds food visually. Exponential reduction in irradiance with depth shortens the distances from which information is received, so that possible reaction distance is also shortened. The short jumps of the deep habitat require much less metabolism than do the long jumps and pursuit swimming needed in the well-lighted surface. In chaetognaths and copepods, prey are mostly detected from vibrations, and the prey of jellyfish simply run into the tentacles. Thus, these animals require the same reaction capability and, hence, metabolism at all depths. According to Childress (1995), this explains the end of the metabolic decline for the sighted animals at about 800 m; once available light is reduced to occasional photons, vision becomes relatively less useful and reaction distance ceases to change with depth. There is contrary evidence: Ikeda *et al.* (2007; see Chapter 7) compiled copepod respiration results, finding a strong reduction in temperature-adjusted respiration rates of mesopelagic copepods. This generated an interesting polemic (Childress *et al.* 2008; Ikeda 2008).

Reproduction in the mesopelagic

Mid-water copepods, like epipelagic ones, divide into free-spawners (e.g. *Gaussia*, *Megacalanus*) and egg carriers (e.g. *Oncaea*, *Euchirella*, *Paraeuchaeta*, *Valdiviella*). Presumably, the former produce more eggs at each spawning that then suffer greater mortality before hatching, although specific data on fecundity and death rates are lacking. The young

of *Paraeuchaeta* molt through all the naupliar stages before beginning to feed, probably a common life-history feature in deep-living copepods. At least some deep-sea chaetognaths (e.g. *Eukrohnia bathypelagica*) also carry their eggs and even early larvae in "marsupial" sacs attached to the female gonopore. Other than that, chaetognath reproduction does not seem to be particularly specialized for the deep. Deep-living *Sagitta* (s.l.) are free-spawners. Among deep-sea euphausiids, females of the genus *Stylocheiron* carry their eggs glued to their thoracic legs, while *Bentheuphausia* and *Thysanopoda* are probably free-spawners. Eggs of the free-spawners tend to sink, but apparently not fast enough to reach the seafloor before hatching. *Stylocheiron* species produce only a few or a few tens of eggs at a time, implying very low mortality rates, while at least *Thysanopoda* females produce hundreds of eggs per clutch, implying high egg and larval mortality. Shrimp also divide into massively fecund free-spawners, for example all of the Sergestidae, and egg carriers, including the most numerous groups of the Caridea. While the eggs of carriers are protected by the mother's swimming and escape capabilities, they are often very numerous. A 7.5 cm *Acanthephyra quadrispinosa*, for example, was carrying 1500 eggs (Aizawa 1974). Nevertheless, the general division into low-fecundity egg protectors and free-spawners with high fecundity extends well down in the water column.

Sexes are separate in the squids and octopods, including those of the deep sea. Mating involves the male attaching a packet of sperm, a spermatophore, inside the lip of the female's mantle. The female then dispenses the sperm from a seminal receptacle organ to fertilize her eggs. Females develop very large ovaries at the expense of body tissue, and after mating they spawn and die. Hatchlings of a few millimeters have a body form like that of adults, but with relatively shorter arms. The larvae usually live in near-surface layers, working down as they grow. Squid growth is often extremely rapid, especially in forms that migrate up to the surface at night, leading to life cycles of only a year or two. Many of the more active, vertically migratory species also undertake mating and spawning migrations along continental coasts. For example, *Todarodes pacificus* spawns in the northeast corner of the East China Sea adjacent to Kyushu and the Tsuchima Strait (Okutani 1983). An adult female of about 50 cm total length (26 cm mantle length) produces masses of eggs in a gelatinous matrix the size of a soccer ball. These blobs are released near the seafloor, and they are dense enough that they do not rise. Females of this size have been estimated to produce up to 470,000 eggs 0.8 mm in diameter in several such blobs. "Nidamental" glands associated with the oviducts produce the gelling agent; it is now extracted by squid fisheries as a by-product and used for thickening ice cream. Hatching larvae are carried north through the Sea of Japan, and along the east coast of Japan inside the Kuroshio, to feeding grounds in the vicinity of Tsugaru Strait between Hokkaido and Honshu. There they feed and grow until the adults nearing mating condition swim back to the spawning grounds. The feeding grounds are the site of a large night-time fishery that uses brilliant lanterns, as visible from space as a large city, to attract the squid to barbed jigs. A similar migratory and reproductive pattern is followed by *Illex illecebrosus* along the US East Coast (O'Dor 1983), spawning over the outer shelf and slope south of Cape Hatteras, then moving north with the Gulf Stream to feed near Newfoundland. Its egg masses also sink. Some non-migratory, mesopelagic species have highly modified larvae; for example, the doratopsis larva of the Chiroteuthidae has odd swimming mechanics and a pronounced metamorphosis to the adult form. Much remains to be learned about reproduction in the fully oceanic and deep-living squid that do not migrate into surface waters at night.

Mesopelagic fish, according to Marshall (1979), are quite uniform in their reproductive patterns; most (all?) produce eggs less dense than their surroundings. He says they do that by excluding salts from the oocytes before spawning, presumably replacing their osmotic effect with ammonium (or something light). Salts are excluded by an outer membrane (chorion) impermeable to them. Data demonstrating this proposed mechanism are hard to find. Most mesopelagic fish eggs also have a small oil globule, about 1/25th of its volume, enough to increase buoyancy slightly. In any case, the eggs and early larvae are buoyant, so they rise to become part of the near-surface plankton. Fish seem to be the only group of nekton in which eggs of most species rise. Their fecundity can be very high. A 3 cm female of *Cyclothone*, probably the most numerous vertebrate genus in the sea, can produce 300 eggs, and a deep-sea gulper of modest size that spawns once and dies (*Eurypharynx pelecanoides*, mature at 60 cm – mostly jaw and whip-like tail) can produce 33,000 or more 1.3 mm eggs (Nielsen *et al.* 1989). Most eggs are eaten as they ascend, and only a few survive to hatch. Larval mortality is also high. Of course, high fecundity to compensate for low egg and larval survivorship is a common strategy in all marine groups. On the whole, the early larvae are very simple, worm-like creatures 1–3 mm long, usually with a persistent, ventral yolk sac. Early feeding, usually on micro-planktonic protozoa and crustacean nauplii, is important to survival in many species, and often begins before the yolk sac is fully absorbed. Later larvae can be very unusual-looking fish indeed. A favorite example among planktologists is the larva of the black dragonfish, *Idiacanthus*, a deep-sea predator with a prey-attracting chin barbel. The larva carries its eyeballs on slender, mobile stalks up to half as long as the body. Processing information about the location of prey from these distant, variably positioned eyes must involve complex proprioceptive as well as visual sensations. Development usually includes a fairly distinct metamorphosis to a juvenile form resembling the adult more closely, accompanied by a downward ontogenetic migration.

"Life in extreme environments" has become a catch-phrase in oceanographic circles, a selling point for the funding of research programs. Meso- and bathypelagic habitats are extreme environments, and the animals making homes there have marvelous and mostly interpretable adaptations. We should class them, along with hot-vent archaea, as *extremophiles*, and give them the attention their interest deserves.

Chapter 13

The fauna of deep-sea sediments

Seafloor habitats are termed *benthic* (adjective), an organism living in or on the bottom is a *benthont* (noun), and the assemblages of organisms are termed *benthos* (noun). The terms apparently are versions of the Greek "bathos" (βαθος), meaning depth. Benthic habitats share characteristics with both pelagic and terrestrial ones. They are (more-or-less) solid substrates, like the land, but they are continuously submerged in seawater. Thus, the basic physiological problems are the same as those for pelagic ocean life, but the more two-dimensional aspect of a land habitat operates as well. Benthic habitats grade downward in a series: intertidal, subtidal, bathyal (continental slope depths), abyssal, and hadal (trenches). The solid Earth has two principal surfaces, the continental shields above sea level, and mostly at the level of steppe or lowland rainforest (~300 m elevation), and the abyssal plains at depths of about 4500 m. This deep-sea zone occupies about 60% of world ocean area (Fig. 1.6). There are some rocky deep-sea sites, particularly at spreading centers, but most of the ocean is underlain by sediment-covered bottom, 2000 to 5500 m below the surface productive layer. Because of this areal dominance, and because sediment is easiest to sample, we will focus on the benthos of deep-sea sediments. Many insights come from comparisons to the more accessible subtidal and bathyal depths, so data from them will be used as well.

On bottoms within the euphotic zone there are often attached seaweeds, such as the kelp forests from Baja California to Alaska. However, those are in a very narrow zone next to the shore. The biota of the deep-sea floor depends upon food descending from above. At the sediment surface, food accumulates in and can be gathered from a narrow stratum. While only less than 1–2% of surface production reaches abyssal bottoms, it still represents a moderately rich resource compared to the water column just above.

Physical conditions are less variable at abyssal depths than in other large habitats. Temperatures are low, less than 3°C, but never low enough to freeze salt water. Salinity is 33–35 PSS but varies even less within oceanic basins. Oxygen concentration is mostly near saturation, greater than 4 ml liter^{-1}, although, in places where oxygen-minimum layers impinge on continental slopes, it can be much less. There is no sunlight, and below about 1200 m even bioluminescence is minimal. Hydrostatic pressure increases by about 1 atmosphere per 10 m depth, so at abyssal depths the pressure is 300 to 600 atmospheres. At any one site it is nearly invariant from a biological perspective. So far as is known, these conditions have been consistent since at least the end of the Wisconsin glaciation (~8000 YBP).

We can predict, before we begin reviewing the observations, that variations among benthic habitats will depend upon:

1 substrate type – rock or sediment, and within sediment there will be habitat variation with particle size, which we characterize as gravel > sand > silt > clay (and much sediment is characterized as diatom or foram "ooze");

2 depth;

3 food supply.

Biological Oceanography, Second Edition. Charles B. Miller, Patricia A. Wheeler.
© 2012 John Wiley & Sons, Ltd. Published 2012 by John Wiley & Sons, Ltd.

Rock supports assemblages of attached fauna. Until recently, with the advent of submersibles and remotely operated vehicles (ROVs), these have been very difficult to study. Sediments support both *infauna*, animals that nestle in or move through the substrate, and *epifauna* that slide or run about on the surface.

Since most of the seafloor is sediment-covered, and sediments are readily sampled by grabs, corers, or dredges, a majority of studies of "the benthos" have been of infauna. In order to see the animals brought up by a grab, say, the sediment must be removed from around them. A sample is placed on a screen and the sediment is gently washed through. Animals not going through with the sediment are those larger than the mesh, so mesh-size divisions define "ecological" categories:

NAME	GENERAL SIZE	SCREEN
Megafauna	>>1 mm	Any practical size
Macrofauna	>1 mm	Held by 0.5 mm
Meiofauna	0.1–1 mm	Passing 0.5 mm, but held by 0.062 mm
Microfauna	<0.1 mm	

These groups are usually constituted of different animal taxa, and since it is usual for one worker to master only one or a few large groups of organisms (say amphipods, annelids, or clams in the macrofauna, *or* nematodes, harpacticoid copepods or foraminifera), they are usually studied by different people. The life stories among these size categories are even considered to be separate problem areas in ecology. Megafauna (sometimes defined as visible in photos or on ROV video) and macrofauna are easier to study (less microscopy), so they are rather better known. Meiofauna are important ecologically, and they are under active study. The microbiota of bacteria and smaller protozoa are much more numerous than in the water column and account for a majority of overall benthic metabolism.

Sampling gear

It is not necessary for all purposes to collect animals. A great deal can be learned from photo- and videographs of the seafloor. Deep-sea cameras and lighting arrangements became available in the 1960s and provided clear looks at the appearance of the seafloor (Plate 13.1). Cameras can be left in place, taking time-series of photos. This has

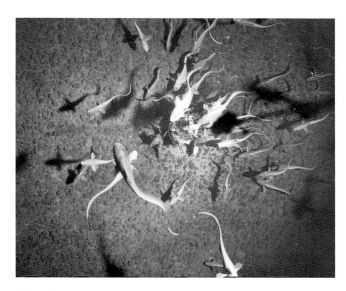

Fig. 13.1 Macrourid fish attracted to a bait anchored just above the seafloor at 5850 m in the northwest Pacific. (From the John Isaacs Papers, Scripps Institute of Oceanography Archives, University of California, San Diego.)

indicated that benthic environments are moderately dynamic, with episodic stirring of sediment by passing sea cucumbers and urchins, with mining and mound building by burrowing deposit feeders. There are intense depositional events when phytodetritus arrives at the end of the spring bloom, or when a large fish or mammal carcass (a *deadfall*) sinks to the bottom. In regard to food deposition, important results have been obtained by deploying cameras to take time-lapse photos of simply the sediment surface or of baits, a tuna body for example, moored at or just above the bottom. Declining blooms rather suddenly create turbid near-bottom phytodetritus layers. Photo sequences of baits (Fig. 13.1) show an arrival sequence of mobile animals.

Early collections of animals from the benthos were made with dredges. A simple dredge is a flat, steel box open at two of the longer sides. One open side is attached to a towing bail, the other to a collecting bag. The bag can be a mesh of chain or string. The dredge is hauled over and partly through the sediment, sieving it and retaining larger animals. Sufficiently rugged designs can be used to collect or scrape over rock. Details of design vary, including epibenthic sleds that slide along, scraping up and sieving the top decimeter or so of sediment. Megafauna can be collected by trawls whose footrope drags along or through the sediment surface. Beam trawls (Fig. 13.2), sometimes with distance-metering wheels alongside, are a modern version of this ancient concept.

For many purposes, it is more useful to collect inhabitants of sediment together with the sediment, doing the

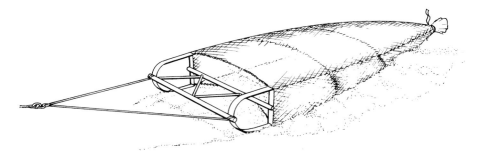

Fig. 13.2 An Agassiz trawl, an early, two-sided version of a beam trawl. (After Gage & Tyler 1991.)

sieving on deck. Thus, the sediment can be characterized and correlations defined between sediment properties (grain size, organic content, porosity) and fauna. A modification of the dredge, a Sanders "anchor" dredge, is sometimes used. Landing either side up on the bottom, it is dragged along for a short distance until the large, non-filtering bag or box (Carey & Hancock 1965) behind is full of sediment. When less disturbance of the sediment is desired, grabs and corers are deployed. Every sort of construction scoop has been used in benthic collecting: draglines, clamshell buckets, and orange-peel grabs. Mostly those depend upon their weight to hold them down as they dig into the sediment. Many grabs are designed expressly for benthic sampling. The simplest designs, for example the van Veen grab, are paired scoops suspended from long lever arms. Deployed by cable open, the arms are released on hitting the bottom so that pulling upward on the cable closes the scoop, collecting a mass of sediment. Mechanically more complex grab designs are numerous, but can be represented by the Smith–McIntyre (Smith–Mac) grab (Fig. 13.3a). Two spring-driven clamshell scoops are hinged into a heavily weighted frame and are cocked before deployment with a lever. Trip pads extending below the frame release the springs on impact with the bottom, and the scoops are driven into the sediment, closing together below a half cylinder of $0.1\,m^2$ surface area. The opening above the enclosed sediment is covered first with a screen, then with a rubber flap. Water passes under the flap during closure, but on retrieval it seals against the screen to prevent washout.

Although widely used, the Smith–Mac and its relatives (Petersen, van Veen, Okean, Campbell, etc., grabs) have been criticized for disturbing the sediment with the pressure wave that they push downward through the water that can blow away the sometimes fluffy sediment surface before impact. They also have substantial failure rates because any small stone or shell bit caught between the scoop edges will allow sediment to wash out on retrieval. Therefore, they have been largely replaced in recent work by box corers (Hessler & Jumars 1974, Fig. 13.3b). Several

cycles of modification have brought these to substantial sophistication, and some lighter designs can be deployed from larger boats as opposed to ships. A stainless-steel box, 25×25 or $50 \times 50\,cm$ and open at both ends, is driven into the sediment by weights. Striking the bottom releases a pin holding the lifting bail, which is also attached to a cable below that passes through pulleys. This cable on retrieval swings a blade down through the sediment, closing round the curved edge of the coring box. At the same time a lid is closed over the top of the core to prevent washing. Because the coring box is open, it has a lesser pressure wave. This is also minimized by very slow final approach to the bottom. Samples appear to be very slightly if at all disturbed and near-bottom water is retained over the top, warmed somewhat during passage through upper layers. Jumars (1975) and others have used box cores to examine the very small-scale patterns of faunal distribution in deep-sea sediment.

Elaborate sledge samplers have been developed for collecting epibenthos and near-bottom plankton just above the seabed. For example, the Macer–GIROQ system is a net with a square-mouth mounted on runners that opens just above the sediment when the sledge touches down. Above that is a plankton net with a shuttered mouth that also opens at arrival on the bottom and closes on lift-off. There are fins to hold the sledge upright (usually) during descent. For detailed description of this and sundry other benthic gear, see Eleftheriou and McIntyre (2005). Very large and mobile fauna must be captured with bigger trawl systems, and some very deep-living and active animals simply are not caught.

Since about 1950 the deep sea floor has been accessible to manned submersibles and these have grown in sophistication and sampling capability. They can be used to deploy sensors, corers, scoops, settlement plates, experimental modifications of the sediment, cameras, and more. More recently, many of these functions can be done with remotely operated vehicles (ROVs) deployed from ships by cable. With suitable propulsion and video relay to the operator on deck, these can visit sites at any depth and serve

(a)

(b)

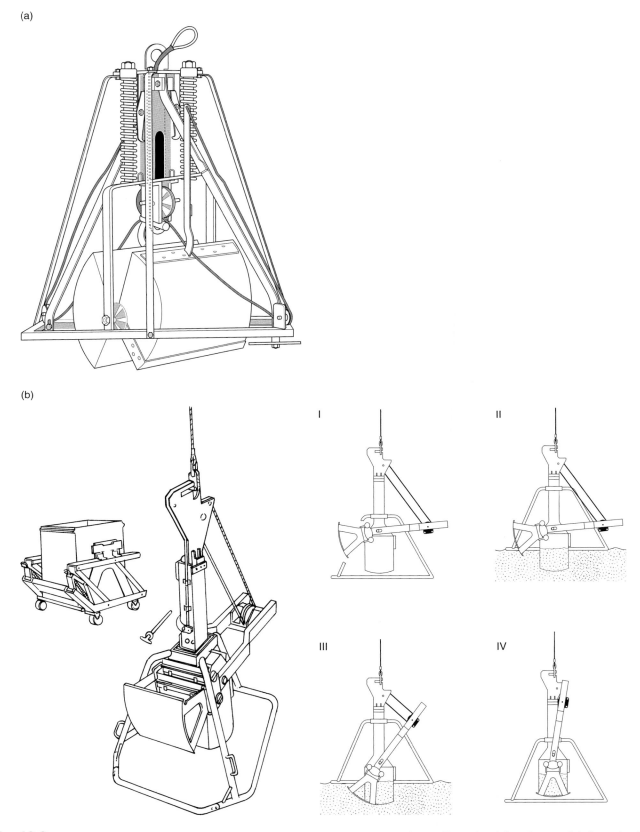

Fig. 13.3 (a) Design of a Smith–McIntyre grab (after Smith & McIntyre 1954). (b) A box corer, shown with core-box dolly and stages of deployment and recovery (I–IV). (After Gage & Tyler 1991.)

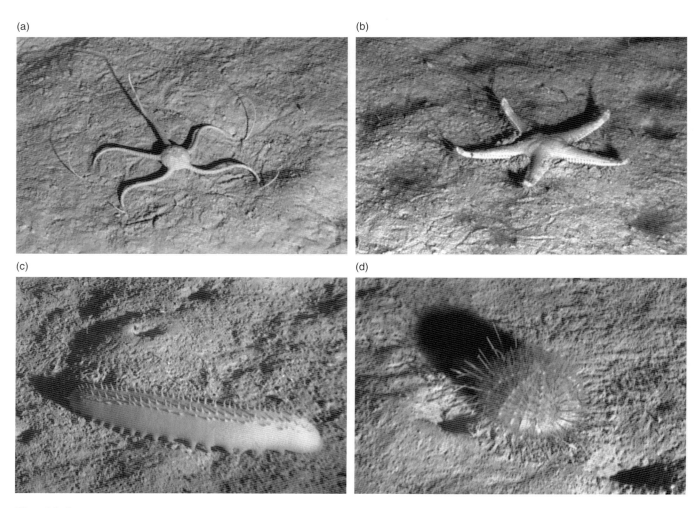

Fig. 13.4 Typical epibenthic echinoderms photographed *in situ*: (a) brittle star or ophiuroid; (b) sea star or asteroid; (c) sea cucumber or holothurian; and (d) sea urchin or echinoid. (All originally by A.L. Rice, courtesy of the UK National Oceanography Centre, and previously published in Gage & Tyler 1991; with permission of Cambridge University Press.)

most of the functions of manned submarines at less expense with risk only to equipment.

The descriptions and pictures that follow of benthic fauna cannot replace actually seeing these animals, particularly if you have a chance to see them alive or shortly after collection. If the chance comes your way to participate in a benthic sampling expedition, say yes.

Megafauna – largest denizens of the deep

Swimming along the deep-sea bottom and sometimes resting on it are specialized fish, particularly the grenadiers (Macrouridae, also called rat-tails) (Fig. 13.1), part of an animal assemblage sometimes termed the "suprabenthos".

Grenadiers have large heads with large eyes (although what they can see in the total dark is a mystery), a wide, deep mouth, and a tall dorsal fin. The body tapers posteriorly, with the tail fin continuous around a narrow, tubular tail with the long, deep ventral fin. Pectoral fins are large; pelvic fins are small. Grenadiers cruise slowly, seeking scent from deadfalls (see ahead) at which they are among the first arrivals. Other fish such as Liparidae, snail fish, lead somewhat similar lives, picking at animals in the sediment. Some deep-sea octopus species move close over the bottom. Particularly at shelf and upper-slope depths, the epifauna or suprabenthos includes crabs and shrimp. Shrimp that rest and feed on the bottom often swim up into the water column.

Trawl catches from all depths include crabs, squat lobsters, and shrimp, but, in the deep sea, the bulk of the megafauna are echinoderms (Figs. 13.4, 13.5d), including brittle stars (Ophiuroidea, as much as two-thirds of the

(a)

(b)

(c)

(d)

Fig. 13.5 Benthic megafauna photographed *in situ*: (a) galathiid crab crawling on one of several siliceous sponges, (b) burrowing anemones, (c) sea pens, and (d) a stalked crinoid. (a, b & c originally by A.L. Rice, courtesy of the UK National Oceanography Centre, d by and courtesy of Craig Young, Oregon Institute of Marine Biology, all previously published by Gage & Tyler 1991; with permission of Cambridge University Press.)

megafauna by numbers), sea stars (Asteroidea), sea cucumbers (Holothuroidea), sea urchins (Echinoidea) and, in places, both stalked (usually attached to rock) and free-living sea lilies and feather stars (Crinoidea). Brittle stars can break away from a limb when it is grabbed by a predator, hence their common name. They are very heavily calcified, both at the surface of the central disk, which bears the mouth ventrally, and in the chains of hard rings armoring the five limbs. The spacing of the rings allows sharp bending of the limbs and great flexibility. Sea stars are also flexible, but body bending is slower and stiffer. The internal organs of sea urchins are enclosed in a strongly calcified case or test, only an array of surface spines and ventral tube feet are movable. Many sea cucumbers are tubular, as their common name implies, but body designs with flaps and rather extended "legs" are also found. A diverse suite of holothurians, more than 20 species, principally in four families (Sars 1867; Miller & Pawson 1990), are documented as regular swimmers. Evidently this capability

evolved repeatedly. All echinoderms move over the seabed by the action of hundreds of tubular feet, tipped with suction cups in some groups, and moved by internal hydraulic pressure. Some ophiuroids can also move by stepping forward with lifted limbs.

Echinoderms have varied diets and feeding modes. Ophiuroids range over the bottom – selecting detritus, ingesting sediment, filtering particles above the surface, and even moving to deadfalls. Sea stars ingest sediment, prey on clams and other animals of low motility, and in one group (Porcellanasteridae) filter water drawn down into a subsurface burrow. Urchins are mostly deposit feeders, ingesting sediment as they move over it. They do move to windrows or masses of detritus and consume them rapidly. Some urchins that have evolved a bilateral symmetry live in burrows and filter water pumped through by movement of their spines. Sea cucumbers ingest sediment as they move across it, leaving tracks behind and depositing fecal blobs modified by mucilage addition and partial digestion of organic matter. Crinoids filter particles from the flow with specialized tube feet, passing the catch to the mouth along the limbs in mucus strands.

A few very large polychaetous Annelida perhaps qualify as megafauna, most of them deep-sea scale worms in the family Polynoidae. Two other worm phyla include large animals: the Hemichordata (including Enteropneusta) and Echiuroidea. These live on or in the bottom, selecting particles from the surface with specialized cephalic structures. Some hemichordates move in oscillating patterns, leaving characteristic fecal traces. Echiurids sit in a burrow, casting their proboscides across the sediment in star patterns (called lick marks) up to a meter in diameter, and depositing their wastes at the bottom of the burrow. They apparently move to new sites fairly frequently, since time-lapse photos will show lick marks for a while, then none. In some sites, usually with fairly persistent currents, there are sessile megafauna, including sponges, anemones, sea pens, crinoids (Fig. 13.5), and gorgonians or soft corals. Sponges found in deeper levels mostly are stiffened by siliceous spicules. They vary in form from flat encrustments to vase shapes. They feed by filtering flow generated by flagellated cells in the walls of internal passages; particles in the flow stick to collar-like structures on these same cells. Sea pens are cnidarians. Their bulbous base buries in the sediment and supports a thick fan of polyps extending into the flow to feed on impinging particles and plankton. Soft corals, that have branching organic skeletons extending from a base fixed to rock, feed in the same way. There are also stony corals in the deep sea with cup-shaped calcite bases several centimeters across, sitting partly buried in the sediment. The sea squirts, phylum Urochordata, are familiar to tide-pool visitors, and some deep-living forms are similar: stalked sacs engaged in mucoid filtering. However, some deep-sea editions of the sea squirt have taken on a predatory format (Havenhand *et al.* 2006) (Fig. 13.6). The incurrent siphon has been modified into a trap, attracting prey from the water then snapping shut around them.

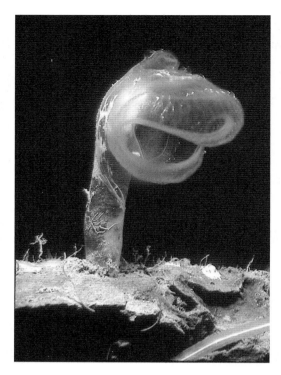

Fig. 13.6 A stalked, predatory sea squirt (*Megalodicopia hians*, Urochordata, Tunicata) attached to the wall of Monterey Canyon. (Photo by Dave Wrobel, © 1995 Monterey Bay Aquarium Research Institute.)

mouth and a seta-less peristomium behind it. The peristomium often bears eye spots, even image-forming eyes, and sundry palps or tentacles that assist in feeding in a variety of ways. In some (e.g. cirratulids), the tentacles reach out and gather sediment or food toward the mouth, in others (e.g. serpulids) they are shaped like feathers and form extensive filtering fans. The anterior esophagus in many families (e.g. glycerids) can be everted to grasp food, then retracted to ingest it. These proboscides can bear stout, recurved spines in predatory forms or be coated with mucus to which organic particles will adhere. Polychaete size varies widely in shallow bottoms, where some species are tiny and the largest forms can exceed a centimeter in diameter and several decimeters in length. In progressively deeper bottoms, however, size diminishes in practically all families.

Jumars (1975) studied a large suite of box cores from abyssal depths in the central Pacific. As is typical, polychaetes were about half the fauna by numbers and bulk. The four most abundant species made up half of the individuals. They were *Chaetozone* sp. (18.5%), *Capitella* sp. (15.8%), *Flabelligella* sp. (11.4%) and *Tharyx* sp. (6%). The first and fourth both come from the family Cirratulidae (Fig. 13.7a), which in shallow water are deposit-feeders living within the sediment, perhaps sorting sediment before ingestion or testing sediment richness with tentacles on the peristomium. Guts of the deep-sea *Chaetozone* and *Tharyx* were filled with sediment. *Capitella* (Fig. 13.7b), of the family

Macrofauna – sieve pickings

Composition of deep-sea macrofauna is characterized according to major zoological groups in Table 13.1. The big players among macrofauna are *polychaetes*, segmented annelid worms with clusters of spines and often flaps (parapodia) along the sides. They are usually over half of both the individuals and the biomass in a screening. Polychaetes belong to more than 80 mostly distinctive families, with many modifications in lifestyle, particularly feeding mode. The simplest polychaetes are very like the familiar earthworm, which is an oligochaete. Modifications on that basic body plan are most pronounced at the head, which is constituted of a conical prostomium segment ahead of the

Table 13.1 The numerical percentage composition of macrofaunal taxa of deep-sea, soft-bottom communities. (After Hessler 1972.)

TAXONOMIC GROUP	NORTHWEST ATLANTIC*		NORTH CENTRAL PACIFIC** 5600 m
	<4000 m	>4000 m	
Porifera	<0.1	0.2	1.1
Cnidaria	0.5	0.5	1.4
Polychaeta	70.4	55.6	54.4
Oligochaeta	0.7	—	2.1
Sipunculida	5.8	4.6	0.4
Echiurida	<<0.1	—	0.4
Priapuloidea/Nemertina/Pogonophora	0.9	—	—
Tanaidacea	1.6	19.3	18.1
Isopoda	1.0	12.2	5.9
Amphipoda	4.1	1.5	—
Cumacea/Misc. Arthropoda	0.1	0.2	—
Aplacophora	0.6	0.3	1.1
Bivalvia	13.0	4.3	7.0
Gastropoda	0.3	0.6	0.4
Scaphopoda	0.5	0.2	2.4
Ophiuroidea	0.3	0.8	0.7
Echinoidea	0.1	0.2	—
Crinoidea/Asteroidea/Holothuroidea	0.3	—	0.4
Ectoprocta	>0.4	—	2.1
Brachiopoda	—	—	0.7
Ascidiacea	<<0.1	—	1.1

*Data for the northwest Atlantic come from anchor dredge samples on the Gay Head–Bermuda transect (Sanders *et al*. 1965). The column for <4000 m averages 10 stations, ranging in depth from 200 m to 2870 m. Seven stations having a depth range of 4436–5001 m were used for the second column.
**The North Central Pacific data are averages of 10 0.25 m² cores, all from the same spot at 28°30′N, 155°20′W at 5497–5825 m depth.

Capitellidae, is a genus found at all depths. They are also deposit-feeders with a simple, earthworm-like shape, but they have a sticky eversible proboscis that probably has some role in selecting better food particles. *Flabelligella* belongs to a group not well represented in shallow waters, and so not well characterized; the deep-sea specimens had pelletized mud in the hindgut, and they are likely also deposit-feeders. Fifth- and sixth-ranked species belonged to families Paraonidae and Spionidae, which also feed on sedi-

ment, although probably more selectively than the others. Thus, the dominant life mode is direct ingestion of sediment, sometimes selecting or rejecting specific particle types.

Rarer on average, but in places abundant, are an array of polychaetes that live in tubes: Ampharetidae (Fig. 13.7c), Terebellidae, Serpulidae, and others. Tubes are made in several variant forms by glands on the body surface. In ampharetids the posterior end of the tube is buried

(a)

(b)

(c)

(d)

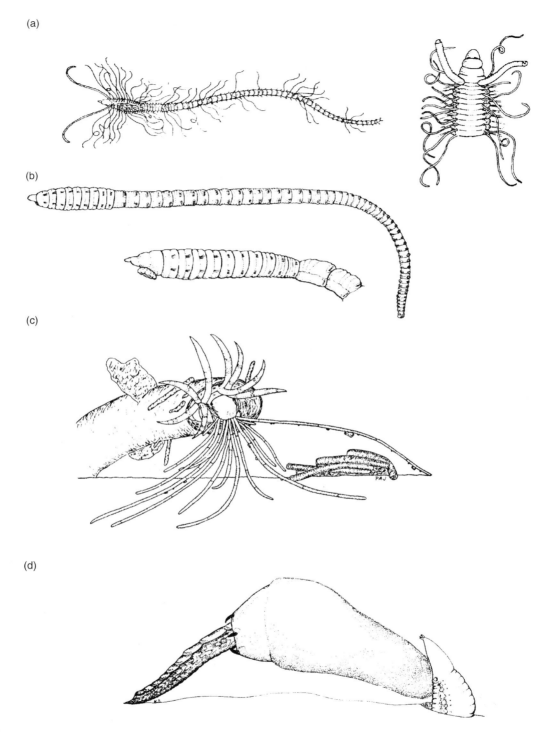

Fig. 13.7 Worms from four families of polychaetous annelids very common in deep-sea muds: (a) Cirratulidae, (b) *Capitella* of the Capitellidae, (c) Ampharetidae, and (d) Glyceridae. (a & b after Day 1967; c & d after Fauchald & Jumars 1979.)

vertically, and the anterior end tips down across the sediment. The worm has tentacles which search about the surface in the vicinity of the tube opening, selecting particles and transferring them to the mouth. Terebellid tubes are nearly fully buried, but the worm partially emerges to search about the surrounding sediment surface with its tentacles. Sabellids filter particles from flow over the sediment surface, as do members of several other families. All are mostly restricted to shallow depths. Predatory polychaetes mostly attack with spinose eversible proboscides, ingesting prey whole. The Glyceridae (Fig. 13.7d) are perhaps the best represented family that feeds in this way in the deep sea.

A group of aberrant annelids, the Pogonophora, lack any gut; they have come strongly to popular attention thanks to *Riftia* and its relatives, the large and colorful worms inhabiting sulfide-rich hydrothermal vent areas. However, much smaller pogonophorans are found in modest numbers in sediments at all depths, particularly where organic matter is abundant. They live oriented vertically, the lower end absorbing sulfide from the reducing layers of the sediment to support chemosynthesis by bacterial symbionts in a trophosome organ. Oxygen to oxidize the sulfide and drive the chemosynthesis is absorbed above the sediment surface by tentacles. DNA sequence studies revealed the close relation of pogonophorans to other annelids, and they superficially resemble the polychaete family Sabellidae in several respects. Formerly a phylum, they have been "reduced" to family status as Siboglinidae.

While polychaetes are dominant at all depths, there are definite changes in their character as sampling descends the continental shelf into the abyss. Shallower forms are larger, more ecologically varied, more colorful. Hartman and Fauchald (1971) describe the change downward in very strong terms:

"The single most conspicuous feature of the abyssal polychaetes . . . is the uniformly small size and the reduced number of body segments at maturity, as compared with their shallow water relatives. The body tends to be linear, plain; parapodia are reduced to small, papillar elevations, with little lamellar development, and armed with smooth, capillary setae which are rarely coarsely serrated or spinose along their free lengths. They lack the highly characteristic modifications uniquely developed in shallow water species. Most species have muted colors if any, or the body is translucent to dusky or black, with no visible pattern. Surface or epithelial modifications are rare . . . Cephalic eyes are absent, reduced or variable in occurrence; they are usually small dark pigment spots on the prostomium." (. . . continuing for several more pages, all emphasizing smallness and simplification).

The arthropod class Crustacea is the second most numerous animal group in deep-sea sediment, mostly the *pericari-dean* forms that brood their eggs and young in special pouches and have sessile (as opposed to stalked) eyes. Included are amphipods, isopods, tanaids, and cumaceans. The egg pouches are constructed from plates on the female's thoracic legs called oostegites. Amphipods are the dominant benthic crustaceans in fresh-water, salt-marsh, shelf, and upper-slope sediments. Typically their bodies are flattened laterally, taking an arched shape (Fig. 13.8a). Downslope they are progressively replaced by isopods and tanaids (Fig. 13.9). The defining differences between amphipods and isopods are technical, but these groups do sort out along the depth gradient. Isopods in shallow habitats most typically are dorso-ventrally flattened and of simple shapes (like the common terrestrial forms known as woodlice or pill bugs), but in the deep sea they take a variety of forms (Fig. 13.8b). For animals living buried in mud, they have amazingly sculptured and decorative shapes. The significance of this level of morphological complexity is not known; it definitely is a trend opposite to that described by Hartman and Fauchald for polychaetes. Tanaids (Fig. 13.8c) are elongate, almost worm-like, but they carry prominent chelae on the first thoracic limbs. Their strong sexual dimorphism often results in separate species names for males and females. Tanaids compose about 60% of deep-sea crustacea, the remainder being isopods and a few amphipods.

Third, or sometimes second, in abundance are the *Pelecypoda* or bivalve mollusks (clams). In shallow waters, the dominant clams are eulamellibranchs, a group with multiply divided gills, which are large in proportion to body size and employed in filtering particles from water above the sediment. At greater depths, eulamellibranchs are progressively replaced (Fig. 13.9) by protobranchs, clams with smaller, simpler gills not involved in feeding. Protobranchs are deposit feeders, sorting sediment with their labial palps and ingesting portions found suitable. Both *Gastropoda* (snails) and *Scaphopoda* (tooth shells) are found in and on sediment at all depths. Some snails deposit feed, others are predators, particularly those that bore into bivalves. Scaphopods can in places be a significant fraction of the deep-sea fauna. Their soft tissue is protected by a gently flaring calcareous tube buried wide end down in the sediment. The head and a muscular, digging foot extend from this lower end. Action of the foot moves the shell and body through the sediment. The head bears a tuft of captacula, thread-like tentacles with sticky, sensory pads at the tips. The foot pushes a cavity into the sediment, and the captacula search along and into the cavity surface for suitable food, which is then transported along the tentacle to the mouth by ciliary bands (Gainey 1972). Most of the diet is foraminifera (Bilyard 1974). *Aplacophora*, a primitive group of mollusks related to chitons and snails, are found in small numbers at abyssal depths.

Several phyla of worms reach significant abundance in deep-sea sediments. Sipunculids (phylum Sipuncula or

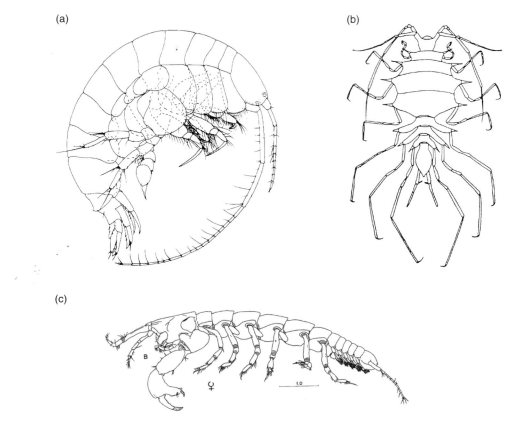

Fig. 13.8 Principal groups of benthic crustacea, all pericaridans: (a) a 6.0 mm amphipod, *Ampelisca* of the Gammaridea (after Myers 1985), (b) a dendrotionid deep-sea isopod (after Hessler *et al.* 1979), and (c) a tanaid, *Neotanais*. (After Gardiner 1975.)

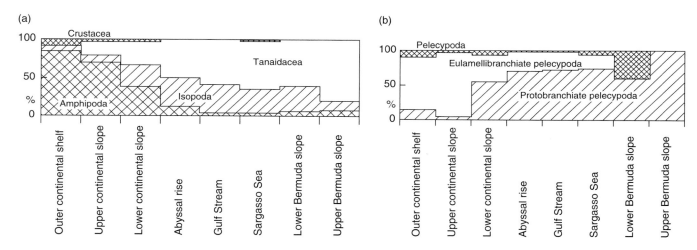

Fig. 13.9 Composition by major groups among (a) crustacea, and (b) pelecypoda in benthic samples collected from shallow to deep stations along the Gay Head to Bermuda transect in the northwest Atlantic. (After Sanders *et al.* 1965.)

Spunculida) look like peanut shells and are fondly called "peanut worms". The body is an elongate sac bearing a tuft of tentacles around the mouth at the narrow anterior end. In some the tentacles are for gas exchange, in others they are active in feeding. Most are non-selective deposit feeders.

Micro- and meiofauna

Compared to the water column above, marine sediments are rich in bacteria (<2 μm) *microfauna* and in a complex assemblage of protozoa and small metazoa termed *meiofauna*. Bacteria adhere to smooth surfaces of clay, silt, and sand grains, and burgeon in cracks and striations on the mineral bits. They are abundant in interstitial water. In surficial, oxygenated layers of the sediment bacteria metabolize organic matter, with particular importance for breakdown of more refractory constituents. They are responsible for a majority of sedimentary oxygen consumption. Their numbers continue to be high down the sedimentary column into anoxic layers where they continue to oxidize organic matter by stripping oxygen from nitrate and, quantitatively more important, from sulfate. The sulfide produced gives sediments and estuarine mud their characteristic, rotten-egg odor. At still deeper levels, bacterial abundance remains high, although their "biogeochemical" activity may not. The proportions that are active metabolically remain an issue; some may be effectively dead but show up in epifluorescent counts of preparations with nuclear stains. It was established long ago that many, sediment-dwelling deep-sea bacteria are obligate cryophiles and barophiles, requiring the cold (<3°C) and hydrostatic pressure of the depths to function. At least some forms that can be cultured do revive and grow (slowly) when recompressed in a laboratory after a trip to the surface in a grab or corer.

Feeding on the bacteria, on microparticulate detritus, possibly on DOM and on each other are the meiofauna. The array of such animals is studied by suspending a sediment sample in a larger volume of seawater and picking through it under a microscope with a pipette. Identification also requires microscopy. A thorough but accessible treatment of what these animals are like, of their biology and ecology, is given by Giere (2009). We only touch on those aspects here. Fully 20 phyla of animals are included. The Gnathostomulida, Kinorhyncha, Loricifera, Gastrotricha, and Tardigrada are found only as meiofauna, some of them also in fresh water, tardigrades even in moist habitats like moss, as well as in marine sediments. The rest are miniscule representatives of phyla usually thought of as much larger animals, for example, gastropods and holothurians. Apart from larvae of macrofauna and with exceptions like Loricifera, small size in metazoans is usually attained by reducing cell numbers, not by reduction of cell size. Rather complex, functional anatomy can be constructed from fewer than 1000 cells. Loricifera, however, pack more than 10,000 cells into ovoid bodies only 0.5 mm long. Typically, organs are reduced in number (one gonad instead of two, for example) and relative size. Numbers of eggs in females ready to spawn are few; Giere shows that ranges of one to four are typical. The low fecundity is partly compensated by (i) keeping larval stages in the sediment rather than as more risky planktonic stages, and (ii) developing very fast to reduce the time until the full adult defensive modes (spines, motility, secretions . . .) are in place. As among zooplankton, many groups are or include hermaphrodites – no individual is lost to either reproductive function – but cross-fertilization remains the rule. Copulatory sperm transfer is most common, but other means of mating include

deposition of sperm packets to be found by conspecific mates ready for female function. Most mate finding and sperm transfer likely involves chemical signaling, but little is known as yet of the specifics.

Many, but not all, meiofauna living in sand (which is mostly restricted to subtidal and slope areas) are very thin and elongate, particularly compared to their relatives in other habitats, adapted to twist through and perhaps anchor in the contorted interstitial spaces of the sediment, and some have a long, skinny tail, likely for the same purpose (Fig. 13.10). Some have adhesive pads or viscous surface secretions for attachment to sediment grains, raising their effective mass and, thus, their ability to stay put as the sand sifts. Some have epidermal spicules or internal stiffeners to enhance abrasion resistance or to enable them to penetrate sediments. In softer silts and clay, the bodies of meiofauna are mostly not exaggeratedly long, but various modes of stiffening in the body wall remain prominent (Fig. 13.11). Mechanisms for moving through sediments tend to emphasize ciliary locomotion in sand; and modes like press, pull back, then advance in muds. In rough order of numerical abundance among meiofaunal groups (although varying greatly with sediment type, water depth, and food availability) are foraminifera, nematodes (and some other worm phyla such as gastrotrichs), copepods (and a few other microcrustacea), flatworms (Platyhelminthes), and microannelids such as *Protodrillus*. There is sufficient convergent evolution in the group that it takes considerable expertise to distinguish, say, miniscule annelids from the pseudo-segmented nematodes and elongated copepods that look like annelids.

Foraminifera ("forams") are amoeba-like protists, moving and feeding by extension of cytoplasmic strands (pseudopodia). They are progressively more important with depth, reaching up to 30% of the benthont biomass below 2000 m (Shirayama & Horikoshi 1989). Most benthic forams have shells of calcite, protein, or adhered sediment particles surrounding their nuclei. Their shells, that are useful in paleoecological reconstructions, are found abundantly as fossils in cores and uplifted deposits. They provide useful contrasts with any planktonic forams preserved with them, and both provide mineral proxies for the composition of seawater in the past (e.g. Cd/Ca ratio as an approximate measure of phosphate concentration at shell deposition; Boyle 1988). There are more than 34,000 extant benthic species described in a still-expanding systematics, plus many extinct species. Reaching out through the sediment with networks of pseudopodia, they feed on bacteria, heterotrophic benthic diatoms (also "meiofauna", actually), detrital particles, and other meiofauna. Other amoeboid protozoa and an array of ciliates also live in sediments as meiofauna or microfauna. At least one foram genus, *Bathysiphon*, reaches a size included in the macrofauna. The cells construct tubular shells of aggregated minerals, usually including sponge spicules, up to 11 cm long

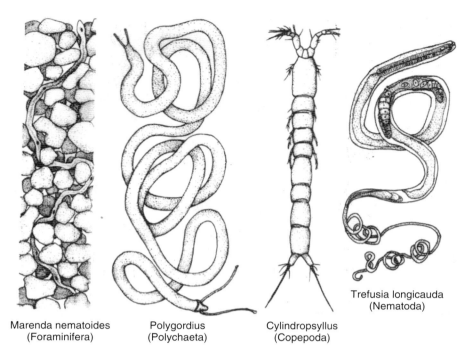

Marenda nematoides
(Foraminifera)

Polygordius
(Polychaeta)

Cylindropsyllus
(Copepoda)

Trefusia longicauda
(Nematoda)

Fig. 13.10 Examples of elongation of body form of sand-dwelling meiofauna, all to the same scale (see the sand grains in the left figure). From left to right: a foram, a polychaete, a harpacticoid copepod, and a nematode. The nematode shows a long, very thin tail, a feature of sand meiofauna in a number of groups. (After Giere 2009.)

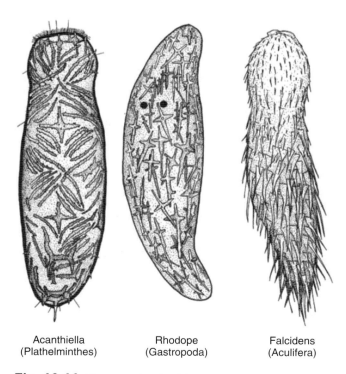

Acanthiella
(Plathelminthes)

Rhodope
(Gastropoda)

Falcidens
(Aculifera)

Fig. 13.11 Three examples of meiofauna from soft sediment, all with epidermal stiffening by spicules. From left to right: a flatworm, snail, and aberrant mollusk (the last is an example of Caudofoveata, a group related according to gene similarities to cephalopods). (After Giere 2009.)

and 2.3 mm in diameter (1.2 mm lumen). About 1 cm is buried in the sediment, the rest stands erect. As tubular networks, they can extend across the sediment to a meter or more. They are abundant at mid-slope depths in zones with very little disturbance. The cytoplasm is loaded with diatom frustules and various organic bits (Gooday *et al.* 1992). Even larger are the foram-related (based on DNA sequences; Pawlowski *et al.*, 2003) Xenophyophorea, multinucleate cells covered with particles. Some are tangles of mineral-covered tubes spread across the sediment surface only millimeters thick to the size of dinner plates; others are smaller balls of tube tangle. Small metazoans are often associated with these massive "protists", in the tube mesh or underneath it (Buhl-Mortensen 2010).

Nematoda are both numerous and remarkably diverse. For example, Lampadariou and Tselepides (2006) found 104 genera (most with numerous species) in just one closely spaced set of stations in the northern Aegean Sea. These small worms are found in virtually all Earth habitats: both free-living and as parasites. The meiobenthic forms move by wriggling against the sediment and feed with an oral complex that varies by diet. Predators may have puncturing stylets, esophageal suction bulbs or circlets of miniscule teeth. Many with simpler mouths ingest whole sediment, digesting from it bacteria, diatoms, detritus, and also absorbing dissolved organic matter. Experiments suggest capability for moving toward preferred foods guided by olfaction (Moen *et al.* 1999). Meiobenthic copepods mostly belong to the order Harpacticoida and occupy the lower limit of copepod adult sizes, about 0.5 mm. They move by digging with their feet and by body flexion. Platyhelminthes ("flat worms", often termed Turbellarians) are significantly abundant and a diverse part of the meiofauna, particularly the order Acoela, with a tiny mouth that opens through the epidermis to a solid column of digestive tissue rather than to an open gut. Acoels are worm-like in smaller sizes, while larger forms are flatter; all of them move through sediment largely by ciliary action.

The complexity of the meiofaunal taxonomic scheme duplicates virtually all of phylum-level zoology from the cnidarians to the chordates, and sediment (or at least very tightly confined spaces) is the only habitat of at least the five "minor" phyla listed above. For extended information, we again recommend Giere (2009).

Gradients with depth and surface productivity

Depth in the sediment

On the continental slope and deeper, the meiofauna and macrofauna within the sediment are found very close to the sediment surface. Macrofauna stir the sediment to depths of at most 10 to 15 cm, with reducing conditions excluding

them from below that level. Exceptions are larger forms that excavate and then irrigate a burrow, so that the oxidative layer can line it for perhaps a centimeter to either side. The bulk of meiofauna are found even closer to the sediment surface (Snider *et al.* 1984), within about 4 cm, but members of the more abundant meiofaunal groups are also found in vastly reduced numbers in deeper layers with significant sulfide concentration. In all groups, there is a strong shift in the genera and species composition at the redox interface. The deeper community, dominated by nematodes, is referred to as the "thiobios" (which includes anaerobic bacteria, of course). Investigation of the necessary physiological adaptations is in progress, hampered of course by the small size of the study subjects.

Abundance versus water-column depth

Trends in faunal numbers and biomass with depth have been treated with statistical sophistication by Rex and Etter (2010), using partial regression to standardize masses of data from the literature to a fictional common suite of collector, sieve-mesh opening and latitude. They provide figures with these comparisons from a wide range of sites, and we show their comparisons between the east and west sides of the North Atlantic (Fig. 13.12), which are statistically indistinguishable. Of course, the standardization has the additional effect of moving plotted points well away from the actual data. However, the general trend with

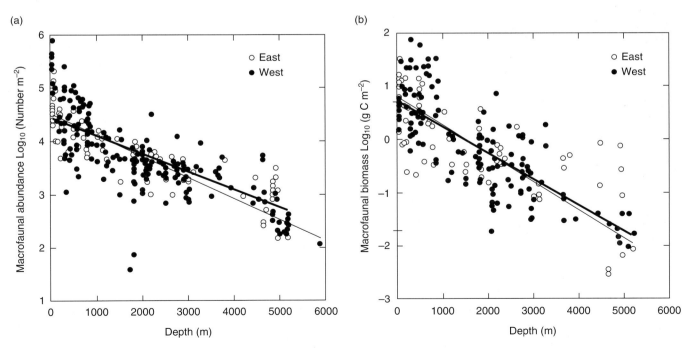

Fig. 13.12 Semi-log plots of (a) numbers of macrofauna vs. depth, and (b) macrofaunal biomass vs. depth for samples from the east and west sides of the North Atlantic. The regression lines (east heavier) are not significantly different. (After Rex & Etter 2010.)

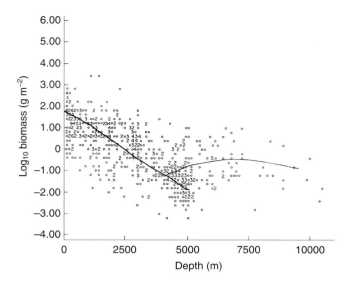

Fig. 13.13 Biomass as wet weight vs. depth from Soviet era oceanographic studies. Trend lines drawn freehand by us. (After Rowe 1983.)

increasing depth could be well represented by almost any transect of samples crossing a continental shelf and slope, extending to the abyss. A compendium (Rowe 1983) of Russian biomass vs. depth results from the extensive cruise work of the Soviet era, mostly results from "ocean" grab samples, shows a similar trend (Fig. 13.13).

Both greater numbers and greater biomass are found nearshore and at shoal depths, decreasing seaward and down. That is because landmasses provide nutrient-rich run-off and/or promote upwelling, both of which enhance surface productivity, and because less of that production is eaten and metabolized before sinking organic matter reaches the shallow bottom. Because trenches with hadal depths are almost entirely found close to island arcs and continents, the trend levels off, even reverses at about 5000 m, just above the abyssal–hadal boundary (Fig. 13.13). Waterlogged organic matter (coconuts appear in trench-floor photographs) supplements the food supply there. There are also recurring sediment slumps and resulting turbidity currents that carry shallow sediment with high carbon content down into trenches.

Production consistently falls off seaward, and the opportunities for consumption on the way to the seafloor are progressively greater. The fitted regressions (Fig. 13.12) show a decrease of numbers by 50-fold from 0 to 5000 m, while biomass decreases by ∼224-fold. In alternate mathematical form, the reductions are exponential at ca. −0.8 $(1000 \, \text{m})^{-1}$ for numbers and ca. −1.1 $(1000 \, \text{m})^{-1}$ for biomass, respectively. The difference is necessarily accounted for by a decrease in individual size. As we discussed particularly for polychaetes, deep-sea benthonts are smaller than related species on the continental slopes and shelves. If the dispersion around the general trends is examined in detail, par-

ticular areas turn out to have higher or lower abundance and biomass as a result of regional variations in conditions, most importantly the supply of organic matter but also sediment type, currents, and temperature. For example, the stocks of every faunal size group offshore from Cape Hatteras, North Carolina, are substantially higher than the typical mid-slope (200–800 m) averages for the North Atlantic. According to Aller *et al.* (2002), "interplay of continental shelf and slope topography with interacting circulation patterns of South Atlantic Bight, Mid-Atlantic Bight, Slope bottom [currents], and Gulf Stream waters near Cape Hatteras, NC, clearly promotes high primary productivity and makes the Hatteras region a likely site for extensive modern carbon export to the neighboring slope." There are other special situations such as off the Peruvian coast: an anoxic zone quite near the shore has almost no macrofaunal biomass, but there are biomass peaks farther offshore where some oxygen is available and food is supplied from nutrient-rich upwelled waters. Rex and Etter's standardization to a fictional fixed latitude removes an important regional distinction. High-latitude seas (save for the Arctic Ocean) tend to have larger standing stocks of benthos than tropical ones. This is at least partly because the large seasonality of primary production (blooms) does not allow it to be so efficiently captured by the pelagic communities as in the more consistent tropics.

Only the decrease in macrofauna is illustrated here, but there are also decreases in all the other categories, with greater rates per meter of added depth for megafauna and progressively smaller rates for meiofauna and bacteria.

Deep-sea species diversity

It has provided a source of endless fascination, not to mention the pleasures of taxonomy, for benthic biologists that *species diversity*, the number of kinds of animal forms, does not decrease dramatically down to considerable depth; the fauna of the upper abyssal floor is about as diverse as that of shallow water, in most comparisons more so. Species diversity is considered by ecologists to have two aspects. An assemblage of organisms, say the macrofauna sieved from an anchor-dredge load of mud, is more diverse if it has a larger total number of species, S. It is also said to be more diverse if the individuals are more equally distributed among the species. Thus, it is more diverse if, as individuals are randomly selected and added to an identified list, the number of species rises rapidly toward S. This aspect of diversity is termed *equitability*. The significance of this is that an animal, say a predator, moving through a low-equitability community, will mostly encounter the same few species over and over. Perhaps a relatively small repertoire of behaviors for interactions will suffice for it. It could afford to specialize on one or very few prey types. A predator moving through a high-equitability community, in contrast, would seldom encounter immediately other animals of the same kind as the one most recently met. It might need a larger repertoire of behaviors. The two aspects of diversity are somewhat tied together, since at least some individuals of every species must be present if S is to be large, creating some equitability, but they are not tightly coupled.

A measure of diversity emphasizing equitability is Simpson's diversity index, L, which is the probability (estimated from a sample of N individuals) that two successive animals encountered at random from an assemblage will belong to the same species:

$$L = \sum_{i=1}^{S} \frac{n_i(n_i - 1)}{N(N - 1)}$$

where the n_i are the numbers in the sample of the ith species. Fifty years ago it was believed that knowing and comparing diversity measures, such as L, among assemblages from different habitats and sites, would provide strong ecological insights. Those mostly did not materialize.

Howard Sanders, the leader of benthic studies along a transect from Gay Head (Martha's Vineyard) to Bermuda, was greatly impressed with the observation that faunal diversity does not decrease with the descent to depth. He expressed assemblage diversity with "rarefaction curves". Such curves (Fig. 13.14) from a variety of sites show that, while benthos of shallow tropical seas (the Bay of Bengal in Sanders' comparison) can be more diverse than Sargasso Sea benthos, most shallow benthic habitats have fewer species per individual than the deep northwest Atlantic. What are these curves? They are "synthetic" species–area curves. Gerlach (1972) developed an actual species–area curve for benthos from the Baltic Sea (Fig. 13.15), which

is a plot of the cumulative number of species as the number of sieved grabs included in the sample is increased, that is, as area sampled is increased. The steepness of such a curve is a measure of its species equitability. Also, the higher its asymptote, that is the more total species, the more diverse it is. If it is steep, then the more likely it is that the next animal that a given worm meets crawling through the mud will be something new, something unexpected and challenging. The higher the curve, the more likely that it will meet something new and unexpected eventually. Rarefaction curves (Sanders 1968, 1969) are just species–area curves modeled statistically from very large faunal analyses. The calculations for increasing sample size, for n specimens up to the actual sample size, N, where N_i is the number of the ith species among S species, go according to:

$$\text{Expected no. species for sample size } n =$$

$$E(S_n) = \sum_{i}^{S} \left[1 - \frac{\left(\dfrac{N - N_i}{n} \right)}{(N / n)} \right].$$

The bracketed quantities in the ratio represent combinations of N items taken in groups of size n. Hurlbert (1971) actually suggested this formula, showing that Sanders' original calculation overestimated expected S at any given n. Comparing whole curves of $E(S)$ vs. n avoids the problem that, since slope (equitability) and asymptote (S) interact, quite different samples can give the same $E(S_n)$ at a given sample size, n. Nevertheless, $E(S_{100})$ and other $E(S_n)$ (occasionally with $n < 15$) have been used to compare diversity

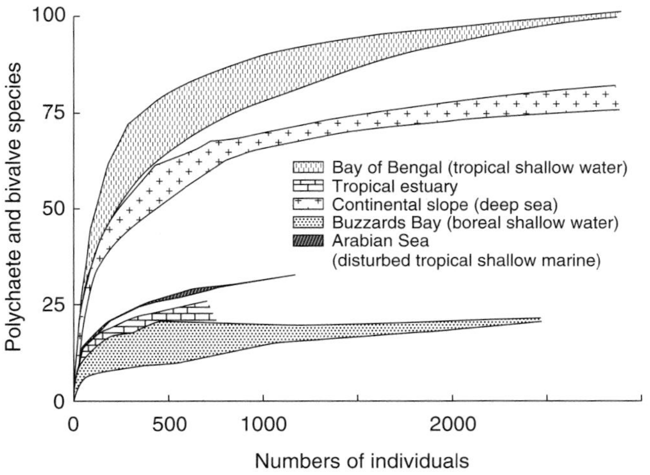

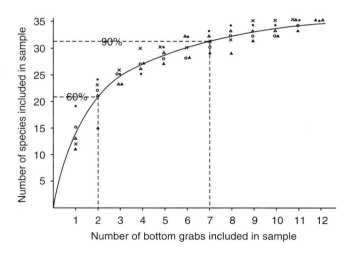

Fig. 13.14 Areas bounding rarefaction curves for polychaetes and pelecypods combined from different marine benthic areas. The deep sea, despite its trophic poverty and nearly invariant conditions, has many species and high species equitability. (After Sanders 1968.)

Fig. 13.15 A species–area curve for benthic infauna from the Helgoland Bight of the Baltic Sea. The fitted curve shows that two grabs will typically capture 60% of the total fauna, seven grabs 90%. (After Gerlach 1972.)

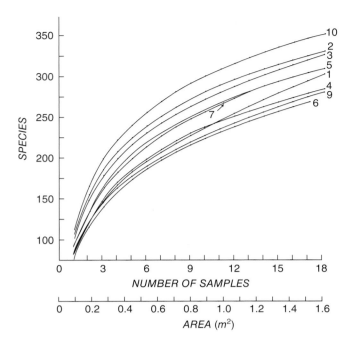

Fig. 13.16 Species–area curves generated by repeatedly adding successively the faunas of 18 randomly selected samples, subsets from the 233 samples collected at 2100 m on the continental slope off New Jersey, USA. The variations between the runs (numbers and the curve ends) represent the effects of multiple scales of patchiness in community composition. (After Grassle & Maciolek 1992.)

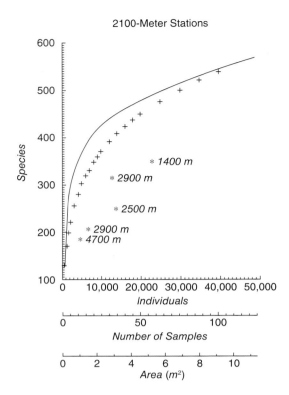

Fig. 13.17 Continuous curve is mean species–area plot calculated from many successive orderings of 125 samples from 2100 m off New Jersey. Plus symbols (+) trace the species rarefaction curve for the same data. (After Grassle & Maciolek 1992.)

between sites. Rarefaction curves remain uniquely popular among benthic ecologists.

One study (Grassle & Maciolek 1992) that came perhaps as close as possible to documenting fully the diversity in the deep sediment of a single region in a narrow range of depth was conducted along the coast of New Jersey, USA. A suite of box cores was taken over ~108 km along the 2100 m isobath to evaluate the habitat in advance of possible oil or gas exploration. Ten stations, augmented by a few both deeper and shallower, were sampled with replication at three seasons in each of two years. An extensive systematic evaluation was made of the macrofauna, identifying >98% of individuals to species, meaning typological species based on morphological variation. Many of them were identifiable to the team of taxonomists as species, but were not species already described. The totals from 233 cores were 798 species in 14 phyla among 90,677 individuals, dominated by 48% annelids, 23% peracarid crustacea, and 13% mollusks.

The 10 dominant species, from 7.1% down to 2.1% of the total, were consistently the dominants and made up quite similar proportions at virtually all stations and seasons. Together they were 35% of individuals. However, only ~20% of all species, including those 10, were found at all stations (in at least one core), whereas 34% occurred at only one station, 11% occurred in only two cores and

28% in only one core. So, the tail of the relative-abundance distribution was very long. There were both a regional core community and a rather wildly varying suite of less-abundant species mixed with them. Grassle and Maciolek calculated species–area curves in several ways: (i) by repeatedly adding their samples in random orders then averaging (Fig. 13.16) and (ii) by the rarefaction formula (Fig. 13.17). The results were a little different, but (i) both showed ~150 species at 1000 specimens (not explicit on Fig. 13.16), (ii) both required progressively more specimens (and sampled area) to add species above 300 or 350 species, but neither was fully asymptotic at 50,000 (or even 90,000) specimens (Fig. 13.17). Separate curves for polychaetes, crustacea, and mollusca were all initially steep (lots of equitability) and had not leveled off for sample sizes in excess of 10,000 individuals (implying large ultimate numbers of species). These curves are not additions of area and individuals from contiguous seabed; several scales contribute to the data, which thus represents a cumulation of species from patches with different histories. They do show that a mosaic of patches can support very large (although impossible to estimate precisely) numbers of species.

The high diversity of deep-sea sedimentary habitats was surprising, when discovered, because it exceeds that of shallower, better-studied habitats. Another extensive sampling and species-identification project, led by Nancy Maciolek, produced one of the better shallow versus deep comparisons, actually calculated by Etter and Mullineaux (2001), between the flanks of Georges Bank offshore from New England, USA (38–167 m) and the deeper continental slope and rise to the southward (250–2180 m). The comparison is not fully fair, since a greater depth range is included in the deeper sample set, but the greater diversity in the deeper sediments is confirmed if only the 1220–1350 m samples are used:

	GEORGES BANK	SLOPE (250–2180 m)	SLOPE (1220–1350 m)
No. of samples	1149	191	63
Ave. species m^{-2}	165	278	319
$E(S_{1000})$	69	156	188

"Synthetic" species numbers vs. sample-area curves (close relatives of rarefaction curves; Fig. 13.18) confirm the greater diversity of the deeper communities. Both evenness (the steeper initial slopes) and total species numbers are greater in the deep sea.

The New England slope study also shows, however, that the downward increase in diversity does not extend indefinitely: $E(S_{100})$ values were 50–65 at 1250 m, and only 40–55 at 2250 m. Similar results have been obtained from many slope to abyss transects. Olabarria (2005), sampling in the Porcupine Sea Bight (NE Atlantic) with an epibenthic sled and trawls, found an increasing number of bivalve mollusk species from 500 to 1200 m, then consistently 41–49 species to 3500 m, and finally a modest decrease onto the adjacent abyssal plain (Fig. 13.19). Svavarsson (1997), also towing sledge samplers to sample upper sediment layers, found a more dramatic rise then decrease in isopod species numbers with depth in the Arctic Ocean north of Iceland (Fig. 13.20). Total species per sample and $E(S_{200})$ both showed that pattern, with $E(S_{200})$ less affected by the number of individuals per sample.

The question that rarefaction curves suggested to Sanders (1968), and which continues to fascinate others (Rex & Etter 2010), is: how can such a homogeneous environment with so few apparent ecological roles retain so many species? Presumably a benthont can: (i) ingest sediment; (ii) filter particulates from the water above; (iii) wait for and move to deadfalls; or (iv) eat other benthonts. This scheme is surely close to the total possible list, especially for infauna. The reason that the question seems important is the "competitive exclusion principle", previously considered when we were puzzling over the high plankton diversity in subtropical gyres. To repeat, it is a sort of ecological "law" derived from a theoretical model and from experiments in jars with paired species of flour beetles or *Daphnia*:

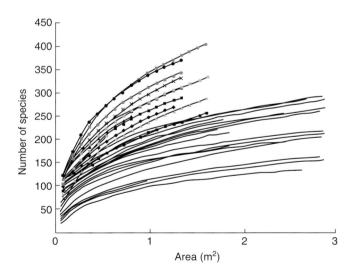

Fig. 13.18 Number of macrofaunal species vs. cumulative sample-area curves based on a large data-set and calculated by a method similar to that of Fig. 13.16 for shelf depths on Georges Bank (no symbols) and for the continental slope (various symbols) to the southeast. Both the initial slopes (equitability) and asymptotes (total species, *S*) are greater for the deeper, slope samples. (Data from Maciolek and others – a US Minerals Management Service project. Figure after Etter & Mullineaux 2001.)

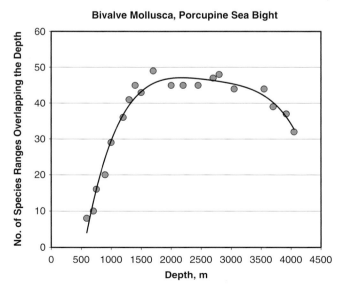

Fig. 13.19 Numbers of bivalve mollusk species, with ranges overlapping depths plotted on the abscissa. Samples were collected with sledges and trawls from the continental slope (500 m) to the abyss in the Porcupine Sea Bight, NE Atlantic. (Data extracted from Olabarria 2005.)

(a)

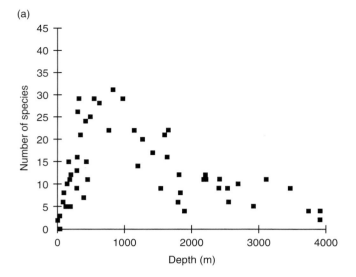

(b)

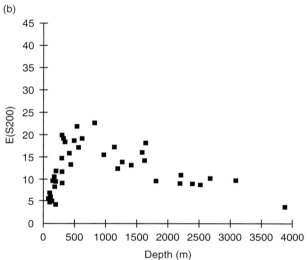

Fig. 13.20 Diversity of isopod species collected with a sediment-skimming sledge along a continental shelf to abyss transect north of Iceland: (a) as total number of species collected and (b) as $E(S_{200})$. (After Svavarsson 1997.)

interactions between species populations competing for limited resources are expected to result eventually in one winner, losers becoming extinct. If that's true, and there are very few distinguishable resources, how can so many deep-sea species persist?

In some habitats, the answer is taken to be that there is great temporal variability, i.e. that continual environmental fluctuations keep the rules of the competition (the "competitive coefficients") shifting. No competition reaches its conclusion before the rules change. In the deep sea, on the other hand, it was once reasonable to think there could be virtually no variations: salinity is the same within 0.1 PSS temperature varies within 0.2°C; food fluctuations should be damped by the transit to the seafloor (or were thought

to be in the 1970s); and so on. In the early 1970s, this great "problem" of deep-sea ecology was the topic of an intense polemic. Scientific disputes are often the most heated over issues that are fundamentally untestable. Explanations of regional or inter-habitat differences in diversity, such as the difference between boreal and tropical forests in number and equitability of tree species, were then being sought by many ecologists, so that there were several classes of hypotheses to choose among. Most of them appeared in the discussion of the deep-sea benthos:

1 Sanders (e.g. 1968, 1969) proposed that high diversity is maintained because deep-sea communities, faced with only mild physical variations, have evolved to ameliorate competition. *Character displacement* produced finer and finer division of the resources among competing species, because even very restricted niches are always present. Evolution of super-specialists paid off in reduced competition and led to co-existence. Character displacement does occur in more readily studied, terrestrial habitats. For example, species of birds competing for food evolve longer and shorter bills respectively, thus separating the levels at which they feed. The character "bill length" is said to be displaced. Sanders called this explanation of high deep-sea diversity the "time-stability hypothesis". The great span of time under stable, unchanging conditions has allowed very refined subdivision of the ecological roles, which he called "biological accommodation". Evidence for character displacement in the deep sea is scanty, however. Determining how four or five species of deposit-feeding, cirratulid polychaetes might be partitioning the uppermost layer of mud is extremely difficult.

2 Deep-sea habitats are not so constant as they seem. Large demersal fish flop about, burying in at times. Sea cucumbers 10 cm wide drag through the mud, digesting everything along swaths that criss-cross everywhere. Sediment miners throw up small volcanoes of sediment. Dead whales, sharks, and pieces of tuna fall (plop) to the seafloor (Smith & Baco 2003), providing huge doses of food to be distributed locally as macrourid, hagfish, and amphipod feces. These events create microstructure in the seemingly homogeneous habitat, and that allows diversity. As the explanation is often phrased, deep-sea benthos is "a spatial mosaic of patches", patches in different phases of a succession that recurs regionally, and generally with greater fidelity along isobaths. Fast-growing, mobile species ("opportunists") can take advantage of empty spots. Whole successions are possible, from early colonists through to permanent "climax" species, or at least permanent until the next disturbance. Overall, the endless cycle of disturbance allows co-existence of a large number of forms on a scale of hundreds of meters. In the deep-sea context, this was suggested by Dayton and Hessler (1972), who sought

alternatives to the time-stability hypothesis. They saw no evidence of real niche separation among deposit feeders; noted that species ranges scatter along the depth gradient such that a given species may live with several different assemblages (which does not suggest tight biological accommodation); and proposed that population density is too low to generate much interaction among individuals. Jumars (1975, 1976) showed that there is indeed small-scale patchiness in deep-sea benthos. Of course, patchiness could be a mechanism of biological accommodation in its own right. Rex and Etter (2010) call this "the principal paradigm for explaining local species coexistence".

3 It can be shown in mathematical models of competition that competitive exclusion will not run its course if predators keep competing species from reaching the population limits set by resources. Given suitably unselective predators ("croppers"), exclusion never acts and diversity is sustained. Dayton and Hessler (1972) brought this under the rubric of disturbance, but really it is different.

4 Rex (1981) and Grassle and Morse-Porteous (1987) invoked an "intermediate disturbance" hypothesis (attributable to Joseph Connell). That is, species diversity will be low where disturbance is frequent and violent, only allowing motile opportunist species to survive, and where there is no disturbance so that competitive exclusion can run its course. At some intermediate level of disturbance, the mosaic of patches – ranging from recently settled opportunists to fully accommodated climax assemblages – will have its greatest complexity and diversity will be highest overall. This is, of course, simply a gloss on explanation "2" above.

There may be other aspects to the explanation, but probably all of at least this set of mechanisms are at work to some degree. Grassle and Sanders (1973) offered an answer to Dayton and Hessler. If you read it these several decades later, you can still feel the intensity. Work on the general issue continues, for example, the work of Wei *et al.* (2010) on the shelf-to-abyss gradient of diversity in the Gulf of Mexico. Rex and Etter (2010) review the data and the arguments in a useful book, extending the discussion to meiofauna and megafauna. Many benthic ecologists have moved on to studies of population biology in individual species and to the interaction of given animals with the habitat. It remains intriguing, and persists in being hard to understand fully, that deep-sea sediments sustain hundreds of species in modest areas with little apparent diversity in the habitat.

Benthic biogeography

This topic is of obvious importance, although its study is hampered by the isolation of the seafloor beneath a thick mantle of water. Moreover, distributions are not necessarily moved about by currents to provide genetic exchange across whole gyres as is the case with epipelagic plankton. Thus, a sufficient spatial density of samples to reveal patterns securely is difficult both to acquire and to work up in good systematic detail. Russian workers have given oceanic, abyssal distributions the most attention, having developed an extensive data-set in their worldwide cruise work of the 1950s to 1980s. Vinogradova (1997) has summarized the results. In agreement with impressions from the early global expeditions (named after the ships: *Challenger, Valdivia, Albatross, Galathea, . . .*), the Russian results showed very widespread, in many cases cosmopolitan distributions of macrofaunal genera and all higher taxa. Of course, genera are subjectively delimited by taxonomists, but at least they are in nearly all cases closely related species. On the other hand, species distributions are much more restricted, with 85% of >1000 species in Vinogradova's analysis occurring in one ocean only, and only 4% were cosmopolitan. Most of the latter have very wide depth ranges, and presumably can extend their distributions across ridges. Of course, these proportions, while typical, vary among taxa. Thus, species endemism is high, which corresponds to individual movement being mostly short range and to isolation occurring readily between basins separated by ridges. Many regional studies agree. For example, Menzies (1965) found that only 22 of 158 isopod species in 22 genera collected from the Argentine and Cape basins of the SW Atlantic are present in both. In contrast, all of those 22 genera are also found in the Pacific. Vinogradova found that Pacific and North Indian faunas are more closely related at the generic and higher levels than to either Atlantic or Antarctic faunas. Both the Pacific and Atlantic faunas have distinctive eastern, western, and northern subdivisions. In the Atlantic, the division line follows the Mid-Atlantic Ridge. The Antarctic fauna is partly distinct between the Atlantic, Pacific, and Indian sectors. Trenches showed very high levels of endemism, half or more of the species in a given trench are found only there.

For abyssal sea bottoms deeper than 3000 m, Vinogradova (1997; basic pattern originally published in 1959) proposed that patterns of faunal similarity (shared species, shared proportions of genera, consideration of many taxonomic groups) separate the world seafloor into three main regions (Fig. 13.21): (1) Pacific and North Indian; (2) Atlantic; and (3) Southern Ocean out to the subtropical convergence.

Each region can be divided into provinces (shown in her figure by varied hatching), the demarcation lines largely following mid-ocean ridges. Arctic abyssal fauna are somewhat distinct, but Vinogradova found a stronger relation with the Atlantic province than with the quite strongly distinctive subarctic Pacific one ("$1A_1$"). The province pattern suggests that the dominant spatial barriers allowing allopatric speciation in deep benthos are bathymetric and are, therefore, shaped quite differently from biogeographical patterns in the plankton.

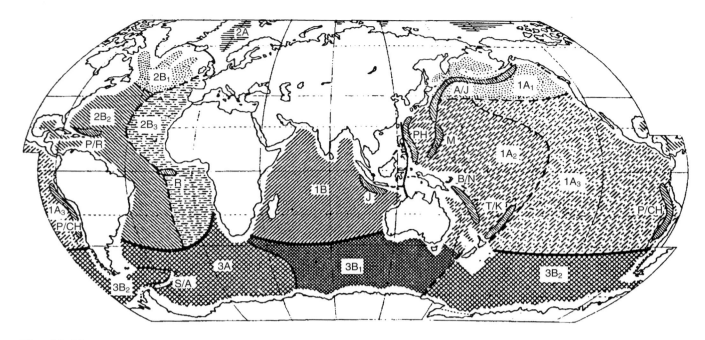

Fig. 13.21 Deep-sea biogeographical regions characterized by distinctive benthic species assemblages defined primarily from Russian sampling in the 1950s to 1980s. (After Vinogradova 1997.)

Kussakin (1973) proposed that the Antarctic abyssal fauna, referring particularly to isopods – his specialty, originated in cold shelf waters and expanded onto deeper bottoms over time, at present reaching close to the limit of equatorward travel of relatively unmixed Antarctic Bottom Water. A division of the circum-Antarctic benthos into sectors has been offered by Griffiths *et al.* (2009) that is similar to Vinogradova's set, each sector south of one of the northward-extending oceans, but with more provinces. If anyone can pull the scattered data together, identification of provinces by an explicit, reproducible procedure (an algorithm) would be useful. Hadal areas, outlined in Fig. 13.21, harbor communities distinct from those on surrounding abyssal bottoms. Exceptions to hadal endemism are species shared between at most two, closely adjacent trenches.

Distributional work on continental shelves and slopes is more detailed, thanks to better accessibility. There have been programs with more and more closely spaced samples. For example, a study by Theroux and Wigley (1998) of macrofaunal distributions on the New England shelf (Gulf of Maine, Georges Bank and southwest to the New Jersey shelf) shows a range of patterns (Fig. 13.22) in every taxonomic group from generalists found everywhere, including some extending far down the slope, to species localized in particular depth zones, in spots washed by specific currents, in topographic features like canyons and in particular sediment types: glacial gravel, sand, and mud. By and large, shelf and slope species have moderately restricted latitudinal ranges, and communities vary at

species level between the sides of ocean basins. A global review was prepared based partly on extensive Russian data by Zezina (1997).

Resources at the seafloor
"Raining" particles

Because most of the seafloor is in continuous darkness, deep benthic communities, apart from those of hydrothermal vent areas, are entirely dependent upon imported nutrition: particulate material sinking (or transported partway by swimmers) from the euphotic zone, deadfalls of large animals like whales and tuna, and to a minor extent waterlogged wood. We know something about the amounts of the falling particles and are still evaluating the importance of the latter two sources. McCave (1975) showed some features of the particle-size distribution in ocean waters. Most of the particles are small; there is an exponential decrease in particle abundance with increasing particle size. Below about 200 m there is generally less than 1 particle ml^{-1} larger than 20 or 30 μm. If the likely sinking rates of these particles are multiplied by their mass, however, the relationship flips. Because the few larger particles fall faster, they carry most of the mass. The particles carrying most of the mass would rarely be collected at all in water bottles. To study supply to the seafloor, something else is needed. That something, developed in the 1970s and deployed in increasing numbers since, is *sediment trapping*.

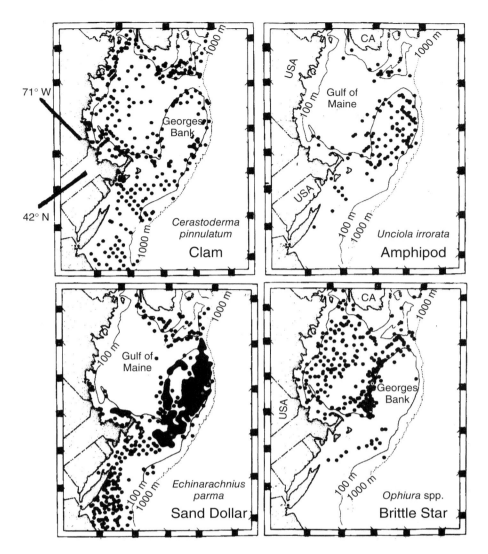

Fig. 13.22 Examples of distinctive distribution patterns of benthic macrofauna on the New England shelf, including Georges Bank and the Gulf of Maine, demonstrated by a dense pattern of Smith–McIntyre grab samples across the region. (Selected from a larger set of maps in Theroux & Wigley 1998.)

Traps come in a variety of shapes and sizes. The first ones were tables that sat on the seafloor with sticky collecting sheets exposed under flow baffles. Those were not effective. Next came various large conical collectors (PARFLUX traps, Dymond traps, etc. (Fig. 13.23a). There are traps based on simple tubes, usually called PITs, particle-interceptor traps (Fig. 13.23b). Tube sizes vary, but typically are small enough to handle readily, 3.5 to 15 cm diameter. There are various schemes for getting time-series of samples, such as carousels at the bottom of the cone which rotate a new collecting cup into place at some interval, often every two weeks. The effort is plagued by biases, and the treatment of that has been to do calibration studies in flumes; to model the collection process so as to calculate correct

fluxes from measured ones; and to seek cool tricks to overcome the problems.

The main flaws are:

1 Lots of zooplankton are attracted to traps and swim in. If the collecting tube contains preservative, then they die and become part of the sample, estimated to be up to four times the actual sinking particulate flux. One solution is to ignore the problem. Another is to have a zooplankton expert pick out animals that have the appearance of having been preserved fresh. Michael Peterson and colleagues invented an indented rotating sphere (IRS) valve to place in the throat of the trap. The plug keeps zooplankton from entering the main trap body so they

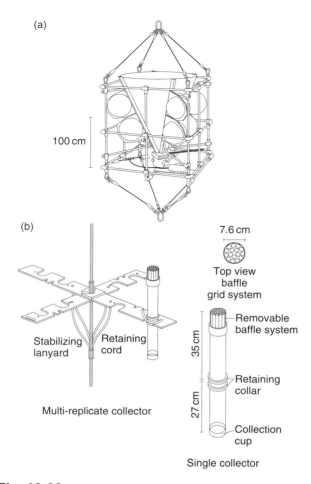

(a)

100 cm

(b)

7.6 cm

Top view
baffle
grid system

Stabilizing
lanyard

Retaining
cord

Multi-replicate collector

Removable
baffle system

35 cm

Retaining
collar

27 cm

Collection
cup

Single collector

Fig. 13.23   (a) Design of the Honjo trap (after Honjo 1982). Mooring
cables attach to upper and lower metal rings. Sediment falling into the cone
works its way down the sides into a cup at the bottom. A spring visible at the
lower right pulls a shutter over the top of the cup before the mooring is
retrieved. (After Honjo *et al.* 1982.) (b) Design of the VERTEX trap and
attachment to the mooring cable. (After Knauer *et al.* 1979.)

can depart again; meanwhile, particles collect in dents
and shallow grooves on the top of the sphere. The dents
are turned down and back up intermittently, dropping
collected particles into the preservative below. Recent
modifications (Peterson *et al.* 2005) include coupling
these collectors to net cones with 1 m diameter mouths
suspended with wave absorbers from surface floats. Data
from these traps are now appearing (e.g. Hernes 2001).
Finally, for deep samples (>2 km), the swimmers are
fewer and assumed to be essentially eliminated by coarse
(1 mm) screening of samples before analysis; marine
snow aggregates primarily break up and pass the screen
(e.g. Honjo *et al.* 2008).

2 Traps are hydrodynamic barriers that force flow around
them into eddies over the top of a trap. This has the
effect of a snow fence on the prairie – greatly enhanced
deposition compared to the actual flux in the region.
This has been modeled in flow flumes and on computers.

Two-fold differences have been found between traps
hung at the same depth in pairs, presumably those leading
and trailing in the flow. The quantitative importance of
snow-fence effects remains uncertain but large. At considerable expense in complexity, mounting traps on floats
neutrally buoyant at a depth of interest may or may not
solve this problem (Stanley *et al.* 2004).

3 In long deployments (and many of the data are from long
deployments) rotting and dissolution of organic matter
sedimented into the traps can be extensive. This causes
enthusiasm for solutions of azide (until ~1985), mercu-
ric chloride or formaldehyde in the trap bottoms, usually
held down by mixing with dense brine. Unfortunately,
dense brine causes osmotic rupture of trapped cells and
animals. The IRS trap mostly solves this, too, retaining
preservative without need for ballasting it with salt
(which can cause osmotic rupture of animals, Peterson
& Dam 1990).

Indented rotating sphere traps have been modified to
determine the distribution of particle sinking speeds
(Peterson *et al.* 2005). Since the rotating valve drops accumulated particles intermittently, a stepping motor at the
bottom of a fairly long tube below the valve changes the
collecting cups at intervals shorter than the rotations, dividing the flux according to sinking speeds (despite some
viscous "wall effects"). Trull *et al.* (2008) applied this
system between 200 and 300 m at both productive and
oligotrophic sites in the North Pacific and in the
Mediterranean. At all sites, they found a geometric distribution of settling speeds, with ~50% of flux sinking between
1000 and 100 m d^{-1}, the remainder descending progressively slower to 1 m d^{-1}. Contradicting a general understanding (to be recited below anyway), the relationship to
particle size was not particularly strong, and, despite the
general consensus that mineral ballasting is critical, the
amounts of inorganic calcium carbonate and opal were not
greater in the faster fractions.

One extensive class of studies employs very large, conical
traps with designs similar to the PARFLUX series (Honjo
et al. 1982): cones, apices down, with sides 14° from vertical and a 1.54 m^2 opening with a honeycomb baffle in the
mouth. They are suspended above mooring anchors,
buoyed up by large Pyrex™ spheres clustered about the
sides. Sedimenting material arriving at the apex cups is
generally preserved by a fixative solution. There are results
from both single-cup versions and time-series versions with
a carousel of cups changed by a stepping motor at suitable
intervals (8.5 days, 21 days, etc.). After a collecting interval,
the whole mooring array, often several of these traps at
different depths, is released from its anchors by a signal to
an acoustically activated hook and comes up in a tangle.
That is sorted out, the cups removed, and the samples
divided into aliquots while wet. Those splits are subjected
to various chemical and biological analyses.

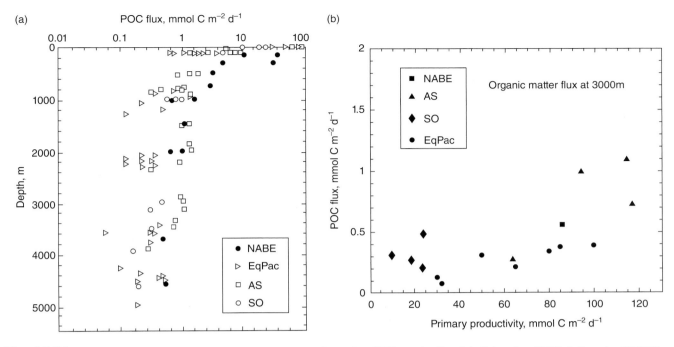

Fig. 13.24 (a) Flux rates of sinking particulate organic matter measured as carbon (POC) as a function of depth based on JGOFS studies using PARFLUX-type traps in various sites: NABE, North Atlantic Bloom Experiment; EqPac, Equatorial Pacific; AS, Arabian Sea; SO, Southern Ocean. (b) Organic-matter flux rates at 3000 m relative to surface primary productivity measured as ^{14}C-uptake rates. (After Berelson 2001.)

There have been several attempts to summarize the results (Lampitt & Antia 1997; Berelson 2001; Lutz *et al.* 2002; Honjo *et al.* 2008). It is not possible to get good estimates of organic matter flux from traps just above the seafloor, because currents near the bottom resuspend sediments and mix them upward for hundreds of meters. However, the rate of resuspension slows sufficiently above those layers that good estimates of the final seafloor input can be obtained from traps at about 3000 meters. Almost all of the results (Fig. 13.24a) fall in the range from 0.1 to 2 mol C m^{-2} d^{-1} (1.2 to 24 mg C m^{-2} d^{-1}). This is a small fraction of the primary production at the surface. Berelson (2001) compared all of the PARFLUX-type trapping studies from the process studies cruises of JGOFS program to the simultaneous production rate estimates (Fig. 13.24b & Fig. 13.25) and found that ~0.5 to 1% of the surface photosynthate generated in the euphotic zone reaches 3000 meters or the bottom. That result is typical for all studies before and since. Consumption in the water column follows a vertical sequence (Fig. 13.25) reasonably well expressed by a function suggested by Martin *et al.* (1987), the "Martin curve":

$$\text{Flux at depth } Z = (\text{POC flux at 100 m}) \times (Z/100)^{-b},$$

where *b* is a fitted parameter. The flux at 100 or 200 m, often termed the "export flux", is measured by a trap there. All of the biases of trapping are extreme above 100 m (or a bit deeper), so that the comparison is made to this some-

what arbitrary standard for the upper level trapping rate, rather than to shallower rates or directly to primary production. As seen in Berelson's figure (Fig. 13.25) the best value of *b* varies from ~0.6 to twice that, and very often the decrease below 2000 m is slower than the fitted equation. The slowing, whether the function fits exactly or not, is due to (i) decrease in the populations of scavengers at depth reprocessing sinking marine snow and fecal matter, and (ii) rather sharp downward acceleration of particles. Sinking rates of typical larger particles (those carrying most of the flux) are 100 to 200 m per day, accelerating by ~50% below that. Mostly, that is attributable to agglomeration of particles, and larger particles sink faster (Stokes 1851). It is widely accepted that sinking of small organic particles is largely due to several forms of agglomeration: (i) in "marine snow" that is at least partly a matrix of polymers secreted by algae and animals, and (ii) fecal pellets of zooplankton (small but dense from copepods, large from salps, etc.) and nekton. The ocean is frequently full of flocculent, filmy stuff, first termed marine snow by William Beebe who observed it from his bathysphere in 1930. Marine snow has been a subject of intense study (e.g. Alldredge 1998; Kiørboe 2000). Much of it is generated by phytoplankton that secrete TEP, standing for *transparent extracellular polysaccharides* (Alldredge *et al.* 1993). The polymers clump into mats. Mucus secreted by zooplankton (mostly by salps, appendicularians, and pteropods) is also an important constituent.

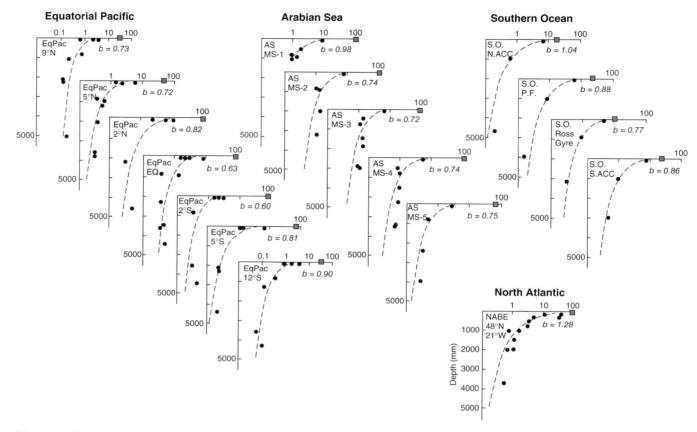

Fig. 13.25 Vertical profiles of POC flux rates estimated from vertical arrays of PARFLUX-type sediment traps deployed in four JGOFS regional study areas. The abscissas are logarithmic scales of POC flux compared to the euphotic-zone primary production rate (the filled squares), both as mol C m^{-2} d^{-1}. The curves are best-fit Martin curves calculated with the b exponents to the right of each curve. (After Berelson 2001.)

Lighter PIT technology with 3.5 cm diameter tubes (Fig. 13.23b) was used by the VERTEX studies in the 1980s led by John Martin. The main trap contents were primarily amorphous mats of organic goo, marine "snow". All sorts of organic particles get stuck in the bits and sheets of snow. Live phytoplankton (microflagellates predominate in numbers, large cells in the flux) are transported to great depths; radiolarians and other protists are common; fecal pellets are present of animals from protozoa (minipellets) to crabs and salps, and all of these are bound in an amorphous mucoid background (e.g. Silver & Gowing 1991). Bacteria are also numerous. In data presented by Urrere and Knauer (1981), pellet fluxes from copepods and other plankton were of the order of 200,000–325,000 m^{-2} d^{-1} in the upper 100 m. That seems a huge number of pellets. However, it is not surprising, given the number of copepods in the upper water column and their rate of defecation. Below 100 m, the numbers progressively fell to 35,000 m^{-2} d^{-1}, then increased again, probably due to midwater consumption and "repackaging" of particles. It is the repeated digestion of the sinking matter that accounts for the reduction downward of organic mass flux. Despite the large number of pellets, they contributed only about 10%

of carbon flux in upper levels and 3% below 1500 m. Snow is the main transport mechanism. All the trappers seem to agree on that.

As a further example, we show the quantification by Berelson (1997), using large conical traps of the fraction of surface production reaching the seafloor along a transect across the equator at 140°W (Fig. 13.26). Primary production was estimated in the surface layer by ^{14}C-uptake (data from Barber 1996, see Fig. 11.34) The effects of equatorial upwelling provided strong gradients of production rates from the more eutrophic equator out into oligotrophic waters at 10°N and 12°S. Flux of organic matter was estimated from trap collections above the seafloor (Fig. 13.26b), and reflects the same gradient. The fraction of primary production reaching the bottom along this transect is about 0.4%, i.e. four parts per thousand remain after filtering down through the 4500 m water column. Benthic respiration rates (see Chapter 14) implied carbon metabolism rates of the same general magnitude (Fig. 13.26c). Sediment traps at a depth of only 105 m produced organic mass flux rates only 3–6% of the primary productivity (Hernes 2001). Most organic matter is consumed very close to the depth at which it is produced by photosynthesis.

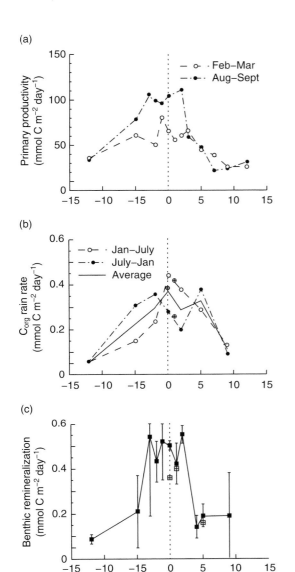

Fig. 13.26 Comparison along a trans-equatorial transect in two seasons (February–March observations were from a period of fully developed El Niño) of (a) primary production, (b) flux of organic carbon above the bottom, and (c) the rate of benthic remineralization based on oxygen consumption (solid squares and open squares with crosses were collected on different expeditions). (After Berelson *et al.* 1997.)

While most of the flux in oceanic areas is of biological origin, most of it is not organic matter as such; biological minerals are dominant. About three-quarters of the mass flux in typical PARFLUX profiles is calcium carbonate: coccoliths, foraminiferan shells, and pteropod shells. Opal (solid, polymerized Si(OH)$_4$: diatoms, silicoflagellates, radiolarians) is another one-eighth or a little more, and carbon in organic matter is about 8%. The remainder is the other elements in organic matter and "lithogenic" particles (mostly clay). The ratio of calcium carbonate to organics increases significantly downward, to at least 1000 m.

Mineral content in sinking agglomerations is important as ballast, accelerating the particles to speeds >100 m d^{-1}.

Wakeham and Lee (1993; Fig. 13.27) assembled the scattered data from trapping studies in oligotrophic areas, comparing at progressively greater depths the downward flux rates of specific components of the organic matter. Much of the consumption and metabolic remineralization of sinking organic matter occurs in the upper 500 m. Below that depth the continued decrease is roughly exponential (somewhat like the Martin curve), which Wakeham and Lee represented by estimating "half-depths", the distances required to reduce the flux by half. Amino acids (protein particles) are removed faster (shorter half depth) than fatty acids, and those are removed faster than the total of all organics. Thus, food quality of sinking organic matter changes downward. That is demonstrated by the substantial shifts in amino-acid composition. Amino acids in sediments (residual after much more metabolic processing) are altered even more. Only moderately refractive organic matter reaches the seafloor to serve as the diet on which an elaborate diversity of bacterial, protist, and metazoan life subsists.

Repeating for emphasis: (i) more organic matter leaves the euphotic zone when the euphotic zone is more productive, and this conclusion applies to all constituents (Wakeham & Lee 1993); (ii) decrease in flux with depth is quasi-exponential, but the rates of decrease are smaller at depth; and (iii) food quantity amounts to <2 mol C m^{-2} d^{-1} in organic matter; and (iv) food quality decreases downward.

Deadfalls and waterlogged wood

The exact quantitative significance of large animal carcasses and waterlogged wood (sticks, logs, coconuts) can only be very crudely approximated. When anchored baits are observed by time-lapse cameras (Fig. 13.1), they are attacked quickly by grenadiers, hagfish, deep-sea sharks, amphipods, and crabs and reduced to skeletons in hours to days. Skeletons of test porpoises are broken up and dispersed in a matter of months (Glover *et al.* 2008). Whale skeletons, however, retain some connections among the bones and can remain intact for several years. For example, a blue- or fin-whale carcass discovered during a submarine investigation in a basin off Southern California in 1987 was still a distinct skeleton in 2005 (Glover *et al.* 2005a), with filamentous mats of the colonial bacterium *Beggiotoa* developed over it. *Bathykurila*, a genus of polynoid scale worms (polychaetes) also found at hydrothermal vents, were grazing on the mats. Thus, whale falls could be "stepping stones", allowing larvae of these worms and other fauna to transit between vents. Whale skeletons can also be densely covered by gutless siboglinid worms similar to vestimentiferans at vents, but belonging to a distinct genus, *Osedax*. They are only recently described from whale falls (Rouse

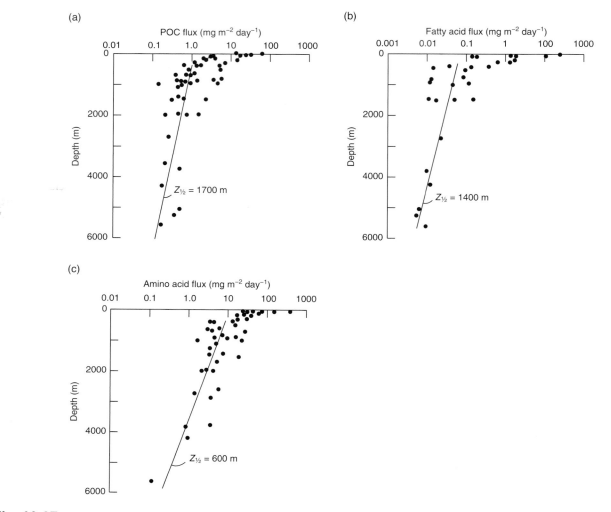

Fig. 13.27 Vertical profiles of (a) flux of particulate organic carbon (POC), (b) fatty acids, and (c) amino acids measured by several trap studies in oligotrophic, oceanic areas. Rates of downward decrease in flux are indicated by depth changes required to reduce flux by half ($Z_{1/2}$). (After Wakeham & Lee 1993.)

2004), but the genus currently has more than eleven species (e.g. Glover 2005b). Rouse reports that nutrition is absorbed through a root-like, posterior structure penetrating the bone and populated by heterotrophic symbiotic bacteria, while Glover suggests that the possibility of chemosynthetic symbionts has not yet been excluded. Whale bone is rich in sulfo-lipids. Whether these several worms are important in the wider economy of the benthos is uncertain but unlikely, whereas the distribution and redistribution of deadfall meat and soft organs may be moderately significant.

Wood is often collected by trawls on the deep seabed. Logs and sticks acquire modestly complex faunas of animals tightly adhering and eating or abrading sockets for themselves, such as limpets. Several groups of clams, shipworms (Teredinidae), and wood-specializing pholads (Xylophagainae), bore into the mass of the wood as they grow (Pailleret *et al.* 2007; Voight 2007). To some extent the importance of the wood is as solid substrate, but given symbiotic cellulose-digesting bacteria for digesting the wood, the case with at least shipworms, it serves also as food.

Seasonal cycling in the deep sea

The 1–2% of primary production that eventually sinks to the deep seafloor has been shown to be almost as strongly seasonal as is surface productivity. Results from an early time-series trap (Deuser 1981) showed the seasonal sequence in the Sargasso Sea of productivity reflected in arrival of materials reaching 3200 m (well above the seafloor). The ~1% fraction reaching depth was found in deep-sea particle flux, and unexpectedly so was the seasonal

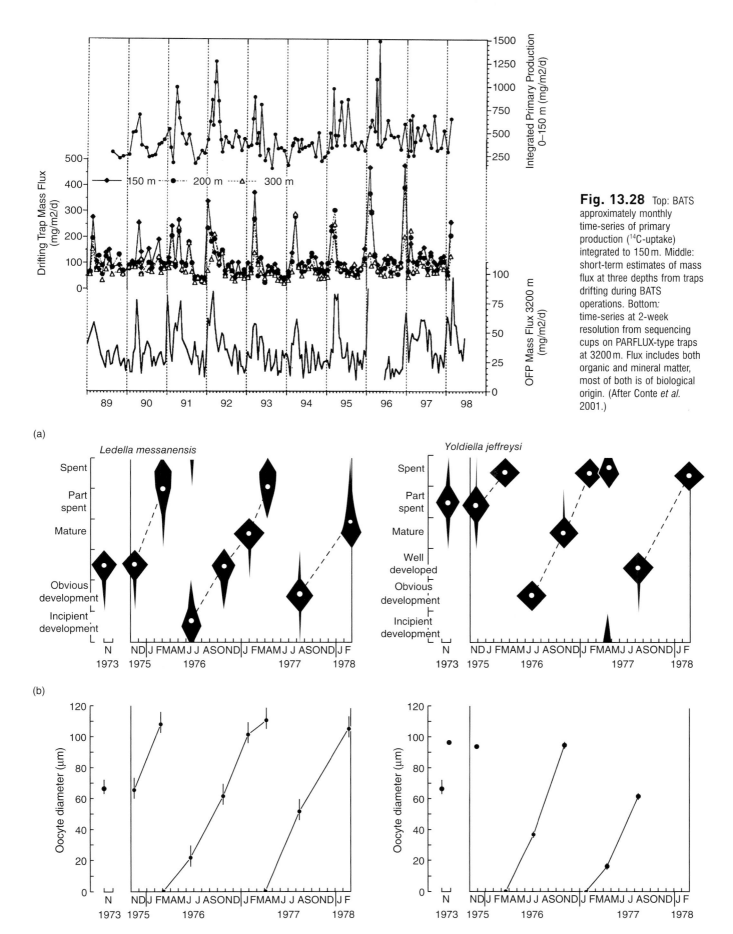

Fig. 13.28 Top: BATS approximately monthly time-series of primary production (^{14}C-uptake) integrated to 150 m. Middle: short-term estimates of mass flux at three depths from traps drifting during BATS operations. Bottom: time-series at 2-week resolution from sequencing cups on PARFLUX-type traps at 3200 m. Flux includes both organic and mineral matter, most of both is of biological origin. (After Conte *et al.* 2001.)

Fig. 13.29 Two aspects of reproductive cycling in two species of clam sampled three times per year in Rockall Trough, northeast Atlantic. In both species, gonad size increases from June to about March, and then decreases again. (a) Each "kite diagram" shows the proportion of individuals in progressively later stages of ovarian development. (b) Oocyte diameter follows the same timing. (After Lightfoot *et al.* 1979.)

variation of productivity. That time-series was extended (Conte *et al.* 2001) near the BATS Sargasso Sea site for 20 years, and with two-week resolution from 1989 to 1998 (Fig. 13.28). The timing of the annual (with exceptions) late winter bloom at this subtropical station is in a general way reproduced by flux at 3200 m (despite the lack of significant time-series cross-correlation according to Conte *et al.*).

The seasonality of deep flux of particles and organic matter is similarly pronounced in the northeast Atlantic, where strong spring blooms occur over oceanic depths, as shown with time-series traps at 48°N, 21°W by Honjo and Manganini (1993). The bloom and flux peaks occurred in May of 1989, later than at BATS, which is typical – the bloom progresses from south to north through spring. When the diatoms deplete available nutrients, the bloom stock "crashes" to the bottom, forming a flocculent layer on the seafloor, and the flux at traps below 2000 m is indeed rich in diatoms at that time.

Tyler (1988) and coworkers have shown that some, but not all, benthic species in the subarctic North Atlantic respond to this cycling of their food supply with cycling of growth and reproduction. For example, the protobranch bivalves *Ledella* and *Yoldiella* are strongly seasonal in reproductive activity (Fig. 13.29) as observed in time-series of oocyte diameter and ovarian state (Lightfoot *et al.* 1979; Tyler *et al.* 1992). Some echinoderms (Tyler *et al.* 1993), brachiopods, and scaphopods show similar cycling. On the whole, the species with reproductive cycling are those with *planktotrophic* larvae – larvae that must feed in the water column to later metamorphose into deep benthos. The *lecithotrophic* types, those with yolky eggs that develop directly to benthic young (in some animals metamorphosing to the benthic phase before hatching), show no ovarian cycling. Thus, isopods are not particularly cyclic in reproduction. This contingency of cycling on reproductive mode suggests that the adaptive aspect is anticipation by spawners of good feeding conditions for their young, possibly on the pulsed fallout of the spring bloom. However, there are few seasonal cues, apart from the pulsing of food, by which to regulate reproductive timing. Perhaps they simply reproduce when resources allow. Along with reproduction, there is strong evidence that bulk metabolism of the deep-sea benthos responds with little delay to the variation of food supply – evidence considered in the next chapter.

Chapter 14

Some benthic community ecology

Ecologists use the term *community* to refer to the suite of organisms constituting the living part of an ecosystem. Unfortunately, the term is burdened with connotations coming from its ordinary, daily use in respect to interacting groups of human beings. A human community functions by a division of labor. Some farm, some butcher, some manage insurance schemes, some are ballerinas, and so on. Some functions are essential, like food gathering and distribution, some less so. Early usage of the term in ecology implied the assumption that natural plant and animal assemblages are also characterized by division of labor. To a degree that is clearly so: plants generate the basic organic matter which allows all biological activity; herbivores convert that to mobile form; carnivores crop the herbivores; decomposers recycle the raw materials.

A question which quickly arises concerns the extent to which specific interactions are obligatory. Are exact or nearly exact combinations of species required to form functional, healthy communities and, thus, ecosystems? Or can most species get along under suitable physical conditions in a variety of combinations with other forms? Schools of thought developed in each direction in late 19th- to early 20th-century plant ecology, opinions eventually associated with the names of Fredric Clements and Henry Gleason. Clements' followers favored the view that most plant species live in quasi-obligatory sets (like the organs in a single organism), while those following Gleason thought the arrays of plants found at most sites were quasi-random assemblages of species that happened to be suitable to the physical habitat, in some cases just those whose seeds first reached patches of open ground. The empirical facts developed down the years fall somewhere between, but generally favor the Gleasonian view. For example, in transects along

habitat gradients, such as up the side of a mountain, plant species are added and removed one at a time, not in matched sets changing at common boundaries. Thus, no species is absolutely required by most of the others. Within the limits of their basic life mode, plants (and animals, too) are sufficiently flexible to deal with a variety of other species as competitors, predators, and associates generally. The same conclusion applies to species of benthic animals proceeding along gradients of depth and organic-matter availability from the intertidal to the deep sea (e.g. Dayton & Hessler 1972). Of course, this is not to say that highly obligate associations never occur. They are particularly frequent for parasite–host combinations.

In benthic ecology, the term *community analysis* is applied to several rather specific approaches to evaluation of faunal assemblages, that is, to interpretation of the combinations of animals found together in samples. One notion of a community, somewhat "Clementsian" in tone, is that they are sets of species that tend to recur wherever conditions are suitable. This notion of recurring assemblages (Fager 1963) has been widely applied, despite the general conclusion that most communities are "Gleasonian". This contradiction will appear in the following examples.

Community analysis: a quantitative approach

In many ecological investigations of particular regions or habitats, there is very little information about the actual life and times of the organisms. All that is available, or even possible to obtain, is a suite of samples representing the

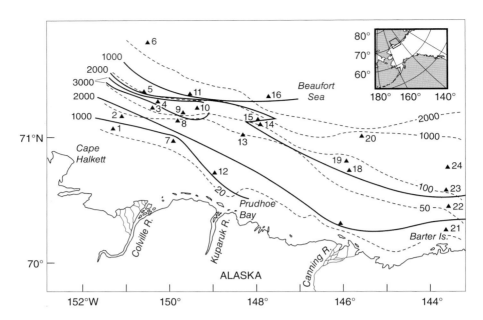

Fig. 14.1 Station positions (▲), bathymetric contours (dashed lines), and numerical abundance of polychaetes (number of worms m⁻², solid lines) on the western Beaufort Sea shelf and slope. (After Bilyard & Carey 1979.)

fauna in different parts of the region, i.e. samples of dead animals preserved in formaldehyde or ethanol and convertible to lists of identified species. The problem is to use those samples to provide information about the character of the habitat. There are important, though limited, ways. A basic assumption is that a spot where an organism is found is a more-or-less suitable habitat for it. Thus, the changes of species composition across a region define differentiable habitats. Once the boundaries of the habitats are defined, they can be examined to determine their characteristic physical, chemical, geological and biological features. It is often possible to hypothesize which features are critical for the well-being of the defining species. As an example, we will examine an old but excellent study by Bilyard and Carey (1979). It has the advantage that questions about the fauna were approached in an ascending order of complexity with useful conclusions emerging at each step. At the final step, they succeeded in showing that species groupings can distinguish habitats that are in some sense more universal than those that might be defined by the presence of a single species.

Bilyard and Carey's problem was to evaluate the number and characteristics of the distinct habitats on the seafloor off the North Slope of Alaska, in the western Beaufort Sea. They had a large suite of samples taken from an ice-breaking ship at positions (Fig. 14.1) ranging from 25 m to 2000 m depth and covering 8° of longitude. Bilyard, who did the sample work-up and analysis, was interested in polychaetes and confined his attention to them. This is a common feature of most such studies; they are limited by the taxonomic expertise of the investigator, and most ecologists can

only deal at a sophisticated level, at the species level, with one or two substantial groups of organisms. As a first pass at a regional analysis, Bilyard counted the annelids in each sample and plotted the results on a map of the isobaths (Fig. 14.1). There was a strong gradient in abundance from west (more) to east (less). That is informative: the distribution of abundance has the same shape as the tongue of summer current that intrudes around Point Barrow from the Bering Strait area, carrying phytoplankton from the rich production over the Bering Sea shelf. This flow is almost certainly a source of enriching organic matter for the underlying sediments, a food source.

The next pass was to identify and count all of the species of polychaetes in each of the samples, generating a book of data-sheets (Fig. 14.2). This part of the work can be a monumental task. Examining animals one by one, sometimes seta by seta, for all 24 samples took over two years. Such an effort is not for those in an overwhelming hurry to publish. Data for any one station are not in themselves very informative. However, comparison of data for all the stations can be. What Bilyard had after his long taxonomic interlude was a large table with two axes, species by stations, filled with abundance estimates. The art of community analysis is to extract a maximum of information from the matrix. It is useful first to simplify it in several ways. Bilyard began by changing all the abundances to "1" for present and "0" for absent, then ordered the species according to presence and absence progressing down the depth gradient (Fig. 14.3).

The ordering, shown for dominant species, is from the narrowest shallow range, through the most inclusive range,

Beaufort Sea Benthos - Polychaeta
Analyst - *Bilyard*

VERIFIED

TRANSECT *E*   STATION 24

No. In (x+1)   No. In (x+1)

Species	No.	In (x+1)	Species	No.	In (x+1)
1. Allia suecica	65	4.19	21. Minuspio cirrifera	13	2.64
2. Allia sp. A	0	0.00	22. Myriochele heeri	5	1.79
3. Amage auricula	14	2.17	23. Nephtys ciliata	1	0.69
4. Anaitides groenlandica	0	0.00	24. Onuphis quadricuspis	7	2.08
5. Antinoella sarsi	0	0.00	25. Ophelina abranchiata	1	0.69
6. Barantolla americana	12	2.56	26. Ophelina cylindricaudatus	0	0.00
7. Capitella capitata	0	0.00	27. Ophelina sp. A	0	0.00
8. Chaetozone setosa	0	0.00	28. Owenia fusiformis	1	0.69
9. Chone murmanica	0	0.00	29. Pholoe minuta	0	0.00
10. Cistenides hyperborea	0	0.00	30. Prionospio steenstrupi	0	0.00
11. Cossura longocirrata	0	0.00	31. Scalibregma inflatum	0	0.00
12. Eclysippe sp. A	35	3.58	32. Scoloplos acutus	0	0.00
13. Eteone longa	2	1.10	33. Sigambra tentaculata	7	2.08
14. Heteromastus filiformis	6	1.95	34. Spiochaetopterus typicus	0	0.00
15. Laonice cirrata	2	1.10	35. Sternaspis fossor	0	0.00
16. Lumbrineris minuta	14	2.71	36. Tauberia gracilis	1	0.69
17. Lumbrineris sp. A	0	0.00	37. Terebellides stroemi	5	1.79
18. Lysippe labiata	0	0.00	38. Tharyx ? acutus	22	3.14
19. Maldane sarsi	346	5.85	39. Typosyllis cornuta	4	1.61
20. Micronephtys minuta	0	0.00			

Fig. 14.2 Actual data-sheet, showing logarithmic data transform, from Gordon Bilyard's analysis of Beaufort Sea polychaete species abundance. (Courtesy of Gordon Bilyard.)

to the most exclusively deep range. Many of the species are seen from this to be generalists; they cover at least the upper 1500 m.

Seven of the 39 species in the figure are found throughout the depth range of the sampling. There are no sharp breaks in the pattern; that is, there are no zones in which both large numbers of forms have their lower boundary *and* large numbers of other forms have their upper boundary. While the fauna changes down slope, there is no sharp habitat break defined by polychaetes that is associated with just depth.

Next Bilyard plotted the number of species found at each station (Fig. 14.4). He had found more species to the east – opposite to the absolute abundance pattern. That may seem surprising, but there it is. Next, he began a comparison of the stations on the basis of their resident faunas. There are many ways to do that. Bilyard first calculated correlation coefficients between the abundances in the list of species for each station pair. Some data standardization is usually employed in this, such as replacing abundance at each station by the number of standard deviations that its estimate falls from the mean for the species across stations. That keeps species from appearing to be correlated just because they are among the more-abundant animals. Joining

those station pairs with the largest correlation coefficients, using straight lines on the map, demonstrated (Fig. 14.5) a strong along-shore alignment pattern. Stations at the same depths tend to have the same species present in roughly the same relative abundances. Clearly, while there are no sharp breaks along the depth gradient in species presence or absence, depth appears to be a major factor affecting the faunal assemblages. Keep in mind, however, that depth is not simply a measure of water pressure. It is usually, as in this North Slope study, correlated with distance from shore and thus with the rate of supply of organic matter from the surface. Production is generally greater near shore, so more food is available to sink. Offshore production is less, and the greater depth gives pelagic organisms more time to find and eat sinking organic matter.

Then Bilyard analyzed the species-by-station table using *clustering* and *ordination* techniques. These are of many kinds, and some importance rides on the choice. Knowing and applying the variety, advantages, drawbacks, and utility of different methods is a discipline in itself (e.g. McCune & Grace 2002). However, virtually all such methods can be referred to a geometric model. Generally, the data must first be standardized, just as for the correlation matrix, s o that all included species, regardless of their absolute

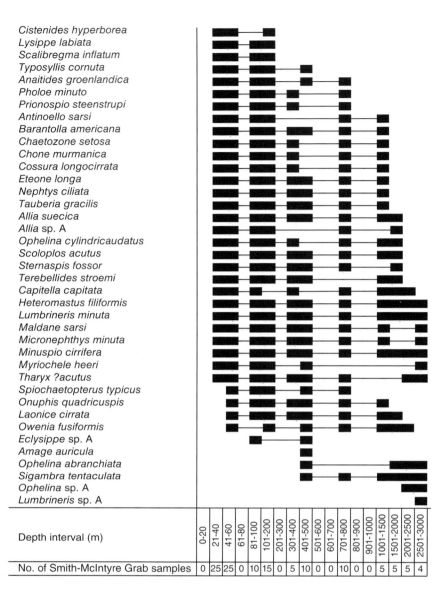

Fig. 14.3 Ordering of dominant polychaete species found in Beaufort Sea sediments according to depth range (intervals at bottom). Species restricted to shallow stations are at the top, generalists are in the center, and species only found at the deepest stations are at the bottom. (After Bilyard & Carey 1979.)

abundance, have roughly equal effect on the outcome. Then the species-by-station table is taken to define either (i) the positions of stations in a space with dimensional axes equal in number to the number of species (S) and each running from zero to the maximum (standardized) abundance of its species, or (ii) the positions of species in a set of axes equal in number to the count (N) of stations and scaled in the same way. In case (i), bunches of station points result from species whose relative abundance behaves in a similar fashion over the station set. In case (ii), bunches define species with similar relative abundances at the different stations. Thus, the definition of the space determines what issue is addressed: which stations have similar species assemblages, or which species behave similarly. Of course,

these two issues are obviously tied together. The term "relative" is needed because standardization of the data removes emphasis from absolute abundance.

Since few studies would have only three species or only three stations, these spaces generally are highly multidimensional and cannot be visualized. Spotting clusters of points in unimaginable space is approached in one of two general modes: *clustering* and *ordination*. Clustering methods can work in several ways. Probably the simplest to understand are agglomerative. Distances are calculated for all possible pairs of, say, stations in species space. The pair with the smallest separation is noted as part of a possible cluster and replaced with a point in the middle between them (in S space). If this removes a small amount of the overall separa-

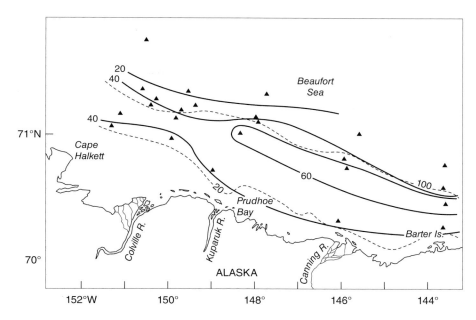

Fig. 14.4 Contour plot of the number of species found at each Beaufort Sea station. For key, see Fig. 14.1. (After Bilyard & Carey 1979.)

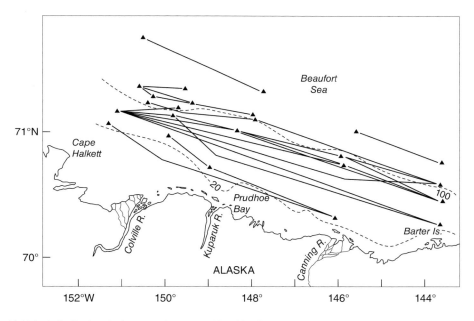

Fig. 14.5 Stations with high similarities in polychaete species composition, identified by high product-moment correlation coefficients (>0.645), connected by solid lines. Similar stations are obviously found at similar depths. For key, see Fig. 14.1. (After Bilyard & Carey 1979.)

tion among stations, then the pairing is given a high rating, if not, then it is rated weak. This procedure is repeated over and over (perhaps weighting new positions according to how many stations are on each side of the new pairing) until all stations are in one cluster. The sequence is examined to find a stopping place that defines a convenient number of clusters, or (better) at which the reductions in remaining intergroup distance jump to large values (as distant clusters are combined). Clusters of stations defined at this stopping

place might then be mapped or examined for commonalities in sediment type or factors likely to differentiate the habitat made "visible" by the species clusters.

Ordination techniques establish the coordinates of, for example, stations in species space, then progressively fit axes through that space with minimum total distance (usually a sum of distances squared, ΣD^2, is minimized) to the station points. In the simplest versions, all related to principal-component analysis, the axes are taken to be at right angles (orthogonal) to each other. Thus, the first axis would be the line through the *S*-dimensional space with minimal ΣD^2, and the second the line at right angles to the first chosen by the same criterion. Those together would define a plane. If the stations define three main clusters in *S*-space, then they will all be near that plane and obvious when their positions are projected onto it and plotted. Lines and planes can be added, showing the positions of additional clusters. Some stations with nearly unique species assemblages will be isolated.

Most workers use methods popular in their organization, with which good familiarity is available locally. On that basis, Bilyard selected a clustering technique called "CLUSB" developed at Oregon State University by D. McIntire and S. Overton. CLUSB puts cluster centers into a geometric model of the data and then assigns stations to centers so as to minimize the summed squares of distances of station points from their cluster centers. Clusterings are tried with more and more centers, stopping when the sum of squares is no longer much reduced by adding another. Bilyard found four station groups that were well separated, and for which no stations included in any cluster were far in species space from their designated center. He plotted the cluster designations A to D at each station location on the isobath map (Fig. 14.6).

The result looks like the species numbers map (Fig. 14.4). That is, the contour of 60 species that appears in the east is attributable to a set of forms that cause identification of a station cluster. Note that the cluster patterns are very much parallel to the depth contours, but that the A cluster extends across the contours in the west. This implies that,

while depth is an important habitat determinant, something else is also important.

Bilyard also applied a canonical correlation analysis, an ordination technique (Box 14.1) involving orthogonal axes. Clusters identified (Fig. 14.7) were the same as those from CLUSB. That is usual. If the clustering is strong, i.e. if data really indicate clusters, then most methods will show that. If it is weak, then different methods will show different clustering. There are now methods for developing a probabilistic evaluation of the "reality" of the clusters (Box 14.1). Bilyard's clusters are tight enough, distinctive enough that such testing would be superfluous. The only data available to Bilyard for seeking habitat variables that might explain the cross-isobath distribution of his A-cluster stations were

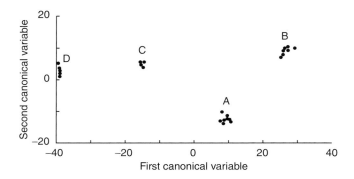

Fig. 14.7 Positions of station clusters (A to D) in a "species space" projected onto a best-fit plane defined by canonical correlation analysis. These clusters are unusually tight for such an analysis. (After Bilyard & Carey 1979.)

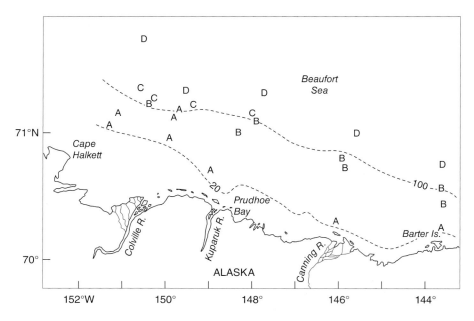

Fig. 14.6 Locations of stations in each of four clusters (A, B, C, and D) identified by both CLUSB and canonical correlation analysis. Stations in the clusters have high similarity in polychaete species composition. (After Bilyard & Carey 1979.)

Box 14.1 Ordination techniques, with the math in words

Ordinations apply a spatial analogy to data tables, which in community ecology are generally lists of the species occurring at each of a suite of sampling stations. Those could be the macrobenthos collected with box corers at locations on a stretch of ocean bottom like Rockall Trough in the northeast Atlantic (Gage *et al.* 2000). The analogy is a "hyperspace" with as many axes, S, as species, and with stations plotted at coordinates set by the abundances of the species represented by the axes: $(n_1, n_2, n_3, \ldots, n_S)$. Of course, this hyperspatial analog cannot be visualized. Ordinations "fit" spaces with fewer dimensions into the original hyperspace, most commonly two or three, and project the station positions onto them, making similarities of species composition among subsets of the stations more obvious.

In *principal-component analysis* (PCA) a line, plane or space with $X < S$ dimensions is placed through the original S-space; we will call it a plane. This is the plane with the smallest possible total of squared direct distances (in S dimensions) from the station points. The points are then projected onto this plane and plotted as station numbers for visualization. Stations with more similar species lists fall closer together.

The mathematics are, obviously(?), to write an equation for the sum of squares of the station distances to the plane, take the derivatives with respect to each axis, set those all to zero (optimization), and solve the equation set for the constants of the plane. The S-space coordinates for each station are fed into that formula to calculate its coordinates on the PC plane. A measure can be generated of how much of the original scatter of stations in S-space is "removed" by fitting principal components. A line (a one-space) removes the most, with lesser contributions from second, third, and higher principal-component axes.

If the raw abundance data are used, the plane will be determined by the very few most abundant species. The distances along their axes will be greatest, and those will be squared. Generally, that is not what is desired, so a nearly consistent first step is to standardize the data, usually by transforming abundances of each species at each station to the distance from the species mean in standard deviation units. Thus, all species have zero mean and similar scales of variation. It is usual to remove from the data-set those species consistently rare and with zeros at many stations. A typical data table can have a great many zeros, which "truncates" the distribution patterns abruptly and does not work out well in the computations (the PCs will be "pulled" toward the origin).

PCA treats the abundances, even after transformation, as linearly related, which is not ecologically realistic.

That may or may not make much difference in revealing groups of related stations by their proximity in the PCA plot. The method can also be used inversely, finding associated groups of species in a space defined by the station axes. All of the PCA computations are fundamentally matrix manipulations, which is how many current algorithms are programmed.

There are many relatives of the PCA approach: factor analysis, principal coordinates, canonical correlation analysis, correspondence analysis, detrended correspondence analysis, canonical correspondence analysis, redundancy analysis, and more. Each has uses in specialized situations and its own drawbacks.

In *non-metric multidimensional scaling* (nMDS or MDS) the original hyperspace is the same. The computations are usually based on some index of distances among the stations in that original S-space, and the indices used generally simplify the data greatly. Sorensen's distance [1 – Bray–Curtis similarity] is popular as is [1 – Jaccard's index]. The Bray–Curtis index is simply the sum for two stations of the smaller of the two proportions of each species. If percentages are used, this is also called a "percent similarity index". If the species proportions are identical, Bray–Curtis = 1 (or 100%). Beware when authors say they are using Jaccard's index, because it has two definitions in the literature that do not produce the same number. Here is one: let A be the number of species occurring just at one station, B the number just at the other, and C the number at both, then $J = C/(A + B + C)$. The proportion of stations with both species present is taken as an index of their similarity. Both Sorensen's and $1 - J$ are distances in a species space with all axes running from 0 to 1. Clearly, the two indices weight the importance of relative abundance differently – important vs. not important.

The computation proceeds by a more efficient method analogous to sliding an arbitrary plane (or X-space) through the S-space, then calculating the sum of squares of all the distances that the points must be moved to sit on it, termed the "stress" for that plane. That sum is stored for later reference and another plane is tried, then another and another. After some sufficiently small stress values are found, the algorithm offers the plane with the lowest value as its result, and it plots the stations where they project onto it (shortest distance). That is the ordination product. The station clusters may or may not be obvious across it. Unlike PCA (and many other schemes), the first axis is arbitrary, not necessarily the one "relieving" the most stress. Thus, if just one axis is fitted, and then independently two, the second may "explain" more of the distance separating

(Continued)

the station points than did the one fitted alone. The so-called first axis is, thus, arbitrary. Three axis nMDS are possible and can be useful if substantially more stress is "relieved". The more-efficient methods start with an arbitrary plane (a PCA plane might be used to start close to a useful result) and then iteratively manipulate the points to determine better locations for the ordination plane. They run the risk that a local minimum of stress can trap the result, requiring that more than one starting plane be tried.

Methods for "testing" the reality of the cluster distinctions appearing in cluster analyses and ordinations have been developed; most are "bootstrap" methods. They depend upon running the clustering or ordination routine many times (say, B [for "bootstrap"] = 1000),

each time randomizing the abundance estimates for each species among the stations. A measure of the strength of clustering, C_{rand}, is calculated for each run to compare with that of the data, C_{data}. If the fraction of measures for which [$C_{rand} < C_{data}$] is smaller than a chosen probability (α), perhaps $\alpha = 0.05$, the clustering is said to be "significant". These probabilities are indicative, but they cannot encompass the entire universe of community-abundance estimates that could be found in the field. To fully test the hypothesis about community structure variation embodied in an ordination or clustering, it is necessary to repeat the entire sampling and taxonomic exercise. That is consistently impractical, and results are judged on the subjective reasonableness of the relation of clusters to environmental variation.

some sediment-particle size analyses. He plotted the positions of the stations, identified by their cluster designations, on a triangle or "ternary" diagram of the proportions of clay, silt, and sand–gravel (Fig. 14.8). Cluster B, with two exceptional stations, is a group of stations with a high proportion of coarse sediment. The two exceptional stations contained some gravel, possibly ice rafted, which may be the habitat requirement of the group. Bilyard's conclusion is not unusual: sediment character is frequently a controlling habitat variable for the composition of benthic fauna. On the other hand, Cluster B appears from the map (Fig. 14.6) to be reasonably well explained by depth alone, since there

is an onshore–offshore gradient in sediment, coarser sediment settling closer to the coast than finer. However, the isobath-crossing pattern of Cluster A is not particularly well explained by sediment characteristics. Cluster A remains unexplained, a potential subject of further research.

It has been decades since this excellent project was completed, and many of the questions that it raised are still unanswered. That is very common in such research. Funds to support the study were available when there was interest in possible offshore oil production and thus the need for background to write environmental impact statements. By the time the study was complete, that interest had waned, because of fear of the impact from massive drifting ice on drilling platforms. A study with similar techniques and goals was carried out by Feder *et al.* (1994) in the region just to the west. It provides a further arctic example of community analysis that we recommend to readers specifically interested. Dunton *et al.* (2006) have contributed a useful food-web analysis (a different form of community evaluation) for the Beaufort Sea benthos. Interest in the arctic shelf benthos has recently accelerated. Because nearshore ice is melting earlier and freezing later, substantial changes can be expected (Carmack & Wassmann 2006).

Benthos, more community ordinations

Benthic studies provide many examples of community analysis. A common application is to discern the range of biological impact from sewage or pollutant outfalls in coastal areas. Similar techniques are applied to plankton, fisheries catches, forest trees, and insects found in wheat fields. The strength of the approach is that organisms tell us what constitutes a suitable habitat for them, and which habitats

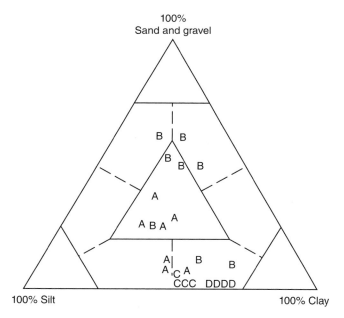

Fig. 14.8 Positions of Beaufort Sea stations identified by polychaete-based cluster designations (A to D) on a ternary diagram of sediment composition. (After Bilyard & Carey 1979.)

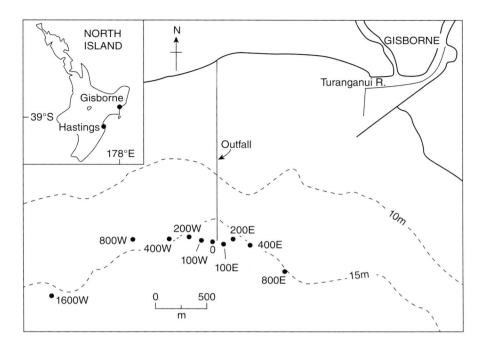

Fig. 14.9 Location map of Gisborne, New Zealand, sewage outfall with positions of benthos sampling stations. (After Roper *et al.* 1989.)

are differentiable by them, without requiring much knowledge of their actual biology. The limit of the method is that it only generates hypotheses about which aspects of a habitat are important to the organisms characteristic of it, not tests of those hypotheses. In benthic ecology, the same hypotheses recur: assemblages of animals change progressively with depth, sediment type, organic richness, and character of the overlying water. Perhaps that would be predicted without even looking at samples, but it is fully confirmed by data from many sites, which is more satisfying.

Radically simplified explanations of several commonly used ordination techniques are provided in Box 14.1. Two books providing detailed explanations of various methods with ecological examples are Legendre and Legendre (1998) and McCune and Grace (2002). Like many other forms of writing (for example, instructions for complex software), descriptions of ordination often assume that the reader knows things that he or she does not. Because the authors know all the details and definitions, they occasionally fail to consider that you might not know one or the other. Also, ordination procedures are feasible largely because computers can carry out the complex calculations, particularly so for nMDS (see Box 14.1). Once the computer program is working, there is a tendency to just send the reader along to try it, assuming that experience will somehow complete the transfer of understanding.

Roper *et al.* (1989) compared the benthic fauna in the near vicinity of city sewer outfalls debouching in the subtidal at about 15 m depth offshore from Gisborne and Hastings on the eastern shore of North Island, New

Zealand. They evaluated the density of a very large number of benthic species at a series of stations close to and variously far from the outfalls (Fig. 14.9 for Gisborne) along the same isobaths as the outfalls. At Gisborne, the analyses included 42 species of polychaetes, 12 mollusks, 27 crustacea, four echinoderms, and four others – a total of 89 species generously represented among 2735 specimens. The Gisborne [species x stations] matrix was evaluated with PCA (Fig. 14.10) and showed distinct community modification near the outfall (in agreement with a cluster analysis also applied). Roper *et al.* then overlaid the positions of the stations in the PCA plot with the amounts of oil/grease in the sediment at the stations, showing that the near-outfall stations (only two close in) were indeed strongly polluted. The Hastings results were similar. Somewhat surprisingly, areas affected by sewage outfalls are of modest size, a few kilometers. They are larger for outfalls from very large cities, like the Hyperion outfall from Santa Monica, California, USA, but still not as large as might be guessed from the tonnages of organic matter coming down the pipes.

Estuaries provide the strongest horizontal gradients in habitat conditions of any marine habitat, save for the intertidal. As fresh water flows in from the upstream river, it tends to ride over water entering from the ocean downstream and moving landward beneath the outflow as a "salt wedge". The vertical salinity gradient between the surface layer and salt wedge can be sharp if freshwater flow is strong and tidal mixing is weak. Conversely, most of the gradient can be along the horizontal axis of the estuary if

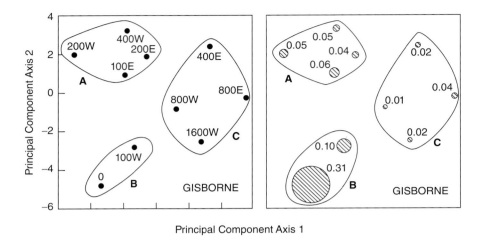

Fig. 14.10 Left: two-dimensional principal-component analysis locating sampling stations around Gisborne outfall in species "space". Right: PCA plot overlaid with the sediment oil and grease content at each sampling location. Sediment community composition close to the outfall is distinct, clearly a response to deposits from sewage. (After Roper *et al.* 1989.)

vertical mixing by tides is a dominant factor. In either case, the salinity along the bottom upstream will rise on flood tides and then fall on the ebbs. Marine benthic animals with their affinities to seaward are defended in several ways from reduced salinity when the tide is low. Oysters, mussels, barnacles, and some other forms living on hard surfaces simply close their shells for the duration, keeping the low-salinity water away from their tissues. Polychaetes, mud shrimp, and others living in the sediment are protected by the increase of salinity downward in their mud sanctuaries. Mixing of fresher water down into sediment and burrows is inhibited by the density gradient established at the highest tides, when the saltiest, most-dense water of the cycle overlies the sediment and sinks into burrows. Thus, estuarine benthos tends to have marine, not freshwater, affinities far inland (e.g. Alexander *et al.* 1935).

San Francisco Bay, on the North American West Coast, extends inland from rocky headlands at the mouth, along an estuary of more than 100 "river km" that drains the Sacramento and San Joaquin Rivers. In addition to effects of a strong salinity gradient and substantial pollution, the estuary's benthic community has been challenged by large numbers of invasive species since the mid-19th century, and dramatically so in the late 1980s by the bottom-paving north Asian clam, *Corbula amurensis*. Peterson and Vayssieres (2010) evaluated a 27-year time-series of seasonal or more frequent benthic fauna analyses of 23 × 23 cm grab samples from four monitoring stations along the estuary (Fig. 14.11). Analyses were combined into mean annual composition tables, and the whole (27 × 4) = 108 sites vs. species table was assessed by nMDS. The distance measures were "Bray–Curtis dissimilarity indices computed from 4th-root transformed [a frequently applied strategy to reduce the effects of differential species abundance] annual

mean invertebrate abundance data". The ordination diagram (Plate 14.1) shows a clear upstream–downstream pattern. The position along the first ordination axis was then plotted vs. salinity (Plate 14.2), showing an extremely tight relationship. Peterson and Vayssieres also showed that:

"Benthic assemblage composition was more sensitive to mean annual salinity than other local physical conditions. That is, benthic assemblages were not geographically static, but shifted with salinity, moving down-estuary in years with high delta outflow, and up-estuary during years with low delta outflow, without strong fidelity to physical habitat attributes such as substrate composition or location in embayment vs. channel habitat." [From their abstract.]

The ordination procedure provided Peterson and Vayssieres with an efficient means for presenting the community gradient, thanks to the reduced dimensionality of the ordination plots. It allowed ready demonstration of its main cause (Plate 14.2). There is more information in Plate 14.1: while all the station clusters from a zone in the estuary fall above the same sector of nMDS axis 1, they separate strongly on axis 2. Another ordination plot coupled with community composition pie diagrams (Plate 14.3) shows convincingly that some of that variability was likely an effect of the *C. amurensis* invasion.

Community analysis – a functional guild approach

There are completely different approaches to community analysis that require extensive biological expertise and,

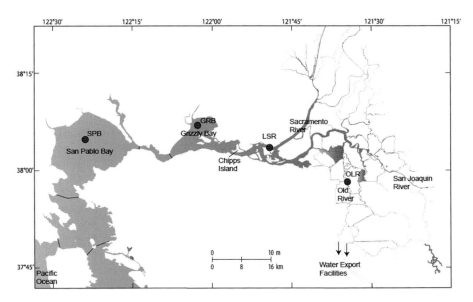

Fig. 14.11 Map of the inner estuary adjacent to San Francisco Bay. Sampling stations were located at the black circles. (After Peterson & Vayssières 2010.)

thus, are much less common despite the insights they generate. Under the punning[1] title, "A Diet of Worms", Fauchald and Jumars (1979) gave an example of how the extreme detail available (and that may become available) in our knowledge of the biology of species and higher taxa of animals can be applied in community analysis. Most ecologists become expert in the biology of some group of organisms, and they apply that biology to enlarge the understanding of how those organisms interact with their habitats. Fauchald and Jumars are both interested in polychaetes. An advantage of polychaetes for such studies (Jumars & Fauchald 1977) is that, in addition to numerical and biomass dominance, they originated in the Precambrian, so their radiation into families is very ancient. Thus, the absence of an extant polychaete life mode in any given habitat is unlikely to be due to insufficient time to disperse to it. Probably this argument applies to most marine groups, so the real advantage is that Jumars and Fauchald know polychaetes.

[1]As history buffs will recall, the Diet of Würms was a meeting in 1521 at Würms of the General Assembly (nobility) of the German states, called by Charles V, Holy Roman Emperor, to review the opinions of Martin Luther. Luther defended himself, ending with the famous words (here in English), "Here I stand. I can do no other. God help me, Amen." The Emperor was not sympathetic, but rather a defender of the faith. Luther disappeared on the way home and was wrongly rumored to have been murdered. The history goes on, but does not really concern us here. The pun, while fun, has no significance for Fauchald and Jumar's paper.

In their 1979 paper, Fauchald and Jumars presented a functional classification, mostly based on feeding mode, for the polychaetes, a class consisting of 81 family groups totaling (up to 1980) roughly 6000 species (now ~9000). Much of the known biology of feeding for the entire class was included in a family-by-family review. They used this to construct a classification of polychaete feeding under three main aspects.

1 Size of particles eaten: microphages vs. macrophages
   • Microphages eat tiny particles and tend to eat them in bulk. Divided again by feeding stratum: burrowers, filter feeders and surface deposit feeders
   • Macrophages eat large chunks either whole or by removing bites. Divided again between herbivores and carnivores
2 Mobility required for feeding
   • Sessile: never moving, sometimes not capable of moving
   • Discretely motile: sometimes moving to get better foraging, but usually not moving while actually feeding
   • Motile: moving while feeding
3 Mode of ingestion (varies between micro- and macrophages)
   • jaws (sometimes based on an eversible pharynx)
   • tentacles
   • pumping
   • "X" – special

The overall classification came out as shown in Table 14.1, which is simplified here from the original, and in

Table 14.1 Summary of a functional classification of polychaetous annelids by Fauchald and Jumars (1979). Acronym designations are followed by families with members included and one or more example genera. H, herbivore; C, carnivore; B, burrower; F, filter feeder; S, surface deposit feeder; J, jawed; T, tentaculate; P, pumping; X, other feeding morphology.

Macrophages

HMX – Non-jawed, motile herbivores that eat large particles are all from families of small-bodied polychaetes. So the particles they eat are only relatively large. The only well-studied family is the Paraonidae – *Paraonis fulgens*. They burrow about in the surface of sediments and take single diatoms and foraminiferans as food.

HMJ – Typical are the Nereidae – *Nereis*. These usually form mucous tubes in the sediment, but can move from them and form new tubes. They emerge from their tubes and explore the neighboring zone of the surface sediment. Most eat bits of drifting plant (Fig. 14.12a).

HDJ – Onuphidae – *Diopatria cuprea*. Build a fibrous tube of mucus and protein, the tip of which sticks up above the sediment surface. The worm decorates it with stones, bits of wood, sediment, or growing algae. Move rarely, but they can, and adults can establish new tubes. Large bits of alga passing the tube are held and chewed (Fig. 14.12b).

CMJ – Other Onuphidae, Nereidae, and Syllidae. Crawl along, sometimes on hard substrates, chewing on sponges, cnidarians, etc. (Fig. 14.12c).

CMX – Amphinomidae – *Hermodice carunculata*. These chew on coral polyps, etc. They may live in coral sand, especially during mid-day. They have toxic spines. There is an eversible, muscular lower lip for rasping and squeezing prey (Fig. 14.12d).

CDJ – Glyceridae – *Glycera* spp. and others. These have enormous, eversible probocides armed with a circlet of spines, which connect in some species to a poison gland. They are ambush predators, sitting quietly in the sand and popping the proboscis out to catch passing prey (Fig. 14.12e). This family is pretty flexible, however, and includes detritivores.

Microphages

Filter-feeders

FDT – Tentaculate, discretely mobile – Sabellidae (e.g. *Sabella* sp.). These are tube-dwelling and only move if displaced. The crown of tentacles makes a current by moving as a whole. Ciliary currents along tentacle filaments move particles to the mouth (Fig. 14.12f).

FDP – Pumping – Arenicolidae – *Arenicola* spp. Worms that live in U-tubes. They move water through the tube by peristaltic pumping with the body. This pulls water through a sand plug ahead of the animal, which is eaten after a period of filtering. Fauchald and Jumars indicate that despite prolonged study the feeding is still not fully understood (Fig. 14.12g).

FST – Sabellaridae. Tubiculous, reef-building polychaetes. Feeding similar to the Sabellidae.

FSP – Chaetopteridae – *Chaetopterus* spp. These live in a U-tube, pumping with swinging movements of large parapodia that move water along the tube from incurrent to excurrent openings. An anterior parapodium produces a mucous net (much like that in salps) which is progressively consumed (Fig. 14.12h).

Surface deposit feeders

SMJ – Motile, jawed – Lumbrinereidae – *Lumbrinereis* spp. These have jaws on an eversible proboscis. They move through the very surface of sediment, opening the path by thrusts of the partially everted probocis, and poking out intermittently. Some ingest sediment; some are carnivores, herbivores, etc. (Fig. 14.12i).

SMT – Flabelligeridae – *Flabella* spp. These move near the sediment surface, collecting particulate food from the interface with large palps (Fig. 14.12j).

SMX – Capitellidae – *Capitella* spp. A group of simple, earthworm-like species that move through the very surface of the sediment, ingesting it whole.

SDJ – Nereidae, Onuphidae (for both, see above).

SDT – Spionidae – *Pygospio* spp. These form a tube from sediment particles. The worm projects above the sediment surface and the tentacles sweep the surrounding area. Particles of interest are transported along the palps and eaten (Fig. 14.12k).

SDX – Some Arenicolidae (see above).

SST – Ampharetidae – much like Spionids, but less motile (Fig. 14.12l).

Burrowing deposit feeders

BMJ – Many mobile, digging carnivores also ingest sediment whole (Nereidae, Nephtydidae, Lumbrinaridae).

BMT – Spionidae have an example.

BMX – Many families, e.g. Pectinaridae (*Pectinaria* spp.). There are many modes for burrowing, for selecting sediment, for ingestion. Some sorts of tentacle are common, but "T" is used for long tentacles and burrowers usually have short tentacles (Fig. 14.12m).

BSX – This seems a contradiction in terms, and Fauchald and Jumars discuss that. The Maldanidae are tubiculous, eating sediment at the ends of the tube system. They are mobile within the tube, and the tubes grow through the sediment. Sediment is exchanged at the ends of the tube by repeated cave-ins (Fig. 14.12n).

Fig. 14.12. While feeding is the basis of the classification, there is strong correlation with other life functions. Mobile and tubiculous forms, for example, have different modes of defense against predators, different requirements for mating, and so on. There are 21 groups with significant "occupation" by kinds of polychaetes, each group identified by an acronym. For example, BMJ is a burrowing motile-jawed subclass of microphagic feeders. Fauchald and Jumars call these "feeding guilds". Ecologists borrowed the notion of guilds from late-medieval history. Guilds were the craft unions of the late Middle Ages, such as the tanner's guild, baker's guild, and so on. These had limited entry by an apprenticeship system; held the secrets of the craft as closely as practice allowed; and struggled artfully to keep prices high by control of supply. There were often social aspects as well, with a guild hall for meetings, weddings, and temporary shelter. Some had simple insurance schemes for members. For ecologists, the term is akin to the Eltonian *niche*, the notion that each species has a role in the ecological community. Members of guilds are known (or often just believed) to have similar functional modes in their habitats. Of course, each modern user of the term has his or her own intentions for it, so some caution is in order when adopting guild schemes from their developers.

In an earlier paper, Jumars and Fauchald (1977) used their classification system to examine the relative occurrence of different feeding-guild members in soft sediments at different depths. For data, they took a species-by-station (30 stations) table generated mostly by the renowned polychaete systematist Olga Hartman for shelf and slope sediments off the southern California coast, adding to it some Pacific slope and abyssal data generated by Jumars. They displayed the data on ternary diagrams (Figs. 14.13 & 14.14), like that of Bilyard, but here the axes refer to the proportional guild composition of the annelid fauna, rather than to the sediment mixture. Proportions were calculated over the total array of individuals in a sample, excluding carnivores (which does give the diagram only three axes, but seems unfortunate; we would like to know the relative importance of carnivores). Each basis of classification (motility, feeding stratum) was analyzed separately. The results show that sessile modes are less common than either motile or discretely motile modes among sediment-dwelling polychaetes of the shelf. Only one sample from the coastal survey showed a predominance of sessile worms. In deep continental-basin sediments all three motility modes were represented (points near the triangle center), while the abyssal sample was evenly split between motile and discretely motile worm species, but almost no species were fully sedentary. In other words, sessile habit is most common at mid-depths. Filter-feeding is a strategy of relatively few individuals in soft sediments at all depths. On coastal bottoms, assemblages vary from all surface deposit-feeding to all burrowing, including all combinations between.

Jumars and Fauchald (1977) discussed the biases of the sampling and sundry statistical niceties at great length. For example, after deciding that most samples in the shelf survey done with grabs probably underestimated burrowers, they eliminated burrowers from the comparisons shown in Figs. 14.13 and 14.14, in order to generate Fig. 14.15. Examining their graph, they hypothesized that sessile worms are rare inshore because unstable sediments and turbulent events, which are frequent to depths as great as 30 m, militate against that lifestyle. Probably it is necessary for successful adults (the ones that we see in samples) to retain the ability to re-establish themselves in the bottom and to return to the sediment surface after slumps, migration of sand waves, and other disturbances. Deeper on the shelf, where sediment stability is higher and food is abundant, sessile life modes (but not filtering above the sediment) are more favored. At the greatest depths, where food is least abundant, Jumars and Fauchald hypothesized that enough food cannot acquired by sitting still. This analysis is very informative, although in itself it does not prove the hypotheses, only suggests them. Jumars and Fauchald were clear about that. More deep-sea data were obviously needed. It is often difficult to commandeer a ship and set off to get needed additional data. It would have been useful to know something about the proportion of carnivorous polychaetes, even if they did not fit conveniently on the triangle graphs.

Guild analysis could be much more extensively applied than it has been. Probably the sheer bulk of biological knowledge required is rarely assembled in one or a few people. Also, the results remain subjective (not just that, they are verbal or only quasi-quantitative), although that could be viewed as an advantage as well as a problem. The first edition of this book stated that, "In the present scientific climate, where natural historic intuitions stated in words are mostly valued as poetry, . . . , it will be difficult to generate a well-funded fashion for this sort of work in science." However, well before that, by 1992, the paper had been cited 245 times, and Fauchald (1992) commented on its success:

> "Many of our conclusions in 'The Diet' are outdated. We were wrong, even spectacularly wrong sometimes. We are now in the second 'post-diet' generation of papers citing the first 'post-diet' generation, and still, sometimes, 'The Diet' itself. . . . 'The Diet' is becoming hidden behind layers of investigations with better results and better theory, in part as consequence of its existence."

That has continued, reaching 1049 citations by the end of 2010, and we now have third- and fourth-generation studies of exactly how benthic animals move and eat; how they accomplish ingestion; how much they eat under

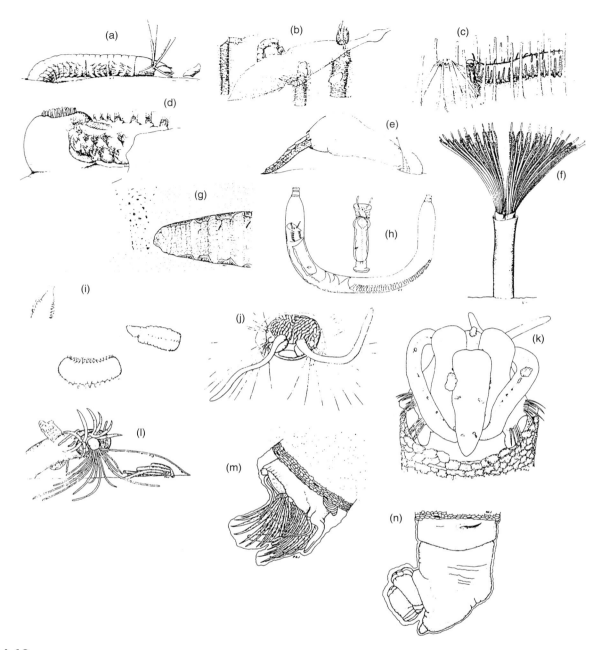

Fig. 14.12 Sketches of the anterior (feeding) ends of polychaete families from different guilds as defined (see Table 14.1) by Fauchald and Jumars (1979). (a) HMJ, Nereidae; (b) HDJ, Onuphidae; (c) CMJ, Syllidae; (d) CMX, Amphinomidae; (e) CDJ, Glyceridae; (f) FDT, Sabellidae; (g) FDP, Arenicolidae; (h) FSP, Chaetopteridae; (i) SMJ; Lumbrineridae; (j) SMT, Flabelligeridae; (k) SDT, Spionidae; (l) SST, Ampharetidae; (m) BMX, Pectinaridae; (n) BSX, Maldanidae. (After Fauchald & Jumars 1979.)

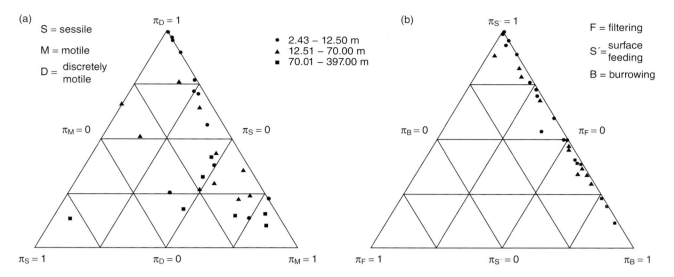

Fig. 14.13 Ternary diagrams for samples from the continental shelf off southern California. (a) Stations are located in the diagram according to the proportions of discretely motile (π_D), motile (π_M), and sessile (π_S) polychaete worms. (b) Stations are located according to proportions of surface feeding ($\pi_{S'}$), burrowing (π_B), and filtering (π_F) polychaetes. (After Jumars & Fauchald 1977.)

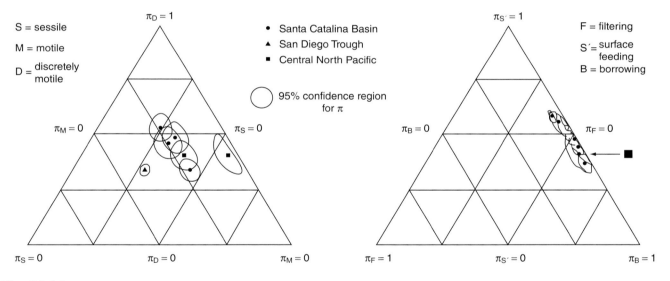

Fig. 14.14 Ternary diagrams like those in Fig. 14.13, only for deep-water samples in the Santa Catalina Channel, San Diego Trough, and central North Pacific. (After Jumars & Fauchald 1977.)

different circumstances; how digestion and assimilation proceed. We must represent the variety of those results by only a few examples.

Benthont movement and feeding

Both burrowing and feeding by macrobenthos have been studied primarily in intertidal animals. They are readily

available and they survive well in the laboratory. Likely the results are in most respects applicable to deep-sea fauna. Burrowing through sand and mud, especially stiff mud, has been assumed to be extremely energy demanding. However, that is a subject of ongoing research, and for several reasons it is still neither established nor refuted. In particular, it is difficult to distinguish metabolic energy required for digging (with associated inefficiencies) from that required by digestion and basal respiration. The mechanics of burrowing, specifically moving forward through sediment, have been reviewed by Dorgan *et al.* (2006). They divide the subject

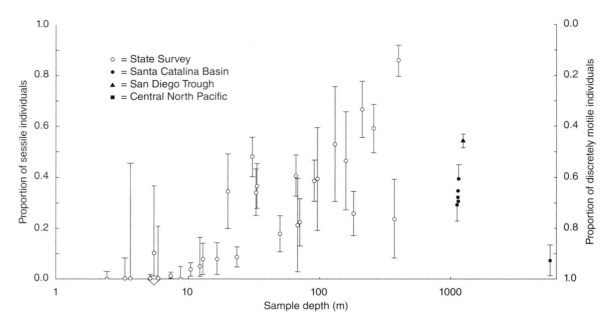

Fig. 14.15 Changes in relative proportions of sessile and discretely motile polychaetes in samples from different depths in several sampling surveys. Confidence limits (95%) are for proportion estimates and are based on the binomial distribution. (After Jumars & Fauchald 1977.)

according to sediment type: moving in sand and progressing through mud work differently.

Sand, both on a wet beach and subtidally, tends to be a stack of grains held in place by gravity, each grain resting at a few points or edges on the grains below and to the sides. The packing usually does not minimize the empty space filled with water between grains. Find a wet beach (nearly flat, fine sand, drained at ebb tide to the sand surface, but with interstitial spaces filled with water) and stand on it. Wiggle your feet and you will disturb the stacking, liquefying the fluid–granular mix as the grains find less space-filling arrangements. You will sink. When you step aside, you will leave a depression holding water that will then drain down into the beach. Sand dwellers use this liquefaction to move. Because the magnitude of the gravitational vector sustaining the stacking is much less effective against horizontal disturbance, much less force is required to liquefy the sediment when burrowing across the beach than down into it.

A worm moving horizontally pushes against the sand ahead, anchoring its length by extending setae into the sand alongside or by hydraulically expanding some specialized segments. Pushing against the sand ahead is by hydraulic extension, a process akin to squeezing a sausage-shaped balloon with both hands. Circular muscles contract, the body narrows and extends forward. Some worms, such as the polychaete *Nephtys*, have an eversible proboscis that pops forward under such increased pressure to push against the sand ahead. It is then drawn back by longitudinal

muscles. In addition, lateral body oscillations can liquefy the sediment all along the body, enabling motion like swimming through a dense but fluid medium. Sand bottoms are mostly shallow enough to be repeatedly restacked by wave action, so that animals moving through them do not permanently convert the bottom to optimum packing of the grains, stopping their movement employing liquefaction. Motion in sand, especially moving to just below the surface, like mole crabs (*Emerita*) repositioning on a beach, can also involve simply digging and then moving into the hole. Liquefaction helps cover the animal's downward retreat. Animals with access to the water above can enhance liquefaction by pumping water into the sand ahead.

Deep-sea mud (and off-channel estuarine deposits) are at the opposite end of the particle-size spectrum with interstitial spaces smaller than even meiofauna. Moreover, the miniscule particles are consistently bound together in a matrix of mucopolymers, forming an elastic, quasi-solid different in behavior from saturated sand. Organic "matrix surrounds all particles, of whatever size" (Watling 1988). Dorgan *et al.* (2005, 2008) use clear gelatins of various penetrabilities to model mud, with the advantage that benthonts moving through gel blocks can be seen and photographed. They have used the polychaete *Nereis* as a model animal that can be induced to move forward in gelatin by orienting them in vertical starter burrows. *Nereis* pushing through such media force the propagation of cracks somewhat ahead of the anterior tip. The force opening the crack, again from hydraulic expansion of the anterior end of the worm (that is, the "prostomium", the anterior sides and in *Nereis* a proboscis) is tensile, pushing perpendicular to the direction of travel and causing the crack to expand in a U-shape ahead. Stress (tensile force per unit area) on the sediment is amplified ahead of the worm by the lever action of the worm on the elastic mud walls behind the opening tip of the crack. Both "O-ring" body expansions and setae may serve as anchors along the length of the worm to assist in pushing the anterior tip forward – a very old notion of how worms, including earthworms, move. The news is the splitting of the sediment well ahead of the prostomium. After establishing a crack, the worm moves into it. Setae are drawn inward as segments elongate and are pulled forward. An important aspect of the cracking process is that the burrow shape is flat and wide (Fig. 14.16; Dorgan *et al.* 2005); thus many polychaetes when actually in the sediment are strongly compressed in the dorso-ventral direction.

Many other benthonts moving beneath the sediment surface also generate cracking. Snails and clams both extend a narrow edge of their foot as a wedge anchored by the shell in the burrow behind, then pull the shell into the crack. Dorgan *et al.* (2008) provide a detailed mechanical description of movement in mucopolymer-bound mud based on the stress-to-strain relationships of "visco-elastic" substances, basically a description of crack propagation. Application to animals other than polychaetes remains sketchy. Possibly, for example, the curved, flattened back of mud-dwelling amphipods serves as a cracking wedge, but detailed observations are required to explain their motility in polymer-stabilized mud. It has been suggested that cracking demands far less energy than pushing through mud, if mud is modeled as a massive and extremely viscous semi-fluid. Perhaps, but the energy costs of burrowing remain to be adequately quantified.

As sediment ahead liquefies or cracks, the newly accessible space ahead of a burrower can be explored by tactile and olfactory senses for particularly desirable food particles: organic-rich and bacteria-laden clots of sediment, nematodes, and other meiofauna are ingested, generally along with plenty of the sediment matrix. Digestion extracts organic matter, but the mass of benthont fecal matter can be nearly indistinguishable from the mass ingested. Nutrition is derived from small fractions of relatively large throughput. On an Oregon (USA) sand beach, with ~6 mg POM per gram of sand, gut throughput for the polychaete *Euzonus mucronata*, which ingests bulk sediment, is mostly mineral. However, uptake by the worms from traces of ^{14}C-labelled organic detritus was sufficient that an apparently realistic 10% assimilation efficiency for all POM could supply its needs (Kemp 1986). Most deposit-feeders consume particles from very near the sediment surface, but there are also specialists that live on organic components considerably deeper, at least toward the bottom of the most active bioturbation (see below) zone. The capitellid polychaete *Heteromastus filiformis* orients head down and feeds on deposits 15 cm and farther below the sediment surface (cm bss), recovering modest fractions of usable organic matter from a largely refractory supply, possibly depending upon bacteria as food. Clough and Lopez (1993), using isotope labeled bacteria from sediment >15 cm bss, estimated their retention efficiency at 8%. Oxygen for metabolism is apparently acquired by the tail of the worm closer to the water column.

The bulk of deposit-feeders, both macrobenthos and megafauna, mostly select and assimilate organic matter quite recently arrived from euphotic zone production. This has been demonstrated for the upper slope offshore of Cape Hatteras (DeMaster *et al.* 2002; samples from 1996) and the upper slope at 64°S in the Bellinghausen Sea (Purinton *et al.* 2008; samples from 2000). The comparisons were of the ^{14}C contents of upper ocean particulates (largely plankton) with that of sediment organic matter. The upper-ocean organic carbon has been enriched by +50

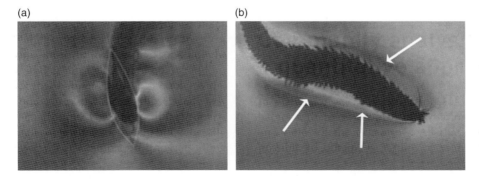

Fig. 14.16 Outlines of a burrow made by *Nereis virens* in a gelatin block photographed with polarized light. (a) Anterior, end-on view of the crack with the worm's dorsal and ventral surfaces against the gelatin. (b) Dorsal view with the outline of the burrow indicated by arrows. (From Dorgan *et al.* 2005.)

to +170 ppm (relative to wood from 1890) by bomb-test ^{14}C, which was increased in the atmosphere during the 1950s and 1960s by +700 ppm. The variability is due to varying proportions of upwelled water with old (low ^{14}C) carbon. Offshore from Cape Hatteras, surface algae (*Sargassum*) using this enriched source have ^{14}C content of ~+109 ppm. Organic carbon in the sediment, however, was quite old, ^{14}C of −41 to −215 ppm, while that in benthic animals (six families of polychaetes and a fish) was +40 to +83 (one worm only was +20), clearly enriched by bomb carbon (DeMaster *et al.* 2002).

Particularly during the open-water season of diatom blooms, the coastal Antarctic is subject to substantial upwelling, so bomb carbon is a much smaller component in surface plankton, with ^{14}C negative relative to standard at −135 ± 10 ppm . There remains, however, a strong difference between that level and carbon in surface sediment at −234 ± 13 ppm. A variety of trawl-collected surface-deposit feeders, mostly echinoderms (sea cucumbers and urchins), had tissue ^{14}C levels of −125 ± 13 ppm, comparable to surface plankton not to bulk sedimentary organic matter (Purinton *et al.* 2008). Clearly, the benthonts are constituted primarily of carbon from recent, near-surface production. They must achieve that by selective feeding on more-nutritious particles and by digestion and assimilation primarily of newly produced organic matter. The latter is partly insured by the refractory character of molecules of greater age; they are resistant to enzymatic attack. For example, Ahrens *et al.* (2001) have shown, with isotope-labeled food, that the polychaete *Nereis* (again, experiments favor readily accessible nearshore forms) absorbs fresh algae with 55–95% efficiency, compared to 5–18% for carbon in bulk sediment. Larger worms showed the greater efficiencies, possibly due to longer gut passage times. Selective ingestion has been observed in experimental work with a wide variety of benthic animals (e.g. Taghon 1982). In Taghon's study with miniscule glass beads, smaller beads were selected over larger, and protein-coated beads over bare ones, by deposit-feeding worms and clams. Either or both "preferences" could result from more favored particles bonding more effectively to mucus on feeding surfaces. To some extent, also, "selection" may simply reflect accelerated feeding when fresh organic matter arrives at the seafloor (Taghon & Jumars 1984).

Bulk benthic processes

As for pelagic ecosystems, processes in benthic habitats can be bundled into bulk effects, and their overall rates measured. Studies in this mode allow placement of the seafloor into general biogeochemical models of the global ecosystem. Many such bulk measures are of sediment mixing and oxygen-utilization rates.

Bioturbation

One bulk effect of benthic biota is sediment mixing, referred to as *bioturbation*. This mixing is most readily demonstrated by profiles of radionuclides in the upper layers of sediment. For example, radioactive carbon (^{14}C) is generated in the atmosphere by cosmic rays (also for a while by bomb tests) and, like other carbon isotopes, it becomes incorporated in organic matter and reaches the sediment in descending organic particles, including marine snow, fecal pellets, and whale carcasses. Then it is mixed downward into the sediment by the burrowing of worms; by the churning and ingestion of passing sea cucumbers; and by the tail swooshes of passing macrourid fish. It eventually reaches a depth in the sediment called the bioturbation limit, L, which varies beneath oxic waters from about ~1.5 to ~20 cm, averaging between 5 (Teal *et al.* 2008) and 10 cm (Boudreau 1998) across all depths. As sediment is progressively added to the pile, the layer being mixed moves up, remaining at the surface. If 1 cm is added at the surface, then 1 cm just below depth L will be protected from further mixing. Throughout its residence in the mixing layer the ^{14}C is reduced by beta decay with a half-life of 5730 years. The amount of reduction can be interpreted in terms of an apparent age, which the mixing continually averages between the most recently arrived carbon and that added at all different ages back to the time of deposition of the layer just at L. However, this physical averaging does *not* produce an accurate mean time since deposition for the layer as a whole. That is because inorganic carbon in the water column ages and loses ^{14}C after sequestration below the sea surface, and carbon isotopes do not fully equilibrate to atmospheric proportions in the surface layers where primary production occurs. Thus, settling particulate organic matter already has an apparent ^{14}C age of hundreds to thousands of years (except that, since bomb testing, it can appear to be younger than zero age). Sediments in the bioturbated layer do increase in apparent age from coastal sites at about 400 years to deep-sea abyssal plains at 12,000 years (Emerson *et al.* 1997; Hedges *et al.* 1999), a function of relative rates of organic matter input and most importantly of overall sedimentation rate.

Below L, there is no further mixing, so ^{14}C decreases exponentially, as expected from radioactive decay, and apparent age increases with depth. Thomson *et al.* (2000) provide an example (Fig. 14.17) from just west of Rockall Bank northwest of the British Isles, which shows $L = 17.4$ cm. The slight increase in ^{14}C in the upper 3 cm is generally attributed to bomb-test contamination. "Conventional years" are ^{14}C ages compared to a tree-ring standard. Sedimentation rate at the Rockall site, estimated from these and other data, is 4.4 cm (1000 yr)$^{-1}$.

Values of L estimated from isotope information are not constant, but depend upon the half-life of the isotope used.

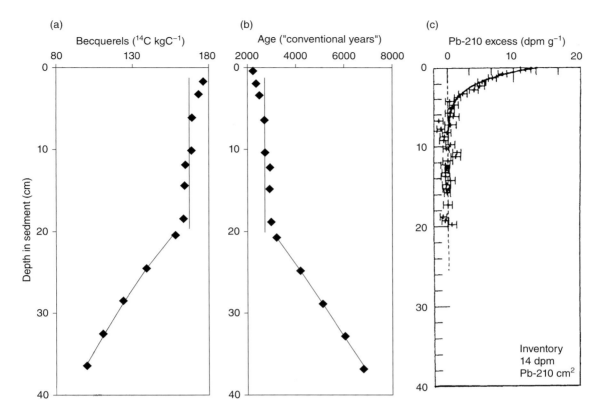

Fig. 14.17 Profiles of bioturbation tracers from a sediment core taken from 1100 m on Rockall Plateau, northeast Atlantic. (a) ^{14}C, (b) ^{14}C age, and (c) ^{210}Pb excess. (a & b from data kindly provided by J. Thomson; c after Thomson *et al.* 2000.)

A very commonly used tracer of sediment mixing is "excess" lead-210 (^{210}Pb). This is an intermediate product in the serial decay of radium; its half-life is 22 years. In sediment, the excess ^{210}Pb is that which has settled from the water column attached to particles and can be differentiated from that sustained by radium in the sediment. This excess is measured by counting its decay products, as well as determining the amount of radium to permit subtraction of "radium supported" ^{210}Pb. Down-core data (Fig. 14.17c) from the same core do not have the same form as the ^{14}C profile; excess ^{210}Pb falls off exponentially from near the sediment surface, reaching zero at 8–10 cm, which is used as its estimate of L. The half-life is short enough that below L there is none. The difference between the two tracers is caused by the difference in their half-life and also by the intermittence of bioturbation. Most of the time the sediment must just be sitting there, its stacking not much disturbed. Once in a while, a worm moves some aside to make a burrow, and surface sediment falls down the hole. Then it sits there again. The mixing events will happen often enough to homogenize a more durable tracer, but not a more ephemeral one, and more durable tracers will eventu-

ally be mixed to greater depths by the more occasional visits of larger animals.

The pattern of ^{210}Pb and several other tracers can be used to estimate a rate of bioturbation using a diffusion model. Waiting times between mixing events are taken in this model to be small relative to the total time available, and a continuous mixing rate D_B is determined as a bulk diffusion coefficient by analogy to Fick's Law:

$$D_{\mathrm{B}} \frac{\mathrm{d}^2 C}{\mathrm{d}z^2} - \lambda C = 0$$

in which $\mathrm{d}C/\mathrm{d}z$ is the concentration gradient of isotope downward in the sediment and λ is the radioactive decay rate. D_{B}, a bioturbative diffusion coefficient, can be estimated from the isotope profile, and for the site near Rockall Bank the result is 0.088 cm^2 yr^{-1}. Like L, D_{B} depends upon tracer half-life, but in the opposite direction of course. Smith *et al.* (1997) provide a comparison (Fig. 14.18) for sites along a Pacific transect near the equator at 140°W, showing that D_{B} for excess thorium-234 (a uranium-238 daughter with half-life of 24 days) is markedly greater than

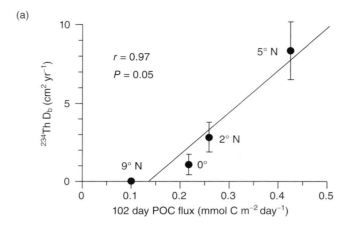

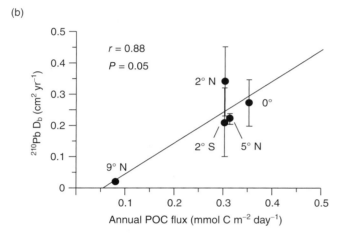

Fig. 14.18 Mean (±SD) sediment mixing coefficients (D_b) vs. POC flux into traps 700 m above bottom at stations along 140°W at the latitudes near the equator as indicated. (a) Excess ^{234}Th. (b) Excess ^{210}Pb. Coefficients are higher for tracers with longer half-life. Bioturbation is faster where more organic matter is supplied to feed sediment-mixing animals. (After Smith *et al.* 1997.)

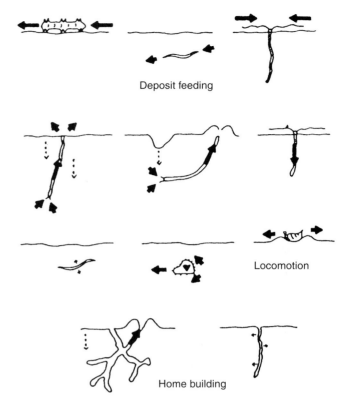

Fig. 14.19 Cartoons of the principal modes of sediment mixing by benthic animals, all largely self-explanatory, but see Wheatcroft *et al.* (1990) for full explanations. (After Wheatcroft *et al.* 1990.)

for ^{210}Pb. The *L* values for ^{234}Th were only 2–3 cm (Pope *et al.* 1996). Estimates of D_B for both isotopes were correlated with organic carbon input as estimated from sediment traps. Apparently, greater food supply leads to more biological activity, which leads to greater rates of bioturbation, a point carefully reviewed by Boudreau (2004).

Profiles are not always as clean as in the ^{210}Pb example from Rockall Trough (Fig. 14.17c). A brief, larger mixing event can occur and disrupt the orderly exponential progression of small mixing steps and radioactive decay. Repeated traverses of the sediment by large echinoderms (Plate 13.1) can introduce breaks and shoulders into the downward curve. Wheatcroft *et al.* (1990) make an extended argument that steady diffusion is not how bioturbation really works, even if diffusion mathematics produces reasonably good fits to the cleaner profiles of isotopes and other tracers. Biological activity in sediment will occur in

stronger and weaker pulses. Much of sediment movement by deposit feeders will be horizontal with very slight vertical displacements, such that most sediment mixing is not really represented by one-dimensional (vertical) diffusivity estimates. Moreover, they point out that vertical transfer of sediment often is not continuous across the mixing layer, but that animals, particularly worms, will usually mine at one level and deposit tailings at the other end of the body. An echiurid, for example, will sweep particles from the sediment surface with its proboscis, ingest them, and then deposit the somewhat changed material as feces at the bottom of its burrow some centimeters down. Sediment in the layer between is not affected. Thus the right model might be advective rather than diffusive. They give a diagram (Fig. 14.19) of different burrowing modes that produce sediment transfers of different kinds.

A tracer with a very short half-life in sediment is oxygen, which diffuses into interstitial water from the water column above. The majority of oxygen profiles determined by inserting microelectrodes slowly down from the sediment surface (Fig. 14.20) show a simple, curving form tending to zero a short distance down, a form explicable from a shifting balance between molecular diffusion from above through pore water (see below) and respiration. Half-life is

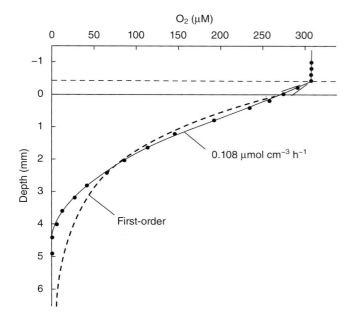

Fig. 14.20 Microelectrode profile of oxygen concentration in sediment at 15 m depth in Aarhus Bay, Denmark. Data are dots. Solid horizontal line is sediment surface; horizontal dashed line is at the top of the diffusive boundary layer (DBL). Solid fitted line is a zero-order model of consumption (no dependence on concentration) at a rate of 0.108 μmol cm^{-3} h^{-1}. Dashed curve is a best-fit first-order model. (After Rasmussen & Jørgensen 1992.)

not the correct representation for oxygen, since its utilization by benthic metabolism is approximately zero order (that is, not dependent upon oxygen concentration). Moreover, oxygen is not a tracer in the same sense as an isotope in the solid phase, since it can diffuse through pore-water spaces independently from bioturbation. It is continuously supplied from above as it is used in the sediment. However, in most places, particularly nearshore, if the supply from above were suddenly capped, the complete utilization times would be of the order of 2 hours. That is why the diffusive penetration is often shallow. Most, but not all, oxygen profiles are unaffected by bioturbation. Intermittent stirring events that homogenize the oxygen profile below the diffusion limit are quickly eradicated by the rapid utilization of oxygen by sediment microorganisms. The low frequency of deeper oxygen penetration, although it does occur, is an indication that most of the sediment is resting between bioturbation events most of the time.

Particle size may make a difference in the rate of downward mixing into the sediment column. Wheatcroft (1992) used a submersible equipped with a large spice shaker to spread smooth, spherical glass beads ranging from 8 to 420 μm diameter over a marked square meter of sediment at 1240 m in the Santa Catalina Basin off California. Cores collected on another visit 997 days later showed (Fig.

14.21) that the beads had spread downward to about 7 cm depth. Because the tracer was inserted as a "plane source" at the sediment surface, the form of the expected distribution is not a simple exponential, but rather a curve as fitted to the data in the figure. Best-fit D_B values decrease with increasing particle size, but inspection of the figure (and others in the paper) suggests the differences are small. All but the largest sizes, which are almost pebbles as seen by small, deep-sea macrofauna, were moved to the same L, and the general form of the curves is the same.

Boudreau (1998) has shown that L is independent of water column depth (Fig. 14.22a). For a large number of isotope profiles, mostly ^{210}Pb, it averages 9.8 ± 4.5 cm (±SD) (but see Teal *et al.* 2008). It also is not affected by sedimentation rate, which is usually correlated with the supply of organic matter and inversely correlated with water column depth. Wheatcroft *et al.* (1990) have speculated that L is set by downward compaction of the sediment (Fig. 14.22b), that it gets progressively harder to move through sediment as its water content decreases downward. This hypothesis of rising cost of deeper burrowing is difficult to test. Boudreau (1998, 2004), in contrast, proposes that bioturbation stops rather close to the sediment surface because most of the non-refractory organic matter is used up in sediments deeper than a few centimeters. There simply is no return to burrowers from seeking food deeper. It should also be remembered that, especially on continental margins where organic input is highest, sediments below a few centimeters are usually anoxic, have negative redox potential, and contain toxic reduced substances including sulfide. Thus, burrowers would encounter difficulty in sustaining activity below a few centimeters. In recent years, complex models of bioturbation rates and depths have been developed, star turns with partial differential equations. Bernard Boudreau, Filip Meysman, Jack Middleburg, and Karline Soetaert are prominent in this work.

Filled burrows in sediment can be seen in X-ray photographs of core slices. Enhanced downward ventilation along burrow walls promotes precipitation of heavy minerals, manganese, and iron solubilized in reducing sediments. These, plus mineral and compaction differences between burrow fill and burrow surround, generate a differential in X-ray absorption, and thus images can be made. There is a definite layering of burrow sizes: mostly tiny ones in the uppermost few centimeters; a middle layer several decimeters thick with larger burrows often visible by color differences as well as X-ray signatures; and a layer of final burial in which burrows fade out (Berger *et al.* 1979).

For many oceanographic purposes, the key point about bioturbation is that *it occurs*; it must not be forgotten or ignored in stratigraphic work to reconstruct faunal or climatic history. Its effect is that the record of depositional events is smeared in time, losing resolution rather dramatically. Oceanic sedimentation rates can be extremely small, as low as 0.1 mm (1000 yr)$^{-1}$ under the subtropical Pacific

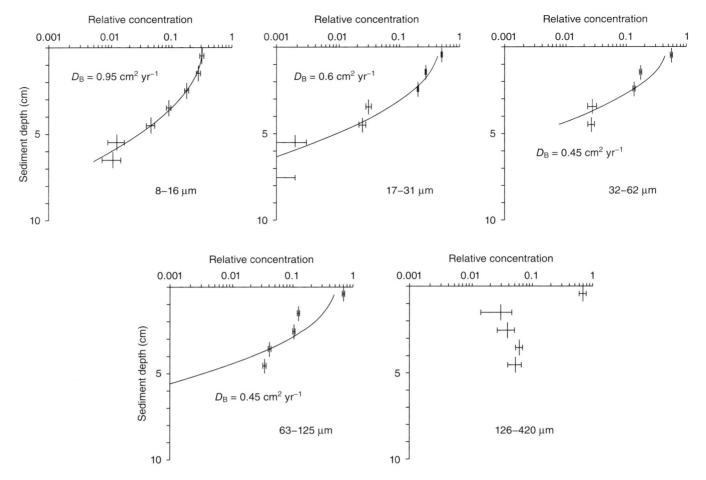

Fig. 14.21 Dispersion curves fitted to progressively larger sizes of tracer particles added to sediment in a basin off southern California. There is a slight tendency for smaller particles to be moved faster and to have larger dispersion coefficients. The largest particles were moved but could not be fitted with a dispersion model. (After Wheatcroft 1992.)

gyres. For sites with such low rates, even with a small L of 4.0 cm, the time at which a particular fossil arrived in the sediment can only be known within about 400,000 years, longer than the periods of the glacial–interglacial cycle. Brief events, for example glacial terminations, are not represented by a plane of abrupt change in the sediment, but by widely dispersed and upward-skewed vertical distributions, as modeled by Guinasso and Schink (1975). Moreover, final storage in the stratigraphic record for events of paleontological or climatic interest only occurs after mixing to the deepest level (L) represented by tracers of the longest half-life, and is influenced by the very smallest, most long-term values of D_B. Of course, stratigraphy is not hopeless. Something can be made of changes downward through strata, provided that sufficient care is exercised. There are sites with much more rapid deposition and sites under nearly anoxic bottom water, such as the Santa Barbara Basin off California, USA, that have no bioturbation and sediments showing annual layers termed *varves*. It is appropri-

ate that sedimentologists have given very close study to these few sites.

Bioturbation has also been credited with a role in the evolution of global biogeochemical processes. The notion is that in the late Precambrian, as autotrophs began to increase oxygen levels in the atmosphere and oceans, some plants (mosses and liverworts by 700 MYA, fungi and lichens a bit later) colonized land, generating enhanced erosion that transferred clay particles to the sea. Charged clay surfaces in water bound organic matter to the particles, increasing the burial of organic carbon, reducing net relative respiration and, thus, increasing global oxygen levels (Kennedy *et al.* 2006). Increased oxygen and buried food supported an adaptive radiation of marine metazoans that initiated bioturbation. Bioturbators broke up sediment-surface microbial mats, extended organic-matter distributions deeper, and eventually established the main features of the modern near-balance between organic-matter production and its metabolic recycling.

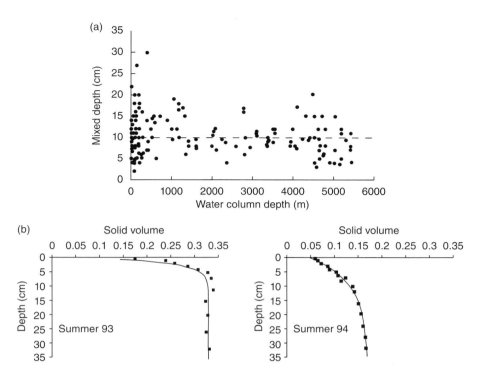

Fig. 14.22 (a) Mixing depths, *L*, in marine sediments around the world, based on excess ^{210}Pb. There is no trend in *L* with depth. (After Boudreau 1998.) (b) Two examples of sediment compaction curves from the continental slope seaward of Nova Scotia. Solid volume $(1 - \varphi)$, where φ is sediment porosity, is plotted vs. depth in the sediment. (After Mulsow & Boudreau 1998.)

Sediment sculpting

Animal activities producing bioturbation also change the shape of the sediment surface. This can be subtle or pronounced. Sediment miners move sediment to the surface, forming mounds, often with volcano-like shapes (Fig. 14.23). Burrowing sea cucumbers, *Molpadia*, make the most dramatic mounds, but many different forms, particularly polychaetes, echiurids, and enteropneusts (acorn worms) make smaller bumps. Pits and grooves are formed in many ways. Ampharetid polychaetes excavate small feeding pits at the ends of their parchment-like tubes. These tubes are partly vertical in the sediment, partly crossing the surface horizontally to the pit. Any upright animal tubes or slender sea pens generate local turbulence close to the sediment and thus increase scouring. Megafaunal echinoderms (brittle stars, sea stars, urchins and sea cucumbers) moving across the bottom produce tracks, often with a flat bottom surface and side ridges raised from slightly to several centimeters above the sediment surface (Plate 13.1). Benthic ecologists refer to the array of animal marks in sediment by the German term *Lebensspuren*, life traces. It was once thought that activity in the deep-sea is slow enough that *Lebensspuren* might have ages of decades or longer. Time-lapse camera studies and repeated submersible visits have made clear that this sculpturing changes and changes again on a time scale of weeks (e.g. Wheatcroft *et al.* 1989).

Sediment can be both stabilized and destabilized by animal secretions and activity. Mucus added to sediment in gut contents can enhance its coherence at least temporarily. In the opposite direction, formation of large numbers of clam pseudofeces, tiny compacted balls of sediment, can make the bottom more fluid. It is believed that burrowing activity in the rich nearshore sediments deposited on steep slopes in submarine canyons can sometimes set off slumping and trigger turbidity currents.

Total benthic metabolism

Measurement

Sediment community oxygen consumption (SCOC) has been extensively studied as a measure of the metabolism of everything living in the seafloor. That is of importance for evaluating the role of the seafloor in the overall economy of organic matter in the sea. Ronnie Glud (2008) has provided a landmark review of the status of this work, including evaluation of technical issues not considered here.

Fig. 14.23 Sediment mounding from burrowing sea cucumbers. (After Gage & Tyler 1991.)

If you wanted to measure the oxygen consumption of all infaunal residents (from bacteria to clams) of an intertidal mudflat, you could take a can open on one end, install a valve on the other and a stirrer inside, wade out when the tide was at mid-thigh height, fill the can with bay water, and insert the open end slowly into the sediment with the valve open to let water out as the can slid into the mud. You would be careful to insert it to a known depth so that the volume inside would also be known. Then you would attach a syringe to the valve, take a water sample, and close the valve. That sample would be analyzed for oxygen content ($[O_2]$) before an incubation period. A series of samples taken over time would show a decline in $[O_2]$. The rate of oxygen use, or total oxygen utilization rate (TOU) would be an estimate of the benthic metabolism for the area under the can over the study interval.

Modern technology allows this to be done with automatic sampling at any depth in the sea, by using canisters attached to benthic landers. Landers, essentially big tripods, descend to the seafloor, set down gently, wait for the dust to clear, and then slowly sink one or several canisters into the sediment below them. Oxygen concentration is determined in a time-series of syringe samples or by a recording electrode. Respiration measures have also been tried by inserting canisters into retrieved box cores. However, return to the surface involves decompression and warming, both of which alter oxygen-uptake rates, generally increasing them.

Consumption of oxygen by the biota of sediments would render the sediments anoxic if oxygen did not diffuse in from the water column above. Oxygen concentration can be profiled in sediments with exquisite vertical precision, using microelectrodes – usually platinum wires embedded in glass micropipettes and compared to silver–silver chloride reference electrodes above the sediment surface. Microelectrodes are also mounted on motorized microprofiling instruments carried by landers (e.g. Reimers 1984). After the dust clears, the electrodes are driven slowly into the sediment, producing profiles (Fig. 14.20). Electrode output can be calibrated with the oxygen concentration (perhaps determined by Winkler titration) of a bottom-water sample and the value several centimeters in the sediment (usually zero). More recently glass optical-fibers have been doped at the tip to produce fluorescent signals convertible to oxygen concentration, creating oxygen "optodes". These, too, have been adapted to penetrate the sediment from landers (Wenzhöfer et al. 2001).

Profiles from oxygen electrodes (Fig. 14.20) and optodes match the deep water column $[O_2]$ well above the water–sediment interface. Descending to within a few millimeters of that surface, approaching the level of no water motion right where water meets mud, the probe enters a diffusive boundary layer (DBL) in which the slope of the $[O_2]$ matches the initial subsurface profile. This slope, $d[O_2]/dz$, can be used to estimate the diffusive oxygen uptake rate, $DOU = D_O \, d[O_2]/dz$. D_O is the molecular diffusivity of oxygen in water, a function of temperature – slower when cold (Armstrong 1979):

Temperature (°C)	D_0 (cm^2 s^{-1} × 10^{-5})
0	0.99
5	1.27
10	1.54
15	1.82
20	2.10
25	2.38
30	2.67

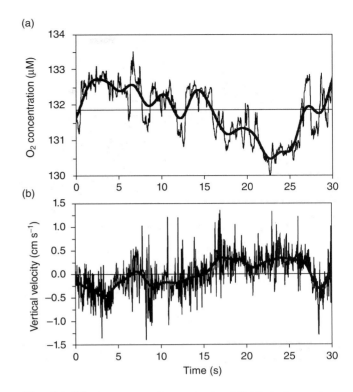

Fig. 14.24 Simultaneous 30-second series of [O$_2$] fluctuations and vertical velocities (+ is up) recorded at 25 Hz at 15 cm above the sediment in Aarhus Bay, Denmark. The horizontal lines are means. Thick lines are from low-pass filtering. (After Berg *et al.* 2003.)

More sophisticated diffusion-reaction models take account of both the fractional interstitial porosity and an effect of particle shape and packing called *tortuosity*, which can be measured from profiles of sediment resistance to electrical current (e.g. Berg *et al.* 1998). In general, these models are "zero order", that is they assume no dependence on oxygen concentration, and, in fact, biological oxygen utilization does not depend upon oxygen availability except at very low concentrations (Fig. 14.20), only on other factors, primarily the availability of organic matter (Cai & Reimers 1995). Typically, TOU is somewhat greater than DOU, because the metabolism of dispersed larger organisms is included and the larger area is more likely to incorporate the metabolic "hot spots" in the microbial microstructure. Macrofauna can generate local sediment ventilation, and electrodes occasionally show effects of this ventilation as subsurface peaks in oxygen (Glud *et al.* 1994).

Neither canisters nor electrodes can be readily pressed into sand or gravel sediments. In part, that problem stimulated development by Peter Berg *et al.* (2003) of a non-penetrating TOU method: eddy correlation. An acoustic Doppler velocimeter (ADV) is suspended from a stand next to an oxygen electrode, both at 10 to 15 cm from the sediment. Records are accumulated of vertical velocity, *w*, and [O$_2$]. This is well above the DBL (Fig. 14.20). Extremely small, but measurable, drops in [O$_2$] due to eddy advection and mixing out of the lower concentration in the DBL correlate with upward velocity pulses ($\sim$0.5 cm s^{-1}); rises correspond to oxygen being mixed down by downward velocity pulses. The magnitudes of these changes are minute but recordable at high frequencies (e.g. 64 Hz with running averages at 8 Hz used to reduce instrument noise). Results for an estuarine deployment (Fig. 14.24) show correlation of quite large fluctuations compared to the deep sea. You are left to think about why the flux (TOU) over a deployment of many minutes is the average of $w'[O_2]'$, the product of the fluctuating components (after subtracting the means from all measures of *w* and [O$_2$] to get w' and $[O_2]'$). A comparison of this technique to canister and electrode profile results at 1450 m in Sagami Bay (eastern Honshu) showed excellent agreement (Berg *et al.* 2009). Eddy cor-

relation equipment is being commercially manufactured, and widespread data will be forthcoming (stay tuned).

Measurement results

An assembly of the global data-set (Glud 2008), eliminating sites with very low [O$_2$], shows an exponential drop-off of both TOU and DOU with depth (Fig. 14.25a). For TOU, the change is 85-fold from 10 m to 4000 m. The scatter does not overwhelm the definite relationships, but it implies strong differences in organic-matter supply among sites. This corresponds to an exponential increase in the depth of oxygen penetration (Fig. 14.25b, from electrode profiles). Where less organic matter is supplied to the sediment to support respiration, the oxygen can diffuse farther into the sediment. Beneath oligotrophic waters and at great depth, oxygen can diffuse down 8 cm or even 20 cm (Wenzhöfer *et al.* 2001). The difference is clearly related to organic-matter availability, that is, to biological oxygen demand. However, the depth of oxygen penetration is strongly variable, even on local scales, as close together as separate probe drops can be made (Fig. 14.26). In a rich coastal site, Sagami Bay at 1450 m, the variation in O$_2$ penetration was from 0.2 to 1.2 cm. This local patchiness is due to local concentrations of organic matter (food), to

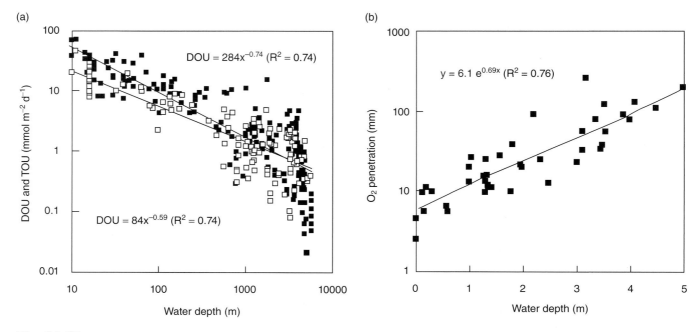

Fig. 14.25 (a) Global assemblage of DOU (open squares) and TOU (closed squares) results versus depth. Log–log plots with regression equations. (b) Oxygen penetration into sediment vs. water column depth. (After Glud 2008.)

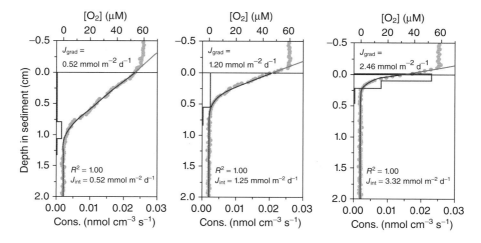

Fig. 14.26 Three sample profiles (the extremes and a median example) of $[O_2]$ versus depth in the sediment of Sagami Bay, Honshu, Japan, at 1450 m water depth. J_{grad} values are flux rates calculated from the slope of the upper profile. Implied oxygen-consumption rates of the sediment community are on the lower abscissa. Similar local variations of the flux and depths of penetration of oxygen are typical. (After Berg *et al.* 2009.)

stirring along the trails of megafauna; to nearby presence or not of burrowers; to bacterial mat present or absent – all resulting in substantial variation in microbial activity on a scale of centimeters. Similar variation occurs in all seabeds.

Wenzhöfer and Glud (2002) have compared DOU between areas of high surface production (e.g. the West African upwelling zone) with that in oligotrophic central waters. On bottoms 1 km deep, the difference is a factor of four; at 3 km it is a factor of two. Overhead production is a major determinant of deep-sea biological activity, but less organic matter reaches the bottom when it must traverse

more water to get down. It is not surprising to find that reflected in SCOC results; they simply agree with trap flux, macrofauna abundance, and so on.

Food supply relative to metabolism

Once TOU and DOU can be measured, the obvious question is: to what extent does it compare with supply rates of organic matter descending to the seafloor? Answering is complicated by the seasonal cycling of both supply and consumption. So, a sufficient comparison requires a time-series of considerable length and at least seasonal resolution. K.L. Smith and colleagues (Baldwin *et al.* 1998; Smith *et al.* 2009) have obtained such a time-series in a study of Station "M" on an abyssal plain at 4100 m depth, 220 km west of Point Conception, California. From 1989 to 1998, they made 36 cruises at roughly seasonal intervals. Additional bursts of sampling have been made through 2010. At each visit they retrieved and replaced conical sediment traps (baffled openings were 0.25 m²) moored at 600 and at 50 m above the bottom (mab). The traps collected downward particle flux in a series of cups filled with preservative and automatically changed every 10 days. The cup-changing mechanisms sometimes jammed, so some data were lost, but intervals in the series are complete. Cruise participants also made a series of benthic observations, most prominently sediment oxygen-consumption rates over two days with canisters on landers. Other studies were conducted as well, including camera sled photography, trawls for megafauna, and submersible dives.

Supply rates of particulate organic carbon (POC) measured by traps at both 600 and 50 mab (Fig. 14.27a & b) showed strong seasonal cycling. The POC time sequences at the two depths were very close, confirming that very little is consumed from sinking organic matter below 2000 m. Rates at 50 mab were about 10% higher than those at 600 mab, which can be attributed to lateral particle transport close to the bottom. Maxima in POC flux followed, with a lag of 2–3 months (relatively immediately), the seasonal highs in expected coastal upwelling based on crude estimates (Bakun indices) of the regional, upwelling-favorable winds. In addition to seasonal cycling, there was variability in total annual flux (Table 14.2; sampling gaps do not allow updating after 1997–1998). POC flux correlated somewhat (Spearman rank correlation = 0.3, $P < 0.001$) with strength of upwelling (Bakun index; BUI) 20 to 60 days earlier, and upwelling strength varies considerably between years.

The reduction in frequency of visits to Station M after 1997–1998 was briefly preceded by the beginning of

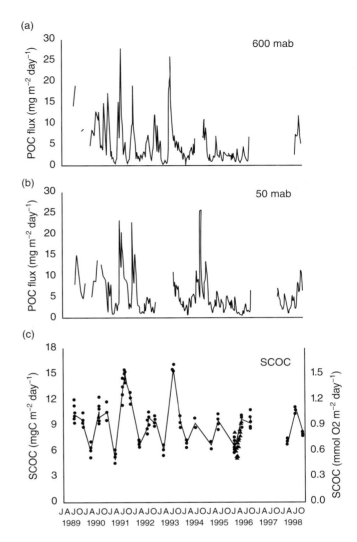

Fig. 14.27 (a and b) Time-series of particulate organic carbon flux into traps at 600 and 50 m above bottom (at 4100 m) 220 km west of Point Conception, California, USA (Station M). (c) Sediment community oxygen consumption at the same station. Left-hand scale is implied organic carbon metabolized if respiratory quotient is 0.85. (After Smith *et al.* 2001.)

Table 14.2 Annual totals of POC flux compared to sediment oxygen consumption (SCOC, converted to units of carbon utilized) at Station "M". (After Smith *et al.* 2001.)

	YEAR								
	89–90	**90–91**	**91–92**	**92–93**	**93–94**	**94–95**	**95–96**	**96–97**	**97–98**
POC flux (g C m⁻² yr⁻¹)	3.18	2.51	1.63	2.32	1.91	1.53	0.67	—	1.45
SCOC (g C m⁻² yr⁻¹)	3.21	3.75	3.34	3.81	3.12	2.94	3.06	—	3.37
POC/SCOC	0.99	0.67	0.49	0.61	0.61	0.52	0.22	—	0.43

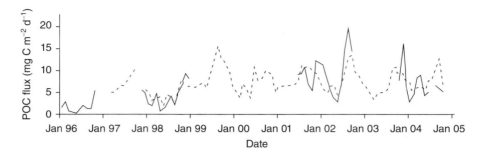

Fig. 14.28 Solid lines: sediment-trap POC flux (50 m above bottom) at Station M. Dashed line: multiple regression model of the data (solid lines) based on satellite chlorophyll (200 km radius), temperature and upwelling indices as independent variables. (After Smith *et al.* 2006.)

availability of satellite chlorophyll data (1996). Smith *et al.* (2006) have modeled a continuous time-series of POC flux to the Station M trap levels for 1996 to 2004 (later updated to 2006; Smith *et al.* 2009). Satellite chlorophyll (Chl) and temperature (*T*) time-series were run though a net production algorithm, and a multiple regression was generated for POC as dependent variable with the Chl, *T*, and BUI series as independent variables. The resulting relationship suggests that POC flux was greater from late 1999 through 2006 than in 1996 through 1998 (Fig. 14.28). These shifts in POC flux (if they did occur) may have had some effects on benthic fauna, and they appear to correspond crudely to variations in some of the climate oscillation indices (Smith *et al.* 2009; in regard to climatic variation indices see Chapter 16).

Smith *et al.* (2001) converted sediment community oxygen consumption (SCOC) rates to measures of organic-carbon oxidation using an assumed respiratory quotient of 0.85 (CO_2 produced per O_2 consumed), appropriate to a mixture of carbohydrate and lipid. Like supply, these consumption rates cycle strongly with season (Fig. 14.27c), being greatest in summer–fall, and lowest in winter, with a mean of roughly $10\,mg\,C\ m^{-2}\,d^{-1}$. Like most ecological time-series, this one is not long enough to fully establish relationships. It appears that summer peaks in SCOC were higher in two of three years with greater POC flux, 1991 and 1993 but not in 1994, than in years with lower flux, 1992, 1995, 1996. A major conclusion from the seasonal series is that response to greater input is rapid, even immediate; organic matter is metabolized soon after it arrives at the seafloor.

On annual average, more organic matter appears to have been consumed than was supplied, roughly 180% more, the discrepancy being greater when supply was smaller, since consumption remained more constant than supply (Fig. 14.27). Moreover, Smith *et al.* (2001) point out that the discrepancy should be assumed to be even greater to account for the roughly 9% of supply that is eventually sequestered deep in the sediment. That proportion sequestered is highly variable worldwide. Reimers *et al.* (1992)

give a general average as 13% of supply for the organic carbon finally sequestered below the bioturbation limit near the continental margin. This "burial efficiency" is much less, probably less than 2%, in bottoms beneath oligotrophic, oceanic waters. For Station M, at the outer fringe of the California Current upwelling zone, 9% is a reasonable number.

Smith *et al.* (2008) used time-series of seafloor photographs to evaluate the importance of large organic aggregates arriving at the seafloor – massive clots of diatoms and radiolaria. These masses were not collected by the traps – the baffles apparently hold them until they are swept off. Aggregates did provide additional pulses of potential food to the benthos, mostly simultaneous with the seasonal peaks of POC flux measured by traps. They were generally not enough to match the rates of sediment metabolism (SCOC). Glud (2008) states that all comparisons on shorter time scales show the same discrepancy: more metabolism implied by SCOC than can be supported by trap-measured POC flux.

Smith's measurement of the standing stock of organic matter at Station M is $150\,g\,C\,m^{-2}$ in the top 3 cm of sediment. While the seasonal cycle of respiration implies that this may be somewhat degraded stuff, it could sustain metabolism of $3–4\,mg\,C\,m^{-2}\,yr^{-1}$ for a considerable time, requiring only very occasional replenishment beyond the usual flux of POC. Traps do not measure the transfer via large food falls (whale bodies, etc.), a food source redistributed over the bottom by scavengers at a rate that cannot at present be estimated. Lateral supply in near-bottom currents is also possible, particularly adjacent to the continental slope where Station M is located. It might be thought, since most of the actual benthic metabolism is attributable to bacteria and other microbes which could assimilate DOC from interstitial water, that this would be an additional source. However, that DOC is dilute (standing stock $\sim 0.16\,g\,C\,m^{-2}$ to 3 cm), and it is unlikely to be replenished from DOC in the overlying water where it is even less concentrated with older ^{14}C ages, implying less food value. It must come from sediment POC, so metabolism of DOC does not help to explain the apparent imbalance of oxygen uptake and POC sedimentation. Andersson *et al.* (2004) take the view that SCOC must represent the real rate of organic-matter flux, estimating a relatively large global rate of transfer to the deep sea. However, a map of DOU and TOU measurement sites (Seiter *et al.* 2005) shows extreme bias toward coastal zones, leaving vast oceanic areas with no measurements and bottom oxygen concentration varying 10-fold.

Community response

Another question of interest is how do macro- and mega-faunal communities respond to the variations in food input. Ruhl (2008) has examined the Station M megafauna from the time-series of camera sled photographs, showing both substantial stability of the community composition and discernible shifts in composition. The pictures reveal that 10 invariant species, all echinoderms, constitute 99% of crawlers on the sediment surface: eight sea cumbers (holothurians), one urchin (echinoid), and one brittle star (actually all ophiuroids lumped). In 37 surveys, the brittle stars always ranked one or two. With one outstanding exception, the rank-order shifts were about those expected from sampling variation. That exception was that the usually dominant cucumber, *Elpidia minutissima*, shifted in 2001–2004 pictures (Plate 14.4)) from consistently ranking number one or two to number 10. In fact, it essentially disappeared. In the same interval, the urchin *Echinocrepis rostrata* shifted in both relative abundance, moving up from fifth rank to third, and increasing by about 10-fold in population density.

The urchin increase clearly involved a recruitment event, since median individual size dropped substantially. Ruhl's (2008) cluster analysis of the varying proportions of these 10 species from 1989 to 2004 showed convincing grouping of the sampling dates into eras, of which the shift in 2001, persisting to 2004, is the most distinctive. Ruhl sees a less-convincing correspondence to blocks of time in the Station M food-input data and approximations (Smith *et al.* 2001, 2006), and suggests that the changed composition of 2001 and later years could have been a response to the severe 1998 El Niño. Changes in macrofauna counted from box cores occurred but were more chaotic (Smith *et al.* 2009, their Fig. 4).

An Atlantic time-series

A similar but less protracted time-series of deep-sea trophic studies was conducted by British ecologists on the Porcupine Abyssal Plain: Station PAP at 4800 m in the North Atlantic subpolar gyre at 49°N, 16.5°E, due west of Brittany, centered between the European slope and the Mid-Atlantic Ridge. Results for 1996 through 2004 are published (Lampitt et al. 2010a). POC flux (Fig. 14.29) was measured with a conical 0.5 m² trap at 3100 m with collecting cups changed every two weeks in intervals of high flux, every eight weeks in periods of low flux. Satellite chlorophyll estimates, averaged for a 200 km radius circle around PAP, became available for comparison from October 1997. Chlorophyll (and thus primary productivity) and flux varied on roughly similar schedules (Lampitt 2010b). However, in some years, 1999 and 2001, peak flux lagged peak chlorophyll by several months, while in other years

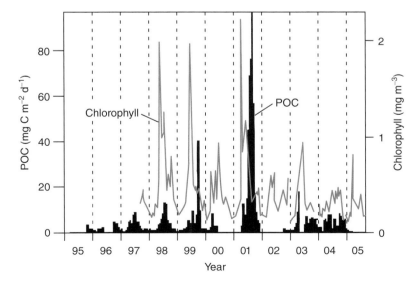

Fig. 14.29 Histogram bars: time-series estimates of particulate organic-carbon flux from traps at 3000 m above the Porcupine Abyssal Plain (49°N, 16°30′W) in the NE Atlantic. Continuous line (mostly above the bars): satellite estimates of chlorophyll concentration averaged for a 200 km radius around the trap station. (After Lampitt *et al.* 2010b).

no lag was obvious. A single year, 2001, had much greater flux, although chlorophyll levels in that year were not markedly more than in 1998 and 1999. In summers of 2002 through 2005, chlorophyll was low compared to 1998 to 2001, and flux was also low (as it was in 1998 and – save for one sample – 1999). The late-summer pulses of POC flux were actually periods of greater ratios of POC flux to mineral flux: there was relatively low flux of particulate calcium carbonate and opal. This aligns with doubts in the literature that mineral ballasting is a particularly significant determinant of rates of food supply to the deep sea (reviewed by Passow & De La Rocha 2006).

Lampitt *et al.* (2009) have suggested, based on continuous plankton recorder data, that late-summer population bursts of phaeodarians, 0.1 to 1 mm protists with siliceous skeletal bits embedded in a protein matrix and that feed with sticky pseudopods along fine opal spines, are likely to be responsible for bringing down large POC flux. Neither phaeodarians nor their spicules are abundant in trap materials, spicules are thought to dissolve during descent, but mini-pellets (protozoan feces) can be abundant in late summer trap samples. They were in 1990. Unfortunately, the 1999 and 2001 samples were not suitable for the necessary microscopic examination. Nevertheless, speculations of this sort do provide hypotheses for later testing.

No SCOC data were gathered with the PAP flux time-series, but a very close examination was made of the variations in abundance and composition of trawl-captured megafauna by Billett *et al.* (2010). Swaths of 6 to 15 hectares were swept with an 8.6 m wide otter trawl in the vicinity of PAP, but spread out enough on the abyssal plain to evaluate the generality of faunal shifts. Sea cucumbers (holothurians) were the dominant group, >60% of individuals and ~90% of biomass. Up to 1995, this dominance was shared by six species. Probably starting in 1995, but not sampled until 1996, there was an increase of the sea cucumber *Amperima rosea*. It had not been present before, but by 1996 it was the numerical dominant. Also increasing were *Ellipinion molle* (holothurian) and *Ophiocten hastatum* (ophiuroid). *Amperima* was present for four years, declined, and then went through a shorter uptick – not shared with *E. molle* – in 2001 to 2003. Then both species dropped to low levels; the "*Amperima*

Event" was over. The high POC flux of 2001 suggests that both events were likely to be driven by food supply. Some of the usual dominants increased only slightly in these intervals, but *Peniagone diaphana* decreased sharply, then came back. Billett *et al.* (2010) point out that the time required to rework the entire surface area of the sediment decreased during the event from >2.5 years to <6 weeks, and throughout that period seasonal accumulations of flocculant detritus were not seen on the PAP seabed. While cause and effect cannot be spelled out in close detail, it is certain that trophic relations, and through them population processes of benthonts, were greatly changed and changed back on a scale of a few years. The frequency of such events cannot be determined, but, taken with faunal shifts seen at Station M, they suggest dynamic and variable effects on benthic fauna driven by shifts of surface productivity and export rates. Similar variations can be found between geographical sites similar in most respects except for food supply from overhead production. For example, Maynou and Cartes (2000) report strong shifts in community composition of epibenthic, bathyal shrimp from adjacent basins with different annual productivity in the Mediterranean.

Closing note

An old *New Yorker* cartoon shows some society ladies on sofas, drinking tea. One of them is saying, "I don't know why I don't care about the bottom of the ocean, but I don't." If you share some sympathy with her sentiment, there are many other subjects in oceanography that should capture your interest. On the other hand, there are those among us fascinated with the mud and its denizens, and new tools are granting them rapid progress. As we continue to drill deeper in the seafloor for oil, with new spills eventually inevitable, and continue to consider bottom waters as a possible repository for carbon dioxide from fossil fuel, it is more than academically important that those benthologists make progress. All Earth's ecosystems are interconnected, so we need to understand the benthos, even though our view is obscured by kilometers of water.

Chapter 15

Submarine hydrothermal vents

Studies in the early 1970s by J. (Jack) B. Corliss (1973) suggested that basalt dredged from crustal spreading centers along mid-ocean ridges had been strongly modified by percolation of very hot seawater next to them. He assembled an impressive array of circumstantial evidence to that effect, much of it derived from isotopic comparisons of rocks from various sites. Without having seen a submarine geyser, he drew a diagram of water moving down through sediments and basalt in the vicinity of a subsurface magma intrusion then entrained upward into a convectively driven spring above it, a submarine geyser. Nearly identical diagrams are used to describe the sub-seafloor circulation around vents now that they are found and well studied. Corliss argued that many terrestrial sulfide-rich ore bodies were in fact precipitated along submarine spreading axes, eventually buried in sediment, subducted with their tectonic plate, and, finally, uplifted with continental mountain ranges. His model for the concentration process was simply stated (Box 15.1).

Corliss and a group of marine geologists and chemists, mostly from Oregon State University, advanced arguments that a submersible should be sent to examine these submarine "geysers", if in fact they existed. Considerable work went into getting the expedition approved, but finally in 1977 the group set sail aboard the Woods Hole submersible support ship R/V *Lulu* to make a number of dives in the submersible *Alvin* to the crest of the Galapagos Ridge, close to the equator in the eastern Pacific. Possible venting sites had been identified by towed thermistors and cameras prior to the dives. Corliss was first down. He promptly found the vents and saw that they were surrounded by a remarkable community of attached and motile benthic animals (Corliss *et al.* 1979; Plate 15.1). The history of this initial discovery of the vents is told by Cone (1991). Ridges, vents, and vent biology are reviewed in books by Humphris *et al.* (1995), Van Dover (2000) and Desbruyères *et al.* (2006).

The expedition was well prepared. Their cameras, rock-collecting equipment (grasping arms and holding baskets), thermometry, and chemical sampling gear allowed an amazingly complete first characterization of the vent systems, including collections of the animals. It was soon evident that the system was very unusual, since large concentrations of organisms whose near relatives are aerobic were living in water supersaturated in sulfide, which is a poison for cytochromes, the electron transfer enzymes fundamental to oxidative metabolism. The source of food, however, was immediately obvious. The vent walls and water were covered with bacterial mats that were assumed to be a food source. The presence of sulfide suggested that these would be chemosynthetic bacteria utilizing sulfide as a source of reducing capacity, a group of bacteria already well known from their presence in sulfide-rich sediments. Biologists who heard the story and saw the movies figured that out long before it was proved. Some (e.g. Enright *et al.* 1981), anxious that this be proved before it was accepted, suggested that enhanced flow along the bottom, driven by warm water rising from the vents, might concentrate particulate food for filter-feeders. There are filtering animals near vents, especially patches of serpulid polychaetes with tentaculate crowns. However, extended observation has shown that their principal food is particles emerging from the vent, not particles approaching from the surrounding seafloor.

Corliss turned the animals over to the invertebrate curators at the Smithsonian Institution, who distributed them to experts on the various groups represented. It took those experts a long while to come out with descriptions and names, but they did an admirable and complete job. What

Biological Oceanography, Second Edition. Charles B. Miller, Patricia A. Wheeler.
© 2012 John Wiley & Sons, Ltd. Published 2012 by John Wiley & Sons, Ltd.

Box 15.1 From "The sea as alchemist" (Corliss 1973)

"Most of the metals of economic interest are the last metals to crystallize in the slowly cooling lavas. These minerals occupy grain boundaries in the rocks and thus are accessible to leaching by the circulating, heated seawater through the formation of soluble chloride complexes . . . Interaction of seawater and basalt leads, at shallow depth, to simple leaching of rocks with little apparent alteration of the major mineral phases. At somewhat greater depths and higher temperatures, the interaction leads to alteration of major mineral phases and conversion of basaltic rocks to greenschists. During such alteration, considerable iron is leached from the rocks along with lesser amounts of manganese, copper, nickel, lead, cobalt, and other metals. The remaining iron is extensively oxidized from its initial reduced state in the magma. This oxidation is coupled with reduction of the sulfate in seawater which initiates the precipitation of the metals from solution as sulfides. The solutions rise and are vented into seawater as submarine hot springs. A reddish-brown precipitate of iron hydroxide then forms, incorporating other metals in the hydrothermal solutions and scavenging elements from the overlying seawater. Most of the precipitate settles, forming a layer of metalliferous sediment around the vent. Some fraction may be more widely dispersed into the seawater . . ."

we have learned since is that most of the sessile fauna around the vents, for example, from the first-visited sites, the tube worms (pogonophorans now known as vestimentiferans, *Riftia pachyptila*), the clams (*Calyptogena magnifica*) and the mussels (*Bathymodiolus thermophilus*) have evolved symbiotic relationships with sulfide-oxidizing chemosynthetic bacteria which provide their nutrition. The mussels also can filter-feed. Other members of the community, the crabs (*Bythograea thermydron* and several galatheid crabs) and fish in particular, are wandering about taking nips off the sessile forms. Others, including zooplankton that swarm in the vent plumes, filter-feed on the bacteria flooding from the vent. The first vents discovered were relatively quiet, without abundant mineral particulates in the flow or violent jets. Subsequent expeditions somewhat farther north along the East Pacific Rise found venting of dense black plumes loaded with minerals from tall chimneys. Those were immediately termed black smokers. Quite similar biota surrounded them.

Nine years after the initial vent discovery, Rona *et al.* (1986) found black smokers on the subtropical section of

the Mid-Atlantic ridge, vents with substantially different fauna. Vents have been explored now on "back-arc" subduction areas in the western Pacific and on the central Indian Ocean spreading ridge. Hydrothermal vents are located along much of the volcanically and tectonically active ridge systems of the world's oceans (Fig. 15.1). Different ridges spread at different rates, 0.1 to 17.0 cm yr^{-1}, depending upon the activity in the subtending magma chambers and their proximity to the ridge surface. Spreading speed is determined from the ages of magnetic reversals in the surrounding seafloor basalt (ages from comparisons to potassium–argon-dated magnetic-field signatures of Earth's polarity reversals in stacked basalts on land). Fast spreading is associated with magma as shallow as 1 km below the seafloor, slowest spreading with deeper and less-continuous sources of heat and molten rock. Greater hydrothermal activity is generally associated with faster spreading and closer proximity of water to actual magma. Vents along back-arc trenches occur where continental crust is moving away from subducting oceanic plate, magma filling between them. The plumbing of vents is not fully known, but upward channels through cracked basalt overlie recent lava intrusions into the upper crust, often at sites right in the axial valleys ("linear calderas") at the centers of spreading ridges. The upward flow is replenished with cooler seawater percolating through subsurface fissures from the sides and eventually from the ocean above. Heating can be sufficient to cause phase separation of brines and supercritical liquid–vapor mixtures at subsurface levels. One vent system discovered in 2005, at 3000 m depth at 5°S on the Mid-Atlantic Ridge, emits steam at up to 464°C (Koschinsky *et al.* 2008). In this supercritical phase, above 407°C, greater pressure does not condense water to liquid. However, water emitted from most vents is at 300°C or less with no obvious steam eruption (bubbles) above the seafloor. The hydrostatic pressure is usually great enough to sustain the liquid phase as temperatures cool on approach to the surface of the crust (Von Damm 1995).

The vents are not only of biological interest. Seawater dissolves copious amounts of practically every element soluble in extremely hot water as it percolates through the hot basaltic rocks of the spreading centers. For many of these elements, the vents have been shown to be the main input term for their budget in seawater. Their concentrations in seawater are what they are because of the input rates at the vents. Since this sort of budgetary chemistry is complex; since we have only known about the vents for 25-odd years; and since the vents can be visited only very occasionally, the consequences of this observation are still being studied. Many of the metallic species dissolved in vent discharge are not soluble at ordinary seafloor temperatures. Thus, when the vent plume enters the cooler surrounding water, the metals precipitate as salts, most often sulfides (e.g. iron pyrite) and sulfates, near the vent mouth. In many cases, particularly for the vents richest in metallic

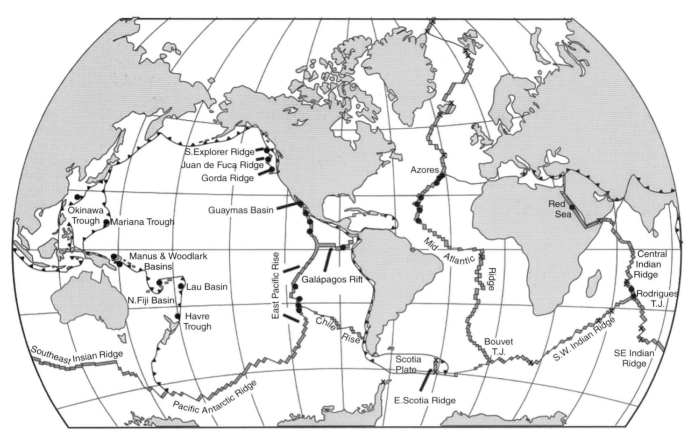

Fig. 15.1 Locations of hydrothermal vents on volcanically active mid-ocean ridges and subduction zones. Spreading centers are shown by parallel lines with offsets at transverse faults. Subduction zones are shown by single lines with arrow heads pointing in the direction of subduction on the side of the overriding plate. Sites sampled for biology are located at the circles; sites at the Xs have been identified from temperature only. (Map modified by Chris German after German & Von Damm, 2003, published with permission from Dr. German and Elsevier.)

solutes, "black smokers", this produces depositional chimneys around the discharge points. These chimneys are rich in exploitable minerals, and interest in them as ores has been considerable, although no deep-sea mining schemes have actually been started.

Black smokers with circulation driven by hot magma are the dominant form of hydrothermal vents along the ridge axes at spreading centers. However, in 2001, Deborah Kelley and others (2005) discovered some tall, white chimneys in the Atlantic, constituted mostly of limestone. These are not in the axis caldera, but well up on the basalt walls to the side. What the discoverers called the Lost City field is in fact on a huge undersea mountain, the Atlantic Massif (30°07′N, 42°07′W), which is rich in faults constituting a percolation system. The mafic (contraction of magnesium + ferric, although the iron is actually ferrous) and ultramafic (more Mg, very low silicate) basalts are subject to serpentinization by seawater moving through the fault cracks. That is, the water oxidizes the iron to the ferric (+3) state, releasing hydrogen and substantial heat that drives the hydrothermal flow and dissolves calcium. The rock left behind is serpentine (which you can look up). The rising water is basic (pH > 10), and when mixed with cool deep-sea water containing carbon dioxide, it deposits calcium carbonate (both aragonite and calcite) in remarkably lovely white towers. At least the initial structures are complex matrices of stony tubes conducting flow through the extending mass (pictures in Kelley *et al*. 2005). Temperatures in the tube systems are 40–90°C.

Lost City towers are not inhabited by abundant macrofauna supported by chemosynthetic bacteria, either free-living or symbiotic with animals. Likely that is because of the near absence of carbon dioxide and low sulfide levels in the venting water. There is a surprisingly rich meiofauna in the pores and cracks of the towers, both actively venting and cold; some larger snails and crustaceans are fairly abundant on the towers, and non-endemic epibenthos are present in the surrounding area. The moderately hot, basic water is loaded with molecular hydrogen and contains some methane that serves as a microbial substrate. The matrix of active Lost City vents harbors a mixture of microbes dominated by archaea of the order

Methanosarcinales, many of which are methanogens of anoxic habitats and some of which can grow by reducing CO_2 with H_2 (one form of chemoautotrophy). Some of the archaeal–bacterial consortium depends upon oxidation of hydrogen and methane for energy, and a fraction of oxidative metabolism by the microbial consortium is supported by sulfate reduction, producing some sulfide. Deborah Kelley is confident that more calcium carbonate tower systems will be found, but, given the rate of new discoveries, they must be relatively few.

Chemosynthesis

Sulfide-driven chemosynthesis was not newly discovered with the finding of the deep-sea vents. Chemosynthetic sulfur bacteria have been known since the 19th century (Winogradsky 1887), and they are particularly active in sediments. Sulfur is abundant in seawater as the sulfate ion, SO_4^{2-}. Wherever oxygen concentration drops to very low levels (<0.1 ml liter^{-1}), bacteria oxidizing organic matter will turn to sulfate and nitrate to obtain oxygen to metabolize organic matter. The products are sulfide, S^{2-}, and ammonium, NH_4^+. These reduced compounds represent a store of energy, of chemical potential. When oxygen is again available, chemosynthetic bacteria can oxidize NH_4^+ and S^{2-}, obtaining energy to drive a sequence of carbon-assimilating reactions producing organic matter, exactly the same reactions as the light-independent reactions of photosynthesis mediated by the Calvin–Benson cycle and by other enzymes. Plants use light to reduce NAD^+ to NADH and to generate ATP. Chemosynthetic bacteria use energy from oxidizing ammonium or sulfide with oxygen to carry out these reactions. Their enzyme, called ATP sulfurylase, mediates production of ATP from sulfide oxidation.

It is the simultaneous requirement for oxygen *and* sulfide that determines which sites will support sulfide chemosynthesis, including sites other than vents. In sediment, for example, these two solutes are available at the interface between anoxic (deeper) and oxic (shallower) layers, usually a rather narrow zone. In some organic-rich salt marshes this interface may be almost exactly at the sediment surface, where sulfur bacteria can form a pink scum. Near deep-sea vents, the zones sustaining such chemosynthesis are where the sulfide-rich vent water mixes with the relatively oxygen-rich deep-sea water. Oxidation of sulfide is thermodynamically favorable with a very low energy barrier, and it proceeds spontaneously in oxic solution. That makes the zones small where both oxygen and sulfide are at sufficient concentrations for chemosynthesis.

The basic reaction is:

$$HS^- + 2\,O_2 \rightarrow SO_4^{2-} + H^+,$$

producing -790 kJ mol^{-1} (free energy of formation).

Only bacteria perform this as a biochemical feat, capturing a fraction of the energy for chemosynthesis. So, for animals to take advantage, they must either graze on those bacteria right at the interface (or as they are swept away from it), or they must harbor the bacteria symbiotically in or on their bodies. In the case of vent animals, symbioses have produced the most dramatic life forms. A few modes of symbiosis, those of some "charismatic" invertebrates, are reviewed below.

In addition to sulfide-based chemoautotrophs, bacterial varieties have been found in hydrothermal vents that utilize other reduced species (often referred to as electron donors because they lose electrons as they are oxidized in energy-yielding reactions), particularly elemental hydrogen, H_2:

$$H_2 + \tfrac{1}{2}\,O_2 \rightarrow H_2O, \text{ producing } -263 \text{ kJ mol}^{-1}.$$

There are also bacteria, "methanotrophs", or "methylophiles", which oxidize methane, obtaining energy from:

$$CH_4 + 2\,O_2 \rightarrow HCO_3^- + H^+ + H_2O,$$
$$\text{producing } -803 \text{ kJ mol}^{-1},$$

and converting the carbon to complex organic matter. Both hydrogen and methane are dissolved in vent effluent. Different bacteria in vent plumes and deposits derive energy from oxidation of metallic ions, particularly $Fe^{2+} \rightarrow Fe^{3+}$ and Mn^{2+} to higher states (Jannasch 1999). It is not established that any of those are chemoautotrophs; the energy return from the iron oxidation is much lower, only -26 kJ mol^{-1}. However, the chemistry of vents exploitable by prokaryotes is remarkably complex and far from fully studied. For example, the chimneys have bacterial coatings that apparently utilize the sulfide in solid pyrite (FeS_2) and other metallosulfide deposits to drive chemoautotrophy (Wirsen *et al.* 1993).

Cary and Giovannoni (1993) have examined the mechanisms by which the young of vent animals obtain their symbionts. They used PCR-amplified DNA encoding SSU rRNA from the symbionts to make probes for characteristic sections of their DNA, then sought that DNA in eggs. The question was, do the mothers install a symbiont inoculum in the egg? The answer for *Riftia* is no. Apparently, young worms must capture their own symbionts from the flow field around a vent. However, for *Bathymodiolus* the answer is yes; the egg is equipped with a complement of viable symbionts.

Biogeography of vent faunas

Because vent fields develop, become populated, then cool, lose flow, and die, their dependent faunas dying with them, their fauna must have reasonably long-range larval transport to populate new sites. Otherwise, species extinction would accompany the death of local vent communities. Given long-range larval transport, one might expect vent faunas to be cosmopolitan, to be about the same everywhere along the entire ridge system. That is not the case. Tunnicliffe *et al.* (1998) reviewed the species then known to occupy vents and concluded that ranges of most species, even major groups, extend only along restricted ridge segments, surely because there are long stretches without vents. "The great majority of species are found at only one site" (Tunnicliffe *et al.* 1998). Ramirez Llodra *et al.* (2007) provide an informal global geography with good photographs of assemblages. Bachraty *et al.* (2009) provide a statistical analysis of regional faunal patterns based on a catalogue they have assembled of 592 species (~80% regionally endemic) in 332 genera from 63 vent fields. They found that vent fields examined to date can be divided, with only small overlaps in faunal composition, into six provinces (Fig. 15.2). That list will likely grow as more sections of the spreading ridge system are examined. For example, Schander *et al.* (2010) have collected fauna from the Mohn Ridge at the north end of the Mid-Atlantic Ridge (71°N, 500–700 m) by ROV and identified 180 species. In contrast to many low-latitude sites, no obvious vent specialists were found, but there is enhanced abundance of regional fauna: sponges, hydroids, anemones, diverse annelids, a gastropod (*Rissoa* sp.), and more. Shander *et al.* found abundant "smoker" bacteria, but no obvious or elaborate symbioses. This suggests that arctic vents (there are also some on the Gakkel Ridge farther north, known only from temperature studies) are relatively new, not old enough for evolution of vent specialists. When the very long ridge system surrounding Antarctica can be fully explored (which the prevailing weather makes less attractive than tropical ridges) more vents and possibly species will be found.

There are vestimentiferans on long sections of the system in the eastern Pacific and in the western Pacific, but not in the Atlantic, or Indian Ocean. They differ at the (admittedly somewhat arbitrary) genus level among the East Pacific Rise (*Tevnia* and *Riftia*), western Pacific (*Alaysia* and *Arcovestia*) and the vents of the Explorer–Juan de Fuca–Gorda ridge system seaward of Canada and the northwest USA (*Ridgea*). The eastern Pacific vent systems lack the provannid snails dominant in the western Pacific and the alvinocaridid shrimp (e.g. *Rimicaris*) of the Mid-Atlantic and Central Indian Ridges.

The Atlantic system lacks the vesicomyid *Calytogena* clams of the east Pacific, but does have related mussels. Bathymodiolid mussels are the most cosmopolitan of the larger fauna. Atlantic sulfide chimneys support swarms of alvinocaridid shrimp, but lack the vestimentiferans and alvinellid worms, a complementary exclusion which is not explained, except by distance and isolation. It would seem that the roughly 40-million-year existence of deeply submerged, oceanic ridges in the Atlantic should have been long enough for some species of vestimentiferan to establish there, but it has not been. The barriers are too extensive. Also, but not likely given the time for evolution to operate, differences in the chemistry of the vent plumes, deriving from differences in rising magmas, may support different strategies for exploitation of chemoautotrophic potential.

Vents in the so-called back-arc systems of the northwestern Pacific are dominated by provannid snails (*Ifremeria* and *Alviniconcha*), as are the hydrothermal areas of the

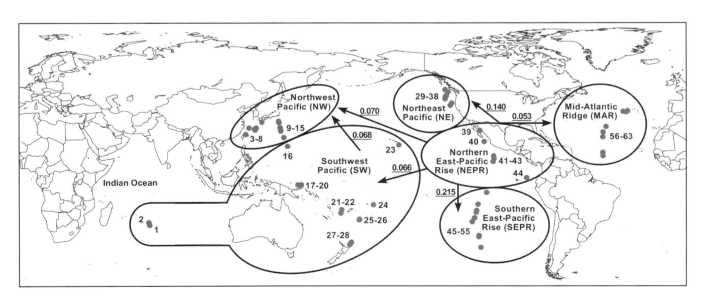

Fig. 15.2 Map showing six biogeographical areas statistically identified for hydrothermal vents by Bachraty *et al.* (2009). The arrows represent the apparent directions of faunal exchange with "coefficients of dispersal" (see the paper). (After Bachraty *et al.* 2009.)

southwest Pacific. These two areas are mostly distinctive at the species level. The vents on the Central Indian Ridge initially studied by Hashimoto *et al.* (2001) share elements (families and genera) with both the southwest Pacific vent province (provannid snails) and the Mid-Atlantic Ridge (distinctive *Rimicaris*). Simplifying a quote from an early paper (Van Dover *et al.* 2001) describing the Indian Ocean vent fields (called Kairei and Edmond) on the east side of its rift valley wall around 24°N:

> "The invertebrate community of the Kairei Field is characterized by an abrupt transition between . . . black smokers dominated by dense swarms of shrimp (*Rimicaris* sp.) in 10° to 20°C fluids and an ambient-temperature (1° to 2°C) peripheral zone dominated by anemones (*Marianactis* sp.) . . . . At the base of shrimp swarms, mussels, hairy gastropods [*Alviniconcha* n. sp.], and "scaly-foot" gastropods [see section on charismatic invertebrates below] occur in separate clusters of up to several hundred individuals. Brachyuran crabs (*Austinograea* n. sp.), turbellarian flatworms, nemerteans, and a second shrimp species (*Chorocaris* n. sp.) occur in sometimes dense but isolated patches in the narrow transition zone separating anemones and swarming shrimp. Other gastropods (limpets and . . . snails), large . . . polychaetes, and clusters of stalked . . . barnacles (*Neolepas* n. sp.) were observed frequently in this transition zone. . . . Vesicomyid shells [like *Calyptogena*] were collected within 1 km of the Kairei Field, but no live clams have been observed." [Bracketed comments are ours.]

As you may or may not imagine, finding a massive diversity of reasonably large animals "new to science" and all abundantly packed in one place (color photos in Van Dover *et al.*) causes an explosion of joy, a dopamine flood, in the mind of a biologist. They get to name new species after colleagues (e.g. *Alviniconchia hessleri* for Robert Hessler) and to study details of their biology. For example, *A. hessleri* has been shown to have endosymbiont epsilonproteobacteria in its gills (Suzuki *et al.* 2005).

There is exchange at long ranges, but it is also limited. Possibly, shallow venting, as in the Azores, and related cold-seep and whale-bone faunas (e.g. *Osedax*) have served as sources of animals adapted for symbioses with chemosynthetic bacteria to new vent fields.

Representative "charismatic" vent invertebrates

The first visit to the Galapagos Rise vents produced sharp color photographs of clusters of invertebrate animals of unusual size for deep-sea benthos. Particularly prominent

were vestimentiferan worms, eventually named *Riftia pachyptila*. The pictures evoked curiosity and extensive research into its specific adaptations. Similar interest has attached to many of the larger vent invertebrates, a few of which have been studied in physiological and molecular detail. Four of those are reviewed here, plus brief notes on some clams, but with no attempt to cover every detail known for them, much less for the entire suite of vent faunas that currently exceeds 500 species.

Riftia pachyptila Jones, 1981

All aspects of vestimentiferan biology have been reviewed by Monika Bright and François Lallier (2010), who provide 258 of the key references. The group name comes from the muscular collar, the vestimentum (Fig. 15.3), that sur-

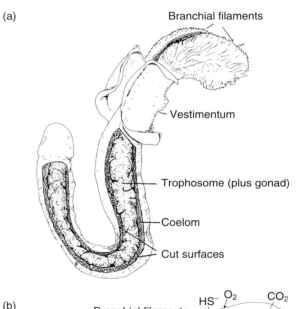

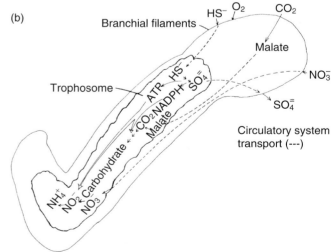

Fig. 15.3 Diagrammatic sections of *Riftia pachyptila*. (a) Anatomical layout. (b) Some biochemical exchanges. (After Felbeck & Somero 1982.)

rounds worms just posterior to the gill plume (Plate 15.1) and that expands to hold the worm at the top of its tube with the plume exposed to the mixture of vent flow and bottom water. Since the discovery of these worms at hydrothermal vents in 1977, over 600 papers (and abundant popularizations) about vestimentiferans have been published. Some of the interest extends to the related genera hosting chemosynthetic bacterial endosymbionts in addition to *Riftia*: *Tevnia* and *Oasisia* living along the East Pacific Rise, *Ridgea piscesae* along the Gorda Ridge System, *Alaysia* spp. at West Pacific back-arc vents, and *Escarpia* spp., *Lamellibrachia* spp., and a few other species living on sulfide-exuding cold-seeps. *Osedax* spp. boring into and obtaining sulfide from whale skeletons are more distantly related. The species most "emblematic" of hydrothermal vents has been *Riftia pachyptila*, still the sole species in its genus. Because of its size, brilliant-red gill plume, and exotic biology, it has attracted a long sequence of submarine dives supporting studies of its morphology, genetics, physiology, symbiosis, reproduction, embryology, and ecology.

Glands in the vestimentum secrete the tube, increasing its length at the upper end. Its wall is leathery, a combination of ~25% chitin and the remainder protein. The bottom end is closed. There can be septa closing off older, unoccupied sections, so probably tube material can be secreted by epidermis well posterior to the vestimentum. Tube growth and worm growth are extremely fast; tube extension was estimated from photographic series by Lutz (1994) as $85\,cm\,yr^{-1}$, and by others from changes in size-modes over a few weeks as twice that. In either case, it is the most rapidly growing large invertebrate. There are two posterior sections of the worm, the trunk and a short, spinose anchor bulb called the opisthosome. If the plume is nipped by a crab or poked by a submarine arm, the plume can be rapidly retracted into the protective tube.

In *Riftia* (Fig. 15.3) and its relatives, symbiotic bacteria live in a large organ in the trunk called the *trophosome*. When the skin is opened, it looks like an elongated bunch of greenish grapes. Each lobule is a colony of cells, of *bacteriocytes*, with a large blood vessel at its core and smaller vessels extending like spokes to the periphery and then opening into a sinus beneath an outer tissue layer. Blood flow is from the periphery toward the central vein. The bacteriocytes are generated in a central epithelium and then repeatedly divide and migrate along the spoke vessels toward the cell layer at the outer surface. While dividing repeatedly in this transit, the symbiotic bacteria change shape: rods, then small cocci, then large cocci. At the periphery, bacteriocytes digest the symbionts, die, and are resorbed (Pflugfelder *et al.* 2009). This recurrent cycling is likely to be part of the harvesting of nutriment from the bacteria by the worm.

Gas exchange at the plume is promoted by its extremely fine surface subdivision, averaging $22\,cm^2\,g^{-1}$ of dry body mass (Anderson *et al.* 2002), a ratio greater than that of all measured gills, except that of another vent annelid, *Paralvinella* ($47\,cm^2\,g^{-1}$). In addition, the diffusion distance through tissue from water to blood is extremely short at the tips of the gill filaments, only 1 to $2\,\mu m$. The blood flow, driven by a heart in the vestimentum section, must transport S^{2-}, O_2, CO_2, and fixed nitrogen (as NO_3^- in *Riftia*, unusual for animals but an adaptation because the deep-sea mostly lacks NH_4^+) from the gill to the trophosome, return the by-products of sulfide oxidation to the gill for removal and distribute nutrition from the symbiosis throughout the body. Substantial complexity for these exchanges, particularly of carbon species, is entailed by the necessity to maintain alkaline internal pH against the inward gradient of H^+ from the acidic vent fluid–seawater mixture. The details of this physiology are reasonably well studied (see Bright & Lallier 2010). A few aspects deserve mention because they illustrate the high level of specialized adaptation achieved by this worm.

The vascular blood and coelomic fluid are rich in specialized hemoglobins, actually two molecular versions in blood (HbV1 at 3600 daltons and HbV2 at 400 daltons) and one in the coelomic fluid (HbC1 at 400 daltons). A complex of linked subunits, HbV1 is large but not close to the size of human, tetraplex hemoglobin (68,000 daltons). *Riftia* hemoglobins are not contained within cells. Both sulfide and oxygen bind to HbV1 in roughly equal amounts from roughly equal external concentrations (Arp *et al.* 1987). Oxygen is bound at the heme group, much as for other oxygen-transport hemoglobins. Sulfide is bound in the protein superstructure of the molecule, possibly at some sectors rich in cysteine and methionine (S-containing amino acids), possibly at ligated zinc ions (Flores & Hourdez 2006), but that remains unresolved. Binding of S^{2-} has been shown for HbC1, likely providing protection for other molecules that might be poisoned by it, such as the heme-centered mitochondrial cytochromes that are key to oxidative electron transport. There is some evidence that cytochromes of *Riftia* have lower affinity for S^{2-} than those of other animals. Hemoglobin-bound S^{2-} in vestimentiferan blood is of the order of 100-fold habitat levels, while S^{2-} dissolved in coelomic fluid is about 10-fold less. Thus, lots of S^{2-} moves to the trophosome, while relatively little is in contact with the animal's own cells. At the trophosome, both oxygen and sulfide are taken off the hemoglobin by concentration gradients, and chemosynthesis proceeds in the bacteria. In many animal–microbe symbiotic relations, the symbiont pays its bills by dumping excess organic product into its vicinity, which the host absorbs and utilizes. Probably symbiosis in *Riftia* functions in the same way, but the cycle of bacteriocyte production, death, and resorption may also enable the worm to acquire bacterial chemosynthate. By-products of the chemosynthesis are sulfate (SO_4^{2-}) and protons, which are carried away in the blood, apparently unbound, and removed at the plume by active transport (metabolic processes reviewed in detail by Childress & Girguis 2011).

An obvious question is how do the young of animals attached to solid substrate, as *Riftia* is, transfer from old active sites to new active sites. Transfer is made more important because vents can be short-lived. For many vent species, it is now known and usually involves planktonic larvae. *Riftia* releases eggs or fertilized zygotes, that eventually hatch into quite ordinary annelid larvae (trochophores). They are wafted along the seafloor in the general flow. Sexes are separate in *Riftia*, with gonopores at the base of the red gill plume, leading in the male to distinctive external channeling that some sources say facilitates copulation. However, Van Dover (1994) has observed external discharges of eggs and sperm (apparently distinctive) in rapidly dispersing plumes, stating that the eggs sink. Zygotes may be fertilized in the distal oviducts by sperm from male discharges, not fertilized in the water after egg discharge. Both ovaries and testes run the length of the trunk adjacent to the trophosome, implying massive reproductive output.

Brooke and Young (2009) collected zygotes from females with pipettes, then reared them in both pressurized, thermally controlled incubators and in plastic chambers moored just above the colonies. They determined that the lipid-rich new zygotes, with sufficient waxy lipids to qualify them as lecithotrophic, rise very slowly, about 2 m d^{-1}, and develop into ciliated larvae in a little over 20 days. Buoyancy declines as the wax is metabolized, so presumably the eggs and larvae stay quite close to the seafloor. Cell division rate and the fraction developing normally fell off rapidly above 4° to 5°C, both *in situ* and in incubators. Zygotes do not develop at all without substantial pressure. Brooke and Young's rearing results were better at 238 atm than at 170 or 102 atm. There was no development at 34 atm. Based on average respiration rates and larval composition, Marsh *et al.* (2001) estimated that larval dispersion can last about 38 days (34 to 44 d). They modeled maximum along-ridge transits in that time of ∼100 km. Other models of along-ridge flow suggest distances over twice that far. In any event, potential transits around extended ventless barriers (as from the East Pacific Rise at 21°N to the Gorda Ridge area) are not possible.

Settlement, observed on plates set on near-vent substrate, begins a sequence of metamorphosis from a somewhat modified trochophore with mouth, gut containing some particles and anus, to a juvenile living in a tube and nourished by a trophosome. The symbionts are either ingested or "infect" through the body wall; there are bacteria in the wall and coelom of these early juveniles that stain with genetic probes from the symbionts. Genetically identical bacteria do live on surfaces and in the water around vents. There have been no observations of *Riftia* releasing symbionts to supply these larvae with an inoculum. The trophosome develops from the vicinity of the foregut, which then disappears (Nussbaumer *et al.* 2006).

Different vestimentiferans harbor different symbionts, but the closely studied species each have only one "phylotype" of internal gammaproteobacteria. The symbionts of all three East Pacific Rise species share "phylotype 2". It has not been cultured, but its genome has been sequenced and a name proposed: *Endoriftia persephone* ("Candidatus" status). The genome of this bacterium, unlike endosymbionts passed from mother to egg and obligately symbiotic, retains a full repertoire of enzymes for independent life. That includes chemoreceptor functions and motility, possibly enabling them to find and "infect" advertising larvae by chemotaxis (Robidart *et al.* 2008). Metabolic studies of *Riftia* in pressurized aquaria showed only sulfide oxidation (via ATP sulfurylase) as the energy source for chemosynthesis. RuBisCO is present, and recent work suggests there are also proteins present for an alternative carbon-fixation pathway (Markert *et al.* 2007).

Riftia are found primarily in crevices conducting warm vent fluids, only 3° to 12°C with occasional higher pulses, with maximum sulfide and typical oxygen in the vicinity of 150 μM and 50 μM, respectively. As flow slows, cools, and becomes more dilute, *Riftia* is often replaced by *Bathymodiolus* or *Calyptogena*.

Alvinella pompejana Desbruyères and Laubier, 1980

Pompeii worms, the polychaete *Alvinella pompejana* and relatives (Fig. 15.4), live in parchment-like (glycoprotein) tubes, secreted from an anterior-ventral gland plate, on the sides of vent chimneys on the East Pacific Rise (9° and 13°N) and Galapagos Ridge. The tubes may direct some of the subsequent deposition of metallic sulfides so that the outer ends of the tubes project from the accreting mineral mass. The largest adults are very large worms for the deep sea: 12 mm diameter by 95 mm length, living in tubes up to 2 cm diameter. Observations of *A. pompejana* behavior vary. Desbruyères *et al.* (1998) reported an activity cycle of 5–10 min down in the tube, followed by short intervals, less than 30 s, with the gill plumes extended into the surrounding flow, whereas Cary *et al.* (1998) reported that they mostly sit with the plumes outside. Most temperature measures in the vicinity of the tube colonies are 20–45°C, but it is much hotter a few centimeters deeper in the chimney wall. Cary *et al.* (1998) measured a gradient within the tubes from 80°C inside at the posterior end to 20°C at the opening, an extreme range for the ends of one animal. Di Meo-Savoie *et al.* (2004) placed narrow (3 mm) thermocouples a few centimeters into many occupied tubes at different East Pacific Rise vents to make extended temperature recordings. Typical results ran 30° to 80°C and in one case 50° to 110°C for over 4 h (Fig. 15.5), including one stretch continuously >95°C for 8 min. A controversy over whether that could possibly be survived by an animal made

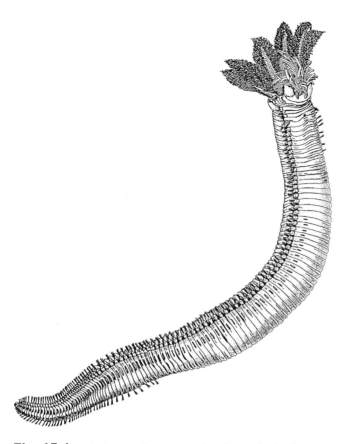

Fig. 15.4 *Alvinella pompejana*, type species of a new family of polychaetes found in association with hydrothermal vents. (After Desbruyères & Laubier 1986.)

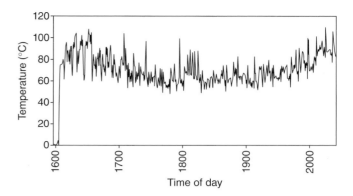

Fig. 15.5 Four-hour temperature time-series from 3 cm inside an occupied tube of *Alvinella pompejana* at 9°51.8′N. (After Di Meo-Savoie *et al.* 2004.)

of lipids and proteins appears to be settled (although some authors still express uncertainty; apparently it is almost too much to believe). Structural collagen molecules have been studied in detail, showing a structure reasonably expected to be especially heat tolerant. Short-term temperature variations are probably caused by variations in vent flow and in mixing rates with surrounding water. In addition, Di Meo-Savoie *et al.* used microelectrodes and ship-board chemistry to show the unusual chemistry of the water surrounding pompeii worms: undetectable oxygen (<5 μM), low levels of free sulfide, but substantial dissolved FeS, seawater concentrations of sulfate (22 to 27 mM), acid conditions (pH = 5.3 to 6.4), and abundant total dissolved iron (up to >700 μM). All of those suggest a chemically harsh environment, especially so at >30°C, and no particular support for sulfide-based chemosynthesis.

The physiology of the extraordinary heat tolerance of *Alvinella* is not yet fully understood. Related *Paralvinella* spp. tolerate somewhat lower maximum temperatures, to 40° or 50°C for a few hours, and have been shown to have substantial amounts of heat-shock proteins (*Hsp*), molecules that protect other proteins from unfolding and denaturing. Increases of *Hsp* can be stimulated in *Paralvinella*

by raising the temperature for a while in pressurized aquaria (Cottin *et al.* 2008). However, the increase takes hours to achieve, so it can only be part of the physiology of surviving rapid temperature shifts (Fig. 15.5). Aspects of the high-temperature enzymology of bacteria associated with *Alvinella* are better studied. Symbiotic bacteria have not been found internally in *Alvinella* spp. However, copious colonies, sometimes described as a "fleece", of filamentous bacteria live on dorsal and lateral epidermal bumps along the edges of each segment. The bumps are glandular with a mucoid secretion evident in surface cells and externally, into which the long (up to 0.6 mm) bacterial colonies are twisted (Desbruyères *et al.* 1985). Thus, the worm's surface is specially elaborated to accommodate these episymbionts. Other chemoautotrophic bacteria live in the intersegmental grooves and line the inner surfaces of the tube. Most endosymbionts at vents are gammaproteobacteria, but Cary *et al.* (1997) have shown that most of the external filamentous forms on *Alvinella* are epsilonproteobacteria (as are those of *Rimicaris*). There are two dominant bacterial "phylotypes" (i.e. with significantly variant SSU RNA) (Campbell *et al.* 2001). One of those has been cultured, named *Nautilia profundicola*, and its complete genome sequenced (Campbell *et al.* 2008) to characterize its metabolism in detail. It is, indeed, a "chemolithotroph", but not a sulfide oxidizer. Rather, according to Campbell *et al.* (2008), the "cells are strictly anaerobic. Chemolithoautotrophic growth occurs with molecular hydrogen or formate as the electron donor and elemental sulfur as the electron acceptor, producing hydrogen sulfide". In the *Alvinella* consortium there are also sulfate or sulfite reducing epsilonproteobacteria that oxidize formate and acetate in laboratory cultures, and others that are sulfur-oxidizing autotrophs (Campbell *et al.* 2001). So, the *Alvinella*-associated bacterial community is phylogenetically restricted, but still trophically complex.

The worms have a feeding apparatus with grooved and ciliated tentacles, comparable to that with which ampharetids, a related polychaete family, gather particulate food. In addition, alvinellids have a peculiar lip structure which may gather bacteria from surfaces in bulk. However, it has not been possible to observe feeding on the bacterial associates directly. Bacteria on the inner tube wall are almost certainly grazed. Those on the dorsum may be grazed, since the worm is very flexible, or they may supply inocula to the colonies on the tube walls. Not all of the behavior that we would like to see is sufficiently accessible. The anterior gut contents are indeed filamentous bacteria, and other bacteria are compressed in the foregut into mucous-bound masses (Desbruyères *et al.* 1985). Solid feces in the posterior gut are mostly elemental sulfur, but also contain glucosamine, an ammoniated sugar found in bacterial cell walls (Saulnier-Michel *et al.* 1990). Thus, the diet is certainly the bacterial associates. The sulfur suggests a significant role for bacteria using sulfate as an oxidant.

Lee *et al.* (2008) isolated and identified genes for two metabolic enzymes in DNA collected from the bacterial fleece of one *A. pompejana* specimen from the East Pacific Rise (9°N). Using standard gene insertion techniques, they created a laboratory bacterium that could generate those enzymes in sufficient quantity to extract for *in vitro* characterization of the effect of temperature on the rates of the reactions that they catalyze. The most dramatic result (Fig. 15.6) was for glutamate dehydrogenase, a metabolic enzyme common and essential to all cells. Lee *et al.* measured the substrate conversion rate at 40°C, then pulsed the temperature to a much higher level and measured the rate over time. Initially, the rate was progressively higher for all temperatures up to 90°C, then fell off over 10 min for temperatures above 60°C. The longer-term increase in rates was less than a doubling for an increase from 40° to 75°C, when most enzyme reaction rates would increase ~10-fold. Lee *et al.* propose that at least some enzymes of such ther-

mophilic organisms have two forms in a shifting equilibrium: active and inactive-undamaged. At sufficiently high temperature, the inactive form will progress at some rate to a denatured (unrecoverable) form. As the temperature rises, more of the enzyme takes the inactive form, stabilizing the overall reaction rate as the substrate turnover by the active form accelerates. A full description of the chemical kinetics of this scheme predicts the form of the rate response to temperature. Since the bacterial surface symbionts have evolved this mode of temperature regulation, perhaps something similar has evolved in the worms. Certainly some quite extraordinary enzymology exists to allow *A. pompejana* to survive temperatures and rapid temperature variation well outside the tolerances of most animals.

Sexes are separate in pompeii worms, and, unlike related annelids, the females have sperm storage chambers (spermathecae). Both sexes have a single pair of gonoducts with the gonopore just behind the tuft of gills. Males have a modified pair of tentacles near the gonopore that are somehow involved in copulation, which has to occur outside adjacent tubes of males and females. Development to maturity is rapid, a few months according to settling plate studies. Both sperm and eggs are of unusual form (reviewed by Padillon & Gaill 2007). Fully grown females can be developing up to 80,000 oocytes in the body cavity with ~3000 eggs ready for spawning in the oviducts. Spawning apparently is not synchronous across a colony. Studies of zygote development (see Padillon & Gaill 2007) comparable to those for *Riftia* give somewhat different results. Deep-sea pressure is required for development, but, unlike *Riftia* zygotes, greater warmth is tolerated and necessary, with optimal development at ~10°C but none at 2°C or >20°C.

The pompeii worm is an example of something else. It is the type species of a new family, Alvinellidae (Desbruyères & Laubier 1986), although initially they seemed satisfactorily placed in the Ampharetidae (Desbruyères & Laubier 1980). There is a tendency to enhance the importance of vent research by elevating the systematic level of the animals. Thus, at one point the vestimentiferan pogonophora were raised to phylum status. Oddly, *all* the pogonophora were later demoted, based on molecular genetic evidence, to family level among the annelids as Siboglinidae, some of which have been known from sedimentary habitats since 1900. That seems extreme, given the dramatic divergence of pogonophora from other annelids. However, devotion to the principle that higher-level nomenclature must reflect phylogeny does dictate the demotion. In the case of Alvinellidae, the separation of a new family is based on careful evaluation of morphological and molecular data. Examination of very large collections from vent sites has shown that there are a number of distinct species, divided at present into two genera: *Alvinella* and *Paralvinella* (Desbruyères & Laubier 1986).

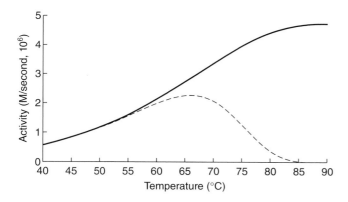

Fig. 15.6 Substrate conversion rate of glutamate dehydrogenase enzyme constructed using its gene taken from an epsilonproteobacterial strain episymbiotic with *Alvinella pompejana*. The black line shows rates immediately after raising temperature from 40°C to that on the abscissa. The dashed line shows rates after 10 min at that higher temperature. (After Lee *et al.* 2008.)

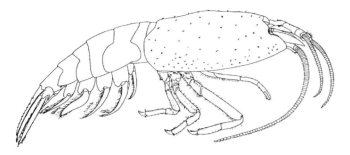

Fig. 15.7 *Rimicaris exoculata*, a bresiliid shrimp from black smoker chimneys on the Mid-Atlantic Ridge. Appearance of the male shrimp. (After Williams & Rona 1986.)

Rimicaris exoculata Williams and Rona, 1986

These are a 6 cm shrimp (Fig. 15.7) discovered swarming in masses on the order of 3000 m^{-2} over the sides of black-smoker chimneys along the Mid-Atlantic Ridge (MAR). Originally *Rimicaris*, now a genus with several species, including *R. kairei* from vents on the Central Indian Ridge, were placed in a subgroup of Caridean shrimps, the Bresiliidae, that mostly live in bore holes on shallow sponges. Later, in line with the tradition of vent taxonomy, they came to represent a distinctive family, the Alvinocarididae, sometimes listed as a group of Bresilioidea. There are, for now, seven genera, including *Rimicaris*, *Alvinocaris*, and *Chorocaris*, that live with similar life patterns on mid-Atlantic and Indian Ocean smokers. The maxillary exopodites (limbs near the mouth) of *R. exoculata* carry fields of short setae, all of which are covered with a fur of filamentous, sulfide-oxidizing, chemosynthetic bacteria. Similar cultures grow on the inner surfaces of the gill chambers. The shrimp achieves gas exchange for this on-board bacterial farm by holding position in turbulent mixing flows along the vent walls. It seeks water at 20 to 40°C, which contains both sulfide and oxygen, harvesting meals by scraping over the maxillae and gill chambers with other mouthparts (Gebruk *et al.* 1993).

The species name *exoculata*, meaning without eyes, is a misnomer. During development, the eyes simplify from a fairly standard form of ommatidia and migrate back below the upper carapace where they look out through transparent chitin (Van Dover *et al.* 1989). The two retinal planes are ~0.25 × 1 cm, a total of ~50 mm^2. All of the optical focusing parts are lost, and the rhodopsin-packed rhabdomes hypertrophy into a thick (100 μm) layer backed by white diffusing cells that return uncaptured photons back to the rhabdome. The absorption spectrum peaks at 500 nm, and White *et al.* (2002) have detected very dim green light emitted from the water emerging from the vents, water hotter than the thermal tolerance of the shrimp. At wave-lengths shorter than 650 nm, this light exceeds the expected blackbody radiation for vent temperatures. At ~10^4 photons cm^{-2} s^{-1} sr^{-1} (sr = steradian), it is not visible to people and is generated by an uncertain mechanism. Light emitted by bubbles collapsing under extreme compression; energy release from crystal formation; and chemiluminescence are suggested as possible sources by White and colleagues. The function of light detection for the shrimp could be avoidance of scalding, finding the way back to vents areas, or something else. It is very likely that different information is gained from this light than would come from the more abundant blackbody irradiance at longer wavelengths. However, it is also possible that evolution of the *Rimicaris* pigment from that of a more ordinary deep-sea shrimp naturally favored the shorter end of the visible spectrum. *Chorocaris* has a similar eye, although not so large and not shifted back into the carapace. In addition to an unusual visual system, likely involved in orientation to and on vents, *Rimicaris* have sulfide-sensing dendrites under pores on the second antenna (Renninger *et al.* 1995). The intensity of activity in these nerves is exponentially proportional to Na$_2$S concentration, and thus suitable for navigating to sulfide sources. Olfactory response to sulfide is not unusual, you can smell it, but Renninger *et al.* did not find it in other shrimp.

Nicole Dubilier and colleagues (Peterson *et al.* 2010) sequenced DNA encoding SSU rRNA from numerous samples of the episymbiont bacteria living on *R. exoculata* collected at four sites along 8500 km of the MAR from 36°N to 4°S, the entire range of the species. They found two groups of bacteria, one of epsilon- and one of gammaproteobacteria. The epsilon group divided into some closely related subgroups that partially sorted along the ridge. All of those were related to known sulfide/sulfuroxidizing chemoautotrophs, including those of *Alvinella*. The gamma group was both less abundant, although found on all shrimp from all sites (well, just three from each site) and less diverse, but their closest metabolically characterized relative is also a sulfur-oxidizing chemoautotroph. While there are other bacteria on the shrimp surfaces, these two proteobacterial groups are the consistent dominants and appear to be obligatory symbionts, at least for the shrimp. The Dubilier team also examined two mitochondrial genes from the shrimp, finding only trivial variation along the 8500 km. They make the point that the chemistry of vent fluids along this stretch is not consistent (Schmidt *et al.* 2008). At the south end, the vents are "basalt hosted", which makes them rich in sulfide and depleted in hydrogen and methane. At the north end, vent fluids are heated in ultramafic rock, which leads to the opposite: low S^{2-}, high H$_2$ and CH$_4$. The epibiont bacteria of the shrimp, however, almost certainly have very similar chemical requirements. This can be termed an obligate relationship for the shrimp or bacteria, which names the observation without explaining it. Possibly, the two dominant proteobacterial strains carried by the shrimp are metabolic virtuosos capable of several modes of chemoautotrophy, but probably not.

Teams led by Bruce Shillito (e.g. Ravaux *et al.* 2003) have directly tested the high-temperature tolerance of *R. exoculata*, which live on the walls of vents conducting water at 350°C into the ocean just a few meters away. Shrimp were collected by a suction system on an ROV, brought to the surface, and promptly returned to 230 atm pressure in sealed, "endoscope"-equipped aquaria. After observation of their survivorship at moderate temperature, the temperature was raised. Death occurred at remarkably low temperatures, 33 to 37°C. In later work, Shillito and colleagues characterized a 70-kilodalton, heat-inducible heat-shock protein (of the *Hsp70* class) that the animal accumulates above only 25°C, implying that its optimal habitat temperature is somewhat less than that. Possibly, at many sites the best sulfide/sulfur and oxygen concentrations are found at that temperature, which is higher than at the East Pacific sites occupied by *Riftia*.

According to limited studies (Ramirez Llorda *et al.* 2000), *Rimicaris* reproduction is similar to that of other caridean shrimp. The male packages sperm in a spermatophore and attaches it to the female gonopore. The sperm are transferred into holding chambers in the female reproductive tract. Oocytes are present in the ovary at all stages of development. Clutches of eggs are produced and she attaches them to her legs for brooding. Very few brooding females have been collected from vent swarms, and Ramirez Llorda *et al.* speculate that they leave the massive swarms during egg development to avoid having the eggs brushed off in traffic. The few broods that have been counted were just fewer than 1000 eggs. Larval *Rimicaris*, identified by molecular genetics (Dixon & Dixon 1996), have ordinary compound eyes and phytoplankton in their guts. They are believed to ascend high in the water column and have a long planktotrophic phase providing extended potential transport. The rate of successful return to vents must be small.

The scaly-foot gastropod

An extraordinary, coiled snail (Plate 15.2), up to 5 cm in diameter, was found clustered around the bases of a vent field on the Central Indian Ridge in 1999 (Van Dover *et al.* 2001). A Linnean binomen has been suggested for it and can be found on the web, but it is not properly published to date. However, Warén *et al.* (2003) give a good description in their on-line supplementary materials, with molecular sequence evidence of its relationship to snails classified in a group termed Neomphalina: *Hirtopelta* and *Peltospira*, snails abundant near western Pacific vents.

The scaly-foot snail is unusual in two respects. First, its foot, that appears not fully retractable, is covered by overlapping plates formed of iron–sulfur minerals: pyrite (FeS_2, which has S–S bonds, as well as Fe–S) and greigite (Fe_3S_4, a mixed-valence compound). These plates can be drawn together, making a closed armor round the foot, and they seem to derive from an extensively modified operculum. The outer layer of the entire shell coil is also covered with a pyrite–greigite layer. Mechanical properties of the armor have been studied in detail (Suzuki *et al.* 2006; Yao *et al.* 2010). The mineral outer coating of the coiled shell is underlain by a thick sheet of resilient protein termed "conchiolin", and the layer next to soft tissue is rather ordinary molluscan periostracum. Apparently, this provides a combination of puncture and compression resistances. There are crabs on the Indian Ridge vents that likely could crush less well-protected snails or nip their feet. Yao *et al.* speculate about designing military armor based on this pattern. It seems warfare is never far from our minds or research budgets. Finding hard parts constituted of iron compounds does occur elsewhere among mollusks, particularly the hematite coatings on chiton radulas. Nevertheless, the scaly-foot minerals are substantially different, a suite strongly favored by the high iron and sulfide content of vent emissions.

Second, the esophagus has large side expansions of soft tissue, the cells of which harbor dense masses of gammaproteobacteria related to thiotrophic (sulfide oxidizing) chemosynthetic forms from other vent endosymbioses (Goffredi *et al.* 2004). The details of the physiology that supplies oxygen and sulfide to these internal structures remains to be learned, as does the means for distributing the product from these glandular tissues to the rest of the snail. Scaly-foot snails live, together with other snails – *Alviniconcha* – around the base of smokers with both sulfide and oxygen available to them. Warén *et al.* (2003) report a very small digestive system compared to typical snails, suggesting strong dependence upon the endosymbionts for nutrition. However, they also show a very elaborate radula structure, which must retain some scraping function. Goffredi *et al.* report abundant filamentous epsilon- and gammaproteobacteria coating the scales, but it is difficult to see how the snail's mouth would reach them for ingestion. Unlike *Alviniconcha* and its relatives, there are no bacteria in the gill.

Not much is known about reproduction of the scaly-foot snail. Warén *et al.* report an unusually anterior position for the ovary, which could mean anything, and they found spermatophores in the female gonoduct. Apparently, males deliver sperm in packages.

Bathymodiolus and *Calyptgena*

Unlike the scaly-foot gastropod, most of the mollusks sustaining chemoautotrophs harbor them in gill filaments. That is true of the mussel genus *Bathymodiolus*, species of which are found in widely dispersed venting areas, and the clam *Calyptogena magnifica* of the East Pacific Rise. Gills are vascularized and expanded for a wide area of contact with external fluid in any animal, thus locating symbionts

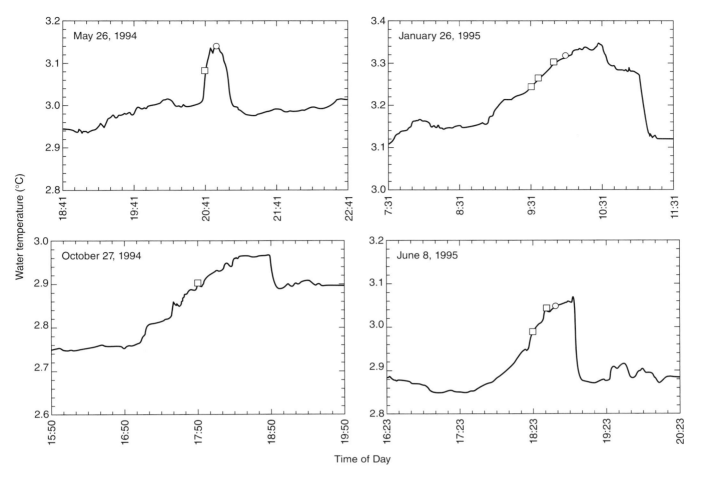

Fig. 15.8 Four examples of sperm release events (squares) by *Calyptogena soyoae*, three of them followed by egg-release events (circles) during slight temperature rises at the cold seep clam bed (1174 m depth) in Sagami Bay, Japan. (After Fujiwara *et al.* 1998.)

there readily achieves high rates of gas exchange. This localization has been shown both by microscopy and by presence in these tissues of the carbon-fixing enzyme RuBisCO (Felbeck 1981). Methanotrophic bacteria have been reported (Cavanaugh *et al.* 1992) to live symbiotically in *Bathymodiolus*, apparently co-existing with sulfide-oxidizing chemoautotrophs (Fisher *et al.* 1993).

These bivalves are distinctive for their dependence on endosymbiotic, chemosynthesizing bacteria, but vesicomyids (*Calyptogena*) are not restricted to hydrothermal systems. They also occur at cold seeps where squeezing of subducting sediments pushes sulfide, hydrogen, and other chemicals into the benthic boundary layer. Fujiwara *et al.* (1998) made an interesting observation about *C. soyoae* and *C. okutani* living in such a seep area in Sagami Bay, Japan. It may or may not apply to vent bivalves. The two clams release both sperm and eggs freely into the water where they must meet to join. This requires nearly simultaneous release. In the cold-seep clams, that is achieved by sensing short-term temperature increases and by olfactory

sensing of gametes in the water. Fujiwara *et al.* set up a video-recording observatory to watch *C. soyoae* over time (1.5 years) while also recording temperature. Temperature increases of only ~0.2°C (Fig. 15.8) induced sperm release, which apparently induced egg release; at least all cases of egg release were preceded by sperm release. Fujiwara *et al.* then induced spawning events using a heater inside a Plexiglass® dome placed with a submersible over the clam bed; again, sperm were produced first, then eggs. The observations of natural spawning have been elaborately repeated (Fujikura *et al.* 2007). Of 213 sperm releases, 90 were followed by egg release. Eggs were never released first. Males spray sperm in an arc by swinging their siphons. Pulses of flow along the turbulence-inducing, rough surface of the clam bed would mix the streams of sperm and eggs together, ensuring high fertilization rates. Similar observations have not been reported from hydrothermal clam beds, but similar signaling should be possible. There are pulses of warmer (and colder) water passing over hot vent fields due to tidal and turbulent flow variations.

Ova of *Bathymodiolus* spp. are planktotrophic, sustained by feeding, on what exactly is uncertain, possibly on the enhanced bacterial content of vent plumes. They are likely relatively long lived, extending transport. Larvae of vesicomyid clams and many vent snails and annelids are more limited to the nutrition supplied in the egg ("lecithotrophic"), and likely must settle sooner. This planktotrophic–lecithotrophic distinction is not very strongly correlated with the degree of genetic variation along the ranges of these animals (Audzijonyte & Vrijenhoek 2010).

Faunal arrangement around vents

Different megafauna around a hydrothermal vent have different requirements for flow rates (exchange) and for specific concentrations of sulfide and oxygen. On the East Pacific Rise, the smaller vestimentiferan *Tevnia jerichonana* is an early colonizer of new vents, settling in vicinities up to 30°C and persisting until they cool to about 5°. *Riftia* arrive later, displacing *Tevnia* and forming colonies in sectors of active vents where temperatures in their clumps are lower. The mussels do not require quite such high sulfide levels and are usually found farther away from the vents, although they may be attached about the bases of the tube worms. The clams appear to prosper in slower flows with lower sulfide concentrations, and they are most often found along horizontal cracks in the basalt, through which hydrothermal flow is more of a seep. They thrust their foot, which is modified for gas exchange, down into the crack. The foot presumably also anchors them. At the periphery of a vent field is an array of filterers, particularly tubiculous serpulid polychaetes. The sulfide is already too dilute and reduced by oxidation to support chemoautotrophy, but sufficient bacteria disperse from the interior walls of the vents to provide a greatly enriched diet compared to the deep-sea background. The spatial layout of an East Pacific Rise vent field community has been described by Hessler *et al.* (1988). An extensive literature now covers faunal layout and succession in other submarine hydrothermal regions.

Longevity of vents and colonization of new vents

Magma-heated venting areas range in longevity from a few years to several decades. Traverses along the axial rift of the Galapagos Rise on the initial 1977 expedition found sites of former vents and animal colonies indicated by deposits of shells, particularly those of *C. magnifica*. Dead chimneys have been found at sites close to active chimneys.

Since vents do not last indefinitely, the animals must have moderately rapid growth rates, especially *Riftia*. *Calyptogena* can be aged fairly accurately by dating of the layers of the shell: the largest specimens are of the order of 25 years old. A section of the East Pacific Rise at 9°N with very active spreading was the site of a lava flow in April 1991, just before a visit by scientists in *Alvin* (Lutz *et al.* 1994). Lava had spread over previously active vent sites that had supported copious fauna. Fauna not actually buried included scorched clusters of tube worms. Hydrothermal venting was already active through fissures in the new lava, and particulate matter clouded the water. Bacteria were abundant as mats up to 5 cm thick on the bottom. On a return visit in March 1992, venting was more localized, and the bacterial mats much less extensive. Possibly, crabs active in the area had eaten back the surface coating of bacteria. Venting fissures were surrounded by populations of *Tevnia*, a small (to 30 cm) vestimentiferan that appears soon after new vents open and is apparently suited to bathing in very hot water. No *Riftia* were present at all. A final visit in December 1993 showed masses of *Riftia* with individuals exceeding 1.5 m in length completely covering sites tagged for relocation that had been bare rock with dispersed *Tevnia* 20 months earlier. Small mussels and small *Calyptogena* were also found around localized areas of active venting. The active sites were supporting complex hydrothermal communities. Thus, the build-up time for the communities is quite short, a few months to a few years. Their demise can be nearly instantaneous due to lava burial or termination of venting.

In a study similar to that by Lutz *et al.*, Tunnicliffe *et al.* (1997) followed the colonization of a completely new vent on the Juan de Fuca Ridge. It was downstream in the poleward mean flow from a new axial volcano on the ridge. The vent stream included some flocculent material, so they called the vent "Floc". The Floc site was barren when discovered in summer 1993. By summer 1994 there were eight metazoan species, and by summer 1995 there were 21, which compares well with 24 at an obviously older site upstream they named "Source". A final visit only slightly later in 1995 found Floc to be dying, and the species count had already dropped to 12. Colonization is relatively rapid; so is denudation after flow stops. Most submersible and ROV visits to ridge areas with vents find both live and dead assemblages scattered along the axis. Dead vents are indicated mostly by cold chimneys, clam shells, and calcified worm tubes.

Like *Riftia* (see above), most vent animals have planktonic dispersal, and settlement on experimental plates has been observed (Craddock *et al.* 1997; Vrijenhoek 1997) for limpets, several worms including alvinellids, and both *Bathymodiolus* and *Calyptogena*. Some larvae reach new vents, certainly vastly more do not. Clearly, many can settle in the vicinity of their parents, which allows large colonies to accumulate. Arriving at new, habitable vent sites must be facilitated to some extent by channeling of near-bottom flow along the axial valleys of spreading centers, where new vents are most likely to erupt. However, the overall loss rates must be higher than for almost any other oceanic, benthic animals. The reproduction required to compensate for these losses must be allowed by the copious food supply, which supports both large body size and high fecundity.

Several workers, particularly Craddock, Vrijenhoek, and colleagues, have examined the degree of genetic differentiation between stocks of vent animals at different distances along and between ridge systems. The tools have been isozyme frequency estimates for various metabolic enzymes and DNA sequence variation. Isozymes are variant forms of enzymes which mediate specific metabolic reactions. They are identified by their distinct migration rates under electrophoresis, with development of the paper electrophoretograms using the metabolic reactions. The results are mixed. Limpets of the genus *Lepetodrilus* were shown by an isozyme analysis to have no greater variation at great distances along the East Pacific Rise and Galapagos Rift than at short distances. Even though features of their larval shells suggest they do not feed much while planktonic, and thus have only short dispersal as larvae, their gene flow is apparently high (Craddock *et al.* 1997). A similar result obtains for *Bathymodiolus thermophilus* over 2370 km along the Galapagos Rift. Its larval shells do suggest (by comparison to better-known shallow clams) a capability for feeding and thus for long-distance dispersal (Craddock *et al.* 1995). According to Vrijenhoek (1997), there are cases of vent animals with planktonic larvae that do show some genetic gradients with distance, suggesting a stepping-stone model of dispersal. On the other hand, vent animals that brood their young, like the amphipod *Ventiella sulfuris*, show strong genetic differentiation between sites, and genetic distinction increases with separation distance (France *et al.* 1992). Gene flow, and by implication migration, between sites is very low in these forms. This species has only been found at vent sites. How they and other brooders reach new sites is an open question. Recently, Audzijonyte and Vrijenhoek (2010) have questioned the statistical adequacy of sampling along ridges to determine genetic homogeneity versus divergence. There is always more to be done.

Site for the origin of life?

Since the mid-19th century, when Darwin established, by thorough review of paleontological results, that life on Earth has gone through a progression of evolutionary phases, the question of how life originally got started has waited for a full and satisfying answer. Some continental thinkers had outlined this problem much earlier (see Wächterhäuser 1997). Thinking about the problem has

been active since the 1920s (Oparin 1924, 1938), when sufficient knowledge of biochemical processes was available to support speculation. Biomolecular knowledge has burgeoned since, so these speculations have started to take on a semblance of actual possibility.

Very shortly after the discovery of deep-sea hydrothermal vents, Sarah Hoffman, a graduate student working at the time with J.B. Corliss (Corliss *et al.* 1981), noticed the similarity between vent systems and an experimental approach to "synthesizing" simple living forms attempted earlier by Sidney Fox (citing Fox 1971; see also Fox & Dose 1972), who was in part inspired by geysers. The actual relation to Fox's experiments was not exact, only intriguing. His experiments had three levels: production of amino acids, formation of protein-like structures ("proteinoids") and formation of "protocells" from the proteinoids. The first step was achieved in the gas phase by passing methane through a concentrated ammonium hydroxide solution, heating it to 1000°C in a bed of sand, lava, or other matrix, then rapidly quenching it in cold ammonia solution. The heating produced a racemic mixture of amino acids, including some not found in biological systems. Polymerization of amino acids was also achieved by heating, but because the reaction is an elimination of water, it was carried out anhydrously, just dry amino acids mixed and heated (175°C) in a tube. Using a mixture of several amino acids was essential; polymerization does not occur with just, say, glycine. This produced fairly long chains of somewhat organized "proteinoid", various conditions giving different but crudely repeatable order and residue composition to the amino acid sequence. Finally, he tossed some hot proteinoid into boiling water, producing little vesicles that he called protocells. These had a membrane-like structure, and a mass of them could inefficiently catalyze some simple metabolic reactions like decarboxylation of pyruvic acid.

The dry, then wet, then dry, then wet sequence of the Fox "synthesis" would be hard to reproduce anywhere, and especially so in hydrothermal vents, where wet would be persistent. Nevertheless, Hoffman argued that small, constituent organic molecules might be generated from reduced species (CH_4, NH_3, H_2, H_2S, all found abundantly in venting waters) deep in the hottest part of the vent (>600°C), convectively lifted and cooled by mixing with surrounding waters to levels (>250°C) where polymerization might be favored, then cooled further to form structures like Fox's protocells. From that point a sort of metabolism would develop, eventually becoming subject to selection for stability, most rapid growth, and division at sufficient sizes as a precursor to reproduction.

Hoffman and Corliss were both experts on Archean geology, on rocks and deposits of the very greatest ages. They noted that a number of structures in rocks greater than 3 billion years in age, structures containing microscopic bits long interpreted as the earliest fossil traces of life (bits that Fox & Dose had noted looked identical to protocells) could best be interpreted as having been formed in once-submerged hydrothermal vents. Thus, the age of the earliest traces of possible microbial life was not greater than ages likely for remains of submerged hydrothermal sites. The overall fit certainly suggested further examination of the possibility. Part of the evidence suggesting an origin in hydrothermal vents is that evolutionary trees (Woese & Fox 1977) developed from DNA sequences for SSU rRNA sequences, phylogenies based mostly on the assumption that evolutionary advance will be accompanied by increased molecular complexity, tend to root among thermophilic Archaea. The most primitive of those microbes are chemoautotrophic and found in or near deep-sea hydrothermal vents (Kandler 1998).

The subsequent history is complex, involving personality conflicts as well as objective evaluations of the chemistry involved. However, hydrothermal vents have persisted as one of the potential candidates for the initial steps in assembling beings using energy from simple reactions to sustain themselves, self-reproducing entities, something alive. Current origin-of-life theories classify in two main groups: "information first", usually based on recent discoveries that RNA can both carry genetic information and act enzymatically (early life as an "RNA world"), and "energy transduction first" leading to an enzyme system somewhat akin to mitochondrial electron transfer. In the latter, given production of a molecular "energy currency" like ATP, perhaps many things eventually become possible. Both schemes usually invoke formation of vesicles ("protocells") by fatty acids emulsifying in water capable of containing suites of "prebiotic" reactions. Proponents of both starting points have come in recent speculations to suppose that the ele-vated temperature and possibly the clay and pyrite or calcite surface chemistries of hydrothermal vents could be part of the process of "abiogenesis".

Dyson (1999) suggested that reproducing, protein-based metabolic systems and nucleic-acid-based replication systems developed separately, possibly at different sites, and later coalesced to provide reproduction with an exact replication mechanism. Their coalescence would carry back to an earlier stage the theme of evolution through symbiotic combinations that gave the eukaryotes mitochondria and chloroplasts. This has at least esthetic appeal.

Even a fully elaborated scheme for *de novo* synthesis of living systems will require a leap of faith to qualify as the accepted story of the origin of life on Earth (or even elsewhere). The best we can hope for is a rational scheme, possibly one duplicated in the laboratory. That story, if it is ever developed, would become part (with the Big Bang, a theory for consciousness and other components) of our scientific origin myth. We leave this topic at that, but provide a few references for those interested: Budin *et al.* (2009); Budin & Szostak (2010); Lane *et al.* (2010); Shrum *et al.* (2010). These recent papers suggest that deep vents heated by serpentinization and sustaining alkaline pH have substantial potential for installing functions in life-initiating protocells.

Sarah Hoffman published important studies in several areas of oceanography and geology. She died in late 2010, succumbing to long-fought bone cancer. Jack Corliss was dismissed by his university for not publishing sufficiently. Two of the most important discoveries (and papers) in the history of science were not enough. He moved to Hungary, where he is still thinking important thoughts.

Chapter 16

Ocean ecology and global climate change

The phrase "global climate change" has long since become a mantra chanted by Earth scientists before international boards and national legislatures as we seek to influence environmental policy and to generate research support. A wide range of issues is involved in this chanting, and their advocates compete for attention and money. Biological processes in the ocean have a part in some climate control mechanisms, and changed climate will affect marine life. Thus, climate change and its effects have become a significant part of biological oceanography. Change has become synonymous with warming, and global warming of the atmosphere and oceans, relative to mid-20th-century temperatures, is definitively upon us. The subject is sufficiently complex that the United-Nations-sanctioned Intergovernmental Panel on Climate Change (IPCC) required four weighty volumes to cover it in their 2007 report, now already outdated. The best we can do in a single chapter is outline how biological processes in the ocean may affect climate on long time scales, show how ocean climate varies on decadal time scales, and give some examples of the plethora of effects of warming.

Climate has never been constant on the Earth. Shifts in mean temperature; temperature-cycle amplitude; precipitation rate; snow versus rain; seasonal snowfall and ice melting; evaporation; and length of growing season, are among the most important driving forces for evolution. At present, during an interglacial period, we are in a reasonably favorable climate compared to much of the Pleistocene era, and even compared to the 17th to 19th centuries. The 20th century saw mostly moderate conditions, gradual warming and long growing seasons compared to the few previous centuries. Since the 1980s, warming has been a source of serious and public concern, and warming promises to remain an important issue for the foreseeable future. Not only is the climate changing, the present distribution of climates over the Earth is coupled to the means for feeding and sheltering ~6.9 billion people (February 2011), more than four times the population of only 100 years ago. It is reasonable to think we are dependent upon climate conditions certain to change.

Thus, we are right to be worrying about the retreat of montane glaciers; melting of the Greenland and antarctic ice sheets; summertime loss of Arctic Ocean sea ice; poleward shift of malarial mosquitoes; enhanced intensity of short-term climate cycles like the El Niño–Southern Oscillation; changes in rainfall quantities and distribution; and other climatic effects of warming. Concern is complicated by our recently strengthened realization that climates cycle at periods from decadal to millennial, as well as at the long-recognized glacial–interglacial period. We are trying to trace shifts in the North Atlantic Oscillation, the Pacific Decadal Oscillation, and other oscillations partially coupled to them worldwide. Fisheries appear to be subject to climate-driven oscillations of as long as 60 years, with different phases termed *regimes*. Several examples of such regime shifts will be considered in Chapter 17. This complex, multi-decadal cycling interacts with and obscures the longer-term, but rapid, general warming currently underway. Even if global climate change has become a mantra, it is essential that we stay with the process of sorting it out, come to understand it as thoroughly as possible, and take mitigating action – particularly reduced use of fossil fuel. Some other proposed mitigations are sited in the oceans and are biological in proposed mechanism.

Biological Oceanography, Second Edition. Charles B. Miller, Patricia A. Wheeler.
© 2012 John Wiley & Sons, Ltd. Published 2012 by John Wiley & Sons, Ltd.

They do not promise moderation of warming commensurate with their difficulty and cost. We must also prepare to live in a warmer world; considerable heating is likely to be inevitable. Parallel to warming attributable to fossil carbon in the atmosphere, the fraction of fossil-fuel CO_2 dissolving in the sea is shifting the carbonate buffering system to greater acidity. That has acquired intense interest more recently, since about 2000.

Biological production and the subsequent redistributions of organic matter in the oceans have a part in climate control, and we will examine that from several perspectives. This is not a complete essay on global climate change issues. The IPCC is attempting to keep abreast of both the data and the current best thinking on the issues. For thorough coverage, see their most recent full report: IPCC (2007; a summary is available on the worldwide web at http://www.ipcc.ch) and the occasional updates posted there.

Global warming and CO_2

The weather is, on broad average (a definition of "climate"), getting warmer (Fig. 16.1). Available data since the mid-1800s are averages of distributed temperature observations, the quality of which was good then and has generally increased steadily (mostly because of added observing stations). In fact, the warming has been going on since before the industrial era, but is now accelerating, albeit in intriguingly stepped fashion. Warming in the 20th century occurred in two episodes: 1905–1940 and 1976–present. Means were fairly steady from 1940 to 1976, followed by more warming. The interval from ~2000 to present was the warmest in the instrumental temperature record (Fig. 16.1),

although it was also intriguingly flat. It is considered by some oceanographers that the heating in this "flat" period has been focused in the ocean, where a great deal of heat ~0.64 W m^{-2} was stored in upper layers from 1993 (mostly from 1998) to 2008 (Fig. 16.2; Lyman *et al.* 2010). Other strong temperature oscillations have occurred in historical times, including the warm medieval optimum and the "little ice age" from 1600–1860, a cold phase in which continental glaciers re-expanded and growing seasons were short. Our current concern with warming comes because it is happening in an era when record-keeping and science are advanced enough to see the process and offer explanations, because its effects will in some ways be unpleasant, and finally because we seem to be causing it. A significant aspect

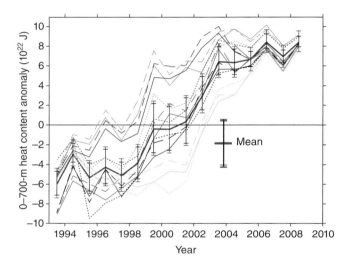

Fig. 16.2 Thirteen estimates with their mean (dark line with std. error bars) of global ocean heat increase plotted as anomalies from 1993 to 2008. The gain was about 13×10^{23} Joules. (After Lyman *et al.* 2010.)

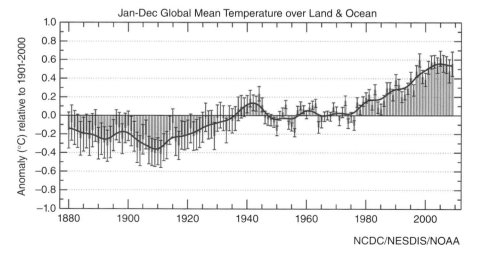

Fig. 16.1 Global average temperatures, 1880–2010. US National Oceanic and Atmospheric Administration (2010). Go to http://www.ncdc.noaa.gov/img/climate/research/2009/global-jan-dec-error-bar.gif.

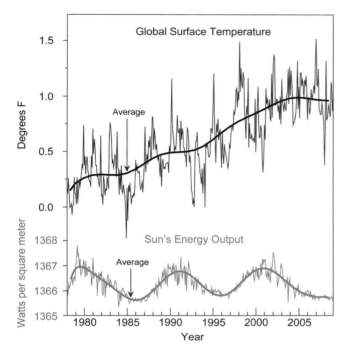

Fig. 16.3 Increasing global average surface atmospheric temperature from 1979–2009 and solar irradiance (W m⁻²) incident at the outside of the atmosphere over the same interval. (NOAA data.)

of warming is that it has been greater at high than at low latitudes (temperature anomaly geography by decades from 1880 to 2009 is shown in sequence at: http://earthobservatory.nasa.gov/Features/WorldOfChange/decadaltemp.php?src=eoa-features.

The most likely explanation of the recent acceleration of warming is the increasing accumulation in the atmosphere of so-called greenhouse gases. To understand the issues surrounding global warming, particularly as affected by these gases, it is necessary to master the notion of the Earth's *radiation balance*, a topic peripheral (but critical) to biological oceanography. So, briefly, the sun shines on the atmosphere, delivering energy to areas perpendicular to its rays at the rate of 1365.6 to 1367.0 watts m⁻² (varying with the sunspot cycle) mostly in the visible part of the spectrum (Fig. 16.3). That mostly passes through the atmosphere and heats the land and the sea. Of course, radiance on any given area depends upon the solar zenith angle. Warmed, the land and sea emit energy as light of longer wavelength, infrared light (IR), back toward space. The Earth warms until its IR emissions to beyond the atmosphere equal, on average, the incoming energy. That is the radiation balance, and the composite of temperatures at which it is achieved depends strongly upon absorption of the IR emissions by gases in the atmosphere. Typical in-bound light is almost entirely in a wavelength band from 350 to 750 nm, while at surface temperatures averaging ~15°C the wavelength of outbound light peaks from 12 to

20 μm (the "15 micron" band, a factor of ~38 longer). Those wavelengths are absorbed by atmospheric gases: transferred into vibrations of their molecular bonds and into increased molecule velocity (higher temperature). That both warms the atmosphere and requires that the surface be warmer to emit sufficient amounts of the "15 μm" IR, and somewhat shorter wavelengths that are not so strongly absorbed, to attain balance. Balance is attained at relatively short time scales (a few days).

The absorbing gases are water vapor, carbon dioxide (which absorbs IR in the 12 to 17 μm band centered at 15 μm), methane, nitrous oxide, and more. Methane and nitrous oxide, while smaller contributors to the air mixture, are also stronger IR absorbers on a per molecule basis than CO_2. Methane in the air and retained in surface ecosystems is mostly biologically produced: fermentation in cattle guts; bacterial methanogenesis in marshes, rice paddies and sediments, etc. It is also a variable but dominant (~95 mole %) component of natural gas, some of which is lost to the atmosphere in production, transport and use. There are several events that could result in large methane releases, including widespread tundra meltdown and degassing (adding a huge positive feedback to global warming) and seafloor disturbances that could release methane–water clathrates buried just below surface sediments of many continental shelves. Nitrous oxide is generated by fossil-fuel burning and naturally by, among other things, phytoplankton metabolism.

For the moment, increasing CO_2 is the main concern. Fossil-fuel burning (changing carbon-isotopic composition assures us that coal, oil, and natural gas are the main sources) has been and is increasing the carbon dioxide concentration in the atmosphere (Fig. 16.4), an increase

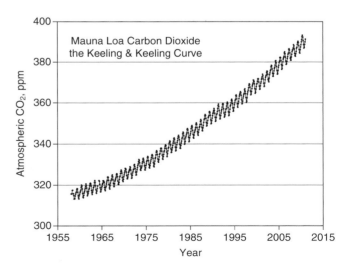

Fig. 16.4 Plot of time-series of Mauna Loa Observatory CO_2 concentration (ppm), 1958–2010. This is the widely disseminated "Keeling Curve" established by C.D. Keeling and maintained by R.F. Keeling (Data provided by R.F. Keeling, Scripps Institution of Oceanography.)

that has strongly accelerated since the middle of the 20th century. There are additions from burning of biomass (wood, etc.) and opening of soils. The overall increase is reduced by a third to half, relative to fuel burned, by dissolution of CO_2 in the ocean, less so by recent reforestation and effects minor in the short term, but significant on time scales of thousands of years, like incorporation in soils and weathering of silicate minerals. Warming from absorption of IR by CO_2 allows the warmer atmosphere to retain more water vapor, which is a more-abundant IR absorber, generating a multiplier effect. A commonly accepted radiative balance model predicts a temperature increase of $0.3°C$ watt m^{-2} if the water-vapor multiplier is neglected, and $0.6°C$ watt^{-1} m^{-2} if it is included (it must be included). It also predicts that doubling of CO_2 should increase the rate of heating over all by 4 watts m^{-2}, so at twice the pre-industrial level of 270 ppmv CO_2 (= 540 ppmv) we might expect warming on the order of $2.4°C$. That is,

$$\Delta T_{\text{double CO2}} = 0.6°C \text{ watt}^{-1} \text{ m}^{-2} \times 4 \text{ watt}^{-1} \text{ m}^{-2} = 2.4°C.$$

A tangle of possible feedback effects (the topic of a vast literature) are subsumed (or ignored) in that. As you worry (that is the right word) about this, keep in mind that light absorption by a component of a fluid mixture is negatively exponential with respect to its concentration. That is, if we double CO_2 and get an increase "X" of IR absorbance or associated warming, it must be doubled again to raise that effect to $2X$. At sufficiently high levels, all IR in the 15 μm band (and some other bands) will convert to heat in the atmosphere, with the temperature rising enough to sustain radiation balance at somewhat shorter wavelengths. In the fairly deep past, during the Eocene, CO_2 was much higher than twice current levels, and life went on. It was warmer on average, sea level was much higher, with inland seas in places like Kansas, USA (essentially no ice was stored on land), and there were tree ferns and crocodiles on Ellesmere Island. No level of CO_2 likely from human activity will be the end of the world, just of the climate as people and all extratropical species are accustomed to it. The focus on CO_2, here and in the worldwide discussion, is because it has one of the longer atmospheric residence times among greenhouse gases and has a large anthropogenic production. Keep in mind, however, that other gases and a complex skein of land and sea processes are involved in climate control. There are abundant details in IPCC (2007, Vol. 1).

Burning of fossil fuel is the largest contributor to CO_2 release, currently greater than the 2007 total of 8.4 GtC yr^{-1} ("GtC" stands for gigatonnes of carbon, or 10^{15} grams. Total use of fossil fuel went up steadily after about 1750, and it increased rapidly from the late 1940s until now (Fig. 16.5). It is popular to say that use in this interval has increased exponentially, but that is only true of some components, not the total. The linear increase is dramatic enough, adding 0.11 GtC per year to the annual CO_2 incre-

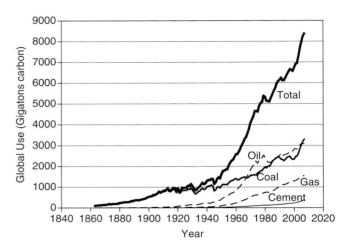

Fig. 16.5 Estimated yearly global amounts of carbon dioxide released to the atmosphere from different sources and in total. The units are carbon in CO_2 in millions of metric tonnes. Data are from CDIAC, Carbon Dioxide Information Analysis Center, US Dept. of Energy (http://cdiac.ornl.gov/trends/emis/meth_reg.html.)

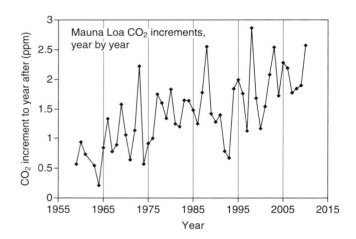

Fig. 16.6 Annual increments to atmospheric CO_2 calculated from the annual average data for Fig. 16.4. (Data provided by R.F. Keeling, Scripps Institution of Oceanography.)

ment since ~1950 (Fig. 16.6), when the post-WWII global economy suddenly caught fire (almost literally). The annual increment is currently four-times that of the immediate post-WWII years. Virtually all of the CO_2 generated by this burning of fossil fuel passes through the atmosphere, and currently about half of it remains there (Fig. 16.4). The annual average atmospheric concentration at the end of 2010 was 390 ppmv (parts/million by volume) of CO_2, and it will pass 400 ppmv very soon. Concentration increases currently at about 1.7 to 2.5 ppmv yr^{-1} (2004–2010 average). The year-to-year variation of the increase (Fig. 16.6) is not because fuel use varies that much. It is caused by complex interactions of weather cycles with partial removals to the ocean. Strongest increases are during El Niño years. Some ocean areas take in CO_2, others release it. Particularly important as a source to the atmosphere is the eastern tropical Pacific. Reduction of upwelling by El Niños there should actually reduce the net transfers from the ocean. Therefore, the positive effect of El Niños on atmospheric CO_2 increase must occur on land. It is believed to be caused by the associated tropical droughts in the western Pacific and Asia that generally reduce terrestrial net primary production and dry the fuel for the tropical fires common during El Niños. Since 1 ppmv = 2.125 Gt of carbon, the increase currently averages ∼4.4 GtC yr^{-1}, about half of estimated burning of fossil-fuel carbon. When it was first noticed, this difference was referred to as "missing fossil-fuel carbon". We now know where most of it goes, and "missing" is often used for unaccounted remainders.

We need a comparison for scaling these numbers of gigatonnes. A good one is an estimate from satellite data (Behrenfeld & Falkowski 1997) of global, oceanic, primary production at ∼44 GtC yr^{-1}. Annual fossil-fuel burning at approximately 8.5 GtC yr^{-1} is about 19% of that, and ocean-sequestered fossil-fuel CO_2 is about 4.5%. Terrestrial photosynthesis is of the same general magnitude as oceanic. To a close approximation, both in the ocean and on land, the carbon newly bound into organic matter each year is equal in amount to that released by respiration, hence the obvious cycling in Fig. 16.4. Thus, additions of CO_2 from burned fossil fuel are equal to a major part of the global carbon cycle each year. They are the main source of net change.

Fossil carbon not retained in the atmosphere on year-to-year time scales is accumulating primarily in two places: dissolved in the oceans and taken up and stored (for a time at least) by terrestrial plants, particularly trees. The air–ocean–land partition of burned carbon has been measured in several ways. Likely the most accurate are estimates from measures of the amounts of oxygen removed from the atmosphere by the burning carbon ($C + O_2 \rightarrow CO_2$), with some additional reduction from oxidizing the hydrogen (producing water) in petroleum and natural gas. It is also necessary to estimate the release of both O_2 and N_2 from the ocean caused by warming on decadal scales. The net change in atmospheric oxygen partial pressure is only a small decrement to the 20.9% in the atmosphere, too small to measure directly, particularly given the variation in actual [mol O_2/volume] driven by varying atmospheric pressure. However, the ratio $O_2 : N_2$ can be very precisely measured (i) by interferometry from the shift in air refractive index as the ratio changes (Keeling *et al.* 1998a), or (ii) by mass spectrometry (Bender *et al.* 1996). The final measurements are very precise, but both techniques involve refined gas-handling procedures because oxygen adsorbs to collector surfaces and valve gaskets, reacts significantly with some metal ducting (e.g. stainless steel) and passes through valves and orifices at different rates than nitrogen. Moreover, standards and reference air must be produced with precise proportions of O_2, N_2, Ar, and CO_2 then protected from interaction with their containers. By extended trials, those problems have been solved sufficiently (e.g. Keeling *et al.* 1998a) to provide useful data.

The use of ratios measured relative to arbitrary standards produces somewhat abstruse arithmetic for evaluating temporal changes in the atmosphere and relating them to the disposition of fossil-fuel carbon dioxide to different global reservoirs – arithmetic well presented by Manning and Keeling (2006). The first-order variable used is:

$$\delta(O_2 / N_2) = \frac{(O_2 / N_2)_{sample} - (O_2 / N_2)_{reference}}{(O_2 / N_2)_{reference}} \times 10^6$$

in units termed "meg", parts per million. The global reduction in the ratio from 1998 to 2002 was ∼21 meg/year (slope of the fitted curve in Fig. 16.7), after removal of smooth and opposite seasonal variations in the northern

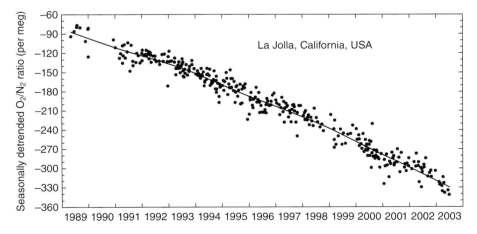

Fig. 16.7 Seasonally adjusted (to the annual means of seasonal data, Fig. 16.8) time-series of O_2/N_2 ratios relative to reference air at La Jolla, California, USA (33°N, 117°W). Units are millionths (meg). (After Manning & Keeling 2006.)

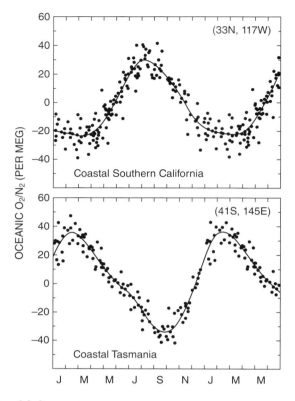

Fig. 16.8 Northern- and southern-hemisphere seasonal cycles of O_2/N_2 ratios showing the inverse seasonality of net production (photosynthesis > respiration) and net metabolism. (After Keeling *et al.* 1998b.)

and southern hemispheres (Fig. 16.8). A budgetary argument for changes in atmospheric CO_2 and O_2 allows estimation of the ocean and land sinks for CO_2:

$$\Delta CO_2 = F - Oc - L \text{ and}$$

$$\Delta O_2 = -\alpha_F F + \alpha_B L + Z,$$

with F = moles fossil C burned, Oc = ocean CO_2 uptake, L = land CO_2 uptake, α_F = O_2:C_F ratio for fossil-fuel burning ($\sim$1.39), α_B = O_2 produced by an *average* mole of terrestrial organic matter produced by photosynthesis ($\sim$1.1) and Z = the air:sea exchange of O_2 (small but toward the air due to upper ocean warming). Quantities on the left are measured, ΔO_2 from $\Delta(O_2/N_2)$ by assuming N_2 is nearly fixed (with approximation for release on warming). On the right, F is reported by governmental agencies tracking coal mining and gas and oil production. The equations are solved for Oc and L. According to Manning and Keeling (2006) the sinks for 1993 to 2003 were: $Oc = 2.2 \pm 0.6$ and $L = 0.5 \pm 0.7\,GtC\,yr^{-1}$. In those years, $F \approx 6.5\,GtC\,yr^{-1}$, so the ocean took up about a third of the fossil-fuel CO_2, and overall the amount left in the air was somewhat over half the fuel input. The uncertainty in L is larger than the mean estimate, apparently including the possibility that

deforestation and soil opening make land a source not a sink.

Thus, the ocean has a major part in modifying the impact of fuel use on the atmospheric CO_2 levels. It is believed that the positive estimate for land uptake, L, in recent years can be accounted for by reforestation in northern hemisphere temperate areas, including eastern North America and Scandinavia. As agriculture has become more intensely focused in the most productive areas, sizeable places such as New England have returned to forest since about 1950. The stock of biomass in these forests is still increasing. Harvest rates have slowed in some temperate, lumber-producing forests, as in the west of the US, allowing some reaccumulation of biomass. Photosynthesis generating that biomass returns oxygen to the atmosphere, causing the lowering of O_2 and O_2/N_2 to be less than expected from the inventory of burned fuel. Photosynthesis in the ocean does not have this effect since, as frequently stated in this text, almost all marine photosynthesis is respired very soon after it is produced, converting photosynthetic oxygen back to CO_2. The budget can be worked in more detail, keeping seasonal-scale track of CO_2 uptake and return (Fig. 16.8). Oxygen released to the atmosphere in spring–summer is taken up again in fall–winter. This must be partly physical and due to the effect of the temperature cycle on solubility. Biology has a part, generating O_2 by photosynthesis and consuming it by metabolism. Data on tropical forests suggest that their biomass continues on balance to be transferred to the atmosphere, since we are cutting them down and burning most of the wood faster than they regenerate. The annual net terrestrial storage of carbon only applies to the global average.

Another fate of some fossil-fuel CO_2 may be chemical conversion to soluble form by weathering of silicate minerals. As CO_2 levels rise, acidity of rain and soil water increases by formation of carbonic acid, H_2CO_3. This can attack a wide array of silicates, for example wollastonite, $CaSiO_3$:

$$CaSiO_3 + H_2CO_3 \rightarrow CaCO_3 + SiO_2 + H_2O.$$

The products are carried to the sea, incorporated in shells, mostly of foraminifera and diatoms, then deposited in sediments. Chemical weathering is believed to have been very important in limiting and sometimes reducing CO_2 levels in the atmosphere over long geological time scales. For example, a reduction is believed to have occurred over the past 12 million years (Myr), because the uplift of the Himalayas, which started 12 Mya ago, exposed great expanses of new rock surface to weathering at a high rate. The importance of enhanced chemical weathering due to current anthropogenic CO_2 increase has been little addressed. Detecting the increase in calcite deposition resulting from gradually progressing CO_2 levels would be difficult, since it could happen anywhere in the global ocean (as coral-reef build-up, foraminiferan shell deposition or inorganic oolites in tropical waters) and only need remove about $1 \, GtC \, yr^{-1}$ to be important. Moreover, the pH lowering in the sea resulting from increased carbonic acid (from fossil-fuel CO_2 in the atmosphere) acts to reduce calcite deposition rates in reefs and coccolithophores (Riebesell *et al.* 2000). Sorting out the magnitudes and interactions of constituent processes in global warming is a considerable challenge, but one being widely addressed.

It is certain that much of the "lost" fossil-fuel carbon ends up in the ocean, but it is less certain exactly what processes mediate its dissolution. According to marine chemists, the general answer is fairly simple: the transfers that matter are atmosphere–ocean exchanges involving CO_2 solubility. The sequestration occurring as we add fossil fuel CO_2 to the atmosphere is essentially a non-equilibrium process. A map of CO_2 invasion and evasion rates (Plate 16.1), based on station data for CO_2 contents (air and water) and exchange rates, shows that much of the influx is in areas of deep-water formation: (i) around the Antarctic, where sinking is driven by chilling coupled to salt release during ice formation, and (ii) the Norwegian Sea and/or the Irminger Sea, where sinking is driven by arctic refrigeration of salty Gulf Stream water. In these areas, the new levels of CO_2 (at the fossil-fuel enhanced pCO_2 of the atmosphere) equilibrate with the cold, polar water that carries an increased concentration to depth as it sinks. Evasion sites, most dramatically the eastern tropical Pacific, Arabian Sea, and deep southwestern basin of the Bering Sea, return the

gas to the atmosphere at lower rates (in the global total) set by the concentrations established at depth over the whole interval that the water has been sequestered, that is up to thousands of years. Thus, return to the atmosphere represents the pre-industrial atmospheric levels. Adding to that non-equilibrium effect, the CO_2 evasion now must work to push the gas into the atmosphere against a reduced concentration gradient, because there is more in the atmosphere. So, until a new equilibrium is established, probably a 3000–6000-year delay, some of the fossil-fuel CO_2 will move quickly into the ocean.

However, the CO_2 concentration at the evasion sites is not the same as that at the invasion sites – it is higher. While water moves along at depth, carbon is added from respiratory oxidation of organic matter sinking into it all along the route of deep-water flow. We assume that factors affecting primary-production rates probably have not changed much over 200 years, so there probably has been no large change in this biological sequestration during the industrial build-up. It is taken to be a large but constant term in the overall exchange. The current consensus (IPCC 2007) for the overall budget (Fig. 16.9) puts the biological contribution at about $11 \, GtC \, yr^{-1}$. That is a large enough component of the budget that, if it does vary, it could account for major shifts in the overall input–output balance for labile carbon in the ocean. "Labile" in this context means not fixed into rock, but free to come and go between gaseous and dissolved carbonate pools and organic matter.

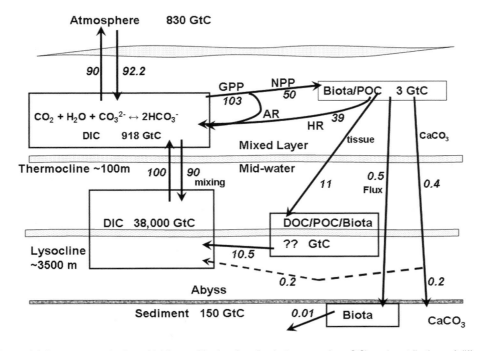

Fig. 16.9 Approximate global ocean reservoir sizes of labile, combined carbon (in gigatonnes carbon, GtC), and contributions of different processes to transfers of CO_2 to and from the ocean and among carbon-bearing constituents in the ocean. Rates in italics are per year. Abbreviations: GPP, gross primary production; NPP, net primary production; AR, autotrophic respiration; HR, heterotrophic respiration; DIC, dissolved inorganic carbon; DOC, dissolved organic carbon. Data from Sarmiento and Gruber (2006), IPCC (2007), and Emerson and Hedges (2008).

The air–sea CO_2-flux map (Plate 16.1) also shows this biological carbon transfer into the sea. Much influx to the ocean interior occurs in regions without deep-water formation: the subarctic North Atlantic, subarctic Pacific and those subantarctic areas just downwind of land. In those regions, the substantial influx of CO_2 to the ocean interior is by the "biological pump" (a term from Longhurst & Harrison 1989). In these areas, surface CO_2 is reduced because photosynthesis removes it from the water, increasing the atmosphere-to-water gradient and the flux into the ocean. Respiration does not balance phytoplankton uptake of CO_2 in surface layers across these regions, because sinking particulates (from tiny fecal pellets to dead whales) and vertical migrations move carbon out of the surface to accumulate at depth. Below the permanent pycnoclines typical of mid-latitudes or the haloclines in subarctic and subantarctic regions, these particles eventually oxidize, adding to the supersaturation of mid-waters with CO_2. Over intervals longer than just the industrial era, changes in atmospheric CO_2, as will be seen for glacial–interglacial cycling, certainly involve changes in operation of both the physical and biological carbon pumps.

The data upon which the flux map is based make possible areal summations of input and output (Gruber *et al.* 2009; Plate 16.2). There are four ways to calculate these sums, represented by the bars of different colors in the plate. While the areas to which the bars apply are not explicit, it is clear that input to the ocean is greater than output, a very large proportion of which is located in the tropical Pacific. Within their errors, the methods agree with the O_2/N_2 method that the net uptake in the period of the surveys (1995) was ∼2 GtC yr^{-1}. Approximate chemical calculations allow separation of anthropogenic carbon from "natural", leading to the conclusion that the physical transfer of CO_2 in the Southern Ocean is now into the ocean, whereas without fossil-fuel carbon it would be out. Elsewhere, the directions of transfers are unchanged, but the amounts have changed greatly, for example the reduction of export by almost half in the tropics. The global carbon budget diagram (Fig. 16.9) shows that net changes are small relative to overall transfers. Thus, both input and output of CO_2 to the ocean are about 90 GtC yr^{-1}, with a difference of about 2.2 GtC yr^{-1}, the anthropogenic input. Again, virtually all primary production is respired on very short time scales, with very little organic matter stored in water or sediment. Thus, the small net changes are widely dispersed across the oceans, making them extremely difficult to measure. It is remarkable that completely different estimates of ocean uptake agree so closely.

The climatic impact of the industrial era increase of atmospheric carbon is estimated by complex models of the circulation of the global atmosphere, the biosphere–atmosphere interaction with respect to CO_2 (also methane, etc.), air–sea energy exchange, and the incoming–outgoing energy budget. Making and running these models is now a regular industry with dozens of practitioners. Apart from numerical problems such as necessarily coarse grid scales, the main difficulties with these models are the large uncertainties deriving from the complexity of the system and from critical processes that are not well quantified. Examples include: (i) the poor characterization of the effect of increased CO_2 on primary productivity, probably small for phytoplankton but modestly important on land; (ii) the impact of warming on cloud formation, altitude and coverage, and thus global albedo; and (iii) the impact of warming on deep-water formation and thus deep-sea ventilation rates.

CO_2 and the glacial–interglacial cycle

At least a part of global climate variation on glacial–interglacial time scales is coupled to this same transfer of CO_2 into (and out of) the deep-sea that accounts for fossil fuel carbon not in the air. This topic, too, is freighted with complexities. Ruddiman (2007a) has sorted some of them out in intuitive fashion and should be consulted for fuller treatment. The ocean below its permanent pycnoclines has the largest reservoir of labile carbon on Earth, some 38,000 GtC, compared to 830 GtC in the atmosphere (Fig. 16.9). If deep-water formation were to slow or stop, as it apparently does at least intermittently in glacial eras, the driving force for both horizontal deep-sea circulation and ventilation of accumulated CO_2 at evasion sites would decrease markedly. During a long re-equilibration interval, there would be a net transfer of CO_2 to the deep sea by the biological pump, and atmospheric CO_2 levels would drop. According to gas analyses of datable ice cores from Antarctica (Fig. 16.10) and Greenland, the atmospheric CO_2 level was repeatedly close to 185 ppmv at the glacial maxima at 350, 260, 140, and 25 kyr BP (i.e. thousands of years before 1950, somewhat approximately determined). With the break-ups of glaciation, atmospheric CO_2 rebounded to the pre-industrial level of 270 ppmv, actually and briefly (a few thousand years) to 290 ppmv in the three earlier interglacials. The declines in atmospheric CO_2 during the four major glaciations did not have exactly the same patterns. They occurred in pulses that paralleled the accumulation of ice as represented by the increase in deuterium and $^{18}O_2$ in glacial water (see below). Return of CO_2 at the major glacial terminations is fast, covering the whole 100 ppmv range in ∼5 kyr. That fits both a slow, or pulsed with gaps, reduction by ocean uptake during growth of glaciation, mostly by biological pumping, and rapid return by onset of ventilation of the deep sea at the end of the glaciation.

It is established that glacial–interglacial cycles are driven in the first instance by oscillations in solar irradiance received by the Earth, oscillations due to cyclic changes in

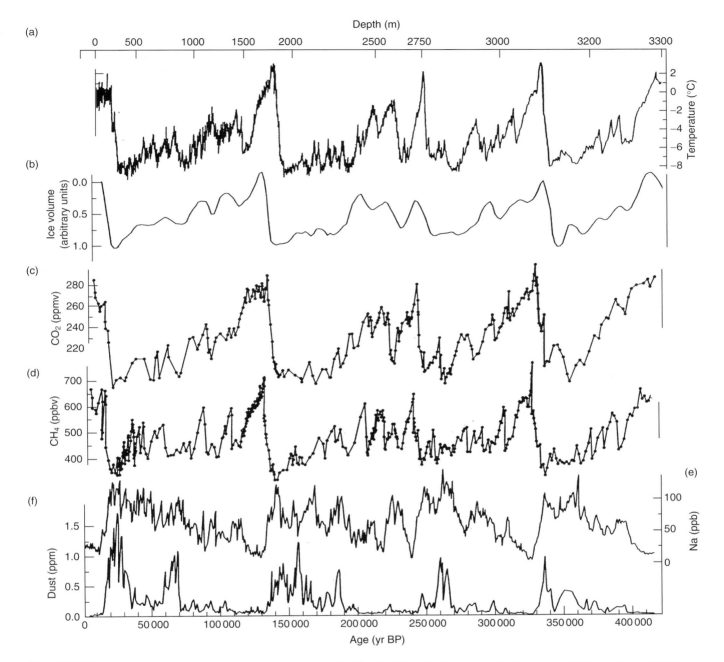

Fig. 16.10 Analytical time-series for constituents of a deep ice core from Vostok station, East Antarctica, and other sources. (a) Temperature variation at Vostok, based primarily on deuterium variation, a proxy for air temperature near the deposition site. (b) Global volume of ice on land estimated from $\delta^{18}O$ in marine core carbonate. (c) CO_2 and (d) methane in air from Vostok ice. (e) Sodium and (f) dust in Vostok ice. Sodium is a proxy for marine storm activity. Dust may be a proxy for iron transport to the Antarctic and thus to the surrounding ocean. (After Petit *et al.* 1999.)

the Earth's rotation and its orbit around the sun. These are termed Milankovitch cycles, after Milutin Milankovitch, a Serbian mathematician who worked on the problem of estimating combined effects of Earth's rotational and orbital variations. Glaciers grow or recede depending upon the balance between winter snow accumulation and summer melting rates. Since glaciers always form in and advance from regions that are cold all winter, every winter, and they receive snow addition every year. The amount is only weakly affected by temperature variation. However, the response of summer melting to temperature variation is strong. A few degrees of temperature increase can switch condtions from leaving some of last winter's snow until fall to melting all of it *plus* some of the previous accumulations before fall. Moreover, since the Antarctic, during at least the past 5 million years, has been continuously cold and always glaciated, the main changes are those of northern, continental ice sheets. Thus, the aspect of insolation cycles that matters most is variation of sunlight during the northern summer near the Arctic Circle, the zone from which ice sheets advance.

Once snow remains unmelted all summer and northern glaciers begin to grow, there are important positive feedbacks. (i) Glaciers have high albedo (90% vs. 15% for vegetation) and reflect light back to space, enhancing cooling. (ii) The great glaciers of the ice ages slowly rose to altitudes of 1–3 km, lowering temperature on their surfaces at the vertical lapse rate of about $-6.5°C\,km^{-1}$, thus further protecting the high core from summer melting. (iii) General cooling reduces the amount of water vapor in the atmosphere, reducing its role as a greenhouse gas and cooling the atmosphere further. (iv) Of main interest here, the reduction of CO_2 by 100 ppmv allowed more 15 μm IR to pass through the atmosphere to space, reducing the mean global temperature needed to sustain radiation balance. The list of feedback suggestions is even more extensive, some speculative, some with counter-evidence.

Summer insolation varies with cycles in four aspects of the Earth–sun rotation:

1 The tilt of the Earth's axis of rotation relative to the ecliptic, the plane of rotation around the sun, varies from 22.25 to 24.25° (currently 23.44°) with a period of about 41 kyr. Summer warming is greater at greater tilt, since the polar regions point more directly at the sun. The value at a given time is termed the *obliquity*. Two degrees difference in tilt seems like a small number, but its effects are large. At 24.25° versus 22.25° the Arctic Circle shifts south 2°, summer daylengths are slightly shorter, and summer noon sunlight spreads over 8% more area.

2 The tilted axis wobbles (axial precession) in the plane of the ecliptic with a period of 25.7 kyr.

3 The ellipse of the Earth's rotation around the sun also rotates around the sun once about every 22 kyr, moving perihelion and aphelion with respect to the seasons. Aspects (2) and (3) combine as "precession of the equinoxes" (and the solstices) with a net period of around 23 kyr. Summer warming is greater when the summer solstice occurs at perihelion (where the ellipse lies closest to the sun), least at aphelion. Perihelion at present is in the northern winter.

4 Ellipse eccentricity ($[a^2 - b^2]^{1/2}/a$, where a and b are the long and short axes) varies between 0.005 and 0.0607 (currently 0.0167), shifting the annual variation in Earth–sun distance, with a period of about 100 kyr. Greater eccentricity produces closer perihelion, and farther aphelion.

The resulting overall range of summer isolation at 65°N is from approximately 410 to 500 watts m⁻², a huge variation. Ellipse eccentricity acts to "modulate" the effect of precession of the equinoxes, which is possibly the reason that eccentricity variation acts as the switch between glacial and interglacial times (see Ruddiman 2007a). Pulses of ice formation and loss are also evident in ice-volume proxies at the obliquity period, 41 kyr.

The time course of the amount of glacial ice piled on the land is best represented among available "fossil variables" (ice-volume "proxies") by changes in the level of oxygen-18 (^{18}O) in the ocean as recorded in glacial ice (Fig. 16.10) and fossil carbonate (foraminifera, coral). Lighter $H_2{}^{16}O$ molecules are more likely to evaporate than $H_2{}^{18}O$ molecules, and more likely to be included in glacial ice. Thus, seawater becomes enriched in ^{18}O as glacial accretion progresses. As ^{18}O level rises in the ocean during ice accumulation, it also rises in water evaporating, so that stacks of glacial ice, old near the bottom and young at the top, record global ice volume as changes in ^{18}O content, determined from the oxygen in the water itself. Measurement is done by mass spectrometry, and the result expressed as a fractional difference from the oxygen in a standard. This fraction is called $δ^{18}O$ ("del-O-18") with units of parts per thousand (‰). Deuterium in water rises by the same mechanism and is measured as δD, which can be more accurately calibrated in antarctic ice as a temperature index than can $δ^{18}O$. Profiles of both variables have been developed for ice cores from central Antarctica and central Greenland – sites where lateral ice movement is minimal and stacking with age is close to vertical – as well as from montane glaciers worldwide. Age is determined from ^{14}C levels and other dating techniques, with special precision when annual layers of dust deposit can be counted.

Equilibration of $^{18}O : {}^{16}O$ between water and dissolved CO_2 will occur in a matter of hours via carbonic acid and bicarbonate formation. Thus, oxygen in carbonate included in new coral or foraminiferan shells will partially record the water values of $^{18}O : {}^{16}O$. Equilibration among different global oxygen pools (water, atmospheric gas, dissolved O_2, carbonate) by exchange on time scales of a few thousand years means that all will record relative removal of ^{16}O into ice. For ocean carbonate, the record will show some lag, and lag will vary with depth of shell deposition (say, benthic forams vs. surface ones) and with latitude, since there are several ^{18}O-enriching cycles of evaporation and precipitation along the route from the equator to the ice sheets.

A composite foraminiferan $δ^{18}O$ "stack" has been generated by Lisiecki and Raymo (2005), extending back 5.3 million years. (Fig. 16.11 presents 3.6 Mya). It shows that throughout that long interval, there were many cycles of glaciation, most of them rather restricted in the volume of ice accumulated and initially with dominance of the 41 kyr obliquity cycle. An interval of transition followed from 1.1 to 0.676 Mya; then came a series of six major glaciations with overall periodicity at (quite roughly) 100,000 years with continuing oscillations of ice volume (Fig. 16.11) corresponding to the shorter 41 and 23 kyr oscillations. It is

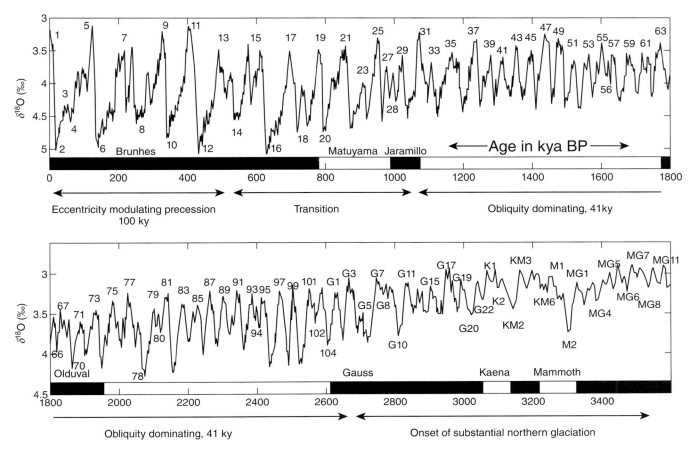

Fig. 16.11 A consensus "stack" of ^{18}O content estimates from carbonate in marine sediment cores. The data, stratigraphically aligned and averaged, are proportional estimates as $\delta^{18}O$ (‰), plotted on an inverted scale to show intervals of low terrestrial ice volume and low ^{18}O content as peaks. Age estimates are from various sources, including ^{14}C in carbonate, magnetic stratigraphy shown as black and white bars (Brunhes, etc.). Primary responses to Earth's orbital variations in different eras are indicated next to duration arrows. The small numerals are designations for "marine isotope stages" (MIS), as for example, MIS-5 is the switch at the end of the interglacial. The "stack", termed LR04 for Lorraine Lisiecki and Maureen Raymo who combined the data in 2004, is often used for comparison with variables in glacial ice cores and new sediment profiles. (After Lisiecki & Raymo 2005.)

thought, although uncertainty remains, that the longer cycling became possible as a result of persistence of extensive parts of the northern ice sheets through the warm phases of the shorter cycles, accelerating glacier recovery with a return of cooler summers via the feedbacks mentioned above. Only when the eccentricity maximum put the summer solstice (and perhaps also the tilt) at the perihelion was summer warming sufficient to ablate the ice sheets almost completely over a period of a few thousand years. Moreover, almost as soon as this combination of orbital and rotational conditions passed, ice sheets re-advanced in somewhat intermittent fashion for nearly the full 100 kyr cycle period. Thus, large and usually growing northern ice sheets have been present through much of the past 675 kyr. Re-advance during the 100 kyr cycles was not slower than the retreats, but advances were more intermittent than the major ablations, taking longer overall.

The orbital control of global ice volume (*y*) is generally demonstrated by the fit of a model based on orbital varia-tions, developed by Imbrie and Imbrie (1980), to data like the $\delta^{18}O$ stack. The model is:

$$\frac{dy}{dt} = \frac{1 \pm b}{T_m}(x - y),$$

where *x* is the orbital forcing value (e.g. from Berger 1977) as the net sum at time *t* of obliquity, precession, and eccen-tricity effects on insolation at 65°N, T_m is the mean time lag of the ice response and *b* is a non-linearity parameter sub-tracted when ice is growing (slowing growth) and added when it is not (speeding ablation). The fit with $T_m = 15$ kyr and $b = 0.6$ (Fig. 16.12, Lourens *et al.* 2010) is mostly very good. The orbital effect pulse at ~150 kyr BP is also not present in ice core δD or $\delta^{18}O$ data; the theory is excellent but does not explain everything. There are also shorter-term, smaller-scale variations, the Dansgaard–Oeschger variations seen in Greenland ice cores, and the Antarctic

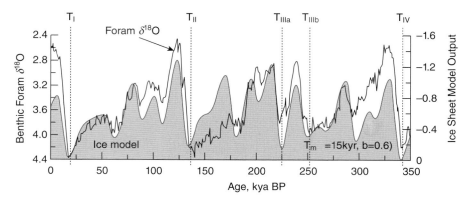

Fig. 16.12 Comparison of an Imbrie and Imbrie model (T_m = 15 kyr, b = 0.6; see text) of global ice volume response to orbital variations (shaded time-series) with the LR04 $\delta^{18}O$ "stack" (Fig. 16.11). Onset times of glacial terminations are T_I, T_{II}, etc. (After Lourens *et al.* 2010.)

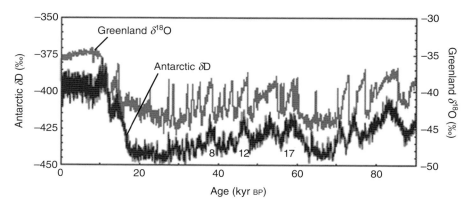

Fig. 16.13 δD (deuterium in glacial ice, ‰, indicative of temperature) in the EPICA core from Dome C in Antarctica (lower curve) plotted with $\delta^{18}O$ from the NorthGRIP ice core in Greenland (upper curve). The series are each on their locally determined time scale without squeezing or stretching to emphasize the matches, but matches are quite obvious among short-term events, mostly Dansgaard–Oeschger events in Greenland and antarctic isotopic maxima (AIM; 8, 2, and 17 are numbered) at Dome C. (After Wolff *et al.* 2009.)

isotopic maxima that appear to be coincident or somewhat offset (Wolff *et al.* 2009; Fig. 16.13). They may be driven by variations in the oceanic deep-ventilation considered next.

Finally, we come back to the interaction of glaciation with CO_2. It is almost proportionally removed from the atmosphere as ice piles up on land (Fig. 16.10). It rapidly and again proportionally reappears in the atmosphere as ice ablates from the major sheets. Its 15 μm IR absorption is clearly another of the key feedbacks that sustains cooling during ice-sheet growth and promotes temperature rise as ice sheets melt. The only likely reservoir of suitably labile CO_2 is the deep ocean. While it remains difficult to document the specific timing, it is clear that ice build-up recurrently reduces the thermohaline circulation of the oceans, an idea closely connected with and promoted by Wallace Broecker. There must be reduced formation of deep water in the North Atlantic. As it becomes ice covered far to the south, the Gulf Stream does not carry high-salinity water beneath very cold arctic winds, so cooling does not raise its density sufficiently to sink it to the seafloor. There can also be intermittent spreading of glacial meltwater across

the Atlantic, enhancing stratification. It has been suggested that formation of North Atlantic Deep Water (NADW) may be replaced by subduction of shallower, less-voluminous "Glacial North Atlantic Intermediate Water". There is also likely to be inhibition of gas exchange by year-around presence of sea ice around Antarctica to much lower latitudes than today. When deep water is not formed or re-exposed at the surface, ventilation is diminished, and organic matter exported to depth and converted to CO_2 remains in the sea. The relatively rapid returns (a few kyr) occur in periods when heating allows resumption of deep-water formation, thus, ventilation. Retreat of antarctic sea ice, allowing ventilation via upwelling in the Southern Ocean, is certainly also significant – perhaps the leading event in warming cycles. Carbonate in antarctic foraminifera showed a strong drop in ^{14}C age during the last glacial termination (Skinner *et al.* 2010), in agreement with return to the surface, and exchange with the atmosphere, of carbon long sequestered at depth. The drops in ^{14}C age actually occurred in several pulses associated with short-term variations in the deglaciation rate known from various stratigraphically documented events long hypothesized to be associated with variations in deep-water formation rates: Heinrich Stadial 1, Bølling–Allerod, and Pre-Boreal–Younger Dryas.

The mechanism for sequestering carbon when glacial-era processes slow ventilation must primarily be the biological pump. We would like to be able to project back to glacial conditions, say via models, to determine the global rate of sequestration. However, that probably cannot be done convincingly, because the annual amounts involved are extremely small. Bringing CO_2 down by $\sim$70 ppmv in $\sim$30 kyr (290 to 220 ppmv, the "rapid" initial drawdown of the Wisconsin glaciation) only required sequestration of $\sim$145 GtC from the atmosphere's interglacial content of $\sim$600 GtC, which is $\approx$3 years' global marine primary productivity. Thus, the rate was less than 0.005 GtC yr^{-1}, a miniscule change readily produced by even small changes in ventilation or export. It would have been immeasurable by any direct means, had we been about with our ships, sediment traps, and thorium methods. In fact, a modest change in the amount of vertical mixing of the oceans by swimming animals (Dewar et al. 2006; Katija & Dabiri 2009) could have changed exchange that much. Comparing 0.005 GtC yr^{-1} to the 2.2 GtC yr^{-1} currently moving into the ocean from fossil fuel gives a strong sense of the magnitude of current human involvement in climate control and ocean chemistry.

It was for a time popular to suppose that something about the glacial eras enhanced primary production in order to generate the organic matter to increase vertical export and sequestration during ice-sheet growth. The primary candidate was enhanced dust delivery to the oceans. It was likely that the mean global wind velocity did increase as the thermal gradients from the equator to the poles intensified, and that would have carried dust farther seaward from sources on land, where more soil open to wind erosion would have resulted from greater aridity. Dust delivery might well have been most significant in the sub-antarctic sector of the Atlantic and western Indian Oceans, with the Argentinian pampas upwind in the westerlies belt to supply the dust. Something like the present-day iron-limitation of subantarctic production might have been alleviated to a degree by that increase in dust. However, look again at Fig. 16.10. The initial drawdown of the Wisconsin (and the prior two glaciations at least) occurred before much dust appeared. Dust does spike at the glacial maxima, but not during the main CO_2 drawdown. That argument is not conclusive, since the ice records available are far inland in Antarctica, isolated by several opposing wind belts from the likely main paths of over-ocean dust transport. Dust flux data from an ice core representing 800 kyr at a site called Dome C (75°S) (Lambert et al. 2008) shows that the pattern of low flux during interglacials and in initial glaciation ($\sim$0.03 mg m^{-2} yr^{-1}) and high dust later (to $\sim$15 mg m^{-2} yr^{-1}) recurred through eight glacial cycles.

There is, however, some confirmation of low dust over the Southern Ocean during initial phases of reglaciation. Study of a single sediment core from the Atlantic subantarctic (Martinez-Garcia et al. 2009) shows that deposited dust and iron were low during the present and previous interglacials and for periods of $\sim$30 Myr afterward (Fig. 16.14), only rising as full glacial development was approached. Eventually, possibly due to iron enrichment, both diatom deposition (not measured by Martinez-Garcia et al., but evident from gamma-ray attenuation data at the associated ODP site 1090) and alkenone deposition (a signature of prymnesiophyte abundance) peaked, but only at the glacial

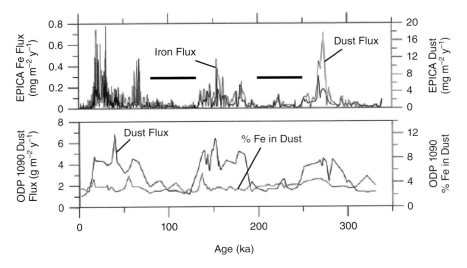

Fig. 16.14 Comparison of dust deposition rates (black lines) with interpretive rescaling as iron supply rates (gray lines and right-hand scales) between the EPICA Dome C ice core and a sediment core from the Atlantic sector of the subantarctic (ODP-1090) for the three most recent cycles of glaciation. The bars indicate periods during re-expansion of terrestrial ice sheets, which were periods of low dust deposition on both Antarctica and the Southern Ocean. (After Martinez-Garcia et al. 2009.)

maxima. As stated by Martinez-Garcia *et al.*, the initial phase of CO_2 drawdown must have been mediated by physics, not dust or its contained iron. The obvious candidate is reduced ventilation. That could have been reduced North Atlantic deep-water formation or expanded, annually prolonged Antarctic sea-ice coverage or both. In contrast to these subantarctic data, results from four late Pleistocene cores *inside* the Antarctic polar front at both Pacific and Atlantic longitudes (Anderson *et al.* 2009) show enhanced diatom sedimentation just during the last deglaciation, then ceasing. Anderson *et al.* suggest this uptick in diatom deposition was caused by increased silicic acid availability during the resumption of ventilation. It is not clear why it would stop after full onset of the interglacial. Data from ODP site 1094, also inside the polar front, show *low* diatom deposition rates (at least based on approximate opal proxies) during the CO_2 drawdowns following the peaks of earlier interglacials.

There are other and earlier data suggesting only a small role for dust fertilization in CO_2 cycling. Elderfield and Rickaby (2000) suggested, based on sediment profiles of cadmium/calcium ratios in planktonic foraminifera from the subantarctic, that there was little difference between modern and ice-age dissolved-phosphate levels in the subantarctic sector of the Southern Ocean. The Cd/Ca reflects phosphate levels because cadmium substitutes for zinc in a tiny fraction of enzymes in which zinc is a cofactor, and dissolved zinc through its role as a nutrient is closely proportional to phosphate. Greater iron availability should have allowed phytoplankton to consume more major nutrients, reducing phosphate (and cadmium) concentration in the upper water column. Like many more recent writers, Elderfield and Rickaby, having downplayed iron effects, speculated that CO_2 was lower in the glacial because area-related and seasonal increase in ice cover prevented some of the ventilation occurring now in the Southern Ocean. They proposed that there was no change in biological productivity involved, at least not in the Southern Ocean. There could have been a switch to diatom production in some latitudes (again, Anderson *et al.* 2009) without a general increase in productivity due to iron.

In addition to the general glacial-interglacial cycling, ice-core and other records show shorter-term variations: the Dansgaard–Oeschger (D–O) variations, mentioned above, during glacial eras and Bond events during the Holocene. The D–O events, most strongly evident in Greenland ice cores (Dansgaard *et al.* 1993), are rapid (~40 years) warmings with much longer cooling phases and total cycle times around 1470 years. The warming phases correspond to Heinrich events, which are depositions of iceberg-rafted rock debris in North Atlantic sediments. Not only temperature proxies change in D–O events, but there are strong increases in ice $\delta^{18}O$, implying strong reductions in glacier volume, freshening of surface layers and cyclic reduction of deep-water formation and

thermohaline circulation. Possibly, they were key factors in the longer-term CO_2 drawdown. Bond events (e.g. Bond 1997) in the Holocene, eight since the end of the Wisconsin glaciations, with the last about 1400 years ago, are identified from rock debris in Atlantic sediments. They have roughly the same frequency as D–O events, but are not obviously correlated with temperature shifts. In contrast, some warming events during the previous interglacial, and evident in Greenland cores, were multi-decade pulses up to 14°C (GRIP Members 1993). Holocene Bond events may generate sea-surface freshening and for a time reduce thermohaline circulation, but that is not established. Neither D–O nor Bond events are fully explained in terms of cause. A likely candidate is cycling between ice build-up and ice-sheet instability. Immediate ecological consequences of these events are not well characterized. The lesson that they offer for present-day concerns is that climate can shift dramatically in very short intervals, whatever the causes.

That covers some of the data and thinking about the glacial cycling process. Data gathering and thinking about process interactions continue. Much of the research is motivated by the hope that fully understanding those mechanisms will give insight about the likely impact of anthropogenic increases in greenhouse gases. Unfortunately, as pointed out to a vast public audience by (former US Vice President) Al Gore, we are now so far beyond the remarkably rigid, glacial cycling between 185 and 290 ppmv, that the main insight may well be that those mechanisms will no longer have the same effects. Ruddiman (2007b) has suggested that we should have seen substantial cooling as long as 8 kyr ago, but we have not. He attributes that to the possible greenhouse effects, starting then, of increasing CO_2 and methane derived from opening of soils and felling of forests for agriculture and development of rice culture in paddies. One effect of enhanced greenhouse gases, comparable to their "feedback" effect in the glacial eras, will be continued melting of glaciers. Both Greenland and Antarctica are in or moving toward net ice loss. Since the bulk of ice in both places sits on land, the rise in sea level that melting can generate may inundate island and coastal areas to depths of tens of meters on decadal to centennial time scales. The possible amounts are recalculated often enough that we do not cite them.

Iron fertilization of the sea to counteract global warming

As discussed elsewhere, primary productivity over substantial stretches of ocean (HNLC areas), particularly growth of larger phytoplankton (>8 μm), is limited by iron availability. Iron, while a major constituent element of the Earth as a whole, is very dilute in seawater. That is because ferric iron (Fe^{3+}, the form absorbable by cells) forms iron hydroxide ($Fe(OH)_3$) in basic solutions such as seawater (once closely buffered to pH $\sim$8.3). Iron hydroxide has a solubility coefficient of the order of 10^{-12} and forms a flocculant precipitate reducing the levels of free ferric iron to subnanomolar levels ($<1 \times 10^{-9}$ moles Fe^{3+} L^{-1}). Ferrous iron (Fe^{2+}) is oxidized to ferric in seawater and joins the precipitates. Iron is required as a cofactor for many enzymes and as a constituent of some pigments, so its limited availability controls the growth of larger phytoplankton. A substantial suite of twelve studies involving direct additions of iron to $\sim$100 km^2 patches of HNLC ocean (eastern tropical Pacific, subarctic Pacific, Southern Ocean) have proved that, indeed, iron is the limiting factor. Reviews of these studies have been provided by de Baar et al. (2005) and Boyd et al. (2007). Even before those "IRON-EX" studies, addition of iron to HNLC areas to sequester CO_2 from the atmosphere was suggested. It continues to be a contentious issue. One of us (Miller, e.g. Strong et al. 2009a) has taken public stands against this "ocean iron fertilization" (OIF) approach to reduction of atmospheric CO_2, and the following reflects that bias. We continue to think opposition to OIF is well founded. That is not to say that global society should do nothing to reduce CO_2 emissions and even CO_2 in the atmosphere. This book is not about what can or should be done, but something must be.

In 1989, with remarkable public notice in 1990 (remarkable considering that rather esoteric science was at issue), the late John Martin both hypothesized the interaction of glacial–interglacial cycling with iron-limitation in HNLC areas and suggested that we might be able to use enhanced iron transport to those areas to remove CO_2 from the atmosphere. The idea was to add iron-rich dust to broad HNLC stretches of the Southern Ocean, promoting growth of large phytoplankton. This regional phytoplankton bloom would more completely strip the waters of major nutrients and (the theory went) would mostly sink, carrying large quantities of organic matter into the deep sea. This would amount to a mitigation of the industrial CO_2 problem, possibly returning the climate to the somewhat cooler conditions prior to, say, 1960.

This created a storm of controversy. The conclusion, which came from a simple calculation by Michael Pilson and from a box model by Peng & Broecker (1991), was that even getting all of the nutrients upwelling in the Antarctic converted to organic matter and then sunk would have a very small impact on atmospheric CO_2. Moreover, most of the sequestered carbon would not be at particularly great depths and would come back to rejoin the atmosphere in only 30–40 years. More elaborate modeling studies suggested that sustaining even a modest impact on atmospheric CO_2 would require continued iron additions forever. The project seemed so unpromising that the environmentalist furor around OIF blew itself out for a time.

Starting about 1994, a sequence of commercial OIF schemes has appeared (Strong et al. 2009b). Various companies have formed, then dissolved, prominently including GreenSea Ventures (led by Michael Markels), Planktos (Russ George) and Climos (Dan Whaley and Margaret Leinen). Their schemes have had different details, but all included the hope of making money by selling "carbon credits". GreenSea proposed a tropical addition using slow-release capsules of major nutrients as well as iron. Climos, while still extant, has mostly gone quiet since late 2008. Climos is/was principally interested in the potential of the unutilized major nutrients of the Southern Ocean. Substantive arguments can be made that OIF would not remove sufficient CO_2 from the atmosphere to be worth the bother.

First, for OIF to work significantly, the gain would come almost entirely from the Southern Ocean, the site of the most significant availability of residual (southern autumn) nitrate and phosphate in the euphotic zone (see NOAA–NODC World Ocean Atlas 2009 April–June surface nitrate map at: http://www.nodc.noaa.gov/OC5/WOA09F/pr_woa09f.html). However, the subantarctic sector does not have significant unused amounts of silicic acid (compare the April–June surface silicate map in the World Ocean Atlas 2009) to match its "left over" NO_3^- and PO_4^{3-}. Two iron-ex studies, the more northerly trial of "SoFex" and the 2009 "LohaFex" study have both shown that chlorophyll increases induced by added iron were not due to diatoms and export was minimal. It is essential for OIF to work that it induce diatom blooms that sink mostly unconsumed by grazers. Without silicate, microflagellates increase but do not sink. That pushes the OIF potential south of 55° to 60°S, depending upon latitude. A more difficult and remote ocean area does not exist.

Second, the potential of OIF for CO_2 sequestration must be evaluated by models. Among the best available is that by Zahariev et al. (2008), a global circulation model (GCM) including vertical mixing and incorporating primary production controlled by spatial variation in temperature, nitrogen, mixing depth, and iron limitation (modeled from the spatial distributions of annual minima of residual nitrate). Alleviating all iron limitation in the model (i.e. using all nitrate everywhere, every year) removes 0.9 GtC in the first year, but that drops progressively to only 0.2 GtC yr^{-1} after 30 years of OIF. The reason for the drop is that vertical resupply would not fully replenish the removed nitrate during the seasons of low production. Actually, the estimate is too large by a factor of $\sim$2, since the model neglected the silicate "problem" in the subantarctic. Sequestering 0.1 GtC yr^{-1} by repeated spreading of iron (say as $Fe_2(SO_4)_2$) across the entire sector south of the polar front every year until fossil fuel runs out or can be replaced, does not seem a promising strategy in the face of >8 GtC yr^{-1} of fossil-fuel carbon now passing through the atmosphere.

Third, a remarkable natural iron-ex study initiated on 7–8 August 2008 has now been evaluated. It was the dusting of the oceanic Gulf of Alaska, an HNLC area, by volcanic ash from an eruption of Mt. Kasatochi in the Aleutian Islands. Fortunately, some oceanographic information (Hamme *et al.* 2010) could be developed from satellite records, moorings, and a research cruise in the area. The ash was carried east in an expanding plume by the then prevailing winds (Plate 16.3a) generating a bloom, primarily of diatoms, of about twice the usual late-summer standing stocks (Plate 16.3b). The extent of the bloom was documented by both SeaWIFS and MODIS satellite sensors (Plate 16.3c), and satellite chlorophyll levels were confirmed by ship-borne observations. Ash particles were identified in the Gulf surface from a sample taken at 50°N, 145°W (Station P) on 21 August. Experiments with fresh ash added to seawater do induce both strong iron release and diatom growth. Both pCO_2 and pH recorded at a Station P mooring shifted sharply down and up, respectively, over a week starting about 12 August (Plate 16.3d), and mixed-layer fluorescence from a seaglider operating nearby roughly doubled (Plate 16.3e). Late August fucoxanthin levels were unusually high, indicating that the bloom included diatoms. Alternative hypotheses for this bloom have been plausibly rejected (Hamme *et al.* 2010).

Hamme *et al.* calculated that the Kasatochi-induced drop in pCO_2 (~25 μatm) implies a drawdown of 0.3–0.7 moles C m^{-2}, given the mixed-layer depths. Extrapolating to the area of the bloom (maximally 2 million km²) suggests a maximum likely drawdown of 0.017 GtC of carbon. Based on that, Hamme *et al.*, applying an unlikely high export rate of 50%, suggest that the entire ash-plume effect might have sequestered <0.01 GtC. The SERIES iron-addition study in the same region estimated a 7% export rate from strongly enhanced production. Thus, a realistic sequestration estimate would be <0.001 GtC. The Kasatochi plume was larger than any intentional iron-addition experiment that could be done, and it would be very costly to exceed the modest likely levels of iron it supplied over such a large region. Thus, "Kasatochi" shows that potential for OIF to reduce atmospheric CO_2 is trivial. Projecting the 0.001 GtC sequestered to the portion of the Antarctic seas with substantial silicate, the sequestration might be 0.01 GtC per treatment (calculation by Phillip Boyd). The oceans are already absorbing ~200 times that much annually. OIF is not going to help us. Unfortunately, volcanoes are not going to help much, either.

Decadal-scale changes in ocean conditions and biota

Climate is changing, and it will change in some direction, probably warming, even if we act significantly and soon to

reduce greenhouse-gas emissions. Since temperature affects every chemical and biological reaction, even moderate warming will change, well, everything. To predict the effects on ocean biota, we must extrapolate from the longest time-series data available. Apart from fossil proxies, those series are less than a century long. In general, the expectations from greater warmth begin with more stable water-column stratification. That means fewer phytoplankton nutrients mixed upward, less primary production, and shorter rations all the way up the food chain and all the way down to the benthos. Stratification variation in fact drives a major part of variation in global ocean productivity (Behrenfeld *et al.* 2006; Fig. 16.15). That is because the dominant factor is productivity in the huge equatorial belt. That drops dramatically when the West Pacific warm pool collapses along the equator during El Niño (see below), overriding the productive eastern tropical waters. The warm layer buries nutrient-rich layers more than 100 m deeper, making them inaccessible to upwelling. Thus, the anomaly in global primary production estimated from chlorophyll and temperature data closely tracks both an index (MEI) of El Niño strength and stratification estimates from

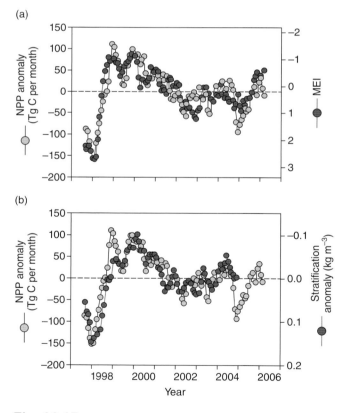

Fig. 16.15 Monthly anomaly of global ocean net primary production (NPP) rate, estimated from satellite chlorophyll (mostly) and temperature data, for 1997 through 2005, compared to (a) the Multivariate ENSO Index (MEI) and (b) global anomaly of stratification (mostly determined by El Niño vs. normal conditions in the equatorial Pacific). (After Behrenfeld *et al.* 2006.)

a data-assimilating model of water-column structure. The enhanced stratification from global warming will operate differently and vary in different modes, but the principle is quite general. Warming increases stratification and reduces upward nutrient mixing, which reduces primary production. Guesses at the magnitude of the effects can be generated by models that consistently carry substantial uncertainties. So, we leave them for others. The responses of marine ecosystems to the longer climate cycles (alternating decades or longer stretches of similar seasonal cycles of weather) may give us a preliminary sense of what is coming.

It has been argued (e.g. Mackas & Beaugrand 2010) that mesozooplankton abundance and species composition are among the more accessible indices of climate response of marine ecosystems. Life cycles are mostly less than a year, hence stocks will be quite responsive to interannual climate changes. Yet they are long enough that modest sampling frequency (say, monthly) can indicate shifts. Some short time-series, like the approximately monthly HOTS zooplankton samplings, show substantial abundance changes. In that subtropical area (Sheridan & Landry 2004) net zooplankton doubled in the eight years from 1994 to 2002 (Fig. 11.32), an increase among small animals remaining near the surface day and night and affecting about equally the summer peaks and winter lows. There was no increase

in night samples of diel migrators. Whether that change is part of a longer cycle or a really long-term change is always an issue with short time-series (and all time-series of direct biological observations are too short).

There are some time-series of population estimates in the sea that extend to decades. Three for zooplankton are the continuous plankton recorder (CPR) survey out of the United Kingdom, which was established about 1952 and fully operational by about 1958, the California Cooperative Oceanic Fisheries Investigations (CalCOFI) survey, which started in 1950, and more than 50 years of collections by Japanese Fishery Agency scientists, the "Odate collection", that are being reanalyzed. The North Atlantic CPR survey pulled a Hardy plankton recorder (Fig. 16.16) from commercial vessels on standard routes, collecting plankton in serial fashion on rolls of gauze. These rolls have been analyzed on a regular basis in a standard way for over 50 years. The program is kept going currently by the private Sir Alister Hardy Foundation for Ocean Science (SAHFOS) located at Plymouth, England. CalCOFI sampled offshore from California on onshore–offshore lines of stations spaced every 40 nautical miles using research vessels and standard ring or bongo nets. Time intervals between CalCOFI surveys have varied from monthly in some early years to quarterly in recent decades, with some longer gaps.

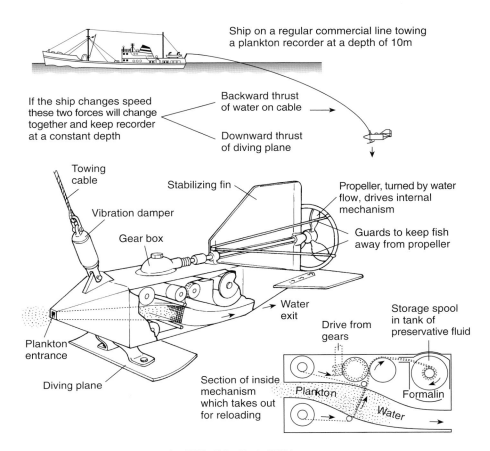

Fig. 16.16 Sketch of a Hardy continuous plankton recorder (CPR). (After Hardy 1970.)

The Japanese collections were 150–0 m vertical hauls with 0.45 m ring-nets with 333 µm mesh, made all around Japan and east to the date line in huge numbers from about 1950 to present. Initially, they were only examined for biomass, a project of Kazuko Odate. Recent re-evaluations have been called "the Odate Project". All three series show long-term changes in stocks of plankton, and consistently the changes follow multi-year sequences of changing wind patterns, ocean temperatures and variations of circulation. The changes are generally termed "oscillations" and character-ized by indices, the North Atlantic Oscillation (NAO), and the Pacific Decadal Oscillation (PDO). There are other named "oscillations".

CPR results: the changing Northeast Atlantic

Early reports from the CPR study were vague, mostly couched in statistical constructs difficult to interpret as specific changes in biology. More-recent work initiated by Frédéric Ibanez and colleagues has clarified things. The situation varies by region (Fig. 16.17). North and east of Britain, the stock of the dominant, large herbivorous copepod, *Calanus finmarchicus*, has varied inversely with the NAO, a measure of wind strength across the North Atlantic (Fig. 16.18a). The NAO is the difference in baro-metric pressure between the Icelandic low and Azores high. When NAO is high, winds pushing the Gulf Stream east are strong. When it is low, the winds are weaker. The whole distribution of wind over the region changes between high NAO and low NAO (Fig. 16.18b & c). In the "low" pattern, flow into northern European waters is from farther north, the locus of the *C. finmarchicus* stock, and is colder and weaker. In the "high" pattern, flow is from farther south and carries more and warmer waters toward Scotland and Norway. The NAO moved from the 1960s to 1990s toward stronger NAO, toward a warmer Europe. The zooplankton along the recorder routes in the northern North Sea and northwest of Scotland became more and more constituted of species with southern affinities. Particularly, in the North Sea, abundance changes of *C. finmarchicus* (northern affini-ties) and *Calanus helgolandicus* (southern) followed those changes in the NAO both in the mean, and with striking interannual detail (Fig. 16.19). So, the plankton respond, partially via advective effects, to the changes in the global atmospheric circulation. In the 2000s the NAO trend leveled, and at the time of writing (2011) has plunged to a low negative value. The published CPR data have not caught up with this change. The dominance of *C. helgolan-dicus* in all of the North Sea persisted through at least 2005 (Fig. 16.20).

We understand these shifts in moderate detail. There is a theory, mostly due to the Scottish worker Michael Heath, for this effect on *Calanus*. The main bulk of the eastern

Atlantic *C. finmarchicus* stock is in the Norwegian Sea. In periods of low NAO, there is more deep overflow in winter from the Norwegian Sea southward across the Iceland–Scotland ridge. This flow, driven by deep-water formation, travels along a narrow, deep channel west of the Faroes Islands. It carries with it the resting stage of *C. finmarchicus*. They mature in February–March and rise to the surface to spawn. Some are then carried by surface flow into the northern North Sea, the CPR data region with the strongest inverse relation of *Calanus* stock to the NAO. The sequence goes:

Low NAO → more deep transport → more mothers → more *C. finmarchicus*.

High NAO → less deep transport → fewer mothers → less *C. finmarchicus*.

Each effect works in the opposite direction on the abun-dance of *C. helgolandicus*, the species found in warmer waters to the south and east. High NAO corresponds with more northward transport of not only *C. helgolandicus*, but the whole eastern boundary current plankton community from off northern Africa and from the Mediterranean shifts northward, raising diversity and reducing mean body size. The shift from one regime to the other is likely to have knock-on effects up the food chain. *Calanus finmarchicus* stocks peak in spring when cod (*Gadus morhua*) spawn and larvae depend upon its nauplii for food, and the juveniles upon the later copepodites. *Calanus helgolandicus* peaks in mid- to late summer, too late to nourish cod reproduction. Euphausiids and the small copepod *Pseudocalanus* are also important in larval–juvenile cod nutrition. By considerably obscure arithmetic, Beaugrand and Kirby (2009) generated from North Sea CPR data an index of availability from March to September of planktonic cod food and compared it to recruitment of "age 1" cod (Fig. 16.21). While the diversity of zooplankton, particularly copepod species, went up after about 1983, the likely food for juvenile cod went down and so did cod recruitment. The pulses of North Sea cod recruitment centered at 1963 and 1978 are termed the "Gadoid outbreak", which possibly ended due to the shift in plankton abundance, species, and seasonal timing.

Along the coasts of Labrador, Nova Scotia, and New England there are actually cyclic shifts in the opposite direction to those in the northeast Atlantic. As *C. helgolan-dicus* moves north, *C. finmarchicus* and even arctic zoo-plankton species tend to shift south and west. This is connected to greater advection out of the Labrador Sea, caused by the same shifts in North Atlantic low- and high-pressure centers recorded by the NAO. The changes are less dramatic, but definite, and are spelled out by Greene and Pershing (2000).

The North Sea and general North Atlantic (Hátun *et al.* 2009) plankton changes are typical of the broad coastal

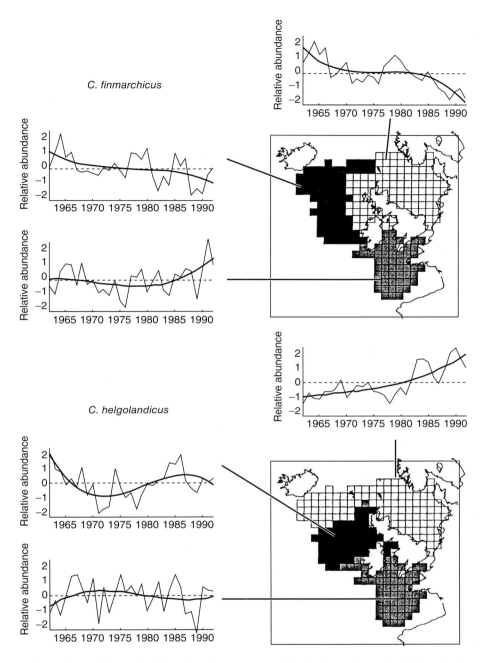

Fig. 16.17 Variation of *Calanus finmarchicus* and *Calanus helgolandicus* abundance in the North Sea, Norwegian Bight (west of southern Norway) and Faeroes–Iceland sector 1958–1996. (After Planque & Ibanez 1997.)

current systems. They involve both changes in conditions locally and changes in the plankton delivered by the prevailing currents. To learn whether the biological changes will prove in the very long term to be predictors of persistent climate changes, we will have to wait. Possibly, higher-latitude plankton will adapt, shifting the mobile, oscillating patterns back to cycles similar to those of available time-series. Possibly, high latitudes will be permanently occupied by low-temperate and even tropical species.

CalCOFI: shifts in the California Current

Plankton samples from the CalCOFI series were rarely counted in much specific detail, apart from the fish larvae. A few years were thoroughly studied, but that was all that the available funding and human energy allowed. However, total zooplankton displacement volume data (drain the sample on a fine mesh, resuspend in a measured volume of

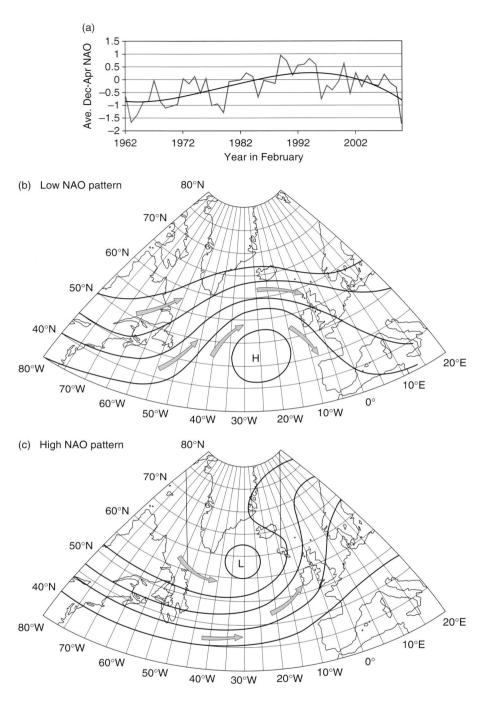

Fig. 16.18 (a) Time-series of the North Atlantic Oscillation (NAO) index estimates for 1950 to 2000. NAO is the atmospheric pressure difference between Iceland and Portugal, plotted here as anomalies. (From NOAA.) General effects of low NAO (b) and high NAO (c) on ocean current patterns in the North Atlantic. (After Mann & Lazier 1991.)

water and measure again) were scrupulously kept. John McGowan and colleagues (e.g. Roemmich & McGowan 1995a, b) have provided us with analyses, and Lavaniegos and Ohman (2007) have offered some reinterpretations. The most complete time-series have been sustained in the area offshore of the Southern California Bight (Fig. 16.22a)

Displacement volume of oblique plankton hauls fell off progressively from approximately 1976 until 1998 (Fig. 16.22b), reaching ~25% of the long-term mean, then sharply rebounded to near the pre-1976 mean in 1999. Of course, emphasizing that trend skips across the strong pulse in 1984–1986 following the extreme 1983 El Niño.

Nevertheless, the decreasing trend corresponded to a sharp upward step in mean temperature (Fig. 16.22c) and a shift in the PDO (Fig. 16.22d). Unlike the NAO, the PDO is derived directly from Pacific temperature fields north of 20°N as the first principal component of the spatial variation (Mantua *et al.* 1997). Many biological variables across that region are correlated with it. Of course, the correlations need not arise from direct effects of temperature. Atmospheric pressure indices like the NAO exist for the Pacific and also correlate at interannual time-scales with the PDO. McGowan *et al.* (2003) have suggested that the effect of the temperature increase on California Current zooplankton was mediated by increased stratification, reducing upward transfer of nutrients by upwelling and mixing. A contingency comparison in color (Plate 16.4) makes their point by visual inspection.

Rebstock (2001a & b) reanalyzed a selected set of southern California CalCOFI samples from 49 years, estimating the abundance of dominant copepod species. She found that species composition did not vary in the long term, although there were short-term composition shifts during six years; three of those were El Niño years. However, the abundance of dominant forms in her data, and thus copepods overall, did decline in correlation with the Roemmich–McGowan pattern, especially in the prolonged period of high ocean temperatures in the early 1990s. The same array of zooplankton species appeared to have sustained substantially smaller stocks.

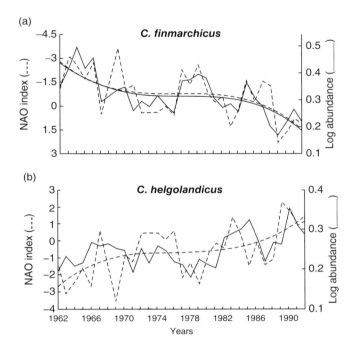

Fig. 16.19 Regional and annual average CPR abundances of (a) *C. finmarchicus* and (b) *C. helgolandicus* plotted with the NAO index (NAO scale reversed for *C. finmarchicus*), 1962–1992. Curves are fitted cubic polynomials. (After Fromentin & Planque 1996.)

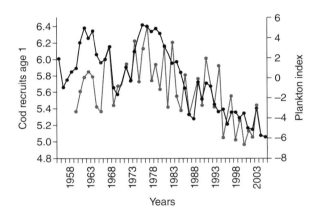

Fig. 16.21 Long-term changes in North Sea zooplankton (1958–2007, upper series at the left), compared to rate of cod recruitment at age 1 lagged one year (1963–2007). (After Beaugrand and Kirby 2010.)

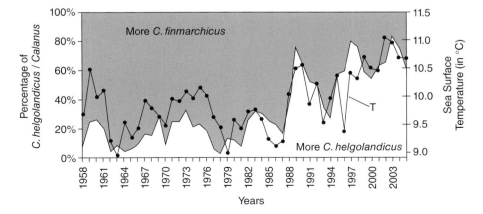

Fig. 16.20 Progressive shift in the relative abundance of *Calanus finmarchicus* and *Calanus helgolandicus* (*C. fin*/total, %) in North Sea CPR surveys compared to change in sea-surface temperature (line joining dots), 1958–2005. (After Beaugrand 2009.)

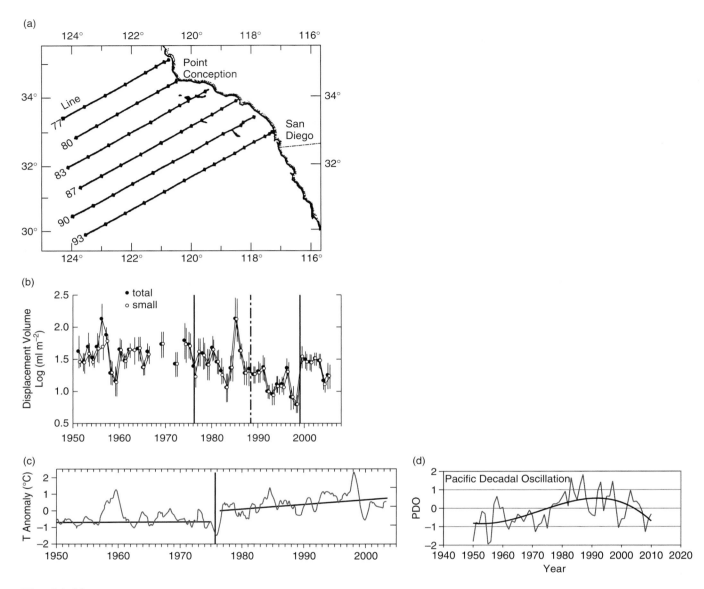

Fig. 16.22 (a) Location of CalCOFI sampling lines in the southern California (SoCal) region (Point Conception to San Diego and seaward to 124°W). (b) Time-series of zooplankton displacement volume estimates from CalCOFI samples in the SoCal region. Total volume (●); volume of animals <5 cm long (○); vertical double lines are standard errors; vertical bars are suggested times of regime shifts, less certain, especially for the California Current in 1989 (dot–dash). (Graph by M. Ohman.) (c) Sea-surface temperature time-series from SoCal, showing the step-like increase of the 1970s. (d) Time-series of annual means of the Pacific Decadal Oscillation index. (b & c from Mackas & Beaugrand 2010; d plotted from tables maintained by Nathan Mantua: http://jisao.washington.edu/pdo/PDO.latest.)

Lavaniegos and Ohman (2007) again revisited samples from the Bight, just those from one spring cruise each year. They combined splits of all night samples from the mid-spring cruises of 45 springs, subsampled that mixture, then counted and measured plankton groups. After converting to carbon biomass by length–mass ratios, they concluded that the only statistically significant drop in abundance was due to salps and doliolids. Possibly, but most of the detailed features of the displacement-volume time-series recur in their total carbon estimates and those for copepods,

euphausiids, and chaetognaths. The year-to-year troughs and spikes mostly correspond, respectively to warm El Niño years with advection of southern plankton, and cold years with more extended advection from the north. The nature and persistence of the stock rebound in the years after 1999 remains to be evaluated.

Shorter time-series of nearshore zooplankton samples from Oregon and Vancouver Island show both faunal and abundance shifts correlated with the PDO, even the PDO variations within single years (Mackas *et al.* 2004). At these more northerly sites, the shifts are caused by changes in the oscillation between southern zooplankton species in winter, northern species in summer.

The Odate project

Chiba *et al.* (2006) report the results from a taxonomic analysis of the Odate collection. A 1433 sample subset of the Odate collection, and some other collections from similar net hauls, was selected representing the Oyashio (sites deeper than 500 m in the vicinity of 42°N and colder than 5°C at 100 m) in months from February through October of each year from 1960 to 2002. The wet-weight biomass (Fig. 16.23) of the whole samples varied by a factor of approximately three almost exactly at the periodicity of the PDO. Copepods were identified and 59 of the more abundant species were evaluated statistically. The species divide (by cluster analysis) into five seasonal "communities": species peaking in April (including *Neocalanus cristatus* and *N. flemingeri*), May–July (*N. plumchrus, Eucalanus bungii, Oithona similis, Pseudocalanus minutus*), July, September, and October. The last three groups have warm-water associations and are likely to represent the frequent summer–fall overwashing of the Oyashio by Kuroshio loops

and rings. Chiba *et al.* characterize the shifts of faunal composition as follows (paraphrased):

The abundance of the spring community gradually increased from 1960 to 2002. Changes of life-history timing coincided with the climate regime shift in the mid-1970s, indicated by the PDO. After the regime shift, the timing of peak spring-community abundance was delayed by one month, from March–April to April–May, whereas the spring–summer community peaked earlier. That resulted in an overlap of the highest biomass period for the two communities in May. Wintertime cooling followed by rapid summertime warming after the mid-1970s was likely responsible for delayed initiation and early termination of the productive season for both groups. Then in the mid-1990s the PDO and phenology shifted back. Warm winters followed by cool summers lengthened the productive season and again separated the life history patterns of the two groups.

Chiba *et al.* suggest that decadal climatic cycles may affect winter–spring and spring–summer differently, such that the combined changes determine annual productivity. Clearly the period of colder winters and rapid summer warming was less conducive to overall biomass production. Exactly

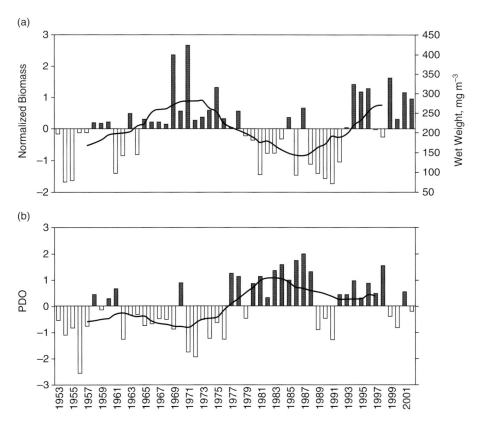

Fig. 16.23 (a) Time-series for the Oyashio region at depths >500 m east of Hokkaido of mean May to July zooplankton wet weight. Bar tips are at observed values in mg m^{-3}, right scale; tips are rescaled as anomalies in standard deviation, left scale). Data are from the Odate collection. (b) The Pacific Decadal Oscillation. The solid lines indicate 10-year running means. Zooplankton biomass is low when the PDO is high and inversely. (After Chiba *et al.* 2006.)

how copepod populations would be affected by the interactions of temperature, flow, illumination, nutrient supply and predators that control the spring-bloom food chain where the Oyashio flows over the slope remains to be fully resolved.

General points about decadal cycling

The equation between ecosystem responses to decadal climate variations and the likely responses to general, long-term climate warming has been made repeatedly (e.g. Richardson 2008). For instance, the distribution and abundance shifts correlated with NAO and PDO occur more rapidly than the latitudinal shifts occurring in the terrestrial realm. Marine plankton shift hundreds of kilometers per year, while land plants and insects move poleward at <10 km yr^{-1}. Clearly the fluid habitat generates advection not operating on land, i.e. advective reach that reverses often and completely. Once the ocean warms overall, more and more of the extended species distribution shifts in the sea are likely to become permanent. Conceivably, some species will be squeezed poleward into rather refugial bands or pockets. Species (like the cod of the North Sea) that do not move are likely to find changed timing of their food resources, and will have to cope with that in the same area where they must compete with invaders at their own trophic level from previously more temperate climes. Accepting all that, the equation is imperfect; the best we can do is leave the best possible time-series data for future generations to determine how ocean ecology varies over time.

Such data are in short supply. We need time-series from all over the oceans so as to get a sense of the global health of marine pelagic ecosystems. There are signs that we will start more such series for our grandchildren to study. The work of collecting, analyzing, and making ongoing scientific sense of such surveys is substantial, and few are willing to undertake it. The CPR program on lines extending in all directions from Britain appears to be stably sustained by SAHFOS. Sonja Batten and SAHFOS (e.g. Batten & Freeland 2007) have initiated a new CPR line from Alaska to the Strait of Juan de Fuca and one from Canada to Japan, which should contribute time-series if there is institutional will to keep them going. There are other short or young zooplankton series here and there, and they recently have been examined (Batchelder *et al.* 2011). The gradually developing Global Ocean Observing System (GOOS) does not promise to include much biology, apart from nutrients, fluorescence, and possibly spectral light absorbance, but it should improve our knowledge of water-column and surface-weather variation at a few points. The ARGOS program, if it can be sustained, will provide volumetric maps of hydrography and possibly variables related to phytoplankton. Thus, there is hope that we will know in some detail what happens to oceans as the climate changes.

Phenology effects

Phenology is the study of the mechanisms that tune an organism's life-history timing to the cycling, generally seasonal cycling, of its ecosystem. The term is also used in the form, "a butterfly's phenology", meaning its particular life-cycle adaptation to the timing of habitat variation. One of the major features of climate change is shifting of seasonal events, in particular winters end earlier with climate warming, springs start earlier, so peak primary production comes earlier and growing seasons can last longer if nutrients are continuously supplied. Peak seasons for 56 of 86 phytoplankton- and zooplankton-species populations recurrently estimated by the CPR survey in the northeast Atlantic have now shifted one to six weeks earlier than in the late 1950s and early 1960s:

> "The general pattern observed for taxa that peak when the water column is mixed or in a transitional state is to show considerable variability in phenology, whereas taxa associated with low turbulent conditions have virtually all advanced in their seasonality (34 out of 37 taxa between May–August)." (Edwards & Richardson 2004).

That suggests that the date of stratification is a key seasonal event. Oddly, peaks of a substantial fraction (26 of the 66 evaluated) shifted to later dates. Marine life will not be wiped out by global warming (or acidification, either, see below), but substantial shifts in the biota and rapid evolution of many populations are likely.

Meanwhile, both plants and animals often use cues more fixed to the calendar, particularly daylength variation, to time the onset and emergence from resting (or diapause) stages, to meet for mating and spawning, to lay up lipid for diapause (all key aspects of phenology). Diapause stages can also require a sufficient duration of cold (or warm) conditions before emergence is enabled, such that significant temperature change can prevent emergence. Thus, rapid shifts in the timing or temperature of the periods of optimum or unsuitable habitat conditions will challenge the evolutionary rates at which phenology can be shifted.

On the other hand, life-history controls vary markedly within populations and thus are subject to rapid selective modification. Probably there is a very long-term population-level payoff for sustaining substantial phenological variability (that is, for "bet hedging"). At least one planktonic copepod appears to be preadapted to respond to shifting timing of the annual production peak. *Neocalanus plumchrus* in the subarctic Pacific enters rest in late spring–early summer, then, within a few months, begins to "meter out" maturation from its resting stage at depth. Males and

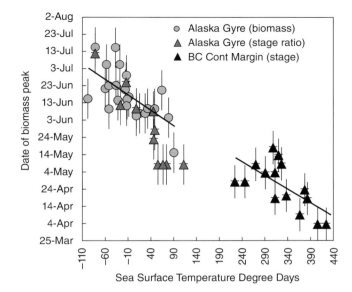

Fig. 16.24 Date of the seasonal maximum of the *Neocalanus plumchrus* population in the northeast Pacific vs. cumulative warmth of the upper ocean during the March–May growing season as "degree days" between when spring warming passes 6°C and the peak. Cumulative degree days depend mostly upon the starting date for spring warming. Peak dates have varied between early April and early July (by 3 months). Stage ratio is the number of fifth copepodites relative to the total of all stages; it corresponds closely to the peak of stock biomass. "Alaska gyre" includes samples from 50°N, 145°W and to the east but well offshore. (After Mackas *et al.* 2007.)

females mature sequentially over the time from about September to February, reproducing at depth over that long period (Miller & Clemons 1988). The nauplii do not need to feed, and they rise to the mixed layer as they develop. Arriving, they may or may not find sufficient food for copepodite development. However, those arriving when the seasonal peak of production is in progress grow and produce the following year's resting stock. Mackas *et al.* (2007) have shown that the date of peak biomass and prevalence of the late copepodite stages ready for diapause varies with the rate of warming of the upper water column from a baseline (6°C) reached every year (Fig. 16.24). Early warming can produce an early April peak; late warming delays development success into July. The prevalence of similarly effective bet-hedging strategies among marine fauna is unknown.

Lessons from El Niño

Similar help with predicting the responses of ocean biota to climate change comes from the response to shorter events than NAO or PDO shifts, events that are part of normal climate variability. Among the most informative are responses to the El Niño–Southern Oscillation cycle,

already considered briefly in regard to its effects on plankton distribution patterns (Chapter 10). El Niño, Spanish for "the boy", gets its name from Christmas, the usual season of its appearance off Peru. The eastern tropical Pacific Ocean warms during these events. Upwelling may or may not cease, but when it continues, fewer nutrients are brought to the surface. There are strong effects on all parts of the ecosystem: phytoplankton, zooplankton, the anchoveta stock, guano birds, and global fish-meal markets.

"Southern Oscillation" refers to the variable atmospheric pressure difference (Fig. 16.25a) along the equator in the Pacific, usually measured between Easter Island or Tahiti and Darwin, Australia. This difference, normally high pressure in the east, low in the west, is a measure of the driving force for the east-to-west trade winds. These persistent winds (20–30 knots night and day, year around) push ocean water west in the equatorial zone, generating an upward slope in sea level from east to west. It is sun-warmed surface waters that move, so the layer of warm water in the west becomes deep, and isopleths of temperature, density, and most other variables slope upward (relative to the surface) to the east. "Permanent" thermoclines are at 350–400 m in the west, only 100–120 m in the east. Equatorial upwelling in the east can thus reach down to rich subsurface layers, making the eastern tropical Pacific a nutrient-rich (apart from iron limitation), relatively high production zone. The western "warm pool" is as oligotrophic as a central gyre. This is the "normal" condition.

Every 3 to 5 years, the balance changes, signaled months in advance by a decrease in the Easter–Darwin pressure difference (Quinn *et al.* 1987). The trade winds subside, the sloping sea surface begins to level, and warm water in the west flows east, beginning an El Niño. The Coriolis effect pushes the flow toward the equator, and it becomes a Kelvin wave traveling east along the equatorial wave guide. It arrives in the eastern tropical Pacific, raising sea level and pushing down the cold layers from which upwelling normally supplies nutrients to the surface. Then flow divides to north and south, continuing along the coasts and held inshore by the Coriolis effect. Actual water displacement can be a thousand kilometers, the displaced mass pushing along water downstream from it and thus warming the coastal ocean as far as Alaska in the north, central Chile in the south. El Niño events vary in intensity. More or less water can move into the east, coupled with extended or shorter poleward effects. Some recent events have been among the most dramatic: 1972, 1983, 1997–1998.

The "double" El Niño of 1982–1983 was notable for the rapid succession of two events, the later one occurring about 6 months from the usual timing with peaks in sea-surface temperature (SST) and sea level in Peru during May–June, not December (Huyer *et al.* 1987). The second wave of that strong event reached far along the North American coast with striking effects. However, strong events have been noted over several centuries in colonial

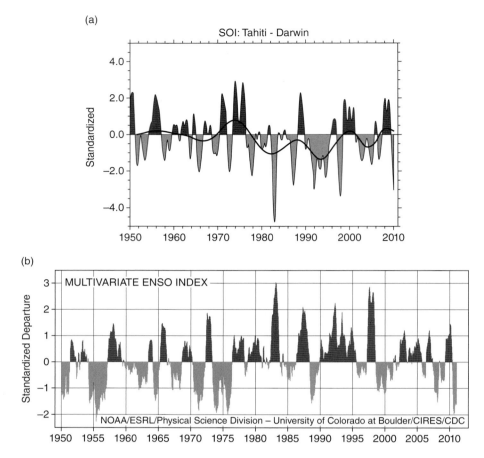

Fig. 16.25 Comparison of (a) a Southern Oscillation Index (SOI, anomaly of a standardized atmospheric pressure difference between Tahiti and Darwin, Australia, that is filtered to remove oscillations shorter than 8 months) and (b) a multivariate ENSO index (MEI). The SOI is from www.cgd.ucar.edu/cas/catalog/climind/soi.html. The smoothing line is a decadal running average. The MEI is based on six variables across the tropical Pacific: sea-level pressure; zonal and meridional components of the surface wind; sea-surface temperature; surface air temperature; and total cloudiness fraction. The MEI has an arbitrary scale, high when SOI is low. It is a NOAA product generated by Klaus Wolter (www.esrl.noaa.gov/psd/people/klaus.wolter/MEI/).

records. Those and other records along the tropical belt, particularly records since 622 AD of the height of the Nile River, measured in adjacent wells of that age (Quinn 1992), show that phenomena related to the Southern Oscillation are long-term features of tropical zone climate. El Niño–Southern Oscillation (ENSO) events correlate with hydrological cycles over much of the Earth. Evaporation from the western Pacific warm pool supplies rainfall over Australia, Indonesia, India, and distantly as far as the steppes of Russia. The El Niño condition reduces this evaporation and can generate droughts of varying severity. The opposite can occur around the eastern Pacific, where El Niño can bring unusually strong low- to mid-latitude rains and flooding.

An El Niño ends when the trade winds resume and slowly push warm surface layers west, raising the eastern Pacific thermocline, allowing upwelling to come from

nutrient-rich layers again. Non-El Niño (in the colder instances called "La Niña") conditions are usually re-established 18–24 months after El Niño onset. This was violated in the early 1990s when the Southern Oscillation Index remained negative (pressure higher at Darwin than Tahiti; Fig. 16.25a), and eastern tropical SST was above average for a half decade. This protracted run of warmth in the eastern Pacific, coupled with a very strong El Niño in 1997–1998 suggests to some that global warming may be increasing the frequency, duration, and intensity of El Niños. But, again, data running back several hundred years do show similar series of very strong and prolonged El Niños in the past.

Impacts of El Niño on marine biota are strong. Those involving plankton, which have short lifetimes and high population growth potential, come and go rapidly with the events. Effects on larger, longer-lived animals, particularly seabirds, involve substantial mortality and long population recovery times. Weakening of the trade winds reduces equatorial upwelling, and deepening of the thermocline and nutricline in the eastern tropical Pacific reduces nutrient supply to the surface from the upwelling that does occur. General effects of El Niños on primary production and higher trophic levels are discussed starting on page 265 of Chapter 11.

El Niño transport along the coast displaces stocks of planktonic algae and animals; they just move poleward with the flow. Latitudinal boundaries of tropical species shift poleward, and corresponding boundaries of subpolar forms retreat upstream in the California Current (Fig. 10.25) and Peru Current. Ochoa and Gómez (1987) found from massive and recurring sampling campaigns before and during the 1982–1983 El Niño that *Protoperidinium obtusum*, a subantarctic species normally dominant all along the Peru coast, was completely replaced by the equatorial species *Ceratium breve*, a shift of 16° of latitude for both. Such complementary shifts of 10° of latitude or farther are commonly associated with strong El Niños. In strong El Niños, summer conditions at a given latitude will match those usual farther equatorward so eastern tropical Pacific and subtropical offshore species displaced poleward will reproduce and their young will grow. They never produce the typical biomass of these eastern boundary currents. However, a single season of normal conditions completely reverses the situation to subpolar faunas with higher biomass. Stocks are partly rebuilt by transport, partly by growth and reproduction along the way. The response of plankton occurs on the same time scales as the El Niño cycle without lasting effects.

Effects on nekton are partly parallel to those of plankton, partly longer lasting, depending upon behavior. During strong El Niño events, subtropical fish like yellowfin tuna with a usual northern limit south of San Diego move northward following the warming, generating landings at Oregon ports. Subtropical species like *Mola mola*, the ocean sunfish, are seen off British Columbia or even Alaska (Pearcy & Schoener 1987). When colder conditions resume, they disappear. Whether these fish die or move south as conditions cool is uncertain, but a return swim is very likely. Coastal anadromous fish like salmon and steelhead trout remain mostly in their usual migratory patterns during El Niños. Smolts of both chinook and coho salmon entering the ocean from coastal rivers between northern California and British Columbia during El Niño conditions suffer strong reductions in survival and growth due to the warmth and low plankton availability.

Effects of El Niño on long-lived, relatively sessile fishes like the rockfish of the shelf (*Sebastes* spp.) are short-term losses in growth potential. It is certain that most older fish weather through El Niño periods, in fact they must do it many times during lives spanning multiple decades.

Recruitment from larval stages can be reduced by warmth and starvation, or possibly favored by unusual retention of larvae inshore (Yoklavich *et al.* 1996). Of course, survival of such fish populations never depends upon successful reproduction in every year. Effects on long-lived benthos are similar, they survive but growth slows. For example, red abalone tagged at about 3 years of age (50–100 mm diameter) around Santa Rosa Island in the Southern California Bight grew by 37 mm at age 5 years in the 1978–1980 normal interval. In the 1981–1983 interval with a strong El Niño, they grew only 30 mm, a modest but significant effect (Haaker *et al.* 1998). There was no measurable difference in mortality rate. Again, relatively sessile animals like red abalone that live 25+ years must be adapted to survive multiple El Niños. Farmed oysters in Washington state showed record lows in the ratio of meat to shell volume during the 1982–1983 El Niño (Schoener & Tufts 1987), exacerbating a longer-term decline in body condition initiated in 1976 coincident with the general shift to warmer conditions. Effects on short-lived and infaunal benthos are little studied.

At the top of the food chain, both seabirds and marine mammals suffer very large mortality during strong El Niños. Persistent counting of dead birds along Oregon beaches shows sharp increases in mortality during warm, low-production periods. Guano birds off Ecuador and Peru (guanay cormorant, Peruvian booby, tropical penguins) can suffer massive mortalities, and flocks can be reduced 60% or more in a single season when anchoveta have moved seaward and deeper than usual during El Niño. It can take several cycles of normal conditions and mild El Niños for stocks of these birds to recover. Seals do not simply shift poleward during El Niños, because of fidelity to particular breeding and pupping sites along the coast. Seal pups off southern California suffer high mortality during warm spells, and growth of larger juveniles is sharply curtailed. El Niño-induced scarcity of fish can render seals so malnourished that they cannot deposit enamel in the layers added annually to their teeth. Thus, El Niño years are recorded in their teeth as abnormally thick dentin deposits.

It can be predicted, from the effects of El Niño on coastal biota, that longer-term warming will produce more permanent shifts in the latitudinal limits of plankton and nekton species. Tropical species will be displaced poleward; so will temperate and boreal species. Probably the warming (accelerating metabolism) and enhanced water-column stability due to warming (reducing nutrient supply from depth) will reduce production over a wider latitudinal range. The richer, more productive temperate-boreal communities will have smaller latitudinal ranges, reducing global marine production overall. Less production means less fish, which means less fishery production. While older, more resident animals like rockfish may be protected from warming by their deep, cool habitat, their young usually must survive a period at the surface. Warming and reduced production will reduce that survival, or shift the zone of larval success poleward. Recent losses of Atlantic cod from sites at the warmer limits of its range, like Georges Bank and the Grand Banks, while at least partly caused by excessive fishing, may become permanent due to ocean warming. Just the loss of a few such cod populations caused dramatic economic and social displacements along the New England and Canadian coasts. If and as warming continues, we can expect more such disruptions to our interaction with marine life.

Ocean acidification

$$CO_2(aq) + H_2O \rightleftharpoons H_2CO_3 \rightleftharpoons H^+ + HCO_3^-$$

$$H^+ + CO_3^{2-} \rightleftharpoons HCO_3^-$$

Unfortunately, this is something more to worry about (see for example, Vol. 22 of *Oceanography*, a 2009 special issue on ocean acidification). Carbon dioxide newly released into the atmosphere dissolves in the ocean and reversibly combines with water, becoming carbonic acid. That dissociates, adding hydrogen ions and bicarbonate to the water chemistry. The hydrogen ions (dashed line in the reaction equations) reversibly combine with free carbonate to generate more bicarbonate. The dissociation constants favor bicarbonate, both raising the $[H^+]$ (lowering pH) and reducing the concentration of free carbonate. The second reaction is both temperature and pressure dependent: more carbonate remains at warmer temperatures, less at cold, and greater pressure pushes the reaction toward bicarbonate. Thus, descending in the ocean (colder, more pressure) carbonate concentration goes down, so carbonate minerals, particularly calcium carbonate, tend to dissolve. As the ocean absorbs more CO_2, the depths at which this effect is pronounced will rise.

Greater acidity will have significant effects on the rather open physiology of marine organisms with exchange membranes exposed to seawater, from phytoplankton and protozoa to fish. At approximately three-fold pre-industrial CO_2 levels, pH will drop to ~7.8 (Feely *et al.* 2009). The present level is already down by about 0.05 pH units globally, but more in some coastal areas, and the general effects are not well characterized or mostly even identified.

There is potential that increased bicarbonate concentration will moderately stimulate phytoplankton photosynthesis and growth, while also shifting carbon physiology. For example, Wu *et al.* (2010) report culture studies of the diatom *Phaeodactylum tricornutum* adapted through many generations to atmospheric CO_2 in culture headspace (pH 8.12) and to three-fold pre-industrial CO_2 (pH 7.8).

Carbon uptake rate of the high-CO_2 culture was 12% greater, but dark respiration was 34% greater as well, so growth only increased by 5%. There was a marked effect on the carbon-uptake system at high CO_2; its K_s value increased 20% (reduced affinity). Similar effects have been found for prymnesiophytes, *Prochlorococcus*, and others, but not all species show these modest increases, and the likely impacts of persistently high DIC and low pH remain obscure.

Carbonate biominerals have received the most attention and concern. Important phytoplankton (coccolithophores), some coralline algae, and animals (foraminifera, pteropods, clams, corals, echinoderms, . . .) secrete their shells, support skeletons, spines, and spicules from $CaCO_3$. There are two mineral forms of $CaCO_3$, calcite (foraminifera, clams, echinoderms) and aragonite (pteropods, corals), and they have different temperature and pressure responses. Aragonite becomes soluble at greater carbonate concentrations than calcite and is susceptible to dissolution at somewhat warmer temperatures and shoaler depths. The levels of DIC accumulated at depth, and deep layer $[H^+]$ are less in the Atlantic, more in the Pacific, and thus shell dissolution occurs at lesser depths in the Pacific. That largely determines the difference in sediment composition: lots of $CaCO_3$ in the Atlantic, even pteropod oozes, mostly opal (diatoms, radiolarians) in the Pacific (the Pacific is also mostly deeper). The levels at which aragonite and calcite readily dissolve are termed their *compensation depths*.

Much remains to be learned about mineralization in marine animals, but their shell-forming organs typically take up calcium and carbonate by active transport and lay down mineral. Some forms can sustain shell structure despite less-basic conditions than "normal" seawater, pH ~8.1, either by keeping actively secretory tissue near the mineral surface or by adding impermeable organic coatings. Other animals, particularly aragonitic forms lose shell even while alive at pH values around 7.6–7.8. A plethora of experimental research is now being reported on the responses of particular species to high levels of CO_2 in aquarium headspaces, or simply to acid added to their water. An early suite of results is reviewed by Doney *et al.* (2009). Calcification is reduced in some coccolithophores but not others, reduced in planktonic foraminifera, mollusks, echinoderms, tropical corals, and red algae, even in formation of the calcium sulfate hexahydrate (no carbonate at all) statoliths of larval jellyfish (Winans & Purcell 2010). Particularly strong effects are observed in larvae such as initial "D-shell" formation in clams and oysters, spicule formation in sea cucumber and brittle-star larvae. Shells of adult pteropods exposed to modestly reduced pH begin to lose mass quickly, become friable, and break off at the edges. Coastal waters can be more heavily impacted than oceanic areas, which is not only an anthropogenic effect. Water upwelling from great depths has substantially lower pH from long-term accumulation of carbonate species. Effects are stronger for high-latitude, cold-water forms, as expected, with attribution of failed spicule formation in antarctic echinoderms to low pH. Oyster hatcheries in Oregon and Washington States (northwest USA) now buffer seawater to a higher pH in their culture systems during the upwelling season in order to sustain D-shell formation.

A closing note

As you contemplate all of this and study climate-change effects in other sources, it is well to keep in mind how very strong the effects of very modest climatic changes can be. Just one example is offered here. Brander (1997) shows (Fig. 16.26) the effect of water-temperature variation on two cod stocks, one off West Greenland, one off the Faroe Islands. The warmer it is in these cold locales (up to about 11°C), the faster cod can grow. The close tracking of size to temperature implies a causal link of some sort, and differences of only 1.0 to 1.5°C in temperature produce a *two-fold* change in the size of 4-year-old cod. Getting bigger cod faster seems to be a good thing. Remember, however, that apparently desirable changes near the colder or warmer limit of a species are likely matched by undesirable changes at the other end of its range. More on the climate-change effects on fish, squid, and fisheries will be found in the next chapter.

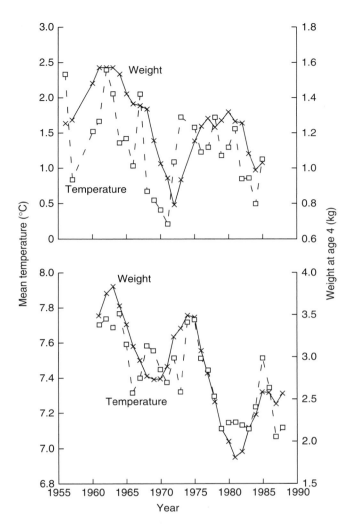

Fig. 16.26 Mean annual temperatures (squares and dashed lines) and mean weight (crosses and solid lines) of 4-year-old-cod landed at (a) West Greenland, and (b) the Faeroes Islands. (After Brander 1997.)

Chapter 17

Fisheries oceanography

Our goal here is to provide an initial look at the mental machinery of fishery biologists. Studies of fisheries are a major part of the practical side of ocean science; so all oceanographers should know something of the ideas in this neighboring field. We will consider the population units ("stocks") that fishery biologists try to discern and evaluate; review the simplest models of stock dynamics and the interaction of fishery production with economics; and briefly examine the overall status of world fisheries. We can only touch the surface of this deep, complex subject.

To understand the data supplied in what follows, you must have mental pictures of a metric tonne (SI unit symbol is "t") and of a *million metric tonnes* (Mt) of fish, crabs, or squid. A million tonnes is also 10^{12} grams (1 teragram = 1 Mt), if that's useful. One metric tonne is the mass of water it would take to fill a cubical box with sides of 1 m. That is about the volume surrounded by a small desk. A metric tonne of fish, then, is an amount that might all be tossed into one large fish box and hoisted from a fishing boat onto a processing plant dock. It is an amount you can visualize. One Mt of fish can be visualized in a similar way. It is about the amount of fish it would take to fill a box sitting over two, side-by-side soccer fields (each 100×50 m) to a height of 100 m. That amount of fish would fill one of the very largest international soccer stadia or the Los Angeles Memorial Coliseum to the top row of seats. In the late 1990s, the world catch of wild fish and other marine animals peaked at about 90 Mt, about 90 stadia full of fish, crabs, shrimp, oysters, and squid. Picture a row of these huge fish hoppers, side by side for about 18 km. It is not an inconceivable amount, just a great deal. The total production of all "fish" is rapidly increasing due to advancing mariculture and aquaculture, which are not always well separated in available data. They bring the sum up to ~160 Mt.

Stocks or "unit" stocks

Some critics (e.g. Gauldie 1991) have claimed that the notion of stocks is outdated; that the impact on all stocks of harvesting one of them makes separating and distinguishing them risky; and that distinguishing stocks without considering their interaction with the economics of fisheries isolates them in an undesirable way. On the other hand, some fishery scientists are thinking that substantially more refined identification of stocks is in order (Hauser & Carvalho 2008). It is likely that the advantages of studying animals with close attention to their species and population relationships transfer completely to fisheries work. It is not fruitful to lump together animals that grow fast with others that grow slowly, or animals which spawn once with those reproducing repeatedly, because each will have different demands for management. That is true even if it is unavoidable that we catch them in the same nets. If fishing for one impacts the other, we just have to account for that. It remains useful to analyze the interaction of fishing with stocks separately from the interaction of fisheries with their markets, and then to tie the two together. Nevertheless, the unit stock concept is often rather wishfully applied, and exploitable organisms grouped together for fishery management are often not the neat interbreeding subpopulations defined in the ideal case.

David Cushing (1995 and earlier books) was the modern champion of the unit stock concept, although thinking on a definition of stocks goes back over a century. According to older definitions, a stock was simply those fish of a given kind, or even of a general class like "flatfish", accessible to the boats fishing out of a given port, or from a cluster of ports grouped for statistical summary of landings. More recently, an ideal *unit stock* has been seen as a reasonably

Biological Oceanography, Second Edition. Charles B. Miller, Patricia A. Wheeler.
© 2012 John Wiley & Sons, Ltd. Published 2012 by John Wiley & Sons, Ltd.

strict breeding group of a biological species, just one kind of fish, squid, or shrimp, etc. The notion that there are such groups arises partly because a large number of species participate in seasonal migrations to restricted spawning locations in the sea that are remarkably consistent from year to year. It is these mating meetings that define subpopulations with a countable (at least in principle) number of individuals, distinctive both in terms of size and age structure, with enough homogeneity in behavior and other aspects of biology to make them manageable under harvest. Such stocks have persistent proportions of variant genotypes (as represented by blood types, other protein variants, or DNA sequences), and they vary in those proportions from stocks with mating migrations at different times or sites. Thus, genetic markers can help to define the limits of unit stocks, keeping always in mind that *proportions* of genotypes in population samples are required to decide if a group of fish (or port landings) is in one stock or another. Similar proportions do not imply any very great level of genetic intermixing, since only a few individuals switching their mating site or time will provide some gene flow.

Cushing offers the term *hydrographic containment* as one explanation for the recurrence across species, even phyla, of mating conventions. The North Sea plaice (*Pleuronectes platessa*), for example, gather over a modest region in the southwest corner of the North Sea (Fig. 17.1a) to spawn. Individuals travel long distances to reach this site. The nearly certain reason that selection has promoted this site as the place for plaice to mate and spawn is that the most suitable juvenile rearing habitat is in the Dutch Waddensee, and the drift of the larvae from the English Channel will bring them to just outside Texel Gate, the entry to the Waddensee, as they reach the juvenile life phase. Matching of spawning sites to subsequent larval-drift paths provides minimal loss of larvae and juveniles to waters unsuitable for survival, and it also provides for arrival at a suitable juvenile habitat. That is the meaning of hydrographic containment.

Gulf of Alaska pollock (*Theragra chalcogramma*) meet in March–April for a mating spree just to the northwest of Kodiak Island over a shelf trench adjacent to Cape Kekurnoi on the Alaskan Peninsula. Eggs sink to about 150 m and are not advected much during 2 weeks of development. On hatching, the larvae ascend to the upper 50 m, drifting southwest in the Alaska coastal current, often in patches and often entrained for a time in back-eddies, as they feed in the plankton-rich Shelikoff Strait, growing at ~2 mm d^{-1} through May. There are other pollock stocks associating at other spawning sites and times, particularly one on the eastern end and seaward edge of the Bering Sea shelf in May.

Atlantic cod (*Gadus morhua*) generally have about a dozen major and many minor spawning sites (Brander 1997; Fig. 17.1a), from Lofoten in Norway to Georges Bank, where breeding groups gather for relatively short spawning seasons. For example, cod from all over and around Georges Bank move to the northeast peak of the bank (Fig. 17.1b) to spawn in late February. In February–March, *Calanus finmarchicus* (copepod) adults that are ready to spawn will accumulate over this site, and their spawning provides copious nauplii for cod larvae to eat. Just as important, drift from this point retains the larvae over the bank until they are ready to swim along the bottom. Stocks at the many mating sites are distinct genetically, although an endless debate continues about the fidelity of individuals to mating groups and the precise statistical evaluation of genetic distinctions.

At least some groups that mix in feeding habitats remain strongly distinct genetically, each group's members taking part in migrations to different breeding sites. For example, Svein-Erik Fevolden and colleagues have studied Atlantic cod in Norwegian waters, long thought on the basis of blood types and otolith distinctions to be two stocks. There are "northeast Arctic cod" (NEAC) that mostly feed as juveniles and adults offshore in the Barents Sea and extending to the Spitsbergen area, and Norwegian coastal cod (NCC) that mostly feed closer inshore and in fjords. In December to January, the NEAC migrate along-shore to breeding areas near offshore banks in the Lofoten vicinity. The hatched larvae are carried in the northward coastal current back to the Barents, where the juveniles grow to maturity. The adults return north as well for the summer–fall feeding season. The NCC both spawn and feed in and near fjords all along the coast, occasionally mixing with NEAC, particularly adjacent to fjords farthest into the Arctic. Sarvas and Fevolden (2005) examined the frequencies of the A and B alleles (there are only two) of the nuclear gene for pantophysin, a membrane protein found in secretory tissues. They developed a PCR technique for rapid identification of the alleles and found more than 90% B in cod caught offshore and in the Barents Sea identifiable as NEAC and 50–100% A in cod captured in fjords and spawning at sites scattered along the Norwegian coast (Fig. 17.2). Selection alone is unlikely to sustain this level of distinction; it derives from breeding structure, although Case *et al.* (2006) cite evidence for a degree of selection. The intermediate gene proportions from some of the areas to the north can be due to either mixing of NCC and NEAC in the collections, or to actual interbreeding.

A complex literature considers the level of genetic distinctions among Atlantic cod in detail, without fully explaining within-population variation in allele proportions (usually compared between years), but establishing with certainty that fish stocks diverge genetically over wide geographical ranges (e.g. Pogson *et al.* 2001). That shows that gene flow is slow. In part, that may simply be distance, but it also can be attributed (particularly for intra-regional cases like the Norwegian-Arctic cod and Norwegian fjord cod) to separation into stocks with spawning-site fidelity, As a further example, Kovach *et al.* (2010) have used the

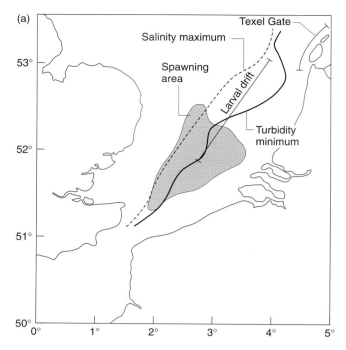

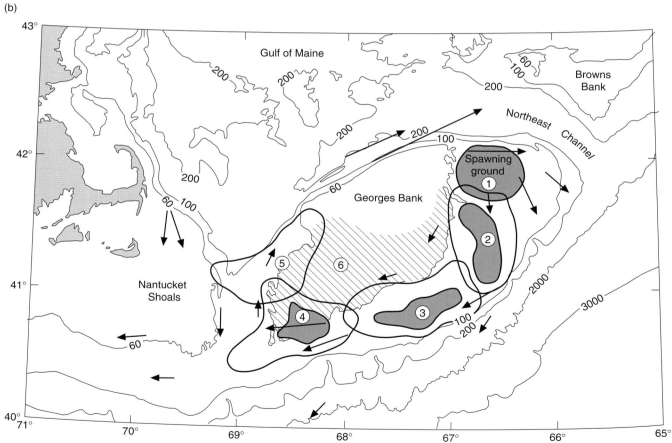

Fig. 17.1 Three examples of spawning migrations to sites providing hydrographic containment. (a) Spawning site of North Sea plaice. The usual larval drift is toward the Waddensee juvenile rearing area entered by Texel Gate. (After Cushing 1995.) (b) Cod and haddock spawning area on the northeast peak of Georges Bank showing typical dispersion of stage 2, 3, and 4 larvae during post-spawning drift. (After GLOBEC NW Atlantic implementation plan.) (c) Spawning sites and typical larval drift patterns for North Atlantic cod. Drift generally carries the young back to feeding sites from which adults migrate to spawning sites. (After Brander 1997.)

(c)

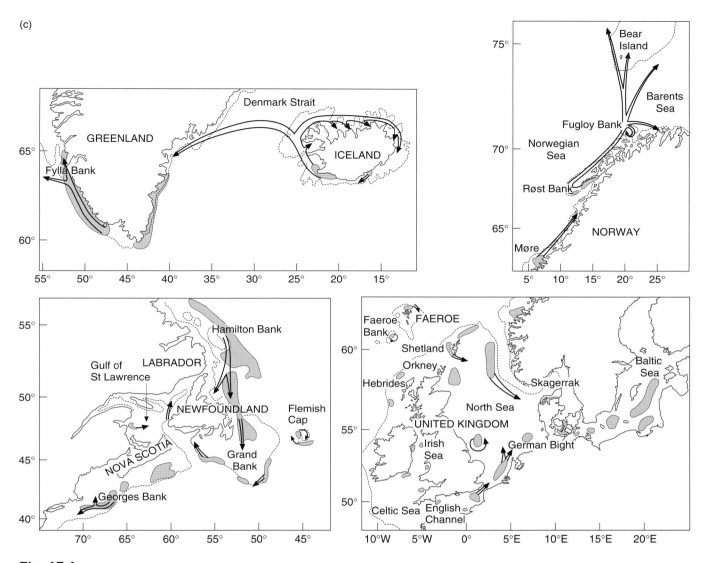

Fig. 17.1 (Continued)

pantophysin alleles to show that the group of cod spawning at the northeast peak of Georges Bank (Fig. 17.1b) are distinct from spring-spawning cod along the southwest coast of the Gulf of Maine, and that both are distinct from those spawning over shallow bottoms around Cape Cod. While catches from all of those stocks would be landed at, say, Plymouth, Massachusetts, USA, their distinct habits might warrant trying to manage them separately (Hauser & Carvalho 2008), although it could not be done easily.

More examples of spawning migrations can be found among the hake, herring (Sinclair 1988), tuna, elasmobranchs, squid, and a wide range of demersal fishes: plaice, flounder, halibut. Even whales migrate, often thousands of miles, switching from subpolar feeding grounds to subtropical areas where they visit special local sites for calving and mating. In their case, it cannot be for hydrographic containment, but the common feature applies that some

sites are best for reproduction, others for feeding. Atlantic and Pacific salmon species provide extreme examples of the phenomenon, with each stream area defining a unit stock. The freshwater provides their larvae ("alevins") hydrographic containment in a sense, and probably more importantly it serves as a refuge from marine predators. Genetic studies, similar to those on Atlantic cod, have been done for many species, generally with similar results. The unit stock and hydrographic containment concepts are useful idealizations for a wide range of marine animals.

While mating sites are many and widespread, there are also large ocean areas and stretches of coastal ocean, even very productive ones, with little spawning by fish with pelagic larvae. That is because young left there would be swept away to unsuitable habitats. Breeding adults do not know this; natural selection installs breeding times and locations as part of a reproductive program that produces

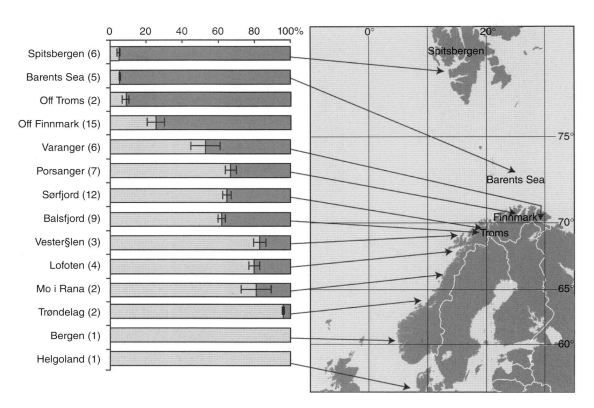

Fig. 17.2 Average ±S.E. frequencies of two pantophysin alleles in samples of post-juvenile (≥1 year) Atlantic cod (*Gadhus morhua*) from different regions along the Norwegian coast. (After Sarvas & Fevolden 2005.)

survivors, a program by which the young arrive in appropriate places at appropriate times. Fish leaving young where they will be carried out of a suitable habitat do not have grandchildren. A prime example of a spawning gap (Parrish *et al.* 1981) is the California coastal zone from Cape Mendocino to Point Conception, where eddying streamers carry newly upwelled water far out to sea (represented by the fat arrow in Fig. 17.3), mixing it eventually into the oligotrophic central gyre. Largely separate stocks of northern anchovy (*Engraulis mordax*) spawn to the north of Cape Blanco (southern Oregon) and to the south of Point Conception (at Santa Barbara), but not much between. Adult anchovy, and sardines also, wander through central Californian water, but they belong to spawning stocks to the north or south. The Pacific hake (*Merluccius productus*) actually migrates back and forth across this long stretch of coast in large schools, moving north to feed off Oregon, Washington, and British Columbia from March into summer, then moving south again starting about October to spawn seaward of northern Baja California in winter (Bailey *et al.* 1982; Fig. 17.4). The exact location of hake spawning is not well known because it is offshore, deep, thought to be broken into patches, and seems to be in different places different years. The eggs, which float upward, and early larvae are carried north and inshore by the seasonal surface flow. The extent to which larvae are retained

inshore after spawning determines subsequent recruitment to the fishery. Thus, an index of "year class strength" (see below) is negatively correlated with the amount of offshore flow estimated from an index of the along-shore wind stress in January (Bailey *et al.* 1982).

Fish and squid that do spawn off central California have different reproductive strategies. Fish like the kelp greenling (*Hexagrammus decagrammus*) and ling cod (*Ophiodon elongatus*) deposit their adhesive eggs on the bottom and the young hatch as rather advanced larvae. Presumably, they stay near the bottom and are not carried away. Surf perches (Embiatocidae) bear live young, large, "precocious larvae", probably capable of enough swimming to remain in the zone very close to shore from which water is not rapidly exchanged offshore during upwelling. The rock fishes (*Sebastes* spp.) are live-bearing, although the young released are small, in yolk-sac stage, and not particularly competent. However, the bulk of releases are in January to early March, when flow near the coast is onshore and northward (Moser & Boehlert 1991). *Sebastes* larvae do get carried seaward, and they are among the most common larvae of demersal fish in the plankton beyond the continental slope. Possibly, they are capable of a very long larval phase, retaining some chance of eventually reaching a suitable settling habitat over the shelf or slope. The very oldest pre-settlement stages of *Sebastes* are mostly captured in the

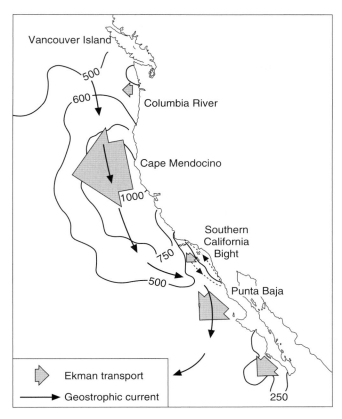

Fig. 17.3 Generalized summer circulation off the west coast of North America. Broad arrows indicate the relative level of offshore Ekman transport, which mostly occurs in jets at topographic projections. Simple arrows show the direction of mean geostrophic current, including the inshore gyre in the Southern California Bight. Contours represent a wind-mixing index ($m^3 s^{-3}$). Pelagic spawning is minimal from Cape Mendocino to Point Conception. (After Bakun 1993.)

neuston, that is, in the upper few centimeters of the sea. This layer responds most readily to the wind, so possibly the fish can navigate shoreward by surfacing when the skin of the sea is carried toward the shelf.

Dynamic methods

Models for stock management are designed in a variety of ways. Most emphasize some aspect of the following general input–output relationships. A catchable fish population, or unit stock, can be considered to have some total mass, B. This mass can be changed by four processes as represented in the following diagram:

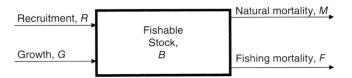

If A is a general age-structure variable, X is fishing pressure, and Y is fishery yield, then:

$$\Delta B/\Delta t = R(B,A) + G(B,A) - M(B,A) - Y(X,B,A)$$

(Eqn. 17.1)

Or, in words:

Stock Change = Recruitment + Growth
        − Natural Mortality − Fishing Mortality.

Parentheses in the symbolic statement indicate "function of" in the usual way. For example, fishing mortality, Y, must be some function of fishing effort, stock size, and stock age structure.

The objective of fishery dynamics is to replace R, G, M, and Y by a system of explicit functions (a model) that realistically mimics the behavior of the stock, thus enabling manipulation of X so that $Y(X,P,A)$ can be maximized on a stable, long-term basis. This has been done with modest success in some fisheries, although the success of the manipulation is often less than that of the models. There are two reasons for this last caveat: (i) The models are adjusted to fit data from the past, and so can be tuned to fit, while manipulation involves prediction of the future, which is more difficult to tune. Moreover, the models generally assume an unchanging dynamics of the stock, while both ocean climate and the interaction of the stock with its predators and prey are nearly certain to shift substantially through time. This point is the main basis of a recent book by Alan Longhurst (2010 – *Mismanagement of Marine Fisheries*). (ii) Managers are often overruled in the politics of fishing (Rosenberg (2003). In some approaches, R and G are explicitly included. In other, simpler approaches, the relations among X, B and $Y(X,B,A)$ are the focus. We will consider the both types briefly. Readers seeking an extensive, fully mathematized treatment could consult Quinn and Deriso (1998).

Often input–output management schemes assume strong *density dependence* of survival and growth. That is, if a stock is drawn down it is expected to respond with better survival and growth, because more resources become available to support that. Of course, some other animal may fill that carrying capacity. For example, in the case of slow-growing fish, the substitute appears often to be fast growing, and thus opportunistic, sometimes squid or jellyfish. In the following, "fish" will simply mean the product of a fishery, which could be clams, shrimp, squid, or, of course, fish.

Stock size (measuring B)

The first-order problem is simply to estimate the biomass and age composition of the stock. As any wildlife manager will tell you, it is not simple to estimate accurately the abundance of deer or geese. Stocks in the ocean are even

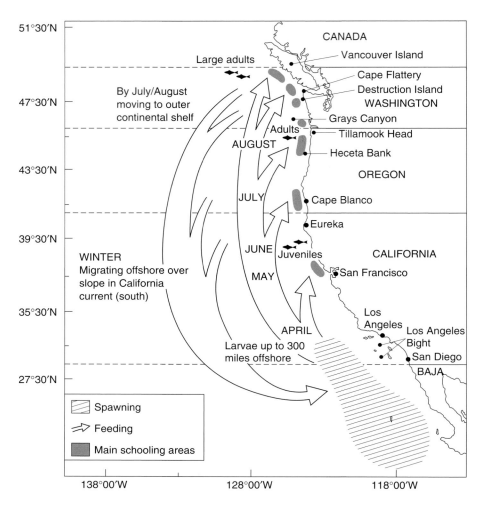

Fig. 17.4 General routes and timing of the spawning and feeding migrations of Pacific hake. (After Bailey *et al.* 1982.)

more difficult, since it is not easy to peer into the ocean. So, the size of stocks is not usually known explicitly (tagging methods – see below – not withstanding). The most common approach is to suppose that the harder it is to catch a fish, the fewer fish there are. That is, the yield, *Y*, per unit fishing effort, *X*, should be less if the stock, *B*, is smaller (and vice versa):

$$Y / X = \text{Catch per unit effort (CPUE)}, \propto B$$
$$Y / X = qB \qquad \text{(Eqn. 17.2)}$$

where *q* is the catchability coefficient, the proportion of the stock removed by one unit of effort. The captured fish are weighed or counted at the dock, and the effort made to catch them is estimated in some suitable units like hook-hours or number of days fishing by standard boats. You get that measure of effort by urging or requiring fishers to keep logs. A government fishery agency might require log submission as a condition of sale. For fisheries that have, over

the course of time, established several approximate equilibria between fishing pressure and population dynamics, this can work; the shape of the *Y/X* vs. *X* curve (*X* on the abscissa) declines monotonically. In the simplest cases the relationship is roughly linear (Fig. 17.5a). If the ordinate intercept is called qP_{natural}, an estimate of catch rate from one unit of effort for the stock size without fishing, we can represent this relationship by:

$$Y/X = qB_{\text{nat}} - qkX, \qquad \text{(Eqn. 17.3)}$$

where *k* is the slope of the relationship. In a *relative* sense, this *Y/X* vs. *X* relation gives us an idea how many fish there are in a stock.

Before using this relationship, we must examine its main assumption, that *Y/X* is proportional to *B*. Note immediately that this is a dangerous equation statistically, since even for two series of random variates, Y_i and X_i, Y_i/X_i will be negatively correlated with X_i. That is, *Y/X* can be

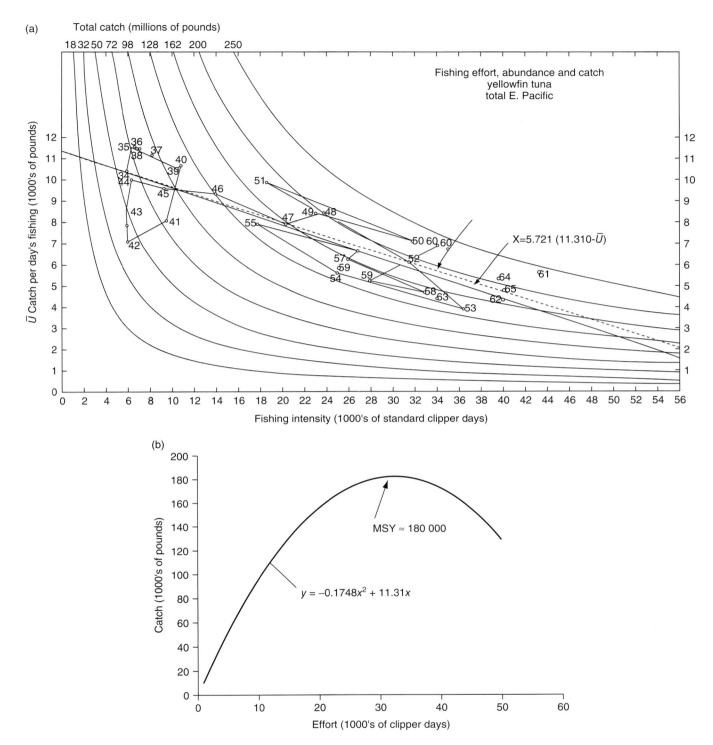

Fig. 17.5 (a) CPUE (catch per unit effort) for yellowfin tuna in the eastern tropical Pacific, 1934–1965. (b) Line in (a) converted to show catch vs. effort, a parabola. (After Schaefer 1967.)

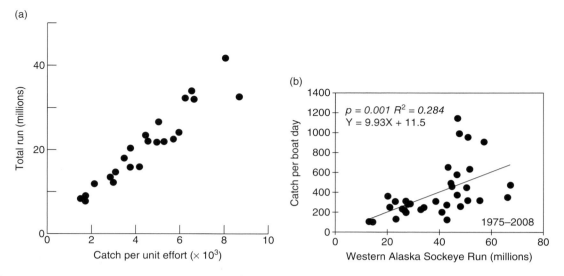

Fig. 17.6 (a) CPUE for Bristol Bay sockeye salmon *versus* total run size from catch plus river counts. (After Tanaka 1962.) (b) Comparison of the CPUE for the June south Peninsula sockeye purse seine fishery and the size of the western Alaska sockeye run, 1975–2008. (After Martin 2009.)

artifactually correlated with *X*. Fortunately, in many cases the relationship is close enough not to be attributed to this artifact alone. One way is to examine *Y*/*X* in a fishery that is somehow associated with a total count of the stock. The Bristol Bay sockeye salmon (*Oncorhynchus nerka*) fishery fits the requirements. The commercial fishery captures returning fish with gill-nets (effort in "fathom-days") in Bristol Bay, but fishing is forbidden in the inlets to the rivers. The yield of the gill-net fishery is measured by counting salmon. Once they are swimming up the rivers, people at the fish ladders and weirs watch and count the fish that escaped the fishery. By including results from multiple years, the correlation of *Y*/*X* to *B* = (*Y* + river counts) can be examined (Fig. 17.6a).

The CPUE estimates appear to be excellent. CPUE estimates are also used for much less exact comparisons of catch rates to stocks. For example, Martin (2009) has compared the June sockeye catch (mixed gear with effort in boat-days) just near the south tip of the Alaska Peninsula with estimates of the total annual sockeye runs for western Alaska (Fig. 17.6b), predictably finding a much less exact fit.

Another test can be applied to fish (or clams) which move infrequently (or very slowly) so that they can be totally fished out of some modest area without influence from migrants moving into the vacated space. Tropical snappers (Lutjanidae) are territorial around reefs, moving only short distances from their base to grab prey (or baits). Over 10 weeks, King (2007) directed fishers in Western Samoa as they fished down the *Etelis coruscans* stock with hook and line on a 12 km² patch of reef not previously fished. A time-series plot (Fig. 17.7a) of CPUE vs. total catch, *Y*sum, has *q* as its slope. Leslie and Davis originally used this technique to estimate rat abundances, in which

case eliminating them was a good idea anyway, so the plot is called a "Leslie plot". Estimates of the original total stock are either the intercept on the *Y*sum axis or the intercept on the CPUE axis, *a*, divided by *q* (which are mathematically equal). It is the linearity of the series that implies the proportion of CPUE to stock size. A similar test has been done for dredging of clams and scallops. In one experiment (P. Rago, unpublished), a patch of bottom (11,613 m²) with substantial numbers of razor clams was repeatedly dredged, each track sieving about 1500 m². The Leslie plot (Fig. 17.7b) of bushels (the unit in the fishery under study, 1 bushel ≈135 clams) captured vs. total clams was again quite linear. It was clearly very difficult to get the last few clams. There are very few such results because a requirement of depletion fishing can be devastation of the stock. But they do show that in general CPUE can be a measure of stock. In most cases, however, there is no way to estimate *q*, so *Y*/*X* must be used as a proxy for actual abundance numbers.

While CPUE is useful in some fisheries, it must be kept in mind that these measures are sensitive to even small shifts in fishing technique. For example, North Pacific halibut (*Hippoglossus stenolepis*) are captured by laying out lines of baited hooks across the bottom, leaving them for 12–24 hours, then reeling in the line and, on most days, fish. Up until 1982–1983, the fishers used "J-hooks", then switched to circle hooks (Fig. 17.8). The capture rate per hook fished rose by a factor of two to three, depending upon fish size (IPHC 1998). An apparently simple change in hook shape raised the hooking rate and reduced escapes dramatically. Sometimes subtle changes in technique occur in a fishery without managers being aware. Since fishers are intensely aware of their own CPUE, these can be

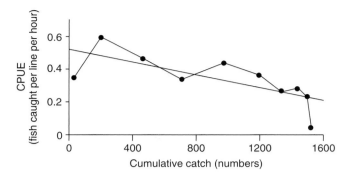

(a)

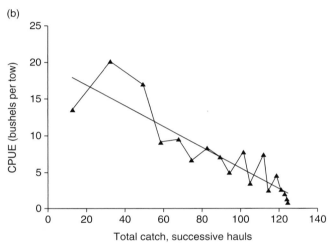

(b)

Fig. 17.7 Tests of CPUE by depletion fishing. (a) CPUE (number per line-hour) plotted against cumulative catch for the snapper *Etelis coruscans* near western Samoa. Fitted q was 0.0002 (fraction of stock caught per line-hour) for the fishing zone, or 0.0023 km^{-2}. (After King 2007.) (b) CPUE for razor clams vs. cumulative catch. q is 0.14 per haul. (Data from Paul Rago, NMFS.)

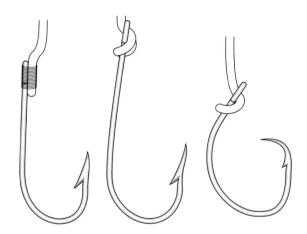

Fig. 17.8 Halibut hooks. The two designs on the left were used before 1982. The circle hook on the right was introduced in 1982–1983. (After IPHC 1998.)

improvements in gear, search efficiency, communication among fishers, focusing fishing at the most productive times, and simply more experience in the work. Changes of CPUE caused by technique modification can appear to managers to be increases in stock size. Whenever stock condition warrants it, managers like to (or legally must) provide fishers higher quotas or longer seasons, so misinterpreted upward CPUE shifts can readily result in over-fishing.

In fact, CPUE as a measure of stock size has received extended criticism early and late, as reviewed by Maunder *et al.* (2006). There turn out to be a many biases. A partial list includes:

1 a tendency to excessively high values early in a fishery as the most susceptible fish are captured, with later CPUE values becoming underestimates;

2 failure of CPUE to fall as stocks decline because fishers focus on the remaining centers of distribution;

3 intentional misreporting of both catches and effort to avoid reaching quotas, to make stocks look strong, to make fishers appear successful (low Y, high Y, low X, and high X are all possible biases to reports under different circumstances);

4 continuous or stepped improvements in the power of a given level of fishing effort, X (at least as measured) to catch fish (also outlined above);

5 shifts in the accessibility of the stock to prevalent fishing methods because of climate or weather variations, prey redistributions or other environmental changes.

To overcome at least (2), (3), and (4), some fisheries are regulated on the basis of management agency research surveys on standard station grids with fixed methods, in recent decades coupling effective sonar biomass assessments with trawling or species-appropriate capture methods. There is a reasonably good track record for such agency surveys as predictors of subsequent catch relative to industry effort. Fishers often view agency surveys as wrong because, in their view, no attention is paid to where the fish are; that is, fishers know where to fish and they do not tell the agencies. That, of course, is the cause of error source (2) above. The struggle to get reliable data, to "see" the fish in the sea is endless. Current recommendations are to manage by "integrated assessment" (Maunder *et al.* 2006) and even more broadly focused "ecosystem-based management", mentioned below. Assessments have CPUE as an input, but age structure; reproductive biology; estimated recruitment; growth dynamics; distribution patterns; predator and prey ecology; shifting oceanography; and more, are taken into account. The better-managed fisheries in the United States, in particular that on the Eastern Bering Sea pollock stock (Ianelli 2005), are regulated in light of fisheries-independent trawl and acoustic surveys, and exploitation is regulated to well below estimates of maximum sustainable yield (considered below). In addition, fishing boats in the pollock and several other US fisheries are required to have agency observers aboard, ensuring greater integrity in catch and bycatch reporting. Elaborate mathematical/statistical models are typical stock assessment tools, albeit requiring both explicit and implicit assumptions.

Recruitment ($\Delta B/\Delta t = R(B,A) + \ldots$)

Fish are considered to be "recruited" to a fishery when they are large enough to be captured by its methods. Thus, size at recruitment sometimes can be chosen by the fishers or by fishery managers (usually by decree). For example, the pots used to capture Dungeness crab on the northwest coast of the USA have an escape port just smaller than the legal size, a minimum width of carapace. Many small crabs visit the pots then leave again, and undersized crabs found by crabbers retrieving their pots are tossed back to the sea. Females are all thrown back, too. Thus, recruits are quite strictly defined as *male* crabs larger than the legal size. For bottom-trawl fisheries, mesh size can be adjusted, and pre-recruits slip through the openings – at least that is the intention. Recruitment rate (numbers of young entering the catchable stock per time) depends upon a list of variables, and understanding how those factors control the rate is a major scientific challenge for fisheries biologists. Variables include the number of eggs spawned, hatching success, survival to recruitment age or size, retention throughout development within a suitable habitat, and the unused carrying capacity of the habitat available for new recruits. Survival of larvae depends upon temperature, speed and direction of currents, availability of sufficiently nutritious food of the right size, and the activity of predators and parasites. Carrying capacity can be mostly filled by large numbers of older individuals, which may inhibit recruitment – sometimes by cannibalism. Survival and growth of potential recruits are reduced by competition with larger and tougher, older individuals.

Usually, recruitment is taken to be a quasi-parabolic function of stock size (Fig. 17.9a), referred to as a "spawner–recruit" curve. At the low end of the scale (often an "experiment" forced by over-fishing, but which can also be a result of climate degradation relative to the requirements of the fish) there are indeed very few recruits when there are very few spawners. Spawner–recruit relations for several herring stocks (Fig. 17.9b & c), illustrate this well. The points on the graphs retreated toward the origin as the stocks were fished down. This is very simply understood; *if there are no parents, there will be no young.* The relationship away from the origin often, but not always (Fig. 17.10) tends to a wild scatter. Intense effort and emotion have gone into choosing and justifying the best mathematical functions to fit spawner–recruit curves. There are curves due to Beverton and Holt (same as the Michaelis–Menten function), Ricker,

Cushing, Shepard, and others for fitting these stock–recruitment relationships. With rare exceptions, the debate is useless, since no one of these deterministic equations looks better as a representation of the scatter than the rest. There are several reasons beside competition for a spawner–recruit curve to drop at the right. For fish with planktonic larvae, it will be rare that larval numbers are so great that they limit their own resources, since larvae are a minor part of the overall pelagic ecosystem, but it could happen. Large numbers of larvae could also attract predator attention, reinforcing successful prey search patterns, raising mortality.

Of course, fisheries should always be managed in light of the necessity to sustain recruitment, that is, with the intent to keep plenty of fish active as parents. That we so often fail to operate under this simple precept is astounding. Sometimes fisheries are managed with the notion of maximizing recruitment. A model is chosen in which recruitment peaks below the maximum stock of spawners, and the stock is then reduced by fishing to the vicinity of that recruitment peak. The best examples are salmon fisheries. These are in a sense ideal for the purpose, especially if salmon are fished in or near their natal river, since spawners for the next cycle are the same fish counted as recruits in the present cycle. They are active in both roles at nearly the same time. The idea is that to maximize the number of salmon returning to a stream (recruits) there should have been some optimal number of spawners in the previous generation (stock). Since the a priori expectation is a domed function, we might guess that substantially fishing down the number of spawning salmon would actually increase the number of returning salmon available for the fishery. The argument is strained because of the separation of the freshwater and marine juvenile phases. It has to be argued that fewer eggs can produce more and healthier seaward-migrating fry or smolts by reducing competition in freshwater, and then that the increase in seaward migrants will improve final adult returns. In some instances, the data are convincing, in others not. Results for the pink salmon (*Oncorhynchus gorbuscha*) stock spawning in the Kodiak Archipelago off Alaska illustrate both cases (Fig. 17.11a & b). If "recruits" are taken to be fry that are headed for the sea, then the data seem to show that reducing the stock to 2–3 million spawners does not reduce recruits significantly, and in some years seems to generate an addition to fry entering the sea of about 1.5-fold. If recruits are taken to be the adults returning as 2-year-olds, the relation (for which there are more years of data) is less clear. Two million spawners can return 3 million or 15 million adults to the combined fishery and spawning stocks. Survival at sea is surely variable and crucial to salmon recruitment.

Application of a spawner–recruit analysis as a management tool assumes in a general way that habitat factors affecting the relationship are close to constant. That is never true. In the case of Kodiak and other Alaskan salmon, there

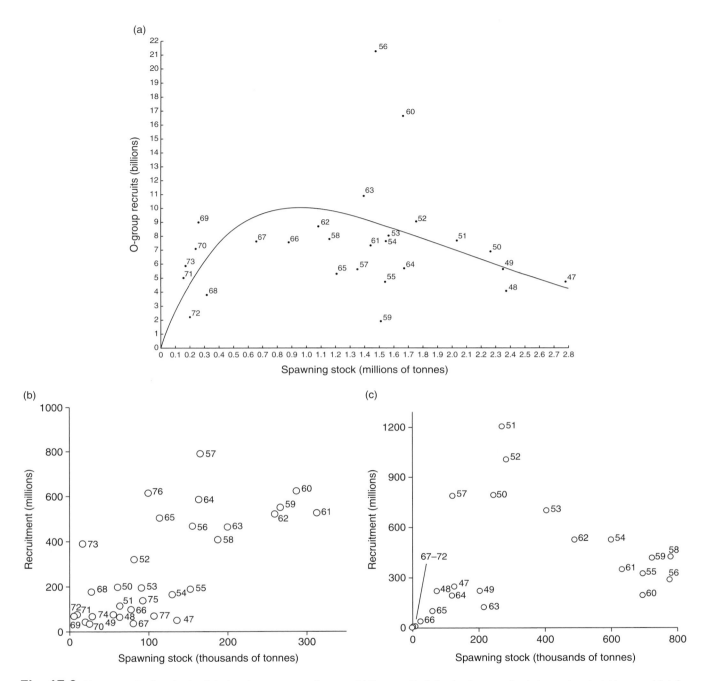

Fig. 17.9 Three examples from herring fisheries of spawner–recruit curves. (a) Zero-age North Sea herring vs. estimated spawning stock biomass of fish 2 years and older, fitted by a Ricker curve. (After Saville & Bailey 1980.) (b) Icelandic summer-spawning herring ("1-ringers" vs. spawning stock). The data from 1947–1961 form a quasi-parabolic spawner–recruit curve. (c) Icelandic spring-spawning herring. (b & c from data in Jakobsson 1980.)

was a habitat change in the 1970s that enhanced ocean survival, and thus the abundance of recruits relative to spawners dramatically increased (Fig. 17.11c & d). Points from before and after the "regime shift" are jumbled together in the spawner–recruit curve, making it of questionable utility. In general, if there are enough data to define a curve, the time-series will be affected by changes in the system.

Most spawner–recruit relationships are affected by the sheer variation from year to year in the success of larvae in reaching recruitment, the *year-class* effect. We cite only one example among a vast assortment. The Alaska pollock (*Theragra chalcogramma*) stock in the Bering Sea is closely monitored by annual fishery surveys that determine the number of fish reaching the age of one year. The variation

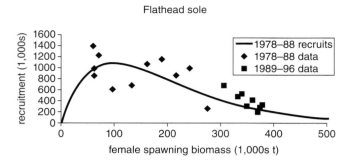

Fig. 17.10 Spawner–recruit relationship for Bering Sea flathead sole (*Hippoglossoides elassodon*) to which a Ricker curve ($R = aS \exp[-bS]$, a and b constants) fits reasonably well. The curve was fitted to data for 1978–1988, and then predicted the data for 1989–1996. While there were no points with very low recruitment, R for $S = 0$ is certain to be zero, anchoring the curve. (From Wilderbuer *et al.* 2002.)

(Ianelli 2005; Fig. 17.12a) is very great and depends, in ways often difficult to specify, on winter weather, predator activity, mesozooplankton production, variation in currents during the larval drift, and more. In this fishery, as in many others, the yields for multiple years can depend upon the availability and growth of a particularly successful year class. The ridge of increasing mean size of pollock landed from 1992 to 1997 (Fig. 17.12b) was caused by the dominance of the yield by the strongly successful 1989 year class. Huge effort has gone into understanding year-class strength variation of different stocks, with varying but always modest success.

Growth ($\Delta B/\Delta t = \ldots + G(B,A) \ldots$)

Fish grow at different rates at different sizes (ages). The rate at which the "average" individual member of a fish

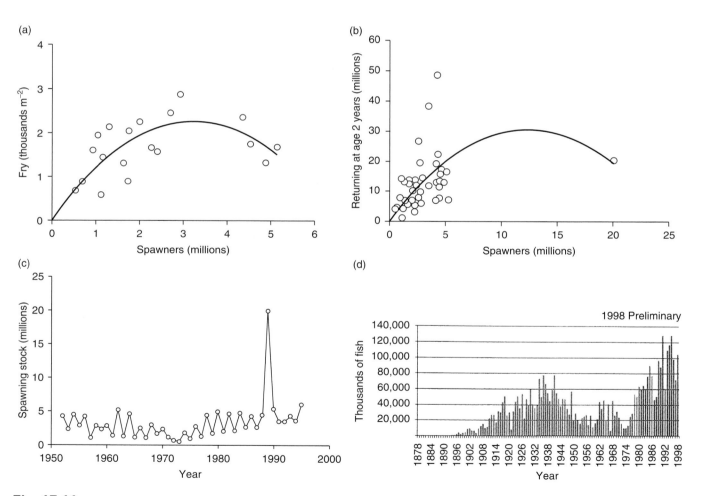

Fig. 17.11 Pink salmon. (a) Relationship of spawner abundance to 1-year fry (still freshwater, ready to migrate) for various streams in the Kodiak Archipelago, Alaska. The curve is a parabola. (Data from Donnelly 1983.) (b) Spawners vs. returning 2-year-old fish relationship for the Kodiak Archipelago, taking all streams together. The curve is a parabola, clearly much influenced by the single large point (1989, fishers weren't prepared to catch the available fish). (Data from Donnelly 1983, and updates in Myers 2001.) (c) Time-series of spawner abundance for the Kodiak Archipelago. Pink salmon from all regions have alternating strong and weak year-classes. (Data from Myers 2001.) (d) Total Alaskan landings of pink salmon, showing rise to very high levels after the mid-1970s. (Data from the Alaska Department of Fish and Game.)

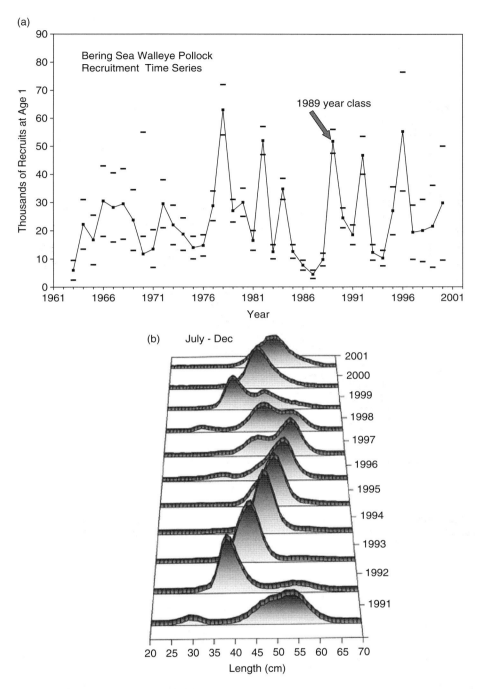

Fig. 17.12 (a) Recruitment variation of age 1 walleye pollock (*Theragra chalcogramma*) in the eastern Bering Sea (EBS) stock as determined by fishery-independent trawl survey. Bars shown for each year are upper and lower recruitment confidence limits from the survey results. Notice the strong recruitment in 1989. (b) Year-on-year size–frequency distributions for the summer catches of the EBS pollock fishery. The ridge from 40 cm in 1992 to 55 cm in 1997 is the 1989 year class. ((a) from data in Ianelli 2005; (b) after Ianelli 2005.)

stock can add biomass gives a fishery manager an idea of how fast biomass can be sustainably removed from a fish population. The relationships of size to age and of growth rate to age typically have the forms shown in Fig. 17.13. There are many considerations affecting the age or size at which it is most desirable to catch any particular kind of fish. It may only be palatable in a limited part of its life span. In fact, the eggs may be the target of the fishery: caviar, herring roe, and urchin ovaries are prized delicacies. A fish may need to reach a given size before it can

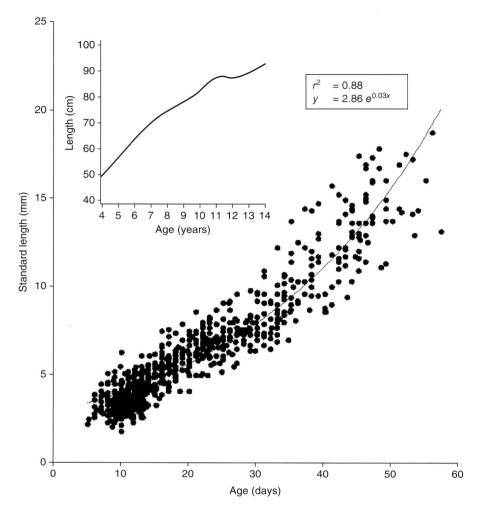

Fig. 17.13 Length at age for larval Atlantic cod (*Gadus morhua*) from Georges Bank (after Green *et al.* 1999). Inset: length of cod at age 4 years and greater that are recruited to the Greenland fishery (after Riget & Engelstoft 1998). There are also moderate differences in growth among year classes.

reproduce significantly, or the most desirable growth rates may be confined to a narrow age span. In the latter case, it might be desirable to remove the fish in older, larger categories so that their slow, inefficient growth does not "waste" fish food. Accounting for growth potential in management means adjusting the size (age) at recruitment so that fish growing at substantial rates remain in the field making meat for eventual harvest. The trade-off can be that more time to grow is coupled with more time for natural mortality, which reduces the final take.

Another factor is that it can be difficult to leave some of the larger individuals behind while taking somewhat smaller specimens. The reason that might be valuable is that most long-lived "fish" (cod, halibut, grouper, and lobster, . . . , but not salmon or squid) grow continuously, albeit with slowing, to great potential sizes, often at ages measured in decades. Reaching those sizes, they are available to fewer predators, so their death rate declines. The benefit of leaving

them in the stock would be that very large females of many species produce disproportionally more eggs than younger, smaller females. For example, Bobko and Berkeley (2004) showed that black rockfish (*Sebastes melanops*) in the California Current double their annual egg output per gram of body tissue from 374 to 549 between age 6 (near earliest reproduction) and 16 years, while the total fecundity rises from 300,000 to a million embryos (their young are released after hatching). Moreover, big, old, fat, fecund female fish (lately called BOFFFF or just BOFF) actually produce bigger and better individual eggs by supplying them with more oil and protein. As demonstrated by Berkeley *et al.* (2004), larvae from those eggs develop and grow faster, are more resistant to starvation (Fig. 17.14) and thus have better survival chances. For some fish and some capture techniques, individuals exceeding a size threshold can be returned to the sea, surviving temporary capture, but for many the ascent to the surface or crushing in trawls is fatal.

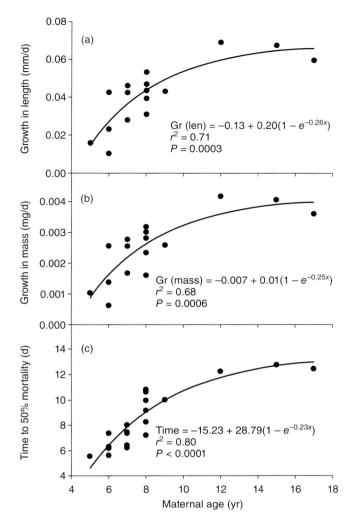

Fig. 17.14 Relationships for black rockfish (*Sebastes melanops*) between the ages of females releasing larvae and (a) larval growth rate in length, (b) growth rate in mass, and (c) median time to starvation when held without food. Similar relationships are found for the size of oil globules in the larval yolk sac. (After Berkeley *et al.* 2004.)

Often the real issue involving growth is to know the size and age structure of the stock. Fishing usually removes the larger, older fish from the population, partly because they are sought by fishers and, thus, usually favored by the technology chosen. They are, of course, less numerous even in the absence of fishing. Thus, onset of intense fishing shifts the age structure to smaller, younger, usually less-fecund fish. Moreover, once a fishery is established and the larger age groups are gone, there is tremendous economic pressure (from people whose jobs and boat payments depend upon the fishery) to allow capture of younger and younger fish. That may simply result from continued fishing. For, example, size composition of the Peruvian anchoveta (*Engraulis ringens*) stock progressively empha-sized smaller, younger fish during the intense fishery of the late 1960s and early 1970s. By 1971, most of the stock was

made up of small individuals. Their egg output was not great enough to overcome the low larval survival caused by the 1972 El Niño, and the fishery crashed. The stock finally came back in the late 1990s and is again being heavily fished. An attempt by managers to close the fishery in response to the 1998 El Niño held for about 10 days before political pressure reopened it.

Natural mortality ($\Delta B/\Delta t = \ldots - M(B,A) \ldots$)

This is usually the least well known of the variables of a fishery, and the least accessible to management. It can some-times be estimated by tagging studies, and from time to time lots of energy goes into the effort. If we tag T fish of the recruitment age (tag part of a "cohort"), we can assume the following proportionality over the remaining life span of the fish of that age class:

$$T : T_R : T_L :: N : Y : M$$

where:

T_R = the number of returned tags over the whole life of the cohort,
T_L = the number of tags never seen again,
N = the number of individuals in the cohort,
Y = the total yield of the cohort to the fishery (number),
M = the number in the cohort that died uncaught.

We have the following necessary relations:

$$T = T_L + T_R, \text{ and } N = Y + M.$$

From the proportionality assumed:

$$M = (T_L/T_R)Y \qquad \text{(Eqn. 17.4)}$$

Hence:

$$N = Y + (T - T_R)Y/T_R. \qquad \text{(Eqn. 17.5)}$$

Evaluation of population sizes, N, by marking and recap-ture is called the Lincoln Index method of population esti-mation. It seems easy. We get an estimate of actual population and of mortality at the same time. It is never so easy. The problems are: (i) both tagged and recaptured samples tend to be trap-prone individuals; (ii) the statistics for return of a very small fraction of the original tagged lot (the usual case) are unfavorable (big estimator variances); (iii) the assumption of a closed population is often violated in the sea; and (iv) tag mortality. Tags kill; there are almost no tags without some tag-associated mortality. Hence, $M < (T_L/T_R)Y$. How much less is extraordinarily hard to estimate.

Regardless of their practical utility, an instructive value of the equations above is that Y and M compete for the available stock: $N = Y + M$. If we don't eat them, something else will. Fishers are well aware of this competitive relationship with other fish consumers. They are sometimes tempted to take such direct actions as shooting seals.

Tags are also useful for study of population dispersal and migration (see Jennings *et al.* 2001). Fish tagged at one site and captured at another have moved by some route between those points. With sufficient tag returns, it is possible to reconstruct typical movement patterns. For example, Pacific halibut is a continental-shelf and upper-slope species whose eastern stock is spread around the perimeter of the Gulf of Alaska. It usually migrates downslope to spawn, and the larvae are transported north and west around the Gulf. Recoveries of tags placed in young adult fish show a general migration back toward the south and east, keeping the population from removal to the west by larval advection. In recent years, tags have become very sophisticated, internally recording temperature, pressure, light levels, and other variables. In some cases, a migration trajectory can be derived from daylength changes or other data, and understanding of large fish behavior is improving rapidly thanks to this data.

Yield and fishery mortality ($\Delta B/\Delta t = \ldots - Y(X,B,A)$)

The yield per unit effort relation (Eqn. 17.3) above,

$$Y / X = qB_{nat} - qkX,\ \text{is readily solved for } Y:$$

$$Y = qB_{nat}X - qkX^2 \qquad \text{(Eqn. 17.6)}$$

That function suggests that there should be a parabolic relationship between yield and effort: the *yield versus effort curve* (Fig. 17.5b). Such a relationship is what economists for several hundred years have called a "Law of Diminishing Returns". Each section of the curve can be explained verbally as follows:

1 The more fishing done at low levels of exploitation, the more fish caught, but at a declining rate of increase.

2 Maximum sustainable yield, called MSY by everybody.

3 At large X, the population gets so depleted that fish become hard to catch and yield falls off.

Usually much of the reproductive stock has been removed in fisheries that are well down the right side of the curve. In that sense the yield vs. effort curve is the recruit vs. spawner stock curve plotted backward.

While MSY is part of every fishery manager's mental machinery, it is no longer explicitly used in many fisheries. It is there, however, always working just under the surface, sometimes reappearing as "reference points" in international agreements about limits to catch or effort. The yellowfin tuna (*Thunnus albacares*) fishery of the eastern

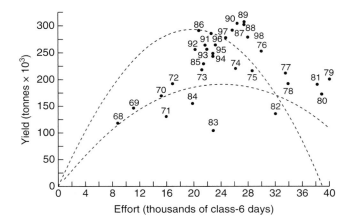

Fig. 17.15 Yield vs. effort curves for yellowfin tuna in the eastern tropical Pacific. Separate curves apply for each of two eras, 1968–1984 and 1985–1998. Distinctive production regimes apparently applied in these periods. (After IATTC 2000.)

tropical Pacific was managed for an interval by an explicit MSY model, and Schaefer's (1967) analysis of catch and effort served as the basis for that management. The fishery has had several major eras. A California coastal fishery for albacore failed in the middle 1930s, so canneries supported development of some refrigerated boats for a distant-water yellowfin fishery in the eastern tropical Pacific. After World War II the fishery extended farther and farther south, joined in those years by boats from Latin American nations. From about 1930 to 1958 the system of fishing was to drive a boat next to a school sighted by a masthead lookout, release some silvery bait fish from a tank, then fish in the ensuing feeding frenzy with silvery hooks. Tuna were heaved into the boat by muscle power. Such boats were called "tuna clippers", and the unit of effort recorded by managers was the "standard clipper day", obtained from a comparison of the fishing effectiveness of clippers of different sizes. In the second era, from 1958 on, the fishery switched fairly quickly to purse seines, enormous curtains of net strung around a tuna school by a small boat dropped off the fishing vessel. When the free end is returned to the ship, the net is closed beneath the school by pulling in a pursing cable. Next the net is pulled back aboard with a hydraulic power block, concentrating the tuna alongside. From there they are lifted out of the water and frozen in the holds. Seiners and clippers were both working for enough years to get a comparison of their fishing power, so the unit of effort remained "standard clipper days".

Schaefer's plot (Fig. 17.5a) of CPUE vs. effort (thousands of clipper days) is reasonably linear. He employed several approximate corrections, trying to overcome the fact that the fishery age structure, recruitment, and yield never had a chance to come to equilibrium with any given level of fishing effort. Rather, effort increased steadily through the years. All that argument can be studied in the original paper.

The plot was readily converted to a yield vs. effort curve (Fig. 17.5b), which suggested an annual MSY around 200,000 pounds of tuna. The Inter-American Tropical Tuna Commission eventually enforced this as an annual quota for the fishery. Ships would radio their catches to headquarters where they were summed. When the quota was reached, a message was sent back to cease fishing and return to port.

The third era began in 1967, when the fishery reached its quota far ahead of its usual schedule. The fishers had discovered that setting their seines around schools of porpoises would capture submerged schools of yellowfin tuna that live beneath them. There is an interaction between porpoises and tuna that is still not fully understood, but the fishers learned to exploit it. Moreover, the tuna beneath porpoises are much larger than the average in surface schools, so the catch rate, the CPUE, increased dramatically, ruining the analysis and basis for management. The Commission more or less threw up its hands and let fishers fish until their boats were submerged to the Plimsoll line. In fact, the entire eastern Pacific yellowfin tuna fishery was subject to no limits from 1980 to 1998. At present, there is again a nominal MSY process for managing this fishery, since the treaty establishing the Inter-American Tropical Tuna Commission (IATTC) requires it. However, entry of Korean and Japanese long-liners, and a much more complex scheme of fishing effort from the Americas, have made effort estimation very complex. A yield vs. effort graph for the period 1968 to 1998 (Fig. 17.15; IATTC 2000) has been developed on the basis of just the CPUE for large ("class-6") seiners. Since those account for most of the eastern Pacific catch, they can represent the whole fishery for our purposes.

Clearly, the graph breaks into two eras, before and after 1985. IATTC analysts fitted separate yield–effort parabolas for the two periods. The earlier period was clearly one of increasing effort, and while total yield remained relatively constant, CPUE went down. Once CPUE got low enough, boats were sold or moved to the Atlantic or western Pacific where fishing was better. The stock rebounded, and yields improved dramatically, soon far exceeding those of the era up to 1984, with less effort expended. Examination of the catch showed higher recruitment rates and faster growth in the later era. This change is not the least unusual; the success of fish stocks changes over time, and usually we cannot confidently identify the causes. It makes obvious the difficulty of managing by this sort of MSY analysis, which assumes that the biomass of the stock will respond in the same way over very long periods to more or less fishing pressure. Evaluation of the eastern Pacific yellowfin tuna fishery and its stock have become so complex, the terminology so arcane (e.g. Maunder & Watters 2001), that explaining it goes beyond our goal in this chapter.

A few other examples of fisheries managed on the basis of MSY may be instructive. Hunter *et al.* (1986) developed two yield vs. effort graphs (Fig. 17.16) for Atlantic yellow-

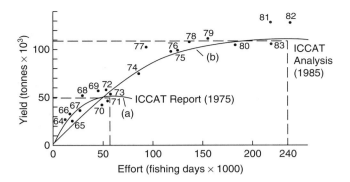

Fig. 17.16 Two yield vs. effort curves for yellowfin tuna in the eastern Atlantic Ocean. Curve (a) was fitted by the management commission after 1973, predicting that MSY had been reached. Increased fishing to 1983 supported (b), a different curve and higher MSY. (After Hilborn & Walters 1992, citing results of Hunter *et al.* 1986.)

fin tuna. One was for data up to 1975, and a fitted parabola suggested an MSY around 50,000 metric tonnes. However, the growth of effort did not stop at the roughly 60,000 fishing days producing that yield, it rapidly increased to over 200,000 days by 1983, leading to a yield curve peaking over 100,000 metric tonnes, and still not obviously trending downward. Hilborn and Walters (1992) draw the excellent caveat from this that (at least in the absence of a detailed population dynamic analysis) "*you cannot determine the potential yield from a fish stock without overexploiting it*". Once that is done, of course, the damage to the stock, even its place in its ecosystem, may well not be repaired simply by reducing X to the level at the apparent MSY. It is possible that no MSY will be evident from effort and yield data. In the northeast Pacific halibut fishery, effort is measured as baskets of baited hooks deployed on the bottom for a day. For many fishing regions, the graph of annual CPUE vs. effort (e.g. Fig. 17.17) is not linear, but has an obvious curve. It turns out that the data in Fig. 17.17 can be fitted with a curve by assuming a constant catch of 24.4 million pounds and dividing it by the effort in thousands of sets. In other words, it doesn't matter how many sets are put out above about 300,000, the catch is about the same. Of course, the CPUE vs. effort function is well defined, even if curved, so it can be converted to a yield vs. effort plot, which for this range of effort is a flat plateau, always about 24 million pounds. The time sequence of effort was, in fact, generated by a regulatory agency, which saw the decline in CPUE from 1925 and had the authority after about 1931 to reduce effort. In this case, CPUE was a useful management tool even when no particular MSY was specified. The agency was also concerned about the age structure of the stock, which had shifted to younger fish.

The MSY argument leads the unwary to assume that a prudent exploiter would seek to establish a level of fishing effort, X, such that the MSY would be realized. There are

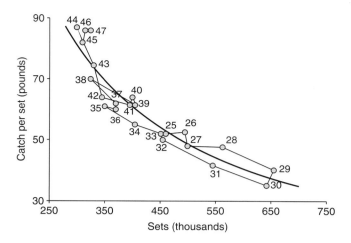

Fig. 17.17 Catch per set (CPUE) vs. sets (effort in baskets of hooks) for North Pacific halibut on the "southern grounds". The curve represents a constant total yield of 24.4 million pounds. (Based on data of the International Pacific Halibut Commission.)

reasons for fishing at somewhat lower rates, and there are severe difficulties in realizing either MSY or the *X* appropriate to MSY. That there should be some MSY against which to design a fishery is a clear ideal. However, because there is oversimplification in arriving at the analysis, it is not a sufficient management tool. It does have many problems; prominent is the assumption that *Y/X* is proportional to *B*, which works better for some stocks and fishing situations than for others. Another is that we cannot study *Y* at large *X* without endangering the stock, as pointed out above. Then there is the assumption that the biomass of the stock is all the same, just a variable responding in a deterministic way to fishing effort. Of course, fishing usually rearranges the age, size, and reproductive structure of the stock, and substantial lag times are involved in its response to changes in fishing mortality. Reducing a stock also gives advantage to its ecological competitors, which may increase. Reversing this replacement may not occur quickly or at all if fishing is scaled back. Finally, MSY assumes the habitat is constant, supporting the stock at the same level consistently. Virtually every stock examined over long intervals shows that this not true, as illustrated above for the eastern Pacific yellowfin tuna. Carmel Finley (2011), has stated well the overall problem with MSY management: "the emphasis tends to fall on *maximum*, not on *sustainable*".

However, clear and generally correct idealizations are very valuable and MSY retains that value. Peter Larkin's (1977) "Epitaph for the Concept of MSY" is appropriate for fishery experts, but for the rest of us the idea still has some useful life:

"Here lies the concept MSY,
It advocated yields too high,
And didn't spell out how to slice the pie.

We bury it with best of wishes,
Especially on behalf of fishes.
We don't know yet what will take its place,
But we hope it's as good for the human race."

Larkin's last line shows even he thought it had its virtues.

Current management techniques (and/or philosophies) include "virtual population analysis" (VPA) and "reasonable and prudent alternatives" (RPA) and "ecosystem based management" (EBM, lately rephrased as "ecosystem approach to management", EAM). A suite of techniques for evaluating stock size, condition, and interaction with fishing, VPA incorporates detailed analyses of size, age, and reproductive status, and applies year-class estimates to take account of the uneven recruitment rates among years. At least in the United States, RPA is a legal mandate to find and apply management schemes that protect stocks, protect other populations affected by fisheries, and prevent undue harm to marine habitats. Ecosystem-based managers attempt to consider everything about exploited stocks and their habitat, with concern for habitat health ranking above concern for yield. The general terms of EBM are spelled out by Pikitch *et al.* (2004). It can at least be mandated, if not readily applied. In all cases, the stated goals are for fisheries to be prudently managed with an eye to long-term environmental health and stock protection. All such efforts are made vastly more complex by their interaction with economic and political factors. In the USA, for example, management is not done by scientists alone, but by regional management councils composed of stakeholders in the fishery – fishers, packers, and scientists. Both the councils and the scientific agencies somehow sustain their work despite recurring interference from the courts, which are brought into play through suits filed by environmental organizations on behalf of fish and by fishing folk trying to protect their livelihood. Thus, the complexity of the scientific terminology and issues and the usually strained economics of fishing (next) have become overlain by the historical arcana rife in our legal system.

Fishery economics

Fishery biology interacts profoundly with fishery economics. If we have a well-behaved and well-understood fishery, we can control it to any of several ends:

- for fishers to make lots of money;
- to get lots of nutritious, tasty food;
- to protect the stock as fully as possible while still exploiting it at a useful rate.

Usually we don't achieve control, for a variety of reasons. Perhaps the most important is that most fisheries are free, unowned resources. Anyone who can obtain a boat can go fishing. Thus, so long as there is a dollar, yen, or krone to be made in a fishery, more fishers will show up to attempt to earn it. Therefore, in most of the world's fisheries, the average profit is very close to nothing. This comes about as follows. The dollar (or yen, ¥) return to a fishery is roughly proportional to yield (although large yield will depress prices). That is, $\$ = kY$. Yield has a parabolic relationship to effort, or at least varies as some curve showing diminishing returns. So, dollar return must also be parabolic with effort. Cost of effort, on the other hand, increases roughly linearly with effort: each additional boat costs about the same as the one before it. These two relationships are shown together as follows:

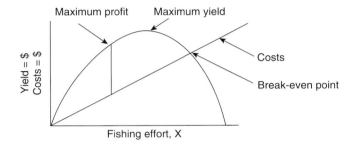

Below break-even point there is return to investment; carpenters are tempted to go fishing | Above break-even point fishers tend to become carpenters

So long as the fishery is below the break-even point, new entrants (buying boats and starting to fish) can make a profit. Approaching the break-even point there is still a profit, but it is divided between more and more boats. At the break-even point, the mean profit is zero. In fact, many or most fisheries move about their break-even point with year-to-year variations in catch and effort. *On average, the return to investment in many fisheries is zero*. If a fishery were to be maintained by regulation near MSY, then too many boats would appear to take that catch, again distributing the profit so thinly that the mean would be near zero. Limiting fishing days or gear type often does this. Thus, fishers (and expensive gear) must sit idle much of the time or use the least-efficient equipment. Fishing industries regulated in these ways are typically low on productivity per unit capital, and labor productivity is abominable. The best solutions are franchise fisheries (where the fisher folk own the fish stock) and limited-entry fisheries. The politics of those are fearsome, but such schemes are being established in some places, notably coastal Japan, British Columbia, Alaska, and even for US West Coast trawl fisheries.

Regime changes

Since the late 1980s, fisheries scientists have been fascinated by what they have come to call "regime changes" (e.g. Hollowed *et al.* 1987); in fisheries these are the responses of stocks to the decadal climate oscillations introduced in the last chapter. An example was given above, the shift in CPUE of yellowfin tuna around 1985. Such substantial differences in stock levels and production between multi-year and multi-decade intervals occur everywhere in the oceans, to all sorts of fish. The best-studied examples are found among the forage fishes (mostly Clupeidae: sardines, anchovies, herrings, capelin, and others), which are subject to industrial fisheries. Bakun (Bakun 1996; Bakun *et al.* 2009) points out that these fish are the "waist" in what he calls "wasp-waisted" pelagic ecosystems. Phytoplankton and zooplankton are usually moderately diverse, with many species, while the schooling forage fish feeding upon them belong to one or two species, one generally being strongly dominant at any given time. Those forage fish are fed upon in turn by a diverse range of fish, squid, birds, and mammals: mackerel, salmon, cod, ommastrephid squid, seals, terns, and whales. Thus, intermediate trophic transfers pass through a low-diversity link, the wasp waist, between more complex levels below and above.

In temperate and eastern tropical neritic seas (out to 100+ nautical miles), the dominant forage species is often a sardine or an anchovy. For example, the dominant schooling fish in the California Current are usually *Sardinops sagax cerulea* (sardine) or *Engraulis mordax* (anchovy). Fishing off the California coast began at the end of the 19th century and became a substantial industry in the 1920s (Fig. 17.18a), generating good data on take. The sardine fishery, particularly off central California, built up to takes over 0.6 Mt in the 1930s, when biomass is believed to have been 3.6 Mt, and sustained that level to the late 1940s. Then it crashed, failing to support any fishery after 1951 with biomass reaching 0.01 Mt in the 1960s. It isn't clear exactly when anchovy stocks appeared to replace these sardines, because no significant fishery for them developed until the 1970s. Development of fisheries requires not only a stock of fish to catch, but also a market and entrepreneurs to invest in processing. Somebody, or sometimes chickens, must eat the catch or convert its oil content to paint. Nevertheless, a fishery did emerge, grew to 0.3 Mt, and then crashed in two steps by 1990. To a slight extent it has been replaced since by renewed sardine fishing, but at rates probably not taking full advantage of the available stock, which is believed to be more than 0.9 Mt (down from a peak of 1.8 Mt in 2000, not yet up to estimates from the 1930s, ~3.6 Mt). Landings have increased and in 2007 were ~0.08 Mt off California, ~0.3 Mt off Mexico. Thus, the harvest of forage fish off California shows a series of multi-decade "regimes".

Similar sequences have affected sardine (*Sardinops melanostictus*) and anchovy (*Engraulis japonicus*) fisheries off Japan (Fig. 17.18b), off western South America (*S. sagax, E. ringens*) (Fig. 17.18c) and off southwest Africa (Namibia). All of these regions show similar fishery regime

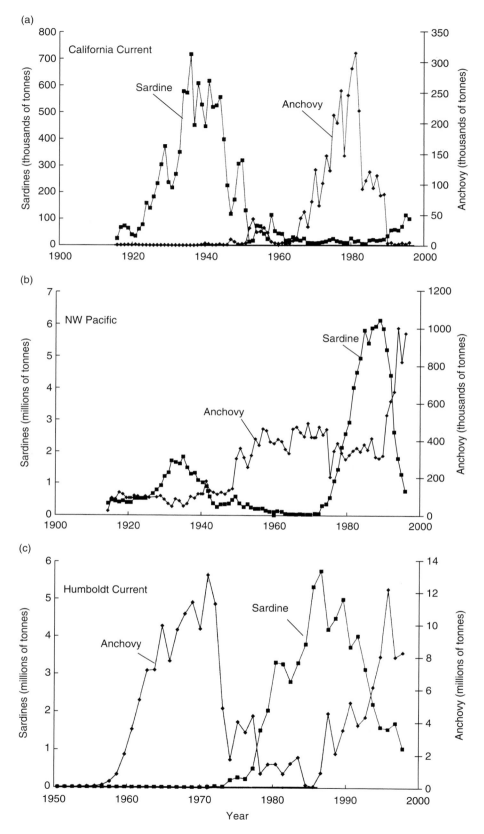

Fig. 17.18 (a) Annual catch from California Current sardine and anchovy stocks. (b) Annual catches of northwest (Asian) Pacific sardine and anchovy. (c) Annual catches of Chilean sardine and Peruvian anchoveta. (Data from Schwartzlose *et al.* 1999.)

shifts. In most of them changes in stock size are accompanied by shifts in geographical distribution. That is well illustrated by the Japanese sardine, which has a much smaller spawning range than feeding range (Schwartzlose *et al.* 1999). Spawning is inshore around southern Japan and central China, a range in which the stock mostly remains to feed when numbers are low. In those times, the catch is mostly confined to the southern Sea of Japan and inshore South China Sea. When the stock is large, however, it expands for feeding into the northern Sea of Japan and eastward into the Pacific, sometimes sustaining fishing east of the dateline. Both the California and Chilean sardines also show range expansions when abundant and contractions when stocks are reduced (e.g. Bakun *et al.* 2010; Fig. 17.19).

Regime shifts off Asia, California, and South America show some degree of simultaneity. That was first commented upon by Kawasaki (1983), based mostly on the sardine changes. Common timing is most evident for the Japanese and Chilean sardines, and is made more evident by normalizing the catch statistics and plotting them over the same date scale (Alheit & Bakun, 2010, Fig. 17.20), and in both cases the sardine pulses were simultaneous with dramatic dips in the associated anchovy stocks. On the other hand, sardines were not strongly fished off Peru and Chile until after the anchoveta collapse of the early 1970s,

so the comparison does not go back very far. Californian sardines did not follow the Japanese and Chilean stocks down in the 1990s. Anchovy cycles are less similar. The northwest Pacific anchovy catches only dropped by a quarter after 1970, not to near zero like those of anchoveta off Peru. The shift in Asian anchovy landings could as well be explained by replacement of anchovy in the marketplace by the newly accessible sardines. There is no question that as sardines declined in both regions after 1990, anchovy stocks and catches boomed in both. At a minimum, regime durations are similar in all three regions, and they do involve a switch-off between the two wasp-waist fish species. Californian and South African clupeid fisheries both have roughly the same periods as the Asian and South American ones, but they are somewhat out of phase.

A few of the hypotheses proposed in respect to regime control of "wasp-waist" forage fish warrant review. While it is reasonable to postulate a role for fisheries in the wasp-waist regime shifts, we also know that they occurred regularly in both directions long before there were any industrial fisheries. The data for that are anchovy and sardine scale-deposition rates in the sediments of anoxic basins, among which the Santa Barbara Basin in the southern California Current is the best studied (Baumgartner *et al.* 1992). This subsidence basin, just southeast of Point Conception, California, has a flat-bottomed central pit at 580 m depth

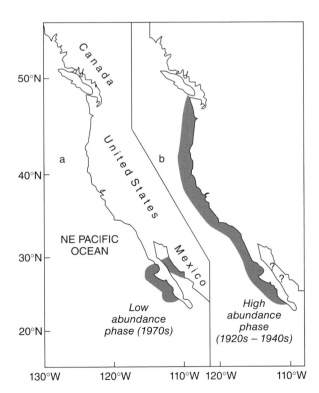

Fig. 17.19 Distributional patterns of California sardines (*Sardinops sagax caeruleus*) when more (right) and less (left) abundant. (After Bakun *et al.* 2010.)

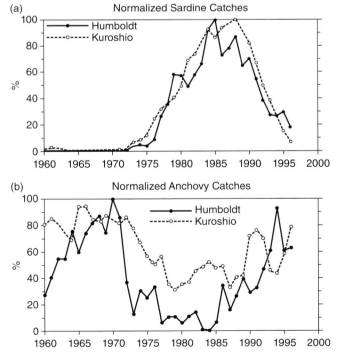

Fig. 17.20 Sardine (a) and anchovy (b) landing rates for the Japanese and Humboldt Current fisheries standardized to percentage of maximum rates. (After Alheit & Bakun 2010, based on data from Schwartzlose *et al.* 1999.)

that is partly filled with sediment. This is only very rarely flushed and generally remains anoxic with almost no benthic macrofauna. Thus, there is little bioturbation, and the sediment retains visible annual layers or "varves". These are terrestrial sediment from winter runoff, alternating with pelagic ooze from summer production. Over the past few thousand years there have been occasional intervals with bioturbation, and there have been some slumps. Thus, exact dating by varve counting becomes problematic below about 1100 yr BP, but various corrections give good approximate dating over several thousand years. There are variations in scale counts between cores, and problems for deeper levels of inadequate sample volumes (low counts and associated higher variances). However, by counting only to 10-year resolution and doing some averaging between cores, the picture emerges (Fig. 17.21) of substantial temporal variability in both species.

It is established for the late 20th century that sediment scale counts from the Santa Barbara Basin are reasonably well correlated with fishery estimates of the regional anchovy and sardine stocks. Most of the scales in the sediment are from fish less than 2 years old, those with highest mortality rates. There is no evidence of scale deterioration or loss from the sediments, in fact counts decrease over the 2000-year series as a whole. Anchovy are generally the dominant fish, two to three times more abundant on average over two millennia. Both species come and go in quasi-periodic fashion, anchovy with more obvious regularity than sardines. Spectral analysis shows a dominant period for anchovy of about 100 years, and for sardines of about 160 years. The spectra show some shorter period "energy" as well, but that isn't obvious in the raw series. Both of these periods are longer than the few cycles of variation apparent in fishery results, suggesting that the system has

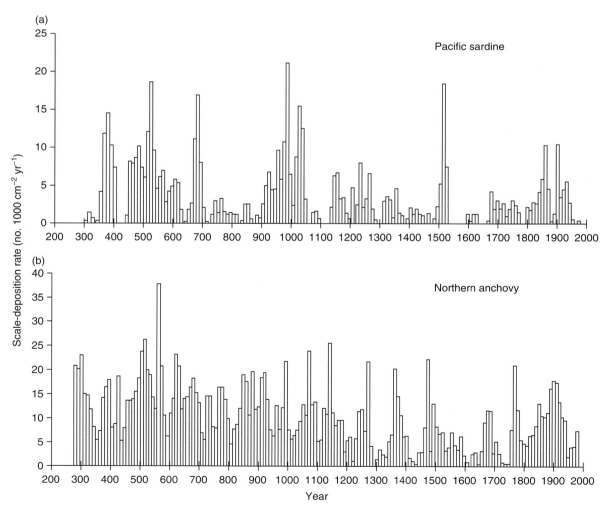

Fig. 17.21 Time-series of (a) sardine and (b) anchovy scale abundance (as deposition rates) in sediments of the anoxic Santa Barbara Basin from *c*. 300 CE to the present. (After Baumgartner *et al.* 1992.)

been changed by some combination of industrial-scale fishing and recent shifts in the basic character of the habitat. In any case, the sediments establish that regime shifts have always existed; we just see them now through the rather distorting lens of fishery data. Shorter time series from sediment under anoxic waters off Peru and Chile (e.g. Díaz-Ochoa *et al.* 2009) also show alternations, with anchoveta scales usually dominant, seconded by myctophid, sardine, or hake scales.

Key factors in all regional cases appear to be differences between sardines and anchovy in respect to temperature preferences and best food-particle sizes. In the Humboldt and California Current systems, the high sardine intervals have been relatively warmer and higher in salinity, likely with less equatorward flow and less coastal upwelling, while periods of anchovy dominance have been colder inshore with stronger upwelling. Both genera feed primarily on zooplankton, but sardines have more closely spaced gill rakers and filter the smaller animals typical of warmer, more offshore waters. Anchovy are primarily visual particle feeders, thriving on somewhat larger zooplankton typical in colder, more productive nearshore upwelling ecosystems (van der Lingen 2006). Thus they are distinctive in both temperature optima and preferred food-types. The Japanese comparison is similar in respect to food type, but reversed in respect to temperature. That is likely to represent distinctive adaptation schemes matching eastern and western boundary current systems.

For the Californian case, Rykaczewski and Checkley (2008) have suggested that the food-size preference combines with the relative strength of two distinct modes of upwelling (Fig. 11.39) to generate regimes favorable and unfavorable to sardines. Coastal upwelling forced by equatorward nearshore winds generates cold, nutrient-rich ecosystems with diatoms supporting abundant larger zooplankton (*Calanus, Euphausia*, . . .). If that is the only prevalent upwelling effect, then anchovy are more likely to thrive. In sardine-favorable intervals, winds are stronger well offshore than close inshore, such that there is strong *wind-stress curl* (a spatial gradient in wind speed). Then, in addition to coastal upwelling, slower, more dispersed upwelling occurs by different physics, by "Ekman pumping" offshore of the main upwelling (Fig. 11.39). That extends the zone of enhanced primary production seaward, but at a less-intense level. This supports greater production of smaller zooplankton (possibly *Paracalanus* and *Oithona*) without strong cooling, setting the scene for increases in sardines. Rykaczewski and Checkley show the effect by comparing levels of curl-driven upwelling (calculated from monthly mean wind-speed maps) for the May–July season of sardine larval development with sardine "surplus production", basically a measure from fisheries and survey data of sardine recruitment, a statistic primarily influenced each year by the abundance of 0-age sardines (Fig. 17.22). The resemblance is close enough that they may well be right. A

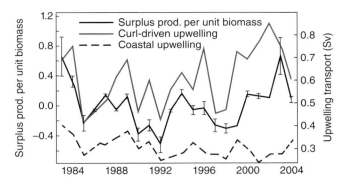

Fig. 17.22 Time-series of relative recruitment rate of California sardine compared to coastal upwelling and curl-driven upwelling May through July off Southern California. Recruitment is uncorrelated with coastal upwelling but significantly ($r = 0.62$, $P \sim 0.005$) correlated with the curl effect. (After Rykaczewski & Checkley 2008.)

similar comparison for the Peru–Chile region will be enlightening.

The timing of Humboldt Current switches between anchovies and sardines has been attributed by Alheit and Bakun (2010) directly to the distinctions between the species in temperature preference and feeding ecology. The rise in sardines starting about 1972 corresponded with a strong and prolonged rise in near-coast SST and salinity. Conditions shifted from predominance of cold coastal water (14–18°C, $S = 35$) to more prevalent subtropical surface conditions (18–27°C, $S = 35.1$–35.7). Large zooplankton became diminished, anchovy recruitment (already hammered by fishery removal of older spawners) fell, and stock abundance plummeted and stayed down. Sardines were favored by the new conditions, their stock coming to support fisheries of 6 Mt yr^{-1}. In the late 1980s, conditions shifted back toward lower temperatures and salinity. Coastal sardines declined and anchovy finally recovered and were again fished at a 12 Mt yr^{-1} rate in 1993. They dropped from that peak, but stocks and yields have remained high, except for 1998. Fishing was actually stopped for a significant interval during the El Niño of that year, and then resumed.

Synchrony of the Humboldt and Japanese sardine "outbursts" of 1973–1993 appears to have to do with at least Pacific-scale atmospheric "teleconnections", but the Japanese case is quite different. The Japanese sardine migrates to spawn in winter where the Kuroshio approaches Japan. The larvae are carried north as they grow, the adults migrating also into the Kuroshio–Oyashio confluence where both juveniles and adults feed and grow. Recruitment is strongest during cooler periods in the spawning grounds, which tend to correspond to greater zooplankton stocks in the spring feeding grounds. This combination appears to be associated with the strength of the Asian winter monsoon, for which of course there is an index, the MOI (high from

1970 to 1990), and low winter values of the Arctic Oscillation (AO, low in the same interval). These are, however, closely enough associated with periods of decadal variations indexed by the PDO and even SOI that the exact atmospheric effects controlling variations remain to be resolved. Unlike the Peru case, the 1980s burst of the Japanese sardine did not greatly diminish Japanese anchovy catches, but more anchovies were taken as the sardine declined. That may have been a stock increase or simply switching of the fishery to the more available alternative. We clearly must wait and observe(!) through at least one more cycling from anchovy to sardines in all of these regions to more fully understand the control mechanisms. More long-term climate variation, including anthropogenic warming, may cause such cycles to shift in duration and in their relationships to atmospheric events.

Regime shifts have major economic and cultural consequences for entire regions along the coast. There was a dramatic regime shift affecting salmon over the past three decades. About 1976 there was a fairly sudden warming across the entire northern reach of the Pacific, at least a 0.5°C increase in mean surface temperatures. This was beneficial to salmon at the northern end of their range, so that catches in the northern Gulf of Alaska and eastern Bering Sea climbed to unheard-of heights. Meanwhile, with warming and probably reduced lower-trophic-level production, the ocean survival of salmon plummeted off Washington and Oregon, USA. Most stocks became inadequate to support commercial fishing. Not only was it hard to catch fish in the Pacific Northwest in the 1980s and early 1990s, but prices were depressed by abundant, and therefore cheap, salmon from Alaska. Boats were sold, processing plants closed and port waterfronts were converted to tourist restaurants. In Alaska, in contrast, boats were purchased, processors expanded, college students got summer jobs cutting fish, and fishers and others made a good living. In about 1999 there was another shift, at least off the Oregon and Washington coasts. Ocean survival, particularly of coho salmon, rebounded, and in 2001 the number of salmon returning to the Columbia River hatcheries set a record (perhaps partly because ocean fishing had not yet rebounded in response). Returns have increased and decreased on a year-to-year basis since. With some one-year exceptions, Alaska catches have remained high.

Terminology notes

An elaborate, often cute, terminology has developed around regime shifts, recruitment variation, and their relationships to fish biology and behavior. Andrew Bakun (2010), who has coined some of these terms, considers them to represent hypotheses that have recurring application in understanding stock changes and species interactions in marine fisher-

ies ecology. For example, we have defined *wasp-waisted ecosystems* sufficiently above. Bakun calls these terms "conceptual templates" and "schematic constructs" (which turn out to be the same thing), basically a tool kit of hypotheses that may (or may not) be adjusted to fit a specific situation. Here are some more that are in current use.

Match–mismatch (David Cushing) refers to the necessity for strong recruitment that spawning of a stock and development of its larval feeding requirements match the timing of food producing events. For cod that might be a match (or not) of their hatching with spawning of *Calanus* to provide the larvae with early and abundant nauplii to eat. *Critical periods* for finding food before starvation are a related conceptual template, one making matches (as opposed to mismatches) a survival necessity.

Those two concepts are closely similar to *Lasker windows*. Reuben Lasker emphasized the need for larvae to find sufficient food and associated that with the need for strong vertical layering of zooplankton prey. Finding suitable concentrations would be a "window" to growth beyond the prey sizes of planktonic predators, to juvenile status and to mobility. Lasker coupled this to the importance of ocean stratification so that layers of prey for larvae would not be disturbed in the larva's critical period.

Connectivity (Michael Sinclair and others) – egg and larval drift must connect planktonic young with suitable locations for juvenile development. As detailed above for the term "hydrographic containment", spawning sites become selected as places usually providing that connectivity. It can fail if currents are unusual (another sort of match–mismatch requirement). This relates to *ocean triads*, Bakun's notion that large stocks can develop in areas where: (i) the ocean is enriched (upwelling, mixing, iron available, etc.), (ii) concentrations are sustained (convergences, fronts, etc.), and (iii) flow connects larvae to juvenile habitat.

P2P loop (Bakun) – "prey to predator" reversed interaction. The notion is that in marine ecosystems forage fish can prey heavily on the eggs and larvae of their predators (e.g. cod, hake, bluefish, . . .), thus protecting themselves because those potential predators do not grow up. Bakun suggests, for example, that the recently booming herring stocks in the northwest Atlantic may be eating cod spawn, slowing cod stock recovery, sustaining their own peak abundance. Quantifying the importance of P2P can be difficult.

Loopholes (Bakun and Kenneth Broad) – places where predation on eggs or larvae is low enough to overcome their disadvantages as habitat. Bakun and Broad (2003) give the example of tuna spawning in regions of minimal production relative to the rest of their adult range, possibly to lower the predation threat to larvae. Emphasis on not being eaten can be as important as eating well.

Predator pits (Bakun) – at low density, of say a forage fish, its predators will ignore it and survival will be high. To reach high density, the stock must cross a level of abundance (a low level, but not the lowest possible, the "pit") in which it is controlled by its attentive predators. If good conditions allow a breakout above the "critical point" at which predation becomes saturated, then the population can have low per capita mortality and reach very high abundance.

School trap (Bakun, again) – a far from dominant species of fish may school with the dominant species, tolerating food or other conditions less than ideal for it, in order to obtain the predator avoidance benefits of large schools. This reduces its potential for increase in one respect, protecting it in another. Mixed schools have been widely observed. Schooling tendency can be strong enough that huge schools cannot act effectively to search the habitat for food (they are "trapped"). Maybe. Bakun modifies this as the *school-mix feedback loop* – dominant fish with inappropriate behavior (say some genotype) can lead whole schools to destruction. Maybe.

Optimal stability window (Ann Gargett) – another window. The general suggestion has been taken from Gargett's (e.g. 1997) notion that an intermediate level of turbulence provides ideal biological conditions. Intermediate wind speeds, wave activity and vertical mixing rates are likely to provide the best conditions for survival.

In addition, a number of explanatory notions have been adopted in fisheries from general ecological terminology; these are also attempts to generate conceptual templates. A *portfolio effect* (coined by Frank Figge) has been claimed by Schindler *et al.* (2010) for lake-spawning sockeye salmon. The variability of the overall stock returning to Bristol Bay, Alaska, is much less that those of the individual runs to different lakes. Like a stock-investment portfolio, diversification of subpopulation behavior provides some stability to the overall system. Others are the "*rivet hypothesis*" – loss or sharp decline in one or a few species ("rivets" in the ecosystem structure) can cause general breakdown; and *functional compensation* – more than one species provides each function in the habitat, and reduction of one can be compensated by others, with redundancy providing stability (the opposite of the popping rivet notion). Like physicists naming quarks, gluons, and charm, fisheries ecologists will continue to coin suggestive (and cute) terminology.

Status of world fisheries

The United Nations Food and Agriculture Organization (FAO) is the world's central data-gathering authority for fisheries. Each year they produce the FAO Fisheries Yearbook, an elaborate statistical summary including data from several years prior to 2 years before publication. The FAO has websites with the majority of their recent graphs and charts. A current review of world fisheries can be found by searching under fisheries at the following site: http://www.fao.org/.

Basic recent capture fisheries data, as well as inland fisheries and aquaculture production figures, can be found by following links at: http://www.fao.org/fishery/statistics/en ("en" for English, also available in French and Spanish)

Use of the data files may require downloading and installation of the FAO "Fishstat" software. Some of the more general tables are also available on the worldwide web: ftp://ftp.fao.org/FI/STAT/SUMM_TAB.HTM.

You should check these sources in addition to the summaries below, since it is best to inform yourself about the status of fisheries with up-to-date results. Even data for years well past are often updated as new reports are received at the FAO from fisheries agencies around the world. For a second opinion on global fisheries status, see Garcia and Rosenberg (2010).

We (the people) fish harder and harder all the time. Whether or not the overall fishery production of the oceans is sustainable remains to be seen (Box 17.1). In about 1973 total production of all marine fisheries stopped increasing at the dramatic quasi-exponential rate, about 6% per year, which had persisted since World War II (Plate 17.1). The last phase of growth at that rate was sustained by development of the Peruvian anchoveta fishery, which peaked around 12 Mt. This temporary ceiling was reached in the 1970s with surprising suddenness as the Peruvian anchoveta fishery crashed in 1972. For a time after 1973, the world total stayed close to 65 Mt live weight, perhaps increasing at an unsteady 1% per year. Continued growth of demersal (near-bottom) fisheries held the total up during the 1970s. This latter phase was accompanied by substantial investment in large, world-ranging trawlers. However, that came to an end in 1978 as the industrial fishing nations (most importantly Japan and the USSR) stretched the world demersal resources to their limit.

Then in the mid-1980s came a series of recurring increases. The trawling industry remained in place and output held steady for a time. Initially, the burst of the mid-1980s was in the top few species: Alaskan pollock, Japanese sardine, capelin, and Chilean jack mackerel. Several of these species, particularly the clupeids, had dramatic increases after the Pacific regime shift of 1976. It was warmer, and production was definitely greater in these stocks. Some fisheries responded immediately (Japan, Chile), while others (herring off New England, the California sardine) did not have the markets to drive a return to intensive harvests. Chilean fisheries took much of the mid-1980s increase, and Chile was the No. 4, even No. 3, fishing nation in those years. That was from fishing on sardine and mackerel stocks, plus fishing along its southern coast on a variety of species. In the 1990s, the Peruvian anchoveta fishery returned to over 9 Mt yr^{-1}, in 1995 12 Mt, which covered some declines in demersal stock production worldwide, actually raising the world total and leading to 1996–1997 marine-capture fishery totals over 86 Mt. Those were the highest totals to date. Total yield dipped in 1998 to about

Box 17.1 Potential fishery production

Several workers (Ryther 1969; Pauly & Christensen 1995) have supposed that it should be possible to predict the rate at which fisheries can remove biomass from the sea by combining a global estimate of primary production with a simple ecological calculation. They approximate the trophic level (T) of "fish" taken by fisheries, apply a general ecological efficiency at each food-chain step leading to T; take account of the need to leave much new biomass in the ocean to support the health and reproduction of stocks; then provide a final estimate by appropriate multiplications. Current estimates of global primary production are approximately 44 GtC per year (Behrenfeld & Falkowski 1997). More productive coastal areas might be roughly 10% of ocean area, but they produce about 50% of fishery output. They also have relatively short food chains with much fishery production from the third and fourth trophic levels, so perhaps $T \approx 3.5$. The remaining seas are more oligotrophic, and food chains are longer, with fishery production at perhaps $T \approx 5.5$.

Ecological efficiency, EE, the fraction of its production that one tropic level transfers on average to its predators, is in fact extremely difficult to estimate. Of course, the value is constrained: it cannot be zero, and it cannot be greater than typical lifetime growth efficiency, that is, (tissue formed)/(food eaten). For many marine animals, growth efficiency is quite high, especially for juveniles, perhaps 30%. Ecological efficiencies are less, since only transfers "upward" to higher-order predators count, not "losses" to the decomposer trophic level. So, like Ryther and all other practitioners of these predictions, let us guess. As a start, let EE $\approx$ 20%.

Expressing the model as an equation and leaving two-thirds of production for stock maintenance (not a strongly prudent number, actually), we have:

$$\text{Potential catch} = 0.33\,(44\ \text{GtC}\,[0.5\text{EE}^{3.5} + 0.5\text{EE}^{5.5}])$$
$$= 0.027\ \text{GtC}.$$

This must be converted from carbon to "fish". Using reasonable, approximate factors (carbon/dry weight = 0.4, dry weight/wet weight = 0.3), we have:

$$\text{Potential catch} = 0.027\ \text{GtC}/(0.3 \times 0.4)$$
$$= 0.227\ \text{Gt wet weight}.$$

And, finally, this must be converted to millions of metric tonnes (Mt):

$$\text{Potential catch} = 0.227 \times 10^3\ \text{Mt/Gt} = 227\ \text{Mt}$$

That is about double the global catch, including discarded bycatch. It is of importance that the world fishery take is of the same order of magnitude as this outcome, whatever its associated uncertainty. It definitely implies that removal of biomass by us, "the people", is a very large factor in marine ecology at higher trophic levels.

If you examine the equation, you will see that it is exquisitely sensitive to the value chosen for EE, a value we must guess. Putting potential catch at current catch, about 125 Mt, and solving for EE (iteratively or by logs), gives 17%, by no means an unlikely mean value for the oceans, given the state of current knowledge of trophic ecology. Of course, the calculation is also sensitive to the trophic level chosen as representative for fishery production, the proportions assigned to coastal and oceanic production, and the accuracy of the grand sum of primary production. You should explore this yourself.

If limitation of potential catch to one-third of production is eliminated, and EE is reduced to 10%, the supposed potential is only 19 Mt, less than 20% of actual. This all but proves that EE around 10%, which was assumed by ecologists a generation ago as standard, and which is still taught in elementary ecology courses, must be too low for at least the oceans.

We may very well be fishing accessible resources as intensely as the stocks can bear, or more intensely in the long term, even if the apparent potential catch is double the actual catch. Accessibility is the issue. Unexploited squid and krill alone represent enough productivity to account for the difference. Other production at the trophic levels of fishery stocks appears in little-exploited jellyfish, in mid-water fishes, and in widespread oceanic stocks not possible to exploit commercially. Thus, the calculation, along with the recently flat catch totals and other considerations, suggests that we are pushing close to the oceans' potential to produce fish for human use.

---

78 Mt as the Peruvians established a partial fishing moratorium during an El Niño. Anchoveta catches rebounded in 1999, taking the global capture total to 84.5 Mt. Also in the 1990s, the demersal trawling fleet decreased as the former USSR, Poland, and East Germany retreated from intense

fishing activity. The retreat was enforced by the installation of capitalist-style accounting for costs, particularly opportunity costs. Earlier, Communist Bloc fishing was inefficient from an economic standpoint. World totals were not greatly reduced by reduction in Eastern Bloc harvesting. The gap was filled by return of the Peruvians to exploitation of the anchoveta. Since 2000, annual capture fisheries totals have been 80 to 84 Mt, consistently led by anchoveta ranging from 7 to 10.7 Mt and Alaskan pollock at 2.7 to 2.9 Mt (latest complete data are for 2008).

In addition to capture fisheries, coastal waters around the world now host substantial mariculture operations (distinct from aquaculture operations in freshwater). Important products are salmon (mostly in Norway and Canada), mollusks (mussels, oysters, clams, and scallops) and shrimp. These activities produced over 12 Mt of these products in 1999 and 19.7 Mt in 2008. Note that FAO reports whole weight of mollusks, including shells, and most shells end up in piles near the shore, not inside diners. Aquaculture has grown even more, from 18.4 Mt in 1999 to 32.9 Mt in 2008. In both cases, and marine-capture landings as well, there has been doubt about totals reported by the Chinese based on the special value of over-reporting in their accounting system (e.g. Brander 2007). That is not accounted for here.

Many fishing operations do not bring to port all that they catch. Captured fish and invertebrates other than the species targeted by a fishing operation are termed "bycatch". Some bycatch is valuable enough to return to port, so is included in the catch totals. But much bycatch has no accessible market, is made up of species closed to fishing by regulations, or just has no commercial value. Fishers shovel that part, most of it killed or injured by the gear and hauling aboard, back into the ocean. Discarded bycatch is a major component of the mortality imposed by fisheries on marine populations. Alverson *et al.* (1994) used observer estimates to calculate a likely world total in 1992 of 27 Mt, or about a third of the FAO landings estimate for that year. Discarded bycatch probably has been about that proportion of landings for a long time (assumed to show the bycatch impact in Plate 17.1), amounts greater than the total clupeid landings.

At the present time, China is the number-one fishing nation, even accounting for likely over-reporting. It has risen to that rank by dramatic increases since 1980, mostly concentrated around its own coast and very broadly distributed among many species. Moreover, China has more than matched its increased fishing effort in expansion of pond (and rice paddy) culture of carp and *Tilapia*. Extending fisheries and aquaculture has made China self-sufficient for good-quality protein.

The capacity of commercial fishery interests to push the total catch upward has been astounding. So far as we know, few new resources exist that might be exploited to increase fishery production. However, it does not pay to say that a ceiling has been reached; leave it to the Japanese fishing masters to find more edible stocks. Mid-water squid populations might be a resource with some potential, if efficient means of capture can be devised. At present, jigging under night-lights for species that migrate to the surface in the dark is the only effective capture method. Japanese vessels fishing worldwide obtain large takes with automated jigging gear. Returns to trawling at depth are not large enough to make it economical. Antarctic krill (*Euphausia superba*) are a resource with potential to produce 50 to 200 Mt per year, and the Japanese, West Germans, and Eastern Bloc nations tried for a while to develop that fishery. However, a decade of research and preliminary commercial development brought the 1982 catch to 0.53 Mt, and it has been less in all years since, e.g. 0.016 Mt in 2008 (not among the top 69 species on the world list). The stock is in the most remote region of the world ocean, fuel costs for exploitation are extreme, and efforts to make krill palatable to people are only marginally successful. Processing is difficult, since the digestive glands of the animals tend to break, spilling enzymes into the tissue which digest it and make it rot almost instantly. Very rapid, specialized preservation methods are required to produce a usable product. As a consequence of these difficulties, krill have not been very promising compared to alternative investments.

Pauly *et al.* (1998) have shown that we are progressively moving closer to the base of the food chain in our exploitation, which they characterize as "fishing down the food chain". They argue that we are reducing stocks of higher-trophic-level fish (e.g. cod, tuna, swordfish) and moving more and more effort to planktivorous fish (pilchards, anchoveta, etc.). It is arguable whether placing our take at higher or lower trophic levels is the best strategy for fishery exploitation, but it is very likely that the trend does not represent a strategy. It simply reflects (i) the recurring reduction in stocks of the predatory fish favored at the table, as noted by Pauly *et al.*, but just as important is (ii) the recurring resurgence and exploitation of forage fish stocks, particularly the Peruvian anchoveta. At times the "mean trophic level" statistic used actually moves up. That occurs when the anchoveta or similar stocks collapse. The overall changes are a small fraction of one trophic level and probably do not mean much.

Ecological impacts

Pollution and fishing impose the largest impacts of human activity on marine environments. The obvious impact of fishing is removal of fish, which disrupts food chains directly by removing biomass from populations of the larger animals. There must be "top-down" effects from removing predators, freeing prey stocks to increase, then to over-exploit their own food sources. There are relatively few documented examples. Perhaps the best case study was the near removal of the blue, fin and other krill-eating whales around Antarctica in the 1960s and 1970s. We have no direct evidence that krill increased, but crab-eater seals, penguins, and other populations of krill-eating animals doubled and redoubled. Such rearrangements of food webs probably always occur when large stocks are heavily exploited, we just don't generate the data to see them.

Overall levels of exploitation of fishery stocks and the degree and significance of habitat modification by fishing activity are currently areas of active study and intense dispute. There is no question that the fauna of the modern ocean, at least above the bathypelagic and abyssal plains is substantially different from what it was (whatever it was) before industrial fishing. Even the very first fishing industries modified ocean biota. An excellent review of the historical development of fisheries was provided by Callum Roberts in *The Unnatural History of the Sea* (2007), and Anthony Koslow in *The Silent Deep* (2007) has covered the more recent effects of trawling in very deep habitats, particularly the destruction of rich soft-coral communities on seamounts in the hunt for orange roughy and other long-lived fish. Orange roughy were rapidly fished to "commercial extinction". Roberts shows that 17th- to 19th-century whaling generated all but permanent endangered-species status for the right whales and sperm whales, although the latter may finally be coming back. Those devastations were accomplished from sailing ships by launching rowing boats armed with hand-thrown harpoons. Virtually no tropical islands have their former assemblage of large carnivorous fish, particularly groupers (Serranidae) and snappers (Lutjanidae). Those have mostly been fished out by sport and artisanal fishers, and sometimes by large-scale attacks with rotenone or explosives. Rosenberg *et al.* (2005) have roughly reconstructed from boat logs and other data a history of the New England and Scotian shelf cod stocks. They put biomass of the lightly fished stock of 1852 at 1.26×10^6 Mt. Industrial fishing and habitat damage in the middle to late 20th century reduced that to "less than 5×10^4 Mt today". Recovery has not been obvious during a near moratorium on cod fishing (they still are taken as bycatch) since 1994. There are too many such sad tales to recount here.

Prominent in the recent controversies was a paper by Ransom Myers and Boris Worm (2003), based partly on the development of the Japanese long-line fishery for tuna and billfish after World War II. Excellent records were kept for this fishery as it expanded east from Japan into the Pacific and eventually across the world, and a time-series of contour maps of its catches per unit effort show high initial catches at the leading edge of the expansion, followed by progressive declines that finally are global. On the basis of CPUE (catch per 100 hooks fished; Fig. 17.23) trends for oceanic areas, Myers and Worm (2003) estimated that residual stocks are ~10% of their mid-20th-century size for this overall species mix. Criticism followed quickly. For example, Maunder *et al.* (2006, discussed above) strongly criticized use of CPUE for estimating stock size, explicitly because it was the basis of Myers and Worm's claims. Long lining has indeed not been a consistent prac-

tice, and different stocks are differentially susceptible to changing depths of the baits and other modifications. Some stocks, particularly yellowfin tuna in the eastern tropical Pacific, have declined and then recovered to an extent, and a case can be made that they are sustainably harvested. However, others, particularly the larger sizes of bluefin tuna (*Thunnus thynnus*), prized for sushi, are stocks *in extremis*. The value of such fish interacts with availability to sustain fishing, even though catches diminish to nearly zero. One large bluefin (377 kg) was sold for \$396,700 at Tsukiji market in Tokyo in January 2011. That much money can buy lots of boat fuel and search time. The value has also led to live harvest of young specimens for pen-rearing off Australia and in the Mediterranean, which has its own problems and may or may not relieve pressure on the wild stock. Ten percent may or may not be the best overall generalization for high-level predatory fish, but the global reductions in only about 50 years surely have strongly modified ocean ecosystems. Because of the shifting baseline effect, even ocean biologists come to think of the new situation as normal.

Prominent among ecosystem rearrangements of recent decades, one attributable in part to fish stock reduction, is the increasing abundance of jellyfish, particularly of scyphozoan medusae. Recall (Chapter 6) that this group alternates between a small (<1 cm), attached polyp stage and very large, dome-shaped medusae trailing masses of tentacles. The polyps "strobilate" producing larval medusae (ephyra) that grow to substantial sizes – from serving platter-size (*Aurelia*) to dinner-table size (*Nemopilema*). Reducing planktivorous fish stocks may reduce predation on ephyrae. Trawling may remove predators on polyps. Seasonal pulses of abundant medusae, like the summer outbursts of *Chrysaora fuscescens* off Oregon, USA, and in the Bering Sea then compete for zooplankton meals and eat fish larvae, stabilizing the community shift. The most dramatic outbursts have been those of the 2 m, 500 kg Nomura's jellyfish (*Nemopilema nomuri*) in waters surrounding Japan (Kawahara *et al.* 2006). Like many scyphozoan polyps, those of this monster are unusually tolerant of suboxia, which recently is prevalent in their benthic habitat beneath the outflow of the Changjiang River draining northeast China. Thus, it is likely they suffer reduced predation in this area where they are abundant on shallow bottoms, increasing production of larvae. The ephyrae grow as they are carried across the East China Sea, and turn north with the springtime Tsushima warm current into the East Korean/Japan Sea. Some years, Nomura's jellyfish are most prominent along the Korean coast, some years along the Japanese coast. Reaching Tsugaru Strait, the coastal flow carries them into the Pacific and south along-shore of Honshu. In many recent years they have been sufficiently numerous to stop all trawling. Netting just one or two fully developed medusae creates a serious problem for fishing machinery. The economic impact on Japanese and Korean

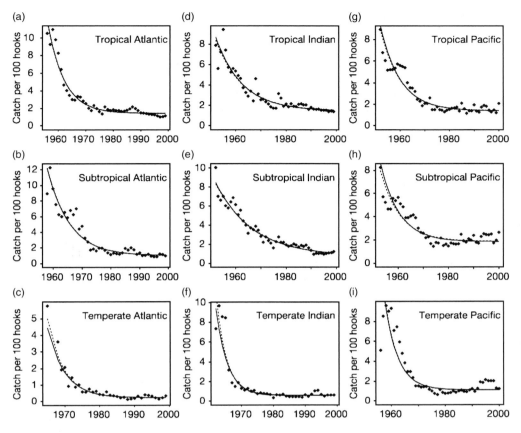

Fig. 17.23 Time-series of CPUE (catch per hundred hooks fished) for different ocean areas fished by the Japanese long-line tuna industry. (After Myers & Worm 2003.)

coastal fisheries is substantial. Richardson *et al.* (2009) provide a general review of the many jellyfish "problems" worldwide.

Fishing quite consistently interacts with climate variations to become over-fishing. We leave this topic to other authors. For a clean statement of how management and politics interact to prevent limitation of fishing to sustainable levels (in so far as the habitat is consistent enough for any semblance of sustainability) see Rosenberg (2003). For a review of the depletion and failure to recover of the northwest Atlantic cod fisheries, see Rose (2004).

The technology of some fisheries is relatively benign, apart from the fish killed. Hauling a trawl or trolling baited hooks through mid-water probably does not damage the biota significantly, apart from the fish targeted and captured. But hauling trawls right against the bottom is another matter entirely. The "tickler chains" of large benthic trawls are hauled though the upper layers of sediment, capturing or breaking masses of invertebrates, plowing the sediment to several decimeters depth, leaving in some habitats a wasteland that takes years to regenerate (e.g. see Jennings *et al.* 2001). Some heavily trawled bottoms, such as Georges

Bank or the eastern English Channel, probably have not approached anything like a natural, untrawled condition in many decades. In fact, into the early 19th century the seabed of the North Sea was entirely covered with oyster beds (Roberts 2007). Early trawlers pulled heavy chains over the whole region to improve the habitat for plaice and sole, and now it is all mud and sand. That is good if you like plaice and sole. In some cases, nearly natural benthic habitat conditions may be required for maintenance of the fish stocks that are targeted by the trawlers that disrupt the bottom. Thus, fishing not only removes fish, but also can hamper stock recovery. The entire issue has come strongly into public attention since the late 1990s, and work is under way to evaluate and mitigate trawl damage. Les Watling, a benthic ecologist, has been a leader on this issue (Watling 2005).

Lost gear of all kinds becomes part of the marine environment for a long time. Crab and lobster traps that have lost their surface floats continue to collect. Latching escape doors with corrodible clasps that last only a little while beyond the anticipated pull date can alleviate this. Synthetic fibers now used almost exclusively in trawls and gill (drift) nets last many years adrift in the ocean. Every mariner has seen seals and whales struggling to swim while stuck in a blanket of old mesh, suffering a slow, tortured death. Lost gill nets go on fishing and fishing, sometimes for years, entangling both fish and mammals. Mammals, particularly the harbor porpoise in coastal regions, are often drowned while entangled in gill nets that are not lost. Recent international conventions have somewhat reduced the net loss problem by requiring shorter, better-marked runs of net, and have reduced bycatch (and possibly catch) by closing some areas. These problems cannot be totally eliminated if we are going to fish. However, the tuna fisheries in the eastern tropical Pacific have achieved great reductions in porpoise mortality, despite continuing to set seines around porpoise schools. Similar reductions in sea-turtle mortality have been achieved in shrimp fisheries by installing turtle-excluding devices in trawl mouths. These and other success stories show that careful design of operations can minimize bycatch and habitat destruction.

A movement now afoot among environmentalists, fishery managers, and even the fishing industry seeks to protect commercial fishery stocks, other populations, and some parts of the marine habitat, by establishing reserves closed to fishing. It is reasonable to expect that refuges could allow some individual organisms to reach large size and high reproductive rates (to become BOFFF); that the young will prosper in undisturbed habitats; and that burgeoning populations will spread from reserves into surrounding fishing zones. Many aspects of reserve design are under study, including physical dimensions, types of bottom to include, and siting that takes advantage of flow patterns. Both spawning sites, where abundant species congregate intermittently, and nursery areas are obvious candidate locations. Evaluation of sites and quantification of reserve effects have become research topics in their own right. Results already available show dramatic success at improving habitat quality and fishing in neighboring zones (Lubchenco *et al.* 2007), sometimes primarily just outside the reserve perimeter.

Last words

If you reached this point by reading the whole book, congratulations and thanks. We, too, are glad to have reached it. Take the day off and go fishing. There are still some fish, if only some.

REFERENCES

Aagaard, K., Andersen, R., Swift, J. & Johnson, J. (2008) A large eddy in the central Arctic Ocean. *Geophysical Research Letters* 35: L09601, doi:10.1029/2008GL033461.

Abbott, M.R., Richman, J.G., Letelier, R.M. & Bartlett, J.S. (2000) The spring bloom in the Antarctic Polar Frontal Zone as observed from a mesoscale array of bio-optical sensors. *Deep-Sea Research II* 47: 3285–3314.

Acuña, J.L., Deibel, D. & Morris, C.C. (1996) Particle capture mechanism of the pelagic tunicate *Oikopleura vanhoeffeni*. *Limnology and Oceanography* 41: 1800–1814.

Agogue, H., Brink, M., Dinasquet, J. & Herndl, G.J. (2008) Major gradients in putatively nitrifying and non-nitrifying Archaea in the deep North Atlantic. *Nature* 456: 788–792.

Ahrens, M.J., Hertz, J. & Lamoureux, E.M. (2001) The effect of body size on digestive chemistry and absorption efficiencies of food and sediment-bound organic contaminants in *Nereis succinea* (Polychaeta). *Journal of Experimental Marine Biology and Ecology* 263: 185–209.

Ailing, V., Humborg, C., Mörth, C.-M., Rahm, L. & Pollehne, F. (2008) Tracing terrestrial organic matter by $\delta^{34}S$ and $\delta^{13}C$ signatures in a subarctic estuary. *Limnology and Oceanography* 53: 2594–2602.

Aizawa, Y. (1974) Ecological studies of micronektonic shrimp (Crustacea, Decapoda) in the western North Pacific. *Bulletin of the Ocean Research Institute, University of Tokyo* 6: 1–84.

Aksnes, D. & Ohman, M.D. (1996) A vertical life table approach to zooplankton mortality estimation. *Limnology and Oceanography* 41: 1461–1469.

Albert, A., Echevin, V., Lévy, M. & Aumont, O. (2010) Impact of nearshore wind stress curl on coastal circulation and primary productivity in the Peru upwelling system. *Journal of Geophysical Research* 115: (13 pp.)

Alexander, W.B., Southgate, B.A. & Bassindale, R. (1935) Survey of the River Tees. II. The estuary – chemical and biological. *Water Pollution Research Technical Paper* No. 5, xiv + 171 pp. Department of Scientific and Industrial Research, United Kingdom.

Alheit, J. & Bakun, A. (2010) Population synchronies within and between ocean basins: apparent teleconnections and implications as to physical–biological linkage mechanisms. *Journal of Marine Systems* 79: 267–285.

Alldredge, A. (1998) The carbon, nitrogen and mass content of marine snow as a function of aggregate size. *Deep-Sea Research I* 45: 529–541.

Alldredge, A.L., Passow, U. & Logan, B.E. (1993) The abundance and significance of a class of large, transparent organic particles in the ocean. *Deep-Sea Research* 40: 1131–1140.

Aller, J.Y., Aller, R.C. & Green, M.A. (2002) Benthic faunal assemblages and carbon supply along the continental shelf/shelf break-slope off Cape Hatteras, North Carolina. *Deep-Sea Research II* 49: 4599–4625.

Aluwihare, L.I., Repeta, D.J. & Chen, R.F. (1997) A major biopolymeric component to dissolved organic carbon in surface seawater. *Nature* 387: 166–169.

Alvain, S., Moulin, C., Dandonneau, Y. & Loisel, H. (2008) Seasonal distribution and succession of dominant phytoplankton groups in the global ocean: a satellite view. *Global Biogeochemical Cycles* 22: GB3001, doi:10.1029/2007GB003154.

Alvariño, A. (1965) Distributional atlas of chaetognatha in the California Current region. *CalCOFI Atlas* No. 3. 299 pp.

Alverson, D.L., Freeberg, M.H., Murawski, S.A. & Pope, J.G. (1994) A global assessment of fisheries bycatch and discards. *FAO Fisheries Technical Papers*, No. 339.

Ambler, J.W. (1986) Effect of food quantity and quality on egg production of *Acartia tonsa* Dana from East Lagoon, Galveston, Texas. *Estuarine, Coastal and Shelf Science* 23: 183–196.

Ambler, J.W. & Miller, C.B. (1987) Vertical habitat partitioning by copepodites and adults of subtropical oceanic copepods. *Marine Biology* 94: 561–577.

Anderson, A.C., Jolivet, S., Claudinot, S. & Lallier, F.H. (2002) Biometry of the branchial plume in the hydrothermal vent tube-worm *Riftia pachyptila* (Vestimentifera;[sic] Annelida). *Canadian Journal of Zoology* 80: 320–332.

Anderson, D.M. & 8 coauthors (2005) *Alexandrium fundyense* cyst dynamics in the Gulf of Maine. *Deep-Sea Research II* 52: 2522–2542.

Anderson, P. & Sorensen, H.M. (1986) Population dynamics and trophic coupling in pelagic microorganisms in eutrophic coastal waters. *Marine Ecology Progress Series* 33: 99–109.

Anderson, R.F., Ali, S., Bradtmiller, L.I., Nielsen, S.H.H., Fleisher, M.Q., Anderson, B.E. & Burckle, L.H. (2009) Wind-driven upwelling in the Southern Ocean and the deglacial rise in atmospheric CO_2. *Science* 323: 1443–1448.

Andersson, J.H., Wijsman, J.W.M., Herman, P.M.J., Middleburg, J.J., Soetaert, K. & Heip, K. (2004) Respiration patterns in the deep ocean. *Geophysical Research Letters* 31: L03304 (4 pp) doi: 10.1029/2003GL018756.

Andrews, K.J.H. (1966) The distribution and life history of *Calanoides acutus* (Giesbrecht). *Discovery Reports* 34: 1–116.

Anonymous (1963) CalCOFI Atlas of 10-meter temperatures and salinities 1949 through 1959. *CalCOFI Atlas* No. 1. California Cooperative Oceanic Fisheries Investigations, La Jolla, California, iv + 296 pp.

Antezana, T., Ray, K. & Melo, C. (1982) Trophic behavior of *Euphausia superba* Dana in laboratory conditions. *Polar Biology* 1: 77–82.

Antoine, D., Morel, A., Gordon, H.R., Banzon, V.F. & Evans, R.H. (2005) Bridging ocean color observations of the 1980s and 2000s in search of long-term trends. *Journal of Geophysical Research* 110: C06009 (22 pp.).

Archer, S.D., Leakey, R.J.G., Burkill, P.H. & Sleigh, M.A. (1996) Microbial dynamics in coastal waters of East Antarctica: herbivory by heterotrophic dinoflagellates. *Marine Ecology Progress Series* **139**: 239–255.

Aristegui, J., Gasol, J.M., Duarte, C.M. & Herndl, G.J. (2009) Microbial oceanography of the dark ocean's pelagic realm. *Limnology and Oceanography* **54**: 1501–1529.

Armstrong, W. (1979) Aeration in higher plants. *Advances in Botanical Research* **7**: 225–332.

Arp, A.J., Childress, J.J. & Vetter, R.D. (1987) The sulphide-binding protein in the blood of the vestimentiferan tube-worm, *Riftia pachyptila*, is the extracellular haemoglobin. *Journal of Experimental Biology* **128**: 139–158.

Arrigo, K.R., van Dijken, G.L. & Bushinsky, S. (2008) Primary production in the Southern Ocean. *Journal of Geophysical Research* **113**: C08004, doi: 10.1029/2007JC004551, 208.

Aruda, A.M., Baumgartner, M.F., Reitzel, A.M. & Tarrant, A.M. (2011) Heat shock protein expression during stress and diapause in the marine copepod *Calanus finmarchicus*. *Journal of Insect Physiology* **57**: 665–675.

Atkinson, A. (1998) Life cycle strategies of epipelagic copepods in the Southern Ocean. *Journal of Marine Systems* **15**: 289–311.

Atkinson, A., Siegel, V., Pakhomov, E.A., Jessopp, M.J. & Loeb, V. (2009) A re-appraisal of the total biomass and annual production of Antarctic krill. *Deep-Sea Research I* **56**: 727–740.

Audzijonyte, A. & Vrijenhoek, R.C. (2010) When gaps really are gaps: statistical phylogeography of hydrothermal vent invertebrates. *Evolution* **64**: 2369–2384.

Azam, F., Fenchel, T., Gray, J.G., Meyer-Reil, L.A. & Thingstad, T. (1983) The ecological role of water-column microbes in the sea. *Marine Ecology Progress Series* **10**: 257–263.

Bachraty, C., Legendre, P. & Desbruyères, D. (2009) Biogeographic relationships among deep-sea hydrothermal vent faunas at global scale. *Deep-Sea Research I* **156**: 1371–1378.

Baier, C.T. & Purcell, J.E. (1997) Effects of sampling and preservation on apparent feeding by chaetognaths. *Marine Ecology Progress Series* **146**: 37–42.

Bailey, K.M., Francis, R.C. & Stevens, P.R. (1982) The life history and fishery of Pacific whiting, *Merluccius productus*. *CalCOFI Reports* **23**: 81–98.

Bailey, S.W. & Werdell, P.J. (2006) A multi-sensor approach for the on-orbit validation of ocean color satellite data products. *Remote Sensing of Environment* **102**: 12–234.

Baines, S.B. & Pace, M.L. (1991) The production of dissolved organic matter by phytoplankton and its importance to bacteria: patterns across marine and freshwater systems. *Limnology and Oceanography* **36**: 1078–1090.

Baker, A.deC., Boden, B.P. & Brinton, E. (1990) *A Practical Guide to the Euphausiids of the World*. Natural History Museum Publications, London, 96 pp.

Bakun, A. (1993) The California Current, Benguela Current, and Southwestern Atlantic shelf ecosystems: a comparative approach to identifying factors regulating biomass yields. In: Sherman, K., Alexander, L.M. & Gold, B.D. (eds.) *Large Marine Ecosystems; Stress, Mitigation and Sustainability*. AAAS Press, Washington, DC, pp. 199–221.

Bakun, A. (1996) *Patterns in the Ocean, Ocean Processes and Marine Population Dynamics*. California Sea Grant, La Jolla, CA, 323 pp.

Bakun, A. (2010) Linking climate to population variability in marine ecosystems characterized by non-simple dynamics: conceptual templates and schematic constructs. *Journal of Marine Systems* **79**: 361–373.

Bakun, A. & Broad, K. (2003) Environmental "loopholes" and fish population dynamics: comparative pattern recognition with focus on El Niño effects in the Pacific. *Fisheries Oceanography* **12**: 458–473.

Bakun, A. & Weeks, S.J. (2008) The marine ecosystem off Peru: what are the secrets of its fishery productivity and what might its future hold? *Progress in Oceanography* **79**: 290–299.

Bakun, A., Babcock, E.A. & Santora, C. (2009) Regulating a complex adaptive system via its wasp-waist: grappling with ecosystem-based management of the New England herring fishery. *ICES Journal of Marine Science* **66**: 1768–1775.

Bakun, A., Babcock, A.E., Lluch-Cota, S.E., Santora, C. & Salvadeo, C.J. (2010) Issues of ecosystem-based management of forage fisheries in "open" non-stationary ecosystems: the example of the sardine fishery in the Gulf of California. *Reviews in Fish Biology and Fisheries* **20**: 9–29.

Balch, W.M., Poulton, A.J., Drapeau, D.T., Bowler, B.C., Windecker, L.A. & Booth, E.S. (2011) Zonal and meridional patterns of phytoplankton biomass and carbon fixation in the Equatorial Pacific Ocean, between 110°W and 140°W. *Deep-Sea Research II* **58**: 400–416.

Baldwin, R.J., Glatts, R.C. & Smith, K.L. Jr (1998) Particulate matter fluxes into the benthic boundary layer at a long time-series station in the abyssal NE Pacific: composition and fluxes. *Deep-Sea Research II* **45**: 643–666.

Ball, E.E. & Miller, D.J. (2006) Phylogeny: the continuing classificatory conundrum of chaetognaths. *Current Biology* **16**: R593–R596.

Ballance, L.T., Pitman, R.L. & Fiedler, P.C. (2006) Oceanographic influences on seabirds and cetaceans of the eastern tropical Pacific: a review. *Progress in Oceanography* **69**: 360–390.

Baltar, F., Aristegui, J., Gasol, J.M., Sintes, E. & Herndl, G.J.(2009) Evidence of prokaryotic metabolism on suspended particulate matter in the dark waters of the subtropical North Atlantic. *Limnology and Oceanography* **54**: 182–193.

Bannister, T.T. (1974) Production equations in terms of chlorophyll concentration, quantum yield, and upper limits to production. *Limnology and Oceanography* **19**: 1–12.

Banoub, M.W. & Williams, P.J.le B. (1973) Seasonal changes in the organic forms of carbon, nitrogen and phosphorus in the English Channel in 1968. *Journal of the Marine Biological Association of the United Kingdom* **53**: 695–703.

Banse, K. (1995) Zooplankton: pivotal role in the control of ocean production. *ICES Journal of Marine Science* **52**: 265–277.

Barber, R.T. & 7 coauthors (1996) Primary productivity and its regulation in the equatorial Pacific during and following the 1992–1993 El Nino. *Deep-Sea Research II* **43**: 933–969.

Barlow, R.G., Mantoura, R.F.C., Gough, M.A. & Fileman, W.T. (1993) Pigment signatures of the phytoplankton composition in the northeastern Atlantic during the 1990 spring bloom. *Deep-Sea Research II* **40**: 459–477.

Basedow, S.L. & Tande, K.S. (2006) Cannibalism by female *Calanus finmarchicus* on naupliar stages. *Marine Ecology Progress Series* **327**: 247–255.

Batchelder, H.P. & Miller, C.B. (1989) Life history and population dynamics of *Metridia pacificus*: results from simulation modelling. *Ecological Modelling* **48**: 113–136.

Batchelder, H.P., Edwards, C.A. & Powell, T.M. (2002) Individual-based models of copepod populations in coastal upwelling regions: implications of physiologically and environmentally influenced diel vertical migration on demographic success and nearshore retention. *Progress in Oceanography* **53**: 307–333.

Batchelder, H.P., Mackas, D.L. & O'Brien, T.O. (2011) Spatial–temporal scales of synchrony in marine zooplankton biomass and abundance patterns: a world-wide comparison. *SCOR Working Group 125 Report*. 54 pp.

Batten, S.D. and Freeland, H.J. (2007) Plankton populations at the bifurcation of the North Pacific Current. *Fisheries Oceanography* **16**: 536–646

Baum, J.K. & Worm, B. (2009) Cascading top-down effects of changing oceanic predator abundances. *Journal of Animal Ecology* 78: 699–714.

Baumgartner, T.R., Soutar, A. & Ferreira-Bartrina, V. (1992) Reconstruction of the history of Pacific sardine and northern anchovy populations over the past two millennia from sediments of the Santa Barbara Basin, California. *CalCOFI Report* 33: 24–40.

Bé, A.W.H. & Tolderlund, D.S. (1971) Distribution and ecology of living planktonic foraminifera in surface waters of the Atlantic and Indian Oceans. In: Funnell, B. & Riedel, W.R. (eds.) *Micropalaeontology of the Oceans*. Cambridge University Press, Cambridge, pp. 105–149.

Bé, A.W.H., MacClintock, C. & Currie, D.C. (1972) Helical shell structure and growth of the pteropod *Cuvierina columella* (Rang) (Mollusca, Gastropoda). *Biomineralisation Forschungsberichte* 4: 47–79.

Beam, C.A. & Himes, M. (1979) Sexuality and meiosis in dinoflagellates. In: Levandowsky, M. & Hutner, S.H. (eds.) *Biochemistry and Physiology of Protozoa*, Vol. 3. Academic Press, New York, pp. 171–206.

Beardall, J. & Morris, I. (1976) The concept of light intensity adaptation in marine phytoplankton: some experiments with *Phaeodactylum tricornutum*. *Marine Biology* 37: 377–387.

Beaugrand, G. (2010) Decadal changes in climate and ecosystems in the North Atlantic Ocean and adjacent seas. *Deep-Sea Research II* 56: 656–673.

Beaugrand, G. (2009) Decadal Changes in climate and ecosystems in the North Atlantic Ocean and adjacent seas. *Deep-Sea Research II* 56: 656–673.

Beaugrand, G. & Kirby, R.R. (2010) Climate, plankton and cod. *Global Change Biology* 16: 1268–1280.

Beckmann, A. & Hense, I. (2009) A fresh look at the nutrient cycling in the oligotrophic ocean. *Biogeochemistry* 96: 1–11.

Behrenfeld, M.J. (2010) Abandoning Sverdrup's critical depth hypothesis on phytoplankton blooms. *Ecology* 91: 977–989.

Behrenfeld, M.J. & Falkowski, P.G. (1997a) Photosynthetic rates derived from satellite-based chlorophyll concentration. *Limnology and Oceanography* 42: 1–20.

Behrenfeld, M.J. & Falkowski, P.G. (1997b) A consumer's guide to phytoplankton primary production models. *Limnology and Oceanography* 42: 1479–1491.

Behrenfeld, M.J. & 9 coauthors (2006) Climate-driven trends in contemporary ocean productivity. *Nature* 444: 752–755.

Béjà, O. & 11 coauthors (2000) Bacterial rhodopsin: evidence for a new type of phototrophy in the sea. *Science* 289: 1902–1906.

Bender, M. & 12 coauthors (1987) A comparison of four methods for the determination of planktonic community metabolism. *Limnology and Oceanography* 32: 1085–1098.

Bender, M., Ellis, T., Tans, P., Francey, R. & Lowe, D. (1996) Variability in the O_2/N_2 ratio of southern hemisphere air, 1991–1994: implications for the carbon cycle. *Global Biogeochemical Cycles* 10: 9–21.

Bender, M., Orchardo, J., Dickson, M.-L., Barber, R. & Lindley, S. (1999) In vitro O_2 fluxes compared with ^{14}C production and other rate terms during the JGOFS Equatorial Pacific experiment. *Deep-Sea Research I* 46: 637–654.

Benner, R. (2002) Chemical composition and reactivity. In: Hansell, D.A. & Carlson, C.A. (eds.) *Biogeochemistry of Marine Dissolved Organic Matter*, Academic Press, San Diego, CA, pp. 59–90.

Benner, R. & 7 coauthors (1993) Measurement of dissolved organic carbon and nitrogen in natural waters: workshop report. *Marine Chemistry* 41: 5–10.

Benoit-Bird, K.J. (2009) Dynamic 3-dimensional structure of thin zooplankton layers is impacted by foraging fish. *Marine Ecology Progress Series* 396: 61–76.

Benoit-Bird, K.J., Cowles, T.J. & Wingard, C.E. (2009) Edge gradients provide evidence of ecological interactions in planktonic thin layers. *Limnology and Oceanography* 54: 1382–1392.

Berelson, W.M. (2001) The flux of carbon into the ocean interior: a comparison of four U.S. JGOFS regional studies. *Oceanography* 14: 59–67.

Berelson, W.M. & 11 coauthors (1997) Biogenic budgets of particulate rain, benthic remineralization and sediment accumulation in the equatorial Pacific. *Deep-Sea Research II* 44: 2251–2282.

Berg, H.C. & Purcell, E.M. (1977) Physics of chemoreception. *Biophysics Journal* 20: 193–219.

Berg, P., Risgaard-Petersen, H. & Rysgaard, S. (1998) Interpretation of measured concentration profiles in sediment pore water. *Limnology and Oceanography* 43: 1500–1510.

Berg, P., Røy, H., Janssen, F., Meyer, V., Jørgensen, B.B., Huettel, M. & de Beer, D. (2003) Oxygen uptake by aquatic sediments measured with a novel non-invasive eddy-correlation technique. *Marine Ecology Progress Series* 261: 75–83.

Berg, P., Glud, R.N., Hume, A., Stahl, H., Oguri, K., Meyer, V. & Kitazato, H. (2009) Eddy correlation measurements of oxygen uptake in deep ocean sediments. *Limnology and Oceanography Methods* 7: 576–584.

Berger, A. (1977) Long-term variation of the earth's orbital elements. *Celestial Mechanics* 15: 53–74.

Berger, W.H., Ekdale, A.A. & Bryant, P.P. (1979) Selective preservation of burrows in deep-sea carbonates. *Marine Geology* 32: 205–230.

Berkeley, S.A., Chapman, C. & Sogard, S.M. (2004) Maternal age as a determinant of larval growth and survival in a marine fish, *Sebastes melanops*. *Ecology* 85: 1258–1264.

Besiktepe, S. & Dam, H. (2002) Coupling of ingestion and defecation as a function of diet in the calanoid copepod *Acartia tonsa*. *Marine Ecology Progress Series* 229: 151–164.

Beversdorf, L.J., White, A.E, Björkman, K.M., Letelier, R.M. & Karl, D.M. (2010) Phosphonate metabolism of *Trichodesmium* IMS101 and the production of greenhouse gases. *Limnology and Oceanography* 55: 1768–1778.

Biddanda, B. & Benner, R. (1997) Carbon, nitrogen and carbohydrate fluxes during the production of particulate and dissolved organic matter by marine phytoplankton. *Limnology and Oceanography* 42: 506–518.

Bidle, K.D. & Azam, F. (2001) Bacterial control of silicon regeneration from diatom detritus: significance of bacterial ectohydrolases and species identity. *Limnology and Oceanography* 46: 1606–1623.

Bieri, R. (1959) The distribution of the planktonic chaetognatha in the Pacific and their relationship to the water masses. *Limnology and Oceanography* 4: 1–28.

Billen, G., Servais, P. & Becquevort, S. (1990) Dynamics of bacterioplankton in oligotrophic and eutrophic aquatic environments: bottom-up or top-down control? *Hydrobiologia* 207: 37–42.

Billett, D.M.S., Bett, B.J., Reid, W.D.K., Boorman, B. & Priede, I.G. (2010) Long-term change in the abyssal NE Atlantic: the "Amperima Even" revisited. *Deep-Sea Reseach II* 57: 1406–1417.

Bilyard, G. (1974) The feeding habits and ecology of *Dentalium entale stimpsoni* Henderson (Mollusca: Scaphopoda). *The Veliger* 17: 126–138.

Bilyard, G. & Carey, A.G. Jr (1979) Distribution of western Beaufort Sea polychaetous annelids. *Marine Biology* 54: 329–339.

Binet, D. & Suisse de Sainte-Claire, E. (1975) Contribution à l'étude du copepode planctonique *Calanoides carinatus*: répartition et cycle biologique au large de la Côte-d'Ivoire. *Cahiers ORSTOM, séries Océanographie* 13: 15–30.

Bishop, J.K.B., Calvert, S.E. & Soon, M.Y.-S. (1999) Spatial and temporal variability of POC in the northeast Subarctic Pacific. *Deep-Sea Research II* 46: 2699–2733.

Bishop, J.K.B., Davis, R.E. & Sherman, J.T. (2002) Robotic observations of dust storm enhancement of carbon biomass in the North Pacific. *Science* **298**: 817–821.

Bissinger, J.E., Montagnes, D.J.S., Sharples, J. & Atkinson, D. (2008) Predicting marine phytoplankton maximum growth rates from temperature: improving on the Eppley curve using quantile regression. *Limnology and Oceanography* **53**: 487–493.

Blachowiak-Samolyk, K. & Angel, M.V. (2008) *An Atlas of Southern Ocean Ostracods*. http://deep.iopan.gda.pl/ostracoda/index.php. (National Oceanography Centre, Southampton, UK).

Bobko, S.J. & Berkeley, S.A. (2004) Maturity, ovarian cycle, fecundity, and age-specific parturition of black rockfish (*Sebastes melanops*). *Fishery Bulletin* **102**: 418–429.

Bochdansky, A.B. & Deibel, D. (1999) Functional feeding response and behavioral ecology of *Oikopleura vanhoeffeni* (Appendiculara, Tunicata). *Journal of Experimental Marine Biology and Ecology* **233**: 181–211.

Bochdansky, A.B., Deibel, D. & Hatfield, E.A. (1998) Chlorophyll-*a* conversion and gut passage time for the pelagic tunicate *Oikopleura vanhoeffeni* (Appendicularia). *Journal of Plankton Research* **20**: 2179–2197.

Bode, A., Cunha, M.T., Garrido, S., Peleteiro, J.B., Porteiro, C., Valdés, L. & Varela, M. (2007) Stable nitrogen isotope studies of the pelagic food web on the Atlantic shelf of the Iberian Peninsula. *Progress in Oceanography* **75**: 115–131.

Bollens, S. & Frost, B.W. (1989a) Zooplanktivorous fish and variable diel vertical migration in the marine planktonic copepod *Calanus pacificus*. *Limnology and Oceanography* **34**: 1072–1083.

Bollens, S. & Frost, B.W. (1989b) Predator-induced diel vertical migration in a planktonic copepod. *Journal of Plankton Research* **11**: 1047–1065.

Bollens, S., Rollwagen-Bollens, G., Quenette, J.A. & Bochdansky, A.B. (2011) Cascading migrations and implications for vertical fluxes in pelagic ecosystems. *Journal of Plankton Research* **33**: 349–355.

Bond, G. & 9 coauthors (1997) A pervasive millennial-scale cycle in North Atlantic Holocene and glacial climates. *Science* **278**: 1257–1266.

Booth, B.C., Lewin, J. & Postel, J.R. (1993) Temporal variation in the structure of autotrophic and heterotrophic communities in the subarctic Pacific. *Progress in Oceanography* **32**: 57–99.

Boss, E. & 7 coauthors (2008) Observations of pigment and particle distributions in the western North Atlantic from an autonomous float and ocean color satellite. *Limnology and Oceanography* **53**: 2112–2122.

Bothe, H., Tripp, H.J. & Zehr, J.P. (2010) Unicellular cyanobacteria with a new mode of life: the lack of photosynthetic oxygen evolution allows nitrogen fixation to proceed. *Archives of Microbiology* **192**: 783–790.

Böttger-Schnack, R. & Huys, R. (2004) Size polymorphism in *Oncaea venusta* Philippi, 1843 and the validity of *O. frosti* Heron, 2002: a commentary. *Hydrobiologia* **513**: 1–5.

Boudreau, B.P. (1998) Mean mixed depth of sediments: the wherefore and the why. *Limnology and Oceanography* **43**: 524–526.

Boudreau, B.P. (2004) What controls the mixed-layer depth in deep-sea sediments? The importance of particulate organic carbon flux. *Limnology and Oceanography* **49**: 620–622.

Boxshall, G.A. & Halsey, S.H. (2004) *An Introduction to Copepod Diversity*. The Ray Society, London. Pt.1, xv + 421 pp.; Pt. 2, vii + pp. 422–966.

Boyd, P.W. & 12 coauthors (2005) The evolution and termination of an iron-induced bloom in the north east subarctic Pacific. *Limnology and Oceanography* **50**: 1872–1886.

Boyd, P.W. & 22 coauthors (2007) Mesoscale iron enrichment experiments 1993–2005: synthesis and future directions. *Science* **315**: 612–617.

Boyd, S.H., Wiebe, P.W. & Cox, J.L. (1978) Limits of *Nematoscelis megalops* in the Northwestern Atlantic in relation to Gulf Stream cold core rings. II. Physiological and biochemical effects of expatriation. *Journal of Marine Research* **36**: 143–159.

Boyle, E.A. (1988) Cadmium: chemical tracer of deepwater paleoceanography. *Paleoceanography* **3**: 471–489.

Boyson Jensen, P. (1919) Valuation of the Limfjord. I. Studies on the fish-food in the Limfjord 1909–1917, its quantity, variation and animal production. *Report of the Danish Biological Station* **26**: 1–44.

Brand, L.E. (1991) Minimum iron requirements of marine phytoplankton and the implications for the biogeochemical control of new production. *Limnology and Oceanography* **36**: 1756–1771.

Brander, K. (1997) Effects of climate change on cod (*Gadus morhua*) stocks. In: Wood, C.M. & McDonald, D.G. (eds.) *Global Warming: Implications for Freshwater and Marine Fish*. Cambridge University Press, New York, pp. 255–278.

Brander, K. (2007) Global fish production and climate change. *Proceedings of the National Academy of Sciences* **104**: 19,709–19,714.

Breitbart, M., Middelboe, M. & Rohwer, F. (2008) Marine viruses: Community dynamics, diversity and impact on microbial processes. In: Kirchman, D.L. (ed.) *Microbial Ecology of the Oceans*, Second Edition, Wiley, New York, pp. 443–479.

Bricaud, A., Babin, M., Morel, A. & Claustre, H. (1995) Variability in the chlorophyll-specific absorption coefficients of natural phytoplankton: analysis and parameterization. *Journal of Geophysical Research* **100**: 13,321–13,332.

Bricaud, A., Claustre, H., Ras, J. & Oubelkheir, K. (2004) Natural variability of phytoplanktonic absorption in oceanic waters: influence of the size structure of algal populations. *Journal of Geophysical Research* **109**: C11010, doi: 10.1029/2004JC002419.

Bright, M. & Lallier, F.H. (2010) The biology of vestimentiferan tube worms. *Oceanography and Marine Biology: an Annual Review* **48**: 213–266.

Brinton, E. (1962) The distribution of Pacific euphausiids. *Bulletin of the Scripps Institute of Oceanography* **8**: 51–270.

Brinton, E. (1967a) Distributional Atlas of Euphausiacea (Crustacea) in the California Current Region, Part 1. *CalCOFI Atlas* No. 5. California Cooperative Oceanic Fisheries Investigations, La Jolla, California, xii + 275 pp.

Brinton, E. (1967b) Vertical migration and avoidance capability of euphausiids in the California Current. *Limnology and Oceanography* **12**: 451–483.

Brinton, E. (1976) Population biology of *Euphausia pacifica* off southern. *California* **74**: 733–762.

Brinton, E., Ohman, M.D., Townsend, A.W., Knight, M.D. & Bridgeman, A.L. (1999, MacIntosh/2000, PC) *Euphausiids of the World Ocean* (CD-ROM expert system). Springer-Verlag, Heidelberg.

Britschgi, T.B. & Giovannoni, S.J. (1991) Phylogenetic analysis of a natural marine bacterioplankton population by rRNA gene cloning and sequencing. *Applied and Environmental Microbiology*, **57**: 1313–1318.

Brock, T.D. (1981) Calculating solar radiation for ecological studies. *Ecological Modelling* **14**: 1–19.

Bronk, D.A. & Glibert, P.M. (1994) The fate of the missing ^{15}N differs among marine systems. *Limnology and Oceanography* **39**: 189–195.

Brooke, S.D. & Young, C.M. (2009) Where do the embryos of *Riftia pachyptila* develop? Pressure tolerances, temperature tolerances, and buoyancy during prolonged embryonic dispersal. *Deep-Sea Research II* **56**: 1599–1605.

Brown, R.M., Herthe, W., Franke, W.W. & Romanovicz, D.K. (1973) The role of the Golgi apparatus in the biosynthesis and secretion of a cellulosic glycoprotein in *Pleurochrysis*. In: Loewus, F. (ed.)

Biogenesis of Plant Cell Wall Polysaccharides. Academic Press, New York, pp. 207–257.

Bruland, K.W. (1980) Oceanographic distributions of cadmium, zinc, nickel and copper in the North Pacific. *Earth and Planetary Science Letters* 47: 176–198.

Brzezinski, M. A. & 16 coauthors (2011) Co-limitation of diatoms by iron and silicic acid in the equatorial Pacific. *Deep-Sea Research II* 58: 493–511.

Buckel, J.A., Steinberg, N.D. & Conover, D.O. (1995) Effects of temperature, salinity, and fish size on growth and consumption of juvenile bluefish. *Journal of Fish Biology* 47: 696–706.

Buckley, T.W. & Miller, B.S. (1994) Feeding habits of yellowfin tuna associated with fish aggregation devices in American Samoa. *Bulletin of Marine Science* 55: 445–459.

Bucklin, A. & Wiebe, P.H. (1998) Low mitochondrial diversity and small effective population sizes of the copepods *Calanus finmarchicus* and *Nannocalanus minor*: possible impact of climatic variation during recent glaciation. *Journal of Heredity* 89: 383–392.

Bucklin, A., Astthorsson, A.S., Gislason, A., Allen, L.D., Smolenack, S.B. & Wiebe, P.H. (2000) Population genetic variation of *Calanus finmarchicus* in Icelandic waters: preliminary evidence of genetic differences between Atlantic and Arctic populations. *ICES Journal of Marine Science* 57: 1592–1604.

Budin, I. & Szostak, J.W. (2010) Expanding roles for diverse physical phenomena during the origin of life. *Annual Reviews of Biophysics* 39: 245–263.

Budin, I., Bruckner, R.J. & Szostak, J.W. (2009) Formation of protocell-like vesicles in a thermal diffusion column. *Journal of the American Chemical Society* 131: 9628–9629.

Buesseler, K.O., Andrews, J.A., Hartman, M.C., Belastock, R. & Chai, F. (1995) Regional estimates of the export flux of particulate organic carbon derived from thorium-234 during the JGOFS EqPac program. *Deep-Sea Research II* 42: 757–776.

Buhl-Mortensen, L. & 8 coauthors (2010) Biological structures as a source of habitat heterogeneity and biodiversity on the deep ocean margins. *Marine Ecology – An Evolutionary Perspective* 31(special issue, 1): 21–50.

Bundy, M.H., Gross, T.F., Vanderploeg, V.A. & Strickler, J.R. (1998) Perception of inert particles by calanoid copepods: behavioral observations and a numerical model. *Journal of Plankton Research* 20: 2129–2152.

Burkholder, J.M. & Glasgow, H.B., Jr (1997) *Pfiesteria piscicida* and other *Pfiesteria*-like dinoflagellates: behavior, impacts, and environmental controls. *Limnology and Oceanography* 42(5, Suppl. 2): 1052–1075.

Burton, R.S., Ellison, C.K. & Harrison, J.S. (2006) The sorry state of F2 hybrids: consequences of rapid mitochondrial DNA evolution in allopatric populations. *The American Naturalist* 168: S14–S24.

Busenberg, S., Kumar, S.K., Austin, P. & Wake, G. (1990) The dynamics of a model of plankton–nutrient interaction. *Bulletin of Mathematical Biology* 52: 677–696.

Buskey, E.J., Lenz, P.H. & Hartline, D.K. (2002) Escape behavior of planktonic copepods in response to hydrodynamic disturbances: high speed video analysis. *Marine Ecology Progress Series* 235: 135–146.

Cai, W.-J. & Reimers, C.E. (1995) Benthic oxygen flux, bottom water oxygen concentration and core top organic carbon content in the deep northeast Pacific Ocean. *Deep-Sea Research I* 42: 1681–1699.

Calbet, A. (2008) The trophic role of microzooplankton in marine systems. *ICES Journal of Marine Science* 65: 325–331.

Calbet, A. & Landry, M.R. (1999) Mesozooplankton influences on the microbial food web: direct and indirect trophic interactions in the oligotrophic open ocean. *Limnology and Oceanography* 44: 1370–1380.

Calbet, A. & Landry, M.R. (2004) Phytoplankton growth, microzooplankton grazing, and carbon cycling in marine systems. *Limnology and Oceanography* 49: 51–57.

Calbet, A. & Saiz, E. (2005) The ciliate–copepod link in marine ecosystems. *Marine Microbial Ecology* 38: 157–167.

Calbet, A., Landry, M.R. & Nunnery, S. (2001) Bacteria–flagellate interactions in the microbial food web of the oligotrophic subtropical North Pacific. *Aquatic Microbial Ecology* 23: 283–292.

Calbet, A. & 8 coauthors (2008) Impact of micro- and nanograzers on phytoplankton assessed by standard and size-fractionated dilution grazing experiments. *Aquatic Microbial Ecology* 50: 145–156.

Calbet, A., Atienza, D., Henriksen, C.I., Saiz, E. & Adey, T.R. (2009) Zooplankton grazing in the Atlantic Ocean: a latitudinal study. *Deep-Sea Research II* 56: 954–963.

Caldwell, D.R. & Chriss, T. M. (1979) The viscous sublayer at the sea floor. *Science* 205: 1131–1132.

Caldwell, D.R. (1978) The maximum-density points of pure and saline water. *Deep-Sea Research* 25: 175–181.

Campbell, B.J., Jeanthon, C., Kostka, J.E., Luther, G.W. & Cary, S.C. (2001) Growth and phylogenetic properties of novel bacteria belonging to the epsilon subdivision of the *Proteobacteria* enriched from *Alvinella pompejana* and deep-sea hydrothermal vents. *Applied and Environmental Microbiology* 67: 4566–4572.

Campbell, B.J. & 10 coauthors (2008) Adaptations to submarine hydrothermal environments exemplified by the genome of *Nautilia profundicola*. *PLOS Genetics* 5: e1000362 (19 pp).

Campbell, L., Nolla, H.A. & Vaulot, D. (1994) The importance of *Prochlorococcus* to community structure in the central North Pacific Ocean. *Limnology and Oceanography* 39: 954–961.

Campbell, R.G., Wagner, M.M., Teegarden, G.J., Boudreau, C.A. & Durbin, E.G. (2001) Growth and development rates of the copepod *Calanus finmarchicus* reared in the laboratory. *Marine Ecology Progress Series* 221: 161–183.

Campbell, R.G., Sherr, E.B., Ashjian, C.J., Plourde, S., Sherr, B.F., Hill, V. & Stockwell, D.A. (2009) Mesozooplankton prey preference and grazing impact in the western Arctic Ocean. *Deep-Sea Research II* 56: 1274–1289.

Cannon, H.G. (1928) On the feeding mechanism of the copepods *Calanus finmarchicus* and *Diaptomus gracilis*. *British Journal of Experimental Biology* 6: 131–144.

Caparroy, P., Pérez, M.T. & Carlotti, F. (1998) Feeding behavior of *Centropages typicus* in calm and turbulent conditions. *Marine Ecology Progress Series* 168: 109–118.

Carey, A.G. & Hancock, D.R. (1965) An anchor-box dredge for deep-sea sampling. *Deep-Sea Research* 12: 983–984.

Carlotti, F. & Nival, S. (1991) Individual variability of development in laboratory-reared *Temora stylifera* copepodites: consequences for the population dynamics and interpretation in the scope of growth and development rules. *Journal of Plankton Research* 13: 801–813.

Carmack, E.C. & Wassmann, P. (2006) webs and physical–biological coupling on pan-Arctic shelves: unifying concepts and comprehensive perspectives. *Progress in Oceanography* 71: 446–477.

Carpenter, E.J. & Capone, D.G. (2008) Nitrogen fixation in the marine environment. In: Carpenter, E.J. & Capone, D.G. (eds.) *Nitrogen in the Marine Environment*, Elsevier, Amsterdam, pp. 141–198. doi:10.1016/B978-0-12-372522-6.00004-9.

Carpenter, E.J. & Guillard, R.R.L. (1971) Intraspecific differences in nitrate half-saturation constants for three species of marine phytoplankton. *Ecology* 52: 183–185.

Carr, M.-E. & Kearns, E. (2003) Production regimes in four Eastern Boundary Current systems. *Deep-Sea Research II* 50: 3199–3221.

Cary, S.C. & Giovannoni, S.J. (1993) Transovarial inheritance of endosymbiotic bacteria in vesicomyid clams found inhabiting deep-sea hydrothermal vent systems. *Proceedings of the National Academy of Sciences* 90: 5695–5699.

Cary, S.C., Cottrell, M.T., Stein, J.L., Camacho, F. & Desbruyères, D. (1997) Molecular identification and localization of filamentous symbiotic bacteria associated with the hydrothermal vent annelid *Alvinella pompejana*. *Applied and Environmental Microbiology* **63**: 1124–1130.

Cary, S.C., Shank, T. & Stein, J. (1998) Worms bask in extreme temperatures. *Nature* **391**: 545–546.

Casciotti, K.L., Trull, T.W., Glover, D.M. & Davies, D. (2008) Constraints on nitrogen cycling at the subtropical North Pacific Station ALOHA from isotopic measurements of nitrate and particulate nitrogen. *Deep-Sea Research II* **55**: 1661–1672.

Case, R.A.J. & 12 coauthors (2006) Association between growth and *Pan I** genotype within Atlantic cod full-sibling families. *Transactions of the American Fisheries Society* **135**: 41–250.

Catton, K.B., Webster, D.R., Brown, J. & Yen, J. (2007) Quantitative analysis of tethered and free-swimming copepodid flow fields. *Journal of Experimental Biology* **210**: 299–310.

Caut, S., Angulo, E. & Courchamp, F. (2009) Variation in discrimination factors (Δ^{15}N and Δ^{13}C): the effect of diet isotopic values and application for diet reconstruction. *Journal of Applied Ecology* **46**: 443–453.

Cavanaugh, C.M., Wirsen, C.O. & Jannasch, H.W. (1992) Evidence for methylotrophic symbionts in a hydrothermal vent mussel (Bivalvia: Mytilidae) from the Mid-Atlantic Ridge. *Applied and Environmental Microbiology* **58**: 3799–3803.

Cavender-Bares, K.K., Karl, D.M. & Chisholm, S.W. (2001) Nutrient gradients in the western North Atlantic Ocean: relationship to microbial community structure and comparison to patterns in the Pacific Ocean. *Deep-Sea Research I* **48**: 2373–2395.

Cermeno, P., Estevez-Blanco, P., Marañon, E. & Fernandez, E. (2005) Maximum photosynthetic efficiency of size-fractionated phytoplankton assessed by ^{14}C uptake and fast repetition rate fluorometry. *Limnology and Oceanography* **50**: 1438–1446.

Chan, A.T. (1978) Comparative physiological study of marine diatoms and dinoflagellates in relation to irradiance and cell size. I. Growth under continuous light. *Journal of Phycology* **14**: 396–402.

Chase, Z., Strutton, P. & Hales, B. (2007) Iron links river runoff and shelf width to phytoplankton biomass along the US west coast. *Geophysical Research Letters* **34**: LO4607. doi:10.1029/2006GL028069.

Chavez, F.P. & Messié, M. (2009) A comparison of eastern boundary upwelling ecosystems. *Progress in Oceanography* **83**: 80–96.

Checkley, D.M. Jr (1980) Food limitation of egg production by a marine, planktonic copepod in the sea off southern California. *Limnology and Oceanography* **25**: 991–998.

Checkley, D.M. Jr & Barth, J.A. (2009) Patterns and processes in the California Current sysytem. *Progress in Oceanography* **83**: 49–64.

Chelton, D.B., Wentz, F.J., Gentemann, C.L., de Szoeke, R.A. & Schlax, M. (2000) Satellite microwave SST observations of transequatorial tropical instability waves. *Geophysical Research Letters* **27**: 1239–1242.

Chen, F., Lu, J.-R., Binder, B.J., Liu, Y. & Hodson, R.E. 2001. Application of digital image analysis and flow cytometry to enumerate marine viruses stained with SYBR Gold. *Applied and Environmental Microbiology* **67**: 539–545.

Chen, J.-Y. & Huang, D.-Y. (2002) A possible Lower Cambrian chaetognath (arrow worm). *Science* **298**: 187.

Chiba, S., Tadokoro, K., Sugisaki, H. & Saino, T. (2006) Effects of decadal climate change on zooplankton over the last 50 years in the western subarctic North Pacific. *Global Change Biology* **12**: 907–920.

Childress, J.J. (1975) The respiratory rates of midwater crustaceans as a function of depth of occurrence and relation to the oxygen minimum layer off southern California. *Comparative Biochemistry and Physiology* **50A**: 787–799.

Childress, J.J. (1995) Are there physiological and biochemical adaptations of metabolism in deep-sea animals? *Trends in Ecology and Evolution* **10**: 30–36.

Childress, J.J. & Girguis, P.R. (2011) The metabolic demands of endosymbiotic chemoautotrophic metabolism on host physiological capacities. *The Journal of Experimental Biology* **214**: 312–325.

Childress, J.J. & Price, M.H. (1978) Growth rate of the bathypelagic crustacean *Gnathophausia ingens* (Mysidacea: Lophogastridae). I. Dimensional growth and population structure. *Marine Biology* **50**: 47–62.

Childress, J.J. & Somero, G.N. (1979) Depth-related enzymic activities in muscle, brain and heart of deep-living pelagic marine teleosts. *Marine Biology* **52**: 273–283.

Childress, J.J., Taylor, S.M., Cailliet, G.M. & Price, M.H. (1980) Patterns of growth, energy utilization and reproduction in some meso- and bathypelagic fishes off Southern California. *Marine Biology* **61**: 27–40.

Childress, J.J., Seibel, B.A. & Thuesen, E.V. (2008) N-specific metabolic data are not relevant to the "visual interactions" hypothesis concerning the depth-related declines in metabolic rates: comment on Ikeda *et al.* (2006). *Marine Ecology Progress Series* **373**: 187–191.

Chisholm, S.W., Olson, R.J., Zettler, E.R., Goericke, R., Waterbury, J.B. & Welschmeyer, N. (1988) A novel free-living prochlorophyte abundant in the oceanic euphotic zone. *Nature* **334**: 340–343.

Chriss, T.M. & Caldwell, D.R. (1984) Turbulence spectra from the viscous sublayer and buffer layer at the ocean floor. *Journal of Fluid Mechanics* **142**: 39–55.

Christensen, V. & Walters, C.J. (2004) Ecopath with Ecosim: methods, capabilities and limitations. *Ecological Modelling* **172**: 109–139.

Church, M., Bidigare, R., Dore, J., Karl, D., Landry, M., Letelier, R. & Lukas, R. (2009) The Ocean is HOT: 20 years of Hawaii Ocean Time-Series research in the North Pacific subtropical gyre. *OCB [Ocean Carbon and Biogeochemistry] News* **2**(1): 2–9.

Claus, C. (1866) Die Copepoden-Fauna von Nizza. Ein Beitrag zur Charakteristik der Formen und deren Abänderungen "im Sinne Darwin's". *Schriften der Gesellschaften sur Beförderung der gesamten Naturwissenschaften zu Marburg* (Suppl 1): 1–34; pls. 1–5.

Claustre, H. & 8 coauthors (2005) Toward a taxon-specific parameterization of bio-optical models of primary production: a case study in the North Atlantic. *Journal of Geophysical Research* **110**: C07S12, doi: 10.1029/2004JC002634.

Clough, L.M. & Lopez, G.R. (1993) Potential carbon sources for the head-down deposit-feeding polychaete *Heteromastus filiformis*. *Journal of Marine Research* **51**: 595–616.

Coale, K. & 18 coauthors (1996) A massive phytoplankton bloom induced by an ecosystem-scale iron fertilization experiment in the equatorial Pacific Ocean. *Nature* **383**: 495–501.

Coale, K. & 47 coauthors (2004) Southern Ocean iron enrichment experiment: carbon cycling in high- and low-Si waters. *Science* **304**: 408–414.

Cohen, D.M. (1964) Suborder Argentinoidea. In: Bigelow, H.B. (ed.) *Fishes of the Western North Atlantic, Part. 4*. Sears Foundation for Marine Research, Memoir No. I, pp. 1–70.

Collin, S.P. (1997) Specialisations of the teleost visual system: adaptive diversity from shallow-water to deep-sea. *Acta Physiologica Scandinavica* **161** (Suppl. 638): 5–24.

Cone, J. (1991) *Fire Under the Sea*. Oregon State University Press, Corvallis, 285 pp.

Conover, S.A.M. (1956) Oceanography of Long Island Sound, 1952–54. IV. Phytoplankton *Bulletin of the Bingham Oceanographic Collection* **15**: 62–112.

Conte, M.H., Ralph, N. & Ross, E.H. (2001) Seasonal and interannual variability in deep ocean particle fluxes at the Oceanic Flux

Program (OFP)/Bermuda Atlantic Time Series (BATS) site in the western Sargasso Sea near Bermuda. *Deep-Sea Research II* **48**: 1471–1505.

Conway Morris, S. (2009) The Burgess Shale animal *Oesia* is not a chaetognath: a reply to Szaniawski (2005) *Acta Palaeontologica Polonica* **54**: 175–179.

Corliss, J.B. (1973) The sea as alchemist. *Oceanus* **17** (Winter 1973–74): 38–43.

Corliss, J.B. & 10 coauthors (1979) Submarine thermal springs on the Galapagos Rift. *Science* **203**: 1073–1083.

Corliss, J.B., Baross, J.A. & Hoffman, S.E. (1981) An hypothesis concerning the relationship between submarine hot springs and the origin of life on Earth. *Oceanologica Acta* **4** (Suppl.): 59–69.

Corno, G., Letelier, R.M., Abbott, M.R. & Karl, D.M. (2006) Assessing primary production variability in the North Pacific Subtropical Gyre: a comparison of fast repetition rate fluorometry and ^{14}C measurements. *Journal of Phycology* **42**: 51–60.

Corno, G., Letelier, R.M., Abbott, M.R. & Karl, D.M. (2008) Temporal and vertical variability in photosynthesis in the North Pacific Subtropical Gyre. *Limnology and Oceanography* **53**: 1252–1265.

Corwith, H.L. & Wheeler, P.A. (2002) El Niño related variations in nutrient and chlorophyll distributions off Oregon. *Progress in Oceanography* **54**: 361–380.

Cottin, D. & 7 coauthors (2008) Thermal biology of the deep-sea vent annelid *Paralvinella grasslei*: in vivo studies. *The Journal of Experimental Biology* **211**: 2196–2204

Cottrell, M.T. & Kirchman, D.L. (2000) Natural assemblages of marine proteobacteria and members of the Cytophaga–Flavobacter cluster consuming low- and high-molecular-weight dissolved organic matter. *Applied and Environmental Microbiology* **66**: 1692–1697.

Cottrell, M.T. & Kirchman, D.L. (2003) Contribution of major bacterial groups to biomass production (thymidine and leucine incorporation) in the Delaware estuary. *Limnology and Oceanography* **48**: 168–178.

Cottrell, M.T., Malmstrom, R.R., Hill, V., Parker, A.E. & Kirchman, D.L. (2006) The metabolic balance between autotrophy and heterotrophy in the western Arctic Ocean. *Deep-Sea Research I* **53**: 1831–1844.

Cowles, D.L., Childress, J.J. & Wells, M.E. (1991) Metabolic rates of midwater crustaceans as a function of depth of occurrence off the Hawaiian Islands: food availability as a selective factor. *Marine Biology* **110**: 75–83.

Cowles, T.J., Desiderio, R.A. & Carr, M.-E. (1998) Small-scale planktonic structure: persistence and trophic consequences. *Oceanography* **11**: 4–9.

Craddock, C., Hoeh, W.R., Lutz, R.A. & Vrijenhoek, R.C. (1995) Extensive gene flow among mytilid (*Bathymodiolus thermophilus*) populations from hydrothermal vents of the eastern Pacific. *Marine Biology* **124**: 137–146.

Craddock, C., Lutz, R.A. & Vrijenhoek, R.C. (1997) Patterns of dispersal and larval development of archaeogastropod limpets at hydrothermal vents in the eastern Pacific. *Journal of Experimental Marine Biology and Ecology* **210**: 37–51.

Crain, J.A. & Miller, C.B. (2001) Effects of starvation on intermolt development in *Calanus finmarchicus* copepodites: a comparison between theoretical models and field studies. *Deep-Sea Research II* **48**: 551–566.

Cullen, J.J. (1999) Iron, nitrogen and phosphorus in the ocean. *Nature* **402**: p. 372.

Cushing, D.H. (1995) *Population Production and Regulation in the Sea: a Fisheries Perspective*. Cambridge University Press, Cambridge, UK, 354 pp.

Cussler, E.L. (1984) *Diffusion, Mass Transfer in Fluid Systems*. Cambridge University Press, Cambridge, UK, 580 pp.

Cuzin-Roudy, J. & Bucholz, F. (1999) Ovarian development and spawning in relation to the moult cycle in Northern krill, *Meganyctiphanes norvegica* (Crustacea: Euphausiacea), along a climatic gradient. *Marine Biology* **133**: 267–281.

Dagg, M. (1977) Some effects of patchy food environments on copepods. *Limnology and Oceanography* **22**: 99–107.

Dagg, M., Strom, S. & Liu, H. (2009) High feeding rates on large particles by *Neocalanus flemingeri* and *N. plumchrus*, and consequences for phytoplankton community structure in the subarctic Pacific Ocean. *Deep-Sea Research I* **56**: 716–726.

Dagg, M.J., Frost, B.W. & Walser, W.E., Jr (1989) Copepod diel migration, feeding and the vertical flux of pheopigments. *Limnology and Oceanography* **34**: 1062–1071.

Dahlgren, U. (1915–1917) The production of light by animals; light production by cephalopods. *Journal of the Franklin Institute* **181**: 525–556. (A series of papers on bioluminescence: 1915–1917, *J.F.I.* 180, 182, 183).

Dale, T., Bagoeien, E., Melle, W. & Kaartvedt, S. (1999) Can predator avoidance explain varying overwintering depth of *Calanus* in different oceanic water masses? *Marine Ecology Progress Series* **179**: 113–121.

D'Alelio, D.D., d'Alcala, M.R., Dubroca, L., Sarno, D., Zingone, A. & Montressor, M. (2010) The time for sex: a biennial life cycle in a marine planktonic diatom. *Limnology and Oceanography* **55**: 106–114.

Daly, K. L. (1990) Overwintering development, growth, and feeding of larval *Euphausia superba* in the Antarctic marginal ice zone. *Limnology and Oceanography* **35**: 1564–1576.

Dam, H.G. & Peterson, W.T. (1988) The effect of temperature on the gut clearance rate constant of planktonic copepods. *Journal of Experimental Marine Biology and Ecology* **123**: 1–14.

Dansgaard, W. & 10 coauthors (1993) Evidence for general instability of past climate from a 250-kyr ice-core record. *Nature* **364**: 218–220.

Darley, M.W., Sullivan, C.W. & Volcani, B.E. (1976) Studies on the biochemistry and fine structure of silica shell formation in diatoms. Division cycle and chemical composition of *Avicula pelliculosa* during light–dark synchronized growth. *Planta (Berl.)* **130**: 159–365.

Davis, C.C. (1977) *Sagitta* as food for *Acartia*. *Astarte Journal of Arctic Biology* **10**: 1–3.

Dawson, P.A. (1973) Observations on the structure of some forms of *Gomphonema parvulum* Kutz. III. Frustule formation. *Journal of Phycology* **9**: 353–365.

Day, J.H. (1967) *A Monograph on the Polychaeta of Southern Africa. Part 2, Sedentaria*. British Museum (Natural History), London, pp. 471–878.

Dayton, P.K. & Hessler, R.R. (1972) The role of disturbance in the maintenance of deep-sea diversity. *Deep-Sea Research* **19**: 199–208.

de Baar, H.J.W. & 33 coauthors (2005) Synthesis of iron fertilization experiments: from the Iron Age in the Age of Enlightenment. *Journal of Geophysical Research* **110**: C09S16, doi:10.1029/2004JC002601 (22 pp).

De Vargas, C., Aubry, M.P., Probert, I. & Young, J. (2007) Origin and evolution of coccolithophores: from coastal hunters to oceanic farmers, In: Falkowski, P.G. & Knoll, A. (eds.) *Evolution of Primary Producers in the Sea*, Academic Press, pp. 251–285.

Décima, M., Landry, M.R. & Rykaczewski, R.R. (2011) Broad scale patterns in mesoplankton biomass and grazing in the equatorial Pacific. *Deep-Sea Research II* **58**: 387–399.

Deibel, D. & Powell, C.V.L. (1987) Ultrastructure of the pharyngeal filter of the appendicularian *Oikopleura vanhoeffeni*: implication for particle size selection and fluid mechanics. *Marine Ecology Progress Series* **35**: 243–250.

del Giorgio, P.A. & Cole, J.J. (2000) Bacterial energetics and growth efficiency. In: Kirchman, D.L. (ed.) *Microbial Ecology of the Oceans*, Wiley, New York, pp. 289–326.

del Giorgio, P.A. & Gasol, J.M. (2008) Physiological structure and single-cell activity in marine bacterioplankton. In: Kirchman, D.L. (ed.) *Microbial Ecology of the Oceans*. Second Edition, Wiley, New York, pp. 243–298.

DeLong, E.F. (1992) Archaea in coastal marine environments. *Proceedings of the National Academy of Sciences*, 99: 10,494–10,499.

DeLong E.F. (2001) Microbial seascapes revisited. *Current Opinion in Microbiology* 4: 290–295.

DeLong, E.F., Taylor, L.T., Marsh, T.L. & Preston, C.M. (1999) Visualization and enumeration of marine planktonic Archaea and Bacteria using polyribonucleotide probes and fluorescent in situ hybridization. *Applied and Environmental Microbiology* 65: 5554–5563.

DeMaster, D.J., Thomas, C.J., Blair, N.E., Fornes, W.L., Plaia, G. & Levin, L.A. (2002) Deposition of bomb 14C in continental slope sediments of the Mid-Atlantic Bight: assessing organic matter sources and burial rates. *Deep-Sea Research II* 49: 4667–4685.

Denman, K.L. & Peña, M.A. (1999) A coupled 1-D biological/physical model of the northeast subarctic Pacific Ocean with iron limitation. *Deep-Sea Research II* 46: 2877–2908.

Denman, K.L., Voelker, C., Peña, M.A. & Rivkin, R.B. (2006) Modelling the ecosystem response to iron fertilization in the subarctic NE Pacific: the influence of grazing, and Si and N cycling on CO_2 drawdown. *Deep-Sea Research II* 53: 2327–2352.

Denton, E.J. (1970) On the organization of reflecting surfaces in some marine animals. *Philosophical Transactions of the Royal Society, London B* 258: 285–313.

Denton, E.J. (1991) Some adaptations of marine animals to physical conditions in the sea. In: Mauchline, J. & Nemoto, T. (eds.) *Marine Biology, Its Accomplishment and Future Prospect*. Hokusen-Sha, Tokyo, pp. 187–193.

Derenbach, J.B., Astheimer, H., Hansen, H.P. & Leach, H. (1979) Vertical microscale distribution of phytoplankton in relation to the thermocline. *Marine Ecology Progress Series* 1: 187–193.

Desbruyères, D. & Laubier, L. (1980) *Alvinella pompejana* gen. et sp. nov., Ampharetidae aberrant des sorces hydrothermales de la ride Est-Pacifique. *Oceanologica Acta* 3: 267–274.

Desbruyères, D. & Laubier, L. (1986) Les Alvinellidae, une famille nouvelle d'annélides polychètes inféodées aux sources hydrothermales sous-marines: systématique, biologie et écologie. *Canadian Journal of Zoology* 64: 2227–2245.

Desbruyères, D., Gaill, F., Laubier, L. & Fouquet, Y. (1985) Polychaetous annelids from hydrothermal vent ecosystems: an ecological overview. *Biological Society of Washington Bulletin* 6: 103–116.

Desbruyères, D. & 17 coauthors (1998) Biology and ecology of the Pompeii worm (*Alvinella pompejana* Desbruyères and Laubier), a normal dweller of an extreme deep-sea environment: a synthesis of current knowledge and recent developments. *Deep-Sea Research II* 45: 383–422.

Desbruyères, D., Segonzac, M. & Bright, M. (eds.) (2006) *Handbook of Deep-Sea Hydrothermal Vent Fauna*, 2nd edn. Denisia 18, Landesmuseum Linz, Austria, 554 pp.

Dessier, A. & Donguy, J.R. (1987) Response to El Niño signals of the epiplanktonic copepod populations in the eastern tropical Pacific. *Journal of Geophysical Research, Oceans* 92: 14,393–14,403.

Deuser, W.G., Ross, E.H. & Anderson, R.F. (1981) Seasonality in the supply of sediment to the deep Sargasso Sea and implications for the rapid transfer of matter to the deep ocean. *Deep-Sea Research* 28: 495–505.

Deutsch, C., Sarmiento, J.L., Sigman, D.M., Gruber, N. & Dunne, J.P. (2007) Spatial coupling of nitrogen inputs and losses in the ocean. *Nature* 445: 163–167. doi:10.1038/nature05392.

Dever, E.P., Dorman, C.E. & Largier, J.L. (2006) Surface boundary-layer variability off Northern California, USA, during upwelling. *Deep-Sea Research II* 53: 2887–2905.

Dewar, W.K., Bingham, R.J., Iverson, R.L., Nowacek, D.P., St. Laurent, L.C. & Wiebe, P.H. (2006) Does the biosphere mix the ocean? *Journal of Marine Research* 64: 541–561.

Diaz, R.J. and Rosenberg, R. (1995) Marine benthic hypoxia: a review of its ecological effects and the behavioral responses of benthic macrofauna. *Oceanography and Marine Biology, An Annual Review* 33: 245–303.

Díaz-Ochoa, J.A., Lange, C.B., Pantoja, S., De Lange, G.J., Gutiérrez, D., Muñoz, P. & Salamanca, M. (2009) Fish scales in sediments from off Callao, central Peru. *Deep-Sea Research II* 56: 1124–1135.

Di Meo-Savoie, C.A., Luther, G.W. III & Cary, S.C. (2004) Physicochemical characterization of the microhabitat of the epibionts associated with *Alvinella pompejana*, a hydrothermal vent annelid. *Geochimica et Cosmochemica Acta* 68: 2055–2066.

Dittmar, W. (1884) Report on researches in the composition of ocean-water, collected by H.M.S. Challenger, during the years 1873–1876. *Report on the Scientific Results of the Voyage of H.M.S. Challenger, during the years 1873–76, Vol. 1*, 247 pp. + 3 plates.

Dixon, D.R. & Dixon, L.J.R. (1996) Results of DNA analyses conducted on vent shrimp postlarvae collected above the Broken Spur vent field during the CD95 cruise, August 1995. *BRIDGE Newsletter* 111: 9–15.

Dodge, J.D. (1972) The fine structure of the dinoflagellate pusule: a unique osmo-regulatory organelle. *Protoplasma* 75: 285–302.

Dodge, J.D. (1979) The phytoflagellates: fine structure and phylogeny. In: Levandowsky, M. & Hutner, S.H. (eds.) *Biochemistry and Physiology of Protozoa*, Vol. 1. Academic Press, New York, pp. 7–57.

Dodge, J.D. (1985) *Atlas of Dinoflagellates*. Blackwell Scientific Publishing, Palo Alto, CA, 119 pp.

Dodge, J.D. & Crawford, R.M. (1970) A survey of thecal fine structure in the Dinophyceae. *Journal of the Linnean Society of London. Botany* 63: 53–67.

Donaghay, P. & Small, L.F. (1979) Food selection capabilities of the estuarine copepod *Acartia clausi*. *Marine Biology* 52: 137–146.

Doney, S.C., Fabry, V.J., Feely, R.A. & Kleypas, J.A. (2009) Ocean acidification: the other CO_2 problem. *Annual Review of Marine Science* 1: 169–192.

Donnelly, R.F. (1983) *Factors affecting the abundance of Kodiak Archipelago Pink Salmon (Oncorhynchus gorbuscha, Walbaum)*. PhD Dissertation, University of Washington, Seattle, vii + 157 pp.

Dore, J.E., Letelier, R.M., Church, M.J., Lukas, R. & Karl, D.M. (2008) Summer phytoplankton blooms in the oligotrophic North Pacific Subtropical Gyre: historical perspective and recent observations. *Progress in Oceanography* 76: 2–38.

Dorgan, K.M., Jumars, P.A., Johnson, B., Boudreau, B.P. & Landis, E. (2005) Burrow extension by crack propagation. *Nature* 433: p. 475.

Dorgan, K.M., Jumars, P.A., Johnson, B.D. & Boudreau, B.P. (2006) Macrofaunal burrowing: the medium is the message. *Oceanography and Marine Biology Annual Review* 44: 85–121.

Dorgan, K.M., Arwade, S.R. & Jumars, P.A. (2008) Worms as wedges: effects of sediment mechanics on burrowing behavior. *Journal of Marine Research* 66: 219–253.

Drebes, G. (1977) Sexuality. In: Werner, D. (ed.) *The Biology of Diatoms*. University of California Press, Berkeley, pp. 250–283.

Droop, M.R. (1968) Vitamin B_{12} and marine ecology, IV: the kinetics of uptake, growth and inhibition in *Monochrysis lutheri*. *Journal of the Marine Biological Association UK* 48: 689–733.

Duarte, C.M. & Regaudie-de-Gioux, A. (2009) Thresholds of gross primary production for the metabolic balance of marine planktonic communities. *Limnology and Oceanography* 54: 1015–1022.

Ducklow, H. (1992) Factors regulating bottom-up control of bacterial biomass in open ocean plankton communities. *Archiv fur Hydrobiologie. Beiheft* **37**: 207–217.

Ducklow, H. (2000) Bacterial production and biomass in the oceans. In: Kirchman, D.L. (ed.) *Microbial Ecology of the Oceans*. Wiley, New York, pp. 47–84.

Ducklow, H.W. & Carlson, C.A. (1992) Oceanic bacterial production. *Advances in Microbial Ecology* **12**: 113–181.

Ducklow, H.W., Carlson, C., Church, M., Kirchman, D., Smith, D. & Stewart, G. (2001) The seasonal development of the bacterioplankton bloom in the Ross Sea, Antarctica, 1994–1997. *Deep-Sea Research II* **48**: 4199–4221.

Dufresne, A. & 20 coauthors (2003) Genome sequence of the cyanobacterium *Prochlorococcus marinus* SS120, a nearly minimal oxyphototrophic genome. *Proceedings of the National Academy of Sciences* **100**: 10,020–10,025.

Dugdale, R.C. & Goering, J. (1967) Uptake of new and regenerated forms of nitrogen in primary production. *Limnology and Oceanography* **12**: 196–206.

Dugdale, R.C., Wilkerson, F.P., Chai, F. & Feely, R. (2007) Size-fractionated uptake measurements in the equatorial Pacific and confirmation of the low Si-high-nitrate low-chlorophyll condition. *Global Biogeochemical Cycles* **21**: GB2005, doi:10.1029./2006GB002722,2007.

Duhamel, S., Dyhrman, S.T. & Karl, D.M. (2010) Alkaline phosphatase activity and regulation in the North Pacific Subtropical Gyre. *Limnology and Oceanography*, **55**: 1414–1425.

Dunton, K.H., Weingartner, T. & Carmack, E.C. (2006) The nearshore western Beaufort Sea ecosystem: circulation and importance of terrestrial carbon in Arctic coast food webs. *Progress in Oceanography* **71**: 362–378.

DuRand, M.D., Olson, R.J. & Chisholm, S.W. (2002) Phytoplankton population dynamics at the Bermuda Atlantic Time-series station in the Sargasso Sea. *Deep-Sea Research II* **48**: 1983–2003.

Durbin, E.G., Durbin, A.G. & Wlodarczyk, E. (1990) Diel feeding behavior in the marine copepod *Acartia tonsa* in relation to food availability. *Marine Ecology Progress Series* **68**: 23–45.

Durbin, E.G., Campbell, R.G., Gilman, S.L. & Durbin, A.G. (1995) Diel feeding behavior and ingestion rate in the copepod *Calanus finmarchicus* in the southern Gulf of Maine during late spring. *Continental Shelf Research* **15**: 539–570.

Durbin, E.G., Casas, M.C., Rynearson, T.A. & Smith, D.C. (2007) Measurement of copepod predation on nauplii using qPCR of the cytochrome oxidase I gene. *Marine Biology* **153**: 699–707.

Dyson, F. (1999) *Origins of Life*, 2nd edn. Cambridge University Press, Cambridge, UK, 100 pp.

Edgar, L.A. & Pickett-Heaps, J.D. (1983) The mechanism of diatom locomotion. I. An ultrastructural study of the motility apparatus. *Proceedings of the Royal Society of London B* **281**: 331–343.

Edgar, L.A. & Pickett-Heaps, J.D. (1984) Diatom locomotion. *Progress in Phycological Research* **3**: 47–88.

Edlund, M.B. & Stoermer, E.F. (1991) Sexual reproduction in *Stephanodiscus niagarae* (Bacillariophyta). *Journal of Phycology* **27**: 780–793.

Edwards, C.A., Batchelder, H.P. & Powell, T.M. (2000) Modeling microzooplankton and macrozooplankton dynamics within a coastal upwelling system. *Journal of Plankton Research* **22**: 1619–1648.

Edwards, E. & Richardson, A.J. (2004) Impact of climate change on marine pelagic phenology and trophic mismatch. *Nature* **430**: 881–884.

Egerton, F.N. (2002) A history of the ecological sciences, part 6: Arabic language science – origins and zoological writings. *Bulletin of the Ecological Society of America*, April 2002: 142–146.

Eissler, Y., Wang, K., Chen, F., Wommack, K.E. & Coats, D.W. (2009) Ultrastructural characterization of the lytic cycle of an intranuclear virus infecting the diatom *Chaetoceros* cf. *wighamii* (Bacillariophyceae) from Chesapeake Bay, USA. *Journal of Phycology* **45**: 787–797.

Elderfield, H. & Rickaby, R.E.M. (2000) Oceanic Cd/P ratio and nutrient utilization in the glacial Southern Ocean. *Nature* **405**: 305–310.

Eleftheriou, A. & McIntyre, A. (2005) *Methods for Study of Marine Benthos*, 3rd Edition. Wiley-Blackwell, Oxford, 218 pp.

Ellis, S.G. & Small, L.F. (1989) Comparison of gut-evacuation rates of feeding and non-feeding *Calanus marshallae*. *Marine Biology* **103**: 175–181.

Elskens, M. & 7 coauthors (2008) Primary, new and export production in the NW Pacific subarctic gyre during the vertigo K2 experiments. *Deep-Sea Research II* **55**: 1594–1604.

Emerson, S. & Stump, C. (2010) Net biological oxygen production in the ocean II: remote in situ measurements of O_2 and N_2 in the subarctic Pacific surface water. *Deep-Sea Research I* **57**: 1255–1265.

Emerson, S., Quay, P., Stump, C., Wilbur, D. & Knox, M. (1991) O_2, Ar, N_2, and ^{222}Rn in surface waters of the subarctic Pacific ocean: net biological O_2 production. *Global Biogeochemical Cycles* **5**: 49–69.

Emerson, S., Stump, C., Grootes, P.M., Stuiver, M., Farwell, G.W. & Schmidt, F.H. (1997) Estimate of degradable organic carbon in deep-sea surface sediments from ^{14}C concentrations. *Nature* **329**: 51–53.

Emerson, S., Stump, C. & Nicholson, D. (2008) Net biological oxygen production in the ocean: remote in situ measurements of O_2 and N_2 in surface waters. *Global Biogeochemical Cycles* **22**: GB3023, doi:10.1029/2007GB003095.

Emerson, S.R. & Hedges, J.I. (2008) *Chemical Oceanography and the Marine Carbon Cycle*. Cambridge University Press, Cambridge, UK, 453 pp.

Endo, Y. & Wiebe, P.H. (2007) Temporal changes in euphausiid distribution and abundance in North Atlantic cold-core rings in relation to the surrounding waters. *Deep-Sea Research I* **54**: 181–202.

Enright, J.T. (1977a) Copepods in a hurry: sustained high-speed upward migration. *Limnology and Oceanography* **22**: 118–125.

Enright, J.T. (1977b) Diurnal vertical migration: adaptive significance and timing. Part 1. Selective advantage: a metabolic model. *Limnology and Oceanography* **22**: 856–872.

Enright, J.T., Newman, J.A., Hessler, R.R. & McGowan, J.A. (1981) Deep-ocean hydrothermal vent communities. *Nature* **289**: 219–221.

Eppley, R.W. (1972) Temperature and phytoplankton growth in the sea. *Fisheries Bulletin* **70**: 1063–1085.

Eppley, R.W. & Thomas, W.H. (1969) Comparisons of half-saturation constants for growth and nitrate uptake of marine phytoplankton. *Journal of Phycology* **5**: 375–379.

Eppley, R.W., Holm-Hansen, O. & Strickland, J.D.H. (1968) Some observations of the vertical migration of dinoflagellates. *Journal of Phycology* **5**: 375–379.

Eppley, R.W., Rogers, J.N., McCarthy, J.J. & Sournia, A. (1971) Light:dark periodicity in nitrogen assimilation of the marine phytoplankters *Skeletonema costatum* and *Coccolithus huxleyi* in N-limited chemostat culture. *Journal of Phycology* **7**: 150–154.

Ericson, D.B. & Wollin, G. (1968) Pleistocene climates and chronology in deep-sea sediments. *Science* **162**: 1227–1234.

Esaias, W.E. & Curl, H.C., Jr (1972) Effect of dinoflagellate bioluminescence on copepod ingestion rates. *Limnology and Oceanography* **17**: 901–906.

Etter, R.J. & Mullineaux, L. (2001) Deep-sea communities. In: Bertness, M.D., Gaines, S.D. & Hay, M.E. (eds.) *Marine Community Ecology*. Sinauer Associates, Inc., Sunderland, MA, pp. 367–394.

Evans, G.T. & Parslow, J.S. (1985) A model of annual plankton cycles. *Biological Oceanography* 3: 327–347.

Evans, W., Strutton, P.G. & Chavez, F.P. (2009) Impact of tropical instability waves on nutrient and chlorophyll distributions in the equatorial Pacific. *Deep-Sea Research I* 56: 178–188.

Fager, E.W. (1963) Communities of organisms. In: Hill, M.N. (ed.) *The Sea*. Vol. 2. Wiley-Interscience, New York, pp. 415–437.

Fahrenbach, W.H. (1963) The sarcoplasmic reticulum of a striated muscle of a cyclopoid copepod. *The Journal of Cell Biology* 17: 629–640.

Falkowski, P.G. (1997) Evolution of the nitrogen cycle and its influence on the biological sequestration of CO_2 in the ocean. *Nature* 387: 272–375.

Falkowski, P.G. & Raven, J.A. (2007) *Aquatic Photosynthesis*, 2nd Edition. Princeton University Press, Princeton, New Jersey, 484 pp.

Falkowski, P.G., Dubinsky, Z. & Wyman, K. (1985) Growth–irradiance relationships in phytoplankton. *Limnology and Oceanography* 30: 311–321.

Falkowski, P.G., Katz, M.E., Knoll, A.H., Quigg, A., Raven, J.A., Schofield, O. & Taylor, F.J.R. (2004) The evolution of modern eukaryotic phytoplankton. *Science* 305: 354–360.

Falk-Peterson, S., Mayzaud, P., Kattner, G. & Sargeant, J.R. (2009) Lipids and life strategy of Arctic *Calanus*. *Marine Biology Research* 5: 18–39.

Fasham, M.J.R. (1995) Variations in the seasonal cycle of biological production in subarctic oceans: a sensitivity analysis. *Deep-Sea Research I* 42: 1111–1149.

Fasham, M.J.R., Ducklow, H.W. & McKelvie, S.M. (1990) A nitrogen-based model of plankton dynamics in the oceanic mixed layer. *Journal of Marine Research* 48: 591–639.

Fauchald, K. (1992) Diet of Worms. *Current Contents* 40: p. 8.

Fauchald, K. & Jumars, P.A. (1979) The diet of worms: a study of polychaete feeding guilds. *Oceanography and Marine Biology Annual Review* 17: 193–284.

Faust, M.A. (1992) Observations on the morphology and sexual reproduction of *Coolia monotis* (Dinophyceae). *Journal of Phycology* 28: 94–104.

Feder, H.M., Naidu, A.S., Jewett, S.C., Hameedi, J.M., Johnson, W.R. & Whitledge, T.E. (1994) The northeastern Chukchi Sea: benthos–environmental interactions. *Marine Ecology Progress Series* 111: 171–190.

Feely, R.A., Doney, S.C. & Cooley, S.R. (2009) Ocean acidification: present conditions and future changes in a high-CO_2 world. *Oceanography* 22: 36–47.

Feinberg, L.R. & Peterson, W.T. (2003) Variability in duration and intensity of euphausiid spawning off central Oregon, 1996–2001. *Progress in Oceanography* 57: 363–379.

Feistel, R. (2005) Numerical implementation and oceanographic application of the Gibbs thermodynamic potential of seawater. *Ocean Science* 1: 9–16.

Felbeck, H. (1981) Chemoautotrophic potential of the hydrothermal vent tube worm, *Riftia pachyptila* Jones (Vestimentifera). *Science* 213: 336–338.

Felbeck, H. & Somero, G.N. (1982) Primary production in deep-sea hydrothermal vent organisms: roles of sulfide-oxidizing bacteria. *Trends in Biochemical Sciences* 7: 201–204.

Fenaux, R. (1976) Cycle vital d'un appendiculaire *Oikopleura dioica* Fol, 1872, description et chronologie. *Annales de l'Institut océanographique, Paris* 52: 89–101.

Fields, D.M. & Yen, J. (1997) The escape behavior of marine copepods in response to a quantifiable fluid mechanical disturbance. *Journal of Plankton Research* 19: 1289–1304.

Fiksen, O. & Giske, J. (1995) Vertical distribution and population dynamics of copepods by dynamic optimization. *ICES Journal of Marine Science* 52: 483–503.

Finkel, Z., Quigg, A., Raven, J.A., Reinfelder, J.R., Schofield, O. & Falkowski, P.G. (2006) Irradiance and the elemental stoichiometry of marine phytoplankton. *Limnology and Oceanography* 51: 2690–2701.

Finley, C. (2011) *All the Fish in the Sea, Maximum Sustainable Yield and the Failure of Fisheries Management*. University of Chicago Press, Chicago, 224 pp.

First, M.R., Miller, H.L. III, Lavrentyev, P.J., Pinckney, J.L. & Burd, A.B. (2009) Effects of microzooplankton growth and trophic interactions on herbivory in coastal and offshore environments. *Aquatic Microbial Ecology* 54: 255–267.

Fisher, C.R., Brooks, J.M., Vodenichar, J.S., Zande, J.M., Childress, J.J. & Burke, R.A.J. (1993) The co-occurrence of methanotrophic and chemoautotrophic sulfur-oxidizing bacterial symbionts in a deep-sea mussel. *Marine Ecology* 14: 277–289.

Fisher, R.A. (1930) *The Genetical Theory of Natural Selection*. Clarendon Press, Oxford, variorum edition, J.H. Bennett (ed.), 2000, 360 pp.

Fitzwater, S.E., Knauer, G.A. & Martin, J.H. (1982) Metal contamination and its effect on primary production measurements. *Limnology and Oceanography* 27: 554–551.

Fleminger, A. & Hulsemann, K. (1974) Systematics and distribution of the four sibling species comprising the genus *Pontellina* Dana (Copepoda, Calanoida). *Fishery Bulletin* 72: 63–120.

Flood, P.R. (1991) Architecture of, and water circulation and flow rate in, the house of the planktonic tunicate *Oikopleura labradoriensis*. *Marine Biology* 111: 95–111.

Flores, J.F. & Hourdez, S.M. (2006) The zinc-mediated sulfide-binding mechanism of hydrothermal vent tubeworm 400 kDa hemoglobin. *Cahiers de Biologie Marine* 47: 371–377.

Fonda Umani, S. & Beran, A. (2003) Seasonal variations in the dynamics of microbial plankton communities: first estimates from experiments in the Gulf of Trieste, Northern Adriatic Sea. *Marine Ecology Progress Series* 247: 1–16.

Fonda Umani, S., Tirelli, V., Beran, A. & Guardiani, B. (2005) Relationship between microzooplankton and mesozooplankton: competition versus predation on natural assemblages of the Gulf of Trieste (northern Adriatic Sea). *Journal of Plankton Research* 27: 973–986.

Fox, S.W. (1971) Self-assembly of the protocell from a self-ordered polymer. In: Kimball, A.P. & Oro, J. (eds.) *Prebiotic and Biochemical Evolution*. North-Holland Publishers Co., Amsterdam, pp. 8–30.

Fox, S.W. & Dose, K. (1972) *Molecular Evolution and the Origins of Life*. W.H. Freeman, San Francisco. [Reprint, 1977, M. Dekker, NewYork] 359 pp.

Frada, M., Probert, I., Allen, M.J., Wilson, W.H. & deVargas, C. (2008) The "Cheshire Cat" escape strategy of the coccolithophore *Emiliana huxleyi* in response to viral infection. *Proceedings of the National Academy of Sciences* 105: 15,944–15,949.

France, S.C., Hessler, R.R. & Vrijenhoek, R.C. (1992) Genetic differentiation between spatially-disjunct populations of the deep-sea hydrothermal vent endemic amphipod *Ventiella sulfuris*. *Marine Biology* 114: 551–559.

Francis, C.A., Roberts, K.J., Beman, J.M., Santoro, A.E. & Oakley, B.B. (2005) Ubiquity and diversity of ammonia-oxidizing archaea in water columns and sediments of the ocean. *Proceedings of the National Academy of Sciences* 102: 14,683–14,688.

Francisco, D.E., Mah, R.A. & Rabin, A.C. (1973) Acridine orange epifluorescence technique for counting bacteria in natural waters. *Transactions of the American Microscopical Society* 92: 416–421.

Franck, V.M., Brzezinski, M.A., Coale, K.H. & Nelson, D.M. (2000) Iron and silicic acid concentration regulate Si uptake north and south of the Polar Frontal Zone in the Pacific sector of the Southern Ocean. *Deep-Sea Research II* **47**: 3315–3338.

Frank, K.T., Petrie, B., Choi, J.S. & Leggett, W.C. (2005) Trophic cascades in a formerly cod-dominated ecosystem. *Science* **308**: 1621–1623.

Frank, K.T., Petrie, B., Shackell, N.L. & Choi, J.S. (2006) Reconciling differences in trophic control in mid-latitude marine ecosystems. *Ecology Letters* **9**: 1096–1105.

Frank, T.A., Porter, M. & Cronin, T.W. (2009) Spectral sensitivity, visual pigments and screening pigments in two life history stages of *Gnathophausia ingens*. *Journal of the Marine Biological Association of the United Kingdom* **89**: 119–129.

Franks, P.J.S., Wroblewski, J.S. & Flierl, G.R. (1986) Behavior of a simple plankton model with food-level acclimation by herbivores. *Marine Biology* **91**: 121–129.

Franz, V. (1907) Die biologische Bedeutung des Silberglanzer in der Fischchaut. *Biologiches Zentralblatt* **27**: 278–285.

Franz, V. (1912) Zur Frage der vertikalen Wanderungen der Planktontiere (Autorreferat). *Archiv für Hydrobiologie und Planktonkunde* **7**: 493–499.

Fréon, P., Barange, M. & Aristegui, J. (2009) Eastern boundary upwelling ecosystems: integrative and comparative approaches. *Progress in Oceanography* **83**: 1–14.

Fromentin, J.-M. & Planque, B. (1996) *Calanus* and environment in the eastern North Atlantic. II. Influence of the North Atlantic Oscillation on *C. finmarchicus* and *C. helgolandicus*. *Marine Ecological Progress Series* **134**: 111–118.

Frost, B.W. (1969) *Distribution of the oceanic, epipelagic copepod genus Clausocalanus with an analysis of sympatry of North Pacific species*. PhD Thesis, Scripps Institute of Oceanography, University of California, San Diego, xxii + 297 pp.

Frost, B.W. (1972) Effects of size and concentration of food particles on the feeding behavior of the marine planktonic copepod *Calanus pacificus*. *Limnology and Oceanography* **17**: 805–815.

Frost, B.W. (1975) A threshold feeding behavior in *Calanus pacificus*. *Limnology and Oceanography* **20**: 263–266.

Frost, B.W. (1989) A taxonomy of the marine calanoid copepod genus *Pseudocalanus*. *Canadian Journal of Zoology* **67**: 525–551.

Frost, B.W. (1993) A modelling study of processes regulating plankton standing stock and production in the open subarctic Pacific Ocean. *Progress in Oceanography* **32**: 17–56.

Frost, B.W. & McCrone, L.E. (1974) Vertical distribution of zooplankton and myctophid fish at Canadian Weather Station P, with description of a new multiple net trawl. *Proceedings of the International Conference on Engineering in the Ocean Environment*, IEEE, **1**: 159–165.

Fry, B. & Sherr, E. (1984) $\delta^{13}C$ measurements as indicators of carbon flow in marine and freshwater ecosystems. *Contributions to Marine Science* **27**: 15–47.

Fu, F.X. & 7 coauthors (2008) Interactions between changing pCO_2, N_2 fixation, and Fe limitation in the marine unicellular cyanobacterium *Crocosphaera*. *Limnology and Oceanography* **53**: 2472–2482.

Fu, Y., O'Kelly, C., Sieracki, M. & Distel, D.L. (2003) Protistan grazing analysis by flow cytometry using prey labeled by in vivo expression of fluorescent proteins. *Applied and Enviromental Microbiology* **69**: 6848–6855.

Fuhrman, J.A. & Azam, F. (1982) Thymidine incorporation as a measure of heterotrophic bacterioplankton production in marine surface waters: evaluation and field results. *Marine Biology* **66**: 109–120.

Fuhrman, J.A. & Noble, R.T. (1995) Viruses and protists cause similar bacterial mortality in coastal seawater. *Limnology and Oceanography* **40**: 1236–1242.

Fuhrman, J.A. & Steele, J.A. (2008) Community structure of marine bacterioplankton: patterns, networks, and relationships to function. *Aquatic Microbial Ecology* **53**: 69–81.

Fuhrman, J.A., McCallum, K. & Davis, A.A. (1992) Novel major archaebacterial group from marine plankton. *Nature* **356**: 148–149.

Fujikura, K. & 7 coauthors (2007) Long-term *in situ* monitoring of spawning behavior and fecundity in *Calyptogena* spp. *Marine Ecology Progress Series* **333**: 185–193.

Fujiwara, Y., Tsukahara, J., Hashimoto, J. & Fjikura, K. (1998) *In situ* spawning of a deep-sea vesicomyid clam: evidence for an environmental cue. *Deep-Sea Research I* **45**: 1881–1889.

Fulton, J.D. (1983) Seasonal and annual variations of net zooplankton at Ocean Station "P", 1956–80. *Canadian Data Reports, Fisheries and Aquatic Science* **374**: i–iii; 1–65.

Gage, J.D. & Tyler, P.A. (1991) *Deep-Sea Biology: A Natural History of Organisms on the Deep-Sea Floor*. Cambridge University Press, Cambridge, UK, 504 pp.

Gage, J.D., Lamont, P.A., Kroeger, K., Paterson, G.L.J. & Gonzalez Vecino, J.L. (2000) Patterns in deep-sea macrobenthos at the continental margin: standing crop, diversity and faunal change on the continental slope off Scotland. *Hydrobiologia* **440**: 261–271.

Gainey, L.F. Jr (1972) The use of the foot and the captacula in the feeding of *Dentalium* (Mollusca: Scaphopoda). *The Veliger* **15**: 29–34.

Gallager, S.M. & Alatalo, P. (n.d.) Swimming and feeding in the Thecosomate pteropod *Limacina retroversa*. Unpublished manuscript.

Galloway, J.N. & 14 coauthors (2004) Nitrogen cycles: past, present, and future. *Biogeochemistry* **70**: 153–226.

Garcia, S.M. & Rosenberg, A.A. (2010) Food security and marine capture fisheries: characteristics, trends, drivers and future perspectives. *Proceedings of the Royal Society, London B* **365**: 2869–2880.

Gardiner, L.F. (1975) The systematics, postmarsupial development, and ecology of the deep-sea family Neotanaidae (Crustacea: Tanaidacea). *Smithsonian Contributions to Zoology* **170**: 1–265.

Gargett, A. E. (1997) The optimal stability "window": a mechanism underlying decadal fluctuations in North Pacific salmon stocks? *Fisheries Oceanography* **6**: 109–117.

Garside, C. (1982) A chemiluminescent technique for the determinations of nanomolar concentrations of nitrate and nitrite in seawater. *Marine Chemistry* **11**: 159–167.

Gaten, E., Shelton, P.M.J. & Herring, P.J. (1992) Regional morphological variations in the compound eyes of certain mesopelagic shrimps in relation to their habitat. *Journal of the Marine Biological Association of the United Kingdom* **72**: 61–75.

Gauldie, R.W. (1991) Taking stock of genetic concepts in fisheries management. *Canadian Journal of Fisheries and Aquatic Science* **48**: 722–731.

Gayral, P. & Fresnel, J. (1983) Description, sexualité et cycle de développement de une nouvelle coccolithophoracée (Prymnesiophyceae): *Pleurochrysis pseudoroscoffensis* sp. nov. *Protistologica* **19**: 245–261.

Gebruk, A.V., Pimenov, N.V. & Savvichev, A.A. (1993) Feeding specialization of bresiliid shrimps in the TAG site hydrothermal community. *Marine Ecological Progress Series* **98**: 247–253.

Gerlach, S.A. (1972) Die Produktionsleistung des Benthos in der Helgoländer Bucht. *Verhandlung der Deutschen Zoologische Gesellschaft* **65**: 1–13.

German, C.R. & Von Damm, K.L. (2003) Hydrothermal processes. In: Elderfield, H. (ed.) *Treatise on Geochemistry, Vol. 6, The Oceans and Marine Geochemistry*. Elsevier, Amsterdam, pp. 181–222.

Giere, O. (2009) *Meiobenthology: The Microscopic Motile Fauna of Aquatic Sediments*. Springer, 422 pp.

Gifford, D.J. (1993) Protozoa in the diets of *Neocalanus* spp. in the oceanic subarctic Pacific Ocean. *Progress in Oceanography* **32**: 223–237.

Gifford, D.J., Bohrer, R.N. & Boyd, C.M. (1981) Spines on diatoms: do copepods care? *Limnology and Oceanography* **26**: 1057–1061.

Gilmer, R.W. & Harbison, G.R. (1986) Morphology and field behavior of pteropod molluscs: feeding methods in the families Cavoliniidae, Limacinidae and Peraclididae (Gastropoda: Thecosomata). *Marine Biology* **91**: 47–57.

Giovannoni, S. & Rappé, M. (2000) Evolution, diversity, and molecular ecology of marine prokaryotes. In: Kirchman, D.L. (ed.) *Microbial Ecology of the Oceans.* Wiley, New York, pp. 47–84.

Giovannoni, S.J. & Stingl, U. (2005) Molecular diversity and ecology of microbial plankton. *Nature* **437**: 343–348.

Giovannoni, S.J. & 13 others (2005) Genome streamlining in a cosmopolitan oceanic bacterium. *Science* **309**: 1242–1245.

Gjøsæter, J. (1984) Mesopelagic fish, a large potential resource in the Arabian Sea. *Deep-Sea Research* **31**: 1019–1035.

Glazier, D.S. (2005) Beyond the "3/4-power law": variation in the intra- and interspecific scaling of metabolic rate in animals. *Biological Reviews* **80**: 611–662

Glibert, P.M. & 7 coauthors (2005) The role of eutrophication in the global proliferation of harmful algal blooms. *Oceanography* **18**: 198–209.

Glover, A.G., Goetz, E., Dahlgren, T.G. & Smith, C.R. (2005a) Morphology, reproductive biology and genetic structure of the whale-fall and hydrothermal vent specialist, *Bathykurila guaymasensis* Pettibone, 1989 (Annelida: Polynoidae). *Marine Ecology* **26**: 223–234.

Glover, A.G., Killstrom, B., Smith, C.R. & Dalgren, T.G. (2005b) World-wide whale worms? A new species of *Osedax* from the shallow north Atlantic. *Proceedings of the Royal Society, London B* **272**: 2587–2592.

Glover, A.G., Kemp, K.M., Smith, C.R. & Dahlgren, T.G. (2008) On the role of bone-eating worms in the degradation of marine vertebrate remains. *Proceedings of the Royal Society, London, B* **275**: 1959–1961.

Glover, D.M. & Brewer, P.G. (1988) Estimates of wintertime mixed layer nutrient concentrations in the North Atlantic. *Deep-Sea Research* **35**: 1525–1546.

Glud, R.N. (2008) Oxygen dynamics of marine sediments. *Marine Biology Research* **4**: 243–289.

Glud, R.N., Gundersen, J.K., Jørgensen, B.B., Revsbech, N.P. & Schulz, H.D. (1994) Diffusive and total oxygen uptake of deep-sea sediments in the eastern South Atlantic Ocean: in situ and laboratory measurements. *Deep-Sea Research I* **41**: 1767–1788.

Goetze, E. (2005) Global population genetic structure and biogeography of the oceanic copepods *Eucalanus hyalinus* and *E. spinifer. Evolution* **59**: 2378–2395.

Goetze, E. (2008) Heterospecific mating and partial prezygotic reproductive isolation in the planktonic marine copepods *Centropages typicus* and *Centropages hamatus. Limnology and Oceanography* **53**: 433–445.

Goetze, E. (2010) Species discovery in marine planktonic invertebrates through global molecular screening. *Molecular Ecology* **95**: 952–967.

Goetze, E. & Bradford-Grieve, J. (2005) Genetic and morphological description of *Eucalanus spinifer* T. Scott 1894 (Calanoida: Eucalanidae), a circumglobal sister species of the copepod *E. hyalinus* s.s. (Claus, 1866). *Progress in Oceanography* **65**: 55–87.

Goffredi, S.K., Warén, A., Orphan, V.J. & Van Dover, C.L. (2004) Novel forms of structural integration between microbes and a hydrothermal vent gastropod from the Indian Ocean. *Applied and Environmental Microbiology* **May 2004**: 3082–3090. doi: 10.1128/AEM.70.5.3082–3090.2004.

Goldman, J.C. (1977) Temperature effects on phytoplankton growth in continuous culture. *Limnology and Oceanography* **22**: 932–936.

Goldman, J.C. & Carpenter, E.J. (1974) A kinetic approach to the effect of temperature on algal growth. *Limnology and Oceanography* **19**: 756–766.

Goldman, J.C., Caron, D.A. & Dennett, M.R. (1987) Regulation of gross growth efficiency and ammonium regeneration in bacteria by substrate C : N ratio. *Limnology and Oceanography* **32**: 1239–1252.

Gómez-Gutiérrez, J., Peterson, W.T. & Morado, J.F. (2006) Discovery of a ciliate parasitoid of euphausiids off Oregon, USA: *Collinia oregonensis* n. sp. (Apostomatida: Colliniidae). *Diseases of Aquatic Organisms* **71**: 33–49.

Gómez-Gutiérrez, J., Feinberg, L.R., Shaw, C.T. & Peterson, W.T. (2007) Interannual and geographical variability of the brood size of the euphausiids *Euphausia pacifica* and *Thysanoessa spinifera* along the Oregon coast (1999–2004). *Deep-Sea Research II* **54**: 2145–2169.

Gómez-Gutiérrez, J., Rodríguez-Jaramillo, C., Ángel-Rodríguez, J.D., Robinson, C.J., Zavala-Hernández, C., Martínez-Gómez, S. & Tremblay, N. (2010) Biology of the subtropical sac-spawning euphausiid *Nyctiphanes simplex* in the northwestern seas of Mexico: interbrood period, gonad development, and lipid content. *Deep-Sea Research II* **57**: 616–630.

Gonzalez, J.M., Sherr, E.B. & Sherr, B.F. (1993) Differential feeding by marine flagellates on growing vs. starving and on motile vs. non-motile prey. *Marine Ecology Progress Series* **102**: 257–267.

Gooday, A.J., Levin, L.A., Thomas, C.L. & Hecker, B. (1992) The distribution and ecology of *Bathysiphon filiformis* Sars and *B. major* DeFolin (Protista, Foraminiferida) on the continental slope of North Carolina. *Journal of Foraminiferal Research* **22**: 129–146.

Goodenough, U.W. & Weiss, R.L. (1978) Interrelationships between microtubules, a striated fiber, and the gametic mating structure of *Chlamydomonas reinhardtii. Journal of Cell Biology* **76**: 430–438.

Gordon, L.I., Codispoti, L.A., Jennings, J.C. Jr, Millero, F.J., Morrison, J.M. & Sweeney, C. (2000) Seasonal evolution of hydrographic properties in the Ross Sea, Antarctica, 1996–97. *Deep-Sea Research II* **47**: 3095–3117.

Gowing, M.M. (1989) Abundance and feeding ecology of Antarctic phaeodarian radiolarians. *Marine Biology* **103**: 107–118.

Grabowski, M.N.W., Church, M.J. & Karl, D.M. (2008) Nitrogen fixation rates and controls at Stn ALOHA. *Aquatic Microbial Ecology* **52**: 175–183.

Gran, H.H. & Braarud, T. (1935) A quantitative study of the phytoplankton in the Bay of Fundy and the Gulf of Maine. *Journal of the Biological Board of Canada* **1**: 279–467.

Grassle, J.F. & Maciolek, N.J. (1992) Deep-sea species richness: regional and local diversity estimates from quantitative bottom samples. *American Naturalist* **139**: 313–341.

Grassle, J.F. & Morse-Porteous, L.S. (1987) Macrofaunal colonization of disturbed deep-sea environments and the structure of deep-sea benthic communities. *Deep-Sea Research* **34A**: 1911–1950.

Grassle, J.F. & Sanders, H.L. (1973) Life histories and the role of disturbance. *Deep-Sea Research* **20**: 643–659.

Green, J., Brownell, S., Jones, R. & Chute, A. (1999) *Age and Growth of Larval Cod and Haddock From the '95 and '96 [Georges Bank] Broad-Scale Program.* http://globec.whoi.edu/globec-dir/reports/siworkshop1999/green.html.

Greene, C.H. & Pershing, A.J. (2000) The response of *Calanus finmarchicus* populations to climate variability in the Northwest Atlantic: basin-scale forcing associated with the North Atlantic Oscillation. *ICES Journal of Marine Science* **57**: 1536–1544.

Greene, C.H. & Pershing, A.J. (2007) Climate drives sea change. *Science* **315**: 1084–1085.

Griffiths, H.J., Barnes, D.K.A. & Linse, K. (2009) Towards a generalized biogeography of the Southern Ocean benthos. *Journal of Biogeography* **36**: 162–177.

GRIP (Greenland Ice-core Project) Members, 40 coauthors (1993) Climate instability during the last interglacial recorded in the GRIP ice core. *Nature* **364**: 203–207.

Gruber, N. & Sarmiento, J. (1997) Global patterns of marine nitrogen fixation and denitrification. *Global Biogeochemical Cycles* **11**: 235–266.

Gruber, N. & 14 coauthors (2009) Oceanic sources, sinks, and transport of atmospheric CO_2. *Global Biogeochemical Cycles* **23**: GB1005 (doi:10.1029/2008GB003349, 21 pages).

Guinasso, N.L. Jr & Schink, D.R. (1975) Quantitative estimates of biological mixing rates in abyssal sediments. *Journal of Geophysical Research* **80**: 3032–3043.

Haaker, P.L., Parker, D.O., Barsky, K.C. & Chun, S.Y.C. (1998) Growth of red abalone, *Haliotis rufescens* (Swainson), at Johnsons Lee, Santa Rosa Island. *California Journal of Shellfish Research* **17**: 747–753.

Haddock, S.H.D. (2004) A golden age of gelata: past and future research on planktonic ctenophores and cnidarians. *Hydrobiologia* **530/531**: 549–556.

Hairston, N.G. & Twombly, S. (1985) Obtaining life table data from cohort analysis: a critique of current methods. *Limnology and Oceanography* **30**: 886–893.

Hales, B., Moum, J.N., Covert, P. & Perlin, A. (2005) Irreversible nitrate fluxes due to turbulent mixing in a coastal upwelling system. *Journal of Geophysical Research* **110**, C10S11. doi:10.1029/2004JC002685.

Halsband-Lenk, C., Pierson, J.J. & Leising, A.W. (2005) Reproduction of *Pseudocalanus newmani* (Copepoda: Calanoida) is deleteriously affected by diatom blooms – a field study. *Progress in Oceanography* **67**: 332–348.

Halsey, K.H., Milligan, A.J. & Behrenfeld, M.J. (2010) Physiological optimization underlies growth rate-independent chlorophyll-specific gross and net primary production. *Photosynthesis Research* **103**: 125–137.

Hama, T., Miyazaki, T., Ogawa, Y., Iwakuma, T., Takahashi, M., Otsuki, A. & Ichimura, S. (1983) Measurement of photosynthetic production of a marine phytoplankton population using a stable ^{13}C isotope. *Marine Biology* **73**: 31–36.

Hamme, R.C. (2010) & 15 coauthors Volcanic ash fuels anomalous plankton bloom in subarctic northeast Pacific. *Geophysical Research Letters* **37**: L19604, doi:10.1029/2010GL044629 (5 pp.).

Hamner, W.M. (1974) Ghosts of the Gulf Stream, blue water plankton. *National Geographic* **146**: 530–545.

Hamner, W.M. (1988) Biomechanics of filter feeding in the Antarctic krill *Euphausia superba*: review of past work and new observations. *Journal of Crustacean Biology* **8**: 149–165.

Hamner, W.M. & Hamner, P.P. (2000) Behavior of Antarctic krill (*Euphausia superba*): schooling, foraging, and antipredatory behavior. *Canadian Journal of Fisheries and Aquatic Sciences* **57** (Suppl. 3): 192–202.

Hannides, C.C.S., Popp, B.N., Landry, M.R. & Graham, B.S. (2009) Quantification of zooplankton trophic position in the North Pacific subtropical gyre using stable nitrogen isotopes. *Limnology and Oceanography* **54**: 30–61.

Hannides, C.C.S., Landry, M.R., Benitez-Nelson, C.R., Styles, R.M., Montoya, J.P. & Karl, D.M. (2009) Export stoichiometry and migrant-mediated flux of phosphorus in the North Pacific Subtropical Gyre. *Deep-Sea Research I* **56**: 73–88.

Hansen, P.J., Bjørnsen, P.K. & Hansen, B.W. (1997) Zooplankton grazing and growth: scaling within the 2–2000 µm body size range. *Limnology and Oceanography* **42**: 687–704.

Harbison, G. R. & McAlister, V. (1979) The filter-feeding rates and particle retention efficiencies of three species of *Cyclosalpa* (Tunicata: Thalicea). *Limnology and Oceanography* **24**: 875–892.

Hardy, A.C. (1924). The herring in relation to its animate environment. Part I. The food and feeding habits of the herring with special reference to the east coast of England. *Fishery Investigations, Series 2*, 7(3): 53 pp.

Hardy, A.C. (1970) *The Open Sea: Its Natural History. Part I: The World of Plankton*, 2nd edn. Collins, London (335 pp.).

Harris, J.E. & Wolfe, U.K. (1955) A laboratory study of vertical migration. *Proceedings of the Royal Society of London, B* **144**: 329–354.

Harrison, P.J., Boyd, P.W., Levasseur, M., Tsuda, A., Rivkin, R.B., Roy, S.O. & Miller, W.L. (editors) (2006). Canadian SOLAS: Subarctic Ecosystem Response to Iron Enrichment (SERIES). *Deep-Sea Research II*, 53(20–22): 2005–2454.

Harrison, W.G., Harris, L.R. & Irwin, B.D. (1996) The kinetics of nitrogen utilization in the oceanic mixed layer: nitrate and ammonium interactions at nanomolar concentrations. *Limnology and Oceanography* **41**: 16–32.

Hartman, O. & Fauchald, K. (1971) Deep-water benthic polychaetous annelids off New England to Bermuda and other North Atlantic areas, Part II. *Allan Hancock Monographs in Marine Biology* **6**: 1–327.

Hartwick, R.F. (1991) Observations on the anatomy, behavior, reproduction and life cycle of the cubozoan *Carybdea sivickisi*. *Hydrobiologia* **216/217**: 171–179.

Harvey, W.H. (1937) Note on selective feeding by *Calanus*. *Journal of the Marine Biological Association of the United Kingdom* **22**: 97–100.

Hashimoto, J. & 9 coauthors (2001) First hydrothermal vent communities from the Indian Ocean discovered. *Zoological Science* **18**: 717–721.

Hasle, G.R. & Syvertsen, E.E. (1997) Marine diatoms. In: Tomas, C.R. (ed.) *Identifying Marine Phytoplankton*, Academic Press, San Diego, pp. 5–385.

Hátun, H. & 8 coauthors (2009) Large bio-geographical shifts in the north-eastern Atlantic Ocean: from the subpolar gyre, via plankton, to blue whiting and pilot whales. *Progress in Oceanography* **80**: 149–162.

Hauser, L. & Carvalho, C.R. (2008) Paradigm shifts in marine fisheries genetics: ugly hypotheses slain by beautiful facts. *Fish and Fisheries* **9**: 333–362.

Hausmann, K., Hülsmann, N., MacHerner, H. & Mulisch, M. (1996) *Protozoology*. Georg Thieme Verlag, Stuttgart, 338 pp.

Havenhand, J.N., Matsumoto, G.I. & Seidel, E. (2006) *Megalodicopia hians* in the Monterey submarine canyon: distribution, larval development, and culture. *Deep-Sea Research I* **53**: 215–222.

Haxo, F.T. (1985) Photosynthetic action spectrum of the coccolithophorid, *Emiliania huxleyi* (Haptophyceae): 19'Hexanoyloxyfucoxanthin as antenna pigment. *Journal of Phycology* **21**: 282–287.

Heath, M.R., Fraser, J.G., Gislason, A., Hay, S.H., Jónasdóttir, S.H. & Richardson, K. (2000) Winter distributions of *Calanus finmarchicus* in the Northeast Atlantic. *ICES Journal of Marine Science* **57**: 1628–1635.

Heath, M.R. & 32 coauthors (2008) Spatial demography of *Calanus finmarchicus* in the Irminger Sea. *Progress in Oceanography* **76**: 39–88.

Hedges, J.I. (1992) Global biogeochemical cycles: progress and problems. *Marine Chemistry* **39**: 67–93.

Hedges, J.I., Hu, F.S., Devol, A.H., Hartnett, H.E., Tsamakis, E. & Keil, R.G. (1999) Sedimentary organic matter preservation: a test for selective degradation under oxic conditions. *American Journal of Science* **299**: 529–555.

Hedges, J.I., Keil, R.G. & Benner, R. (1997) What happens to terrestrial organic matter in the ocean? *Organic Geochemistry* **27**: 195–212.

Hedgpeth, J.W. (1957) Classification of marine environments. In: Hedgpeth, J.W. (ed.), *Treatise on Marine Ecology and Paleoecology,*

Volume 1, Ecology, Geological Society of America, Memoir, 67, pp. 17–27.

Heintzelman, M.B. (2006) Cellular and molecular mechanics of gliding locomotion in eukaryotes. *International Review of Cytology* **251**: 79–129.

Helfenbein, K.G., Fourcade, H.M., Vanjani, R.G. & Boore, J.L. (2004) The mitochondrial genome of *Paraspadella gotoi* is highly reduced and reveals that chaetognaths are a sister group to protostomes. *Proceedings of the National Academy of Sciences* **101**: 10,639–10,643.

Helmkampf, M., Bruchhaus, I. & Hausdorf, B. (2008) Multigene analysis of lophophorate and chaetognath phylogenetic relationships. *Molecular Phylogenetics and Evolution* **46**: 206–214.

Hernández-León, S. & Ikeda, T. (2005) A global assessment of mesozooplankton respiration in the ocean. *Journal of Plankton Research* **27**: 153–158.

Hernández-León, S., Portillo-Hahnenfeld, A., Almeida, C., Bécognée, P. & Moreno, I. (2001) Diel feeding behaviour of krill in the Gerlache Strait, Antarctica. *Marine Ecology Progress Series* **223**: 235–242.

Hernández-León, S., Franchy, G., Moyano, M., Menéndez, I., Schmoker, C. & Putzeys, S. (2010) Carbon sequestration and zooplankton lunar cycles: could we be missing a major component of the biological pump? *Limnology and Oceanography* **55**: 2503–2512.

Hernes, P.J., Peterson, M.L., Murray, J.M., Wakeham, S.G., Lee, C. & Hedges, J.J. (2001) Particulate carbon and nitrogen fluxes and composition in the central equatorial Pacific. *Deep-Sea Research I* **48**: 1999–2023.

Heron, G.A. (2002) *Oncaea frosti*, a new species (Copepoda: Poecilostomatoida) from the Liberian coast and the Gulf of Mexico. *Hydrobiologia* **480**: 145–154.

Hessler, R.R. (1972) The structure of deep benthic communities from central oceanic waters. In: Miller, C. (ed.) *The Biology of the Oceanic Pacific*. Oregon State University Press, Corvallis, pp. 79–93.

Hessler, R.R. & Jumars, P. (1974) Abyssal community analysis from replicate box cores in the central North Pacific. *Deep-Sea Research* **21**: 185–209.

Hessler, R.R., Wilson, G.D. & Thistle, D. (1979) The deep-sea isopods: a biogeographic and phylogenetic overview. *Sarsia* **64**: 67–75.

Hessler, R.R., Smithey, W.M., Boudrias, M.A., Keller, C.H., Lutz, R.A. & Childress, J.J. (1988) Temporal change in megafauna at the Rose Garden hydrothermal vent (Galapagos Rift; Eastern Tropical Pacific). *Deep-Sea Research* **36**: 1681–1709.

Heymans, S.J.J., Guénette, S. & Christensen, V. (2007) Evaluating network analysis indicators of ecosystem status in the Gulf of Alaska. *Ecosystems* **10**: 488–502.

Heymans, S.J.J., Sumaila, U.R. & Christensen, V. (2009) Policy options for the northern Benguela ecosystem using a multispecies, multifleet ecosystem model. *Progress in Oceanography* **83**: 417–425.

Heymans, S.J.J. & Sumaila, U.R. (2007) Updated ecosystem model for the northern Benguela ecosystem, Namibia. In: Le Quesne, W.J.F., Arreguín-Sánchez, F. & Heymans, S.J.J. (eds.) *INCOFISH ecosystem models: transiting from Ecopath to Ecospace*. University of British Columbia Fisheries Center Research Reports 15(6), pp. 25–70.

Higgins, M.J., Molino, P., Mulvaney, P. & Wetherbee, R. (2003) The structure and nanomechanical properties of the adhesive mucilage that mediates diatom substrate adhesion and motility. *Journal of Phycology* **39**: 1181–1193.

Hilborn, R. & Walters, C.J. (1992) *Quantitative Fisheries Stock Assessment: Choice, Dynamics and Uncertainty*. Chapman and Hall, New York, 570 pp.

Hill, R.S., Allen, L.D. & Bucklin, A. (2001) Multiplexed species-specific PCR protocol to discriminate four N. Atlantic *Calanus* species, with a mtCOI gene tree for ten *Calanus* species. *Marine Biology* **139**: 279–287.

Hirst, A.G. & Bunker, A.J. (2003) Growth of marine planktonic copepods: global rates and patterns in relation to chlorophyll a, temperature, and body weight. *Limnology and Oceanography* **48**: 1988–2010.

Hirst, A.G. & Lampitt, R.S. (1998) Towards a global model of in situ weight-specific growth in marine planktonic copepods. *Marine Biology* **132**: 247–257.

Hirst, A.G., Peterson, W.T. & Rothery, P. (2005) Errors in juvenile copepod growth rate estimates are widespread: problems with the moult rate method. *Marine Ecology Progress Series* **298**: 268–279.

Ho, T.-Y., Quigg, A., Finkel, Z.V., Milligan, A.J., Wyman, K., Falkowski, P.G. & Morel, F.M.M. (2003) The elemental composition of some phytoplankton. *Journal of Phycology* **39**: 1145–1159.

Hobbie, J.E., Daley, R.J. & Jaspar, S. (1977) Use of nucleopore filters for counting bacteria by fluorescent microscopy. *Applied and Environmental Microbiology* **33**: 1225–1228.

Hoch, M.P., Snyder, R.A., Cifuentes, L.A. & Coffin, R.B. (1996) Stable isotope dynamics of nitrogen recycled during interactions among marine bacteria and protists. *Marine Ecology Progress Series* **132**: 229–239.

Hollowed, A.S., Bailey, K.S. & Wooster, W.S. (1987) Patterns in recruitment of marine fishes in the Northeast Pacific Ocean. *Biological Oceanography* **5**: 99–131.

Honjo, S. & Manganini, S.J. (1993) Annual biogenic particle fluxes to the interior of the North Atlantic Ocean; studied at 34°N 21°W and 48°N 21°W. *Deep-Sea Research I.* **40**: 587–607.

Honjo, S., Manganini, S.J. & Cole, J.J. (1982) Sedimentation of biogenic matter in the deep ocean. *Deep-Sea Research* **29**: 609–626.

Honjo, S., Dymond, J., Collier, R. & Manganini, S.J. (1995) Export production of particles to the interior of the equatorial Pacific Ocean during the 1992 EqPac experiment. *Deep-Sea Research II* **42**: 831–870.

Honjo, S., Manganini, S.J., Krisfield, R.A. & Francois, R. (2008) Particulate organic carbon fluxes to the ocean interior and factors controlling the biological pump: a synthesis of global sediment trap programs since 1983. *Progress in Oceanography* **76**: 217–285.

Hopcroft, R.R., Roff, J.C., Webber, M.K. & Witt, J.D.S. (1998) Zooplankton growth rates: the influence of size and resources of tropical marine copepodites. *Marine Biology* **132**: 67–77.

Huggett, J., Verheye, H., Escribano, R. & Fairweather, T. (2009) Copepod biomass, size composition and production in the Southern Benguela: spatio-temporal patterns of variation, and comparison with other eastern boundary upwelling systems. *Progress in Oceanography* **83**: 197–207.

Humphris, S.E., Zierenberg, R.A., Mullineaux, L.S. & Thomson, R.E. (eds.) (1995) Sea floor hydrothermal systems: physical, chemical, biological, and geological interactions. *American Geophysical Union, Geophysical Monograph* **91**, 466 pp.

Hunter, J.R., Argue, A.W., Bayliff, W.H., Dizon, A.E., Fonteneau, A., Goodman, D. & Seckel, G.R. (1986) The dynamics of tuna movement: an evaluation of past and future research. *FAO Fisheries Technical Paper* **277**.

Huntley, M.E. (1996) Temperature and copepod production in the sea: a reply. *American Naturalist* **148**: 407–420.

Huntley, M.E. & Boyd, C. (1984) Food-limited growth of marine zooplankton. *American Naturalist* **124**: 455–478.

Huntley, M.E. & Lopez, M.D.G. (1992) Temperature-dependent production of marine copepods: a global synthesis. *American Naturalist* **140**: 201–242.

Hurlbert, S.H. (1971) The nonconcept of species diversity: a critique and alternative parameters. *Ecology* **52**: 577–586.

Hurtt, G.C. & Armstrong, R.A. (1999) A pelagic ecosystem model calibrated with BATS and OWSI data. *Deep-Sea Research, I* **46**: 27–61.

Hutchins, D.A. & Fu, F.-X. (2008) Linking the oceanic biogeochemistry of iron and phosphorus with the marine nitrgen cycle. In: Capone, D.G., Bronk, D.A. & Mulholland, M.R. (eds.) *Nitrogen in the Marine Environment*, Academic Press, Burlington, MA, pp. 1627–1666.

Huyer, A., Smith, R.L. & Paluszkiewicz, T. (1987) Coastal upwelling off Peru during normal and El Niño times, 1981–84. *Journal of Geophysical Research, Oceans* **92**: 14,297–14,307.

Huys, R. & Boxshall, G.A. (1991) *Copepod Evolution*. The Ray Society, London, 468 pp.

Ianelli, J. (2005) Assessment and fisheries management of eastern Bering Sea walleye pollock: is sustainability luck? *Bulletin of Marine Science* **76**: 321–335.

IATTC (Inter-American Tropical Tuna Commission) (2000) *Annual Report of the IATTC, 1998*. IATTC, La Jolla, California.

Ikeda, I. & Dixon, P. (1982) Body shrinkage as a possible overwintering mechanism of the Antarctic krill *Euphausia superba* Dana. *Journal of Experimental Marine Biology and Ecology* **62**: 143–151.

Ikeda, T. (1974) Nutritional ecology of marine zooplankton. *Memoirs of the Faculty of Fisheries, Hokkaido University* **22**: 1–97.

Ikeda, T. (1985) Metabolic rates of epipelagic marine zooplankton as a function of body mass and temperature. *Marine Biology* **85**: 1–11.

Ikeda, T. (2008) Metabolism in mesopelagic and bathypelagic copepods: reply to Childress *et al.* (2008). *Marine Ecology Progress Series* **373**: 193–198.

Ikeda, T., Sano, F. & Yamaguchi, A. (2007) Respiration in marine pelagic copepods: a global bathymetric model. *Marine Ecology Progress Series* **339**: 215–219.

Iles, E.J. (1961) The appendages of Halocypridae. *Discovery Reports* **31**: 299–626.

Imai, K., Nojiri, Y., Tsurushima, N. & Saino, T. (2002) Time series of seasonal variation of primary productivity at station KNOT (44°N, 155°E) in the subarctic western North Pacific. *Deep-Sea Research II* **49**: 5395–5408.

Imbrie, J. & Imbrie, J.Z. (1980) Modeling the climatic response to orbital variations. *Science* **207**: 943–953.

Ingalls, A.E., Shah, S.R., Hansman, R.L., Aluwihare, L.I., Santos, G.M., Druffel, E.R.M. & Pearson, A. (2006) Quantifying archaeal community autotrophy in the mesopelagic ocean using natural radiocarbon. *Proceedings of the National Academy of Sciences* **103**: 6442–6447.

Intergovernmental Panel on Climate Change (2007) *Working Group I Report "The Physical Science Basis"*, Oxford University Press, 996 pp. (Also on-line: http://www.ipcc.ch/publications_and_data/publications_ipcc_fourth_assessment_report_wg1_report_the_physical_science_basis.htm)

IOCCG (International Ocean-Colour Coordinating Group) (2009) Partition of the ocean into ecological provinces: role of ocean-colour radiometry. *IOCCG Report* No. 9, 98 pp. (available on-line).

IPHC, International Pacific Halibut Commission (1998) The Pacific halibut. Biology, fishery and management. *IPHC Technical Report* **40**: 1–63.

Ivlev, V.S. (1945) Biologicheskaya produktivnost' vodoemov (The biological productivity of waters). *Uspekhi Sovremennoi Biologii* **19**: 98–120. Translated in Ricker, W. E. (1966) *Journal of the Fisheries Research Board of Canada* **23**: 1707–1759.

Jacobson, D.M. & Anderson, D.M. (1992) Ultrastructure of the feeding apparatus and myonemal system of the heterotrophic dinoflagellate *Protoperidinium spinulosum*. *Journal of Phycology* **28**: 69–82.

Jahnke, R.A. (2010) Global synthesis. In: Liu, K.K., Atkinson, L., Quiñones, R. & Talaue-McManus, L.(eds) *Carbon and Nutrient Fluxes in Continental Margins, Global Change – The IGBP Series*, Springer-Verlag, Berlin, Heidelberg, pp. 597–615..

Jakobsson, J. (1980) Exploitation of the Icelandic spring- and summer-spawning herring in relation to fisheries management, 1947–77. *Rapports et procès-verbaux des réunions/Conseil permanent international pour l'exploration de la mer* **177**: 23–42.

Jannasch, H.W. (1999) Biocatalytic transformations of hydrothermal fluids. In: Cann, J.R., Elderfield, H. & Laughton, A. (eds.) *Mid-Ocean Ridges: Dynamics of Processes Associated with Creation of New Ocean Crust*. Cambridge University Press, Cambridge, UK, pp. 281–292.

Jannasch, H.W. & Jones, G.E. (1959) Bacterial populations in sea water as determined by different methods of determination. *Limnology and Oceanography* **4**: 128–139.

Jansen, S. (2008) Copepods grazing on *Coscinodiscus wailesii*: a question of size? *Helgoland Marine Research* **62**: 251–255.

Jassby, A.D. & Platt, T. (1976) Mathematical formulation of the relationship between photosynthesis and light for phytoplankton. *Limnology and Oceanography* **21**: 540–547.

Jennings, S., Kaiser, M.J. & Reynolds, J.D. (2001) *Marine Fisheries Ecology*. Blackwell Science, Oxford, 417 pp.

Jeong, H.J., Yoo, Y.D., Kimi, J.S., Kang, N.S., Kim, T.H. & Kim, J.H. (2004) Feeding by the marine planktonic ciliate *Strombidinopsis jeokjo* on common heterotrophic dinoflagellates. *Aquatic Microbiology and Ecology* **36**: 181–187.

Johnson, C.L., Leising, A.W., Runge, J.A., Head, E.J.H., Pepin, P., Plourde, S. & Durbin, E.G. (2008) Characteristics of *Calanus finmarchicus* dormancy patterns in the Northwest Atlantic. *ICES Journal of Marine Science* **65**: 339–350.

Johnson, J.K. (1980) Effects of temperature and salinity on production and hatching of dormant eggs of *Acartia californiensis* (Copepoda) in an Oregon estuary. *Fishery Bulletin* **77**: 567–584.

Johnson, J.K. (1981) Population dynamics and cohort persistence of *Acartia californiensis* (Copepoda: Calanoida) in Yaquina Bay, Oregon. Ph.D. Dissertation, Oregon State University, Corvallis, Oregon, USA, 305 pp.

Johnson, K.S., Riser, S.C. & Karl, D.M. (2010) Nitrate supply from deep to near-surface waters of the North Pacific subtropical gyre. *Nature* **465**: 1062–1065.

Johnson, M.W. & Brinton, E. (1963) Biological species, water masses and currents. In: Hill, M. (ed.) *The Sea*, Vol. 2. Interscience, New York, pp. 381–414.

Johnson, P.W. & Sieburth, J.M. (1979) Chroococcoid cyanobacteria in the sea: a ubiquitous and diverse phototrophic biomass. *Limnology and Oceanography* **24**: 928–935.

Johnson, P.W. & Sieburth, J.M. (1982) *In situ* morphology and occurrence of eucaryotic phototrophs of bacterial size in the picoplankton of estuarine and oceanic waters. *Journal of Phycology* **18**: 318–327.

Johnson, S. (2001) Hidden in plain sight: the ecology and physiology of organismal transparency. *Biological Bulletin* **201**: 301–318.

Jørgensen, E.G. (1964) Adaptation to different light intensities in the diatom *Cyclotella menenghiniana* Kutz. *Physiologia Plantarum* **17**: 136–145.

Jørgensen, E.G. (1969) The adaptation of plankton algae IV. Light adaptation in different algal species. *Physiologia Plantarum* **22**: 1307–1315.

Jumars, P. (1975) Environmental grain and polychaete species diversity in a bathyal benthic community. *Marine Biology* **30**: 253–266.

Jumars, P. (1976) Deep-sea species diversity: does it have a characteristic scale? *Journal of Marine Research* **20**: 643–659.

Jumars, P.A. & Fauchald, K. (1977) Between-community contrasts in successful polychaete feeding strategies. In: Coull, B. (ed.) *Ecology of Marine Benthos*. University of South Carolina Press, Columbia, pp. 1–20.

Kaiser, K. & Benner, R. (2008) Major bacterial contribution to the ocean reservoir of detrital organic carbon and nitrogen. *Limnology and Oceanography* 53: 99–112.

Kampa, E.M. & Boden, B.P. (1954) Submarine illumination and the twilight movements of a sonic scattering layer. *Nature* 174: 869–873.

Kandler, O. (1998) The early diversification of life and the origin of the three domains: a proposal. In: Wiegel, J. & Adams, M.W.W. (eds.) *Thermophiles: the Keys to Molecular Evolution and the Origin of Life?* Taylor and Francis, London, pp. 19–31.

Kapp, H. (2000) The unique embryology of Chaetognatha. *Zoologischer Anzeiger* 239: 263–266.

Karaköylü, E.M., Franks, P.J.S., Tanaka, Y., Roberts, P.L.D. & Jaffe, J.S. (2009) Copepod feeding quantified by planar laser imaging of gut fluorescence. *Limnology and Oceanography Methods* 7: 33–41.

Karl, D.M. & 7 coauthors (2001) Ecological nitrogen-to-phosphorus stoichiometry at station ALOHA. *Deep-Sea Research II* 48: 1529–1566.

Karner, M.B., DeLong, E.F. & Karl, D.M. (2001) Archaeal dominance in the mesopelagic zone of the Pacific Ocean. *Nature* 409: 507–510.

Katija, K. & Dabiri, J.O. (2009) A viscosity-enhanced mechanism for biogenic ocean mixing. *Nature* 460: 624–626 (methods on line) (doi:10.1038/nature08207)

Kaupp, L.J., Measures, C.I., Selph, K.E. & MacKenzie, F.T. (2011) The distribution of dissolved Fe and Al in the upper waters of the Eastern Equatorial Pacific. *Deep-Sea Research II* 58: 296–310.

Kawachi, M., Inouye, I., Maeda, O. & Chihara, C. (1991) The haptonema as a food-capturing device; observations on *Chrysochromulina hirta*. *Phycologia* 30: 563–573.

Kawahara, M., Uye, S.-I., Ohtsu, K. & Iizumi, H. (2006) Unusual population explosion of the giant jellyfish *Nemopilema nomurai* (Scyphozoa: Rhizostomeae) in East Asian waters. *Marine Ecology Progress Series* 307: 161–173.

Kawasaki, T. (1983) Why do some pelagic fishes have wide fluctuations in their numbers? Biological basis of fluctuation from the viewpoint of evolutionary ecology. *FAO Fisheries Report* 291: 1065–1080.

Keeling, P.J. (2010) The endosymbiotic origin, diversification and fate of plastids. *Philosophical Transactions of the Royal Society B* 365: 729–748.

Keeling, R.F., Manning, A.C., McEvoy, E.M. & Shertz, S.R. (1998a) Methods for measuring changes in atmospheric O_2 concentration and their application in southern hemisphere air. *Journal of Geophysical Research* 103: 3381–3397.

Keeling, R.F., Stephens, B.B., Najjar, R.G., Doney, S.C., Archer, D. & Heimann, M. (1998b) Seasonal variations in the atmospheric O_2/N_2 ratio in relation to the kinetics of air–sea gas exchange. *Global Biogeochemical Cycles* 12: 141–163.

Keister, J.E. & Peterson, W.T. (2003) Zonal and seasonal variations in zooplankton community structure off the central Oregon coast, 1998–2000. *Progress in Oceanography* 57: 341–361.

Kelley, D.S. & 25 coauthors (2005) A serpentinite-hosted ecosystem: the Lost City hydrothermal field. *Science* 307: 1428–1434.

Kemp, P.F. (1986) Direct uptake of detrital carbon by the deposit-feeding polychaete *Euzonus mucronata* (Treadwell). *Journal of Experimental Marine Biology and Ecology* 99: 49–61.

Kennan, S.C. & Flament, P.J. (2000) Observations of a tropical instability vortex. *Journal of Physical Oceanography* 30: 2277–2301.

Kennedy, M., Droser, M., Mayer, L.M., Pevear, D. & Mrofka, D. (2006) Late Precambrian oxygenation: inception of the clay mineral factory. *Science* 311: 1446–1449.

Kimmerer, W.J. (1983) Direct measurement of the production:biomass ratio of the subtropical calanoid copepod *Acrocalanus inermis*. *Journal of Plankton Research* 5: 1–14.

Kimmerer, W.J. & McKinnon, A.D. (1987) Growth, mortality, and secondary production of the copepod *Acartia tranteri* in Westernport Bay, Australia. *Limnology and Oceanography* 32: 14–28.

Kimmerer, W.J., Hirst, A.G., Hopcroft, R.R. & McKinnon, A.D. (2007) Estimating juvenile copepod growth rates: corrections, inter-comparisons and recommendations. *Marine Ecology Progress Series* 366: 187–202.

King, M. (2007) *Fisheries Biology: Assessment and Management*, 2nd Edition. Blackwell Science, Oxford, 400 pp.

Kiørboe, T. (2000) Colonization of marine snow aggregates by invertebrate zooplankton: abundance, scaling, and possible role. *Limnology and Oceanography* 45: 479–484.

Kiørboe, T. (2008) *A Mechanistic Approach to Plankton Ecology*. Princeton Univ. Press, Princeton NJ, 209 pp.

Kiørboe, T. & Bagøien, E. (2005) Motility patterns and mate encounter rates in planktonic copepods. *Limnology and Oceanography* 50: 1999–2007.

Kiørboe, T. & Sabatini, M. (1994) Reproductive and life cycle strategies in egg-carrying cyclopoid and free-spawning calanoid copepods. *Journal of Plankton Research* 16: 1353–1366.

Kiørboe, T., Andersen, A., Langlois, V.J., Jakobsen, H.H. & Bohr, T. (2009) Mechanisms and feasibility of prey capture in ambush-feeding zooplankton. *Proceedings of the National Academy of Sciences* 106: 12,394–12,399.

Kipp, N.G. (1976) New transfer function for estimating past sea-surface conditions from sea-bed distribution of planktonic foraminiferal assemblages in the North Atlantic. In: Cline, R. & Hays, J.D. (eds.) *Investigations of Late Quaternary Paleoceanography and Paleoclimatology*. Geological Society of America Memoirs 145: 3–41.

Kirchman, D.L., K'Nees, E. & Hodson, R. (1985) Leucine incorporation and its potential as a measure of protein synthesis by bacteria in natural waters. *Applied and Environmental Microbiology* 49: 599–607.

Kirchman, D.L. (1992) Incorporation of thymidine and leucine in the subarctic Pacific: application to estimating bacterial production. *Marine Ecology Progress Series* 82: 301–309.

Kirchman, D.L. (2000) Uptake and regeneration of inorganic nutrients by marine heterotrophic bacteria. In: Kirchman, D.L. (ed.) *Microbial Ecology of the Oceans*. Wiley, New York, pp. 261–288.

Kirchman, D.L., Ducklow, H. & Mitchell, R. (1982) Estimates of bacterial growth from changes in uptake rates and biomass. *Applied and Environmental Microbiology* 44: 1296–1307.

Kirchman, D.L., Moran, X.A.G. & Ducklow, H. (2009) Microbial growth in the polar oceans – role of temperature and potential impact of climate change. *Nature Reviews/Microbiology* 7: 451–459.

Knauer, G.A., Martin, J.H. & Bruland, K.W. (1979) Fluxes of particulate carbon, nitrogen and phosphorus in the upper water column of the northeast Pacific. *Deep-Sea Research* 26A: 97–108.

Knight, M.D. (1984) Variation in larval morphogenesis within the Southern California Bight population of *Euphausia pacifica* from winter through summer, 1977–1978. *CalCOFI Reports* 25: 87–99.

Koblizek, M. (2011) Role of photoheterotrophic bacteria in the marine carbon cycle. In: Jiao, N., Azam, F. & Sanders, S. (eds.) *Microbial Carbon Pump in the Ocean*, American Association for the Advancement of Science, Washington D.C., pp. 49–51.

Koehl, M. & Strickler, R. (1981) Copepod feeding currents: food capture at low Reynolds number. *Limnology and Oceanography* 26: 1062–1073.

Kolber, Z.S., Prasil, O. & Falkowski, P.G. (1998) Measurements of variable chlorophyll fluorescence using fast repetition rate techniques: defining methodology and experimental protocols. *Biochimica et Biophysica Acta* **1367**: 88–106.

Kolber, Z.S., Van Dover, C.L., Niederman, R.A. & Falkowski, P.G. (2000) Bacterial photosynthesis in surface waters of the open ocean. *Nature* **407**: 177–179.

Koschinsky, A., Garbe-Schönberg, D., Sander, S., Schmidt, K., Gennerich, H.-H. & Strauss, H. (2008) Hydrothermal venting at pressure-temperature conditions above the critical point of seawater, 5°S on the Mid-Atlantic Ridge. *Geology* **36**: 615–618.

Koslow, J.A. (2007) *The Silent Deep: the Discovery, Ecology and Conservation of the Deep Sea*. University of Chicago Press, Chicago, 270 pp.

Kostadinov, T.S., Siegel, D.A. & Maritorena, S. (2010) Global variability of phytoplankton functional types from space: assessment via the particle size distributions. *Biogeosciences* **7**: 3239–3257.

Kovach, A.I., Breton, T.S., Berlinsky, D.L., Maceda, L. & Wirgin, I. (2010) Fine-scale spatial and temporal genetic structure of Atlantic cod off the Atlantic coast of the USA. *Marine Ecology Progress Series* **410**: 177–195.

Kristiansen, S., Farbrot, T. & Naustvoll, L.-J. (2000) Production of biogenic silica by spring diatoms. *Limnology and Oceanography* **45**: 472–478.

Kröger, N., Deutzmann, R. & Sumper, M. (1999) Polycationic peptides from diatom biosilica that direct silica nanosphere formation. *Science* **286**: 1129–1132.

Kruse, S. (2009) Population structure and reproduction of *Eukrohnia bathypelagica* and *Eukrohnia bathyantarctica* in the Lazarev Sea, Southern Ocean. *Polar Biology* **32**: 1377–1387.

Kudela, R.M. & 8 coauthors (2008) New insights into the controls and mechanisms of plankton productivity in coastal upwelling waters of the northern California Current system. *Oceanography* **21**: 46–59.

Kudo, I., Noiri, Y., Imai, K., Nojiri, Y., Nishioka, J. & Tsuda, A. (2005) Primary productivity and nitrogenous nutrient assimilation dynamics during the subarctic Pacific iron experiment for ecosystem dynamics study. *Progress in Oceanography* **64**: 207–221.

Kudo, I., Noiri, Y., Cochlan, W.P., Suzuki, K., Aramaki, T., Ono, T. & Nojiri, Y. (2009) Primary productivity, bacterial productivity and nitrogen uptake in response to iron enrichment during SEEDS II. *Deep-Sea Research II* **56**: 2755–2766.

Kussakin, O.G. (1973) Peculiarities of the geographical and vertical distribution of marine isopods and the problem of deep-sea fauna origin. *Marine Biology* **23**: 19–34.

Lalli, C.M. & Gilmer, R.W. (1989) *Pelagic Snails, The Biology of Holoplanktonic Gastropod Mollusks*. Stanford University Press, Stanford, CA, 259 pp.

Lambert, F. & 9 coauthors (2008) Dust–climate couplings over the past 800,000 years from the EPICA Dome C ice core. *Nature* **452**: 616–619.

Lampadariou, A. & Tselepides, A. (2006) Spatial variability of meiofaunal communities at areas of contrasting depth and productivity in the Aegean Sea (NE Mediterranean). *Progress in Oceanography* **69**: 19–36.

Lampitt, R.S. & Antia, A.N. (1997) Particle flux in deep seas: regional characteristics and temporal variability. *Deep-Sea Research I* **44**: 1377–1403.

Lampitt, R.S., Salter, L. & Johns, D. (2009) Radiolaria: major exporters of organic carbon to the deep ocean. *Global Biogeochemical Cycles* **23**: GB1010 (9 pp).

Lampitt, R.S., Billett, D.S.M. & Martin, A.P. (2010a) The sustained observatory over the Porcupine Abyssal Plain (PAP): insights from time series observations and process studies. *Deep-Sea Research II* **57**: 1267–1271.

Lampitt, R.S., Salter, I., de Cuevas, B.A., Hartman, S., Larkin, K.E. & Pebody, C.A. (2010b) Long-term variability of downward particle flux in the deep northeast Atlantic: causes and trends. *Deep-Sea Research II* **57**: 1346–1361.

Landry, M.R. (1976) *Population dynamics of the planktonic marine copepod,* Acartia clausi *Giesbrecht, in a small temperate lagoon*. Ph.D. Dissertation, Univ. Washington, Seattle, 167 pp.

Landry, M.R. (1978) Population dynamics and production of a planktonic marine copepod, *Acartia clausii*, in a small temperate lagoon on San Juan Island, Washington. *International Revue der gesamten Hydrobiologie* **63**: 77–119.

Landry, M.R. & Calbet, A. (2004) Microzooplankton production in the oceans. *ICES Journal of Marine Science* **61**: 501–517.

Landry, M.R. & Hassett, R.P. (1982) Estimating the grazing impact of marine micro-zooplankton. *Marine Biology* **67**: 283–288.

Landry, M.R., Constantinou, J., Latasa, M., Brown, S.L., Bidigare, R.R. & Ondrusek, M.E. (2000) Biological response to iron fertilization in the eastern equatorial Pacific (IronEx II). III. Dynamics of phytoplankton growth and microzooplankton grazing. *Marine Ecology Progress Series* **201**: 57–72.

Landry, M.R., Ohman, M., Goericke, R., Stukel, M.R. & Tsyrklevich, K. (2009) Lagrangian studies of phytoplankton growth and grazing relationships in a coastal upwelling ecosystem off Southern California. *Progress in Oceanography* **83**: 208–216.

Landry, M.R., Selph, K.E., Taylor, A.G., Décima, M., Balch, W.M. & Bidigare, R.R. (2011) Phytoplankton growth, grazing and production balances in the HNLC equatorial Pacific. *Deep-Sea Research II Marine Ecology Progress Series* **58**: 524–535.

Lane, N., Allen, J.F. & Martin, W. (2010) How did LUCA make a living? Chemiosis in the origin of life. *BioEssays* **32**: doi: 10.1002/bies.201090012.

Lang, B.T. (1965) *Taxonomic review of the copepod genera Eucalanus and Rhincalanus in the Pacific Ocean*. Ph.D. Dissertation, Scripps Institute of Oceanography, University of California, San Diego, xvi + 251 pp.

Larkin, P.A. (1977) An epitaph for the concept of maximum sustained yield. *Transactions of the American Fisheries Society* **106**: 1–11.

Larson, R.J. (1980). The Medusa of *Velella velella* (Linnaeus, 1758) (Hydrozoa, Chondrophorae). *Journal of Plankton Research* **2**: 183–186.

Latz, M.I., Frank, T.J. & Case, J.F. (1988) Spectral composition of bioluminescence of epipelagic animals from the Sargasso Sea. *Marine Biology* **98**: 441–446.

Lavaniegos, B.E. & Ohman, M.D. (2007) Coherence of long-term variations of zooplankton in two sectors of the California Current system. *Progress in Oceanography* **75**: 42–69.

Lavaniegos, B.E. (1995) Production of the euphausiid *Nyctiphanes simplex* in Vizcaino Bay, Western Baja California. *Journal of Crustacean Biology* **15**: 444–453.

Lawrence, J.E. (2008) Furtive foes: algal viruses as potential invaders. *ICES Journal of Marine Science* **65**: 716–722.

Laws, R.M. (1984) Seals. In: Laws, R.M. (ed.) *Antarctic Ecology*. Vol. 2. Academic Press, London, pp. 621–715.

Lazier, J.R.N. & Mann, K.H. (1989) Turbulence and diffusive layers around small organisms. *Deep-Sea Research* **36**: 1721–1733.

Le Borgne, R., Barber, R.T., Delcroix, T., Inoue, H.Y., Mackey, D.J. & Rodier, M. (2002) Pacific warm pool and divergence: temporal and zonal variations on the equator and their effects on the biological pump. *Deep-Sea Research II* **49**: 2471–2512.

Lebourges-Dhaussy, A., Marchal, E., Menkes, C., Champalbert, G. & Biessy, B. (2000) *Vinciguerria nimbaria* (micronekton) environment and tuna: their relationships in the eastern tropical Atlantic. *Oceanologica Acta* **23**: 515–528.

Lee, C.K., Cary, S.C., Murray, A.E. & Daniel, R.M. (2008) Enzymatic approach to eurythermalism of *Alvinella pompejana* and its endosymbionts. *Applied and Environmental Microbiology* **74**: 774–782.

Lee, H.-W., Ban, S., Ikeda, T. & Matsuishi, T. (2003) Effect of temperature on development, growth and reproduction in the marine copepod *Pseudocalanus newmani* at satiating food condition. *Journal of Plankton Research* **25**: 281–271.

Lee, R.E., Kugrens, P. & Mylnikov, A.P. (1991) Feeding apparatus of the colorless flagellate *Katablepharis* (Cryptophyceae). *Journal of Phycology* **27**: 725–733.

Lee, S. & Fuhrman, J.A. (1987) Relationships between biovolume and biomass of naturally-derived marine bacterioplankton. *Applied and Environmental Microbiology* **52**: 1298–1303.

Legendre, P. & Legendre, L. (1998) *Numerical Ecology*, 2nd English Edition. Elsevier Science, Amsterdam, xv + 853 pp.

Leising, A.W., Gentleman, W.C. & Frost, B.W. (2003) The threshold feeding response of microzooplankton within Pacific high-nitrate low-chlorophyll ecosystem models under steady and variable iron input. *Deep-Sea Research II* **50**: 2877–2894.

Leising, A.W., Pierson, J.J., Halsband-Lenk, C., Horner, R. & Postel, J. (2005) Copepod grazing during spring blooms: does *Calanus pacificus* avoid harmful diatoms? *Progress in Oceanography* **67**: 384–405.

Lenz, P.H., Hartline, D.K. & Davis, A.D. (2000) The need for speed. I. Fast reactions and myelinated axons in copepods. *Journal of Comparative Physiology, A* **186**: 337–345.

Lenz, P.H., Hower, A.E. & Hartline, D.K. (2004) Force production during pereiopod power strokes in *Calanus helgolandicus*. *Journal of Marine Systems* **49**: 133–144.

Lessard, E.J. (1991) The trophic role of heterotrophic dinoflagellates in diverse marine environments. *Marine Microbial Food Webs* **5**: 49–58.

Letelier, R.M., Dore, J.E., Winn, C.D. & Karl, D.M. (1996) Seasonal and interannual variations in photosynthetic carbon assimilation at Station ALOHA. *Deep-Sea Research II* **43**: 467–490.

Lewin, J.C. (1955) Silicon metabolism in diatoms. III. Respiration and silicon uptake in *Navicula pelliculosa*. *Journal of General Physiology* **39**: 1–10.

Lewin, R.A. & Withers, N.W. (1975) Extraordinary pigment composition of a prokaryotic alga. *Nature* **256**: 735–737.

Lewis, M.R., Warnock, R.E., Irwin, B. & Platt, T. (1985) Measuring photosynthetic action spectra of natural phytoplankton populations. *Journal of Phycology* **21**: 310–315.

Li, X., McGillicuddy, Jr, D.J., Durbin, E.G. & Wiebe, P.H. (2006) Biological control of *Calanus finmarchicus* on Georges Bank. *Deep-Sea Research II* **53**: 2632–2655.

Lightfoot, R.H., Tyler, P.A. & Gage, J.D. (1979) Seasonal reproduction in deep-sea bivalves and brittlestars. *Deep-Sea Research* **26**: 967–973.

Lin, G., Banks, T. & O'Reilly-Sternberg, L.daS.L. (1991) Variation in $\delta^{13}C$ values in *Thalassia testudinum* and its relation to mangrove carbon. *Aquatic Botany* **40**: 333–341.

Lisiecki, L.E. & Raymo, M.E. (2005) A Pliocene–Pleistocene stack of 57 globally distributed benthic $\delta^{18}O$ records. *Paleoceanography* **20**: PA1003. doi:10.1029/2004PA001071 (17 pp.).

Litaker, R.W., Vandersea, M.W., Kibler, S.R., Madden, V.J., Noga, E.J. & Tester, P. (2002) Life cycle of the heterotrophic dinoflagellate *Pfiesteria piscicida* (Dinophyceae). *Journal of Phycology* **38**: 442–463.

Liu, H., Nolla, H.A. & Campbell, L. (1997) *Prochlorococcus* growth rate and contribution to primary production in the equatorial and subtropical North Pacific Ocean. *Aquatic Microbial Ecology* **12**: 39–47.

Liu, H., Probert, I., Uitz, J., Claustre, H., Aris-Brosou, S., Frada, M., Not, F. & deVargas, C. (2009) Extreme diversity in non-calcifying haptophytes explains a major pigment paradox in open oceans. *Proceedings of the National Academy of Sciences* **106**: 12,803–12,808.

Lochte, K., Ducklow, H.W., Fasham, M.J.R. & Stienen, C. (1993) Plankton succession and carbon cycling at 47°N, 20°W during the JGOFS North Atlantic Bloom Experiment. *Deep-Sea Research II* **40**: 91–114.

Locket, N.A. (1985) The multiple bank rod fovea of *Bajacalifornia drakei*, an alepocephalid deep-sea teleost. *Proceedings of the Royal Society of London, B* **224**: 7–22.

Longhurst, A.R. (1985) Relationship between diversity and the vertical structure of the upper ocean. *Deep-Sea Research* **85**: 1535–1570.

Longhurst, A.R. (2006) *Ecological Geography of the Sea* (2nd Edn.). Academic Press, Amsterdam, 542 pp.

Longhurst, A.R. (2010) *Mismanagement of Marine Fisheries*. Cambridge University Press, 320 pp.

Longhurst, A.R. & Harrison, W.G. (1989) The biological pump: profiles of plankton production and consumption in the upper ocean. *Progress in Oceanography* **22**: 47–123.

Longhurst, A.R., Bedo, A.W., Harrison, W.G., Head, E.J.H. & Sameoto, D.D. (1990) Vertical flux of respiratory carbon by oceanic diel migrant biota. *Deep-Sea Research I* **37**: 685–694.

Loose, C.J. (1993) *Daphnia* diel vertical migration behavior: response to vertebrate predator abundance. *Archiv für Hydrobiologie, Beiheft Ergebnisse der Limnologie* **39**: 29–36.

Lotka, A.J. (1925) *Elements of Physical Biology*. Williams & Wilkins Co., Baltimore, 465 pp.

Lourens, L.J., Becker, J., Bintanja, R., Hilgen, F.J., Tuenter, E., van de Wal, R.S.W. & Ziegler, M. (2010) Linear and non-linear response of Neogene glacial cycles to obliquity forcing and implications for the Milankovitch theory. *Quaternary Science Reviews* **29**: 352–365.

Lubchenco, J. & Partnership for Interdisciplinary Studies of Coastal Oceans (2007) *The Science of Marine Reserves*. 2nd Edition, International Version. www.piscoweb.org. 22 pp.

Lutz, M., Dunbar, R. & Caldeira, K. (2002) Regional variability in the vertical flux of particulate organic carbon in the ocean interior. *Global Biogeochemical Cycles* **16**(3), 1037, doi: 10.1029/2000GB001383.

Lutz, R.A., Shank, T.M., Fornari, D.J., Haymon, R.M., Lilley, M.D., Von Damm, K.L. & Desbruyères, D. (1994) Rapid growth at deep-sea vents. *Nature* **371**: 663–664.

Lyman, J.M. & 7 coauthors (2010) Robust warming of the global upper ocean. *Nature* **465**: 334–337.

Lynch, D.R, Ip, J.T.C., Naimie, C.E. & Werner, F.E. (1996) Comprehensive coastal circulation model with application to the Gulf of Maine. *Coastal and Shelf Science* **16**: 875–906.

MacDonald, J.D. (1869) On the structure of the diatomaceous frustule, and its genetic cycle. *Annals and Magazine of Natural History* **4**(3): 1–8.

MacIntyre, H.L., Kana, T.M., Anning, T. & Geider, R.J. (2002) Photoacclimation of photosynthesis irradiance response curves and photosynthetic pigments in microalgae and cyanobacteria. *Journal of Phycology* **38**: 17–38.

Mackas, D.L & Beaugrand, G. (2010) Comparisons of zooplankton time series. *Journal of Marine Systems* **79**: 286–304.

Mackas, D. & Bohrer, R. (1976) Fluorescence analysis of zooplankton gut contents and an investigation of diel feeding patterns. *Journal of Experimental Marine Biology and Ecology* **25**: 75–85.

Mackas, D.L., Peterson, W.T. & Zamon, J.E. (2004) Comparisons of interannual biomass anomalies of zooplankton communities along the continental margins of British Columbia and Oregon. *Deep-Sea Research II* **51**: 875–896.

Mackas, D.L., Strub, P.T., Thomas, A.C. & Montecino (2005) Eastern ocean boundaries; pan-regional overview. In: Robinson, A.R. & Brink, K.H. (eds.) *The Global Ocean, Interdisciplinary Regional Studies and Syntheses. The Sea*, vol. 14, Part A. Harvard University Press, Cambridge, MA, pp. 21–59.

Mackas, D.L., Batten, S. & Trudel, M. (2007) Effects on zooplankton of a warmer ocean: Recent evidence from the Northeast Pacific. *Progress in Oceanography* **75**: 223–252.

Mackey, K.R.M., Paytan, A., Grossman, A.R. & Bailey, S. (2008) A photosynthetic strategy for coping in a high-light, low-nutrient environment. *Limnology and Oceanography* **53**: 900–913.

MacLulich, D.A. (1937) Fluctuations in the number of the varying hare (*Lepus americanus*). *University of Toronto Studies in Biology*, Series No. 43, Univ. Toronto Press.

Madin, L.P. & Harbison, G.R. (1978) *Thalassocalyce inconstans*, new genus and species, an enigmatic ctenophore representing a new family and order. *Bulletin of Marine Science* **28**: 680–687

Madin, L.P. & 7 coauthors (1996) Voracious planktonic hydroids: unexpected predatory impact on a coastal marine ecosystem. *Deep-Sea Research II* **43**: 1823–1829.

Mann, D.G. (1984) Structure, life history and systematics of *Rhoicosphenia* (Bacillariophyta). V. Initial cell and size reduction in *Rh. curvata* and a description of the Rhoicospheniaceae Fam. Nov. *Journal of Phycology* **20**: 544–555.

Mann, K.H. & Lazier, J.R.N. (1991) *Dynamics of Marine Ecosystems: Biological–Physical Interactions in the Oceans*, 1st edn. Blackwell Science, Oxford, 466 pp.

Mann, K.H. & Lazier, J.R.N. (2006) *Dynamics of Marine Ecosystems*, 3rd Edn., Blackwell, Oxford, 496 pp.

Manning, A.C. & Keeling, R.F. (2006) Global oceanic and land biotic carbon sinks from the Scripps atmospheric oxygen flask sampling network. *Tellus* **58B**: 95–116.

Mantua, N.J., Hare, S.R., Zhang, Y., Wallace, J.M. & Francis, R.C. (1997) A Pacific decadal climate oscillation with impacts on salmon. *Bulletin of the American Meteorological Society* **78**: 1069–1079.

Marañon, E., Holligan, P.M., Varela, M., Mourino, B. & Bale, A.J. (2000) Basin-scale variability of phytoplankton biomass, production, and growth in the Atlantic Ocean. *Deep-Sea Research I* **47**: 825–857.

Marchetti, A., Maldonado, M.T., Lane, E.S. & Harrison, P.J. (2006a) Iron requirements of the pennate diatom *Pseudo-nitzschia*: comparison of oceanic (high-nitrate, low-chlorophyll waters) and coastal species. *Limnology and Oceanography* **51**: 2092–2101.

Marchetti, A., Sherry, N.D., Kiyosawa, H., Tsuda, A. & Harrison, P.J. (2006b) Phytoplankton processes during a mesoscale iron enrichment in the NE subarctic Pacific: Part I – Biomass and assemblage. *Deep-Sea Research II* **53**: 2095–2113.

Marcus, N.H. (1982) Photoperiodic and temperature regulation of diapause in *Labidocera aestiva* (Copepoda: Calanoida). *Biological Bulletin* **162**: 45–52.

Markert, S. & 11 coauthors (2007) Physiological proteomics of the uncultured endosymbiont of *Riftia pachyptila*. *Science* **441**: 247–250.

Marlétaz, F. & 11 coauthors (2006) Chaetognath phylogenetics: a protostome with deuterostome-like development. *Current Biology* **16**: R577–R578.

Marlow, C.J. & Miller, C.B. (1975) Patterns of vertical distribution and migration of zooplankton at Ocean Station "P". *Limnology and Oceanography* **20**: 824–844.

Marra, J.F. (2009) Net and gross: weighing in with 14C. *Aquatic Microbial Ecology* **56**: 123–131.

Marsh, A.G., Mullineaux, L.S., Young, C.M. & Manahan, D.T. (2001) Larval dispersal potential of the tube worm *Riftia pachyptila* at deep-sea hydrothermal vents. *Nature* **411**: 77–80.

Marshall, N.B. (1979) *Developments in Deep-Sea Biology*. Blandford, Poole, UK. 566 pp.

Marshall, S.M. & Orr, A.P. (1934) On the biology of *Calanus finmarchicus*. V. Distribution, size, weight and chemical composition in Loch Striven in 1933, and their relation to phytoplankton. *Journal of the Marine Biological Association of the United Kingdom* **19**: 793–827.

Martin, J.H. & Fitzwater, S. (1988) Iron deficiency limits phytoplankton growth in the northeast Pacific subarctic. *Nature* **331**: 341–343.

Martin, J.H. & 43 coauthors (1994) Testing the iron hypothesis in ecosystems of the equatorial Pacific Ocean. *Nature* **371**: 123–129.

Martin, J.H., Knauer, G.A., Karl, D.M. & Broenkow, W.W. (1987) VERTEX: carbon cycling in the northeast Pacific. *Deep-Sea Research* **34**: 267–285.

Martin, J.H., Gordon, R.M., Fitzwater, S. & Broenkow, W.W. (1989) VERTEX: phytoplankton/iron studies in the Gulf of Alaska. *Deep-Sea Research* **36**: 7649–7680.

Martin, P.C. (2009) Do sea surface temperatures influence catch rates in the June South Peninsula, Alaska, salmon fishery? *North Pacific Anadromous Fish Commission Bulletin* **5**: 147–156.

Martinez, J.M., Schroeder, D.C., Larsen, A., Bratbak, G., & Wilson, W.H. (2007) Molecular dynamics of *Emiliana huxleyi* and cooccurring viruses during two separate mesocosm studies. *Applied and Environmental Microbiology* **73**: 554–562.

Martinez-Garcia, A., Rosell-Melé, A., Geibert, W., Gersonde, R., Masqué, P., Gaspari, V. & Barbante, C. (2009) Links between iron supply, marine productivity, sea surface temperature, and CO_2 over the last 1.1 Ma. *Paleoceanography* **24**: PA1207 (14 pp.); doi:10.1029/2008PA001657.

Martiny, A.C., Kathuria, S. & Berube, P.M. (2009) Widespread metabolic potential for nitrite and nitrate assimilation among *Prochlorococcus* ecotypes. *Proceedings of the National Academy of Sciences* **106**: 10,787–10,792.

Mathews, C.K., van Holde, K.E. & Ahern, K.G. (2000) *Biochemistry*. Benjamin/Cummings, Menlo Park, CA, xxvii + 1159 pp.

Mauchline, J. (1988) Growth and breeding of meso-and bathypelagic organisms of the Rockall Trough, northeastern Atlantic Ocean, and evidence of seasonality. *Marine Biology* **98**: 387–393.

Mauchline, J. (1998) The Biology of Calanoid Copepods. *Advances in Marine Biology* **33**: 1–710.

Maunder, M.N. & Watters, G.M. (2001) Status of yellowfin tuna in the eastern Pacific Ocean. *Inter-American Tropical Tuna Commission, Stock Assessment Report* **1**: 5–86.

Maunder, M.N., Sibert, J.R., Fonteneau, A., Hampton, J., Kleiber, P. & Harley, S.J. (2006) Interpreting catch per unit effort data to assess the status of individual stocks and communities. *ICES Journal of Marine Sciences* **63**: 1373–1385.

Maynou, F. & Cartes, J.E. (2000) Community structure of bathyal crustaceans off south-west Balearic Islands (Western Mediterranean) season and regional patterns in zonation. *Journal of the Marine Biological Association of the United Kingdom* **80**: 789–798.

Mayzaud, P. & Poulet, S. (1978) The importance of the time factor in the response of zooplankton to varying concentrations of naturally occurring particulate matter. *Limnology and Oceanography* **23**: 1144–1154.

McCarthy, J.J., Taylor, W.R. & Taft, J.L. (1975) The dynamics of nitrogen and phosphorus cycling in the open waters of Chesapeake Bay. In: Church, T.M. (ed.) *Marine Chemistry in the Coastal Environment*. American Chemical Society Symposium Series 18: 664–681.

McCave, I.N. (1975) The vertical flux of particles in the ocean. *Deep-Sea Research* **22**: 491–502.

McClatchie, S. (1985) Time-series feeding rates of the euphausiid *Thysanoessa raschii* in a temporally patchy food environment. *Limnology and Oceanography* **31**: 469–477.

McClelland, J.W. & Montoya, J.P. (2002) Trophic relationships and the nitrogen isotopic composition of amino acids in plankton. *Ecology* **83**: 2173–2180.

McCune, B. & Grace, J.B. (2002) *Analysis of Ecological Communities*. MjM Software, Gleneden Beach, Oregon, 300 pp.

McGillicuddy, D., Anderson, D.M., Lynch, D.R. & Townsend, D.W. (2005) Mechanisms regulating large-scale seasonal fluctuations in *Alexandrium fundyense* populations in the Gulf of Maine: results from a physical–biological model. *Deep-Sea Research II* **52**: 2698–2714.

McGowan, J.A. (1963) Geographical variation in *Limacina helicina* in the North Pacific. In: Harding, J.P. & Tebble, N. (eds.) *Speciation in the Sea*. Systematics Association, London, Publication No. 3, pp. 109–128.

McGowan, J.A. (1968) The Thecosomata and Gymnosomata of California. *Veliger* **3** (suppl.): 103–130.

McGowan, J.A. & Walker, P.W. (1979) Structure in the copepod community of the North Pacific Central Gyre. *Ecological Monographs* **49**: 195–226.

McGowan, J.A. & Walker, P.W. (1985) Dominance and diversity maintenance in an oceanic ecosystem. *Ecological Monographs* **55**: 103–118.

McGowan, J.A., Bograd, S.J., Lynn, R.J. & Miller, A.J. (2003) The biological response to the 1977 regime shift in the California Current. *Deep-Sea Research II* **50**: 2567–2582.

McIntyre, A. & 7 coauthors (1976) Glacial North Atlantic 18,000 years ago: a CLIMAP reconstruction. In: Cline, R. & Hays, J.D. (eds.) *Investigations of Late Quaternary Paleoceanography and Paleoclimatology*. Geological Society of America Memoir **145**: 43–76.

McLaughlin, F.A., Carmack, E.C., Macdonald, R.W. & Bishop, J.K.B. (1996) Physical and geochemical properties across the Atlantic/Pacific water mass front in the southern Canadian Basin. *Journal of Geophysical Research, Oceans* **101**: 1183–1197.

McQuoid, M.R. & Hobson, L.A. (1996) Diatom resting stages. *Journal of Phycology* **32**: 889–902.

McQuoid, M.R., Godhe, A. & Nordberg, K. (2002) Viability of phytoplankton resting stages in the sediments of a coastal Swedish fjord. *European Journal of Phycology* **37**: 191–201.

Ménard, F. & Marchal, E. (2003) Foraging behavior of tuna feeding on small schooling *Vinciguerria nimbaria* in the surface layer of the equatorial Atlantic Ocean. *Aquatic Living Resources* **16**: 231–238.

Menkes, C.E. & 9 coauthors (2002) A whirling ecosystem in the equatorial Atlantic. *Geophysical Research Letters* **29**: 10.1029/2001GL014576.

Menzies, R.J. (1965) Conditions for the existence of life on the abyssal sea floor. *Oceanography and Marine Biology Annual Review* **3**: 195–210.

Messié, M., Ledesma, J., Kolber, D.D., Michisaki, R.P., Foley, D.G. & Chavez, F.P. (2009) Potential new production estimates in four eastern boundary upwelling ecosystems. *Progress in Oceanography* **83**: 151–158.

Michael, E.L. (1911) Classification and vertical distribution of the Chaetognatha of the San Diego region. *University of California Publications in Zoology* **8**: 21–186.

Michaels, A.F. & 8 coauthors (1996) Inputs, losses and transformations of nitrogen and phosphorus in the pelagic North Atlantic Ocean. *Biogeochemistry* **35**: 181–226.

Middleboe, M. & Jorgensen, N.O.G. (2006) Viral lysis of bacteria: an important source of dissolved amino acids and cell wall compounds. *Journal of the Marine Biological Association of the United Kingdom* **86**: 805–612.

Miller, C.B. (ed.) (1993) Pelagic ecodynamics in the Gulf of Alaska, results from the SUPER Program. *Progress in Oceanography* **32**: iv + 358 pp.

Miller, C.B. & Clemons, M.J. (1988) Revised life history analysis for large grazing copepods in the subarctic Pacific Ocean. *Progress in Oceanography* **20**: 293–313.

Miller, C.B., Frost, B.W., Wheeler, P.A., Landry, M.R., Welschmeyer, N. & Powell, T.M. (1991a) Ecological dynamics in the subarctic Pacific, a possibly iron-limited ecosystem. *Limnology and Oceanography* **36**: 1600–1615.

Miller, C.B., Lynch, D.R., Carlotti, F., Gentleman, W. & Lewis, C.V.W. (1998) Coupling of an individual-based population dynamical model for stocks of *Calanus finmarchicus* to a circulation model for the Georges Bank region. *Fishery Oceanography* **7**: 219–234.

Miller, J.E. & Pawson, D.L., (1990) Swimming sea cucumbers (Echinodermata: Holothuroidea): a survey, with analysis of swimming behavior in four bathyal species. *Smithsonian Contributions to the Marine Sciences* **35**: iii + 18 pp.

Miralto, A. & 10 coauthors (1999) The insidious effect of diatoms on copepod reproduction. *Nature* **402**: 173–176.

Mobley, C.T. (1987) Time-series ingestion rate estimates on individual *Calanus pacificus* Brodsky: interactions with environmental and biological factors. *Journal of Experimental Marine Biology and Ecology* **114**: 199–216.

Moen, T., Verbeeck, L., de Maeyer, A., Swings, J. & Vincx, M. (1999) Selective attraction of marine bacterivorous nematodes to their bacterial food. *Marine Ecology Progress Series* **176**: 165–178.

Mohr, W., Grosskopf, T., Wallace, D.W.R. & LaRoche, J. (2010) Methodological underestimation of oceanic nitrogen fixation fates. *PLoS ONE* **5**(9): e12583 (7 pp.)

Moisander, P.H. & 7 coauthors (2010) Unicellular cyanobacterial distributions broaden the oceanic N_2 fixation domain. *Science* **327**: 1512–1514.

Moku, M., Kawaguchi, K., Watanabe, H. & Ohno, A. (2000) Feeding habits of three dominant myctophid fishes, *Diaphus theta, Stenobrachius leucopsarus* and *S. nannochir*, in the subarctic and transitional waters of the western North Pacific. *Marine Ecology Progress Series* **207**: 129–140.

Monger, B.C., Landry, M.R. & Brown, S.L. (1999) Feeding selection of heterotrophic marine nanoflagellates based on the surface hydrophobicity of their picoplankton prey. *Limnology and Oceanography* **44**: 1917–1927.

Montagnes, D.J.S. (1996) Growth responses of planktonic ciliates in the genera *Strombilidium* and *Strombidium*. *Marine Ecology Progress Series* **130**: 241–254.

Moore, C.M., Mills, M.M., Langlois, R., Milne, A., Achterberg, E.P., LaRoche, J. & Geider, R.J. (2008) Relative influence of nitrogen and phosphorus availability on phytoplankton physiology and productivity in the oligotrophic sub-tropical Atlantic Ocean. *Limnology and Oceanography* **53**: 291–305.

Moore, J.K. & Abbott, M.R. (2000) Phytoplankton chlorophyll distributions and primary production in the Southern Ocean. *Journal of Geophysical Research, C, Oceans* **105**: 28,709–28,722.

Moreira, D., von der Heyden, S., Bass, D., López-García, P., Chao, E. & Cavalier-Smith, T. (2007) Global eukaryote phylogeny: combined small- and large-subunit ribosomal DNA trees support monophyly of Rhizaria, Retaria and Excavata. *Molecular Phylogenetics and Evolution* **44**: 255–266.

Morel, A. (1991) Light and marine photosynthesis: a spectral model with geochemical and climatological implications. *Progress in Oceanography* **26**: 263–306.

Morel, F.M.M. (1987) Kinetics of nutrient uptake and growth in phytoplankton. *Journal of Phycology* **23**: 137–150.

Morel, F.M.M., Kustka, A.B. & Shaked, Y. (2008) The role of unchelated Fe in the iron nutrition of phytoplankton. *Limnology and Oceanography* **53**: 400–404.

Morris, R.M., Nunn, B.L., Frazar, C., Goodlett, D.R., Ting, Y.S. & Rocap, G. (2010) Comparative metaproteomics reveals ocean-scale shifts in microbial nutrient utilization and energy transduction. *The ISME Journal* **4**: 673–685.

Morton, J.E. (1954) The biology of *Limacina retroversa*. *Journal of the Marine Biological Association of the United Kingdom* **33**: 297–312.

Moser, H.G. & Boehlert, G.W. (1991) Ecology of pelagic larvae and juveniles of the genus *Sebastes*. *Environmental Biology of Fishes* **30**: 203–224.

Moulton, F.R. 1942. *Liebig and after Liebig*. AAAS Publ. No. 16, Washington, D.C.

Moutin, T., Broeck, N.V.D., Beker, B., Dupouy, C., Rimmelin, P. & Bouteliller, A.L. (2005) Phosphate availability controls *Trichodesmiun* spp. biomass in the SW Pacific Ocean. *Marine Ecology Progress Series* **297**: 15–21.

Mullin, M.M. & Brooks, E.R. (1970) Production of the planktonic copepod *Calanus helgolandicus*. *Bulletin of the Scripps Institution of Oceanography* **17**: 89–103.

Mullin, M.M. & Brooks, E.R. (1976) Some consequences of distributional heterogeneity of phytoplankton and zooplankton. *Limnology and Oceanography* **21**: 784–796.

Mullins, T.D., Britschgi, T.B., Krest, R.L. & Giovannoni, S.J. (1995) Genetic comparisons reveal the same unknown bacterial lineages in Atlantic and Pacific bacterioplankton communities. *Limnology and Oceanography* **40**: 148–158.

Mulsow, S. & Boudreau, B. (1998) Bioturbation and porosity gradients. *Limnology and Oceanography* **43**: 1–9.

Mundy, C.J. & 13 coauthors (2009) Contribution of under-ice primary production to an ice-edge upwelling phytoplankton bloom in the Canadian Beaufort Sea. *Geophysical Research Letters* **36**: L17601, doi:10.1029//2009GL038837.

Munn, C.B. (2006) Viruses as pathogens of marine organisms – from bacteria to whales. *Journal of the Marine Biological Association of the United Kingdom* **86**: 453–467.

Muntz, W.R.A. (1976) On yellow lenses in mesopelagic animals. *Journal of the Marine Biological Association of the United Kingdom* **56**: 963–976.

Murray, J.W., Johnson, E. & Garside, C. (1995) A US JGOFS process study in the Equatorial Pacific (EqPac): introduction. *Deep-Sea Research II* **42**: 275–293.

Murray, J.W., Le Bourgne, R. & Dandonneau, Y. (1997) JGOFS studies in the equatorial Pacific. *Deep-Sea Research II* **44**: 1759–1763.

Myers, A.A. (1985) Shallow-water, coral reef and mangrove amphipoda (Gammaridea) of Fiji. *Records of the Australian Museum, Supplement* **5**: 143 pp.

Myers, R. (2001) *Stock Recruitment Database*. http://www.mscs.dal.ca/~myers/welcome.html.

Myers, R.A. & Worm, B. (2003) Rapid worldwide depletion of predatory fish communities. *Nature* **423**: 280–284.

Nagasawa, S. (1984) Laboratory feeding and egg production in the chaetognath *Sagitta crassa* Tokioka. *Journal of Experimental Marine Biology and Ecology* **76**: 51–65.

Nagura, M. & McPhaden, M.J. (2010) Wyrtki Jet dynamics: seasonal variability. *Journal of Geophysical Research* **115**: C07009 (17 pp.).

Nejstgaard, J.C. & 7 coauthors (2008) Quantitative PCR to estimate copepod feeding. *Marine Biology* **153**: 565–577.

Nelson, D.M. & Brand, L.E. (1979) Cell division periodicity in 13 species of marine phytoplankton on a light:dark cycle. *Journal of Phycology* **15**: 67–75.

Nelson, D.M. & Dortch, Q. (1996) Silicic acid depletion and silicate limitation in the plume of the Mississippi River: evidence from kinetic studies in spring and summer. *Marine Ecology Progress Series* **136**: 163–178.

Nelson, D.M. & Landry, M.R. (2011) Regulation of phytoplankton production and upper ocean biogeochemistry in the eastern Equatorial Pacific: introduction to results of the Equatorial Biocomplexity project. *Deep-Sea Research II* **58**: 277–283.

Nelson, D.M. & Smith, W.O. Jr (1991) Sverdrup revisited: critical depths, maximum chlorophyll levels and the control of Southern Ocean productivity by the irradiance-mixing regime. *Limnology and Oceanography* **36**: 1650–1661.

Nelson, D.M., Tréguer, P., Brzezinski, M. A., Leynaert, A. & Quéguiner, B. (1995) Production and dissolution of biogenic silica in the ocean: revised global estimates, comparison with regional data and relationship to biogenic sedimentation. *Global Biogeochemical Cycles* **9**: 359–372.

Nelson, D.M., Brzezinski, M.A., Sigmon, D.E. & Frank, V.E. (2001) A seasonal progression of Si limitation in the Pacific sector of the Southern Ocean. *Deep-Sea Research II* **48**: 3973–3995.

Neuer, S. & 10 coauthors (2007) Biogeochemistry and hydrography in the eastern subtropical North Atlantic gyre. Results from the European time-series station ESTOC. *Progress in Oceanography* **72**: 1–29.

Neuheimer, A.B., Gentleman, W.C., Galloway, C.L. & Johnson, C.L. (2009) Modeling larval *Calanus finmarchicus* on Georges Bank: time-varying mortality rates and a cannibalism hypothesis. *Fisheries Oceanography* **18**: 147–160.

Neuheimer, A.B., Gentleman, W.C., Pepin, P. & Head, E.J.H. (2010) How to build and use individual-based models (IBMs) as hypothesis testing tools. *Journal of Marine Systems* **81**: 122–133.

Neveux, J., Dupouy, C., Blanchot, J., Le Bouteiller, A., Landry, M.R. & Brown, S. (2003) Diel dynamics of chlorophylls in high-nutrient, low-chlorophyll waters of the equatorial Pacific (180°): interactions of growth, grazing, physiological responses, and mixing. *Journal of Geophysical Research – Oceans* **108**: C12,8240, doi:10.1029/2000JC000747.

Newell, C. L. & Cowles, T. J. (2006) Unusual gray whale *Eschrichtius robustus* feeding in the summer of 2005 off the central Oregon coast. *Geophysical Research Letters* **33**: L22S11 doi: 10.1029/2006GL027189.

Nicol, S. (2006) Krill, currents, and sea ice: *Euphausia superba* and its changing environment. *BioScience* **56**: 111–120.

Nicol, S. (2010) BROKE-West, a large ecosystem survey of the South West Indian Ocean sector of the Southern Ocean, 30°E–80°E (CCAMLRDivision58.4.2). *Deep-Sea Research II* **57**: 693–700.

Nicol, S., De la Mare, W.K. & Stolp, M. (1995) The energetic cost of egg production in Antarctic krill (*Euphausia superba* Dana). *Antarctic Science* **7**: 25–30.

Niehoff, B. (2000) Effect of starvation on the reproductive potential of *Calanus finmarchicus*. *ICES Journal of Marine Science* **57**: 1764–1772.

Niehoff, B., Klenke, U., Hirche, H.-J., Irigoien, X., Head, R. & Harris, R. (1999) A high frequency time series at Weathership M, Norwegian Sea, during the 1997 spring bloom: the reproductive biology of *Calanus finmarchicus*. *Marine Ecological Progress Series* **176**: 81–82.

Nielsen, J.G., Bertelsen, E. & Jespersen, Å. (1989) The biology of *Eurypharynx pelecanoides* (Pisces, Eurypharyngidae). *Acta Zoologica (Stockholm)* **70**: 187–197.

Nilsson, D.-E. (1989) Optics and design of the compound eye. In: Hardie, R.C. & Stavenga, D.G. (eds.) *Facets of Vision*, Springer-Verlag, Berlin, pp. 30–73.

Nishida, S. (1985) Taxonomy and distribution of the family Oithonidae (Copepoda, Cyclopoida) in the Pacific and Indian Oceans. *Bulletin of the Ocean Research Institute, University of Tokyo* **20**: 167 pp.

Nishida, S. & Ohtsuka, S. (1991) Midgut structure and food habits of the mesopelagic copepods *Lophothrix frontalis* and *Scottocalanus securifrons*. *Bulletin of the Plankton Society of Japan, Special Volume* **1991**: 527–534.

Nussbaumer, A.D., Fisher, C.R. & Bright, M. (2006) Horizontal endosymbiont transmission in hydrothermal vent tubeworms. *Nature* **441**: 345–348.

Oakley, B.R. & Dodge, J.D. (1976) Mitosis and cytokinesis in the dinoflagellate *Amphidinium carterae*. *Cytobios* **17**: 35–46.

Ochoa, N. & Gómez, O. (1987) Dinoflagellates as indicators of water masses during El Niño, 1982–83. *Journal of Geophysical Research* 92: 14,355–14,367.

O'Dor, R.K. (1983) *Illex illecebrosus*. In: Boyle, P.R. (ed.) *Cephalopod Life Cycles*, Vol. 1. Academic Press, London, pp. 175–199.

Ohman, M.D. & Wood, S.N. (1996) Mortality estimation for planktonic copepods: *Pseudocalanus newmani* in a temperature fjord. *Limnology and Oceanography* 41: 126–135.

Ohman, M.D., Frost, B.W. & Cohen, E.B. (1983) Reverse diel vertical migration: an escape from invertebrate predators. *Science* 220: 1404–1406.

Ohman, M.D., Bradford, J.M. & Jillett, J.B. (1989) Seasonal growth and lipid storage of the circumgiobal, subantarctic copepod, *Neocalanus tonsus* (Brady). *Deep-Sea Research* 36: 1309–1326.

Ohman, M.D., Eiane, K., Durbin, E.G., Runge, J.A. & Hirche, H.-J. (2004) A comparative study of *Calanus finmarchicus* mortality patterns at five localities in the North Atlantic. *ICES Journal of Marine Science* 61: 687–697.

Ohman, M.D., Durbin, E.A., Runge, J.A., Sullivan, B.K. & Field, D.B. (2008) Relationship of predation potential to mortality of *Calanus finmarchicus* on Georges Bank, northwest Atlantic. *Limnology and Oceanography* 53: 1643–1655.

Okutani, T. (1983) *Todarodes pacificus*. In: Boyle, P.R. (ed.) *Cephalopod Life Cycles*, Vol. 1. Academic Press, London, pp. 201–214.

Olabarria, C. (2005) Patterns of bathymetric zonation of bivalves in the Porcupine Seabight and adjacent Abyssal plain, NE Atlantic. *Deep-Sea Research I* 52: 15–31.

Olli, K. & 10 coauthors (2007) The fate of production in the central Arctic Ocean – top-down regulation by zooplankton expatriates? *Progress in Oceanography* 72: 84–113.

Olsen, G.J., Lane, D.L., Giovannoni, S.J., Pace, N.R. & Stahl, D.A. (1986) Microbial ecology and evolution: a ribosomal RNA approach. *Annual Review of Microbiology* 40: 337–366.

Olson, D.B. (2001) Biophysical dynamics of western transition zones: a preliminary synthesis. *Fisheries Oceanography* 10: 133–150.

Olson, D.B. & Hood, R.R. (1994) Modelling pelagic biogeography. *Progress in Oceanography* 34: 161–205.

Olson, R.J. & Boggs, C.H. (1986) Apex predation by yellowfin tuna (*Thunnus albacares*): independent estimates from gastric evacuation and stomach contents, bioenergetics, and cesium concentrations. *Canadian Journal of Fisheries and Aquatic Sciences* 43: 1135–1140.

Olson, R.J., Chisholm, S.W., Zettler, E.R., Altabet, M.A. & Dusenberry, J.A. (1990) Spatial and temporal distributions of prochlorophyte picoplankton in the North Atlantic Ocean. *Deep-Sea Research* 37: 1033–1051.

Olson, R.J. & 11 coauthors (2010) Food-web inferences of stable isotope patterns in copepods and yellowfin tuna in the pelagic eastern Pacific Ocean. *Progress in Oceanography* 86: 124–138.

Oparin, A.I. (1938) *The Origin of Life*. Macmillan, New York. [Russian original, Proiskhozhdenie Zhizny, 1924.], viii + 270 pp.

Orsi, A.J., Whitworth, T. & Nowlin, W.D. (1995) On the meridional extent and fronts of the Antarctic Circumpolar Current. *Deep-Sea Research I* 42: 641–673.

Ouverney, C.C. & Fuhrman, J.A. (1999) Marine planktonic archaea take up amino acids. *Applied and Environmental Microbiology* 66: 4829–4833.

Pabi, S., van Dijken, G.L. & Arrigo, K.R. (2008) Primary production in the Arctic Ocean, 1998–2006. *Journal of Geophysical Research* 113: C08005, doi:10.1029/2007JC004578, 2008.

Pace, N.R. (2006) Time for a change. *Nature* 441: p. 289.

Pace, N.R. (2009) Mapping the tree of life: progress and prospects. *Microbiology and Molecular Biology Reviews* 73: 565–576.

Padillon, F. & Gaill, F. (2007) Adaptation to deep-sea hydrothermal vents: some molecular and developmental aspects. *Journal of Marine Science and Technology* (Taiwan) 15 (Special Issue): 7–53.

Paffenhöfer, G.-A. (1971) Grazing and ingestion rates of nauplii, copepodids and adults of the marine planktonic copepod *Calanus helgolandicus*. *Marine Biology* 11: 286–298.

Paffenhöfer, G.-A. (1984) Food ingestion by the marine planktonic copepod *Paracalanus* in relation to abundance and size distribution of food. *Marine Biology* 80: 323–333.

Paffenhöfer, G.-A. & Lewis, K.D. (1990) Perceptive performance and feeding behavior of calanoid copepods. *Journal of Plankton Research* 12: 933–946.

Pailleret, M., Haga, T., Petit, P., Privé-Gill, C., Saedlou, N., Gaill, F. & Zbinden, M. (2007) Sunken wood from the Vanuatu Islands: identification of wood substrates and preliminary description of associated fauna. *Marine Ecology* 28: 233–241.

Palenik, B. & 37 coauthors (2007) The tiny eukaryote *Ostrecoccus* provides genomic insights into the paradox of plankton speciation. *Proceedings of the National Academy of Sciences* 104: 7705–7710.

Papetti, C., Zane, L., Bortolotto, E., Bucklin, A. & Patarnello, T. (2005) Genetic differentiation and local temporal stability of population structure in the euphausiid *Meganyctiphanes norvegica*. *Marine Ecology Progress Series* 289: 225–235.

Papillon, D., Perez, Y., Caubit, X. & Le Parco, Y. (2006) Systematics of Chaetognatha under the light of molecular data, using duplicated 18S DNA sequences. *Molecular Phylogenetics and Evolution* 38: 621–634.

Park, T. (1993) Taxonomy and distribution of the calanoid copepod family Euchaetidae. *Bulletin of the Scripps Institute of Oceanography* 29: 1–203.

Parker, M.S., Mock, T. & Armbrust, E.V. (2008) Genomic insights into marine microalgae. *Annual Review of Genetics* 42: 619–645.

Parrish, R.H., Nelson, C.S. & Bakun, A. (1981) Transport mechanisms and reproductive success of fishes in the California Current. *Biological Oceanography* 1: 175–203.

Parsons, T.R., Maita, Y. & Lalli, C.M. (1984) *A Manual of Chemical and Biological Methods for Seawater Analysis*. Pergamon Press, Oxford, 173 pp.

Partridge, J.C., Archer, S.N. & Oostrum, J. van (1992) Single and multiple visual pigments in deep-sea fishes. *Journal of the Marine Biological Association of the United Kingdom* 72: 113–130.

Passow, U. & De La Rocha, C. (2006) Accumulation of mineral ballast on organic aggregates. *Global Biogeochemical Cycles* 20: GB1013 (7 pp.).

Pauly, D. & Christensen, V. (1995) Primary production required to sustain global fisheries. *Nature* 374: 255–257.

Pauly, D., Christensen, V., Dalsgaard, J., Froese, R. & Torres, F. Jr (1998) Fishing down marine food webs. *Science* 279: 860–863.

Pawlowski, J., Holzmann, M., Fahrni, J. & Richardson, S.L. (2003) Small subunit ribosomal DNA suggests that the xenophyophorean *Syringammina corbicula* is a foraminiferan. *Journal of Eukaryotic Microbiology* 50: 483–487.

Pearce, I., Davidson, A.T., Thomson, P.G., Wright, S. & van den Enden, R. (2010) Marine microbial ecology off East Antarctica (30–80°E): rates of bacterial and phytoplankton growth and grazing by heterotrophic protists. *Deep-Sea Research II* 57: 849–862.

Pearcy, W.G. & Schoener, A. (1987) Changes in the marine biota coincident with the 1982–83 El Niño in the northeastern subarctic Pacific Ocean. *Journal of Geophysical Research* 92: 14,417–14,428.

Pearcy, W.G., Meyer, S.L. & Munk, O. (1965) A "four-eyed" fish from the deep-sea. *Nature* 207: 1260–1262.

Pearre, S. Jr (1973) Vertical migration and feeding in *Sagitta elegans* Verrill. *Ecological Monographs* **54**: 300–314.

Pearre, S. Jr (1980) Feeding by chaetognatha: the relation of prey size to predator size in several species. *Marine Ecology Progress Series* **3**: 125–134.

Pearre, S. Jr (1981) Feeding by chaetognatha: energy balance and importance of various components of the diet of *Sagitta elegans*. *Marine Ecology Progress Series* **5**: 45–54.

Pearson, A., McNichol, A.P., Benitez-Nelson, B.C., Hayes, J.M. & Eglinton, T.I. (2001) Origins of lipid biomarkers in Santa Monica Basin surface sediment: a case study using compound specific $\Delta^{14}C$ analysis. *Geochimica et Cosmochimica Acta* **65**: 3123–3137.

Peijnenburg, K.T.C.A., Breeuwer, J.A.J., Pierrot-Bults, A.C. & Menken, S.B.J. (2004) Phylogeography of the planktonic chaetognath *Sagitta setosa* reveals isolation in European seas. *Evolution* **58**: 1472–1487.

Peijnenburg, K.T.C.A., Fauvelot, C., Breeuwer, J.A.J. & Menken, S.B.J. (2006) Spatial and temporal genetic structure of the planktonic *Sagitta setosa* (Chaetognatha) in European seas as revealed by mitochondrial and nuclear DNA markers. *Molecular Ecology* **15**: 3319–3338.

Peng, T.-H. & Broecker, W.S. (1991) Factors limiting the reduction of atmospheric CO_2 by iron fertilization. *Limnology and Oceanography* **36**: 1919–1927.

Penry, D.L. & Frost, B.W. (1991) Re-evaluation of the gut-fullness (gut fluorescence) method for inferring ingestion rates of suspension-feeding copepods. *Limnology and Oceanography* **35**: 1207–1214.

Pérez, V. & 12 coauthors (2005) Latitudinal distribution of microbial plankton abundance, production, and respiration in the Equatorial Atlantic in autumn 2000. *Deep-Sea Research I* **52**: 861–880.

Perry, R.I., Cury, P., Brander, K., Jennings, S., Möllmann, C. & Planque, B. (2010) Sensitivity of marine systems to climate and fishing: concepts, issues and management responses. *Journal of Marine Systems* **79**: 427–435.

Peters, F. & Marrasé, C. (2000) Effects of turbulence on plankton: an overview of experimental evidence and some theoretical considerations. *Marine Ecology Progress Series* **205**: 291–306.

Peterson, B.J. (1980) Aquatic primary productivity and the ^{14}C -CO_2 method: a history of the productivity problem. *Annual Review of Ecology and Systematics* **11**: 359–385.

Peterson, H.A. & Vayssieres, M. (2010) Benthic assemblage variability in the upper San Francisco Estuary: a 27-year retrospective. *San Francisco Estuary and Watershed Science*, 8(1): 1–27.

Peterson, J.M., Ramette, A., Lott, C., Cambon-Bonavita, M.-A., Zbinden, M. & Dubilier, N. (2010) Dual symbiosis of the vent shrimp *Rimicaris exoculata* with filamentous gamma- and epsilonproteobacteria at four Mid-Atlantic Ridge hydrothermal vent fields. *Environmental Microbiology* **12**: 2204–2218.

Peterson, M.L., Wakeham, S.G., Lee, C., Askea, M.A. & Miquel, J.C. (2005) Novel techniques for collection of sinking particles in the ocean and determining their settling rates. *Limnology and Oceanography Methods* **3**: 520–532.

Peterson, W.T. & Dam, H.R. (1990) The influence of copepod "swimmers" on pigment fluxes in brine-filled vs. ambient seawater-filled sediment traps. *Limnology and Oceanography* **35**: 448–455.

Peterson, W.T. & Miller, C.B. (1977) Seasonal cycle of zooplankton abundance and species composition along the central Oregon coast. *Fishery Bulletin* **75**: 717–724.

Petit, J.R. & 18 coauthors (1999) Climate and atmospheric history of the past 420,000 years from the Vostok ice core, Antarctica. *Nature* **399**: 429–436.

Pflugfelder, B., Cary, S.C. & Bright, M. (2009) Dynamics of cell proliferation and apoptosis reflect different life strategies in hydrothermal vent and cold seep vestimentiferan tubeworms. *Cell and Tissue Research* **337**: 149–165.

Phleger, F.B., Parker, F.L. & Pierson, W.J. (1953) North Atlantic Foraminifera. *Report of the Swedish Deep-Sea Expedition* **7**: 1–122.

Pickett-Heaps, J.D., Schmid, A.-M.M. & Edgar, L.A. (1990) The cell biology of diatom valve formation. *Progress in Phycological Research* **7**: 1–168.

Pierson, J.J., Frost, B.W., Thoreson, D., Leising, A.W., Postel, J.R. & Nuwer, M. (2009) Trapping migrating zooplankton. *Limnology and Oceanography Methods* **7**: 334–346.

Pikitch, E. & 16 coauthors (2004) Ecosystem-based fishery management. *Science* **305**: 346–347.

Pilson, M.E.Q. (1998) *An Introduction to the Chemistry of the Sea*. Prentice Hall, Upper Saddle River, NJ, USA. 431 pp.

Pinchuk, A.I. & Hopcroft, R.R. (2006) Egg production and early development of *Thysanoessa inermis* and *Euphausia pacifica* (Crustacea: Euphausiacea) in the northern Gulf of Alaska. *Journal of Experimental Marine Biology and Ecology* **332**: 206–215.

Planque, B. & Batten, S.D. (2000) *Calanus finmarchicus* in the North Atlantic: the year of *Calanus* in the context of interdecadal change. *ICES Journal of Marine Science* **57**: 1528–1535.

Planque, B. & Ibanez, F. (1997) Long-term time series in *Calanus finmarchicus* abundance – a question of space? *Oceanologica Acta* **20**: 159–164.

Planque, B., Hays, G.C., Ibanez, F. & Gamble, J.C. (1997) Large-scale variations in the seasonal abundance of *Calanus finmarchicus*. *Deep-Sea Research* **44**: 315–326.

Platt, T. and Jassby, A.D. (1976) The relationship between photosynthesis and light for natural assemblages of coastal marine phytoplankton. *Journal of Phycology* **12**: 421–430.

Platt, T., Bird, D.F. & Sathyendranath, S. (1991) Critical depth and marine primary productivity. *Proceedings of the Royal Society of London B* **246**: 205–217.

Platt, T., Gallegos, C.L. & Harrison, W.G. (1980) Photoinhibition of photosynthesis in natural assemblages of marine phytoplankton. *Journal of Marine Research* **38**: 687–701.

Plourde, S. & Runge, J.A. (1993) Reproduction of the planktonic copepod *Calanus finmarchicus* in the lower St. Lawrence estuary: relation to the cycle of phytoplankton production and evidence for a *Calanus* pump. *Marine Ecology Progress Series* **102**: 217–227.

Pogson, G.H., Taggart, C.T., Mesa, K.A. & Boutilier, R.G. (2001) Isolation by distance in the Atlantic cod, *Gadus morhua*, at large and small geographic scales. *Evolution* **55**: 131–146.

Pointer, M.A., Carvalho, L.S., Cowing, J.A., Bowmaker, J.K. & Hunt, D.M. (2007) The visual pigments of a deep-sea teleost, the pearl eye *Scopelarchus analis*. *The Journal of Experimental Biology* **210**: 2829–2835.

Pomeroy, L.R. (1974) The ocean's food web: a changing paradigm. *BioScience* **24**: 499–504.

Poorvin, L., Hutchins, D.A. & Wilhelm, S.W. (2004) Viral release of iron and its bioavailability to marine plankton. *Limnology and Oceanography* **49**: 1734–1741.

Pope, R.H., DeMaster, D.J., Smith, C.R. & Seltmann, H. Jr (1996) Rapid bioturbation in equatorial Pacific sediments: evidence from excess 234-Th measurements. *Deep-Sea Research* **43**: 1339–1364.

Popp, B.N. & 7 coauthors (2007) Insight into the trophic ecology of yellowfin tuna, *Thunnus albacares*, from compound-specific nitrogen analysis of proteinaceous amino acids. In: Dawson, T.E. & Siegwolf, R.T.W. (eds) *Stable Isotopes as Indicators of Ecological Change*. Elsevier/Academic Press, Terrestrial Ecology Series, San Diego, pp. 173–190.

Post, D.M. (2002) Using stable isotopes to estimate trophic position: models, methods, and assumptions. *Ecology* **83**: 703–778.

Poulet, S.A. & Oullet, G. (1983) Role of amino acids in swarming and feeding of copepods. *Journal of Plankton Research* **4**: 341–346.

Poulet, S.A., Ianora, A., Miralto, A. & Meijer, L. (1994) Do diatoms arrest embryonic development? *Marine Ecology Progress Series* 111: 79–86.

PrasannaKumar, S. & 9 coauthors (2006) Bay of Bengal Process Studies (BOBPS) Final Report: http://drs.nio.org/drs/bitstream/2264/535/3/Report_BOBPS_July2006.p.pdf

Price, H.J. & Paffenhöfer, G.A. (1984) Effects of feeding experience in the copepod *Eucalanus pileatus*: a cinematographic study. *Marine Biology* 84: 35–40.

Provan, J., Beatty, G.E. & Keating, S.L. (2010) High dispersal potential has maintained long-term population stability in the North Atlantic copepod *Calanus finmarchicus*. *Proceedings of the Royal Society, B* 276: 301–307.

Purcell, E.M. (1977) Life at low Reynolds number. *American Journal of Physics* 45: 3–11.

Purcell, J.E. & Madin, L.P. (1991) Diel patterns of migration, feeding, and spawning by salps in the subarctic Pacific. *Marine Ecology Progress Series* 73: 211–217.

Purinton, B.L., DeMaster, D.J., Thomas, C.J. & Smith, C.R. (2008) ^{14}C as a tracer of labile organic matter in Antarctic benthic food-webs. *Deep-Sea Research II* 55: 2438–2450.

Quetin, L.B. & Ross, R.M. (2001) Environmental variability and its impact on the reproductive cycle of Antarctic krill. *American Zoologist* 41: 74–89.

Quigg, A. & 8 coauthors (2003) The evolutionary inheritance of elemental stoichiometry in marine phytoplankton. *Nature* 425: 291–294.

Quinn, T.J. & Deriso, R.B. (1998) *Quantitative Fish Dynamics*. Oxford University Press, Oxford, 560 pp.

Quinn, W.H. (1992) A study of Southern Oscillation-related climatic activity for A.D. 622–1900 incorporating Nile River flood data. In: Diaz, H.F. & Markgraf, V. (eds.) *El Niño: Historical and Paleoclimatic Aspects of the Southern Oscillation*. Cambridge University Press, Cambridge, UK, pp. 110–149.

Quinn, W.H., Neal, V.T. & de Mayolo, S.E.A. (1987) El Niño occurrences over the past four and a half centuries. *Journal of Geophysical Research* 92: 14,449–14,461.

Quinones, R.A. (2010) Eastern boundary current systems. In: Liu, K.K., Atkinson, L., Quiñones, R. & Talaue-McManus, L. (eds.) *Carbon and Nutrient Fluxes in Continental Margins, Global Change – The IGBP Series*, Springer-Verlag, Berlin, pp. 25–120.

Ramirez Llodra, E., Tyler, P.E. & Copley, J.T.P. (2000) Reproductive biology of three caridean shrimp, *Rimicaris exoculata*, *Chorocaris chacei* and *Mirocaris fortunate* (Caridea: Decapoda), from hydrothermal vents. *Journal of the Marine Biological Association of the United Kingdom* 80: 473–484.

Ramirez Llodra, E., Shank, T.M. & German, C.R. (2007) Biodiversity and biogeography of hydrothermal vent species: thirty years of discovery and investigations. *Oceanography* 20: 30–41.

Rappé, M.S., Connon, S.A., Vergin, K.L. & Giovannoni, S.J. (2002) Cultivation of the ubiquitous SAR11 marine bacterioplankton clade. *Nature* 418: 630–633.

Rasmussen, H. & Jørgensen, B.B. (1992) Microelectrode studies of seasonal oxygen uptake in a coastal sediment: role of molecular diffusion. *Marine Ecology Progress Series* 81: 289–303.

Ravaux, J., Gaill, F., Le Bris, N., Sarradin, P.-M., Jollivet, D. & Shillito, B. (2003) Heat-shock response and temperature resistance in the deep-sea vent shrimp *Rimicaris exoculata*. *The Journal of Experimental Biology* 206: 2345–2354.

Raven, J.A. (1985) Physiological consequences of extremely small size for autotrophic organisms in the sea. In: Platt, T. & Li, K.W. (eds.) *Photosynthetic Picoplankton*, Canadian Bulletin of Fisheries and Aquatic Science 214: 1–70.

Razouls, C., de Bovée, F., Kouwenberg, J. & Desreumaux, N. (2005–2011) *Diversity and Geographic Distribution of Marine Planktonic Copepods*. http://copepodes.obs-banyuls.fr/en

Rebstock, G.A. (2001a) *Long-term changes in the species composition of calanoid copepods off southern California*. PhD Dissertation, Scripps Institution of Oceanography, University of California, San Diego.

Rebstock, G.A. (2001b) Long-term stability of species composition in calanoid copepods off southern California. *Marine Ecological Progress Series* 215: 213–224.

Redfield, A.C., Ketchum, G.H. & Richards, F.A. (1963) The influence of organisms on the composition of sea water. In: Hill, M.N. (ed.) *The Sea*. Wiley-Interscience, New York, pp. 26–77.

Reimers, C.E. (1984) An *in situ* microprofiling instrument for measuring interfacial pore water gradients: methods and oxygen profiles from the North Pacific Ocean. *Deep-Sea Research* 24: 2019–2035.

Reimers, C.E., Jahnke, R.A. & McCorkle, D. (1992) Carbon fluxes and burial rates over the continental slope and rise off central California with implications for the global carbon cycle. *Global Biogeochemistry and Cycles* 6: 199–224.

Renninger, G.H. & 7 coauthors (1995) Sulfide as a chemical stimulus for deep-sea hydrothermal vent shrimp. *Biological Bulletin* 189: 59–76.

Renz, J., Mengedoht, D. & Hirche, H.-J. (2008) Reproduction, growth and secondary production of *Pseudocalanus elongatus* Boeck (Copepoda, Calanoida) in the southern North Sea. *Journal of Plankton Research* 30: 511–528.

Reuter, J.G. (1988) Iron stimulation of photosynthesis and nitrogen fixation in *Anabaena* 7120 and *Trichodesmium* (Cyanophyceae). *Journal of Phycology* 24: 249–254.

Rex, M.A. (1981) Community structure in the deep-sea benthos. *Annual Review of Ecology and Systematics* 12: 331–353.

Rex, M.A. & Etter, R.J. (2010) *Deep-Sea Biodiversity, Pattern and Scale*. Harvard University Press, Cambridge, Massachusetts, 354 pp.

Richardson, A.J. (2008) In hot water: zooplankton and climate change. *ICES Journal of Marine Science* 65: 279–295.

Richardson, A.J. & Verheye, H.M. (1998) The relative importance of food and temperature to copepod egg production and somatic growth in the southern Benguela upwelling system. *Journal of Plankton Research* 20: 2379–2399.

Richardson, A.J., Bakun, A., Hays, G.C. & Gibbons, M.J. (2009) The jellyfish joyride: causes, consequences and management responses to a more gelatinous future. *Trends in Ecology and Evolution* 24: 312–322.

Richardson, K., Beardall, J. & Raven, J.A. (1983) Adaptation of unicellular algae to irradiance: an analysis of strategies. *New Phytologist* 93: 157–191.

Richman, S., Heinle, D. & Huff, R. (1977) Grazing by adult estuarine calanoid copepods of the Chesapeake Bay. *Marine Biology* 42: 69–84.

Ricker, W.E. (1946) Computation of fish production. *Ecological Monographs* 16: 373–391.

Riebesell, U., Zondervan, I., Rost, B., Tortell, P.D., Zeebe, R.E. & Morel, F.M.M. (2000) Reduced calcification of marine plankton in response to increased atmospheric CO_2. *Nature* 407: 364–367.

Riget, F. & Engelstoff, J. (1998) Size-at-age of cod (*Gadus morhua*) off West Greenland, 1952–92. *North Atlantic Fisheries Organization, Science Council Studies* 31: 1–12.

Riley, G.A. (1946) Factors controlling phytoplankton populations on Georges Bank. *Journal of Marine Research* 6: 54–73.

Riley, G.A., Stommel, H. & Bumpus, D.F. (1949) Quantitative ecology of the plankton of the western North Atlantic. *Bulletin of the Bingham Oceanographic Collection* 12: 1–169.

Rizzo, P.J. & Nooden, L.D. (1973) Isolation and partial characterization of dinoflagellate chromatin. *Biochimica et Biophysica Acta (Amsterdam)* 349: 402–414.

Roberts, C. (2007) *The Unnatural History of the Sea*. Island Press, Washingon, D.C., 435 pp.

Robertson, J.E. & 7 coauthors (1994) The impact of a coccolithophore bloom on oceanic carbon uptake in the northeast Atlantic during summer 1991. *Deep-Sea Research I* **41**: 297–314.

Robidart, J.C. & 7 coauthors (2008) Metabolic versatility of the *Riftia pachyptila* endosymbiont revealed through metagenomics. *Environmental Microbiology* **10**: 727–737.

Robinson, A.R. & 11 coauthors (1993) Mesoscale upper ocean variabilities during the 1989 JGOFS bloom study. *Deep-Sea Research II* **40**: 9–35.

Robinson, C. & 11 coauthors (2006) The Atlantic Meridional Transect (AMT) Programme: a contextual view 1995–2005. *Deep-Sea Research II* **53**: 1485–1515.

Robinson, C. & 16 coauthors (2010) Mesopelagic zone ecology and biogeochemistry – a synthesis. *Deep-Sea Research II* **57**: 1504–1518.

Robinson, C., Holligan, P., Jickells, T. & Lavender, S. (2009) The Atlantic Meridional Transect (AMT) Programme (1995–2012). *Deep-Sea Research II* **56**: 895–898.

Robinson, G.A. (1970) Continuous plankton records: variation in the seasonal cycle of phytoplankton in the North Atlantic. *Bulletin of Marine Ecology* **6**: 333–345.

Robison, B.H. & Connor, J. (1999) *The Deep Sea*. Monterey Bay Aquarium Natural History Series, 80 pp.

Rocap, G. & 23 coauthors (2003) Genome divergence in two *Prochlorococcus* ecotypes reflects oceanic niche differentiation. *Nature* **424**: 1042–1047.

Roe, P. & Norenburg, J.L. (1999) Observations on depth distribution, diversity, and abundance of pelagic nemerteans from the Pacific Ocean off California and Hawaii. *Deep-Sea Research I* **46**: 1201–1220.

Roemmich, D. & McGowan, J.A. (1995a) Climatic warming and the decline of zooplankton in the California Current. *Science* **267**: 1324–1326.

Roemmich & McGowan (1995b) Sampling zooplankton: correction. *Science* **268**: 352–353.

Roman, M.R., Dam, H.G., Gauzens, A.L., Urban-Rich, J., Foley, D.G. & Dickey, T.D. (1995) Zooplankton variability on the Equator at 140°W during JGOFS EqPac Study. *Deep-Sea Research II* **42**: 673–693.

Roman, M.R., Adolf, H.A., Landry, M.R., Madin, L.P., Steinberg, D.K. & Zhang, X. (2002) Estimates of oceanic mesozooplankton production: a comparison using the Bermuda and Hawaii time-series data. *Deep-Sea Research II* **49**: 175–192.

Rona, P.A., Klinkhammer, G., Nelson, T.A., Tefry, J.H. & Elderfield, H. (1986) Black smokers, massive sulfides and vent biota at the Mid-Atlantic Ridge. *Nature* **321**: 33–37.

Roper, D.S., Smith, D.G. & Read, G.B. (1989) Benthos associated with two New Zealand coastal outfalls. *New Zealand Journal of Marine and Freshwater Research* **23**: 295–309.

Rose, G.A. (2004) Reconciling overfishing and climate change with stock dynamics of Atlantic cod (*Gadus morhua*) over 500 years. *Canadian Journal of Fisheries and Aquatic Sciences* **61**: 1553–1557.

Rosenberg, A.A. (2003) Managing to the margins: the overexploitation of fisheries. *Frontiers in Ecology and the Environment* **1**: 102–106.

Rosenberg, A.A., Bolster, W.J., Alexander, K.E. & Leavenworth, W.B. (2005) The history of ocean resources: modeling cod biomass using historical records. *Frontiers in Ecology the Environment* **3**: 78–84.

Ross, R.M. & Quetin, L.B. (1984) Spawning frequency and fecundity of the Antarctic krill *Euphausia superba*. *Marine Biology* **77**: 201–205.

Ross, R.M., Daly, K.L. & English, T.S. (1982) Reproductive cycle and fecundity of *Euphausia pacifica* in Puget Sound, Washington. *Limnology and Oceanography* **27**: 304–315.

Rothschild, B.J. & Osborne, T.R. (1988) Small-scale turbulence and plankton contact rates. *Journal of Plankton Research* **10**: 465–474.

Round, F.E., Crawford, R.M. & Mann, D.G. (1990) *The Diatoms, Biology and Morphology of the Genera*. Cambridge University Press, Cambridge, UK, 747 pp.

Rouse, G.W., Goffredi, S.K. & Vrijenhoek, R.C. (2004) *Osedax*: bone-eating marine worms with dwarf males. *Science* **305**: 668–671.

Rowe, G.T. (1983) Biomass and production of the deep-sea macrobenthos. In: Rowe, G.T. (ed.) *The Sea*, Volume 8: *Deep-Sea Biology*. Wiley-Interscience, New York, pp. 97–121.

Ruddiman, W.F. (2007a) *Earth's Climate: Past and Future*, 2nd Edn., W.H. Freeman, San Francisco, 388 pp.

Ruddiman, W.F. (2007b) The early anthropogenic hypothesis: challenges and responses. *Reviews of Geophysics* **45**: 2006RG000207R.

Ruhl, H.A. (2008) Community change in the variable resource habitat of the abyssal northeast Pacific. *Ecology* **89**: 991–1000.

Runge, J.A. (1980) Effects of hunger and season on the feeding behavior of *Calanus pacificus*. *Limnology and Oceanography* **25**: 134–145.

Runge, J.A. (1984) Egg production of the marine, planktonic copepod *Calanus pacificus* Brodsky: laboratory observations. *Journal of Experimental Marine Biology and Ecology* **74**: 53–66.

Runge, J.A. (1985) Relationship of egg production of *Calanus pacificus* to seasonal changes in phytoplankton availability in Puget Sound, Washington. *Limnology and Oceanography* **30**: 382–396.

Russell, F. (1939) Hydrographical and biological conditions in the North Sea as indicated by plankton organisms. *Journal du conseil/ Conseil international pour l'exploration de la mer* **14**: 171–192.

Rykaczewski, R.R. & Checkley, D.M. Jr (2008) Influence of ocean winds on the pelagic ecosystem in upwelling regions. *Proceedings of the National Academy of Sciences* **105**: 1965–1970.

Ryther, J.H. (1969) Photosynthesis and fish production in the sea. *Science* **166**: 72–76.

Saito, H. & Tsuda, A. (2000) Egg production and early development of the subarctic copepods *Neocalanus cristatus*, *N. plumchrus* and *N. flemingeri*. *Deep-Sea Research I* **47**: 2141–2158.

Saiz, E. & Calbet, A. (2007) Scaling of feeding in marine copepods. *Limnology and Oceanography* **52**: 668–675.

Sakshaug, E. (2004) Primary and secondary production in the Arctic seas. In: Stein, R. & MacDonald, R.W. (eds.) *The Organic Carbon Cycle in the Arctic Ocean*, Springer-Verlag, Berlin, pp. 57–81.

Sambrook, J., MacCallum & Russell, D.W. (2006) *The Condensed Protocols from Molecular Cloning: a Laboratory Manual*. Cold Spring Harbor Laboratory Press, Woodbury, New York, 800 pp.

Sanders, H.L. (1968) Marine benthic diversity: a comparative study. *American Naturalist* **102**: 243–282.

Sanders, H.L. (1969) Benthic marine diversity and the stability–time hypothesis. *Brookhaven Symposia in Biology* **22**: 71–80.

Sanders, H.L., Hessler, R.R. & Hampson, G.R. (1965) An introduction to the study of deep-sea benthic faunal assemblages along the Gay Head–Bermuda transect. *Deep-Sea Research* **12**: 845–867.

Sarmiento, J.L. & Gruber, N. (2006) *Ocean Biogeochemical Dynamics*, Princeton University Press, Princeton NJ, 503 pp.

Sars, M. (1867) Om nogle Echinodermer og Coelenterater fra Lofoten. *Oversigt Over del Kongeligt Danske Videnskabernes Selskabs Forhandlingar, Christiana* **1867**: 1–7.

Sarvas, T.H. & Fevolden, S.E. (2005) Pantophysin (*Pan* I) locus divergence between inshore *v.* offshore and northern *v.* southern populations of Atlantic cod in the north-east Atlantic. *Journal of Fish Biology* **67**: 444–469.

Saulnier-Michel, C., Gaill, F., Hily, A., Alberic, P. & Cosson-Mannevy, M.A. (1990) Structure and function of the digestive tract of *Alvinella pompejana*, a hydrothermal vent polychaete. *Canadian Journal of Zoology* **68**: 722–732.

Savidge, G. & 7 coauthors (1992) The BOFS 1990 spring bloom experiment: temporal evolution and spatial variability of the hydrographic field. *Progress in Oceanography* 29: 235–281.

Saville, A. & Bailey, R.S. (1980) The assessment and management of the herring stocks in the North Sea and to the west of Scotland. *Rapports et procès-verbaux des réunions/Conseil permanent international pour l'exploration de la mer* 177: 112–142.

Sazhina, L.I. (1968) Hibernating eggs of marine Calanoida. *Zoologicheskii Zhurnal* 47: 1554–1556 (in Russian).

Scanlan, D.J. and 9 coauthors (2009) Ecological genomics of marine picocyanobacteria. *Microbiology and Molecular Biology Reviews* 73: 249–299.

Schaefer, M.B. (1967) Fishery dynamics and present status of the yellowfin tuna population of the eastern Pacific Ocean. *Inter-American Tropical Tuna Commission Bulletin* 12: 89–136.

Schander, C. & 23 coauthors (2010) The fauna of hydrothermal vents on the Mohn Ridge (North Atlantic). *Marine Biology Research* 6: 155–171.

Schindler, D.E., Hilborn, R., Chasco, B., Boatright, C.P., Quinn, T.P., Rogers, L.A. & Webster, M.S. (2010) Population diversity and the portfolio effect in an exploited species. *Nature* 465: 609–612.

Schmid, A.M. (1984) Wall morphogenesis in *Thalassiosira eccentrica*: comparison of auxospore formation and the effects of MT-inhibitors. In: Mann, D.G. (ed.) *Proceedings of the 7th International Symposium on Living and Fossil Diatoms*, O. Koeltz, Koenigstein, pp. 47–70.

Schmidt, C., Vuillemin, R., Le Gall, C., Gaill, F. & Le Bris, N. (2008) Geochemical energy sources for microbial primary production in the environment of hydrothermal vent shrimps. *Marine Chemistry* 108: 18–31.

Schnack-Schiel, S.B. (2001) Aspects of the study of the life cycles of Antarctic copepods. *Hydrobiologia* 453/454: 9–24.

Schnack-Schiel, S.B., Thomas, D.N., Haas, C., Dieckmann, D.S. & Alheit, R. (2001) The occurrence of the copepods *Stephos longipes* (Calanoida) and *Drescheriella glacialis* (Harpacticoida) in summer sea ice in the Weddell Sea, Antarctica. *Antarctic Science* 13: 150–157.

Schoener, A. & Tufts, D.F. (1987) Changes in oyster condition index with El Niño Southern Oscillation events at 46°N in an eastern Pacific bay. *Journal of Geophysical Research* 92: 14,429–14,435.

Schwartzlose, R.A. & 20 coauthors (1999) Worldwide large-scale fluctuations of sardine and anchovy populations. *South African Journal of Marine Science* 21: 289–347.

Scott, T. (1894) Report on the Entomostraca from the Gulf of Guinea. *Transactions of the Linnean Society of London, Series 2* 6(1): 1–161.

Sebastian, M. (1966) Euphausiacea from Indian Seas: systematics and general considerations. *Symposium on Crustacea, Marine Biological Association of India* 1: 233–254.

Seibel, B.A., Thuesen, E.V., Childress, J.J. & Gorodezky, L.A. (1997) Decline in pelagic cephalopod metabolism with habitat depth reflects differences in locomotory efficiency. *Biological Bulletin* 192: 262–278.

Seibel, B.A., Thuesen, E.V. & Childress, J.J. (2000) Light-limitation on predator–prey interactions: consequences for metabolism and locomotion of deep-sea cephalopods. *Biological Bulletin* 198: 284–298.

Seiter, K., Hensen, C. & Zabel, M. (2005) Benthic carbon mineralization on a global scale. *Global Biogeochemical Cycles* 19: GB1010 (26 pp.).

Serret, P., Fernández, E., Robinson, C., Woodward, E.M.S. & Pérez, V. (2006) Local production does not control the balance between plankton photosynthesis and respiration in the open Atlantic Ocean. *Deep-Sea Research II* 53: 1611–1628.

Sheridan, C.C. & Landry, M.R. (2004) A 9-year increasing trend in mesozooplankton biomass at the Hawaii Ocean Time-series Station ALOHA. *ICES Journal of Marine Science* 61: 457–463.

Sherr, E. & Sherr, B. (2000) Marine microbes, an overview. In: Kirchman, D.L. (ed.) *Microbial Ecology of the Oceans*. Wiley, New York, pp. 47–84.

Sherr, E.B. & Sherr, B.F. (2007) Heterotrophic dinoflagellates: a significant component of microzooplankton biomass and major grazers of diatoms in the sea. *Marine Ecology Progress Series* 352: 187–197.

Sherr, E.B. & Sherr, B.F. (2009) Capacity of herbivorous protists to control initiation and development of mass phytoplankton blooms. *Aquatic Microbial Ecology* 57: 253–262.

Sherr, B.F., Sherr, E.B. & Fallon, R.D. (1987) Use of monodispersed, fluorescently labeled bacteria to estimate in situ protozoan bacterivory. *Applied and Environmental Microbiology* 53: 958–965.

Sherr, E.B., Sherr, B.F. & Fessenden, L. (1997) Heterotrophic protists in the central Arctic Ocean. *Deep-Sea Research II* 44: 1665–1682.

Sherr, E.B., Sherr, B.F. & Cowles, T.J. (2001) Mesoscale variability in bacterial activity in the Northeast Pacific Ocean off Oregon, USA. *Aquatic Microbial Ecology* 25: 21–30.

Sherr, E.B., Sherr, B.F., Wheeler, P.A. & Thompson, K. (2003) Temporal and spatial variation in the stocks of autotrophic and heterotrophic microbes in the upper water column of the central Arctic Ocean. *Deep-Sea Research I* 50: 557–571.

Sherr, E.B., Sherr, B.F. & Hartz, A.J. (2009) Microzooplankton grazing impact in the western Arctic Ocean. *Deep-Sea Research II* 56: 1264–1273.

Shimeta, J., Jumars, P.A. & Lessard, E.J. (1995) Influences of turbulence on suspension feeding by planktonic protozoa; experiments in laminar shear fields. *Limnology and Oceanography* 40: 845–859.

Shinn, G.L. (1997) Chaetognatha. In: Harrison, F.W. & Ruppert, E.E. (eds.) *Microscopic Anatomy of Invertebrates*, Vol. 15. *Hemichordata, Chaetognatha, and the Invertebrate Chordates*. Wiley-Liss, New York, pp. 103–220.

Shiomoto, A., Tadokoro, K., Nagasawa, K. & Ishida, Y. (1997) Trophic relations in the subarctic North Pacific ecosystem: possible feeding effect from pink salmon. *Marine Ecology Progress Series* 150: 75–85.

Shirayama, Y. & Horikoshi, M. (1989) Comparison of the benthic size structure between sublittoral, upper-slope and deep-sea areas of the western Pacific. *International Revue Gesamtes Hydrobiologie* 74: 1–13.

Shrum, J.P., Zhu, T.F. & Szostak, J.W. (2010) The origins of cellular life. *Cold Spring Harbor Perspectives in Biology*. doi:10.1101.cshperspect.a002212 (15 pp.) [no volume number exists].

Sieberth, K. McM., Smetacek, V. & Lenz, J. (1978) Pelagic ecosystem structure: heterotrophic compartments of the plankton and their relationship to plankton size fractions. *Limnology and Oceanography* 23: 1256–1263.

Siegel, D.A., Karl, D.M. & Michaels, A.F. (2001) HOTS and BATS: interpretations of open ocean biogeochemical processes. *Deep-Sea Research II* 48: 1403–1404.

Sieracki, M.E., Verity, P.G. & Stoecker, D.K. (1993) Plankton community response to sequential silicate and nitrate depletion during the 1989 North Atlantic spring bloom. *Deep-Sea Research II* 40: 213–225.

Sieracki, C.K., Sieracki, M.E. & Yentsch, C.S. (1998) An imaging-in-flow system for automated analysis of marine microplankton. *Marine Ecology Progress Series* 168: 285–296.

Sigman, D.M., Granger, J., DiFiore, P.J., Lehmann, M.M., Ho, R., Cane, G. & van Geen, A. (2005) Coupled nitrogen and oxygen isotope measurements of nitrate along the eastern North Pacific margin. *Global Biogeochemical Cycles* **19**: GB4022.doi:10.1029/2005GB002458.

Silver, M.W. & Gowing, M.M. (1991) The "particle" flux: origins and biological components. *Progress in Oceanography* **26**: 75–113.

Simard, Y., Lacroix, G. & Legendre, L. (1985) *In situ* twilight grazing rhythm during diel vertical migrations of a scattering layer of *Calanus finmarchicus*. *Limnology and Oceanography* **30**: 598–606.

Sinclair, M. (1988) *Marine Populations, an Essay on Population Regulation and Speciation.* University of Washington Press, Seattle.

Skinner, L.C., Fallon, S., Waelbroeck, C., Michel, E. & Barker, S. (2010) Ventilation of the deep Southern Ocean and deglacial CO_2 rise. *Science* **328**: 1147–1151.

Slawyk, G., Collos, Y. & Auclair, J.C. (1977) The use of the ^{13}C and ^{15}N isotopes for the simultaneous measurement of carbon and nitrogen turnover rates in marine phytoplankton. *Limnology and Oceanography* **22**: 925–932.

Slemons, L.O., Murray, J.W., Resing, J., Paul, B. & Dutrieux, P. (2010) Western Pacific sources of iron, manganese and aluminum to the Equatorial Undercurrent. *Global Biogeochemical Cycles* **24**: GB3024, doi:10.1029/2009GB003693.

Smayda, T.J. (1976) Phytoplankton processes in mid-Atlantic near-shore and shelf waters and energy-related activities. In: Manowitz, B. (ed.) *Effects of Energy-Related Activities on the Atlantic Continental Shelf.* Brookhaven National Laboratory Report No. 50484, pp. 70–95.

Smayda, T.J. (1980) Phytoplankton species succession. In: Morris, I. (ed.) *The Physiological Ecology of Phytoplankton*, University of California Press, Berkeley, pp. 493–570.

Smetacek, V. (1985) Role of sinking in diatom life-history cycles: ecological, evolutionary and geological significance. *Marine Biology* **84**: 239–251.

Smetacek, V. (2009) (temporary citation: http://www.awi.de/en/news/press_releases/detail/item/lohafex_provides_new_insights_on_plankton_ecology_only_small_amounts_of_atmospheric_carbon_dioxide/?cHash=1eb5f2e233)

Smetacek, V. & Passow, U. (1990) Spring bloom initiation and Sverdrup's critical-depth model. *Limnology and Oceanography* **35**: 228–234.

Smith, C.R. & Baco, A.R. (2003) Ecology of whale falls at the deep-sea floor. *Oceanography and Marine Biology* **41**: 311–354.

Smith, C.R., Hoover, D.J., Doan, E.E., Pope, R.H., DeMaster, D.J., Dobbs, F.C. & Altabet, M.A. (1996) Phytodetritus at the abyssal seafloor across 10° of latitude in the central equatorial Pacific. *Deep-Sea Research II* **43**: 1309–1338.

Smith, C.R., & 7 coauthors (1997) Latitudinal variations in benthic processes in the abyssal equatorial Pacific: control by biogenic particle flux. *Deep-Sea Research II* **44**: 2295–2317.

Smith, K.L. Jr, Kaufmann, R.S., Baldwin, R.J. & Carlucci, A.F. (2001) Pelagic–benthic coupling in the abyssal eastern North Pacific: an 8-year time-series study of food supply and demand. *Limnology and Oceanography* **46**: 543–556.

Smith, K.L. Jr, Baldwin, R.J., Glattis, R.C. & Kaufmann, R.S. (2006) Climate effect on food supply to depths greater than 4000 meters in the northeast Pacific. *Limnology and Oceanography* **51**: 166–176.

Smith, K.L. Jr, Ruhl, H.A., Kaufmann, R.S. & Kahru, M. (2008) Tracing abyssal food supply back to upper-ocean processes over a 17-year time series in the northeast Pacific. *Limnology and Oceanography* **53**: 2655–2667.

Smith, K.L. Jr, Ruhl, H.A., Bett, B.J., Billett, D.S.M., Lampitt, R.S. & Kaufmann, R.S. (2009) Climate, carbon cycling, and deep-ocean ecosystems. *Proceedings of the National Academy of Sciences* **106**: 19,211–19,218.

Smith, R.C. & Baker, K.S. (1981) The bio-optical state of ocean waters and remote sensing. *Limnology and Oceanography* **23**: 247–259.

Smith, S.L. & 8 coauthors (1998) Seasonal response of zooplankton to monsoonal reversals in the Arabian Sea. *Deep-Sea Research II* **45**: 2369–2403.

Smith, W. & McIntyre, A.D. (1954) A spring-loaded bottom-sampler. *Journal of the Marine Biological Association of the United Kingdom* **33**: 257–264.

Smith, W.O. Jr & Anderson, R.F. (eds.) (2000a) U.S. Southern Ocean JGOFS Program (AESOPS). *Deep-Sea Research II* **47**: 3073–3548.

Smith, W.O. Jr & Anderson, R.F. (eds.) (2000b) U.S. Southern Ocean JGOFS Program (AESOPS) – Part II. *Deep-Sea Research II* **48**: 3883–4383.

Snider, L.J., Burnett, B.R. & Hessler, R.R. (1984) The composition and distribution of meiofauna and nanobiota in a central Pacific deep-sea area. *Deep-Sea Research* **31**: 1225–1249.

Sochard, M., Wilson, D., Austin, B. & Colwell, R. (1979) Bacteria associated with the surface and gut of marine copepods. *Applied and Environmental Microbiology* **37**: 750–759.

Sommer, F.. Hansen, T., Feuchtmayr, H., Santer, B., Tokle, N. & Sommer, U. (2003) Do calanoid copepods suppress appendicularians in the coastal ocean? *Journal of Plankton Research* **25**: 869–871.

Sorokin, Y. (1964) A quantitative study of the microflora in the central Pacific Ocean. *Journal du conseil/Conseil international pour l'exploration de la mer* **29**: 25–40.

Sosik, H.M. & Olson, R.J. (2007) Automated taxonomic classification of phytoplankton sampled with imaging-in-flow cytometry. *Limnology and Oceanography Methods* **5**: 204–216.

Sowell, S.M. & 7 coauthors (2008) Proteomic analysis of stationary phase in the marine bacterium "Candidatus *Pelagibacter ubique*". *Applied and Environmental Microbiology* **74**: 4091–4100.

Spero, H.J. (1982) Phagotrophy in *Gymnodinium fungiforme* (Pyrrophyta): the peduncle as an organelle of ingestion. *Journal of Phycology* **18**: 356–360.

Spitz, Y.H., Moisan, J.R. & Abbott, M.R. (2001) Configuring an ecosystem model using data from the Bermuda Atlantic Time Series (BATS). *Deep-Sea Research II* **48**: 1733–1768.

Stanley, R.H.R., Buesseler, K.O. & Manganini, S.J. (2004) A comparison of major and minor elemental fluxes collected in neutrally buoyant and surface-tethered sediment traps. *Deep-Sea Research I* **51**: 1387–1395.

Steeman-Nielsen, E. (1952) The use of radioactive carbon (^{14}C) for measuring organic production in the sea. *Journal du conseil/Conseil international pour l'exploration de la mer* **18**: 117–140.

Steinberg, D.K., Carlson, C.A., Bates, N.R., Johnson, R.J., Michaels, A.F. & Knap, A.H. (2001) Overview of the US JGOFS Bermuda Atlantic Time-Series Study (BATS): a decade-scale look at ocean biology and biogeochemistry. *Deep-Sea Research II* **48**: 1405–1447.

Steinberg, D.K., Van Mooy, B.A.S., Buesseler, K.O., Boyd, P.W., Kobari, T. & Karl, D.M. (2008) Bacterial vs. zooplankton control of sinking particle flux in the ocean's twilight zone. *Limnology and Oceanography* **53**: 1328–1338.

Stenseth, N.C., Falck, W., Bjørnstad, O.N. & Krebs, C.J. (1997) Population regulation in snowshore hare and Canadian lynx: asymmetric food web configurations between hare and lynx. *Proceedings of the National Academy of Sciences* **94**: 5147–5152.

Stoecker, D.K., Johnson, M.D., de Vargas, C. & Not, F. (2009) Acquired phototrophy in aquatic protists. *Aquatic Microbial Ecology* **57**: 279–310.

Stokes, G.G. (1851) *Transactions of the Cambridge Philosophical Society* **9**: 8. (Stokes: Mathematical and Physical Papers **3**: 1.)

Straile, D. (1997) Gross growth efficiencies of protozoan and metazoan zooplankton and their dependence on food concentration, predator–prey weight ratio, and taxonomic group. *Limnology and Oceanography* **42**: 1375–1385.

Stramska, M. & Dickey, T. (1993) Phytoplankton bloom and the vertical thermal structure of the upper ocean. *Journal of Marine Research* 51: 819–842.

Strom, S.L. (2000) Bacterivory: interactions between bacteria and their grazers. In: Kirchman, D.L. (ed.) *Microbial Ecology of the Oceans.* Wiley, New York, pp. 351–386.

Strom, S.L. & Buskey, E.J. (1993) Feeding, growth, and behavior of the thecate heterotrophic dinoflagellate *Oblea rotunda. Limnology and Oceanography* 38: 965–977.

Strom, S.L. & Morello, T.A. (1998) Comparative growth rates and yields of ciliates and heterotrophic dinoflagellates. *Journal of Plankton Research* 20: 571–584.

Strom, S.L., Benner, R., Ziegler, S. & Dagg, M.J. (1997) Planktonic grazers are a potentially important source of marine dissolved organic carbon. *Limnology and Oceanography* 42: 1364–1374.

Strom, S.L., Miller, C.B. & Frost, B.W. (2000) What sets lower limits to phytoplankton stocks in high-nitrate, low-chlorophyll regions of the open ocean? *Marine Ecology Progress Series* 193: 19–31.

Strom, S.L., Brainard, M.A., Holmes, J.L. & Olson, M.B. (2001) Phytoplankton blooms are strongly impacted by microzooplankton grazing in coastal North American waters. *Marine Biology* 138: 355–368.

Strom, S.L., Wolfe, G.V. & Bright, K.J. (2007) Responses of marine planktonic protists to amino acids: feeding inhibition and swimming behavior in the ciliate *Favella* sp. *Aquatic Microbiology and Ecology* 47: 107–121.

Strong, A., Chisholm, S., Miller, C. & Cullen, J. (2009a) Ocean fertilization: time to move on. *Nature* 461: 347–348.

Strong, A.L., Chisholm, S.W. & Cullen, J.J. (2009b) Ocean fertilization: science, policy, and commerce. *Oceanography* 22: 236–261.

Strutton, P.G. & Chavez, F.P. (2000) Primary productivity in the equatorial Pacific during the 1997–1998 El Nino. *Journal of Geophysical Research, Oceans* 105: 26,089–26,101.

Strutton, P.G. & 7 coauthors (2011) The impact of equatorial Pacific tropical instability waves on hydrography and nutrients: 2004–2005. *Deep-Sea Research II* 58: 284–295.

Strzepek, R.F. & Harrison, P.J. (2004) Photosynthetic architecture differs in coastal and oceanic diatoms. *Nature* 43: 689–692.

Suggett, D.J., MacIntyre, H.L., Kana, T.M. & Geider, R.J. (2009) Comparing electron transport with gas exchange: parameterizing exchange rates between alternative photosynthetic currencies for eukaryotic phytoplankton. *Aquatic Microbial Ecology* 56: 147–162.

Sullivan, B.K. (1980) *In situ* feeding behavior of *Sagitta elegans* and *Eukrohnia hamata* (Chaetognatha) in relation to the vertical distribution and abundance of prey at Ocean Station "P". *Limnology and Oceanography* 25: 317–326.

Sullivan, B.K. & McManus, L.T. (1986) Factors controlling seasonal succession of the copepods *Acartia hudsonica* and *A. tonsa* in Narragansett Bay, Rhode Island: temperature and resting egg production. *Marine Ecology Progress Series* 28: 121–128.

Sullivan, B.K., Miller, C.B., Peterson, W.T. & Soeldner, A.H. (1975) A scanning electron microscope study of the mandibular morphology of boreal copepods. *Marine Biology* 30: 175–182.

Sun, S., De La Mare, W. & Nicol, S. (1995) The compound eye as an indicator of age and shrinkage in Antarctic krill. *Antarctic Science* 7: 387–392.

Sunda, W.G. & Huntsman, S.A. (1995) Iron uptake and growth limitation in oceanic and coastal phytoplankton. *Marine Chemistry* 50: 189–206.

Sunda, W.G. & Huntsman, S.A. (1997) Interrelated influence of iron, light and cell size on marine phytoplankton growth. *Nature* 390: 389–392.

Sunda, W.G., Swift, D.G. & Huntsman, S.A. (1991) Low iron requirement for growth in oceanic phytoplankton. *Nature* 351: 55–57.

Suttle, C.A. (2007) Marine viruses – major players in the global ecosystem. *Nature Reviews Microbiology* 5: 801–812.

Suzuki, Y. & 7 coauthors (2005) Novel chemoautotrophic endosymbiosis between a member of the epsilonproteobacteria and the hydrothermal-vent gastropod *Alviniconcha* aff. *hessleri* (Gastropoda: Provannidae) from the Indian Ocean. *Applied and Environmental Microbiology* 71: 5440–5450.

Suzuki, Y. & 15 coauthors (2006) Sclerite formation in the hydrothermal vent "scaly-foot" gastropod – possible control of iron sulfide biomineralization by the animal. *Earth and Planetary Science Letters* 242: 39–50.

Svavarsson, J. (1997) Diversity of isopods (Crustacea): new data from the Arctic and Atlantic Oceans. *Biodiversity and Conservation* 6: 1571–1579.

Sverdrup, H.U. (1953) On conditions for the vernal blooming of phytoplankton. *Journal du Conseil international pour l'exploration de la mer* 18: 287–295.

Sverdrup, H.U., Johnson, M.V. & Fleming, R.H. (1942) *The Oceans, Their Physics, Chemistry and Biology.* Prentice Hall, Englewood Cliffs, NJ, 1060 pp.

Taghon, G.L. (1982) Optimal foraging by deposit-feeding invertebrates: roles of particle size and organic coating. *Oecologia* 52: 295–304.

Taghon, G.L. & Jumars, P.A. (1984) Variable ingestion rate and its role in optimal foraging behavior of marine deposit feeders. *Ecology* 65: 549–558.

Takahashi, T. & 30 coauthors (2009) Climatological mean and decadal change in surface ocean pCO_2, and net sea–air CO_2 flux over the global oceans. *Deep-Sea Research* 56: 554–577.

Tanaka, S. (1962) On the salmon stocks of the Pacific coast of the United States and Canada. *International North Pacific Fisheries Commission Bulletin* 9: 69–84.

Tang, E.P.Y. (1996) Why do dinoflagellates have lower growth rates? *Journal of Phycology* 32: 80–84.

Tarrant, A.M., Baumgartner, M.F., Verslycke, T. & Johnson, C.L. (2008) Differential gene expression in diapausing and active *Calanus finmarchicus* (Copepoda). *Marine Ecology Progress Series* 355: 193–207.

Taylor, A.G., Landry, M.R., Selph, K.E. & Yang, E.J. (2011) Biomass size structure and depth distribution of the microbial community in the eastern equatorial Pacific. *Deep-Sea Research II* 58: 342–357.

Taylor, F.J.R. (1976) Flagellate phylogeny: a study in conflicts. *Journal of Protozoology* 3: 28–40.

Taylor, F.J.R. (1987) *The Biology of Dinoflagellates.* Biological Monographs, Vol. 21. Blackwell, Oxford, UK, 785 pp.

Teal, L.R., Bulling, M.T., Parker, E.R. & Solan, M. (2008) Global patterns of bioturbation intensity and mixed depth of marine soft sediments. *Aquatic Biology* 2: 207–218.

Tett, P. (1990) The photic zone. In: Herring, P.J., Campbell, A.K., Whitfield, M. & Maddock, L. (eds) *Light and Life in the Sea.* Cambridge University Press, Cambridge, pp. 58–87.

Tett, P. & Barton, E.D. (1995) Why are there about 5000 species of phytoplankton in the sea? *Journal of Plankton Research* 17: 1693–1704.

Theroux, R.B. & Wigley, R.L. (1998) Quantitative composition and distribution of the macrobenthic invertebrate fauna of the continental shelf ecosystems of the northeastern United States. *NOAA Technical Report NMFS* 140: 1–240.

Thibault, D., Head, E.J.H. & Wheeler, P.A. (1999) Mesozooplankton in the Arctic Ocean in summer. *Deep-Sea Research I* 46: 1391–1415.

Thingstad, T.F. (2000) Control of bacterial growth in idealized food webs. In: Kirchman, D.L. (ed.) *Microbial Ecology of the Oceans.* Wiley, New York, pp. 229–260.

Thompson, G.A., Alder, V.A., Boltovskoy, D. & Brandini, F. (1999) Abundance and biogeography of tintinnids (Ciliophora) and associated microzooplankton in the Southwest Atlantic Ocean. *Journal of Plankton Research* **21**: 1265–1298.

Thomson, J., Brown, L., Nixon, S., Cook, G.T. & MacKenzie, A.B. (2000) Bioturbation and Holocene sediment accumulation rates in the north-east Atlantic Ocean (Benthic Boundary Layer experiment sites). *Marine Geology* **169**: 21–39.

Thor, P. & Wendt, I. (2010) Functional response of carbon efficiency in the pelagic calanoid copepod *Acartia tonsa* Dana. *Limnology and Oceanography* **55**: 1779–1789.

Thuesen, E.V. & Childress, J.J. (1993) Enzymatic activities and metabolic rates of pelagic chaetognaths: lack of depth-related declines. *Limnology and Oceanography* **38**: 935–948.

Thuesen, E.V. & Childress, J.J. (1994) Oxygen consumption rates and metabolic enzyme activities of oceanic California medusae in relation to body size and habitat depth. *Biological Bulletin* **187**: 84–98.

Thuesen, E.V., Kogure, K., Hashimoto, K. & Nemoto, T. (1988) Poison arrow worms: a tetrodotoxin venom in the marine phylum Chaetognatha. *Journal of Experimental Marine Biology and Ecology* **116**: 249–256.

Thuesen, E.V., Miller, C.B. & Childress, J.J. (1998) An ecophysiological interpretation of oxygen consumption rates and enzymatic activities of deep-sea copepods. *Marine Ecological Progress Series* **168**: 95–107.

Torres, J.J., Belman, B.W. & Childress, J.J. (1979) Oxygen consumption rates of midwater fishes off California. *Deep-Sea Research* **26A**: 185–197.

Townsend, D.W., Cammen, L.M., Holligan, P.M., Campbell, D.E. & Pettigrew, N.R. (1994) Causes and consequences of variability in the timing of spring phytoplankton blooms. *Deep-Sea Research* **41**: 747–765.

Trees, C.C., Clark, D.K., Bidigare, R.R., Ondrusek, M.E. & Mueller, J.L. (2000) Accessory pigments versus chlorophyll *a* concentrations within the euphotic zone: a ubiquitous relationship? *Limnology and Oceanography* **45**: 1130–1143.

Tremblay, J.-É. & 9 coauthors (2006) Trophic structure and pathways of biogenic carbon flow in the eastern North Water Polynya. *Progress in Oceanography* **71**: 402–425.

Triemer, R.E. & Brown, R.M. (1977) Ultrastructure of meiosis in *Chlamydomonas reinhardtii*. *British Phycological Journal* **12**: 23–44.

Trull, T.W., Bray, S.G., Buesseler, K.O., Lamborg, C.H., Manganini, S., Moy, C. & Valdes, J. (2008) *In situ* measurement of mesopelagic particle sinking rates and the control of carbon transfer to the ocean interior during the Vertical Flux in the Global Ocean (VERTIGO) voyages in the North Pacific. *Deep-Sea Research II* **55**: 1684–1695.

Tsuda, A. (ed.) (2005) Results from the subarctic Pacific Iron Experiment for Ecosystem Dynamics Study (SEEDS). *Progress in Oceanography* **64**: 91–324.

Tsuda, A. & Miller, C.B. (1998) Mate-finding behaviour in *Calanus marshallae* Frost. *Proceedings of the Royal Society of London, B* **353**: 713–720.

Tunnicliffe, V., Embley, R.W., Holden, J.F., Butterfield, D.A., Massoth, G.J. & Juniper, S.K. (1997) Biological colonization of new hydrothermal vents following an eruption on Juan de Fuca Ridge. *Deep-Sea Research* **44**: 1627–1644.

Tunnicliffe, V., McArthur, A.G. & McHugh, D. (1998) A biogeographical perspective of the deep-sea hydrothermal vent fauna. *Advances in Marine Biology* **34**: 353–442.

Turner, J.T. (2004) The importance of small planktonic copepods and their roles in pelagic marine food webs. *Zoological Studies* **43**: 255–266.

Tyler, P.A. (1988) Seasonality in the deep-sea. *Oceanography and Marine Biology Annual Review* **26**: 227–258.

Tyler, P.A., Harvey, R., Giles, L.A. & Gage, J.D. (1992) Reproductive strategies and diet in deep-sea nuculanid protobranchs (Bivalvia:

Nuculoidea) from the Rockall Trough. *Marine Biology* **114**: 571–580.

Tyler, P.A., Gage, J.D., Paterson, G.J.L. & Rice, A.L. (1993) Dietary constraint on reproductive periodicity in two sympatric deep-sea seastars. *Marine Biology* **115**: 267–277.

Tyrrell, T. (1999) The relative influences of nitrogen and phosphorus on oceanic primary production. *Nature* **400**: 525–531.

Tyrrell, T., Marañon, E., Poulton, A.J., Bowie, A.R., Harbour, D.S. & Woodward, E.M.S. (2003) Large-scale latitudinal distribution of *Trichodesmium* spp. in the Atlantic Ocean. *Journal of Plankton Research* **25**: 405–416.

Uematsu, M., Tsuda, A., Wells, M. and Saito, H. (2009) Introduction to Subarctic iron Enrichment for Ecodynamics Study II (SEEDS II). *Deep-Sea Research II* **56**: 2731–2732.

Uitz, J., Claustre, H., Morel, A. & Hooker, S.B. (2006) Vertical distribution of phytoplankton communities in open ocean: an assessment based on surface chlorophyll. *Journal of Geophysical Research* **111**: C08005, doi:10.1029/2005JC003207.

Uitz, J., Huot, Y., Bruyant, F., Babin, M. & Claustre, H. (2008) Relating phytoplankton physiological properties to community structure on large scales. *Limnology and Oceanography* **53**: 614–630.

Uitz, J., Claustre, H., Gentili, B. & Stramski, D. (2010) Phytoplankton class-specific primary production in the world's oceans: seasonal and interannual variability from satellite observations. *Global Biogeochemical Cycles*, **24**(GB3016): 1–19, doi:10.1029/2009GB 003680.

Urrere, M. & Knauer, G.A. (1981) Zooplankton fecal pellet fluxes and vertical transport of particulate matter in the marine pelagic environment. *Journal of Plankton Research* **3**: 369–387.

Uye, S. (1982) Population dynamics and production of *Acartia clausi* Giesbrecht (Copepoda: Calanoida) in inlet waters. *Journal of Experimental Marine Biology and Ecology* **57**: 55–83.

Uye, S. (1996) Induction of reproductive failure in the planktonic copepod *Calanus pacificus* by diatoms. *Marine Ecology Progress Series* **133**: 89–97.

Van den Hoek, C., Mann, D.G. & Jahns, H.M. (1995) *Algae, an Introduction to Phycology*. Cambridge University Press, Cambridge, UK, 623 pp.

van der Lingen, C.D., Hutchings, L. & Field, J.G. (2006) Comparative trophodynamics of anchovy (*Engraulis encrasicolus*) and sardine *Sardinops sagax* in the southern Benguela: are species alternations between small pelagic fish trophodynamically mediated? *African Journal of Marine Science* **28**: 465–477.

Van Dover, C.L. (1994) *In situ* spawning of hydrothermal vent tubeworms (*Riftia pachyptila*). *Biological Bulletin* **186**: 134–135.

Van Dover, C.L. (2000) *The Ecology of Deep-Sea Hydrothermal Vents*. Princeton University Press, Princeton, NJ, 424 pp.

Van Dover, C.L., Szuts, E.Z., Chamberlain, S.C. & Cann, J.R. (1989) A novel eye in "eyeless" shrimp from hydrothermal vents of the Mid-Atlantic Ridge. *Nature* **337**: 458–460.

Van Dover, C.L., Kaartvedt, S., Bollens, S.M., Wiebe, P.H., Martin, J.W. & France, S.C. (1992) Deep-sea amphipod swarms. *Nature* **358**: 25–26.

Van Dover, C.L. & 26 coauthors (2001) Biogeography and ecological setting of Indian Ocean hydrothermal vents. *Science* **294**: 818–823.

van Duren, L.A., Stamhuis, E. & Videler, J.J. (2003) Copepod feeding currents: flow patterns, filtration rates and energetics. *Journal of Experimental Biology* **206**: 255–267.

Van Haren, H., Mills, D.K. & Wetsteyn, L.P.M.J. (1998) Detailed observations of the phytoplankton spring bloom in the stratifying central North Sea. *Journal of Marine Research* **56**: 655–680.

Van Mooy, B.A.S. & Devol, A.H. (2008) Assessing nutrient limitation of *Prochlorococcus* in the North Pacific tropical gyre by using an RNA capture method. *Limnology and Oceanography* **53**: 78–88.

Van Mooy, B.A.S. & 11 coauthors (2009) Phytoplankton in the ocean use non-phosphorus lipids in response to phosphorus scarcity. *Nature* **458**: 69–72.

Vanderklift, M.A. & Ponsard, S. (2003) Sources of variation in consumer-diet $\delta^{15}N$ enrichment: a meta-analysis. *Oecologia* **136**: 169–182.

Vanderploeg, H. & Paffenhöfer, G.-A. (1985) Modes of algal capture by the freshwater copepod *Diaptomus sicilis* and their relation to food-size selection. *Limnology and Oceanography* **30**: 871–885.

Vaqué, D., Gazelle, J.M. & Maoris, C. (1994) Grazing rates on bacteria: the significance of methodology and ecological factors. *Marine Ecology Progress Series* **109**: 263–274.

Varela, M.M., van Aken, H.M., Sintes, E. & Herndl, G.J. (2008) Latitudinal trends of Crenarchaeota and Bacteria in the meso- and bathypelagic water masses of the Eastern North Atlantic. *Environmental Microbiology* **10**: 110–124.

Vaske, T., Jr, Vooren, C.M. & Lessa, R.P. (2003) Feeding strategy of yellowfin tuna (*Thunnus albacares*), and wahoo (*Acanthocybium solandri*) in the Saint Paul and Saint Peter Archipelago, Brazil. *Boletim do Instituto de Pesca, São Paulo* **29**: 173–181.

Venrick, E.L. (1974) The distribution and significance of *Richelia intracellularis* Schmidt in the North Pacific Central Gyre. *Limnology and Oceanography* **19**: 437–445.

Venrick, E.L. (1982) Phytoplankton in an oligotrophic ocean: observations and questions. *Ecological Monographs* **52**: 129–154.

Venrick, E.L. (1999) Phytoplankton species structure in the central North Pacific, 1973–1996: variability and persistence. *Journal of Plankton Research* **21**: 1029–1042.

Venter, J.C. & 22 coauthors (2004) Environmental genome shotgun sequencing of the Sargasso Sea. *Science* **304**: 66–74.

Vichi, M., Pinardi, N. & Masina, S. (2007a) A generalized model of pelagic biogeochemistry for the global ocean ecosystem. Part I: Theory. *Journal of Marine Systems* **64**: 89–109.

Vichi, M., Masina, S. & Navarra, A. (2007b) A generalized model of pelagic biogeochemistry for the global ocean ecosystem. Part II: Numerical simulations. *Journal of Marine Systems* **64**: 110–134.

Vidal, J. (1980) Physioecology of zooplankton. 1. Effects of phytoplankton concentration, temperature, and body size on the growth rate of *Calanus pacificus* and *Pseudocalanus* sp. *Marine Biology* **56**: 111–134.

Villareal, T.A. & Carpenter, E.J. (2003) Buoyancy regulation and the potential for vertical migration in the oceanic cyanobacterium *Trichodesmium*. *Microbial Ecology* **45**: 1–10.

Vinogradov, M.E. (1968) Vertikal'noe Raspredelenie Okeanicheskogo Zooplanktona. *Izdatel'stvo Nauka, Moskow*. [Translated 1970, "Vertical Distribution of the Oceanic Zooplankton". Israel Program of Scientific Translations, 339 pp.]

Vinogradov, M.E., Volkov, A.F. & Semenova, T.N. (1996) *Hyperiid Amphipods (Amphipoda, Hyperiidea) of the World Oceans*. Translated from Russian edition (1982) by Siegel-Causey, D. Science Publishers, Inc., Lebanon, New Hamphire, USA, 632 pp.

Vinogradova, N.G. (1997) Zoogeography of the abyssal and hadal zones. *Advances in Marine Biology* **32**: 325–387.

Visser, A.W. (2007) Biomixing of the oceans? *Science* **316**: 838–839.

Vogel, S. (1996) *Life in Moving Fluids*, 2nd Edn., Princeton University Press, Princeton NJ, 484 pp.

Voight, J.R. (2007) Experimental deep-sea deployments reveal diverse northeast Pacific wood-boring bivalves of Xylophagainae (Myoida: Pholadidae). *Journal of Molluscan Studies* **73**: 377–391.

Volterra, V. (1926a) Fluctuations in the abundance of a species considered mathematically. *Nature* **118**: 558–60.

Volterra, V. (1926b) Variazioni e fluttuazioni del numero d'individui in specie animali conviventi. *Memorie Royale Accademia Nazionale dei Lincei, Series VI* **2**: 31–113.

Von Damm, K.L. (1995) Controls on the chemistry and temporal variability of sea floor hydrothermal fluids. In: Humphris, S.E.,

Zierenberg, R.A., Mullineaux, L.S. & Thomson, R.E. (eds.) *Sea Floor Hydrothermal Systems: Physical, Chemical, Biological, and Geological Interactions*. American Geophysical Union, Geophysical Monographs 91: 222–247.

Von Stosch, H.A. (1973) Observations on vegetative reproduction and sexual life cycle of two freshwater dinoflagellates, *Gymnodinium pseudopalustre* Schiller and *Woloszynskia apiculata* sp. nov. *British Phycological Journal* **8**: 105–134.

Vrieling, E.G., Gieskes, W.W.C. & Beelen, T.P.M. (1999) Silicon deposition in diatoms: control by the pH inside the silicon deposition vesicle. *Journal of Phycology* **35**: 548–559.

Vrijenhoek, R.C. (1997) Gene flow and genetic diversity in naturally fragmented metapopulations of deep-sea hydrothermal vent animals. *Journal of Heredity* **88**: 285–293.

Wächterhäuser, G. (1988a) Before enzymes and templates: theory of surface metabolism. *Microbiological Review* **52**: 452–484.

Wächterhäuser, G. (1997) The origin of life and its methodological challenge. *Journal of Theoretical Biology* **187**: 483–494.

Wade, I. & Heywood, K.J. (2001) Acoustic backscatter observations of zooplankton abundance and behaviour and the influence of oceanic fronts in the northeast Atlantic. *Deep-Sea Research II* **48**: 899–924.

Wajih, S. & Naqvi, A. (2008) The Indian ocean. In: Carpenter, E.J. & Capone, D.G. (eds.) *Nitrogen in the Marine Environment*, Elsevier, Amsterdam, pp. 631–681.

Wakeham, S.G. & Lee, C. (1993) Production, transport, and alteration of particulate organic matter in the marine water column. In: Engel, M.H. & Macko, S.A. (eds.) *Organic Geochemistry: Principles and Applications*, Plenum Press, New York, pp. 145–169.

Walker, C.B. & 23 coauthors (2010) *Nitrosopumilus maritimus* genome reveals unique mechanisms for nitrification and autotrophy in globally distributed marine crenarchaea. *Proceedings of the National Academy of Sciences* **107**: 8818–8823.

Walsh, I., Chung, S.P., Richardson, M.J. & Gardner, W.D. (1995) The diel cycle in the integrated particle load in the equatorial Pacific: a comparison with primary production. *Deep-Sea Research II* **42**: 465–477.

Ward, P., Shreeve, R.S. & Cripps, G.C. (1996) *Rhincalanus gigas* and *Calanus simillimus*: lipid storage patterns of two species of copepod in the seasonally ice-free zone of the Southern Ocean. *Journal of Plankton Research* **18**: 1439–1454.

Warén, A., Bengtson, S., Goffredi, S.K. & Van Dover, C.L. (2003) A hot-vent gastropod with iron sulfide dermal sclerites. *Science* **302**: 1007–1007.

Warner, J.A., Latz, M.I. & Case, J.F. (1979) Cryptic bioluminescence in a midwater shrimp. *Science* **203**: 1109–1110.

Warrant, E.J. & Locket, N.A. (2004) Vision in the deep sea. *Biological Reviews* **79**: 671–712.

Warren, B.A. (1983) Why is no deep water formed in the North Pacific? *Journal of Marine Research* **41**: 327–347.

Wassmann, P. & 11 coauthors (2004) Particulate organic carbon flux to the Arctic Ocean sea floor. In: Stein, R. & Macdonald, R.W. (eds.) *The Organic Carbon Cycle in the Arctic Ocean*, Springer, Berlin, pp. 101–138.

Waterbury, J.B., Watson, S.W., Guillard, R.R.L. & Brand, L.E. (1979) Widespread occurrence of a unicellular, marine, planktonic cyanobacterium. *Nature* **277**: 293–294.

Watkins-Brandt, K.S., Letelier, R.M., Spitz, Y.H., Church, M., Bottjer, D. & White, A.E. (2011) Addition of inorganic and organic phosphorus enhances nitrogen and carbon fixation in the oligotrophic North Pacific. *Marine Ecology Progress Series* **432**: 17–29.

Watling, L. (1988) Small-scale features of marine sediments and their importance to the study of deposit feeding. *Marine Ecology Progress Series* **47**: 135–144.

Watling, L. (2005) The global destruction of bottom habitats by mobile fishing gear. In: Crowder, L.B. & Norse, E.A. (eds.) *Marine*

Conservation Biology. Island Press, Washington D.C., pp. 198–210.

Wei, C.-L. & 11 coauthors (2010) Bathymetric zonation of deep-sea macrofauna in relation to export of surface phytoplankton production. *Marine Ecology Progress Series* **399**: 1–14.

Weinbauer, M.G. (2004) Ecology of prokaryotic viruses. *FEMS Microbiology Reviews* **28**: 127–181.

Weinbauer, M.G. & Suttle, C.A. (1997) Comparison of epifluorescence microscopy and transmission electron microscopy for counting viruses in natural marine waters. *Aquatic Microbial Ecology* **13**: 225–232.

Weinbauer, M.G., Winter, C. & Hofle, M.G. (2002) Reconsidering transmission electron microscopy based estimates of viral infection of bacterioplankton using conversion factors derived from natural communities. *Aquatic Microbial Ecology* **27**: 103–110.

Weissburg, M.J., Doall, M.H. & Yen, J. (1998) Following the invisible trail: kinematic analysis of mate-tracking in the copepod *Temora longicornis*. *Proceedings of the Royal Society of London, B* **353**: 701–712.

Welch, H.E., Crawford, R.E. & Hop, H. (1993) Occurrence of Arctic Cod (*Boreogadus saida*) schools and their vulnerability to predation in the Canadian high Arctic. *Arctic* **46**: 331–339.

Welschmeyer, N. (1993) Primary production in the subarctic Pacific Ocean: Project SUPER. *Progress in Oceanography* **32**: 101–135.

Wenzhöfer, F. & Glud, R.N. (2002) Benthic carbon mineralization in the Atlantic: a synthesis based on *in situ* data from the last decade. *Deep-Sea Research I* **49**: 1255–1279.

Wenzhöfer, F., Holby, O. & Kohls, O. (2001) Deep penetrating benthic oxygen profiles measured *in situ* by oxygen optodes. *Deep-Sea Research I* **48**: 1741–1755.

West, G.B., Brown, J.H. & Enquist, B.J. (1997) A general model for the origin of allometric scaling laws in biology. *Science* **276**: 122–126.

Wetz, M.S. & Wheeler, P.A. (2007) Release of dissolved organic matter by coastal diatoms. *Limnology and Oceanography* **52**: 798–807.

Wetz, M.S., Wheeler, P.A. & Letelier, R.M. (2004) Light-induced growth of phytoplankton collected during the winter from the benthic boundary layer off Oregon, USA. *Marine Ecology Progress Series* **280**: 95–104.

Wheatcroft, R.A. (1992) Experimental tests for particle size-dependent bioturbation in the deep ocean. *Limnology and Oceanography* **37**: 90–104.

Wheatcroft, R.A., Smith, C.R. & Jumars, P.A. (1989) Dynamics of surficial trace assemblages in the deep-sea. *Deep-Sea Research* **36A**: 71–91.

Wheatcroft, R.A., Jumars, P.A., Smith, C.R. & Nowell, A.R.M. (1990) A mechanistic view of the particulate biodiffusion coefficient: step lengths, rest periods, and transport directions. *Journal of Marine Research* **48**: 177–207.

Wheeler, P.A. (1993) New production in the subarctic Pacific Ocean: net changes in nitrate concentrations, rates of nitrate assimilation and accumulation of particulate nitrogen. *Progress in Oceanography* **32**: 137–161.

Wheeler, P.A. & Kokkinakis, S.A. (1990) Ammonium recycling limits nitrate uptake in the oceanic subarctic Pacific. *Limnology and Oceanography* **35**: 1267–1278.

Wheeler, P.A., Watkins, J.M. & Hansing, R.L. (1997) Nutrients, organic carbon and organic nitrogen in the upper water column of the Arctic Ocean: implications for the sources of dissolved organic carbon. *Deep-Sea Research II* **44**: 1571–1592.

White, A.E., Spitz, Y.H. & Letelier, R.M. (2006) Modeling carbohydrate ballasting by *Trichodesmium* spp. *Marine Ecology Progress Series* **323**: 35–45.

White, A.E., Karl, D.M., Björkman, K.M., Beversdorf, L.J. & Letelier, R.M. (2010) Production of organic matter by *Trichodesmium* IMS101 as a function of phosphorus source. *Limnology and Oceanography* **55**: 1755–1767.

White, S.N., Chave, A.D., Reynolds, G.T. & Van Dover, C.L. (2002) Ambient light emission from hydrothermal vents on the Mid-Atlantic Ridge. *Geophysical Research Letters* **29**: doi: 10.1029/2002GL014977 (4 pp.).

Whitehill, E.A.G., Franks, T.A. & Olds, M.K. (2009) The structure and sensitivity of the eye of different life history stages of the ontogenetic migratory *Gnathophausia ingens*. *Marine Biology* **156**: 1347–1357.

Whitman, W.B. (2009) The modern concept of the prokaryote. *Journal of Bacteriology* **191**: 2000–2005.

Wilderbuer, T.K., Hollowed, A.B., Ingraham, W.J., Spencer, P.D., Conners, M.E., Bond, N.A. & Walters, G.E. (2002) Flatfish recruitment response to decadal climatic variability and ocean conditions in the eastern Bering Sea. *Progress in Oceanography* **55**: 235–247.

Wiebe, P.H. & Benfield, M.C. (2003) From the Hensen net toward four-dimensional biological oceanography. *Progress in Oceanography* **56**: 7–136.

Wiebe, P.H. & Boyd, S.H. (1978) Limits of *Nematoscelis megalops* in the Northwestern Atlantic in relation to Gulf Stream cold core rings. I. Horizontal and vertical distributions. *Journal of Marine Research* **36**: 119–142.

Wiebe, P.H., Flierl, G.R., Davis, C.S., Barber, V. & Boyd, S.H. (1985a) Macrozooplankton biomass in Gulf Stream warm-core rings: spatial distribution and temporal changes. *Journal of Geophysical Research* **90**: 8885–8901.

Wiebe, P.H. & 7 coauthors (1985b) New developments in the MOCNESS, an apparatus for sampling zooplankton and micronekton. *Marine Biology* **87**: 313–323.

Wilderbuer, T.K., Hollowed, A.B., Ingraham, W.j., Jr., Spencer, P.D., Conners, M.E., Bond, N.A. & Walters, G.E. (2002) Flatfish recruitment response to decadal climatic variability and ocean conditions in the eastern Bering Sea. *Progress in Oceanography* **55**: 235–247.

Wilhelm, S.W. & Suttle, C.A. (1999) Viruses and nutrient cycles in the sea. *BioScience* **49**: 781–788.

Williams, A.B. & Rona, P.A. (1986) Two new caridean shrimps (Bresiliidae) from a hydrothermal field on the Mid-Atlantic Ridge. *Journal of Crustacean Biology* **6**: 446–462.

Williams, P.J.leB. (1981) Incorporation of microheterotrophic processes into the classical paradigm of the food web. *Kieler Meeresforschungen Sonderheft* **5**: 1–29.

Williams, P.J.leB. (1995) Evidence for the seasonal accumulation of carbon-rich dissolved organic material, its scale in comparison with changes in particulate material and the consequential effect on net C/N assimilation ratios. *Marine Chemistry* **51**: 17–29.

Williams, P.M., Bauer, J.E., Robertson, K.J., Wolgast, D.M. & Occelli, M.L. (1993) Report on DOC and DON measurements made at Scripps Institution of Oceanography. *Marine Chemistry* **41**: 271–281.

Winans, A.K. & Purcell, J.E. (2010) Effects of pH on asexual reproduction and statolith formation of the scyphozoan, *Aurelia labiata*. *Hydrobiologia* **645**: 39–52.

Winogradsky, S. (1887) Über Schwefelbakterien. *Botanische Zeitgung* **45**: [In sections from 489 to 610.]

Winter, C., Herndl, G.J. & Weinbauer, M.G. (2004) Diel cycles in viral infection of bacterioplankton in the North Sea. *Aquatic Microbial Ecology* **35**: 207–216.

Wirsen, C.O., Jannasch, H.W. & Molyneaux, S.J. (1993) Chemosynthetic microbial activity at Mid-Atlantic Ridge hydrothermal vent sites. *Journal of Geophysical Research* **98**: 9693–9703.

Woese, C.R. & Fox, G.E. (1977) Phylogenetic structure of the prokaryote domain: the primary kingdoms. *Proceedings of the National Academy of Sciences* **75**: 5088–5090.

Wolff, E.W., Fischner, H. & Röthlisberger, R. (2009) Glacial terminations as southern warmings without northern control. *Nature Geoscience* 2: 206–209. (doi:10.1038/NGEO442).

Wood, S.N. (1994) Obtaining birth and mortality patterns from structured population trajectories. *Ecological Monographs* 64: 23–44.

Woods, L.P. & Sonoda, P.M. (1973) Order Berycomorphi (Beryciformes). In: Cohen, D.M. & 9 coeditors, *Fishes of the Western North Atlantic*, Part 6. Memoir Number 1 of the Sears Foundation for Marine Research, pp. 263–396.

Worden, A.Z. & Binder, B.J. (2003) Application of dilution experiments for measuring growth and mortality rates among *Prochlorococcus* and *Synechococcus* populations in oligotrophic environments. *Aquatic Microbial Ecology* 30: 159–174.

Worden, A.Z. & Not, F. (2008) Ecology and diversity of picoeukaryotes. In: Kirchman, D.L. (ed.) *Microbial Ecology of the Oceans, Second Edition*, Wiley-Blackwell, New Jersey, pp. 159–196.

Wu, J., Sunda, W., Boyle, E.A. & Karl, D.M. (2000) Phosphate depletion in the western North Atlantic Ocean. *Science* 289: 759–762.

Wu, S.M. & Rebeiz, C.M. (1988) Chlorophyll biogenesis: molecular structure of short wavelength chlorophyll-*a* (E432: F662). *Phytochemistry* 27: 353–356.

Wu, Y., Gao, K. & Riebesell, U. (2010) CO_2-induced seawater acidification affects physiological performance of the marine diatom. *Phaeodactylum tricornutum* Biogeosciences 7: 2915–2923.

Wuchter, C., Abbas, B., Coolen, M.J.L., Herfort, L. & van Bleijswijk, J. (2006) Archaeal nitrification in the ocean. *Proceedings of the National Academy of Sciences* 103: 12,317–12,322.

Wuchter, C., Schouten, S., Boschker, H.T.S. & Sinninghe Damste, J.S. (2003) Bicarbonate uptake by marine Crenarchaeota. *FEMS Microbiology Letters* 219: 203–207.

Wyllie, J.G. & Lynn, R.J. (1971) Distribution of temperature and salinity at 10 meters, 1960–1969 and mean temperature, salinity and oxygen at 150 meters, 1950–68. *CalCOFI Atlas no. 15*. California Cooperative Oceanic Fisheries Investigations, La Jolla, California, xi + 189 pp.

Yamaguchi, A. & Ikeda, T. (2000) Vertical distribution, life cycle, and developmental characteristics of the mesopelagic calanoid copepod *Gaidius variabilis* (Aetideidae) in the Oyashio region, western North Pacific Ocean. *Marine Biology* 137: 99–109.

Yao, H. & 7 coauthors (2010) Protection mechanisms of the iron-plated armor of a deep-sea hydrothermal vent gastropod. *Proceedings of the National Academy of Sciences* 107: 987–992.

Yen, J., Rasberry, K.D. & Webster, D.R. (2008) Quantifying copepod kinematics in a laboratory turbulence apparatus. *Journal of Marine Systems* 69: 283–294.

Yentsch, C. (1980) Light attenuation and phytoplankton photosynthesis. In: Morris, I. (ed.) *The Physiological Ecology of Phytoplankton*. University of California Press, Berkeley, pp. 95–127.

Yoder, J.A., Aiken, J., Swift, R.N., Hoge, F.E. & Stegmann, P.M. (1993) Spatial variability in near-surface chlorophyll *a* fluorescence measured by the Airborne Oceanographic Lidar (AOL). *Deep-Sea Research II* 40: 37–53.

Yoklavich, M.M., Loeb, V.J., Nishimoto, M. & Daly, B. (1996) Nearshore assemblages of larval rockfishes and their physical environment off central California during an extended El Niño event, 1991–93. *Fisheries Bulletin* 94: 766–782.

Young, J.R., Andruleit, H. & Probert, I. (2009) Coccolith function and morphogenesis: insights from appendage-bearing coccolithophores of the family Syracosphaeraceae (Haptophyta). *Journal of Phycology* 45: 213–226.

Young, R.E. & Mencher, F.M. (1980) Bioluminescence in mesopelagic squid: diel color change during counterillumination. *Science* 208: 1286–1288.

Young, R.E. & Roper, C.F.E. (1977) Intensity regulation of bioluminescence during countershading in living midwater animals. *Fisheries Bulletin* 75: 239–252.

Young, R.E., Roper, C.F.E. & Walters, J.F. (1979) Eyes and extraocular photoreceptors in midwater cephalopods and fishes: their roles in detecting downwelling light for counterillumination. *Marine Biology* 51: 371–380.

Zahariev, K., Christian, J.R. & Denman, K.L. (2008) Preindustrial, historical, and fertilization simulations using a global ocean carbon model with new parameterizations of iron limitation, calcification, and N_2 fixation. *Progress in Oceanography* 77: 56–82.

Zehr, J.P. & Kudela, R.M. (2009) Photosynthesis in the open ocean. *Science* 326: 945–946.

Zehr, J.P. & 8 coauthors (2007) Experiments linking nitrogenase gene expression to nitrogen fixation in the North Pacific subtropical gyre. *Limnology and Oceanography* 52: 169–183.

Zeldis, J. (2001) Mesozooplankton community composition, feeding and export production during SOIREE. *Deep-Sea Research II* 48: 2615–2634.

Zentara, S.-J., Kamykowski, D. (1981) Geographic variations in the relationship between silicic acid and nitrate in the South Pacific Ocean. *Deep-Sea Research* 28A: 455–465.

Zezina, O.N. (1997) Biogeography of the bathyal zone. *Advances in Marine Biology* 32: 389–426.

Zhaxybayeva, O., Doolittle, W.F., Papke, R.T. & Gogarten, J.P. (2009) Intertwined evolutionary histories of marine *Synechococcus* and *Prochlorococcus marinus*. *Genome, Biology and Evolution* 1: 325–339.

Zhou, M. (2006) What determines the slope of a plankton biomass spectrum? *Journal of Plankton Research* 28: 437–448.

INDEX

Note: page numbers followed by b, f, or t refer to Boxes, Figures, or Tables

Biological Oceanography, Second Edition. Charles B. Miller, Patricia A. Wheeler.
© 2012 John Wiley & Sons, Ltd. Published 2012 by John Wiley & Sons, Ltd.